Lecture Notes in Computer Science   7428

Commenced Publication in 1973
Founding and Former Series Editors:
Gerhard Goos, Juris Hartmanis, and Jan van Leeuwen

T0223464

Emmanuel Prouff   Patrick Schaumont (Eds.)

Cryptographic Hardware and Embedded Systems – CHES 2012

14th International Workshop
Leuven, Belgium, September 9-12, 2012
Proceedings

 Springer

Volume Editors

Emmanuel Prouff
French Network and Information Security Agency (FNISA)
51 boulevard de La Tour-Maubourg, 75700 Paris, France
E-mail: e.prouff@gmail.com

Patrick Schaumont
Virginia Polytechnic Institute
The Bradley Department of Electrical and Computer Engineering
302 Whittemore Hall, Blacksburg, VA 24060, USA
E-mail: schaum@vt.edu

ISSN 0302-9743   e-ISSN 1611-3349
ISBN 978-3-642-33026-1   e-ISBN 978-3-642-33027-8
DOI 10.1007/978-3-642-33027-8
Springer Heidelberg Dordrecht London New York

Library of Congress Control Number: 2012945398

CR Subject Classification (1998): E.3, D.4.6, K.6.5, E.4, C.2, G.2.1, J.1

LNCS Sublibrary: SL 4 – Security and Cryptology

Typesetting: Camera-ready by author, data conversion by Scientific Publishing Services, Chennai, India

Printed on acid-free paper

Springer is part of Springer Science+Business Media (www.springer.com)

Preface

The 14th International Workshop on Cryptographic Hardware and Embedded Systems (CHES 2012) was held at the Katholieke Universiteit Leuven, Belgium, during September 9–12, 2012. The workshop was sponsored by the International Association for Cryptologic Research.

CHES 2012 received 120 submissions from 22 countries. The 42 members of the Program Committee were assisted by more than 150 external reviewers. In total, they delivered 498 reviews. Each submission was reviewed by at least four Program Committee members. Submissions by Program Committee members received at least five reviews. The review process was double-blind, and conflicts of interest were carefully handled. The review process was handled through an online review system that supported discussions among Program Committee members. Over the entire review period, more than 660 messages were exchanged between Program Committee members. Eventually, the Program Committee selected 32 papers (a 27% acceptance rate) for publication in the proceedings.

CHES 2012 used, for the first time, an author rebuttal. After four weeks of individual review, and two weeks of initial online discussions, the reviews were forwarded to the submitting authors. The authors were invited to provide a text-only rebuttal of no more than 500 words. CHES 2012 received 110 rebuttals (a 91% response rate). The rebuttals were then included in the online discussion system, to guide the paper decision process in two additional weeks of online discussion.

The program also included two invited talks, by Stephen Murdoch from the University of Cambridge, UK, and by Christof Tarnovsky from Flylogic Engineering. For the first time, the program included two tutorials on cryptographic engineering aimed at newcomers in CHES. The tutorials were given by Junfeng Fan from the Katholieke Universiteit Leuven, Belgium, and by Diego Aranha from the Universidad de Brasil, Brazil.

The Program Committee also identified the best submissions from CHES for their scientific quality, their originality, and their clarity. After deliberation, an ad-hoc committee with no conflict of interest to these submissions evaluated each nomination and selected one of them for award. The CHES 2012 Best Paper Award went to Andrew Moss, Elisabeth Oswald, Dan Page, and Michael Tunstall. Their paper, "Compiler Assisted Masking," discusses the use of compiler techniques to create side-channel countermeasures.

Many people contributed to CHES 2012. We thank the authors for contributing their excellent research, and for participating so enthusiastically in the rebuttal process. We thank the Program Committee members, and their external reviewers, for making a significant effort over an extended period of time to select the right papers for the program. We particularly thank Lejla Batina and Ingrid Verbauwhede, the General Co-chairs, who took care of many practical details of the event. We also thank Thomas Eisenbarth for organizing the poster session at CHES 2012. We are very grateful to Shai Halevi, who wrote the review software, and helped adapt the system to support a rebuttal phase. The website was maintained by Dusko Karaklajic and by Jens Peter Kaps; we appreciate their support throughout CHES. Finally, we thank our sponsors for supporting CHES financially: Cryptography Research, Sakura, Technicolor, Riscure, Infineon, Telecom ParisTech, NXP, and Intrinsic ID.

CHES 2012 collected truly exciting results in cryptographic engineering, from concepts to artifacts, from software to hardware, from attack to countermeasure. We feel priviledged for the opportunity to develop the CHES 2012 program. We hope that the papers in these proceedings will continue to inspire, guide, and clarify your academic and professional endeavors.

July 2012   Emmanuel Prouff
                                                               Patrick Schaumont

CHES 2012

Workshop on Cryptographic Hardware and Embedded System
Leuven, Belgium, 9–12 September, 2012.

Sponsored by the *International Association for Cryptologic Research*

General Co-chairs

Lejla Batina   Radboud University, The Netherlands, and
     Katholieke Universiteit Leuven, Belgium

Ingrid Verbauwhede   Katholieke Universiteit Leuven, Belgium

Program Co-chairs

Emmanuel Prouff   ANSSI, France

Patrick Schaumont   Virginia Tech, USA

Program Committee

Paulo Barreto   University of Sao Paulo, Brazil

Daniel J. Bernstein   University of Illinois at Chicago, USA

Guido Bertoni   STMicroelectronics, Italy

Swarup Bhunia   Case Western University, USA

Zhimin Chen   Microsoft, USA

Dipanwita Roy Chowdhury   Indian Institute of Technology Kharagpur,
     India

Jean-Sébastien Coron   University of Luxembourg, Luxembourg

Hermann Drexler   Giesecke & Devrient, Germany

Thomas Eisenbarth   Florida Atlantic University, USA

Kris Gaj   George Mason University, USA

Catherine Gebotys   University of Waterloo, Canada

Benedikt Gierlichs   Katholieke Universiteit Leuven, Belgium

Christophe Giraud   Oberthur Technologies, France

Louis Goubin   University of Versailles, France

Sylvain Guilley   TELECOM ParisTech, France

Tim Güneysu   Ruhr University Bochum, Germany

Naofumi Homma   Tohoku University, Japan

Marc Joye   Technicolor, France

Kirstin Lemke-Rust   University of Applied Sciences
     Bonn-Rhein-Sieg, Germany

External Reviewers

Nicolas Gama
Berndt Gammel
Laurie Genelle
Benoît Gérard
Martin Goldack
Sezer Goren
Johann Groszschädl
Keisuke Hakuta
Mike Hamburg
Mark Hamilton
Laszlo Hars
Ludger Hemme
Stefan Heyse
Lars Hoffmann
Ekawat Homsirikamol
Philippe Hoogvorst
Yohei Hori
Gabriel Hospodar
Michael Hutter
Kouichi Itoh
Tetsuya Izu
Josh Jaffe
Eliane Jaulmes
Takahashi Junko
Hyunho Kang
Saffija Kasem-Madani
Markus Kasper
Toshihiro Katashita
Hyunmin Kim
Mario Kirschbaum
Paris Kitsos
Ilya Kizhvatov
Boris Koepf
François Koeune
Thomas Korak
Sebastian Kutzner
Tanja Lange
Marc Le Guin
Victor Lomné
Robert Lorentz

Pedro Maat Massolino
Roel Maes
Abhranil Maiti
Benjamin Martin
Marcel Medwed
Filippo Melzani
Oliver Mischke
Amir Moradi
Cédric Murdica
Seetharam Narasimhan
Svetla Nikova
David Oswald
Toru Owada
Ozen Ozkaya
Dan Page
Pascal Paillier
Andrea Palomba
Somnath Paul
Eric Peeters
Chris Peikert
Geovandro Pereira
Christiane Peters
John Pham
Gilles Piret
Thomas Plos
Thomas Poeppelmann
Jürgen Pulkus
Francesco Regazzoni
Christof Rempel
Mathieu Renauld
Francisco
    Rodríguez-Henríquez
Marcin Rogawski
Franck Rondepierre
Vladimir Rožić
Minoru Saeki
Muhammet Sahinoglu
Kazuo Sakiyama
Laurent Sauvage
Mike Scott

Martin Seysen
Rabia Shahid
Kiyomoto Shinsaku
Endo Sho
Youssef Souissi
Mehmet Soybali
François-Xavier
    Standaert
Rainer Steinwandt
Takeshi Sugawara
Ruggero Susella
Daisuke Suzuki
Robert Szerwinski
Yannick Teglia
Adrian Thillard
Enrico Thomae
Michael Tunstall
Malik Umar Sharif
Hasan Unlu
Gilles Van Assche
Ihor Vasylstov
Ingrid Verbauwhede
Frederik Vercauteren
Vincent Verneuil
Ingo Von Maurich
Xinmu Wang
Erich Wenger
Carolyn Whitnall
Marcin Wojcik
Jun Yajima
Tolga Yalcin
Panasayya Yalla
Teruyoshi Yamaguchi
Li Yang
Gavin Yao
Xin Ye
Ramazan Yeniceri
Hirotaka Yoshida
Yu Zheng

Invited Talk I

Banking Security: Attacks and Defences

Steven Murdoch

University of Cambridge, Cambridge, UK

Steven.Murdoch@cl.cam.ac.uk

Abstract. Designers of banking security systems are faced with a difficult challenge of developing technology within a tightly constrained budget, yet which must be capable of defeating attacks by determined, well-equipped criminals. This talk will summarise banking security technologies for protecting Chip and PIN/EMV card payments, online shopping, and online banking. The effectiveness of the security measures will be discussed, along with vulnerabilities discovered in them both by academics and by criminals. These vulnerabilities include cryptographic flaws, failures of tamper resistance, and poor implementation decisions, and have led not only to significant financial losses, but in some cases unfair allocation of liability. Proposed improvements will also be described, not only to the technical failures but also to the legal and regulatory regimes which are the underlying reason for some of these problems not being properly addressed.

Table of Contents

Intrusive Attacks and Countermeasures

Masking

Improved Fault Attacks and Side Channel Analysis

Leakage Resiliency and Security Analysis

Physically Unclonable Functions

Efficient Implementations

Lightweight Cryptography

We Still Love RSA

Hardware Implementations

3D Hardware Canaries

Sébastien Briais[4], Stéphane Caron[1], Jean-Michel Cioranesco[2,3],
Jean-Luc Danger[5], Sylvain Guilley[5], Jacques-Henri Jourdan[1],
Arthur Milchior[1], David Naccache[1,3], and Thibault Porteboeuf[4]

[1] École normale supérieure, Département d'informatique
{stephane.caron,jacques-henri.jourdan,arthur.milchior,
david.naccache}@ens.fr
[2] Altis Semiconductor
jean-michel.cioranesco@altissemiconductor.com
[3] Sorbonne Universités – Université Paris II
jean-michel.cioranesco@etudiants.u-paris2.fr
[4] Secure-IC
{sebastien.briais,thibault.porteboeuf}@secure-ic.com
[5] Département Communications et Electronique
Télécom-ParisTech
{jean-luc.danger,sylvain.guilley}@telecom-paristech.fr

Abstract. 3D integration is a promising advanced manufacturing process offering a variety of new hardware security protection opportunities. This paper presents a way of securing 3D ICs using Hamiltonian paths as hardware integrity verification sensors. As 3D integration consists in the stacking of many metal layers, one can consider surrounding a security-sensitive circuit part by a wire cage.

After exploring and comparing different cage construction strategies (and reporting preliminary implementation results on silicon), we introduce a "hardware canary". The canary is a spatially distributed chain of functions F_i positioned at the vertices of a 3D cage surrounding a protected circuit. A correct answer $(F_n \circ \ldots \circ F_1)(m)$ to a challenge m attests the canary's integrity.

1 Introduction

3D integration is a promising advanced manufacturing process offering a variety of new hardware security protection opportunities. This paper presents a way of securing 3D ICs using Hamiltonian paths[1] as integrity verification sensors. 3D integration consists in the stacking of many metal layers. Hence, one can consider surrounding a security-sensitive circuit part by a wire cage, for instance a Hamiltonian path connecting the vertices of a cube (Fig. 1). In this paper, different algorithms to construct cubical Hamiltonian structures are studied; those ideas can be extended to other forms of sufficiently dense lattices.

Since 3D integration is based on the vertical stacking of different dies, a Hamiltonian cage can surround the whole target and protect its content from physical

---

[1] A Hamiltonian circuit (hereafter "cage" or simply "path" for the sake of conciseness) is an undirected path passing once through all the vertices of a graph.

E. Prouff and P. Schaumont (Eds.): CHES 2012, LNCS 7428, pp. 1–22, 2012.

Fig. 1. Hamiltonian cycle passing through the vertices of a $4 \times 4 \times 4$ cube

attacks. 3D ICs are relatively hard to probe due to the tight bonding between layers [11]. Moreover, the 3D path can even penetrate the protected circuit and connect points in space between the protected circuit's transistors.

A path running through different metal layers and different dies can thus serve as a *digital* integrity verification sensor allowing the sending and the collecting of signals. In addition, the wire can be used to fill gaps in empty circuit parts to increase design compactness and make reverse-engineering harder.

Such a protection proves challenging in terms of design as it requires devising new manufacturing and synthesis tools to fit the technology used [1, 2]. However the resulting structures prove very helpful in protecting against active probing.

Throughout this paper n will represent the number of points forming the edge of a cubical Hamiltonian structure. We will focus our study on cubical structures, but the algorithms and concepts that are presented hereafter can in principle be extended to many types of sufficiently dense lattices of points.

2 Generating Random 3D Hamiltonian Paths

2.1 General Considerations

The problem of finding a Hamiltonian path in arbitrary graphs (HAMPATH) is NP-complete. Membership in NP is easy to see (given a candidate solution, the solution's correctness can be verified in quasi-linear time). We refer the reader to [3] for more information on HAMPATH.

A quick glance reveals that a cube's n^3 vertices, potentially connectable by a mesh of $3n^2(n-1)$ edges, break-down into four categories, illustrated in Fig. 2[2]:

- $(n-2)^3$ vertices corresponding to the cube's innermost edges (*i.e.* not facing the outside) can be potentially connected in any of the possible 3D directions (right, left, up, down, front, rear).
- $6(n-2)^2$ vertices, facing the cube's outside in exactly one direction, can be potentially connected in five possible directions.

---

[2] The depicted cube is shown as a solid opaque object for the sake of clarity.

- 12$(n-2)$ vertices, facing the cube's outside in exactly two directions, can be potentially connected in four possible directions.
- 8 extreme corner vertices can be connected in only three possible manners.

Indeed: $(n-2)^3 + 6(n-2)^2 + 12(n-2) + 8 = ((n-2)+2)^3 = n^3$

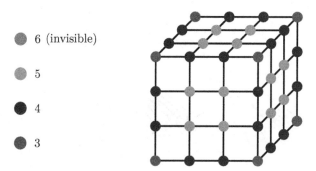

6 (invisible)

5

4

3

Fig. 2. Potential edge connectivity

We observe that for HAMPATH to be solvable in a cube, n must be even. If we depart from point the $(0,0,0)$ and reach a point of coordinates (x,y,z) after visiting i vertices, then $x+y+z$ and i have the same parity. Given that the path must collect all the cube's vertices, the cube size must necessarily be even.

2.2   Odd Size Cubes

The above observation excludes the existence of odd-size cubes unless one skips in such cubes an edge (x,y,z) such that $x+y+z \equiv 1 \mod 2$. To extend the construction to odd $n = 2k+1$ while preserving symmetry, we *arbitrarily* decide to exclude the central vertex (*i.e.* at coordinate (k,k,k)) when n is odd.

Assume that we color vertices in black and white alternatingly (the cube's 8 extreme vertices being black) with black corresponding to even-parity $x+y+z$ and white corresponding to odd parity $x+y+z$. Here $0 \leq x,y,z \leq 2k$. In other words, a $(2k+1)$-cube has $4k^3 + 6k^2 + 3k$ white vertices and $4k^3 + 6k^2 + 3k + 1$ black vertices.

The coordinate of the cube's central vertex is (k,k,k) which parity is identical to the parity of k. When k is even, vertex (k,k,k) is black and when k is odd vertex (k,k,k) is white. If we remove vertex (k,k,k) it appears that:

- When k is even, (*i.e.* $n = 2k+1 = 4\ell+1$) we have as many black and white vertices (namely $4k^3 + 6k^2 + 3k$).
- When k is odd, we have $4k^3+6k^2+3k+1$ black vertices and $4k^3+6k^2+3k-1$ white vertices.

Noting that each edge causes a color switch, we see that Hamiltonian paths in cubes of size $4\ell+3$ cannot exist. Note that if one extra black vertex is removed[3] then (the now asymmetric) construction becomes possible for all k.

---

[3] *e.g.* one of the cube's extreme edges which is necessarily black.

It remains to prove that cubes of size $n = 4\ell + 1$ exist for all $\ell \neq 0$ (Fig. 3). We refer the reader to the extended version of this paper on the IACR ePrint server for further details.

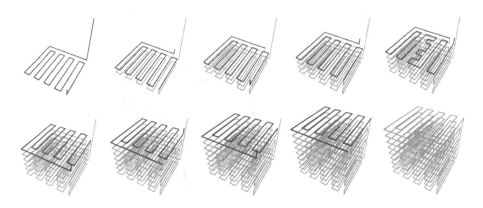

Fig. 3. Constructive proof that cubes of size $n = 4\ell + 1$ exist for all $\ell \neq 0$

3   A Toolbox for Generating 3D Hamiltonian Cycles

3.1   From Two to Three Dimensions

We start by presenting a first algorithm for constructing random[4] Hamiltonian cycles in graphs having a minimum degree equal to at least half the number of their vertices.

Our application requires an efficient algorithm that outputs cycles passing through a very large number of vertices. The first algorithm reduces the problem's complexity by using smaller cycles that we will progressively merge to form the final bigger cycle. Consider the elementary Hamiltonian cycle forming a simple 2×2 square. To combine two such squares all we need are two parallel edges. Merging (denoted by the operator ⟿) can be done in two ways as shown Fig. 4. Note that this association not only preserves Hamiltonicity but also extends it.

Fig. 4. Association of squares along the x axis (leftmost figure), or the y axis (rightmost figure)

In other words, at each step two different Hamiltonian cycles in adjacent graphs are merged, and a new Hamiltonian cycle is created. The process is repeated until only one Hamiltonian cycle remains. We implemented this process in C. As explained previously, our program cannot find Hamiltonian cycles for odd cardinality values simply because such cycles do not exist (see Algorithm 1). The code starts by filling the lattice with 2×2 squares, and then associates them randomly. The program ends when only one cycle is left (Fig. 5).

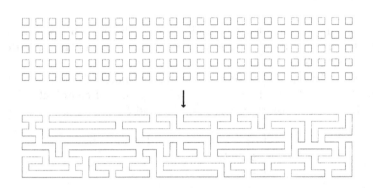

Fig. 5. Rewriting 125 squares filling a 50×10 lattice as a Hamiltonian cycle using Algorithm 1

Algorithm 1. Cycle Merging

1: **Input** $p, q \in 2\mathbb{N}$.
2: let $Q = Q_1, ..., Q_v$ be the $v = \frac{pq}{4}$ squares of size 2 filling the lattice of $p \times q$ points.

3: **while** $\mathrm{Card}(Q) \neq 1$ **do**
4:     choose randomly $\{a, b\} \in Q^2$ with $a \neq b$.
5:     **if** a and b have at least one couple of neighbouring parallel edges **then**
6:         Break a randomly chosen couple of parallel neighbouring edges, verify that they form a single Hamiltonian circuit and merge $c = a \leadsto b$.
7:         **let** $Q = Q \cup \{c\} - \{a, b\}$
8:     **else**
9:         **goto** line 4
10:    **end if**
11: **end while**

The algorithm is pretty fast, and we were able to build Hamiltonian cycles of 10^5 points using a laptop[5] within few seconds. For some p and q values, we observed some runtime spikes in single measurements due to convergence issues. Fig. 6 shows the average runtime over 100 measurements as well as the standard deviation at each point in red.

---

[5] MacBook Air 1.8 GHz Intel Core i7.

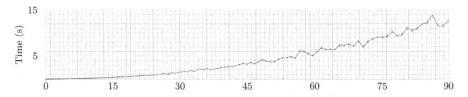

Fig. 6. Cycle Merging runtime as a function of the number of points $\times 10^3$ (average over 100 measurements)

To transform a rectangular 2D Hamiltonian cycle into a 3D one, we run Algorithm 1 for $\{p, q\} = \{p, p^2\}$ to get a $p \times p^2$ rectangle $\mathcal{L}$ similar in nature to the one shown in Fig. 5.

Then, letting (x_i, y_i) denote the Cartesian coordinates of points in $\mathcal{L}$, with the first point being $(0, 0)$, we fold $\mathcal{L}$ into a 3D structure of coordinates (x'_i, y'_i, z'_i) using the following transform where $j = \lfloor \frac{x_i}{p} \rfloor$ and $\ell \equiv j$ mod 2:

$$\varphi = \begin{cases} x'_i = (-1)^\ell (x_i - jp) + \ell(p - 1) \\ y'_i = y_i \\ z'_i = j \end{cases}$$

The result is shown in Fig. 24 (Appendix A).

It remains to destroy the folded nature of the construction while preserving Hamiltonicity. This is done as follow: Identify anywhere in the generated structure the red pattern shown at the leftmost part of Fig. 7 where at positions a, b, c, d edges take any of the blue positions. Iteratively apply this rewriting rule *along any desired axis* until the resulting structure gets "mixed enough" to the designer's taste. Evidently, this is only one possible rewriting rule amongst several.

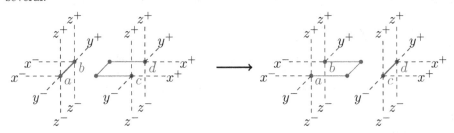

Fig. 7. Rewriting rule

Note that the zig-zag folding φ is only one among many possible folding options as φ may be replaced by any 2D (preferably random) plane-filling curve of size $p \times p$ (e.g. a Peano curve [8]).

3.2 Random Cube Association

Another approach consists in generalizing Algorithm 1 to the associating of elementary 3D cubes. As shown in Fig. 26, one can fill the target lattice by a random sampling of six elementary Hamiltonian cubes (Fig. 25), associate them randomly and further randomize the resulting structure by rewriting.

The algorithm proves very efficient (Fig. 8) and takes a few seconds[6] to compute a random Hamiltonian cube of size 50 (125 000 points).

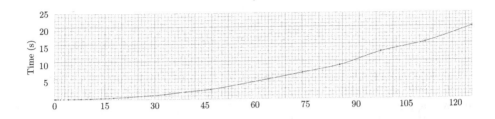

Fig. 8. Random Cube Association runtime as a function of the number of points $\times 10^3$ (average over 100 measurements)

The algorithm picks random parallel edges from different Hamiltonian cycles and attempts to associate them in one new structure. By opposition of the 2D case, the 3D case presents a new difficulty which is that in some cases associable parallel edges suddenly cease to exist. To force termination we abort and restart from scratch if the number of iterations executed without finding a new association exceeds the upper bound p^3. To compute structures over huge lattices (e.g. $n = 100$), one might need to introduce additional association rules (e.g. the rule shown in Fig. 9) to avoid such deadlocks.

Fig. 9. An additional association rule (example)

3.3 Cycle Stretching

Our third algorithm maintains and extends a set of edges E initialized with the four edges defined by the square of vertices $(0, 0, 0)$, $(0, 1, 0)$, $(1, 1, 0)$ and $(1, 0, 0)$. At each iteration, the algorithm selects a random edge $e \in E$ and one of the four extension directions shown in Fig. 10. If such an extension is possible (in other words, by doing so we do not bump into an edge already in E) then E is extended by replacing e by three new edges (one parallel to e and two orthogonal to e in the chosen extension direction). If e cannot be replaced, i.e. none of the four extensions is possible, we pick a new $e' \in E$ and try again.

---

[6] MacBook Air 1.8 GHz Intel Core i7.

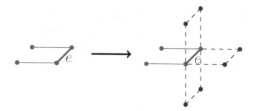

Fig. 10. Extension options

The algorithm keeps track of a subset of E, denoted B, interpreted as the set of potentially stretchable edges of E. B avoids trying to stretch the same e over and over again.

At each stretching attempt the algorithm picks a random $e \in B$. As the algorithm tries to stretch e, e is removed from B (no matter if the stretching attempt is successful or not). If stretching succeeded, e is also removed from E and three new edges replacing e are added to B and E.

The algorithm halts when $B = \varnothing$. If upon halting $|E| = n^3 - (n \bmod 2)$ then the algorithm succeeds, otherwise the algorithm fails and has to be re-launched. Since at most $3n^2(n-1)$ vertices can be added to B, the algorithm will eventually halt.

A non-optimized implementation running on a typical PC found a solution for $n = 6$ in about a minute and a solution for $n = 8$ in 30 hours. The same code was unable to find a solution for $n = 10$ in three weeks. An empirical human inspection of the obtained cubes shows that the resulting structures seem very irregular. Hence, an interesting strategy consists in generating a core cube of size $n = 8$ by cycle stretching, surrounding it by elementary size 2 cubes and proceeding by random cube association and rewriting.

Algorithm 2. Edge Stretching

1: **let** $E =$ the four vertices defined by the square $(0,0,0)$, $(0,1,0)$, $(1,1,0)$, $(1,0,0)$.
2: **let** $B = E$.
3: **while** $B \neq \varnothing$ **do**
4:     **let** $e \in_R B$, we denote the vertices of e by $e = [e_1, e_2]$.
5:     **let** $B = B - \{e\}$
6:     **let** dir $= \{\leftarrow, \rightarrow, \uparrow, \downarrow, \nearrow, \swarrow\}$
7:     **while** dir $\neq \varnothing$ **do**
8:         **let** $d \in_R$ dir
9:         **let** dir $=$ dir $- \{d\}$
10:         **if** d and e are not aligned and stretching is possible **then**
11:             $E = E - \{e\}$.
12:             $E = E \cup \{[e_1, v_1], [v_1, v_2], [v_2, e_2]\}$.
13:             **break**
14:         **end if**
15:     **end while**
16: **end while**

In the above algorithm the sentence "*stretching is possible*" is formally defined as the fact that no edges in E pass through the two vertices v_1, v_2 such that the segment $[v_1, v_2]$ is parallel to e in direction d. Arrows represent right, left, up, down, front and backwards directions, *i.e.* $\leftarrow \uparrow \nearrow \swarrow \downarrow \rightarrow$

3.4 Constraining Existing Hamiltonian Pathfinding Algorithms

A fourth experimented approach consisted in adapting existing HAMPATH solving strategies. (Dharwadker) [4] presents a polynomial time algorithm for finding Hamiltonian paths in certain classes of graphs. Assuming that the graphs that we are interested in are in such a class, we tweaked [4]'s C++ code to find Hamiltonian cycles in cubes. The resulting code succeeded in finding solutions, but these had a too regular appearance and had to be post-processed by rewriting.

We hence constrained the algorithm by working in a randomly chosen subgraph E of the full n^3 cube. We define a *density factor* $\gamma \leq 1$ allowing to control the number of edges in E to which we apply [4]. The ratio of edges in E and n^3 is expected to be approximately γ. Note that because of the heuristic corrective step (9), meant to reduce the odds that certain points remain unreachable, E's density is expected to be slightly higher than γ. The corresponding algorithm is:

Algorithm 3. Edges Selection Routine

1: $E = \varnothing$
2: **for** each vertex $v = (x, y, z)$ of the full cube **do**
3:    **for** each move $dv = (dx, dy, dz)$ in $\{(1,0,0), (0,1,0), (0,0,1)\}$ **do**
4:       **generate** a random $r \in [0, 1]$
5:       **if** $r < \gamma$ and $(0,0,0) \leq v + dv \leq (n-1, n-1, n-1)$ **then**
6:          **add** edge $[v, v + dv]$ to E
7:       **end if**
8:    **end for**
9:    **if** loop 3 didn't add to E any edge having v as en extremity **then**
10:       **goto** line 3
11:    **end if**
12: **end for**

Practical experiments show indeed that as γ diminishes, the generated Hamiltonian cycles seem increasingly irregular (for high (*i.e.* $\simeq 1$) γ values the algorithm fills the cube by successive "slices"). Finding solutions becomes computationally harder as γ diminishes, but using a standard PC, it takes about a second to generate an instance for $\{\gamma = 0.8, n = 6\}$ and an hour to generate a $\{\gamma = 0.86, n = 10\}$ one. The reader is referred to the IACR ePrint version of this paper for visual illustrations of experimental results.

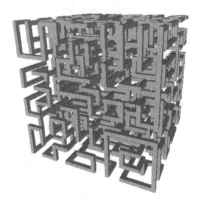

Fig. 11. A $n = 10$ Hamiltonian cycle obtained by a modified version of Dharwadker's algorithm [4]

3.5 Branch-and-Bound

Another experimented approach was the use of branch-and-bound: Using a recursive function, we can try all different cycles. Given a connected portion of a potential Hamiltonian path, this function tries to add all the possible new edges and calls itself recursively. If the function is called with a complete path, the job is done.

We added several heuristic improvements to this method:

1. If the set of vertices unlinked by the current path is disconnected, it is clear that we won't be able to find any Hamiltonian path, and thus we can stop searching.
2. If this set is not connected to the extremities of the current path, we can also halt.
3. The existence of an Hamiltonian path containing a given sub-path only depends on the extremities and on the set of vertices in the path. We can hence use a dynamic programming approach to avoid redundant computations.
4. We tried multiple heuristics to chose the order of recursive calls.

However, those approaches proved much slower than cycle stretching: it appears that the branch-and-bound algorithm makes decisive choices at the beginning of the path without being able to re-consider them quickly. We tried to count all the Hamiltonian cycles when $n = 4$ using this algorithm, but the code proved too slow to complete this task in a reasonable time.

Those results suggest a meta-heuristic approach that would be intermediate between branch-and-bound and stretching: we can make a cycle evolve using meta-heuristics until we obtain an Hamiltonian cycle. Using this method (that we did not implement) we should be able to re-consider any previous choice without restarting the search process.

3.6   Rewriting 3D Moore Curves

Finally, one can depart from a know regular 3D cycle (*e.g.* a 3D Moore curve as shown in Fig.12) and rewrite it. Moore curves are particularly adapted to such a strategy given that the maze entrance and exit are two adjacent edges. However, as shown in Fig.12c (a top-down view of Fig.12b), Moore curves are inherently regular and must be re-rewritten to gain randomness.

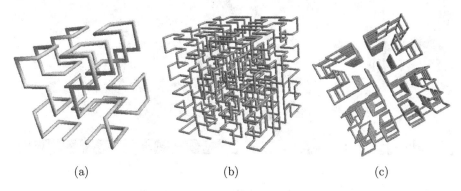

(a)   (b)   (c)

Fig. 12. Example of Moore Curves [5]

4   Silicon Experiments

To test manufacturability in silicon we created a first passive cage meant to protect an 8-bit register. We notice that the compactness of the cage provides a very good reverse-engineering protection.

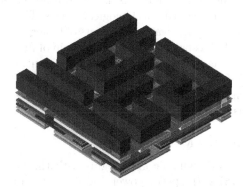

Fig. 13. 3D layout of a cage of size 6 (130nm, 6 Metal Layers Technology)

The implemented structure (Fig. 13) is a $6 \times 6 \times 6$ Hamiltonian cube stretching over six metal layers, the first four metal layers are copper ones, and the last two metal layers are thicker and made of aluminum (130nm RF technology, Fig. 14). The cube is 26μm wide and covers an 8 bit register.

As will be explained in the next section, this first prototype is not dynamic, the Hamiltonian path is not connected to transistors. The implementation of a simplified dynamic structure as described in section 5 is underway and does not seem to pose insurmountable technological challenges. Moreover, all layers of the prototype are processed in one side of the silicon, so this implementation does not prevent backside attack. Backside metal deposit and back to back wafer stacking must thus be investigated to thwart backside attacks as well.

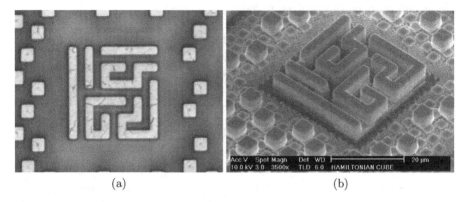

(a)   (b)

Fig. 14. Top layer view (a) and tilted SEM view (b) of a 26μm wide $6 \times 6 \times 6$ cage implemented in a 130nm technology ($\times 2500$)[7]

5   Dynamically Reconfigurable 3D Hamiltonian Paths

A canary is a binary constant placed between a buffer and stack data to detect buffer overflows. Upon buffer overflow, the canary gets corrupted and an overflow exception is thrown. The term "canary" is inherited from the historic practice of using canaries in coal mines as toxic gas biological alarms. The dynamic structures presented in this section are hardware equivalents of biologic canaries: our "hardware canary" is formed of a spatially distributed chain of functions F_i positioned at the vertices of a 3D cage surrounding a protected circuit. In essence, a correct answer $(F_n \circ \ldots \circ F_1)(m)$ to a challenge m will attest the canary's integrity. The device described in this section relies on a library of paths precomputed using the toolbox of algorithms described in the previous section.

5.1   Reconfigurable 3D Mazes

The construction of a 3D dynamic grid begins with the description of a Network On Silicon (NOS) with speed, power and cost constraints [7, 12]. As described in [6, 9], metal wires are shared, or made programmable, by introducing *switch-boxes*, that serve as the skeleton of the dynamic Hamiltonian path. Each switch-box is an independent cryptographic cell that corresponds to a vertex of the graph. The switch-boxes are reconfigurable and receive reconfiguration information as messages flow through the Hamiltonian path during each session c.

---

[7] The structure implemented in silicon is surrounded by fill shapes used as a gaps filler, due to manufacturing constraints (polishing).

All boxes are clocked[8], and able to perform basic cryptographic operations. Six cell-level parameters are used to define each switch-box:

- A *coordinate identifier* i is a positive integer representing the ordinal number of the box's Cartesians coordinates: *i.e.* $i = x + ny + n^2 z$.
- A *session identifier* c is an integer representing the box's configuration: this value is incremented at each new reconfiguration session.
- A *key* k_i shared with the protected processor located inside the cage.
- A *routing configuration* $w_{i,c}$ chosen between the thirty possible routing positions of a 3D bi-directional switch (Fig. 15)[9].
- A *state variable* $s_{i,c}$ computed at each clock cycle from the incoming data $m_{i,c}$ (see hereafter) and the preceding state, $s_{i,c-1}$. The state $s_{i,c}$ is stored in the switch-box's internal memory[10].

$$\begin{cases} m_{i+1,c} = F(m_{i,c}, k_i, w_{i,c}, s_{i,c}) \\ s_{i,c+1} = G(m_{i,c}, k_i, w_{i,c}, s_{i,c}) \end{cases} \tag{1}$$

The output data $m_{i+1,c}$ is computed within box i using the input data $m_{i,c}$ and an integrated cryptographic function F, serving as a lightweight MAC. The final output $m_{n^3,c}$ attests the cage's integrity during session c.

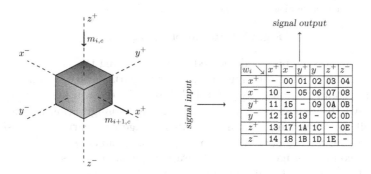

Fig. 15. Example of a 3D switch-box programmed with a routing configuration $w_i = $ 0x13

Each switch-box comprises five logic parts (Fig. 16) that serve to route the integrity attestation signal through the box's six IOs and successively MAC the input values $m_{i,c}$:

- Two multiplexers routing IOs, with three state output buffers to avoid short-circuits during re-configuration.
- A controller commanding the two multiplexers' configuration.

---

[8] We denote by t the clock counter.
[9] For switch-boxes depicted in red, blue and green (Fig. 2) the number of possible configurations drops to (respectively) 6, 12 and 20.
[10] Upon reset $s_{i,0} = 0$ for all i.

– A MAC cell for processing data and a register for storing results.
– A register storing the state variable $s_{i,c}$, the key k_i, the present configuration $w_{i,c}$, the next box configuration $w_{i,c+1}$ and the clock counter t.

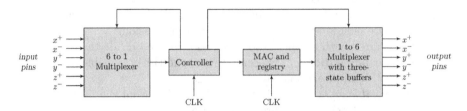

Fig. 16. Logic diagram of a 3D switch-box

The input message $m_{0,c}$, sent through the Hamiltonian path, is composed of two parts serving different goals (Fig. 17):

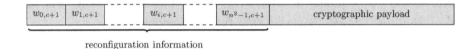

Fig. 17. Structure of message $m_{0,c}$

– The first message part is dedicated to reconfiguring the grid. For a cube of size n, the reconfiguration information has n^3 parts, each containing the next routing configuration $w_{i,c+1}$ of switch-box i. As the routing information of each switch-box can be coded on 5 bits, the reconfiguration information is initially $5n^3$ bits long[11]. Basically, this message part carries the position of all switches for the next Hamiltonian path of session $c + 1$.
– The second message part (cryptographic payload) is used to attest the circuit's integrity, the 64-bits payload will be successively MACed by all switch-boxes and eventually compared to a digest computed by the protected circuit. If possible, one should select a function F that simplifies after being composed with itself to reduce the protected circuit's computational burden.

5.2 Description of the Dynamic Grid and the Integrity Verification Scheme

Upon reset, each switch-box is in a default configuration $w_{i,0}$ corresponding to an initial predefined hardwired Hamiltonian path for session $c = 0$. The input and the output boxes (S_0 and S_{n^3-1}) are only partially reconfigurable; namely, the routing of S_0's input and the routing of S_{n^3-1}'s output cannot be changed. To clarify the reconfiguration dynamics, we denote by t the number of clock ticks elapsed since

---

[11] Note that the reconfiguration information part of the $m_{i,c}$'s gets shorter and shorter as i increases, $i.e.$ as the message approaches the last switch-box.

system reset assuming a one bit per clock tick throughput; given that 5 bits are dropped at each "station", a full reconfiguration route (session) claims

$$5 \sum_{j=0}^{n^3-1} (n^3 - j) = \frac{5}{2} n^3 (n^3 + 1)$$

clock ticks, which is the time needed for the reconfiguration information to flow through all n^3 switch-boxes *i.e.* the number of clock ticks elapsed between the entry of the first bit of $m_{0,c}$ into S_0 and the exit of the last bit of $m_{n^3,c}$ from S_{n^3-1}. Note that this figure does not account for the time necessary for payload transit[12].

At $t = 0$: A new session c starts and the first bit of $m_{0,c}$ is received by S_0 form the protected processor.

For $5 \sum_{j=0}^{i-1} (n^3-j) = \frac{5}{2}i(2n^3+1-i) \leq t \leq 5 \sum_{j=0}^{i} (n^3 - j) - 1 = \frac{5}{2}(i+1)(2n^3 - i) - 1$: All switch-boxes except S_{i-1} and S_i are inactive (dormant). S_{i-1} sends the message $m_{i-1,c}$ to S_i which performs the following operations:

- Store the reconfiguration information $w_{i,c+1}$, for the next Hamiltonian route of session $c + 1$.
- Compute $m_{i+1,c}$ and update $s_{i,c+1}$ as defined in formula (1).

At $t = 5 \sum_{j=0}^{n^3-1} (n^3 - j) = \frac{5}{2}n^3(n^3 + 1)$: The first bit of message $m_{n^3,c}$ emerges from the grid (from S_{n^3-1}) and all switch-boxes re-configure themselves to the new Hamiltonian path $c+1$. $m_{n^3,c}$ is received by the protected processor who compares it to a value computed by its own means. At the next clock tick a new message $m_{0,c+1}$ is sent in, and the process starts all over again for a new route representing session number $c + 1$.

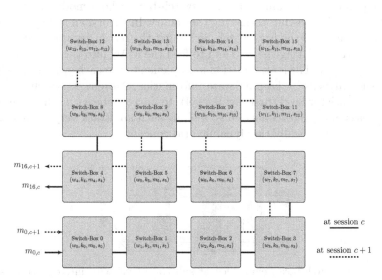

Fig. 18. 4×4 dynamic switch-box grid routed at c and $c + 1$ (illustration)

---

[12] $p(n^3 + 1)$ where p is the payload size in bits.

If one of the switch-boxes is compromised then the digest output by the path will be altered with high probability and the fault will be detected by the mirror verification routine implemented in the protected processor (Fig. 19). The device could then revert to a safe mode, and sanitize sensitive data.

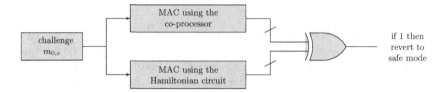

Fig. 19. Device integrity verification scheme

The verification circuit's size essentially depends on the MAC's size and complexity. Note that the XOR gate is a weak point: if it is bypassed the entire canary becomes pointless. Luckily, the XOR is spatially protected by the Hamiltonian path that surrounds it.

5.3   Vulnerability to Focused Ion Beam (FIB) Attacks

The proposed dynamic structure complies with the Read-Proof Hardware requirements described in [10]: the structure is easy to evaluate, relatively cheap (in some case no additional masks would be required) and can't be easily removed without damaging the chip.

Even though an attacker might modify some switch-box interconnections using FIB equipment, one cannot bypass a switch-box without modifying the digest computation logic and thus triggering the canary. In theory, an attacker may microprobe the input of the first switch-box to get the reconfiguration path, feed it into an FPGA simulating the grid and re-feed the MAC into the target, thus bypassing the canary. The state function s_i implemented in each switch-box should prevent such attacks by keeping state information. Moreover, switch-boxes are defined at transistor level (first metal level): to microprobe each cell the attacker has to bypass many interconnections, making such an attack very complex. Fig. 20 describes schematically the dynamic grid concept.

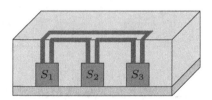

Fig. 20. Three switch-boxes embedded at substrate level with interconnections over the top layers

The successive grid configurations are precomputed by an external Hamiltonian path generator using the strategies described in Section 3. This configuration data should be stored in a non-volatile memory located under the cage.

6   Perspectives and Open Problems

Hardware canaries present an advantage with respect to analog integrity protection such as PUFs and sensors: being purely digital, hardware canaries can be integrated at the HDL-level design phase be portable across technologies. The proposed solution would, indeed, increase manufacturing and testing complexity but, being purely digital, would also increase reliability in unstable physical conditions, a common problem encountered when implementing analog sensors and PUFs.

The previous sections raise several sophistication ideas. For instance, instead of having the processor simply pick a reconfiguration route in a pre-stored table, the processor may *also re-write* the chosen route before configuring the canary with it. Devising more rewriting rules and developing lightweight heuristics to efficiently identify where to apply such rules is an interesting research direction.

Another interesting generalization is the interleaving in space of several disjoint Hamiltonian circuits. Interleaved canaries will force the attacker to overcome several spatial barriers. It is always possible to interleave a cube of size $n - 1$ in a cube of size n without having the two cubes intersect each other[13] as illustrated in Fig. 21.

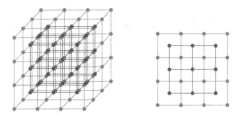

Fig. 21. A size 4 cube interleaved with a size 3 cube (3D and front view)

Fig. 22 shows the result of such a (laborious!) physical interleaving for a cube of size 4 and a cube of size 5. Note that interleaving remains compatible with a dynamic evolution of *both cubes* as canaries do not touch each other nor share any hardware (edges or vertices).

Finally, functions F for which the evaluation of $F(x) = (F_{n^3-1} \circ \ldots \circ F_0)(x)$ is faster than n^3 individual evaluations of F_i are desirable for efficiency reasons. XOR, bit permutation, addition, multiplication and exponentiation (*e.g.* modulo 251) all fall into this category[14]. Note that $F_i(x) = k_i \times x^{k_i'} \mod p$ works as well.

---

[13] Remove the (k, k, k) point from the center of the odd cube as explained before.

[14] Evidently, input should be nonzero for multiplication, nonzero and $\neq 1$ for exponentiation etc.

Fig. 22. Interleaving a Hamiltonian cube of size 4 and a Hamiltonian cube of size 5

In the first dynamic prototype the F_i's will be formed of XORs and bit permutations. Devising computational shortcuts taking into account an evolving internal state $s_{i,c}$ are also desirable.

Appendix

A   Circuit Folding

Fig. 23. 10×100 Hamiltonian rectangle $\mathcal{L}$ prepared to be folded

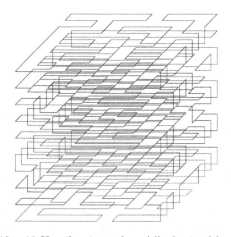

Fig. 24. $10 \times 10 \times 10$ Hamiltonian cube $\varphi(\mathcal{L})$ obtained by folding Fig. 23

B   Random Cube Association

Five elementary cubes in Fig. 26 are shown in red to underline that all cubes forming Fig. 26 are still disjoint.

Fig. 25. The six elementary Hamiltonian cubes of size 2

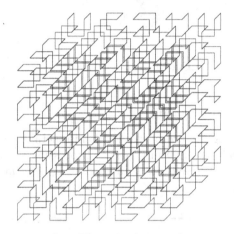

Fig. 26. Elementary 2×2 cubes filling the lattice of points forming a cube of size $n = 10$

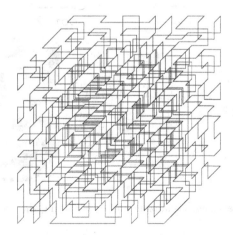

Fig. 27. An $n = 10$ Hamiltonian path obtained by randomly associating Fig. 26

C   Experimental Pre-silicon Models

Having obtained several construction plans, we decided to try and construct concrete examples using copper supplies before migrating to silicon. We used an industrial robot to cut 12mm⌀ copper segments of various sizes. A measurement of the dimensions of off-the-shelf right angle connectors (Fig. 28) revealed that if a 1-unit segment is h millimeters long, then an i-unit segment has to measure $(h + 16) \times i - 16$ millimeters.

Fig. 28. Angle connector

C.1   Visualizing and Layering

Layering and visualizing the prototypes (and chip metal layers) was done using an *ad-hoc* software suite written in C and in Processing[15]. The software allows decomposing a 3D structure into layers and rotating it for inspection.

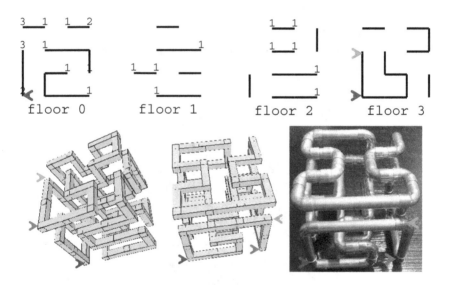

Fig. 29. Layering, visualizing and constructing the prototypes

---

[15] http://processing.org/

C.2   Assembly Options

Segments were assembled using several techniques ranging from soldering to super-glue. The disadvantage of welding was the risk of unsoldering an angle connector while soldering the nearby one (and this indeed happened at times). Super-glue happened to be less risky but called for dexterity as the glue would harden in a couple of seconds and thereby make any further correction impossible. All in all super-glue was preferred and allowed the generation of a variety of experimental pre-silicon cubes shown in Fig. 30. 3D printing using stereolitography or thermoplastic extrusion (fused deposition modeling) were considered as well.

Fig. 30. Experimental pre-silicon cubes

References

[1] Ababei, C., Feng, Y., Goplen, B., Mogal, H., Zhang, T., Bazargan, K., Sapatnekar, S.: Placement and Routing in 3D Integrated Circuits. IEEE Design and Test of Computers 22(6), 520–531 (2005)

[2] Alexander, A.J., Cohoon, J.P., Colflesh, J.L., Karro, J., Peters, E., Robins, G.: Placement and routing for three-dimensional FPGAs. In: Fourth Canadian Workshop on Field-Programmable Devices, pp. 11–18 (May 1996)

[3] Bollobás, B.: Graph Theory: An Introductory Course, p. 12. Springer, New York (1979)

[4] Dharwadker, A.: The Hamiltonian Circuit Algorithm. Proceedings of the Institute of Mathematics, 32 (2011)

[5] Dickau, R.: Hilbert and Moore 3D Fractal Curves. The Wolfram Demonstrations Project, http://demonstrations.wolfram.com/HilbertAndMoore3DFractalCurves

[6] Goossens, K., van Meerbergen, J., Peeters, A., Wielage, P.: Networks on Silicon: Combinig Best-Effort and Guaranteed Services. In: Proceedings of Design Automation and Test Conference (DATE), pp. 423–425 (2002)

[7] Kim, J., Verbauwhede, I., Chang, M.-C.F.: Design of an Interconnect Architecture and Signaling Technology for Parallelism in Communication. IEEE Trans. VLSI Syst. 15(8), 881–894 (2007)

[8] Moore, E.H.: On Certain Crinkly Curves. Trans. Amer. Math. Soc. 1, 72–90 (1900)

[9] Rijpkema, E., Goossens, K.G.W., Radulescu, A., Dielissen, J., van Meerbergen, J., Wielage, P., Waterlander, E.: Trade offs in the design of a router with both guaranteed and best-effort services for networks on chip. In: Proceedings of Design, Automation and Test Conference in Europe (DATE), pp. 350–355 (March 2003)

[10] Tuyls, P., Schrijen, G.-J., Škorić, B., van Geloven, J., Verhaegh, N., Wolters, R.: Read-Proof Hardware from Protective Coatings. In: Goubin, L., Matsui, M. (eds.) CHES 2006. LNCS, vol. 4249, pp. 369–383. Springer, Heidelberg (2006)

[11] Valamehr, J., Huffmire, T., Irvine, C., Kastner, R., Koç, Ç.K., Levin, T., Sherwood, T.: A Qualitative Security Analysis of a New Class of 3-D Integrated Crypto Co-processors. In: Naccache, D. (ed.) Cryphtography and Security: From Theory to Applications. LNCS, vol. 6805, pp. 364–382. Springer, Heidelberg (2012)

[12] Verbauwhede, I., Chang, M.-C.F.: Reconfigurable interconnect for next generation systems. In: Proceedings of the Fourth IEEE/ACM International Workshop on System-Level Interconnect Prediction (SLIP 2002), April 6-7, pp. 71–74 (2002)

Breakthrough Silicon Scanning Discovers Backdoor in Military Chip

Sergei Skorobogatov[1] and Christopher Woods[2]

[1] University of Cambridge, Computer Laboratory, Cambridge, UK
sps32@cam.ac.uk
[2] Quo Vadis Labs, London, UK
chris@quovadislabs.com

Abstract. This paper is a short summary of the first real world detection of a backdoor in a military grade FPGA. Using an innovative patented technique we were able to detect and analyse in the first documented case of its kind, a backdoor inserted into the Actel/Microsemi ProASIC3 chips for accessing FPGA configuration. The backdoor was found amongst additional JTAG functionality and exists on the silicon itself, it was not present in any firmware loaded onto the chip. Using Pipeline Emission Analysis (PEA), our pioneered technique, we were able to extract the secret key to activate the backdoor, as well as other security keys such as the AES and the Passkey. This way an attacker can extract all the configuration data from the chip, reprogram crypto and access keys, modify low-level silicon features, access unencrypted configuration bitstream or permanently damage the device. Clearly this means the device is wide open to intellectual property (IP) theft, fraud, re-programming as well as reverse engineering of the design which allows the introduction of a new backdoor or Trojan. Most concerning, it is not possible to patch the backdoor in chips already deployed, meaning those using this family of chips have to accept the fact they can be easily compromised or will have to be physically replaced after a redesign of the silicon itself.

Keywords: Hardware Assurance, silicon scanning, side-channel analysis, silicon Trojans and backdoors, PEA.

1 Introduction

With the globalisation of semiconductor manufacturing, integrated circuits become vulnerable to malevolent activities in the form of Trojan and backdoor insertion [1]. An adversary can introduce Trojans into the design during a stage of fabrication by modifying the mask at a foundry or fab. It can also be present inside third parties' modules or blocks used in the design. Backdoors could be implemented by malicious insiders at the design house. From the attacker's point of view there is not much difference between Trojans and backdoors, because in most cases the device would be analysed as a black box with very limited information provided by the manufacturer. Neither would it be known who inserted

E. Prouff and P. Schaumont (Eds.): CHES 2012, LNCS 7428, pp. 23–40, 2012.

the undocumented features of additional capabilities or at what stage of the process this occured.

In a search for the ideal target we decided to test the Actel/Microsemi ProA-SIC3 A3P250 device because of its high security specifications and wide use in military and industrial applications. According to the chip manufacturer: *"Low power flash devices are unique in being reprogrammable and having inherent resistance to both invasive and noninvasive attacks on valuable IP"* [2].

In this paper we demonstrate how a deliberately inserted backdoor and additional functionalities can be found in the 'highly secure' Actel/Microsemi ProA-SIC3 Flash FPGA (field-programmable gate array) chip used in both military and sensitive industrial applications. Actel, who developed ProASIC3 devices, market them as chips which *"offer one of the highest levels of design security in the industry"* [3]. These FPGAs are unique by being low-power, live on power-up and inherently secure as no configuration data is stored outside the device: *"In contrast to SRAM-based FPGAs, ProASIC3 configuration files cannot be read back via JTAG or any other method"* [4].

All our experiments were carried out on the Actel/Microsemi ProASIC3 A3P250 Flash FPGA device [2]. Fabricated with a 0.13 μm process with 7 metal layers, this chip incorporates 1,913,600 bits of bitstream configuration data. According to the manufacturer's documentation on this chip: *"Even without any security measures (such as FlashLock with AES), it is not possible to read back the programming data from a programmed device. Upon programming completion, the programming algorithm will reload the programming data into the device. The device will then use built-in circuitry to determine if it was programmed correctly"* [2]. Our research revealed that there is some hidden functionality inside the JTAG controller of this chip and one of the functions is the covert access to the configuration data. The JTAG controller itself is a part of the silicon design as in all FPGA chips and cannot be changed after the chip is manufactured.

To our knowledge, this is the first documented case of finding a backdoor inserted in a real world chip. When we talk about backdoors we treat them differently from other undocumented features most chip manufacturers insert in their devices for factory testing and debug purposes. The difference between a backdoor and an undocumented command is the ability to gain access to the programmed user IP when it is not supposed to be possible. Undocumented commands are known to exist in JTAG for failure analysis or for debugging but they are not designed to circumnavigate the security scheme of the device. The dictionary gives the following definition: *"backdoor – an undocumented way to get access to a computer system or the data it contains"* [5]. This is exactly what we found in the third generation of Actel/Microsemi Flash FPGA chips. The same approach can be used to find Trojans, altering the way the scanning is performed slightly.

Several Trojan detection approaches have been proposed in recent years. These can be divided into three major categories. One is full reverse engineering of the chip which gives an in-depth analysis of the chip [6]. However, this has some drawbacks – it is an extremely expensive and time consuming operation, and it will not work for cases where the Trojan is present only in a small fraction of chips. The second category is an attempt to activate the Trojan by applying test vectors and comparing the responses with expected responses [7,8,9]. This might not work in situations where the Trojan is activated under rare conditions. For modern complex circuits it is close to impossible to verify all states. In addition, this approach will not detect Trojans designed to leak the information rather than take control of the hardware [10]. The final category uses side-channel analysis to detect Trojans by measuring circuit parameters such as power consumption, electro-magnetic emissions and timing analysis. These methods can be used against golden samples [11,12] or within the same integrated circuit (IC) to minimise the variations between samples [13]. However, the effectiveness of side-channel analysis methods greatly depends on the sensitivity of the measuring equipment [14].

One of the most widely used approaches in Trojan and backdoor detection is to employ differential power analysis (DPA) techniques [15] to detect any abnormalities in the device operation. However, due to the latency introduced by the setup and the substantial noise of the acquisition equipment, it normally takes a very long time to scan silicon chips. With modern devices such as FPGAs, it could be infeasible to detect any Trojans or backdoors with DPA techniques. We used a new sensing technique which detects tiny variations in the device operation and is thus able to detect small variations which are well below the noise level in a standard DPA setup.

If a bug is found in firmware programmed into an FPGA then it can be rectified by a firmware update. However, if the Trojan or backdoor is present in the silicon itself, then there is no way to remove the bugs other than replacing all the affected silicon chips, as has happened several times with bugs found in Intel CPUs. The cost of such an operation is enormous and can seriously affect an organisation's revenue.

If a potential attacker takes control of the FPGA device, he can cause a lot of damage to the device. For example, he can erase or even physically destroy the FPGA by uploading a malicious bitstream that will cause a high current to pass through the device and burn it out. Using the backdoor, an attacker can extract the IP from the device and make some changes to the firmware, inserting new Trojans into its configuration. That would provide a wide range of capabilities in carrying out more sophisticated attacks at a later stage.

Greater danger could come from a new and disturbing possibility of a large scale remote attack via a network or the Internet on the silicon itself. If the key is known, commands can be embedded into a worm to scan for JTAG, then to attack and re-program the firmware remotely. This cannot be excluded since the manufacturer specifically designed such remote access feature for ProASIC3 and other Flash FPGA devices: *"Military ProASIC3/EL devices with AES-based security allow for secure, remote field updates over public networks such as the Internet, and ensure that valuable IP remains out of the hands of system overbuilders, system cloners, and IP thieves. The contents of a programmed device cannot be read back, although secure design verification is possible"* [16].

This paper is organised as follows. Section 2 gives a brief introduction into chip access and scanning approaches, and FPGA security. Section 3 introduces the experimental setup, while Section 4 sets out our results. Section 5 discusses limitations and possible improvements. The impact of the research is discussed in the concluding section.

2   Background

With the growing complexity of integrated circuits the importance of post production testing and functional verification is growing. This is necessary to address the issues in failure analysis and to perform design verification for correctness, and to eliminate inevitable bugs [17]. The majority of chip manufacturers use the JTAG (Joint Test Action Group) interface as a standard port for IC testing [18]. However, until recently it was primarily used for boundary scan testing rather than internal IC testing. In the early 2000s the JTAG specification was expanded with programming abilities and security features to meet the FPGA market demands [19]. However, even before then chip manufacturers were using the expanded JTAG usually referred as IEEE 1149.x. This expansion was not standardised and for most chips was kept confidential. In that respect, the knowledge of the test interface being a JTAG did not give any advantage to the outsider over a proprietary test interface. However, this allowed chip manufacturers to use standard JTAG implementation libraries without compromising on the security of their chips. It was important for manufacturers to use undocumented or disguised commands for granting access to the JTAG or test interface, because in some chips it provided access to the internal memory, usually holding the end user IP and secret data [20].

The JTAG interface is operated via test access port (TAP) pins which control the state machine (Fig. 1a). It has two registers – IR (instruction register) and DR (data register) into which the serial data can be shifted and then executed. The IR registers must be selected first and then, depending on the command, DR data shifted in. The length of the IR register varies from chip to chip and normally lies between 4 and 32 bits. Some commands do not involve the DR register, for others its length could be many thousands of bits.

For many chips, and especially for secure microcontrollers and secure FPGAs, the commands and data fields of JTAG registers are not documented. However, an inquisitive attacker can gain most of this information from development kits supplied by a particular device's chip manufacturer. Even when the availability of such kits is restricted by the manufacturer, their clones can be found in the third world. For FPGA chips the task of gathering more information about JTAG commands was simplified by the introduction of a special high level test language called STAPL (Standard Test and Programming Language) [21]. All the commands and data fields in the programming file compiled by design tools are easily identifiable with both subroutines and meaningfully named variables. An example of security state verification procedure is given in the Appendix.

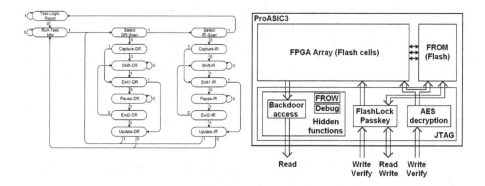

Fig. 1. (a)JTAG TAP state machine, (b)Simplified ProASIC3 security

Knowing all the JTAG commands is not sufficient to search for backdoors. Firstly, the obtained list could be incomplete because the STAPL file is compiled only with commands which serve a particular task. Secondly, although subroutines, functions and variables are meaningfully named, the IR level commands are not explained and usually remain as numbers. That complicates the reverse engineering of the JTAG functionality. What adds to the complexity is the sequence of commands. For complex devices it will not be just one command executed for a particular function, but a series of commands mixed with data. Each command could be not solely IR or IR+DR, but an endless list of possible combinations such as IR+IR, IR+DR+DR, IR+DR+IR+DR and so forth.

Searching for Trojans could represent an easier task, because in that case the design is known as well as its likely implementation in silicon. This operation is usually performed by the chip manufacturer or its subcontractors. However, from an attacker's point of view, there is not much difference between Trojans and backdoors as he is looking for any potential vulnerability within the silicon chip.

As a target for our experiments we chose the Actel/Microsemi ProASIC3 A3P250 device [4] for many reasons. Firstly, it has high security specifications and is positioned as the device with highest security protection in the industry. Actel who developed ProASIC3 chips market them as devices which *"provide the most impenetrable security for programmable logic designs"* [16,22]. Secondly, ProASIC3 chips are widely used in military and industrial applications especially in critical systems. Therefore, without doubt, ProASIC3 devices posed suitable challenges for this research. Any outcome occurring from analysing this device will have a greater impact and will be more useful compared to the results obtained from low-end security chips such as normal microcontrollers or standard FPGAs.

ProASIC3 devices have several levels of security protection. When the new part is shipped to the customer there is no security activated. Even at that level the security protection is high. This is because *"Even without any security measures (such as FlashLock with AES), it is not possible to read back the pro-gramming data from a programmed device"* [2]. The manufacturer claims that the readback function for the FPGA Array is not physically implemented thus making these devices inherently secure: *"Low power flash devices do not support read-back of FPGA core programmed data; however, the FlashROM contents can selectively be read back (or disabled) via the JTAG port based on the security set-tings established by the Microsemi Designer software"* [2]. Higher security level offers activation of a special user key that protects rewriting of any security set-tings: *"Designers have the ability to use a FlashLock Pass Key to prohibit any write or verification operations on the device"* [4].

For remote updates of the device both the configuration bitstream and internal Flash memory can be encrypted with the AES device master key. This function can only be used to decrypt the data being sent to the chip for writing and verification. There is no way to pass the internal data back to the outside world even in an encrypted state.

The highest level of protection turns the device into no-longer-programmable chip and should be considered with caution as in case of any bug found, the chip will have to be physically replaced: *"The purpose of the permanent lock feature is to provide the benefits of the highest level of security to IGLOO and ProASIC3 devices. If selected, the permanent FlashLock feature will create a permanent barrier, preventing any access to the contents of the device. This is achieved by permanently disabling Write and Verify access to the array, and Write and Read access to the FlashROM. After permanently locking the device, it has been effectively rendered one-time-programmable"* [2].

The backdoor we found allows readback access to the configuration data. There are some other hidden JTAG functions which give low-level control over the internal shadow memories and allow modification of hidden registers. How-ever, we do not consider them a backdoor because they are not directly associated with undocumented access to the protected data. The simplified outlook of the ProASIC3 security is presented in Fig. 1b.

We evaluated all levels of protection in ProASIC3 devices and were able to circumvent the security at each level. Table 1 summarises the security protection levels in the ProASIC3 devices according to our research findings. The Passkey offers the highest level of security for reprogrammable chip, while the Permanent Lock should be used as the last resort and will turn the device into a one-time programmable (OTP) chip. However, despite it being a seemingly ultimate protection mechanism, the Permanent lock has some physical security flaws. We found it vulnerable to some fault injection attacks, but as this does not fall within the scope of this paper, we have not gone into further details.

Although with the backdoor we found it is possible to extract the IP blocks – Array configuration from the FPGA, there are other ways to extract the config-uration bitstream IP. One was published in 2010 and uses special type of optical fault injection attacks called 'bumping attacks' [23]. Another method uses the vulnerability of the AES implementation and in particular the message authenti-cation code (MAC) used to protect the integrity of the encrypted bitstream [24].

Table 1. Security protection levels in ProASIC3

Secure area	Read Access	Verify Access	Write Access	Secure Lock	AES Encryption	Expected Security	Attack Time
FROM (Flash)	Yes	Yes	Yes	Yes	Yes	Medium	Seconds
FPGA Array	No	Yes	Yes	Yes	Yes	High	Days
AES Key	No	Yes	Yes	Yes	No	Medium	Seconds
FlashLock Passkey	No	Yes	Yes	Yes	No	Very High	Hours
Backdoor Key	No	Yes	Yes	Yes	No	Very High	Hours
Permanent Lock	No	No	Yes	No	No	Ultra High	Minutes

By compromising the AES key in ProASIC3 the IP could be extracted even without access to the encrypted bitstream. Although there is no readback access for AES-encrypted bitstream configuration, verification is allowed and can be brute forced given small number of unknown bits. An attacker can pass authentication, then write a mask configuration file containing all zeros but a small number of ones per each row, e.g. 16 bits in a 832-bit row. Writing '1' over '0' into the Flash configuration memory has no effect, while writing '0' over '1' changes its state. Since each row can be verified independently in 2 ms time, he can brute force unknown bits row by row. With 50 samples we successfully extracted full IP from A3P250 in 1 week. There is a MAC security feature to prevent arbitrary writing in AES mode through validation of data. We broke it by figuring out that it uses feedback-shift register with just 4 bits of uncertainty per AES CBC (cipher-block chaining) block and easily bruteforceable off-line. Moreover, we managed to disable the MAC verification by modifying few lines in the controlling STAPL file, thus making any arbitrary writing seamless.

3   Experimental Method

Initially, we analysed the chip with standard design tools from Actel – Libero IDE and FlashPro. The sample of A3P250 device was connected to a standard Actel FlashPro3 programmer. All of the JTAG operations are undocumented for the ProASIC3, however, using Actel development software we were able to generate a series of STAPL files which we analysed to determine the commands used for different operations. Once we learned the JTAG communication we moved onto exploring the field of undocumented features. For that we built a special test board with a master JTAG interface and simple functions controlled by PC software via an RS-232 interface for convenience (Fig. 2a). The ProASIC3 chip was placed into a ZIF socket for easier handling. During this stage we gathered information about the command field and data registers.

The next step was to determine which commands have data fields and measure the size of the DR registers. We then used a classic DPA setup to analyse the side-channel emission from the ProASIC3 devices during decryption and to access operations as well as other undocumented commands. We constructed a simple

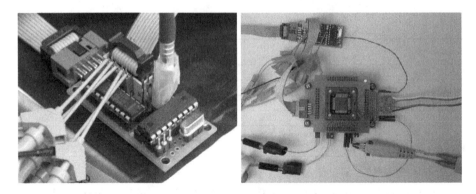

Fig. 2. Test setup: (a)control board, (b)DPA analysis

prototype board with a ZIF socket for the A3P250 device (Fig. 2b) and connected it to our test board which was providing some additional triggering functions for the oscilloscope. The power consumption was measured via a 20 Ω resistor in the V_{CC} core supply line with the Agilent 1130A differential probe and acquired with the Agilent MSO8104A digital storage oscilloscope. Then the waveforms were analysed using MatLab software with our own proprietary program code.

We tried all available JTAG command fields in different combinations and observed all the traces scanned with DPA. In this way we were able to separate commands with different functions. The unknown commands were then tested with different data fields, while we observed the response and tried to understand their function. DPA is a good approach to find normal commands; however it can hardly help in understanding their functionality because of high noise and the number of traces required.

In the next set of experiments we used PEA technology (described in our paper [24]) to achieve an improved signal-to-noise ratio (SNR) in an attempt to better understand the functionality of each unknown command. Some operations were found to have robust silicon level DPA countermeasures. For example, the Passkey is documented as another layer of security protection on top of the AES encryption in the ProASIC3 to prevent IP cloning. Some DPA countermeasures found in the Passkey protection include very good compensation of any EM leakage and broadband spectrum spreading of side-channel emissions for the bit comparison leakage; internal unstable clock; high noise resulting in SNR below −20 dB. The prototype sensor setup we used is presented in Fig. 3a.

The system consists of a control interface that can be represented by a personal computer, a remote control with embedded processor or other human interface (Fig. 3b). The test algorithm is either present inside the test generator or it is supplied via the control interface. Each device under test (DUT) requires its own test algorithm, which is a part of a standard device operation and consists of a list of commands to run the DUT in the way required by the tester; for example, to establish an authentication or to decrypt the data. The test signal generator

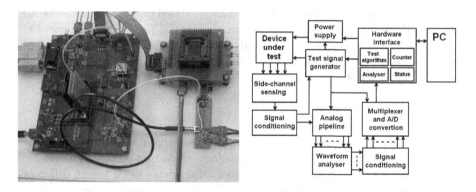

Fig. 3. (a)Prototype board with our sensor, (b)Block diagram of PEA setup

produces sets of test patterns according to the programmed algorithm. One part of the algorithm is fixed while the other is changing. The power supply of the DUT is provided by programmable power supply. The clock of the AC source can be synchronised to the external clock provided by the signal generator, which in turn can be synchronised to the device's internal clock. This could be done by injecting the clock signal from the generator into the DUT power supply line. That allows a significant improvement over the existing measurement equipment setup by significantly reducing the jitter influence on the measurement results.

As the DUT performs a requested operation it leaks some information via side channels as a side effect of the device operation. Those side-channel responses are measured with dedicated sensors specific for each type of side-channel emission, in our case the resistor in the core supply line. The sensors output the signals in analog form which are then put through a signal conditioning circuit to amplify the signal and reduce the noise by applying various filters. The signal is then delayed by one clock period, determined by the test signal generator. The purpose of the delay is to be able to compare the device side-channel response to different input test data. The pipeline delivers its delayed output to a waveform analyser which compares the new signal with the delayed signal for the determined number of points and provides an output, which is the difference between them. The signal from the analyser is conditioned using amplifiers and filters to meet the requirement of the acquisition system, which then converts it in a multiplexed way into a digital form. The output from the multiplexer is then transferred to the hardware interface. The response analyser makes the decision on the reply based on the predetermined decision making patterns and updates the status register, which is checked by the control system. Our invention of this new analysis technique is covered by a patent which is available to the public [25]. Our improvement comes from: real-time attack with no latency associated with an oscilloscope hardware/software, network and memory; lower noise with better probe design, analog signal processing and efficient filtering.

In the end we used a silicon scanning technique based on PEA combined with a classic DPA setup (resistor in power line, differential probe, oscilloscope, PC with MatLab). Communication was analysed with DPA, while the functions of control registers were tested with both DPA and PEA. This is because for some operations DPA is not sensitive enough.

4   Results

Scanning the JTAG command field for any unknown commands by checking the length of the associated DR register revealed an interesting picture. There were plenty of commands for which the associated DR register has a length different from one, hence, used by the JTAG engine. Fig. 4a shows some of these registers learned from the STAPL file analysis as well as newly discovered ones. Not only that, but some registers were impossible to update with a new data (Fig. 4b). Most of these registers were representing a Read-Only memory, referred to as FROW in the STAPL file. Only one row was actually read, but three address bits allowed eight rows to be accessed. All those hidden and non-updatable registers were found to be imprinted into certain locations in FROW memory. However, every single ProASIC3 chip has unique values stored in FROW and, hence, in hidden registers suggesting that this memory was initialised at a factory and then locked against overwriting. Now we knew for sure that there is some hidden functionality in the ProASIC3 chips.

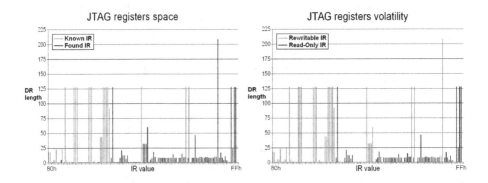

Fig. 4. JTAG scanning results: (a)hidden DR registers, (b)non-volatile DR registers

Although they do not have any specialised DPA countermeasures, the ProA-SIC3 devices are at least 100 times longer to attack using DPA than non-protected conventional microcontrollers such as PIC, AVR, MC68HC, MSP430 etc. The robust hardware design features are complemented with the total lack of information about JTAG engine operation, hardware implementation and commands. That makes any attacks on the ProASIC3 chips quite a challenging task. Fig. 5a shows the result obtained by comparing single traces for different input data. Averaging over 4,096 traces gives a low-noise result, but takes a couple of minutes to acquire (Fig. 5b). As can be seen, for single traces the noise overshadows any useful signal with SNR being below −20 dB. The FFT spectrum of the power trace does not have any characteristic peaks (Fig. 6a) and filtering will not be very effective for substantially improving DPA results.

The noise can be reduced by using a frequency locking technique. There are publications on the successful use of these techniques on FPGAs [26]. That way the timing jitter between traces can be reduced to approximately ten degrees of the phase shift at 19.7 MHz. However, on the other hand this injects a strong carrier frequency which needs to be filtered out to avoid any influence on the power analysis results. Despite good synchronisation and triggering results we did not observe any improvements compared to the standard DPA setup because of a very strong presence of a 19.7 MHz signal in the power trace which we were unable to eliminate of completely.

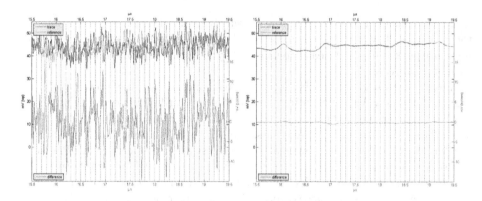

Fig. 5. Power analysis results: (a)single trace difference, (b)average of 4,096 traces

Various DPA techniques were attempted to extract the Passkey, however, we were unable to get even a single bit in two weeks time using our off-the-shelf DPA equipment (oscilloscope with differential probe and PC with MatLab). The traces that appeared using DPA accounted for many functions including memory access, AES, Passkey and other, yet to be learned, functions. Even for an unprotected implementation of AES encryption it would require at least 256 traces to be averaged to reduce the noise and get a reliable correlation with key bits (Fig. 5). The PEA approach allowed the key bits to be guessed at in real time and with a very good correlation with the key bits. The outstanding sensitivity of the PEA is owed to many factors. One of which is the bandwidth of the analysed signal, which for DPA, stands at 200 MHz while in PEA at only 20 kHz. This not only results in much lower electronic noise, which is proportional to the square root of the bandwidth, but the cost of the acquisition hardware becomes several orders of magnitude lower. This also impacts on the latency thus allowing real-time analysis, because the signal produced for the analysis has almost 100% correlation with the key bits guesses

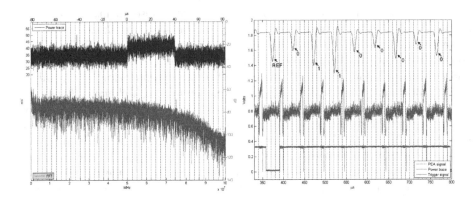

Fig. 6. Power analysis of ProASIC3: (a)FFT spectrum, (b)PEA scan for AES key

(Fig. 6b). This makes extraction time extremely fast. All that needs to be done in the end for the key extraction is to demodulate the signal and compare it with the reference peak. This can be easily performed by a simple one-dollar microcontroller with on-chip ADC.

Initially we analysed all the active JTAG commands using power analysis. Fig. 7a shows how AES authentication and Passkey verification traces look, while Fig. 7b shows traces of Array verification and Flash FROM reading commands. With the analysis of JTAG commands, one particular function was requesting a 128-bit key with the similar low-leakage DPA resistance property as the Passkey. It also had robust countermeasures that proved to be DPA resistant. In addition to an unstable internal clock and high noise from other parts of the circuit, the Passkey and backdoor access verification had their side-channel leakage considerably reduced compared to AES operation. This was likely to be achieved through using a well compensated silicon design together with ultra-low-power transistors instead of standard CMOS library components. In addition, the useful leakage signal has a spread spectrum with no characteristic peaks in frequency domain, thus making narrow band filtering useless. We used the similar PEA approach to extract both the Passkey and the backdoor key by looking for any notable changes in the response from our sensor for correct and incorrect guesses. However, due to much more robust DPA countermeasures it took us approximately one day to achieve this using simple PEA hardware. For the classic DPA setup, in order to achieve at least 0.1 mV difference detectable by an oscilloscope, at least 32 consecutive bits of the key must match. Given the input noise of probe+oscilloscope of 1mV at least 64 synchronous or 1024 asynchronous averages must be performed. It takes about 15 seconds to average the signal on MSO8104A to get a positive SNR. Finding all the unknown bits of the key with DPA would take 2^{32} times longer or approximately 2,000 years. Further investigation of the backdoor key operation revealed that it unlocks many of the undocumented functions, including reprogramming of secure memory areas and IP access.

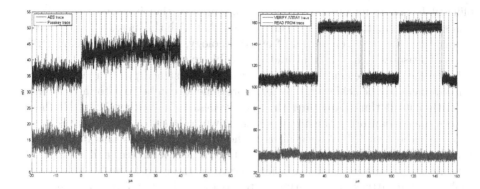

Fig. 7. Examples of JTAG power traces: (a)AES vs Passkey, (b)Array vs FROM

At this point we went back to those JTAG registers which were non-updatable as well as FROW to check whether we could change their values. Once the backdoor feature was unlocked, many of these registers became volatile and the FROW was reprogrammable as a normal Flash memory. Actel has a strong claim that *"configuration files cannot be read back via JTAG or any other method"* in the ProASIC3 and in their other latest generation Flash FPGAs [4]. Hence, they claim, they are extremely secure because the readback access is not implemented. We discovered that in fact Actel did implement such an access, with a special key used for activation.

Alongside this backdoor there is another layer of security in the guise of data permutation to obscure information and make IP extraction less feasible. This can also be dealt with using a simple brute force attack, because permutation functions do not withstand differential cryptanalysis as every single-bit change at the input results in a single-bit change at the output. Our experiments showed how some information can be found via systematic testing of device operations. Through this method, for example, we found the correspondence between bits in the 832-bit verification data and bits in the data bus.

5   Implications and Further Improvements

Many countermeasures are designed to defeat high end oscilloscopes and their known noise, latency and signal issues. These countermeasures prevent themselves from being broken in an affordable time through suppressing the signal or by bringing it to a higher noise level. Our approach through the use of bespoke hardware and the removal of the oscilloscope from the testing process, is designed to have the sensitivity to detect even the smallest variation in signal, which then allows more detailed analysis. The setup with which we achieved these eye-opening results is in its most basic form, employing a single pipeline (one channel).

Having taken this technology to proof of concept, we would like to develop it by building a multi-pipeline system consisting of 100 channels as well as new, more efficient hardware for our probes, with the aim of further improving sensitivity and speed by a factor of 10. We firmly believe that with the increase in capability planned for the next generation of our technology, defeating more secure DPA countermeasures is a very real and achievable expectation. Using a low-noise side-channel measurement setup with a carefully designed probe a 10× further improvement can be achieved. Further improvements can be done to the scanning algorithm itself thus improving the effectiveness by a further 10×. All these improvements can bring the analysis time down to hours or even minutes.

We noticed that FPGA security relies heavily on obscurity. This ranges from the lack of any documentation on the JTAG access interface, and absence of information on the internal operations, down to the data formats. This works well unless an attacker is determined to discover all this information on their own. Alternatively, more information can be gained through the analysis of the development tools and programming files for some chips. That certainly raises a concern about the amount of information a potential attacker can gain through development kits.

Some DPA and design cloning countermeasures might be ineffective in light of efficient silicon scanning techniques. For example, Intrinsic ID offers a software level solution for secure storage of crypto keys [27]. However, for an attacker who has full access to the chip through a backdoor and is capable of extracting the bitstream, localising and defeating the protection mechanism will be trivial. He will still have to understand the proprietary bitstream encoding, however, this can be achieved in several ways from reverse engineering the development software, through active attacks on chips, to reverse engineering the FPGA chip itself. Therefore, solutions with silicon-level fingerprinting using physical unclonable functions (PUF) will be ineffective in the presence of backdoors.

One could possibly argue that the backdoor we discovered is a bug or something overlooked by the developers. However, this is not the case as we performed intensive investigation into this problem and found proof that the backdoor was deliberately inserted and even used as a part of the overall security scheme. The backdoor feature was designed as a part of the JTAG security protection mechanism and traces can be found in the Actel's Libero FPGA design software. Anyone with this free software installed on their Microsoft Windows machine can go to the Search option in the Start menu and search for one of the fuse names taken from Actel generated STAPL file. For example, search for the word ULUWE in all files. This will return all STAPL files together with templates and algorithm description files. Inside some of those files there is a proof of the designed backdoor feature.

At the same time, other hidden features could be used to assist even more deeper recovery of the information from the chip. For example, if someone overproduced pre-programmed ProASIC3 chips for some important design and then decided on using another product; they can erase the chips and sell them on the grey market. No one would expect any security flaw because the chips are erased and no longer hold any useful data. This is not quite true and in fact there are still some tiny traces of the information left deep inside the Flash cells [28].

With the help of some hidden registers we found during the JTAG scan we were able to adjust the reference voltage of the read-sense amplifiers used by the backdoor and successfully extracted the bitstream configuration from the erased ProASIC3 chip. If you use ProASIC3 in a sensitive military system that had a destructive wipe as a security feature in case of the systems capture, erasing the chip would not be sufficient to prevent the readback of the user IP. This was a good example of combining the backdoor with hidden and undocumented JTAG debug functions.

6   Conclusion

Our experiments had achieved the affordable time for scanning of two weeks. As a result we were able to locate and exploit an undocumented backdoor in the Actel/Microsemi ProASIC3 chip positioned as the industry's highest security device. To our knowledge this is the first documented case of a backdoor inserted in a real world device with critical applications. Not only can a poorly protected AES key be extracted from the ProASIC3 chips in no time and with minimal effort, but the Passkey which was believed to be unbreakable and which was robust against DPA attacks can also be extracted.

The discovery of a backdoor in a military grade chip raises some serious questions about hardware assurance in the semiconductor industry. When you use and buy an embedded system or computer it is assumed, wrongly in our opinion, that the hardware is completely devoid of any vulnerabilities. We investigated the ProASIC3 backdoor problem through Internet searches, software and hardware analysis and found that this particular backdoor is not a result of any mistake or an innocent bug, but is instead a deliberately inserted and well thought-through backdoor that is crafted into, and part of, the ProASIC3 security system. We analysed other Actel/Microsemi products and found they all have the same backdoor. Those products include, but are not limited to: Igloo, Fusion and SmartFusion. The ProASIC3 is heavily marketed to the military and industry and resides in some very sensitive and critical products. From Google searches alone we have found that the ProASIC3 is used in military products such as weapons, guidance, flight control, networking and communications. In industry it is used in nuclear power plants, power distribution, aerospace, aviation, public transport and automotive products. This permits a new and disturbing possibility of a large scale Stuxnet-type attack via a network or the Internet on the silicon itself. If the key is known, commands can be embedded into a worm to scan for JTAG, then to attack and reprogram the firmware remotely. The backdoor is close to impossible to fix on chips already deployed because, unlike software bugs in a PC Operating System, you cannot issue a patch to fix this. Instead one has to replace all the hardware which could be extremely expensive. It may simply be a matter of time before this backdoor opportunity, which has the potential to impact on many critical systems, is exploited.

The chip manufacturer suggests the possibility of performing remote upgrade of the firmware in ProASIC3 devices via TCP/IP: *"The system containing the low power flash device can be assigned an IP address when deployed in the field.*

*When the device requires an update (core or FlashROM), the programming in-
structions along with the new programming data (AES-encrypted cipher text) can
be sent over the Internet to the target system via the TCP/IP protocol. Once the
MCU receives the instruction and data, it can proceed with the FPGA update.
Low power flash devices support Message Authentication Code (MAC), which
can be used to validate data for the target device"* [2]. However, as we have already mentioned, the implementation of the MAC is very insecure and trivial to
break. That only increases concerns about the remote updates and taking over
the control of systems based on the ProASIC3 and other Actel/Microsemi Flash
FPGAs.

Having a security related backdoor on a silicon chip jeopardises any efforts of
adding software level protection. This is because an attacker can use the underlying hardware to circumvent the software countermeasures. Using PUFs is not
likely to offer much help as the firmware that calculates them could be extracted
and then reverse engineered to defeat the protection layer. Using encryption as
an additional protection layer does not always help. Moreover, it could make
things worse, as in the ProASIC3, where the AES key can be extracted in less
than a second's time [24] compared to hours required for Passkey extraction.

Most of the current DPA-based research into silicon chips is centred on looking
for Trojans. There has been little research conducted into comparing legitimate
chips with counterfeits, using DPA. This is primarily because standard DPA
equipment generates many terabytes of data which requires complex analysis,
resulting in a long lead time and a high cost. As PEA offers superior sensitivity
and high speed analysis, it would be possible to check all chips to be used in
assembly, not just a sample batch. Currently this is not possible with any other
comparison methods whether by DPA or reverse engineering.

A debug port, factory test interface or JTAG can all potentially be used as
points to scan the silicon chip for backdoors or Trojans. Most chips manufactured
these days have at least one of these features present. Until the development of
the efficient silicon scanning techniques, it has been unfeasible to test real silicon
chips for Trojans or backdoors. Using a low-cost system it becomes possible to
independently test silicon for backdoors and Trojans in a matter of weeks. It
would take many years to perform the same task using standard DPA. Most
silicon chips are now designed and made abroad by third parties. Is there any
independent way to evaluate these products that are used in critical systems?

References

1. Tehranipoor, M., Koushanfar, F.: A survey of hardware Trojan taxonomy and
   detection. IEEE Design and Test of Computers (2010)
2. Military ProASIC3/EL FPGA Fabric User's Guide. Microsemi (2011),
   http://www.actel.com/documents/Mil_PA3_EL_UG.pdf
3. Design Security in Nonvolatile Flash and Antifuse FPGAs, Security Backgrounder,
   http://www.actel.com/documents/DesignSecurity_WP.pdf
4. Actel ProASIC3/E Production FPGAs, Features and Advantages (2007),
   http://www.actel.com/documents/PA3_E_Tech_WP.pdf

5. The Free Dictionary. Backdoor, http://www.thefreedictionary.com/backdoor
6. Torrance, R., James, D.: The State-of-the-Art in IC Reverse Engineering. In: Clavier, C., Gaj, K. (eds.) CHES 2009. LNCS, vol. 5747, pp. 363–381. Springer, Heidelberg (2009)
7. Jha, S., Jha, S.K.: Randomization Based Probabilistic Approach to Detect Trojan Circuits. In: Proc. 11th IEEE High Assurance System Engineering Symp., pp. 117–124 (2008)
8. Banga, M., Hsiao, M.: A Region based Approach for the Identification of Hardware Trojans. In: IEEE Int. Workshop on Hardware-Oriented Security and Trust, HOST, pp. 40–47 (2008)
9. Wolff, F., Papachristou, C., Bhunia, S., Chakraborty, R.S.: Towards Trojan-free Trusted ICs: Problem Analysis and Detection Scheme. In: Design, Automation and Test in Europe, DATE 2008, March 10-14, pp. 1362–1365 (2008)
10. Wang, X., Tehranipoor, M., Plusquellic, J.: Detecting Malicious Inclusions in Secure Hareware: Challenges and Solutions. In: IEEE Int. Hardware-Oriented Security and Trust, HOST (2008)
11. Agrawal, D., Baktir, S., Karakoyunlu, D., Rohatgi, P., Sunar, B.: Trojan Detection using IC Fingerprinting. In: IEEE Symp. on Security and Privacy, SP, pp. 296–310 (2007)
12. Jin, Y., Makris, Y.: Hardware Trojan Detection using Path Delay Fingerprint. In: IEEE Int. Workshop on Hardware-Oriented Security and Trust, HOST (2008)
13. Du, D., Narasimhan, S., Chakraborty, R.S., Bhunia, S.: Self-referencing: A Scalable Side-Channel Approach for Hardware Trojan Detection. In: Mangard, S., Standaert, F.-X. (eds.) CHES 2010. LNCS, vol. 6225, pp. 173–187. Springer, Heidelberg (2010)
14. Rad, R., Tehranipoor, M., Plusquellic, J.: A Sensitivity Analysis of Power Signal Methods for Detecting Hardware Trojans under Real Process and Environmental Conditions. IEEE. Trans. in VLSI 18, 1735–1744 (2009)
15. Kocher, P., Jaffe, J., Jun, B.: Differential Power Analysis. In: Wiener, M. (ed.) CRYPTO 1999. LNCS, vol. 1666, pp. 388–397. Springer, Heidelberg (1999)
16. Military ProASIC3/EL Low Power Flash FPGAs Datasheet. Microsemi (2012), http://www.actel.com/documents/Mil_PA3_EL_DS.pdf
17. Tehranipoor, M., Wang, C.: Introduction to Hardware Security and Trust. Springer (2011)
18. JTAG Boundary scan. IEEE Std 1149.1-2001
19. JTAG Programming specification. IEEE 1532-2002
20. Da Rolt, J., Di Natale, G., Flottes, M.-L., Rouzeyre, B.: New security threats against chips containing scan chain structures. In: IEEE Int. Workshop on Hardware-Oriented Security and Trust, HOST, pp. 110–115 (2011)
21. Actel, ISP and STAPL, Application Note AC171, http://www.actel.com/documents/ISP_STAPL_AN.pdf
22. ProASIC3 Frequently Asked Questions, Actel Corporation, Mountain View, CA 94043-4655 USA, http://www.actel.com/documents/pa3_faq.html
23. Skorobogatov, S.: Flash Memory 'Bumping" Attacks. In: Mangard, S., Standaert, F.-X. (eds.) CHES 2010. LNCS, vol. 6225, pp. 158–172. Springer, Heidelberg (2010)
24. Skorobogatov, S., Woods, C.: In the blink of an eye: There goes your AES key. IACR Cryptology ePrint Archive, Report 2012/296 (2012), http://eprint.iacr.org/2012/296
25. Integrated Circuit Investigation Method and Apparatus. Patent number WO2012/046029 A1

26. Skorobogatov, S.: Synchronization method for SCA and fault attacks. Journal of Cryptographic Engineering (JCEN) 1(1), 71–77 (2011)
27. Intrinsic ID, Quiddikey on ProASIC3 FPGAs, http://www.intrinsic-id.com/quiddikey_on_Actel_FPGA.html
28. Skorobogatov, S.: Data Remanence in Flash Memory Devices. In: Rao, J.R., Sunar, B. (eds.) CHES 2005. LNCS, vol. 3659, pp. 339–353. Springer, Heidelberg (2005)

Appendix

A STAPL Example Code

```
PROCEDURE IS_SECOK USES GV,DO_EXIT;
IF ( ! (SECKEY_OK==0) ) THEN GOTO SECOK;
STATUS = -35;
PRINT "Error, pass key match failure";
CALL DO_EXIT;
SECOK:
LABEL_SEPARATOR = 0;
ENDPROC;
PROCEDURE DO_CHECK_R USES GV,DO_EXIT,DO_READ_SECURITY;
CALL DO_READ_SECURITY;
IF ( ! (ULARE==0) ) THEN GOTO Label_70;
STATUS = -37;
PRINT "FPGA Array Encryption is not enforced.";
PRINT "Cannot guarantee valid AES key present in target device.";
PRINT "Unable to proceed with Encrypted FPGA Array verification.";
CALL DO_EXIT;
Label_70:
IF ( ! (ULARD==1) ) THEN GOTO SKIPRCHK1;
STATUS = -30;
PRINT "FPGA Array Verification is protected by pass key.";
PRINT "A valid pass key needs to be provided.";
CALL DO_EXIT;
SKIPRCHK1:
IF ( ! (ULARD==0) ) THEN GOTO Label_71;
CHKSEC = 0;
Label_71:
LABEL_SEPARATOR = 0;
ENDPROC;
```

Simple Photonic Emission Analysis of AES

Photonic Side Channel Analysis for the Rest of Us

Alexander Schlösser[*,1], Dmitry Nedospasov[*,2], Juliane Krämer[2],
Susanna Orlic[1], and Jean-Pierre Seifert[2]

[1] Optical Technologies, Technische Universität Berlin, Germany
{schloesser,orlic}@opttech.tu-berlin.de
[2] Security in Telecommunications, Technische Universität Berlin, Germany
{dmitry,juliane,jpseifert}@sec.t-labs.tu-berlin.de

Abstract. This work presents a novel low-cost optoelectronic setup for
time- and spatially resolved analysis of photonic emissions and a corresponding methodology, Simple Photonic Emission Analysis (SPEA).
Observing the backside of ICs, the system captures extremly weak photo-
emissions from switching transistors and relates them to program running in the chip. SPEA utilizes both spatial and temporal information
about these emissions to perform side channel analysis of ICs. We successfully performed SPEA of a proof-of-concept AES implementation and
were able to recover the full AES secret key by monitoring accesses to the
S-Box. This attack directly exploits the side channel leakage of a single
transistor and requires no additional data processing. The system costs
and the necessary time for an attack are comparable to power analysis
techniques. The presented approach significantly reduces the amount of
effort required to perform attacks based on photonic emission analysis
and allows AES key recovery in a relevant amount of time.

Keywords: Photonic side channel, emission analysis, optical, temporal
analysis, spatial analysis, AES, full key recovery.

1 Introduction

Most side channel attacks focus on system-wide information leakage. However,
photonic side channels also allow selective in-depth analysis of specific parts
of the hardware. Leakage can even be extracted from single transistors within
an integrated circuit (IC). This selectivity has important implications for side
channel analysis. Potentially, signals can be captured that consist entirely of
leakage and are not impeded by side effects originating from the rest of the
system.

Photonic side channel analysis can be considered far more powerful than other
side channel analysis techniques in use today. Recovering side channel leakage
across large areas of an IC or logic becomes unnecessary. Instead, by investing
some time and effort into spatial photonic analysis of an IC's layout, potential

---

[*] Equal contribution.

E. Prouff and P. Schaumont (Eds.): CHES 2012, LNCS 7428, pp. 41–57, 2012.

weaknesses of the implementation can be efficiently identified and exploited. In this fashion, attacks targeting single transistors become reality. Moreover, targeting specific elements of a chip's logic results in significantly better Signal-to-Noise-Ratios (SNR). Subsequent analysis of such signals can be as simple as a binary evaluation of the traces, akin to Simple Power Analysis (SPA).

The main contributions of this paper are as follows:

A Low-Cost Photonic Emission Analysis System. The system is a low-cost solution to capture photonic emissions of ICs. At approximately the price of a mid-range oscilloscope, it is specifically tailored for photonic side channel analysis. This system also cuts down on measurement times when compared to other state-of-the-art photonic emission analysis methods, see Section 2.4.

A Novel Methodology: Simple Photonic Emission Analysis. With this methodology we are able to recover signals that consist entirely of side channel leakage. By carefully identifying potential targets, we can exploit the spatial separation of logic circuits to eliminate noise within the measurement, circumvent countermeasures and eliminate other potential side effects stemming from the rest of the cryptosystem. Signals recovered by this methodology require little to no additional analysis or post-processing and the key can be recovered directly by simply observing the traces akin to SPA.

Results of a Successful SPEA of AES. Using SPEA in combination with our photonic system, we were able to correctly recover the full secret key of a Proof-of-Concept (PoC) AES-128 implementation running on a common microcontroller, the ATmega328P. The process technology of the ATmega328P, approximately 250 nm, is the technology used in most smartcard deployments today. We exploited the photonic side channel leakage of the row address decoders to monitor accesses to the AES S-Box and were able to recover the full AES secret key. This attack works in software and hardware, and even in the presence of hardware countermeasures, such as memory scrambling and encryption.

Advantages and Limitations of Our System and Methodology. In addition to presenting the system, the methodology and practical results, we present an overview of several potential countermeasures to this kind of attack. Also, we explain how such an attack could be extended to AES implementations using compressed tables and to alternative hardware implementations, including other volatile and non-volatile memories.

Organization. The rest of this work is structured as follows: In Section 2 we present additional background information on photonic emissions in CMOS, detection techniques, the AES algorithm and related work. Section 3 describes the optoelectronic system used in this work. In Section 4 we detail our attack against a PoC AES implementation. Section 5 presents additional considerations for our system and methodology and also includes several potential countermeasures. Finally, we conclude in Section 6.

2   Background

2.1   Photonic Emissions in CMOS

In CMOS technology, carriers gain kinetic energy in a transistor's conductive channel as they are accelerated by the source-drain electric field. At the drain edge of the channel where the field is most intense, this energy is released in radiative transitions, generating photons [21]. This hot-carrier luminescence is dominant in n-type transistors due to the higher mobility of electrons as compared to holes. Consequently, optical emissions of CMOS logic show a data-dependent behaviour similar, but not equal to power consumption. The photon generation rate is proportional to the supply voltage and the transistor switching frequency.

Since multiple interconnect layers of modern IC designs prevent generated photons from escaping the IC on the frontside, hot-carrier luminescence is best observed from the backside. In this case emitted photons have to pass through the silicon substrate, which is absorptive for wavelengths shorter than the bandgap energy, leaving only very few Near-Infrared (NIR) photons for analysis. To reduce absorption the substrate can be mechanically thinned with standard backside polishing machines or alternative techniques [12].

2.2   Detection Techniques

Detection of hot-carrier luminescence from the backside, i.e., single near-infrared photons, must overcome two issues. Firstly, charge coupled devices (CCD) exhibit high spatial resolution, but only allow slow frame rates; single pixel detectors like Photo Multiplier Tubes (PMT), Avalanche Photo Diodes (APD) and Superconducting Single Photon Detectors (SSPD) offer picosecond timing resolution, but only for one small detection area. Secondly, readily available and affordable Si-based detectors only cover a fraction of the relevant NIR spectral range. Thus, for efficient photonic emission analysis more complex and expensive solutions, such as InGaAs-based detectors, are necessary. This is especially true for analysis of modern ICs with small feature sizes, as the emission spectrum shifts further to the infrared with decreasing transistor gate length.

One of the most complex detector technologies in use today is Picosecond Image Circuit Analysis (PICA), which is based on gated multi-channel plates with NIR-sensitive cathode materials. It was developed explicitly for failure analysis of semiconductors [20,10]. PICA delivers both spatial and temporal resolution, but offers only very limited NIR-sensitivity. Additionally, integrated PICA systems have a starting price of around one million Euros in 2012. In contrast to our optoelectronic system, it is unlikely that these systems will ever become a commodity.

2.3   The AES Algorithm

The Advanced Encryption Standard (AES) is a secret key encryption algorithm based on the Rijndael cipher [4]. AES has a fixed block size of 128 bits and operates on a 4×4 matrix of bytes, named the state. Depending on the length of the key, which is 128, 192, or 256 bits, the cipher is termed AES-128, AES-192, or AES-256. The algorithm is specified as a number of identical rounds (except for the last one) that transform the input plaintext into the ciphertext. AES consists of 10, 12 and 14 rounds for 128-, 192- and 256-bit keys, respectively.

Since our attack exploits the leakage obtained during the beginning of the first round of AES, we present only the two operations that are executed until then, namely AddRoundKey and SubBytes. In the AddRoundKey step, each byte of the state is combined with a byte of the round key using the exclusive or operation ($\oplus$). The round key is derived from the original secret key using Rijndael's key schedule; each roundkey is the same size as the state, i.e., 128 bits. The first AddRoundKey operation uses the original secret key, or the first 128 bits of the secret key for AES-192 and AES-256, respectively. In the SubBytes step, each byte in the state is replaced with its entry in a fixed 8-bit lookup table, denoted the S-Box. This is the only operation that provides non-linearity in the algorithm.

2.4   Related Work

In the failure analysis community hot-carrier luminescence has been used primarily to characterize implementation and manufacturing faults and defects [6,15]. In this field the technology of choice to perform backside analysis is PICA [1] and superconducting single photon detectors [17]. Both technologies are able to capture photonic emission with high performance in their respective field, but carry the downside of immense cost and complexity. One of the first uses of photonic emissions in CMOS in a security application was presented in [7], where the authors utilize PICA to spatially recover information about exclusive or operations related to the AddRoundKey operation of AES. Employing PICA in this manner, led to enormous acquisition times. This is especially true considering the size of the executed code. It took the authors 12 hours to recover a single potential key byte [7]. In the same time our system recovers all 16 bytes of the 128-bit AES key twice. In [16] low-cost equipment was used to capture photonic emissions via backside analysis and gain basic information about the operations executed on an IC. Even though the author presented low-cost solutions to both spatially and temporally resolved photonic emission analysis, no attacks using temporal information were demonstrated. Most recently an integrated PICA system and laser stimulation techniques were used to attack a DES implementation on an FPGA [5]. The authors showed that the optical side channel can be used for differential analysis and partly recovered the secret key using temporally resolved measurements. As the authors noted, the use of equipment valued at more than two million Euros does not make such analysis particularly relevant. Additionally, the analysis strongly relied on a specific implementation of DES in which inputs were zeroed. The results required differential analysis and full key recovery was not presented.

In the field of electromagnetic side channel analysis, location-dependent leakage was successfully exploited in an attack on an elliptic curve scalar multiplication implementation on an FPGA using a near-field EM probe [9]. The authors scanned the die surface and collected EM traces at every point. They demonstrated that location-dependent leakage can be used in a template attack and countermeasures against system-wide leakage thus can be circumvented.

Side channels based on monitoring memory accesses have also been researched in the field of cache attacks. In 2004, Bernstein conducted a known-plaintext memory-access timing attack on the OpenSSL AES implementation that uses precomputed tables [2]. The mathematical analysis of our attack is very similar.

3   Experimental Setup

To increase the relevance of semi-invasive optical vulnerability analyses and attacks, the experimental system was constructed with off-the shelf components and employs readily available technical solutions. As neither Si- nor InGaAs-based detectors can deliver both spatial and temporal resolution in the NIR range for an affordable price, our system combines the inherent advantages of both detector technologies in an integrated system. The overall complexity and cost of this system is considerably lower than common semiconductor failure analysis and even power analysis systems, as its price is comparable to that of a mid-range oscilloscope.

3.1   Hardware

The experimental setup consists of two detectors optically and electrically connected to the Device Under Test (DUT) via a custom-built near-infrared microscope and an FPGA-based controller, see Figure 1(a). The DUT is soldered onto a custom printed circuit board and mounted on lateral travel stages. The microscope itself uses finite conjugate reflection type objectives with high numerical aperture and gold plated mirrors to achieve maximum throughput and a spectrally flat transmission curve. After passing the objective, the hot-carrier luminescence spectrum is split by a dichroic mirror and each part directed to the relevant detector. A single Si-CCD serves as the primary detector and captures NIR photons below the silicon bandgap energy. Its mega-pixel deep depletion sensor is back illuminated and thermoelectrically cooled to ensure optimal NIR sensitivity and low dark current rates. This detector can take dark-field reflected-light as well as emission images through the substrate silicon with a diffraction-limited spatial resolution below 1 μm. The acquisition time necessary for adequate emission images ranges from a few seconds to many minutes. It depends strongly on the supply voltage of the DUT, the switching frequency of the transistors under observation and the substrate thickness. Optimized software implementations can increase the execution loop frequency and thereby the switching frequency, which often reduces acquisition times to seconds [11].

The secondary detector is a single InGaAs/InP Avalanche Photo Diode (APD) commonly found in telecom applications (Telcordia GR-468-CORE). It is operated in Geiger mode and thermoelectrically cooled. Increased dark current

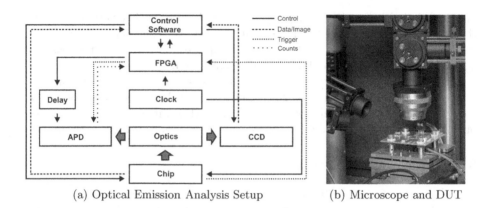

(a) Optical Emission Analysis Setup   (b) Microscope and DUT

Fig. 1. The NIR microscope connects the DUT to the detectors. The two detectors are controlled via an FPGA-based controller, which handles gate synchronization and delay control as well as time-to-amplitude conversion and multichannel counting, see Figure 1(a). The DUT is mounted upside down on a custom printed circuit board underneath the microscope objective, see Figure 1(b).

and afterpulsing, common to InGaAs/InP-APDs, are reduced by gated operation and extensive quenching circuits. The diode is coupled to the microscope via an optical fiber. The fiber's aperture can be freely positioned in the image plane and its object plane size varied by changing the position and magnification of the fiber coupler. Areas of interest, identified in an emission image, can thus be selected for temporal analysis with high spatial selectivity. Because of its spectral sensitivity above $1\,\mu m$, unlike the CCD, this detector does not require a thinned DUT substrate, as silicon is transparent in this spectral range. Hence, if spatial orientation relative to the IC's layout can be obtained by other means, substrate thinning can be omitted completely. This is especially true when applying our methodologies across multiple samples of an identical IC, as only a single sample has to be prepared.

In gated operation the APD is rendered sensitive only for a short window in time, the detection gate, in every signal cycle. To reconstruct the complete signal temporally, the detection gate has to be synchronized and shifted relative to the signal with every signal cycle iteration, similar to a sampling oscilloscope. Provided the gate delay can be controlled with high resolution, the time resolution and inversely the measurement time depends only on the minimal gate width. To implement this detection scheme we use an FPGA-based controller phase-locked to the DUT clock. As the DUT executes the target program code, the phase-locked FPGA digitally delays and triggers the APD detection gate. Detection events are sent back to the FPGA and counted in the corresponding time bins. An additional analog delay can be employed for fine delay control. The absolute time resolution of our system is jitter-limited to approximately $1\,ns$.

The measurement time to reconstruct the extremely weak photoemission signals can be immense: Hundreds of thousands of samples may be necessary to achieve an adequate SNR. To drastically reduce the measurement times, the FPGA triggers hundreds of APD detection gates per execution of the IC. This results in interleaved measurements.

For our practical evaluation the DUT board consisted of an inversely soldered ATmega328P supplied with 5 V operating voltage, a 16 MHz quartz oscillator as well as decoupling capacitors and an I/O header for external communications and programming. To shorten the emission image acquisition times, the backside of the ATmega328P was mechanically polished with an automated backside sample preparation machine. The remaining substrate was approximately 25 μm thick.

3.2   Software

The PoC implementation running on the ATmega328P microcontroller consisted of a software AES implementation. To increase the frequency of the execution, only the first `AddRoundKey` operation and part of the first round of the AES algorithm were computed on the chip after which the input was reset and the measurement restarted. Specifically, the implementation computed only the first `AddRoundKey` and `SubBytes` operations, see Section 2.3. Most notably, the AES S-Box, see Table 2 in the appendix, was implemented in the microcontroller's data memory, i.e. the SRAM. Both stack- and heap-based AES implementations were tested and resulted in offsets within SRAM as described in Section 4.1.

4   Practical Results

This section details our practical results in which we applied SPEA to mount a practical attack against a Proof-of Concept (PoC) AES implementation.

4.1   Monitoring SRAM Access

Address logic is implemented similarly across all platforms and memories. Hence, it is a particularly interesting and relevant target. In our implementation the S-Box is contained within the SRAM, which led us to consider possible side channels that exist within this memory. Specifically, we considered how SRAM is accessed in general and how the S-Box would be accessed in the PoC AES implementation. Memory is structured in rows and columns. In the case of the ATmega328P, each 512 kilobyte SRAM bank is made up of 64 rows and 64 columns. Thus, each row stores a total of 64 bits, or 8 bytes. An SRAM cell read begins with assertion of the word line [13]. When any cell within a row of memory is accessed, the word line for the entire row is asserted. This row select signal is driven by an inverter that is part of the row decode logic [22].

We used the Si-CCD detector and techniques introduced in [11] to analyze the emissions of these memory accesses for different addresses and values. With spatial resolution, accesses between even two adjacent rows of SRAM can be clearly differentiated, see Figure 2(a) and 2(b). The same kind of analysis can also be applied to the column addresses.

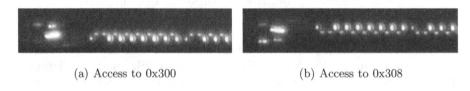

(a) Access to 0x300   (b) Access to 0x308

Fig. 2. 120 s emission images of memory accesses to two adjacent memory rows obtained with the Si-CCD detector

By studying the emission images of our PoC AES implementation we identified the S-Box within memory, see Figure 3 and Table 3 in the appendix. The emissions of the row drivers are clearly visible to the left of the individual memory lines. The emission image Figure 3 reveals that the S-Box spans 33 rows of the SRAM and not 32 as expected. Since the S-Box was implemented as an array of bytes within data memory, its address depends on how many other variables are allocated within memory. It is therefore unlikely to be aligned to the beginning of a memory row. During our experiments, the S-Box always exhibited an offset unless the address was set explicitly.

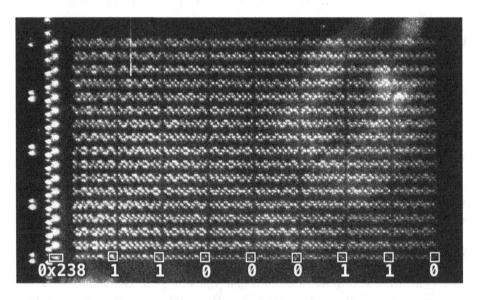

Fig. 3. Optical emission image of the S-Box in memory. The 256 bytes of the S-Box are located from 0x23F to 0x33E, see Table 3 in the appendix. The address 0x23F is the seventh byte of the 0x238 SRAM line, i.e. the S-Box has an offset of 7 bytes. The emissions of the row drivers are clearly visible to the left of the memory bank. The image allows direct readout of the bit-values of the stored data. The first byte for example, as shown in the overlay, corresponds to $01100011_2 = 63_{16}$, the first value of the AES S-Box.

4.2 Key Recovery Using the Photonic Side Channel

By observing time-resolved access patterns to a specific row within the S-Box, the set of key candidates can be greatly reduced. In this attack on the first round of AES, we observe accesses to a single fixed row r over 256 input messages, in which all bytes of the input are equal. That is, for every text byte $p_i = i$, $i \in \{0, \ldots, 255\}$, there is an input message m_i that consists of 16 concatenated p_i's. Thus, after the exclusive or of the first AddRoundKey operation, every element of the S-Box is accessed exactly once for every processed byte $b \in \{1, \ldots, 16\}$ in the first AES round. However, we only measure accesses to rows, i.e., we use the access patterns to a given row r to eliminate certain key candidates. If potential offsets are also taken into consideration, the number of key candidates that can be identified using this attack are shown in Table 1.

Table 1. Number of candidates per key byte and unresolved bits of the whole key, depending on the offset and row. For an offset of 0, there are only 32 rows.

Offset	remaining candidates per key byte		unresolved bits of the whole key	
	$r = 1$ or $r = 33$	$r \in \{2, ..., 32\}$	$r = 1$ or $r = 33$	$r \in \{2, ..., 32\}$
0	8	8	48	48
1	1	1	0	0
2	2	2	16	16
3	1	1	0	0
4	4	8	32	48
5	1	1	0	0
6	2	2	16	16
7	1	1	0	0

Note, uneven offsets always result in unique access patterns, allowing for full key recovery directly.

An attack against our PoC implementation was thus implemented in three parts: (1) an offline precomputation of potential access patterns, (2) an online measurement of emissions over all possible input bytes, and (3) an evaluation resulting in a reduced set of key candidates.

Offline Precomputation. For the first round of AES and independent of the value b, to determine which row of the S-Box is accessed by a plaintext byte p_i, key byte $k_j = j$, $j \in \{0, \ldots, 255\}$, and for a given offset $o \in \mathbb{N}^{\geq 0}$ and width of the rows $w \in \mathbb{N}^{\geq 1}$, we introduce the following function $row : \{0, \ldots, 255\} \times \{0, \ldots, 255\} \to \mathbb{N}$:

$$row(p_i, k_j) = \lfloor (p_i \oplus k_j) + o)/w \rfloor + 1$$

Using this function, we can generate the array A, storing the sets of plaintexts accessing row $r_l = l$, $l \in \{1, \ldots, \lfloor (256 + o)/w \rfloor + 1\}$ for key byte k_j:

$$A(r_l, k_j) = \{p_i | r_l = row(p_i, k_j)\}$$

Time-Resolved Emission Measurement. Expecting input-dependent accesses to the memory rows, we set our APD detector to measure the photonic emissions from the row driver inverter corresponding to a fixed memory row with S-Box elements. Emission traces for all 256 inputs were captured, with an acquisition time of 90 seconds each. A row access resulted in a clearly defined photon detection peak with a high SNR, see Figure 4. Empirically defining a threshold level l determines if a fixed row r of the S-Box was accessed within the measurement. For a period of time t_b in which the b-th byte is processed, and for the measured photonic emission intensity I_r of a row r:

$$M_r(t_b) = \{p_i | max(I_r(p_i, t_b)) > l\}$$

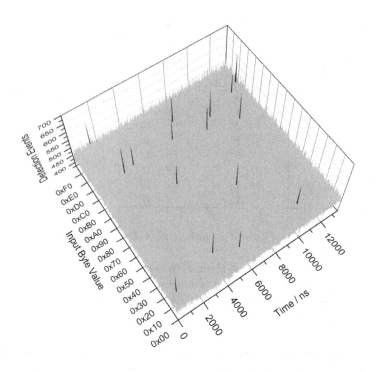

Fig. 4. Time-resolved photonic emission traces of the PoC implementation over all possible input bytes. All 16 peaks, corresponding to the bytes of the AES-128 key, are clearly identifiable. Specifically, results are shown for the following key, [0x10, 0xF1, 0xB3, 0xB7, 0x1E, 0x81, 0x12, 0xBA, 0xD1, 0x56, 0xAD, 0xBB, 0x17, 0xA2, 0xCA, 0xD5]. Note the SNR, comparing peaks to the noise floor of traces that are not accessing the SRAM. A two dimensional plot, Figure 5, can be found in the Appendix.

Full Key Recovery. Finally, we can identify the resulting set of candidates for the b-th key byte by analyzing, which key bytes could have caused the measured accesses, using the precomputed array A:

$$K(b) = \{k_j | A(r, k_j) = M_r(t_b)\}$$

For an S-Box with an uneven offset this results in a complete key recovery. In the case of an even offset a maximum of eight candidates per key byte remain. This set can be easily reduced by a second set of measurements selecting another row or cross-correlation with measurements from column decoder logic.

The measurement time to fully capture the photonic emission signal over time and all possible inputs, as seen in Figures 4 and 5, amounted to just over six hours. However, since accesses to the S-Box occur at constant points in time in every execution, these are the only points that need to be observed in subsequent measurements. This cuts down the necessary measurement time immensely. After initial analysis, any subsequent attacks on identical implementations can therefore recover the key in less than 45 minutes.

5   Discussion

The effectiveness and relevance of the presented attack depends on and therefore highlights the importance of preliminary spatial analysis. It demonstrates how the measurement time can be considerably reduced, while greatly boosting the SNR, by targeting the leakage of specific transistors directly. This basic methodology can be applied to many different attack vectors. Effectively, every transistor exhibiting data-dependent behaviour becomes a potential target.

The initial spatial analysis is necessary to allow for at least a basic understanding of the chip's functionality and the identification of potential points of interest. In our approach we use a Si-CCD, which operates with very few impeding photons at the edge of the spectral range to which silicon is sensitive. This low cost approach requires DUT substrate thinning. However, more expensive InGaAs-cameras can also be used, which are sensitive to photons above $1\,\mu m$ wavelength. In this spectral range silicon is transparent and substrate thinning therefore becomes unnecessary. It is worth noting that many modern security ICs, such as smartcards, have a far thinner substrate than common general purpose microcontrollers. As a result, many security ICs do not have to be thinned at all for semi-invasive optical backside analysis. In contrast, modern ICs generate less and less photonic emissions due to lower supply voltage. However, recent works have demonstrated that emission analysis of modern ICs is possible [18,19], especially if NIR-sensitive detectors are employed. An example of such a device is the InGaAs/InP-APD used in this work.

In the presented attack, full key recovery was achieved for an S-Box with an uneven offset. For a fully-aligned S-Box implemented in 8-byte-rows of memory, the set of potential key candidates can still be greatly reduced to approximately 2^{48}, see Table 1. However, by performing additional temporal measurements of alternative points of leakage on the chip, the set of potential key candidates can nevertheless be reduced to a single key candidate. It is feasible, for example, to exploit the leakage of the column decode logic and reveal the exact address of memory accesses using cross-correlation. SPEA can potentially be extended to any addressable memory. It can also easily be adopted to AES-192 and AES-256 and to alternative implementations, which use compressed tables. If additional measurements can be captured in a reasonable amount of time, the attack could even be implemented as an unknown-plaintext attack, as demonstrated for a cache timing attack by [8].

It is worth noting, that many industry standard countermeasures, in fact, do not prevent photonic emission attacks at all. Shields and meshes generally only protect against attacks from the frontside. Memory encryption and scrambling [14] may protect against probing attacks, but have no effect on the optical emissions. For SPEA, the memory access patterns would be unaffected. However, memory encryption would make the initial spatial analysis more cumbersome. It would obfuscate the values in memory, preventing the memory from being read out optically. Memory scrambling would, on the other hand, potentially make the attack far easier. Since the goal of memory scrambling is to obfuscate the layout of memory in terms of addresses, it increases the likelihood that a single S-Box element may be isolated in a row of memory. If the positions of memory are spatially obfuscated, accessing a single line would reveal all of its elements.

Nevertheless, several potential hardware and software countermeasures to our attack do exist. Specifically, hardware and software delays, masking and dummy rounds can make such attacks vastly more difficult. Any form of randomization forces longer measurement iteration times, thus greatly increasing trace acquisition times. The effects of such countermeasures are identical to the effects on power analysis, as described in [3] and could be minimized with more advanced, e.g. differential, signal analysis techniques. On the other hand, at least in this specific attack randomization techniques could be thwarted by employing APD detection gates long enough to encompass any and all randomization clock cycles. Since accesses to the S-Box occur at vastly different points in time, the resulting temporal resolution would still be sufficient to yield all S-Box accesses.

It has also been argued, that shrinking structure sizes will eventually defeat optical emission analysis and thus photonic side channels. As already mentioned, recent works show that shrinking feature sizes do not eliminate optical emission [17,18,19] and photonic detection techniques continue to improve rapidly. Also, in practice structure sizes only apply to the smallest structures on an IC. In every IC there are plenty data-dependent transistors that have far larger channels than that of the smallest logic on the chip. One such example is the very row driver exploited in this work, which is sized up to cope with the large capacitances of SRAM memory rows.

As a countermeasure that prohibits any optical emission attacks we would like to propose an active shield or mesh on the backside of the IC. While a single metal layer on the backside can trap all photons generated within the chip, active integrity checks prevent the removal of such a layer. Further countermeasures are under development with our partners.

6 Conclusion

Optoelectronic systems are currently the only systems capable of recovering leakage from single transistors directly. Despite this fact, the photonic side channel attacks presented so far failed to utilize both temporal and spatial information in the attacks [5,7,16]. By combining temporal measurements of an InGaAs/InP-APD detector with the spatial resolution of a Si-CCD, potential targets are identified quickly and easily. Also, unprecedented levels in terms of SNR and therefore information leakage over acquisition time are achieved. The optoelectronic system employed in this work outperforms many state of the art optoelectronic systems, for a fraction of the cost of the system utilized in [5]. The necessary acquisition time for the attack presented is comparable to power analysis attacks.

In our *Simple Photonic Emission Analysis* of AES, we initially evaluated emission images and subsequently performed temporal measurements. As a result we were able to mount a successful attack that utilizes both spatial and temporal information retrieved from photonic emissions. Provided the spatial information from the preliminary evaluation of emission images, we were able to focus the temporal measurements directly on a single transistor of the IC. The information leakage was so high that no additional analysis was necessary to recover all 16 bytes of the AES-128 secret key, akin to SPA and for a similarly low price in terms of equipment. The optoelectronic system employed in this work costs approximately the same as a mid-range oscilloscope, yet because it is spatially selective, offers vastly better characteristics in terms of leakage SNR. To the best of our knowledge this is the first work to combine temporal and spatial photonic side channel analysis. It demonstrates that even the leakage of a single transistor can be exploited directly to recover the full AES secret key.

Acknowledgements. The authors acknowledge support by the German Federal Ministry of Education and Research in the project PhotonDA through grant number 01IS10029A and the Helmholtz Research School on Security Technologies. Also, the authors would like to thank our project partners at NXP Semiconductors Germany for their insight and cooperation, the Semiconductor Devices research group at TU Berlin for sample preparation and our colleagues Enrico Dietz, Sven Frohmann, Collin Mulliner and Cristoph Bayer for helpful discussions and feedback.

References

1. Bascoul, G., Perdu, P., Benigni, A., Dudit, S., Celi, G., Lewis, D.: Time Resolved Imaging: From logical states to events, a new and efficient pattern matching method for VLSI analysis. Microelectronics Reliability 51(9-11), 1640–1645 (2011), http://dx.doi.org/10.1016/j.microrel.2011.06.043
2. Bernstein, D.: Cache-timing attacks on AES (2004), http://cr.yp.to/papers.html#cachetiming

3. Clavier, C., Coron, J.-S., Dabbous, N.: Differential Power Analysis in the Presence of Hardware Countermeasures. In: Koç, Ç.K., Paar, C. (eds.) CHES 2000. LNCS, vol. 1965, pp. 252–263. Springer, Heidelberg (2000),
http://dx.doi.org/10.1007/3-540-44499-8_20

4. Daemen, J., Rijmen, V.: The design of Rijndael: AES – the Advanced Encryption Standard. Springer, Heidelberg (2002)

5. Di-Battista, J., Courrege, J.-C., Rouzeyre, B., Torres, L., Perdu, P.: When Failure Analysis Meets Side-Channel Attacks. In: Mangard, S., Standaert, F.-X. (eds.) CHES 2010. LNCS, vol. 6225, pp. 188–202. Springer, Heidelberg (2010),
http://dx.doi.org/10.1007/978-3-642-15031-9_13

6. Egger, P., Grutzner, M., Burmer, C., Dudkiewicz, F.: Application of time re-solved emission techniques within the failure analysis flow. Microelectronics Re-liability 47(9-11), 1545–1549 (2007),
http://dx.doi.org/10.1016/j.microrel.2007.07.067

7. Ferrigno, J., Hlaváč, M.: When AES blinks: introducing optical side channel. In-formation Security, IET 2(3), 94–98 (2008),
http://dx.doi.org/10.1049/iet-ifs:20080038

8. Gullasch, D., Bangerter, E., Krenn, S.: Cache games – bringing access-based cache attacks on AES to practice. In: 2011 IEEE Symposium on Security and Privacy, pp. 490–505 (2011), http://dx.doi.org/10.1109/SP.2011.22

9. Heyszl, J., Mangard, S., Heinz, B., Stumpf, F., Sigl, G.: Localized Electromag-netic Analysis of Cryptographic Implementations. In: Dunkelman, O. (ed.) CT-RSA 2012. LNCS, vol. 7178, pp. 231–244. Springer, Heidelberg (2012),
http://dx.doi.org/10.1007/978-3-642-27954-6_15

10. Kash, J., Tsang, J.: Dynamic internal testing of CMOS circuits using hot lumines-cence. IEEE Electron Device Letters 18(7), 330–332 (1997),
http://dx.doi.org/10.1109/55.596927

11. Nedospasov, D., Schlösser, A., Seifert, J., Orlic, S.: Functional integrated circuit analysis. In: 2012 IEEE International Symposium on Hardware-Oriented Security and Trust, HOST (2012)

12. Nohl, K., Evans, D., Starbug, S.: Reverse-engineering a cryptographic RFID tag. In: 17th USENIX Security Symposium, pp. 185–193 (2008),
http://www.usenix.org/event/sec08/tech/full_papers/nohl/nohl_html/

13. Rabaey, J.M., Chandrakasan, A.: Digital Integrated Circuits. A Design Prespective, 2nd edn. Pearson Education (2003)

14. Rankl, W., Effing, W.: Smart Card Handbook, 4th edn. Wiley (2010)

15. Selmi, L., Mastrapasqua, M., Boulin, D., Bude, J., Pavesi, M., Sangiorgi, E., Pinto, M.: Verification of electron distributions in silicon by means of hot carrier lumi-nescence measurements. IEEE Transactions on Electron Devices 45(4), 802–808 (1998), http://dx.doi.org/10.1109/16.662779

16. Skorobogatov, S.: Using Optical Emission Analysis for Estimating Contribution to Power Analysis. In: 2009 Workshop on Fault Diagnosis and Tolerance in Cryptog-raphy (FDTC), pp. 111–119 (2009), http://dx.doi.org/10.1109/FDTC.2009.39

17. Song, P., Stellari, F., Huott, B., Wagner, O., Srinivasan, U., Chan, Y., Rizzolo, R., Nam, H.J., Eckhardt, J., McNamara, T., Tong, C.L., Weger, A., McManus, M.: An advanced optical diagnostic technique of IBM z990 eServer microprocessor. In: Proceedings of the IEEE International Test Conference, ITC 2005, pp. 1227–1235 (2005), http://dx.doi.org/10.1109/TEST.2005.1584091, doi:10.1109/TEST.2005.1584091

18. Tosi, A., Stellari, F., Pigozzi, A., Marchesi, G., Zappa, F., Heights, Y.: A Challenge for Emission Based Testing and Diagnostics. Reliability Physics, 595–601 (2006), http://dx.doi.org/10.1109/RELPHY.2006.251284

19. Tsang, J.C., Fischetti, M.V.: Why hot carrier emission based timing probes will work for 50 nm, 1V CMOS technologies. Microelectronics Reliability, 1465–1470 (2001), http://dx.doi.org/10.1016/S0026-2714(01)00194-9

20. Tsang, J.C., Kash, J.A., Vallett, D.P.: Picosecond imaging circuit analysis. IBM Journal of Research and Development 44(4), 583–603 (2000), http://dx.doi.org/10.1147/rd.444.0583

21. Villa, S., Lacaita, A., Pacelli, A.: Photon emission from hot electrons in silicon. Physical Review B 52(15), 10993–10999 (1995), http://www.dx.doi.org/10.1103/PhysRevB.52.10993

22. Weste, N.H.E., Harris, D.: CMOS VLSI Design: A Circuits and Systems Perspective, 4th edn. Addison Wesley (2010)

Appendix

Table 2. The AES S-Box, used during SubBytes operation, in hexadecimal representation. The first 4 bits of the input determine the row, the last 4 bits determine the column.

	0	1	2	3	4	5	6	7	8	9	a	b	c	d	e	f
0	63	7c	77	7b	f2	6b	6f	c5	30	01	67	2b	fe	d7	ab	76
1	ca	82	c9	7d	fa	59	47	f0	ad	d4	a2	af	9c	a4	72	c0
2	b7	fd	93	26	36	3f	f7	cc	34	a5	e5	f1	71	d8	31	15
3	04	c7	23	c3	18	96	05	9a	07	12	80	e2	eb	27	b2	75
4	09	83	2c	1a	1b	6e	5a	a0	52	3b	d6	b3	29	e3	2f	84
5	53	d1	00	ed	20	fc	b1	5b	6a	cb	be	39	4a	4c	58	cf
6	d0	ef	aa	fb	43	4d	33	85	45	f9	02	7f	50	3c	9f	a8
7	51	a3	40	8f	92	9d	38	f5	bc	b6	da	21	10	ff	f3	d2
8	cd	0c	13	ec	5f	97	44	17	c4	a7	7e	3d	64	5d	19	73
9	60	81	4f	dc	22	2a	90	88	46	ee	b8	14	de	5e	0b	db
a	e0	32	3a	0a	49	06	24	5c	c2	d3	ac	62	91	95	e4	79
b	e7	c8	37	6d	8d	d5	4e	a9	6c	56	f4	ea	65	7a	ae	08
c	ba	78	25	2e	1c	a6	b4	c6	e8	dd	74	1f	4b	bd	8b	8a
d	70	3e	b5	66	48	03	f6	0e	61	35	57	b9	86	c1	1d	9e
e	e1	f8	98	11	69	d9	8e	94	9b	1e	87	e9	ce	55	28	df
f	8c	a1	89	0d	bf	e6	42	68	41	99	2d	0f	b0	54	bb	16

Table 3. AES S-Box in 32×8 implementation with an offset of 7. The sum of row and column index yields the entry's index, each entry denotes the corresponding input and output value.

	0	1	2	3	4	5	6	7
0x338	f9 (99)	fa (2d)	fb (0f)	fc (b0)	fd (54)	fe (bb)	ff (16)	
0x330	f1 (a1)	f2 (89)	f3 (0d)	f4 (bf)	f5 (e6)	f6 (42)	f7 (68)	f8 (41)
0x328	e9 (1e)	ea (87)	eb (e9)	ec (ce)	ed (55)	ee (28)	ef (df)	f0 (8c)
0x320	e1 (f8)	e2 (98)	e3 (11)	e4 (69)	e5 (d9)	e6 (8e)	e7 (94)	e8 (9b)
0x318	d9 (35)	da (57)	db (b9)	dc (86)	dd (c1)	de (1d)	df (9e)	e0 (e1)
0x310	d1 (3e)	d2 (b5)	d3 (66)	d4 (48)	d5 (03)	d6 (f6)	d7 (0e)	d8 (61)
0x308	c9 (dd)	ca (74)	cb (1f)	cc (4b)	cd (bd)	ce (8b)	cf (8a)	d0 (70)
0x300	c1 (78)	c2 (25)	c3 (2e)	c4 (1c)	c5 (a6)	c6 (b4)	c7 (c6)	c8 (e8)
0x2F8	b9 (56)	ba (f4)	bb (ea)	bc (65)	bd (7a)	be (ae)	bf (08)	c0 (ba)
0x2F0	b1 (c8)	b2 (37)	b3 (6d)	b4 (8d)	b5 (d5)	b6 (4e)	b7 (a9)	b8 (6c)
0x2E8	a9 (d3)	aa (ac)	ab (62)	ac (91)	ad (95)	ae (e4)	af (79)	b0 (e7)
0x2E0	a1 (32)	a2 (3a)	a3 (0a)	a4 (49)	a5 (06)	a6 (24)	a7 (5c)	a8 (c2)
0x2D8	99 (ee)	9a (b8)	9b (14)	9c (de)	9d (5e)	9e (0b)	9f (db)	a0 (e0)
0x2D0	91 (81)	92 (4f)	93 (dc)	94 (22)	95 (2a)	96 (90)	97 (88)	98 (46)
0x2C8	89 (a7)	8a (7e)	8b (3d)	8c (64)	8d (5d)	8e (19)	8f (73)	90 (60)
0x2C0	81 (0c)	82 (13)	83 (ec)	84 (5f)	85 (97)	86 (44)	87 (17)	88 (c4)
0x2B8	79 (b6)	7a (da)	7b (21)	7c (10)	7d (ff)	7e (f3)	7f (d2)	80 (cd)
0x2B0	71 (a3)	72 (40)	73 (8f)	74 (92)	75 (9d)	76 (38)	77 (f5)	78 (bc)
0x2A8	69 (f9)	6a (02)	6b (7f)	6c (50)	6d (3c)	6e (9f)	6f (a8)	70 (51)
0x2A0	61 (ef)	62 (aa)	63 (fb)	64 (43)	65 (4d)	66 (33)	67 (85)	68 (45)
0x298	59 (cb)	5a (be)	5b (39)	5c (4a)	5d (4c)	5e (58)	5f (cf)	60 (d0)
0x290	51 (d1)	52 (00)	53 (ed)	54 (20)	55 (fc)	56 (b1)	57 (5b)	58 (6a)
0x288	49 (3b)	4a (d6)	4b (b3)	4c (29)	4d (e3)	4e (2f)	4f (84)	50 (53)
0x280	41 (83)	42 (2c)	43 (1a)	44 (1b)	45 (6e)	46 (5a)	47 (a0)	48 (52)
0x278	39 (12)	3a (80)	3b (e2)	3c (eb)	3d (27)	3e (b2)	3f (75)	40 (09)
0x270	31 (c7)	32 (23)	33 (c3)	34 (18)	35 (96)	36 (05)	37 (9a)	38 (07)
0x268	29 (a5)	2a (e5)	2b (f1)	2c (71)	2d (d8)	2e (31)	2f (15)	30 (04)
0x260	21 (fd)	22 (93)	23 (26)	24 (36)	25 (3f)	26 (f7)	27 (cc)	28 (34)
0x258	19 (d4)	1a (a2)	1b (af)	1c (9c)	1d (a4)	1e (72)	1f (c0)	20 (b7)
0x250	11 (82)	12 (c9)	13 (7d)	14 (fa)	15 (59)	16 (47)	17 (f0)	18 (ad)
0x248	09 (01)	0a (67)	0b (2b)	0c (fe)	0d (d7)	0e (ab)	0f (76)	10 (ca)
0x240	01 (7c)	02 (77)	03 (7b)	04 (f2)	05 (6b)	06 (6f)	07 (c5)	08 (30)
0x238								00 (63)

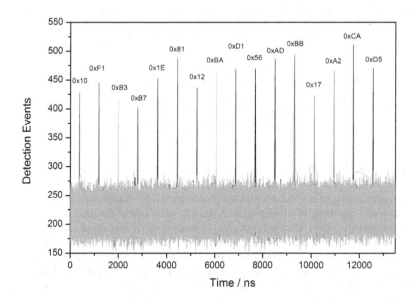

Fig. 5. Time-resolved photonic emission traces of the PoC implementation over all possible input bytes as a 2D representation. All 16 peaks, corresponding to the 16 Bytes of the AES key, are clearly identifiable and show a width of 20 ns. This corresponds to the employed detection gate width.

Table 4. Amount of unresolved bits of the complete key for AES-192 and AES-256, depending on the offset and row. For an offset of 0, there are only 32 rows. Attacking AES-192 or AES-256 requires measuring the photonic emissions of **SubBytes** during the first two rounds.

Offset	unresolved bits of the whole AES-192 key		unresolved bits of the whole AES-256 key	
	$r = 1$ or $r = 33$	$r \in \{2, ..., 32\}$	$r = 1$ or $r = 33$	$r \in \{2, ..., 32\}$
0	72	72	96	96
1	0	0	0	0
2	24	24	32	32
3	0	0	0	0
4	48	72	64	96
5	0	0	0	0
6	24	24	32	32
7	0	0	0	0

Compiler Assisted Masking

Andrew Moss[1], Elisabeth Oswald[2], Dan Page[2], and Michael Tunstall[2]

[1] School of Computing, Blekinge Institute of Technology,
Karlskrona, Sweden
andrew.moss@bth.se

[2] Department of Computer Science, University of Bristol,
Merchant Venturers Building, Woodland Road,
Bristol BS8 1UB, United Kingdom
{eoswald,page,tunstall}@cs.bris.ac.uk

Abstract. Differential Power Analysis (DPA) attacks find a statistical correlation between the power consumption of a cryptographic device and intermediate values within the computation. Randomization via (Boolean) masking of intermediate values breaks this statistical dependence and thus prevents such attacks (at least up to a certain order). Especially for software implementations, (first-order) masking schemes are popular in academia and industry, albeit typically not as the sole countermeasure. The current practice then is to manually 'insert' Boolean masks: essentially software developers need to manipulate low-level assembly language to implement masking. In this paper we make a first step to automate this process, at least for first-order Boolean masking, allowing the development of compilers capable of protecting programs against DPA.

Keywords: Compiler assisted cryptography, masking, DPA.

1 Introduction

Cryptographic software provides a challenging target for software engineering. High-level languages improve programmer productivity by abstracting unnecessary details of the program execution and freeing the programmer to concentrate on the correctness of their implementation. Unfortunately, the details that are generally abstracted away are the behavioural properties of programs, in order to focus on their functional results. In cryptography, the way in which a value is computed may lead to observational differences that an attacker could use to compromise security. If values within the computation that must remain secret, such as cryptographic keys, influence the observational behaviour then information will leak and may render the system insecure. If a compiler is allowed to handle the low-level decisions for a given implementation then it must also take how information may leak into account.

State-of-the-art compilers can rival the efforts of a human in producing high performance code. For example, effective methods of register allocation, instruction selection and scheduling often depend on knowledge of the operational details of memory latency and pipeline behaviour. Extensions to the execution

E. Prouff and P. Schaumont (Eds.): CHES 2012, LNCS 7428, pp. 58–75, 2012.

models targeted by C compilers such as GCC and VisualStudio allow counter-measures to be automatically applied against buffer overflow attacks [1]. A more detailed execution model (e.g. stack frame layout) allows the compiler back-end to perform program transformation that is aware of security constraints.

The increasingly complex threat of physical (e.g. fault and side-channel) attacks on cryptographic implementations offers an interesting extension of the above security case. Automatic resolution of said threat is now an emerging research theme and, alongside more theoretical results in this area (e.g. [2]), a range of concrete compilation systems exist. For example, Molnar et al. [3] construct a binary translation (i.e., compilation) tool that resolves control-flow based leakage using the Program Counter Model (PCM) formalism; Lux and Starostin [4] describe a tool which detects and eliminates timing side-channels in Java programs (demonstrating the tool by highlighting an attack against the FlexiProver implementation of IDEA). Likewise, suitable EDA tool-chains [5] can, given some HDL model, automatically implement countermeasures against power-analysis attacks: a back-end which processes some logical netlist can replace standard cells with a secure logic style equivalent (e.g. WDDL [6]) before producing a physical netlist.

Set within this general context, we focus on a specific challenge: given a source program, the goal is to automatically apply Boolean masking. We therefore introduce a simple type system and make use of static analysis to determine whether statements (and associated variables) leak in a source program, with the aim to automatically transform an insecure program into one that is secure against (first-order) DPA. Our approach currently supports Boolean masking and hence can be used to secure any program (e.g. AES, DES, Present, etc.) that can be masked in this way.

2   Background

At execution time, a value in a program is a particular bit-pattern. The meaning ascribed to that pattern is dependent upon the context around the code. This basic property of computers makes them flexible, as one pattern of bits can represent many different values depending on the program being executed. However, it can also be a source of error as the meaning of the value is not denoted in the executable code, but rather in the source-level description. Type systems are a method of reducing potential errors by denoting the kind (or type) of value that a particular variable represents. Compilers can then use this type information statically (when compiling the program) to rule out erroneous behaviour. In this way types can be seen as a static guarantee of safety. Although type theory is a long establish field within the languages and compilers community, the authors believe there is no previous work on using types to describe masking countermeasures.

Conventionally lattices are used within static analysis of programs to produce conservative results. A lattice is a partially ordered set of values in which every pair of values has a well-defined supremum and infimum. The analysis is guaranteed to be sound (if inexact) as a conservative approximation may use a bound to over-approximate unknown values during analysis. Information flow analysis annotates program values as high, or low, security and prevents data-flow between high-security and low-security values. Non-interference was introduced [7] as a property that can be proven from the program semantics by allowing the erasure of a high-security region without producing any observable difference in the low-security region (under the assumption that erasing a secret value does not leak).

Our work shares some similarities to information flow; secret values are annotated by the programmer and their secrecy is treated as a value in a lattice allowing the compiler to propagate secrecy information through the program. The main difference is the role of the adversary in the system. In information flow the adversary is considered to be on the "edge" of the computation, while execution of code within the high-security region is not observable. Power analysis can be used to make observations at every point in a program; all secure information must be hidden by masking, but the adversary has the chance to observe all masked operations. While previous work is analytical, a decision is made if a program is secure, this work takes a (potentially) broken program and converts it to a functionally equivalent program that meets the behavioural definition in the model.

Compilers typically operate on an Intermediate Representation (IR) of a program. Input text is parsed into an Abstract Syntax Tree (AST) that represents the structure of program. The AST is then converted to an IR that more closely resembles the execution of instructions on the target machine. During this process temporary variables are introduced to store the intermediate results in computing expressions. A 3-address form represents each instruction in the program as two input operands, an opcode and an output operand, e.g. $r \leftarrow a$ xor b. When the output operand is unique for each instruction this form is called Static Single Assignment (SSA). Multiple write operations to the same variable are renamed to separate instances to ensure this property so that a sequence of the form $x \leftarrow a$ xor $b; x \leftarrow x$ xor y becomes the sequence $x_1 \leftarrow a$ xor $b; x_2 \leftarrow x_1$ xor y.

The uniqueness of each target operand implies that loop-free programs form a directed acyclic graph with instructions and variables as vertices, and denoting usage by edges between those vertices. This graph is conventionally termed a *dataflow* representation of the program. In such a graph each vertex v has a set of *ancestors* defined as every vertex where a path exists that reaches v.

3   DPA Attacks and Mask-Based Countermeasures

In conducting a DPA attack, an attacker will try to recover information about a secret (typically a cryptographic key) by using information about the power consumption of a cryptographic device while it is manipulating the secret in cryptographic operations. To perform such a DPA attack, an attacker selects a so-called intermediate value: e.g. in the specific example that we use to illustrate our work later in this article, the attacker might select the input or output of the AES SubBytes operation when applied to the first byte of the AES state.

This intermediate value only depends on a small part of the secret key (in our example only eight bits), which allows an attacker to predict this intermediate value (using knowledge of the input data) for all possible values of that small part of the key. Next an attacker uses a leakage model for the device under attack to map these predicted intermediate values to hypothetical power consumption values: assuming the leakage model is reasonably correct, only the set of hypothetical power consumption values that are related to the correct key guess will match those power consumption values that an attacker can observe from the device itself. Several statistical tools can be used to 'match' hypothetical and real data, e.g. Pearson's correlation coefficient, distance-of-means test, etc.

3.1   Masking to Prevent DPA

As can be inferred from the previous description, DPA attacks can only be successful if an attacker can define an intermediate value (based on a suitably small part of the secret key) that is somehow related to the instantaneous power consumption of the device. Thus, DPA attacks can be prevented by making it impossible for an attacker to predict the intermediate values used on the device. A popular method to this purpose is referred to as 'masking'. Using AES to illustrate the central principle: instead of holding the AES state and AES key 'as they are' one applies a random value to them. For example, the first byte a of the AES state is then represented as a pair (a_m, m), with m being the so-called mask which is a number chosen at random from a suitable uniform distribution, such that $a = a_m \oplus m$. Equivalently, the first byte of the first AES round key is then represented as pair (k_n, n), with $k = k_n \oplus n$, and n is chosen at random from a suitable uniform distribution.

In the encryption process itself these masked values need to be processed correctly and securely. For example, if two masked bytes are exclusively-ored, we need to ensure that the result is masked again: $a_m \oplus b_n$ may be carried out but $a_m \oplus b_m$ would result in $a \oplus b$ being vulnerable to DPA and must not happen. Similarly, table look-ups must be executed such that both inputs and outputs are masked.

Previous work on masking schemes has explored various options for the efficient computation of various cryptographic functions, e.g. the efficient and secure masking of the AES SubBytes operation has been extensively discussed. We make use of the work in [8] and [9] by extracting some necessary properties of secure masking schemes. A useful observation made in these previous works was that 'secure against' DPA attacks is synonymous to the concept of statistical independence between variables, i.e. two variables $a_m = a \oplus m$ and a are statistically independent, if the distribution of a_m is independent of the choice of a (for independently chosen uniformly distributed m).

This can be related to some elementary operations involving Boolean variables. Clearly, if a is arbitrary and m is chosen uniformly at random then $a_m = a \oplus m$ is uniformly distributed (and hence its distribution is the same irrespective of the choice of a). Furthermore, $a_{m_a} \times b_{m_b}$, $a_{m_a} \times m_b$, $(a_{m_a})^2$, $p \times a_{m_a}$ (p a constant), and $\sum a_i \oplus m$ can also be shown to be independent of the unmasked values a and b (see [9]). It follows directly that we can guarantee the independence of the output of any operation involving two masked input operands as long as the inputs are independently masked.

We note that we have the implicit assumption that only computation leaks, i.e. masks do not contribute to the leakage of the device when only stored in memory.

3.2 Masked Variables as Type Annotations

In order to allow static checking of secrecy we annotate each type with information to record if it may be revealed publically, or if it should be hidden, and if so which mask will be used. Rather than a variable a : int, this produces two alternatives, namely a : public int, and a : secret$\langle m \rangle$ int. The programmer's choice indicates to the compiler which state must remain hidden and which set of masks will be used to do so.

Hence, this additional type annotations allow a compiler to keep track of the 'flow' of masks and intermediate values, to check whether our basic masking rule holds, and if necessary to 'backtrack' variables if there is a problem and add masks to intermediate variables such that the basic masking rule applies. In other words, if a programmer implements a description of an algorithm without any masking, but declares variables related to key and/or state as secret, we can provide the rest of the masking automatically and hence relieve the programmer of that burden.

3.3 Assumptions

We make the following assumptions about the attacker, the programmer, and the device in the remainder of this article. The attacker has the ability to execute the program repeatedly, and on each execution run an observation (in form of the power consumption) is made on the values computed within the program. The attacker also has access to the inputs and output data of the encryption algorithm, only the masks and keys are hidden from the attacker. The programmer must mark every value as either secret or public: a valid compilation requires that all secret values have at least one mask. Both programmer declared variables and temporary variables inserted during compilation must meet this requirement.

public values are already known to the attacker. At no stage can a public value (statistically) depend upon a secret value in a computation. This is partially analogous to information flow (in the dependence constraint, also called non-interference).

secret values must be masked with random values which are chosen randomly from a suitable uniform distribution.

The device on which the cryptographic algorithm is implemented supplies (pseudo)random numbers which are uniformly distributed. In each execution run a new set of random numbers is selected and used as masks.

The goal of formalizing a model of countermeasures into a mechanically checkable procedure is not to prove that programs are leakage-free. Although such a goal is desirable with the current state of modelling the complexity of the power consumption characteristics of modern cryptographic devices it is not tractable. Rather we seek to automate the checking of necessary conditions that must be fulfilled in order for a program to be leakage-free. Although the specific characteristics of a particular device may still cause the program to leak information, the automation of the process enables further study of the specific issues.

4   Algorithm

The algorithm operates directly on an intermediate representation of the program. Our system initially parses the source into an Abstract Syntax Tree (AST), and then converts the AST into a list of instructions in 3-operand form. During conversion all constant bounded loops are statically unrolled and function calls are inlined. The programs that interest an attacker are ciphers with simple control-flow that are converted to straight-line code by this process. The result is a list of instructions and a set of initial variable declarations. Some of the declarations made by the programmer will have security annotations, none of the temporary variables introduced when converting expressions will be annotated.

The algorithm is designed to imitate the process used by a human engineer. The first step is inferring what is known about the security of each value in the program. Our system represents the security annotation as part of the type signature of each variable in the system; the secrecy of a value can be inferred from the secrecy of the operands and the kind of instruction used to create it analogously to the propagation of type information. We refer to this propagation phase as type inference, described in Section 4.1. After a single type inference pass two outcomes are possible:

1. Inference successfully checked the security of every value in the program and detected no leakages.
2. Inference operated to a point where it detected an error; a type was inferred that showed a leakage of information.

The first case is a successful conclusion and the algorithm terminates by outputting the program in the target assembly syntax. In the second case the algorithm has a record of the particular control point at which leakage occurred. The second phase of the algorithm attempts to prevent leakage using a set of program transformations that model the techniques an engineer currently uses in the same situation. The repair phase is described in Section 4.2.

4.1   Type Inference

Our prototype used in the experiments operates on a simplified version of the CAO type system [10]. In principle there should be no barriers to implementing the algorithm over the full set of CAO types. The algorithm maintains a security annotation for each type in the system, expressed as algebraic data-types these annotations are:

```
mask := Wildcard id | Named n
ann  := public | secret [mask]
```

Every type is either public or secured by a list of masks. Each mask is either named by the programmer or inserted by the compiler. Masks that have been named are used to specify contracts with external pieces of software (i.e. the caller of the routine). Wildcard masks are removed when possible by the compiler. The removal is via substitution of another mask and the process is guarded by the condition that no value can be reduced from more than zero masks to zero masks. In the example these annotations are attached to the following types:

```
type := byte ann | vector n ann | map ann ann
```

Individual byte variables have their own annotation (and hence set of masks), while vectors are assumed to be masked by the same set. Maps describe functions in which the input and output can be masked separately and are used to denote lookup tables such as S-boxes. We only demonstrate types related to byte-values as our target architecture is only 8-bit, although this formulation could be applied to values of any fixed size. Our conversion from AST to 3-address form unrolls loops statically, inlines function calls and converts to an SSA form. As each variable has only a single definition, the program type inference operates in a single forward pass in which the annotation of each variable is inferred from the operation in the instruction and the previously computed annotations of the source operands. The cases for type inference can thus be defined as rules that produce the type on the left when the pattern on the right matches:

```
public                      ← public  xor public
secretx̄                     ← secretx̄ xor public
secretȳ                     ← public  xor secretȳ
secret(x̄ ∪ ȳ) \ (x̄ ∩ ȳ) ← secretx̄ xor secretȳ
```

These rules can be verified from the definition that a value k with an annotation of $\text{secret}\bar{x} = \{x_1, \ldots, x_n\}$ is defined as $k \oplus x_1 \oplus \ldots x_n$, and the same mask in both source operands will cancel under two applications of xor. If any annotation is computed to be secret $\emptyset$ then the inference stops and an error is generated at that control point.

The rules for load and store operations are simpler as essentially we only need to preserve the masking for secret types.

```
secretx̄ ← load secretx̄ public
```

The second operand is the index (offset) in memory. After the loop unfolding during conversion these values are constant and thus known to the attacker. The case for a map is slightly more general (note that maps essentially implement masked table look-up operations):

```
secretȳ ← load (secretx̄ → ȳ) public
```

This assumes that the map has type $\text{secret}\bar{x} \to \bar{y}$. A consequence of these minimal definitions is that any case not included as a valid rule will cause the inference to fail with an error. This is commonly referred to as a 'closed world assumption'.

4.2   Repair Heuristics

Each of our repair rules is designed to function generally on any supplied input program. However, the set of rules is certainly not complete and requires expansion based on the study of other test cases. As a result of this incompleteness, we will refer to these rules as heuristics, although we emphasize that each rule is sound and guaranteed to preserve the security of the program being rewritten. While the compiler ensures the secrecy of variables that are annotated, and infers any necessary conditions on dependent variables (over-approximating where necessary) it cannot diagnose problems in the specification of secrecy that it is given. For example, if the programmer incorrectly labels a secret part of the state as `public` then the program will remain insecure upon compilation.

The repair phase operates after an inference pass and transforms a fixed set of leakages into secure operations. The inference pass handles all forward propagation of information through the acyclic data-flow graph. The type of each temporary value is infered by applying the inference rules to the instruction (and its operands) that produced it. As the program is in SSA form this propagation pushes information through the use-def chains in the program; the SSA form is defined implicitly in terms of the definition and uses of values. The repair phase looks for inconsistent triples of operand types, given the inference rule associated with the instruction:

Store violations occur when a secret value is stored in a public vector. The algorithm forms the repair by introducing a new copy of the vector protected by a fresh wildcard mask.

Map violations occur when a secret value is used as an index in a public map (e.g. if a key derived value indexes an S-box).

Mask collisions occur during a store operation when the mask set for the source operand does not equal the mask set for the target vector (maps are read-only).

Revelations occur when an instruction with secret operands produces a public value, e.g. the mask sets cancel out.

Both of the first two cases occur because the annotation of the structure in memory is less secure than the indexing value. In the case of the vector, a new copy is synthesized in which the elements are covered under a fresh wildcard mask. In the case where the map is a random shuffle applied to create a secure copy. The shuffle is defined by an input and an output mask: $S_{m \to n}[i] = S[i \text{ xor } m] \text{ xor } n$. In both cases substitution is used to convert the program to the secure form: for every following instruction both read and write accesses to the insecure structure are rewritten to use the secure version. A shuffling operation is synthesized to copy the public version into the secret version and inserted directly before the instruction causing the error. This operation is expensive as it requires remasking of the entire table. Positioning this operation before the first use places it before the beginning of the unrolled loop, affecting loop-hoisting of the expensive code.

Both of the second two cases occur because the propagation of the mask sets according to the rules defined in the preceding section have yielded a value that is insecure. In these cases the problem cannot be fixed where it is observed and the algorithm must find a source for the error that can be fixed. For each operand in the error-causing instruction the algorithm considers the set of ancestor values. For each ancestor the algorithm examines the effects of flipping (i.e. adding / removing) a single mask at a time in the mask-set of the ancestor. These single mask flips correspond to the effect of inserting one `xor` instruction on the ancestor value and rewriting the subsequent parts of the chain to use the altered value. In each case the algorithm checks if the problematic value is fixed, and whether any other values are revealed. If no successful repairs are found, the algorithm then considers pairs of flips amongst the ancestor values, triples etc. When the set of successful flips is non-empty the algorithm uses the number of inserted flips as a simple metric to choose the least-cost solution.

4.3   Combined Process

The two phases described are executed in alternating order.

1. Infer the types of all values starting from programmer declarations.
2. If an error occurred then perform a repair action on the program.
3. Repeat until no errors are found or a repair cannot be performed.

5   Worked Example

Algorithm 1 is sufficiently complex to necessitate a demonstration using a suitable example. The AES block cipher has been extensively used in previous work to demonstrate the working principle of (Boolean) masking schemes, and we apply our algorithm to a simplified version of it: we reduce it to a single round operating on a single column of the state.

The input language for our prototype compiler is derived from CAO. The rich type system of CAO is especially suitable for analysis [10] and previous work has shown that the collection types are of benefit in compiling block ciphers [11]. For the MixColumns stage our prototype requires a small number of data-types and so the input language is a subset of CAO.

```
Sbox, xtime : Byte -> Byte
key : secret<a> vector of Byte(4)
def mixcols( in:public vector of Byte(4) ) : secret<X> vector of Byte(4)
{
out : secret<X> vector of Byte(4) )
temp : vector of Byte(4)
 for i in range(4)
   temp[i] := Sbox[ in[i]^key[i] ]
 for i in range(4)
   out[i] := xtime[temp[i]] + temp[(i+1)%4] + xtime[temp[(i+1)%4]] +
             temp[(i+2)%4] + temp[(i+3)%4]
 return out
}
```

The type Byte is used to represent concrete data, while the higher-order type vector is used to indicate logical grouping. In contrast to C the use of an aggregate type does not imply anything about the representation in memory, and is simply a convenience for the programmer [11]. Each type is annotated by a security level. If the variable is already known to the attacker and can be freely revealed the annotation is public. When the variable must remain hidden a set of masks is specified with the secret annotation. In the example each declared set is a singleton although larger sets are inferred for temporary variables during compilation. The programmer has specified the existence of two masks in the source-code:

Algorithm 1. The Automatic Masking Algorithm

---

procedure repairMapViolation(pos,inst)

  $m = $ new Wildcard

  $n_T = $ secret $\{m\} \to o_T$ **where** origMap $= i_T \to o_T$

  substitute every use of original map with *mapM* from *pos* onwards

  insert instructions at *pos* to compute $mapM[i] := orig[i$ **xor** $o_T]$ **xor** m

procedure repairStoreViolation

  $m = $ new Wildcard

  rewrite vector type in *declarations* to secret $\{m\}$

procedure repairAncs

  $ancs := \{anc \mid anc \in \text{UDC}(operand),\ operand \in inst\}$

  $worklist := 2^{ancs}$ (sorted in increasing size and computed lazily)

  for each *ancset* in *worklist* **do**

    Choose one mask in each ancestor in *ancset*

    Flip the mask in each maskset and rerun the inference

    if no new values are made insecure and the problem value is made secure **then**

      Append (*mask, ancset*) to *results*

    end if

  end for

  Sort *results* by size of *ancset*

  if length *results* > 0 **then**

    Insert flip operations into program

  else

    Abort with an error

  end if

procedure topLevel

  while not finished **do**

    bindings := *declarations*

    for each *inst, pos* in *prog* **do**

      extract r, a, b, *operation* from *inst*

      a_T, b_T := lookup a, b in *bindings*

      r_T := infer from *operation*, a_T, b_T

      if not r in *bindings* **then**

        store $r \to r_T$ in *bindings*

      else if *operation* = store and $r_T \in$ vectors and $a_T < r_T$ **then**

        try repairStoreViolation

      else if *operation* = load and $r_T \in$ maps and $b_T < in(r_T)$ **then**

        try repairMapViolation

      else if $r_T \neq$ lookup r in *bindings* **then**

        try rewriting wildcard masks with declared masks to unify masksets

        if rewrite not possible **then**

          Abort with an error

        end if

      end if

      if $r_T <$ max a_T, b_T **then**

        try repairAncs

      end if

    end for

  end while

1. The key is covered by a mask **a**, as this masking operation must have occurred prior to the execution of the mixcols procedure this named mask forms part of an interface with the calling code.
2. The return value is covered by a mask **X**, again this forms part of an interface with the calling code.

Any variable without a security annotation is initially assumed to be public. If this assumption causes an error during the inference stage then it will be rewritten with a more secure annotation.

Both the `Sbox` and `xtime` functions are declared as public mappings from `Byte` to `Byte`. This leaves some flexibility in their definition, previous work [11] shows how declarative definitions can be provided and memorized into lookup-tables by a compiler, or the program can supply a constant array of bytes to encode the mapping.

The algorithm operates according to the process in Algorithm 1. We now illustrate some of the steps involved in iterating the inference and repair processes. Although the algorithm operates on the low-level 3-op form of the code this description will proceed at a source-level for reasons of space and clarity (including presenting the unrolled loops in a rolled form).

5.1   Map Violation Detected in `Sbox`

Type inference fills in intermediate types until it encounters the expression `Sbox[in[i]^key[i]]`. The type of `in[i]^key[i]` is inferred to be `secret<a>`, while the declaration of `Sbox` is of type `Byte -> Byte`. The compiler can "repair" this type error by synthesizing a new copy of `Sbox`, which we will call `Sboxm`. As this expression includes a part with a secret tag the minimum type annotation for `Sboxm` is given by `secret<x> Byte -> secret<y> Byte` for some secure x and y such that $x \neq y$. As the index expression is masked under **a** we can insert a new mask to produce an annotation of `secret<a> -> secret<b>` for some fresh mask **b**. This mask is called a wildcard as we may merge it with other masks later to reduce the number of random values required.

The new table `Sboxm` must be generated at runtime from the original `Sbox` table and the masks. The compiler inserts the following code:

```
Sboxm : secret<a> Byte -> secret<b> Byte
b : fresh Byte;
for i in range(256) :
 Sboxm[i] := Sbox[i^a] ^ b
```

5.2   Store Violation Detected in `temp`

As the programmer did not specify a security annotation for `temp` it defaulted to `public`. The output of the `Sbox` map is annotated by `secret<b>`. This causes an error in the inference as a secret value cannot be stored in a public variable. The compiler fixes this error by altering the declared type of `temp` to be `secret<b>`. This step is valid as it is always sound to increase the security of a variable. The compiler now expands temporaries in the expression evaluation and converts the access to use the masked table:

```
temp : secret<b> vector of Byte(4)
t    : secret<a> Byte
t2 : secret<b> Byte
for i in range(4) :
 t        := in[i] ^ key[i]
 t2       := Sboxm[t]
 temp[i] := t2
```

5.3   Revelation Detected in Second Loop

The algorithm proceeds into the second loop where it tries the following inference until it reaches an error:

```
t3 : secret<c> Byte     // Compiler inserted
t4 : Byte               // Compiler inserted
for i in range(4)
 t3 := xtime[temp[i]]
 t4 := t3 ^ temp[(i+1)%4]
```

The error arises because the type of t3 is declared to be the same type inferred for the expression temp[(i+1)%4]. The inference rules for an xor operation cancel out masks that appear on both sides producing the public annotation for t4. As a variable predecessor in the user defined code (UDC) is annotated secret this constitutes a revelation error. The compiler uses the process described in Section 4.2 to decide upon a repair. As one predecessor temp is a vector it would be more costly to flip the masks uniformly in each element, rather than simply flip the masks on t3. As flipping the existing masks does not produce a solution the compiler inserts a new wildcard mask b.

```
t3,t5 : secret<c,b> Byte
t4,t6 : secret<b>    Byte
out : secret<c,b> Vector of Bytes(4)
for i in range(4)
 t3 := xtime[temp[i]] ^ b              // Inserted flip operation
 t4 := t3 ^ temp[(i+1)%4]
 t5 := t4 ^ xtime[temp[(i+1)%4]]
 t6 := t5 ^ temp[(i+2)%4]
 out[i] := t6 ^ temp[(i+3)%4]
```

5.4   Mask Collisions Detected

Two subsequent iterations detect inequalities in the masking sets. These are resolved by unifying the wildcard mask c with the declared output mask X and inserting a flip operation to remove the mask b from the final result.

```
t3,t5 : secret<X,b> Byte
t4,t6 : secret<b>   Byte
t7 : secret<X,b>
out : secret<X> Vector of Bytes(4)
for i in range(4)
 t3 := xtime[temp[i]] ^ b
 t4 := t3 ^ temp[(i+1)%4]
 t5 := t4 ^ xtime[temp[(i+1)%4]]
 t6 := t5 ^ temp[(i+2)%4]
 t7 := t6 ^ temp[(i+3)%4]
 out[i] := t7 ^ b;
```

6 Application to Practice

Our discussion above demonstrates the working principle of Alg. 1. It explains how this algorithm transforms an insecure (i.e. unmasked) program (in our example this was AES reduced to `SubBytes` and `MixColumns` for the sake of brevity) into secure code (i.e. masked). We now briefly discuss practical aspects such as how the prototype compiler was implemented and the performance overhead resulting from automated masking.

6.1 Prototype Implementation

A prototype compiler was implemented in Haskell that reads the program source, applies Alg. 1, and then and outputs ARM assembly compatible with the Crossworks tools [12]. The use of a declarative language such as Haskell makes the implementation of a rule-based type-checker particularly simple. Haskell in particular is suited to embedding experimental languages due to the presence of monad transformers and their ability to add new forms of control flow.

An extract of this conversion, including annotation of masks as comments denoted a and b, is given below:

```
PUSH   {R3-R12,R14}
LDR    R5,  =in
LDR    R6,  =key
LDR    R7,  =Sbox_M
LDR    R8,  =temp
LDR    R9,  =xtime_M
LDR    R10, =out

LDRB   R0,  [R5, #0]     // [[]], [[]], ?
LDRB   R2,  [R6, #0]     // [[a]], [[a]], ?
EOR    R3,  R0, R2       // [[a]], [[]], [[a]]
LDRB   R0,  [R7, R3]     // [[b]], [[a],[b]], [[a]]
STRB   R0,  [R8, #0]     // [[b]], ?, [[b]]
```

This code snippet shows the sequence of assembly instructions from pushing some registers onto the stack when the function is called, to loading the first byte of the AES state and key, exclusive-oring these two bytes and using them as index for the SubBytes operation. As described before, a masked SubBytes table must be generated each time the AES code is executed (to facilitate readability this is however not included in the code shown here). Then the result of the SubBytes operation is stored. The code is annotated with comments that show how each register is masked for each instruction, where the mask is given provided between the brackets [[ ]]. As we assume that the plaintext is public (i.e. unmasked) and the key is secret (i.e. masked), the first line which refers to loading the plaintext shows an empty masking set. The second line which refers to the loading of the key shows that the mask a is used. In the third line where input and key are exclusive-ored, the result inherits the mask from the key. The fourth line, which refers to the SubBytes operation, show that this operation has a masked input (mask a is used) and maps this input to a value which is masked differently (mask b is used).

6.2 Performance

Performance comparisons are typically highly context specific, in our case it is useful to bear in mind that different strategies such as loop unrolling lead to very different code sizes and execution times. Consequently, we provide two more code snippets, the left-hand one showing an implementation which was hand-coded and uses loops, the right-hand one showing an implementation which was hand-coded and unrolls these loops:

```
        PUSH   {R3-R12,R14}
        BL     SubBytes
// --------------------------
// SubBytes
// Input : R1 - pointer to data
// Output : @R1
// --------------------------
SubBytes:
        MOV    R5, #4
        LDR    R6, =acAESsbox
SubBytes:
        SUB    R5, R5, #1
        LDRB   R7, [R1, R5]
        LDRB   R8, [R6, R7]
        STRB   R8, [R1, R5]
        CMP    R5, #0
        BNE    SubBytes
        BX     LR
```

```
// -------------------
// Macros
// -------------------

.macro  Msub    i=0
        LDRB   R5, [R0, \i]
        LDRB   R5, [R4, R5]
        STRB   R5, [R0, \i]
.endm




        PUSH   {R3-R12,R14}
        LDR    R4, =acAESsbox
Ssub_s0: Msub    #0
```

The clear difference in coding styles leads to different performance figures. The hand-coded assembly version with loops requires 147 clock cycles to compute the

SubBytes and MixColumns function (for one column of the state only), whereas an unrolled version (hand-coded) requires 52 clock cycles. Our algorithm that automatically adds the masking to an unmasked implementation produces code that requires 76 clock cycles.

Another aspect to consider is how many masks are introduced. This choice also depends on performance considerations: MixColumns can be securely masked using four masks but fewer masks are possible if the performance overhead for remasking is acceptable. Our algorithm in general first draws from the set of already existing masks and only adds a new mask if the resulting errors cannot be resolved otherwise. Consequently, our algorithm will lead to an implementation that requires the least possible amount of randomness.

7   Discussion and Outlook

In this paper we detail an algorithm for the automated generation of code that is resistant to first-order DPA, and illustrated the working principle on a concrete and relevant example. While the source code needs to be written in a particular format in which a developer can indicate what needs to be protected against leakage (i.e. for example the cryptographic key), a developer does not need to have any further knowledge about Boolean masking or even the assembly language of a given microprocessor. Indeed, given that our compiler produces code that is comparable to assembly code written by a human, one could use the same source for numerous platforms reducing development cost considerably.

The current version assumes that the target microprocessor leaks information independently for each instruction executed. Some devices may leak information in a different model, where the information leakage depends on consecutive instructions. This may impose a further restriction on the compiler, i.e. that variables masked with the same mask cannot be manipulated in adjacent instructions, and subsequent iterations of our compiler will seek to address this issue.

Our compiler was designed to produce code that would be resistant to first-order DPA. Clearly, higher-order masking schemes as recently reported in the literature ([13], [14]) necessitate a wider range of schemes and operations than what our compiler currently supports. Implementing a wider range of operations and schemes could be achieved using a domain specific language, similar to what has been recently suggested for computing on encrypted data in ([15]). Our current approach is then interesting as our central contribution, which is the static analysis of types w.r.t. information leakage, could complement such a language definition and allow minimising the overall number of masks without compromising the security of the implementation.

Acknowledgments. This work has been supported in part by EPSRC via grants EP/I005226/1 and EP/H001689/1, and also been supported in part the European Commission through the ICT Programme under Contract ICT-2007-216676 ECRYPT II.

References

1. Cowan, C., Pu, C., Maier, D., Hinton, H., Walpole, J., Bakke, P., Beattie, S., Grier, A., Wagle, P., Zhang, Q.: Stackguard: Automatic adaptive detection and prevention of buffer-overflow attacks. In: USENIX Security Symposium, pp. 63–78 (1998)

2. Agat, J.: Type based techniques for covert channel elimination and register allocation. PhD thesis, Chalmers University of Technology (2001)

3. Molnar, D., Piotrowski, M., Schultz, D., Wagner, D.: The Program Counter Security Model: Automatic Detection and Removal of Control-Flow Side Channel Attacks. In: Won, D.H., Kim, S. (eds.) ICISC 2005. LNCS, vol. 3935, pp. 156–168. Springer, Heidelberg (2006)

4. Lux, A., Starostin, A.: A tool for static detection of timing channels in Java. In: Constructive Side-Channel Analysis and Secure Design (COSADE), pp. 126–140. CASED (2011)

5. Regazzoni, F., Cevrero, A., Standaert, F.-X., Badel, S., Kluter, T., Brisk, P., Leblebici, Y., Ienne, P.: A Design Flow and Evaluation Framework for DPA-Resistant Instruction Set Extensions. In: Clavier, C., Gaj, K. (eds.) CHES 2009. LNCS, vol. 5747, pp. 205–219. Springer, Heidelberg (2009)

6. Tiri, K., Verbauwhede, I.: A digital design flow for secure integrated circuits. IEEE Trans. on CAD of Integrated Circuits and Systems 25(7), 1197–1208 (2006)

7. Zdancewic, S.A.: Programming Languages for Information Security. PhD thesis, Cornell University (August 2002)

8. Blömer, J., Guajardo, J., Krummel, V.: Provably Secure Masking of AES. In: Handschuh, H., Hasan, M.A. (eds.) SAC 2004. LNCS, vol. 3357, pp. 69–83. Springer, Heidelberg (2004)

9. Oswald, E., Mangard, S., Pramstaller, N., Rijmen, V.: A Side-Channel Analysis Resistant Description of the AES S-Box. In: Gilbert, H., Handschuh, H. (eds.) FSE 2005. LNCS, vol. 3557, pp. 413–423. Springer, Heidelberg (2005)

10. Barbosa, M., Moss, A., Page, D., Rodrigues, N., Silva, P.F.: A domain-specific type system for cryptographic components. In: Fundamentals of Software Engineering, FSEN (2011)

11. Moss, A., Page, D.: Bridging the gap between symbolic and efficient AES implementations. In: Gallagher, J.P., Voigtländer, J. (eds.) Partial Evaluation and Program Manipulation (PEPM), pp. 101–110. ACM (2010)

12. Crossworks for ARM, http://www.rowley.co.uk/arm/

13. Fumaroli, G., Martinelli, A., Prouff, E., Rivain, M.: Affine Masking against Higher-Order Side Channel Analysis. In: Biryukov, A., Gong, G., Stinson, D.R. (eds.) SAC 2010. LNCS, vol. 6544, pp. 262–280. Springer, Heidelberg (2011)

14. Genelle, L., Prouff, E., Quisquater, M.: Thwarting Higher-Order Side Channel Analysis with Additive and Multiplicative Maskings. In: Preneel, B., Takagi, T. (eds.) CHES 2011. LNCS, vol. 6917, pp. 240–255. Springer, Heidelberg (2011)

15. Bain, A., Mitchell, J., Sharma, R., Stefan, D., Zimmerman, J.: A domain-specific language for computing on encrypted data. Cryptology ePrint Archive, Report 2011/561 (2011), http://eprint.iacr.org/

16. ARM7TDMI technical reference manual, http://infocenter.arm.com/help/index.jsp?topic=/com.arm.doc.ddi0406b/index.html

A   Results of a DPA

Whilst we have argued that our algorithm ensures the necessary conditions for masking to be secure in practice one can be reluctant to accept this without some 'concrete' evidence from at least for one 'practical' device.

In the remainder of this section we describe some experiments that were conducted on an ARM7TDMI microprocessor [16] using the example described in Section 5. We use a simple experimental board on which a microprocessor is mounted to acquire power traces using a 'standard' setup: a differential probe is used to acquire power traces. We use a suitable sampling frequency and have an artificially generated trigger point which eases the alignment of traces.

The first experiments focused on compiling the **unmasked** code described in Section 5 using a standard C compiler. In our case this was Crossworks for ARM and the code fragment required 183 clock cycles to execute. We acquired 2000 traces showing the power consumption during the execution of the code, for a constant secret key and a randomly generated input, and performed some standard DPA attacks on those traces (targeting the SubBytes output since this is known to be a good target for AES).

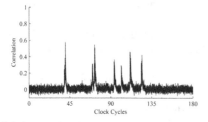 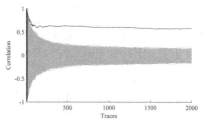

(a) A correlation trace using 2000 traces.
(b) The maximum correlation plotted against the number of traces. The correct hypothesis plotted in black and the incorrect hypotheses in gray.

Fig. 1. Results of a DPA attack on the unmasked implementation which give clear evidence for the vulnerability of such an implementation on the target platform

The correlation trace for the correct hypothesis is shown in the left panel of Fig. 1, where numerous peaks show at what points in time the result of the substitution table is processed by the microprocessor. The right panel shows that as little as 100 acquisitions would be necessary to reveal the correct key byte.

We performed the same analysis on the code **masked** by our algorithm and translated into ARM assembly which requires 76 clock cycles to execute. Fig. 2 shows the results. There are no distinctive peaks in the correlation trace for the correct key hypothesis (left panel). The right panel confirms that any peaks in traces of incorrect key hypotheses are equally significant which means that

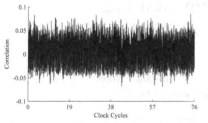 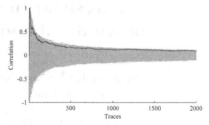

(a) A correlation trace using 2000 traces. (b) The maximum correlation plotted against the number of traces. The correct hypothesis plotted in black and the incorrect hypotheses in gray.

Fig. 2. Results of a DPA attack on the masked implementation demonstrating that no information leaks

no distinction between key hypotheses is possible. Given the strong leakage signals present in the unmasked implementation, even with very few traces, this is a practical confirmation that the masked code is resistant to DPA attacks on this platform (since even a large number of traces produces no distinctive correlation).

Threshold Implementations
of All 3 × 3 and 4 × 4 S-Boxes[*]

Begül Bilgin[1,3], Svetla Nikova[1], Ventzislav Nikov[4],
Vincent Rijmen[1,2], and Georg Stütz[2]

[1] KU Leuven, Dept. ESAT/SCD-COSIC and IBBT, Belgium
[2] Graz University of Technology, IAIK, Austria
[3] University of Twente, EEMCS-DIES, The Netherlands
[4] NXP Semiconductors, Belgium

Abstract. Side-channel attacks have proven many hardware implementations of cryptographic algorithms to be vulnerable. A recently proposed masking method, based on secret sharing and multi-party computation methods, introduces a set of sufficient requirements for implementations to be provably resistant against first-order DPA with minimal assumptions on the hardware. The original paper doesn't describe how to construct the Boolean functions that are to be used in the implementation. In this paper, we derive the functions for all invertible 3×3, 4×4 S-boxes and the 6×4 DES S-boxes. Our methods and observations can also be used to accelerate the search for sharings of larger (e.g. 8×8) S-boxes. Finally, we investigate the cost of such protection.

Keywords: DPA, masking, glitches, sharing, nonlinear functions, S-box, decomposition.

1 Introduction

Side-channel analysis exploits the information leaked during the computation of a cryptographic algorithm. The most common technique is to analyze the power consumption of a cryptographic device using differential power analysis (DPA). This side-channel attack exploits the correlation between the instantaneous power consumption of a device and the intermediate results of a cryptographic algorithm.

Several countermeasures against side-channel attacks have been proposed. Circuit design approaches try to balance the power consumption of different data values [31]. Another method is to randomize the intermediate values of an algorithm

[*] This work was supported in part by the Research Council KU Leuven: GOA TENSE (GOA/11/007) and by the European Commission under contracts ICT-2007-216646 (ECRYPT II). V. Nikov was supported by the European Commission (FP7) within the Tamper Resistant Sensor Node (TAMPRES) project with contract number 258754 and the Internet of Things - Architecture (IoT-A) project with contract number 257521.

E. Prouff and P. Schaumont (Eds.): CHES 2012, LNCS 7428, pp. 76–91, 2012.

by masking them. This can be done at the algorithm level [1,5,12,24], at the gate level [13,27,32] or even in combination with circuit design approaches [25].

Many of these approaches result in very secure *software* implementations. However, it has been shown that *hardware* implementations are much more difficult to protect against DPA [17]. The problem of most of these masking approaches is that they underestimate the amount of information that is leaked by hardware, for instance during glitches or other transient effects. The security proofs are based on an idealized hardware model, resulting in requirements on the hardware that are very expensive to meet in practice. The main advantages of the threshold implementation approach are that it provides provable security against first-order DPA attacks with minimal assumptions on the hardware technology, in particular, it is also secure in the presence of glitches, and that the method allows to construct realistic-size circuits [20,22,23].

1.1 Organization and Contributions of This Paper

The remainder of this paper is organized as follows. In Section 2 we introduce the notation and provide some background material. Section 2.6 contains our first contribution: a classification of S-boxes which simplifies the task to find implementations for all S-boxes. In Section 3 we present our second contribution: a method to decompose permutations as a composition of quadratic ones. We prove that all 4-bit S-boxes in the alternating group can be decomposed in this way. We extend the sharing method in Section 4 and show that all 3×3, 4×4 and DES 6×4 S-boxes can be shared with minimum 3 and/or 4 shares. We investigate the cost of an HW implementation of the shared S-boxes in Section 5. Some ideas for further improvements will be provided in the full version of the paper [2]. Finally, we conclude in Section 6.

2 Preliminaries

We consider n-bit permutations sometimes defined over a vector space $\mathcal{F}_2^n$ or over a finite field $GF(2^n)$. The degree of such a permutation F is the algebraic degree of the (n, n) vectorial Boolean function [6] or also called n-bit S-box. Any such function $F(x)$ can be considered as an n-tuple of Boolean functions $(f_1(x), \ldots, f_n(x))$ called the coordinate functions of $F(x)$.

2.1 Threshold Implementations

Threshold implementations (TI), are a kind of side-channel attack countermeasures, based on secret sharing schemes and techniques from multiparty computation. The approach can be summarized as follows. Split a variable x into s additive shares x_i with $x = \sum_i x_i$ and denote the vector of the s shares x_i by $\mathbf{x} = (x_1, x_2, \ldots, x_s)$. In order to implement a function $a = F(x, y, z, \ldots)$ from $\mathcal{F}_2^m$ to $\mathcal{F}_2^n$, the TI method requires a *sharing*, i.e. a set of s functions F_i which together compute the output(s) of F. A sharing needs to satisfy three properties:

Correctness: $a = F(x, y, z, \dots) = \sum_i F_i(\mathbf{x}, \mathbf{y}, \mathbf{z}, \dots)$ for all $\mathbf{x}, \mathbf{y}, \mathbf{z}, \dots$ satisfying $\sum_i x_i = x, \sum_i y_i = y, \sum_i z_i = z, \dots$

Non-completeness: Every function is independent of at least one share of the input variables x, y, z. This is often translated to "F_i should be independent of $x_i, y_i, z_i, \dots$."

Uniformity (balancedness): For all $(a_1, a_2, \dots, a_s)$ satisfying $\sum_i a_i = a$, the number of tuples $(\mathbf{x}, \mathbf{y}, \mathbf{z}, \dots) \in \mathcal{F}^{ms}$ for which $F_j(\mathbf{x}, \mathbf{y}, \mathbf{z}, \dots) = a_j, 1 \leq j \leq s$, is equal to $2^{(s-1)(m-n)}$ times the number of $(x, y, z, \dots) \in \mathcal{F}^m$ for which $a = F(x, y, z, \dots)$. Hence, if F is a permutation on $\mathcal{F}^m$, then the functions F_i define together a permutation on $\mathcal{F}^{ms}$. In other words, the sharing preserves the output distribution.

This approach results in combinational logic with the following properties. Firstly, since each F_i is completely independent of the unmasked values, also the sub-circuits implementing them are, even in the presence of glitches. Because of the linearity of the expectation operator, the same holds true for the average power consumption of the whole circuit, or any linear combination of the power consumptions of the subcircuits. This implies perfect resistance against all first-order side-channel attacks [23]. The approach was recently extended and applied to Noekeon [23], Keccak [4], Present [26] and AES [19]. Whereas it is easy to construct for any function a sharing satisfying the first two properties, the uniformity property poses more problems. Hence reasonable questions to ask are: which functions (S-boxes) can be shared with this approach, how many shares are required and how can we construct such sharing?

A similar approach was followed in [28], where Shamir's secret sharing scheme is used to construct hardware secure against dth-order side-channel attacks in the presence of glitches. Instead of constructing dedicated functions F_i, they propose a general method which replaces every field multiplication by $4d^3$ field multiplications and $4d^3$ additions, using $2d^2$ bytes of randomness. While the method is applicable everywhere, in principle, there are cases where it may prove too costly.

2.2   Decomposition as a Tool to Facilitate Sharing

In order to share a nonlinear function (S-box) with algebraic degree d, at least $d + 1$ shares are needed [20, Theorem 1]. Several examples of functions shared with 3 shares, namely quadratic Boolean function of two and three variables, multiplication on the extension field $GF(2^{2m})/GF(2^m)$ (e.g. multiplication in $GF(4)$), and the Noekeon S-box have been provided [20, 22, 23]. A realization of the inversion in $GF(16)$ with 5 shares was given in [20]. Since the area requirements of an implementation increase with the number of shares, it is desirable to keep the number of shares as low as possible.

The block ciphers Noekeon and Present have been designed for compact hardware implementations. They have S-boxes, which are not very complex 4×4 cubic permutations. Realizations for these two block ciphers have been presented for Noekeon in [22, 23] and in [26] for Present. In order to decrease the algebraic degree of the functions for which sharings need to be found, these three realizations decompose the S-box into two parts. For the Present S-box, decompositions $S(x) = F(G(x))$ with $G(0) = 0$ have been found where $F(x)$ and $G(x)$ are quadratic permutations [26]. By varying the constant term $G(0)$ the authors found all possible decompositions of $S(X) = F(G(X))$. Both S-boxes $F(x), G(x)$ have been shared with three shares (F_1, F_2, F_3) and (G_1, G_2, G_3) that are correct, non-complete and uniform.

When the AES S-box (with algebraic degree seven) is presented using the tower field approach, the only nonlinear operation is the multiplication in $GF(4)$, which is a quadratic mapping [19]. This observation has lead to a TI for AES with 3 shares. In order to guarantee the uniformity, re-sharing (also called re-masking) has been used four times. Re-sharing is a technique where fresh uniform and random masks/shares are added inside a pipeline stage in order to make the shares follow an uniform distribution again.

A novel fault attack technique against several AES cores including one claimed to be protected with TI method has been proposed in [18]. But as the authors pointed out, contrary to the AES TI implementation in [19], their targeted core has been made without satisfying the non-completeness and uniformity properties by "sharing" the AND gates with 4 shares formula from [19, 20]. Since the used method does not satisfy the TI properties it should not be called a TI implementation of AES. In addition, the TI method was never claimed to provide protection against fault attacks.

2.3   Equivalence Classes for $n = 2, 3, 4$

Definition 1 ([8]). *Two S-boxes $S_1(x)$ and $S_2(x)$ are* affine/linear equivalent *if there exists a pair of invertible affine/linear permutation $A(x)$ and $B(x)$, such that $S_1 = B \circ S_2 \circ A$.*

Every invertible affine permutation $A(x)$ can be written as $\mathrm{A} \cdot x + a$ with a an n-bit constant and A an $n \times n$ matrix which is invertible over $GF(2)$. It follows that there are $2^n \times \prod_{i=0}^{n-1}(2^n - 2^i)$ different invertible affine permutations.

The relation "being affine equivalent" can be used to define equivalence classes. We now investigate the number of classes of invertible $n \times n$ S-boxes for $n = 2, 3, 4$. Note that the algebraic degree is affine invariant, hence all S-boxes in a class have the same algebraic degree.

It is well known that all invertible 2×2 S-boxes are affine, hence there is only one class. The set of invertible 3×3 S-boxes contains 4 equivalence classes [8]: 3 classes containing quadratic functions, and one class containing the affine functions. We will provide a table with a representative of each class in the full version of the paper [2].

The maximal algebraic degree of a balanced 4-variable Boolean function is 3 [7, 16]. De Cannière uses an algorithm to search for the affine equivalent classes which guesses the affine permutation A for as few input points as possible, and then uses the linearity of A and B to follow the implications of these guesses as far as possible. This search is accelerated by applying the next observation, which follows from linear algebra arguments (change of basis):

Lemma 1 ([15]). *Let S be an $n \times n$ bijection. Then S is affine equivalent to an S-box $\tilde{S}$ with $\tilde{S}(0) = 0$, $\tilde{S}(1) = 1$, $\tilde{S}(2) = 2$, ..., $\tilde{S}(2^{n-1}) = 2^{n-1}$.*

In the case $n = 4$, this observation reduces the search space from $16! \approx 2^{44}$ to $11! \approx 2^{25}$.

De Cannière lists the 302 equivalence classes for the 4×4 bijections [8]: the class of affine functions, 6 classes containing quadratic functions and the remaining 295 classes containing cubic functions.[1] We will list the classes in the full version of the paper [2]. The numbering of the classes is derived from the lexicographical ordering of the truth tables of the S-boxes. In order to increase readability, we introduce the following notation $\mathcal{A}_i^n$, $\mathcal{Q}_j^n$, $\mathcal{C}_k^n$ to denote the Affine class number i, Quadratic class number j and Cubic class number k of permutations of $\mathcal{F}_2^n$.

2.4 Order of a Permutation

All bijections from a set X to itself (also called permutations) form the *symmetric group* on X denoted by S_X. A transposition is a permutation which exchanges two elements and keeps all others fixed. A classical theorem states that every permutation can be written as a product of transpositions [29], and although the representation of a permutation as a product of transpositions is not unique, the number of transpositions needed to represent a given permutation is either always even or always odd. The set of all even permutations form a normal subgroup of S_X, which is called the *alternating group* on X and denoted by A_X. The alternating group contains half of the elements of S_X. Instead of A_X and S_X, we will write here A_n and S_n, where n is the size of the set X.

2.5 Known S-Boxes and Their Classes

There are only few cryptographically significant 3×3 S-boxes: the Inversion in $GF(2^3)$, the PRINTcipher, the Threeway and the Baseking S-boxes. They all belong to Class 3. There are many cryptographically significant 4×4 S-boxes. To mention some of them: Twofish, Gost, Serpent, Lucifer, Clefia, HB1, HB2, mCrypton, Klein, Khazad, Iceberg, Puffin, Present, Luffa, Hamsi, JH, Noekeon, Piccolo.

2.6 The Inverse S-Box

Note that S^{-1}, the inverse S-box, is not necessarily affine equivalent to S and in this case may not have the same algebraic degree. We know however, that the inverse of an affine permutation is always an affine permutation. In the case of

---

[1] Independent of [8, 15], Saarinen classified the 4×4 S-boxes using a different equivalence relation [30].

3 × 3 S-boxes it follows that the inverse of a quadratic permutation is again a quadratic permutation. Moreover, it can be shown that the 3 quadratic classes in S_8 are self-inverse, i.e. S^{-1} belongs to the same class as S. In the case $n = 4$, we can apply the following lemma.

Lemma 2 ([6]). *Let F be a permutation of $GF(2^n)$, then $deg(F^{-1}) = n - 1$ if and only if $deg(F) = n - 1$.*

Since the inverse of an affine S-box is affine, and, when $n = 4$, the inverse of a cubic S-box is cubic, it follows that in this case the inverse of a quadratic S-box is quadratic. The Keccak S-box ($n = 5$) is an example where the algebraic degree of the inverse S-box (3) is different from the algebraic degree of the S-box itself (2) [3].

We have observed that there are 172 self-inverse classes in S_{16}. The remaining 130 classes form 65 pairs, i.e., any S-box S of the first class has an inverse S-box S^{-1} in the second class (and vice versa). We will provide the list of the pairs of inverse classes in the full version of the paper [2].

3   Decomposition of 4 × 4 S-Boxes

In this section we consider all 4 × 4 bijections, and investigate when a cubic bijection from S_{16} can be decomposed as a *composition of quadratic bijections*. We will refer to the minimum number of quadratic bijections in such a decomposition as *decomposition length*. Recall that the Noekeon S-box is cubic but defined as a composition of two quadratic S-boxes in $\mathcal{F}_2^4$: $S(x) = S_2(S_1(x))$. Similarly the Present S-box is cubic but has also been shown to be decomposable in two quadratic S-boxes.

Lemma 3. *If an S-box S can be decomposed into a sequence of t quadratic S-boxes, then all S-boxes which are affine equivalent to S can be decomposed into a sequence of t quadratic S-boxes.*

Lemma 4 ([33]). *For all n, the $n \times n$ affine bijections are in the alternating group.*

Lemma 5. *All 4 × 4 quadratic S-boxes belong to the alternating group A_{16}.*

Proof. Since all invertible affine transformations are in the alternating group (the previous Lemma), two S-boxes which are affine equivalent, are either both even or both odd. We have taken one representative of each of the 6 quadratic classes $\mathcal{Q}_i^4$ for $i \in \{4, 12, 293, 294, 299, 300\}$ [8] and have verified that their parities are even.   □

Now we investigate which permutations we can generate by combining the affine and the quadratic permutations. We start with the following lemma.

Lemma 6. *Let Q_i be 6 arbitrarily selected representatives of the 6 quadratic classes $\mathcal{Q}_i^4$. (Hence $i \in \{4, 12, 293, 294, 299, 300\}$.) Then all cubic permutations*

S that have decomposition length 2, are affine equivalent to one of the cubic permutation that can be written as

$$\tilde{S}_{i\times j} = Q_i \circ A \circ Q_j, \tag{1}$$

where A is an invertible affine permutation and $i, j \in \{4, 12, 293, 294, 299, 300\}$.

It follows that we can construct all cubic classes of decomposition length 2 by running through the 36 possibilities of $i \times j$ and the 322560 invertible affine transformations in (1). This approach produces 30 cubic classes. In the remainder, we will denote the S-boxes $\tilde{S}_{i\times j}$ by $i \times j$ and refer to them as the *simple solutions*. In the full version of the paper [2] we provide the list of the simple solutions for all 30 decompositions with length 2. Note that if $Q_i \circ A \circ Q_j = S$, i.e. S can be decomposed as a product of $i \times j$, then $Q_j^{-1} \circ A^{-1} \circ Q_i^{-1} = S^{-1}$. Since for $n = 4$ all quadratics are affine equivalent to their inverse, it follows that S^{-1} is decomposed as a product of $j \times i$. Thus any self-inverse class has decomposition $i \times j$ and $j \times i$ as well. For the pairs of inverse classes we conclude that if $i \times j$ belongs to the first class then $j \times i$ belongs to the second class.

To obtain all decompositions with length 3 we use similar approach as for length 2 but the first permutation Q_i is cubic (instead of quadratic) and belongs to the already found list of cubic classes decomposable with length 2. It turns out that we can generate in this way the 114 remaining elements of A_{16}.

Summarizing, we can prove the following Theorem and Lemma (stated without proof in [9]).

Theorem 1. *A 4×4 bijection can be decomposed using quadratic bijections if and only if it belongs to the alternating group A_{16} (151 classes).*

Proof. ($\Rightarrow$) Let S be a bijection which can be decomposed with quadratic permutations say $Q_1 \circ Q_2 \circ \ldots \circ Q_t$. Since all $Q_i \in A_{16}$ (Lemma 5) and the alternating group is closed it follows that $S \in A_{16}$.
($\Leftarrow$) Lemma 3, Lemma 6 and the discussion following it imply that we can generate all elements of the alternating group using quadratic permutations. $\square$

The left-hand-side columns of Table 1 list the decompositions of all 4×4 S-boxes. Theorem 1 implies that the classes which are not in the alternative group i.e. in $S_{16} \setminus A_{16}$, can't be decomposed as a product of quadratic classes. Now we make the following simple observation:

Lemma 7. *Let $\tilde{S}$ be a fixed permutation in $S_{16} \setminus A_{16}$ then any cubic permutation from $S_{16} \setminus A_{16}$ can be presented as a product of $\tilde{S}$ and a permutation from A_{16}.*

4 Sharing with 3, 4 and 5 Shares

In this section we focus first on the permutations which can be shared with 3 shares, i.e. all S-boxes in $\mathcal{F}_2^3$ and half of the S-boxes in $\mathcal{F}_2^4$. Next we focus on those functions that can be shared with 4 shares, i.e. the other half of the S-boxes in $\mathcal{F}_2^4$. Then, we will show how to share all of these S-boxes in $\mathcal{F}_2^4$ with 5 shares without need of a decomposition.

4.1   A Basic Result

Theorem 2. *If we have a sharing for a representative of a class, then we can derive a sharing for all S-boxes from the same class.*

Proof. Let S be an $n \times n$ S-box which has a uniform, non-complete and correct sharing $\bar{S}$ using s shares S_i. Denote the input vector of S by x, and the shares by x_i. Each S_i contains n coordinate shared functions depending on at most $(s-1)$ of the x_i, such that the noncompleteness property is satisfied. We denote by $\mathbf{x}_i$ the vector containing the $s-1$ inputs of S_i.

   We now construct a uniform, non-complete and correct sharing for any S-box $\tilde{S}$ which is affine equivalent to S. By definition, there exist two $n \times n$ invertible affine permutations A and B s.t. $\tilde{S} = B \circ S \circ A$. In order to lighten notation, we give the proof for the case that A and B are linear permutations. We define $\bar{A}, \bar{B}$ as the $ns \times ns$ permutations that apply A, respectively B, to each of the shares separately:

$$\bar{A}(x_1, x_2, \ldots x_s) = (A(x_1), A(x_2), \ldots A(x_s)),$$
$$\bar{B}(x_1, x_2, \ldots x_s) = (B(x_1), B(x_2), \ldots B(x_s)).$$

Denote $y_i = A(x_i), 1 \leq i \leq s$ and define $\mathbf{y}_i$ as the vector containing the $s-1$ shares y_i that we need to compute S_i. Consider $\bar{S}(\bar{A}(x_1, x_2, \ldots, x_s)) = (S_1(\mathbf{y}_1), S_2(\mathbf{y}_2), \ldots S_s(\mathbf{y}_s))$. By slight abuse of notation we can write $\mathbf{y}_i = \bar{A}(\mathbf{x}_i)$ and see that the noncompleteness of the $\bar{S}_i$ is preserved in $\bar{S} \circ \bar{A}$. Since $\bar{A}$ is a permutation, it preserves the uniformity of the input and since $\bar{S}$ is uniform so will be the composition $\bar{S} \circ \bar{A}$. The correctness follows from the fact that $\bar{S}$ is a correct sharing and that

$$y_1 + y_2 + \cdots + y_s = A(x_1) + A(x_2) + \cdots + A(x_s) = A(x_1 + x_2 + \ldots x_s) = A(x).$$

Consider now $\bar{B}(\bar{S}(A(x))) = (B(S_1(\mathbf{y}_1)), B(S_2(\mathbf{y}_2)), \ldots, B(S_s(\mathbf{y}_s)))$. Since $\bar{B}$ is a permutation, it preserves uniformity of the output and since $\bar{S}$ is uniform, the composition $\bar{B} \circ \bar{S}$ is uniform. The composition is non-complete since the $\bar{S}_i$ are non-complete and $\bar{B}$ doesn't combine different shares. Correctness follows from the fact that $\bar{S}$ is a correct sharing and hence

$$B(S_1(\mathbf{y}_1)) + B(S_2(\mathbf{y}_2)) + \cdots + B(S_s(\mathbf{y}_s))$$
$$= B(S_1(\mathbf{y}_1) + S_2(\mathbf{y}_2) + \cdots + S_s(\mathbf{y}_s)) = B(S(A(x))). \qquad \square$$

4.2   Direct Sharing

The most difficult property to be satisfied when the function is shared is the uniformity. Assume that we want to construct a sharing for the function $F(x, y, z)$ with 3 shares. Then it is easy to produce a sharing which satisfies the correctness and the non-completeness requirements and is rotation symmetric, by means of a method that we call the *direct sharing method*, and that we now describe.

First, we replace every input variable by the sum of 3 shares. The correctness is satisfied if we ensure that

$$F_1 + F_2 + F_3 = F(x_1 + x_2 + x_3, y_1 + y_2 + y_3, z_1 + z_2 + z_3).$$

In order to satisfy non-completeness, we have to divide the terms of the right hand side over the three F_j in such a way that F_j doesn't contain a term in x_j. We achieve this by assigning the linear terms containing an index j to F_{j-1}, the quadratic terms containing indices j and $j+1$ to F_{j-1} and the quadratic terms containing indices j only to F_{j-1}. For example,

$$F(x, y, z) = x + yz, \quad \text{gives:}$$
$$F_1 = x_2 + z_2 y_2 + z_2 y_3 + z_3 y_2$$
$$F_2 = x_3 + z_3 y_3 + z_3 y_1 + z_1 y_3$$
$$F_3 = x_1 + z_1 y_1 + z_1 y_2 + z_2 y_1.$$

Note that the uniformity of sharing produced in this way is not guaranteed. It has to be verified separately. The method can easily be generalized for larger number of shares.

Direct sharing has been used in [26] for the decomposition of the quadratic permutations F and G of the Present S-box S and similarly for Noekeon [23], Keccak [4].

With the direct sharing method we were able to find sharings respecting the uniformity condition for all 1344 permutations of $\mathcal{Q}_1^3$, but none of $\mathcal{Q}_2^3$ and $\mathcal{Q}_3^3$. We were also able to find sharings for all 322560 permutations of $\mathcal{Q}_4^4$, $\mathcal{Q}_{294}^4$ and $\mathcal{Q}_{299}^4$, but none of $\mathcal{Q}_{12}^4$, $\mathcal{Q}_{293}^4$ and $\mathcal{Q}_{300}^4$. So, unfortunately half of the quadratic S-boxes can't be shared directly with length 1 but we still can find a sharing with length 2 by decomposing them as a composition of the already shared quadratic S-boxes. Thus, if we use only direct sharing we will be able to find sharings for all S-boxes in the alternating group but at the cost of longer path.

4.3 Correction Terms

Since direct sharing not always results in an uniform sharing the use of *correction terms* (CT) has been proposed [20, 22]. Correction terms are terms that can be added in pairs to more than one share such that they satisfy the non-completeness rule. Since the terms in a pair cancel each other, the sharing still satisfies the correctness.

By varying the CT one can obtain all possible sharings of a given function. Consider a Boolean quadratic function with m variables (1 output bit), which we want to share with 3 shares. Note that the only terms which can be used as CT are x_i or $x_i y_i$ (or higher degree) for $i = 1, 2, 3$. Indeed terms like $x_i y_j$ for $i \neq j$ can't be used in the i-th and j-th share of the function because of the non-completeness rule and therefore such a term can be used in only 1 share, hence it can't be used as a CT.

Thus counting only the linear and quadratic CT and ignoring the constant terms, which will not influence the uniformity, for a quadratic function with m variables we obtain that there are $3(m + \binom{m}{2})$ CT. Taking into account all possible positions for the CT we get $2^{3(m+\binom{m}{2})}$ sharings. For example, for a quadratic function of 3 variables there are 2^{18} possible CT and therefore for a 3 × 3 S-boxes the search space will be 2^{54}. This makes the exhaustive search (to find a single good solution) over all CT unpractical, even for small S-boxes. For sharing with 4 shares even more terms can be used as CT.

4.4 A Link between the 3 × 3 S-Boxes and Some Quadratic 4 × 4 S-Boxes

Lemma 8. *There is a transformation which expands $\mathcal{Q}_1^3$, $\mathcal{Q}_2^3$ and $\mathcal{Q}_3^3$ into $\mathcal{Q}_4^4$, $\mathcal{Q}_{12}^4$ and $\mathcal{Q}_{300}^4$ correspondingly.*

Proof. Starting from a 3 × 3 S-box S and adding a new variable we can obtain a 4 × 4 S-box $\tilde{S}$. Namely, the transformation is defined as follows: let $S(w, v, u) = (y1, y2, y3)$ and define $\tilde{S}(x, w, v, u) = (y1, y2, y3, x)$. It is easy to check that this transformation maps the first 3 classes into the other 3 classes.   □

The relation from Lemma 8 explains why if we have a sharing for a class in $\mathcal{F}_2^3$ we also obtain a sharing for the corresponding class in $\mathcal{F}_2^4$ and vice versa, i.e., if we can't share a class the corresponding class also can't be shared. The results we have obtained with 3 shares are summarized in Table 1 (middle columns).

Recall that if we use only direct sharing we will be able to share with 3 shares all S-boxes in the alternating group but at the cost of longer path than the one obtained by decomposition. However using CT we found sharing for classes: $\mathcal{Q}_1^3$, $\mathcal{Q}_2^3$, $\mathcal{Q}_4^4$, $\mathcal{Q}_{12}^4$, $\mathcal{Q}_{293}^4$, $\mathcal{Q}_{294}^4$ and $\mathcal{Q}_{299}^4$. So all quadratic classes except $\mathcal{Q}_3^3$ and $\mathcal{Q}_{300}^4$ can be shared with 3 shares and without decomposition. We want to pose an open question: find sharing without decomposition to classes $\mathcal{Q}_3^3$ and $\mathcal{Q}_{300}^4$ or show why they can't be shared with 3 shares in that way.

4.5 Sharing Using Decomposition

As an alternative to the search through a set of correction terms, we can also construct sharings after using decomposition: we try to decompose S-boxes into S-boxes for which we already have sharings. This decomposition problem is more restrained than the basic problem discussed in Section 3 for sharing with 3 shares, since we can use only the quadratic S-boxes for which we already have a sharing. It turns out that this extra requirement sometimes increases the decomposition length by one. For example, decomposition for $\mathcal{Q}_3^3$ is 1 × 2 and 2 × 1, i.e., we obtain a sharing for $\mathcal{Q}_3^3$ at the cost of length 2 (instead of length 1). Similarly $\mathcal{Q}_{300}^4$ can be decomposed as 4 × 12, 4 × 293, 12 × 4, 12 × 294, 293 × 4, 293 × 294, 294 × 12 and 294 × 293 so, again we obtain a sharing with length 2. Table 1 (right columns) gives the results.

Recall that one can't find a sharing with 3 shares for cubic functions outside the alternating group. Thus, 4 shares will be required in this case. Using direct sharing with 4 shares we obtain slightly better results for the quadratic S-boxes compared to 3 shares since we were able to share also class $\mathcal{Q}_{300}^4$ (and therefore $\mathcal{Q}_3^3$ too). The sharing of class $\mathcal{Q}_{300}^4$ has further improved the sharings of $\mathcal{C}_{130}^4$, $\mathcal{C}_{131}^4$ and $\mathcal{C}_{24}^4$ which have sharing with shorter length for 4 shares than for 3 shares. We have also found sharings with 4 shares for the cubic classes $\mathcal{C}_1^4$, $\mathcal{C}_3^4$, $\mathcal{C}_{13}^4$ and $\mathcal{C}_{301}^4$ from $S_{16} \setminus A_{16}$ using direct sharing. By using Lemma 7 we obtain sharings with 4 shares for all 4×4 S-boxes. Observe that the total length of the sharing depends on the class we use ($\mathcal{C}_1^4$, $\mathcal{C}_3^4$, $\mathcal{C}_{13}^4$ and $\mathcal{C}_{301}^4$) and also on the class from the alternating group, which is used for the decomposition. For example, class $\mathcal{C}_7^4$ can be decomposed using $\mathcal{C}_1^4$ with length 4 but with classes $\mathcal{C}_3^4$ and $\mathcal{C}_{13}^4$ it can be decomposed with length 3. Note also that the number of solutions differ. We have found 10, 31 and 49 solutions when using $\mathcal{C}_1^4$, $\mathcal{C}_3^4$ and $\mathcal{C}_{13}^4$ classes, correspondingly. Surprisingly for the classes in the alternating group we have only slight improvement with 4 shares compared to 3 shares and only a few classes in $S_{16} \setminus A_{16}$ have direct sharing with 4 shares. However with 5 shares all classes can be shared directly without decomposition which is a big improvement compare to the situation with 4 shares.

Table 1. Overview of the numbers of classes of 4×4 S-boxes that can be decomposed and shared using 3 shares, 4 shares and 5 shares. The numbers are split up according to the decomposition length of the S-boxes (1, 2, 3, or 4), respectively their shares.

unshared			3 shares				4 shares			5 shares	remark
1	2	3	1	2	3	4	1	2	3	1	
6			5	1			6			6	quadratics
	30			28	2			30		30	cubics in A_{16}
		114			113	1			114	114	cubics in A_{16}
–			–				4	22	125	151	cubics in $S_{16} \setminus A_{16}$

An open question is why for all S-boxes the sharing with 4 shares does not improve significantly the results compared to 3 shares and suddenly with 5 shares we can share all classes with length 1.

Recall that for the Present S-box, decompositions $S(x) = F(G(x))$ have been found in [26]. The authors also made an observation that exactly $\frac{3}{7}$ sharings out of the decompositions automatically satisfy the uniformity condition (i.e. without any correction terms). Recall that with the direct sharing method without CT we (as well as the authors of [26]) were able to share only 3 quadratic classes: $\mathcal{Q}_4^4$, $\mathcal{Q}_{294}^4$ and $\mathcal{Q}_{299}^4$. The Present S-box belongs to $\mathcal{C}_{266}^4$ and has 7 simple solutions but only 3 of them can be shared namely 294×299, 299×294, 299×299, which explains the authors' observation.

In the full version of the paper [2] we provide a complete list for the sharings with 3 and with 4 shares with their lengths. Recall that all classes can be shared with 5 shares with length 1 and that for the S-boxes in $S_{16} \setminus A_{16}$ no solution with 3 shares exist. Note that the DES 6×4 S-boxes can be considered as an affine

2×2 selection S-box with four 4×4 S-boxes attached. Since we have sharings for both 2×2 and 4×4 S-boxes we conclude that we have sharings for the DES 6×4 S-boxes as well.

5   HW Implementation of the Sharings

In this section, our aim is to provide a fair comparison and prediction what the cost (ratio of area to a NAND gate referred to as GE) will be for a protected S-box in a specified library. For our investigations we used the TSMC 0.18μm standard cell library in the Synopsis development tool.

Quadratic classes and cubic classes with length 1 form the basis to all our implementations. Therefore, we concentrated our efforts on these classes. While considering 3×3 S-boxes we synthesized 840 affine equivalent S-boxes for each class. However the number of S-boxes in a class increases to more than 322560 as we move to 4×4 S-boxes. In that case, we choose 1000 S-boxes per class to synthesize.

Table 2. S_8: Quadratic S-boxes sharing

3×3 S-boxes Class # in S_8	Sharing Length (L)	Original S-box	Unshared Decomposed L reg	Shared 3 shares L reg	Shared 4 shares 1 reg	Shared 5 shares 1 reg
$\mathcal{Q}_1^3$ Min	1	27.66	-	98.66	138.00	148.00
$\mathcal{Q}_1^3$ Max		29.66		121.66	150.00	185.66
$\mathcal{Q}_2^3$ Min	1	29.00	-	116.66	174.00	180.00
$\mathcal{Q}_2^3$ Max		29.66		155.00	226.66	220.33
$\mathcal{Q}_3^3$ Min	2	30.00	50.00	194.33	140.00	167.00
$\mathcal{Q}_3^3$ Max		32.00	51.00	201.00	194.33	228.66

Table 3. A_{16}: Quadratic S-boxes sharing

4×4 S-boxes Quadratic Class # in S_{16}	Sharing Length (L)	Original S-box	Unshared Decomposed L reg	Shared 3 shares L reg	Shared 4 shares 1 reg	Shared 5 shares 1 reg
$\mathcal{Q}_4^4$ Min	1	37.33	-	121.33	168.33	186.33
$\mathcal{Q}_4^4$ Max		44.00		223.33	258.00	309.00
$\mathcal{Q}_{12}^4$ Min	1	36.66	-	139.33	204.00	218.00
$\mathcal{Q}_{12}^4$ Max		48.00		253.33	290.33	340.66
$\mathcal{Q}_{293}^4$ Min	1	39.33	-	165.33	194.33	235.00
$\mathcal{Q}_{293}^4$ Max		48.66		297.33	313.00	358.33
$\mathcal{Q}_{294}^4$ Min	1	40.00	-	141.33	170.33	210.33
$\mathcal{Q}_{294}^4$ Max		49.66		261.00	240.00	255.00
$\mathcal{Q}_{299}^4$ Min	1	40.33	-	174.33	211.00	247.00
$\mathcal{Q}_{299}^4$ Max		48.00		298.00	295.33	294.66
$\mathcal{Q}_{300}^4$ Min	2	33.66	58.00	207.33	209.66	249.33
$\mathcal{Q}_{300}^4$ Max		52.66	70.00	346.00	295.00	342.33

Table 4. S_{16}: Cubic S-boxes sharing

4×4 S-boxes Cubic Class # in S_{16}		Sharing Length (L, L')	Original S-box	Unshared Decomposed L' reg	Shared 3 shares L reg	Shared 4 shares L' reg	Shared 5 shares 1 reg
$\mathcal{C}_1^4 \in S_{16} \setminus A_{16}$	Min	1,1	39.66		–	213.66	273.66
	Max		40.33		–	378.00	464.66
$\mathcal{C}_3^4 \in S_{16} \setminus A_{16}$	Min	1,1	40.33		–	230.33	286.33
	Max		43.00		–	413.66	500.66
$\mathcal{C}_{13}^4 \in S_{16} \setminus A_{16}$	Min	1,1	40.33		–	260.00	319.00
	Max		41.33		–	423.00	502.66
$\mathcal{C}_{301}^4 \in S_{16} \setminus A_{16}$	Min	1,1	39.33		–	289.33	350.33
	Max		59.33		–	526.33	605.66
$\mathcal{C}_{150}^4 \in A_{16}$		2,2	46.33	71.66	305.33	430.66	414.33
$\mathcal{C}_{151}^4 \in A_{16}$		2,2	47.33	69.66	286.00	410.00	390.00
$\mathcal{C}_{130}^4 \in A_{16}$		3,2	48.00	97.33	393.00	375.66	442.66
$\mathcal{C}_{131}^4 \in A_{16}$		3,2	50.00	99.00	386.00	363.33	435.66
$\mathcal{C}_{24}^4 \in A_{16}$		4,3	48.33	151.33	674.00	616.66	734.66
$\mathcal{C}_{204}^4 \in S_{16} \setminus A_{16}$		2,2	49.00	80.33	-	413.00	501.33
$\mathcal{C}_{257}^4 \in S_{16} \setminus A_{16}$		2,2	47.66	73.66	-	486.00	594.00
$\mathcal{C}_{210}^4 \in S_{16} \setminus A_{16}$		3,3	47.66	119.33	-	602.00	695.33

In tables 2, 3 and 4 we show the implementation results for each class only the S-box with the *minimum* GE from the result of our original S-box synthesis (over the class), as well as the S-box with the *maximum* GE. However, note that the *Min* and *Max* values should only be taken as indications.

The area results listed in the column *original S-box* for an $n \times n$ S-box include one n-bit register. If a decomposition is necessary for a correct, non-complete and uniform sharing, then we included registers in between every pipelining operation as required [23] which increases the cost as expected.

For classes with decomposition length more than 1, we randomly choose a class representative i.e. an S-box. Then we implement the smallest amongst all possible decompositions of this S-box, namely the one which gives minimum GE. We saw that, classes $\mathcal{Q}_3^3$, $\mathcal{Q}_{300}^4$, $\mathcal{C}_{150}^4$, $\mathcal{C}_{151}^4$, $\mathcal{C}_{130}^4$, $\mathcal{C}_{131}^4$, $\mathcal{C}_{24}^4$, $\mathcal{C}_{204}^4$, $\mathcal{C}_{257}^4$ and $\mathcal{C}_{210}^4$ give relatively small results when implemented as 2×1, 12×4, 12×293, 293×12, $12 \times 4 \times 299$, $299 \times 12 \times 4$, $299 \times 12 \times 4 \times 299$, 3×294, 3×12 and $3 \times 293 \times 12$ respectively. The area figures for $\mathcal{C}_{204}^4$ and $\mathcal{C}_{257}^4$ differ significantly. Closer inspection reveals that this is due to the fact that their decompositions use different S-boxes from $\mathcal{C}_3^4$; the S-box used in the decomposition of $\mathcal{C}_{204}^4$ is smaller than the one in the decomposition of $\mathcal{C}_{257}^4$.

6   Conclusions

In this paper we have considered the threshold implementation method, which is a method to construct implementations of cryptographic functions that are secure against a large class of side-channel attacks, even when the hardware technology is not glitch-free.

We have analyzed which basic S-boxes can be securely implemented using 3, 4 or 5 shares. We have constructed sharings for all 3×3, 4×4 S-boxes and 6×4 DES S-boxes. Thus we have extended the threshold implementation method to secure implementations for any cryptographic algorithm which uses these S-boxes. Note that the mixing layer in the round function of a block cipher is a linear operation and thus it is trivially shared even with 2 shares. Finally, we have implemented several of the shared S-boxes in order to investigate the cost of the sharing as well as the additional cost due to the pipelining stages separated by latches or registers.

Table 5. Range for the ratio $\frac{\text{area of the } \textit{Shared with length L} \text{ S-box}}{\text{area of the } \textit{Original} \text{ S-box}}$

3 shares				4 shares			5 shares	remark
1	2	3	4	1	2	3	1	
3.6–5.2	6.3–6.5	–	–	5.0–7.6	–	–	5.4–7.4	quadratics in S_8
3.3–6.2	6.2–6.6	–	–	4.3–6.4	–	–	5.1–7.4	quadratics in S_{16}
–	6.0–6.6	7.7–8.2	13.9	–	7.3–9.3	12.8	8.2–15.2	cubics in A_{16}
–	–	–	–	5.4–10.2	8.4–10.2	12.6	10.2–14.6	cubics in $S_{16} \backslash A_{16}$

Our results summarized in Table 5 show that such secure implementation can also be made efficient. Note that we consider the cost of *sharing with L registers* which is the total price for the sharing (since it includes the sharing logic plus registers). Observe that the increase of the cost for sharing with 3 shares of a quadratic S-box is similar for $n = 3$ and $n = 4$. As expected, the longer length a sharing has, the more costly it becomes (for 3 and 4 shares). It can be seen that sharings with 4 and 5 shares cost up to 50% more than sharings with 3 shares. However, there are several cases when using 4 or 5 shares reduces the cost by up to 30%, respectively 10%, compared to 3 shares with longer sharing length. For certain S-boxes using 5 shares may be even beneficial compared to 4 shares (up to 4%) but in general 5 shares are up to 30% more expensive than 4 shares.

An obvious conclusion is that the cost of the TI method heavily depends on the class the given S-box belongs to as well as the chosen number of shares and the associated sharing length. Therefore, in order to minimize the implementation cost the number of shares have to be carefully chosen. For all tested S-boxes we were able to find a sharing with cost ranging from 3.3 till 12.8 times the area of the original S-box. However, note that the area numbers are based on a few implementations from each class. The ratios may change significantly if the smallest/biggest S-boxes are found for every class.

Acknowledgements: We would like to thank Christophe De Cannière for the fruitful discussions and for sharing with us his toolkit for affine equivalent classes.

References

1. Akkar, M.L., Giraud, C.: An Implementation of DES and AES, Secure against Some Attacks. In: Koç, Ç.K., Naccache, D., Paar, C. (eds.) CHES 2001. LNCS, vol. 2162, pp. 309–318. Springer, Heidelberg (2001)
2. Bilgin, B., Nikova, S., Nikov, V., Rijmen, V., Stütz, G.: Threshold Implementations of all 3×3 and 4×4 S-boxes. Cryptology ePrint Archive, Report 2012/300, http://eprint.iacr.org/
3. Bertoni, G., Daemen, J., Peeters, M., Van Assche, G.: Keccak specifications. NIST SHA3 contest 2008 (2008)
4. Bertoni, G., Daemen, J., Peeters, M., Van Assche, G.: Building power analysis resistant implementations of Keccak. Round 3 finalist of the Cryptographic Hash Algorithm Competition of NIST (2010)
5. Blömer, J., Guajardo, J., Krummel, V.: Provably Secure Masking of AES. In: Handschuh, H., Hasan, M.A. (eds.) SAC 2004. LNCS, vol. 3357, pp. 69–83. Springer, Heidelberg (2004)
6. Boura, C., Canteaut, A.: On the influence of the algebraic degree of F^{-1} on the algebraic degree of $G \circ F$. e-print archive 2011/503
7. Carlet, C.: Vectorial Boolean Functions for Cryptography (to appear)
8. De Cannière, C.: Analysis and Design of Symmetric Encrytption Algorithms. Ph.D. thesis (2007)
9. De Cannière, C., Nikov, V., Nikova, S., Rijmen, V.: S-box decompositions for SCA-resisting implementations. Poster Session of CHES 2011 (2011)
10. Daemen, J., Vandewalle, J.: A New Approach Towards Block Cipher Design. In: Anderson, R. (ed.) FSE 1993. LNCS, vol. 809, pp. 18–33. Springer, Heidelberg (1994)
11. Daemen, J., Peeters, M., Van Assche, G.: Bitslice Ciphers and Power Analysis Attacks. In: Schneier, B. (ed.) FSE 2000. LNCS, vol. 1978, pp. 134–149. Springer, Heidelberg (2001)
12. Golic, J.D., Tymen, C.: Multiplicative Masking and Power Analysis of AES. In: Kaliski Jr., B.S., Koç, Ç.K., Paar, C. (eds.) CHES 2002. LNCS, vol. 2523, pp. 198–212. Springer, Heidelberg (2003)
13. Ishai, Y., Sahai, A., Wagner, D.: Private Circuits: Securing Hardware against Probing Attacks. In: Boneh, D. (ed.) CRYPTO 2003. LNCS, vol. 2729, pp. 463–481. Springer, Heidelberg (2003)
14. Knudsen, L., Leander, G., Poschmann, A., Robshaw, M.J.B.: PRINTCIPHER: A Block Cipher for IC-Printing. In: Mangard, S., Standaert, F.-X. (eds.) CHES 2010. LNCS, vol. 6225, pp. 16–32. Springer, Heidelberg (2010)
15. Leander, G., Poschmann, A.: On the Classification of 4 Bit S-Boxes. In: Carlet, C., Sunar, B. (eds.) WAIFI 2007. LNCS, vol. 4547, pp. 159–176. Springer, Heidelberg (2007)
16. Lidl, R., Niederreiter, H.: Finite Fields. Encyclopedia of Mathematics and its Applications, vol. 20. Addison-Wesley (1983)
17. Mangard, S., Pramstaller, N., Oswald, E.: Successfully Attacking Masked AES Hardware Implementations. In: Rao, J.R., Sunar, B. (eds.) CHES 2005. LNCS, vol. 3659, pp. 157–171. Springer, Heidelberg (2005)
18. Moradi, A., Mischke, O., Paar, C., Li, Y., Ohta, K., Sakiyama, K.: On the Power of Fault Sensitivity Analysis and Collision Side-Channel Attacks in a Combined Setting. In: Preneel, B., Takagi, T. (eds.) CHES 2011. LNCS, vol. 6917, pp. 292–311. Springer, Heidelberg (2011)

19. Moradi, A., Poschmann, A., Ling, S., Paar, C., Wang, H.: Pushing the Limits: A Very Compact and a Threshold Implementation of AES. In: Paterson, K.G. (ed.) EUROCRYPT 2011. LNCS, vol. 6632, pp. 69–88. Springer, Heidelberg (2011)
20. Nikova, S., Rechberger, C., Rijmen, V.: Threshold Implementations Against Side-Channel Attacks and Glitches. In: Ning, P., Qing, S., Li, N. (eds.) ICICS 2006. LNCS, vol. 4307, pp. 529–545. Springer, Heidelberg (2006)
21. Nikova, S., Rijmen, V., Schläffer, M.: Using Normal Bases for Compact Hardware Implementations of the AES S-Box. In: Ostrovsky, R., De Prisco, R., Visconti, I. (eds.) SCN 2008. LNCS, vol. 5229, pp. 236–245. Springer, Heidelberg (2008)
22. Nikova, S., Rijmen, V., Schläffer, M.: Secure Hardware Implementation of Non-linear Functions in the Presence of Glitches. In: Lee, P.J., Cheon, J.H. (eds.) ICISC 2008. LNCS, vol. 5461, pp. 218–234. Springer, Heidelberg (2009)
23. Nikova, S., Rijmen, V., Schläffer, M.: Secure Hardware Implementation of Nonlinear Functions in the Presence of Glitches. J. Cryptology 24(2), 292–321 (2011)
24. Oswald, E., Mangard, S., Pramstaller, N., Rijmen, V.: A Side-Channel Analysis Resistant Description of the AES S-Box. In: Gilbert, H., Handschuh, H. (eds.) FSE 2005. LNCS, vol. 3557, pp. 413–423. Springer, Heidelberg (2005)
25. Popp, T., Mangard, S.: Masked Dual-Rail Pre-charge Logic: DPA-Resistance Without Routing Constraints. In: Rao, J.R., Sunar, B. (eds.) CHES 2005. LNCS, vol. 3659, pp. 172–186. Springer, Heidelberg (2005)
26. Poschmann, A., Moradi, A., Khoo, K., Lim, C.W., Wang, H., Ling, S.: Side-Channel Resistant Crypto for less than 2,300 GE. J. Cryptology 24(2), 322–345 (2011)
27. Rivain, M., Prouff, E.: Provably Secure Higher-Order Masking of AES. In: Mangard, S., Standaert, F.-X. (eds.) CHES 2010. LNCS, vol. 6225, pp. 413–427. Springer, Heidelberg (2010)
28. Prouff, E., Roche, T.: Higher-Order Glitches Free Implementation of the AES Using Secure Multi-party Computation Protocols. In: Preneel, B., Takagi, T. (eds.) CHES 2011. LNCS, vol. 6917, pp. 63–78. Springer, Heidelberg (2011)
29. Rotman, J.: An introduction to the theory of groups. Graduate texts in mathematics. Springer (1995)
30. Saarinen, M.-J.O.: Cryptographic Analysis of All 4 × 4-Bit S-Boxes. In: Miri, A., Vaudenay, S. (eds.) SAC 2011. LNCS, vol. 7118, pp. 118–133. Springer, Heidelberg (2012)
31. Tiri, K., Verbauwhede, I.: A Logic Level Design Methodology for a Secure DPA Resistant ASIC or FPGA Implementation. In: DATE 2004, pp. 246–251. IEEE Computer Society (2004)
32. Trichina, E., Korkishko, T., Lee, K.H.: Small Size, Low Power, Side Channel-Immune AES Coprocessor: Design and Synthesis Results. In: Dobbertin, H., Rijmen, V., Sowa, A. (eds.) AES 2005. LNCS, vol. 3373, pp. 113–127. Springer, Heidelberg (2005)
33. Wernsdorf, R.: The Round Functions of RIJNDAEL Generate the Alternating Group. In: Daemen, J., Rijmen, V. (eds.) FSE 2002. LNCS, vol. 2365, pp. 143–148. Springer, Heidelberg (2002)

How Far Should Theory Be from Practice?
Evaluation of a Countermeasure

Amir Moradi and Oliver Mischke

Horst Görtz Institute for IT Security, Ruhr University Bochum, Germany
{moradi,mischke}@crypto.rub.de

Abstract. New countermeasures aiming at protecting against power analysis attacks are often proposed proving the security of the scheme given a specific leakage assumption. Besides the classical power models like Hamming weight or Hamming distance, newer schemes also focus on other dynamic power consumption like the one caused by glitches in the combinational circuits. The question arises if with the increasing downscale in process technology and the larger role of static leakage or other harder to model leakages, the pure theoretical proof of a countermeasure's security is still good practice. As a case study we take a new large ROM-based masking countermeasure recently presented at CT-RSA 2012. We evaluate the security of the scheme both under the leakage assumptions given in the original article and using a more real-world approach utilizing collision attacks. We can demonstrate that while the new construction methods of the schemes provide a higher security given the assumed leakage model, the security gain in practice is only marginal compared to the conventional large ROM scheme. This highlights the needs for a closer collaboration of the different disciplines when proposing new countermeasures to provide better security statements covering both the theoretical reasoning and the practical evaluations.

1 Introduction

Security-enabled devices like smartcards play a larger and larger role in our everyday lives. From a mathematical point of view these can easily be protected by modern ciphers which are secure in a black-box scenario where only the inputs and outputs can be observed. Unfortunately, with the discovery of side-channel attacks in the late 90s the security of a device no longer relies only on the use of a secure cryptographic algorithm, but especially on how this algorithm is implemented. In unprotected implementations sensitive information like encryption keys can be recovered by observing so called side channels.

Many different kinds of countermeasures have since been proposed either for protection of software and/or hardware platforms (see [11] for instance). While the masking countermeasures for software are relatively limited mainly to the algorithmic level, dedicated hardware circuits further allow the use of special logic styles and gate-level countermeasures. Preventing side-channel leakage in hardware is especially intricate since glitches in the circuit can cause otherwise

E. Prouff and P. Schaumont (Eds.): CHES 2012, LNCS 7428, pp. 92–106, 2012.

theoretically secure schemes to leak [12]. Because of their wide versatility the community has shown a huge interest to different aspects of masking counter-measures, e.g., [5,6,16,17,19,20]. More recent schemes trying to take the afore-mentioned problem of glitches into account are schemes relying on multi-party computation, e.g., [17,19]. Most articles dealing with side-channel countermea-sures either propose a new scheme and try proving its security in theory under a given leakage model [19] or evaluating the scheme in practice [15]. Because these tasks normally require different backgrounds – math vs. engineering – there have only been few attempts at a joint approach providing theoretically proven secu-rity in addition to practical investigations.

In this work we focus on the practical evaluation of a masking scheme re-cently presented at CT-RSA 2012 [10]. It is based on boolean masking and large Look-Up-Tables (LUTs), and tries to avoid some known shortcomings caused by updating the mask and masked data at the same time. The idea behind the scheme is to store and update the masks in a way that the leakage caused by updating the mask and masked data are not directly related. According to the given proofs the scheme should prevent any kind of univariate leakage if the leakage characteristics of the target device fits to a Hamming distance (HD) model. Taking AES Rijndael in hardware as our case study, we compare the effi-ciency and side-channel leakage of this scheme with the conventional way of using global look-up-tables (GLUT) [18]. Following the guidelines in [10], we have im-plemented the SubBytes transformation using the target masking scheme while facing difficulties when dealing with the linear parts especially MixColumns.

Using an FPGA-based platform we practically examine and compare the side-channel leakage of an exemplary design made considering different masking schemes including the conventional one and those proposed in [10]. We show that the proposed constructions indeed increase the resistance against power analysis attacks, but only when restricting the attack to the model assumed in the original article. Analyzing the resistance of the scheme under more realistic assumptions about the leakage model shows that the new constructions are as vulnerable as the conventional one.

We should stress that our practical results do not show any shortcomings of the scheme considering the assumed leakage model. However, we show that the model taken into account when designing and proving the security of the proposed scheme does not comprehensively consider all the possible leakages which are available in practice.

In Section 2 we define the notations and explain the basics of the target masking schemes. Our design architectures and adaptions to fit to the schemes' assumptions are described in Section 3, and the corresponding evaluation results are provided in Section 4. Finally, Section 5 concludes our research.

2   Preliminaries

When using boolean masking as a side-channel countermeasure, a secret value x, which contributes to the computations of a cryptographic device, is represented by a randomized variable z as $x \oplus \bigoplus_{i=1}^{d} r_i$, where each r_i is an independent random variable with a uniform distribution. Each of z, r_1, ...,r_d is considered as a separate share in this secret sharing scheme, which is called d^{th}-order boolean masking. Since each share contributes to the computations separately, the scheme is supposed to provide security at most against d^{th}-order power analysis attacks. In the literature there exist two distinct definitions for what is the order of an attack. Some previous work define the order via the number of different leakage points considered simultaneously mainly because of the sequential processing in software. Others define the order via the statistical moment applied. In this work we use the following definition:

- An attack which combines v different time instances – usually in v different clock cycles – of each power trace is called v-variate attack.
- The order of an attack – regardless of v – is defined by the order of the statistical moments which are considered in the attack.

For instance, a CPA [4] which combines two points of each power trace by summing them up is a bivariate 1st-order attack, and a CPA which applies the squared values of each point of each trace is a univariate 2nd-order attack. Those attacks where no specific statistical moment is applied, e.g., mutual information analysis (MIA) [2], are distinguished only by v like univariate or bivariate MIA [7].

The focus of this article is 1st-order boolean masking by considering one random value for each secret. The main goal of different hardware implementations of these schemes is to counteract univariate attacks of any order. Therefore, we consider only these attacks in our evaluations.

While the linear operations of a cryptographic algorithm are easy to adjust to the underlying boolean masking, providing solutions to adjust the non-linear parts of different algorithms suitable either for software or hardware platforms still is not trivial. This has indeed taken huge interest by the community and several schemes and techniques have been proposed. One which is the focus of this article is Global Look-up Table (GLUT) introduced in [18]. In the following we briefly explain the scheme w.r.t to the specific case considered in [10].

Suppose that the desired non-linear part of the target algorithm is a bijection and denoted by an $n \times n$ S-box $S : \mathbb{F}_2^n \mapsto \mathbb{F}_2^n$. The secret value x represented by two shares $z = x \oplus r$ and r is mapped to a shared representation of $S(x)$, i.e., $z' = S(x) \oplus r'$ and r'. This mapping is done by a pre-computed look-up table T as $(z, r) \mapsto (z', r')$, where output mask r' is made by a deterministic function U over input mask r. Therefore, the look-up table T maps a $2n$-bit input to a $2n$-bit output and is of a size of $2n \cdot 2^{2n}$ bits. This scheme, which is a usual way of realizing a masked S-box if the table fits into the memory space, is called "conventional" scheme in the rest of this article.

The problem which is observed in [10], is a univariate leakage caused when the registers containing the look-up table input (z, r) are updated by its output (z', r'). The bit flips in the registers are represented by $(\Delta z, \Delta r) = (x \oplus S(x) \oplus r \oplus r', r \oplus r')$. Therefore, a univariate MIA or a univariate 2nd-order CPA can reveal the relation between the bit flips and x. In order to overcome this problem a new scheme, which is shortly restated below, has been introduced in [10].

In the new scheme, x is represented by two shares $z = x \oplus F(r)$ and r, where the bit length of r (denoted by $p < 2n$) is no longer equal to n, and F is a deterministic function from $\mathbb{F}_2^p$ to $\mathbb{F}_2^n$. The look-up table T^* maps a $(n + p)$-bit input (z, r) to an n-bit output $z' = S(x) \oplus F(r')$, and the output mask is computed easily as $r' = r \oplus \alpha$, where α is a non-zero p-bit constant. The table T^* needs $n \cdot 2^{n+p}$ storage bits and still is comparable to the size of T for values p close to n. In this case the register update – same scenario as before – leads to $(\Delta z, \Delta r) = (x \oplus S(x) \oplus F(r) \oplus F(r'), r \oplus r')$. The mask difference is constant, i.e., $\Delta r = \alpha$, but in order to appropriately choose the function F two constructions have been proposed in [10] (we simplified the conditions for clarity):

- $p = n+1$, $\alpha = (\{1\}, \{0\}^n)$, and $F(r) = G(r)$ if $r \in \{0\} \times \mathbb{F}_2^n$ and $F(r) = \{0\}^n$ otherwise. G is an arbitrary bijective function from $\{0\} \times \mathbb{F}_2^n$ to $\mathbb{F}_2^n$.
- $p = n + n'$, and for $r = (r_h, r_l) \in \mathbb{F}_{2^{n'}} \times \mathbb{F}_{2^n}$, $F(r_h, r_l) = G(r_h) \bullet r_l$, where $\bullet$ denotes multiplication over the finite field $\mathbb{F}_{2^n}$. G is an arbitrary injective function from $\mathbb{F}_2^{n'}$ to $\mathbb{F}_2^n - \{0\}$. The constant α is also made by an arbitrary non-zero constant $\alpha' \in \mathbb{F}_2^{n'}$ as $(\alpha', \{0\}^n)$.

Both constructions satisfy the required conditions, i.e., Δr to be constant and distribution of $F(r) \oplus F(r')$ to be uniform. However, in the first construction $F(r)$ is zero for half of its input space. In other words, in half of the cases the secret x is represented by $z = x$ leading to a very high bias in the distribution of z for a given x and uniformly distributed random r. This issue is not seen in the second construction, and the function F is uniformly distributed over $\mathbb{F}_2^n$. We refer to these two constructions as "first CT-RSA" and "second CT-RSA" schemes in the rest of this article.

3   Case Study: AES

Hereafter we consider AES Rijndael as the target algorithm, and try to realize the encryption function using the aforementioned masking scheme. The first step toward our goal is to make the masked S-box. Because of the byte-wise computations ($n = 8$) $16 \cdot 2^{16} = 1M$ bits storage space is required to realize the table T of the conventional scheme. For the first CT-RSA scheme, the same amount of space i.e., $8 \cdot 2^{17} = 1M$ bits, is required to make the T_1^* table. The second CT-RSA scheme also leads to an $8 \cdot 2^{16+n'}$-bit table T_2^*, which is still possible in practice for some small n'.

Therefore, the SubBytes transformation can be easily realized. However, a question arises when trying to rewrite the linear parts of the algorithm under either the first or the second CT-RSA scheme. If the key is not masked, as is assumed in [10], the implementation of AddRoundkey is straightforward. On the other hand, if the key should also be masked this would push the space requirements to $(8 \cdot 2^{26}) = 512M$-bit for the look-up table to map the xored data and key and both their masks to a masked result. This problem becomes worse

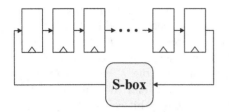

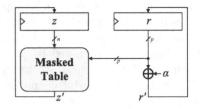

Fig. 1. The serialized design technique   **Fig. 2.** CT-RSA masking architecture [10]

when trying to maintain the masking schemes while implementing mixcolums e.g., by a T-table approach.

In order to follow the scheme presented in [10] we suppose that only the SubBytes transformation is performed using the CT-RSA masking schemes, and for instance – by applying the F function – the masks are transformed to the conventional scheme to perform the rest of the encryption operations. In the following we discuss on issues arising when designing a circuit to solely perform the SubBytes transformation.

3.1 Our Design

When e.g., because of area constraint, there is only one instance of a circuit, e.g., S-box, in a design, the hardware designers usually take advantage of the serialized design methodology. In this technique, as shown in Fig. 1, a rotate shift register with an S-box circuit as the feedback function is employed. As a reference we can mention [3,8,9,21] where this design methodology is used.

In either conventional or CT-RSA masking schemes the look-up tables, i.e., T or T^*, are quite large that integrating more than one table does not seem to be practical. However, the CT-RSA scheme has been designed and its security has been analyzed according to an assumption depicted in Fig. 2. It is assumed that both shares of a state byte are simultaneously replaced by the corresponding shares of the substitution value. This means that if a serialized architecture is considered for implementation, our target masking scheme does not provide the desired proven security. One solution is to use several multiplexers to select each state byte as the S-box input and save its output in the same register. This is usually not a designers' preferred choice because of its higher area overhead and slightly bigger control unit compared to the serialized one. The solution that we have applied to realize the serialized architecture and satisfy the assumption of the CT-RSA masking scheme is shown in Fig. 3. The rotate byte-wise shift register is active between each two table lookups (S-box). Therefore, the S-box output is saved at the same register which contains its input. Compared to both aforementioned architectures this design leads to time overhead since we have separated the shift and the save operations. While there are ways to improve the throughput in this scenario, we deliberately chose to keep it in this simple way since we are mainly interested in evaluating the side-channel leakage of

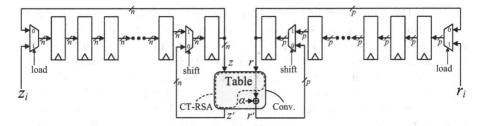

Fig. 3. Our exemplary design to examine conventional and CT-RSA masking schemes

the register updates and there is no need to risk introducing unwanted leakage sources for throughput reasons.

The target platform we selected to implement the schemes is a Virtex-5 FPGA (XC5VLX50) embedded in a SASEBO-GII board [1]. We selected three cases of masking schemes in our experiments to make the corresponding look-up tables:

- **Conventional**, look-up table T needs – as stated before – 1M bits space. The deterministic mask-update function U (see Section 2) is selected as $r' = U(r) = r^4 \oplus 56_h$ in $\mathbb{F}_{2^8}$ and using the Rijndael irreducible polynomial.
- **First CT-RSA**, look-up table T_1^* also needs a 1M-bit space. $\alpha = 100_h$ and the G function is randomly selected (see Appendix for a table representation).
- **Second CT-RSA**, where $n' = 1$, i.e., look-up table T_2^* also needs 1M bits space. $\alpha = 100_h$, $G(0) = b2_h$, and $G(1) = 5f_h$.

There are a couple of different ways to implement a large look-up table in FPGAs. We selected two versions:

- **LUT**, a combination of 6-input 1-output small look-up tables (LUT6 [22]) which allows realizing a large ROM, and
- **BRAM**, a combination of 18k-bit block RAMs (RAMB18 [22]) and a few number of LUT6 which allows implementing a large RAM.

In order to make the T table in the **LUT** version, we required 1365 LUT6 instances in six depth levels for each output bit, i.e., in sum 21 840 LUT6 instances which perfectly fits to the number of available LUT6 instances in our target FPGA, i.e., 28 800. Making each the T_1^* and T_2^* tables similarly we needed 2731 LUT6 instances in seven depth levels for each output bit and in sum 21 848.

The **BRAM** version of table T needs four BRAM18 and one LUT6 for each output bit, i.e., 64 BRAM18 and 16 LUT6 for whole of the T table. Each of tables T_1^* and T_2^* also needs eight BRAM18 and three LUT6 and in sum 64 BRAM18 and 24 LUT6. We should note that 96 BRAM18 instances are available in our target FPGA, and $n' = 1$ for the second CT-RSA scheme is the only option which could fit into the available resources either in the **LUT** or **BRAM** version.

We should emphasize that we omitted using the *Architecture Wizard IP* tool of Xilinx to make the aforementioned look-up tables. Instead, we hard-instanced the BRAM18 and LUT6 instances with our desired contents preventing any optimizations by the synthesizer. Also important to mention is the architectural difference between the **LUT** and **BRAM** versions. The tables made by LUT6 can be seen as a combinational circuit (clockless) which provides output for any value that appears at its input. However, the block RAMs need one clock cycle to provide the desired output of the given input. Therefore, our design (Fig. 3) in **LUT** version needs one clock cycle to save the table output before each shift. It means 32 clock cycles for whole of the SubBytes transformation. But one more clock cycle per state byte is required in the **BRAM** version leading to 48 clock cycles in total.

4   Practical Results

As stated before, we used a SASEBO-GII board as the evaluation platform, and implemented all our experimental designs on its target FPGA (XC5VLX50) running at a frequency of 3MHz. We also measured power consumption traces of the target FPGA using a LeCroy WP715Zi 1.5GHz oscilloscope at the sampling rate of 1GS/s. A 1Ω resistor in the VDD path, a DC blocker, a passive probe, an amplifier, and restricting the bandwidth of the oscilloscope to 20MHz helped to obtain clear and low-noise measurements.

We provided 6 design profiles made as **Conventional**, **First CT-RSA**, and **Second CT-RSA** each in both **LUT** and **BRAM** versions. Each design profile gets 16 plaintext bytes $p_{i\in\{1,\ldots,16\}}$ and according to the target masking scheme makes a masked plaintext byte p_i' of each by means of 16 independent random values $r_{i\in\{1,\ldots,16\}}$ (each 8-bit for the **Conventional** and 9-bit for the **CT-RSA** profiles). 16 secret key bytes $k_{i\in\{1,\ldots,16\}}$, which are fix inside the design, are each XORed with the corresponding masked plaintext byte as $z_{i\in\{1,\ldots,16\}} = p_i' \oplus k_i$. In 16 clock cycles z_i and r_i are serially given to the design (see Fig. 3) to completely fill the shift registers. Depending on the profile after 32 or 48 clock cycles the SubBytes transformation is completed.

We provided a clear trigger signal for the oscilloscope which indicates the start and end of the SubBytes transformation, thereby perfectly aligning the measured power traces. We also restricted the measurements to cover only the S-box computations. We fixed the number of measurements for all profiles to 1 000 000. In the experiments shown below we kept the secret key bytes fix and randomly selected the input plaintext bytes. Moreover, we made sure of the uniform distribution of internal random values r_i.

The technique we used to evaluate the side-channel leakage of these profiles is the *correlation-collision attack* [14]. This attack examines the leakage of one circuit instance that is used in different time instances. It originally considers only the first-order leakage, but according to [13] it can be adapted to use higher-order moments. Since our design profiles realize different 1st-order masking schemes, we restrict our evaluations to consider only the first- and second-order univariate leakage of the profiles. We should again emphasize the goal of the **CT-RSA**

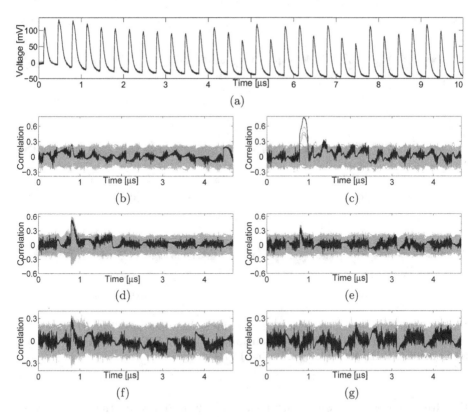

Fig. 4. LUT version, (a) sample power trace, collision attack results by register update model, (left) first-order (right) second-order: (b) and (c) **Conventional**, (d) and (e) **First CT-RSA**, (f) and (g) **Second CT-RSA** profiles, each using 1 000 000 traces

profiles which is preventing the univariate side-channel leakage of any order given the register update leakage model.

We start our evaluations with the **LUT** version of the **Conventional** profile. An exemplary power trace, which shows the S-box table lookup of the first few bytes, is depicted in Fig. 4(a). We consider the leakage caused by register updates when the S-box input is overwritten by its output, i.e., $v_i = (p_i \oplus k_i) \oplus S(p_i \oplus k_i)$ (not considering the masks in the formula). It is indeed the same model which the security of CT-RSA schemes are based on. In order to perform the aforementioned collision attack we need to compare the corresponding leakages of register updates of two different state bytes. We consider the second and the third state bytes, i.e., v_2 and v_3, and search for the correct (k_2, k_3) in a 2^{16} space by comparing each of the first- and second-order univariate leakages (means and variances) of corresponding parts of the measured power traces. The result of these attacks, which are not unexpected, are shown by Fig. 4(b) and Fig. 4(c). The same scenario is considered for the **First CT-RSA** profile, and performing the same attack with the same settings led to the results shown in Fig. 4(d) and

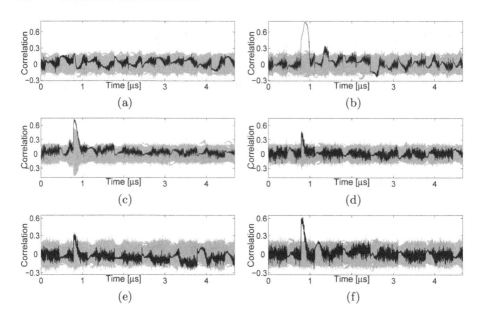

Fig. 5. LUT version, collision attack results by S-box input model, (left) first-order (right) second-order: (a) and (b) **Conventional**, (c) and (d) **First CT-RSA**, (e) and (f) **Second CT-RSA** profiles, each using 1 000 000 traces

Fig. 4(e). Comparing those attack results which are based on the second-order moments (Fig. 4(c) vs. Fig. 4(e)) shows the efficiency of the first CT-RSA scheme to counteract those attacks which use the register update model. The same holds for the **Second CT-RSA** profile, and the attack results depicted in Fig. 4(f) and Fig. 4(g) confirm the efficiency of the second CT-RSA scheme as well.

However, the register update (usually simplified by the HD model) is not the unique source of leakage in hardware. The value of e.g., S-box input or its output also affects the power consumption of the device and hence contributes in information leakage. In our designs the masked S-box input $(p_{i \in \{1,...,16\}} \oplus k_i) \oplus F(r_i)$ and the mask r_i at the same time appear at the look-up table input, and the distribution of leakages which depend on the masked input and the mask is not independent of the unmasked input $p_i \oplus k_i$. Therefore, considering e.g., the S-box input a univariate attack is expected to be successful.

In order to consider such model in our attacks we take the second and the third S-box inputs, $p_2 \oplus k_2$ and $p_3 \oplus k_3$, and search for correct key difference $k_2 \oplus k_3$ amongst 2^8 candidates by comparing each first- and second-order moments of the corresponding parts of the power traces. The result of all six attacks are depicted in Fig. 5, where for each of the profiles exists a successful first- and/or second-order attack. It in fact confirms our claim that the register update is not the sole source of leakage, and in contrast to what is argued in Section 3.2 of [10] the leakage of the combinational logic – including look-up tables – cannot be separated from the leakage of the register update. Indeed, the leakage

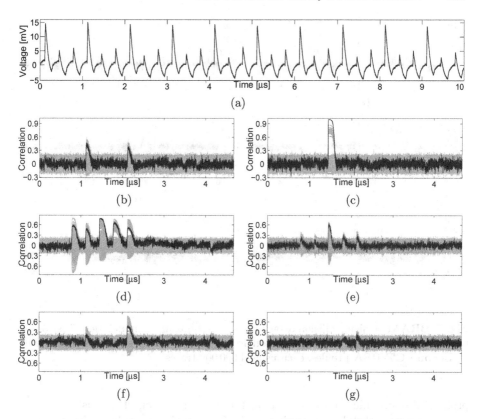

Fig. 6. BRAM version, (a) sample power trace, collision attack results by register update model, (left) first-order (right) second-order: (b) and (c) **Conventional**, (d) and (e) **First CT-RSA**, (f) and (g) **Second CT-RSA** profiles, each using 1 000 000 traces

which can be observed by currently available measurement setups is a mixture of both leakages caused by inherent low-pass filters of the device internals, PCBs, measurement tools, etc [11].

We also should stress the difference between the first-order leakage of the **First CT-RSA** and the **Second CT-RSA** profiles (Fig. 5(c) vs. Fig. 5(e)). The **First CT-RSA** profile has clear first-order leakage in contrast to the other profile. The reason behind this – as stated in Section 2 – is the F function of the first CT-RSA scheme, where the actual mask used to mask the secret, i.e., $F(r)$, is zero for half of the space of r. This results in having the S-box input – and consequently its output – unmask in the computations with the probability of 50%.

Repeating the same scenario of considering the register update as well as S-box input on profiles in the **BRAM** version led to the same results as depicted in Fig. 6 and Fig. 7. Although it needs much less power compared to the **LUT** version (Fig. 6(a) vs. Fig. 4(a)), the **First CT-RSA** and **Second CT-RSA**

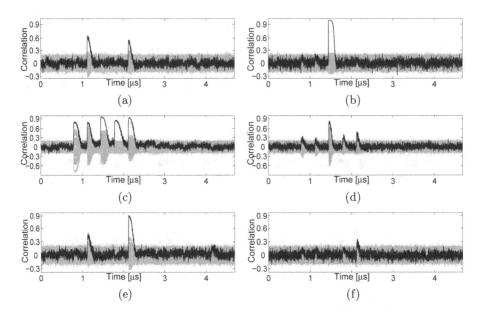

Fig. 7. **BRAM** version, collision attack results by S-box input model, (left) first-order (right) second-order: (a) and (b) **Conventional**, (c) and (d) **First CT-RSA**, (e) and (f) **Second CT-RSA** profiles, each using 1 000 000 traces

profiles also provide robustness against the attack using the register update model. However, they both – similar to the **Conventional** profile – show vulnerability against a collision attack utilizing a straightforward S-box input model.

4.1 Discussions

As mentioned before, the profiles of the **LUT** version can be seen as a huge combinational circuit which sees a masked value and the corresponding mask at its input signals. Therefore, the glitches happening inside the combinational circuit – similar to the results reported in [12] and [14] – are the main source of leakage. Their dependency to the unmasked values cause the designs to be vulnerable.

We should explain an architectural difference between the profiles of the **LUT** and the **BRAM** versions. As stated before, the profiles of the **BRAM** version need one more clock cycle per state byte compared to the **LUT** version. However, according to the results of the **BRAM** version (Fig. 7) the leakage, which depends on one S-box table lookup, appears at more than one clock cycle. Figure 7(a) shows that the input-output of a table lookup affects the power consumption not only at the clock cycle in which the block RAM is active but also at the next time when the block RAM is activated for the next table lookup (a distance of 3 clock cycles in our design profiles). Moreover, the biased distribution of the masks in the **First CT-RSA** profile becomes more problematic in

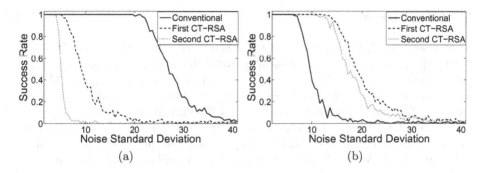

Fig. 8. Success rate of collision attacks by S-box input model in presence of noise using 1 000 000 traces, (a) **LUT** version and (b) **BRAM** version of all profiles

the **BRAM** version, where the leakage related to a table lookup appears in five consecutive clock cycles (see Fig. 7(c)).

The internal architecture of block RAMs of our target FPGA is not publicly available, and in contrast to LUT6 it cannot be simply guessed. Therefore, we can only speculate on the actual reasons behind the strange leakage appearing in the profiles of the **BRAM** version. For instance, there exist additional input and output registers in the block RAMs which can be activated or bypassed. Also, each block RAM contains some cascading registers to be used when combining several small block RAMs to a bigger one. It is ambiguous whether all these registers still get triggered when they are bypassed in the settings. Additionally, the data and address bitwidth of each block RAM can be arbitrarily selected by the settings. This means that there exist several multiplexers and additional logics to provide all possible options. All these unclear issues prevent us from providing a certain reason for the observed leakage in the block RAMs.

At the end we compare the vulnerability of all profiles in presence of noise. Since only the attacks using the S-box input model are successful, we omitted the register update model in this evaluation. With a certain standard deviation we artificially added Gaussian random noise to the specific points of all the 1 000 000 measured power traces, and performed the same attacks as before. We repeated the noise addition and the attack 200 times for each step of the noise standard deviation, and reported the average of the attack success rate in Fig. 8. According to curves shown in Fig. 8(a), the **CT-RSA** profiles of the **LUT** version make the attacks harder compared to the **Conventional** profile. The threshold of the noise standard deviation for a successful attack on **CT-RSA** profiles is considerably lower than that of the **Conventional** one. However, a similar experiment on the profiles of the **BRAM** version shows different results (see Fig. 8(b)). The attack on the **Conventional** profile can be unsuccessful while with the same amount of noise the **CT-RSA** profiles of the **BRAM** version are still vulnerable to the aforementioned attack. The reason is most likely related to the obscure internal architecture of the block RAMs.

5   Conclusions

In this work we have implemented the scheme recently proposed in CT-RSA 2012 [10], and have evaluated its security under the given leakage assumption in the original paper as well as using an approach more close to a real-world scenario. We pointed out the practical issues when realizing this masking schemes for linear functions. Moreover, we addressed the difficulties of the GLUT technique caused by their extremely large resource consumption on FPGAs. For instance, two thirds of the available BRAMs or three quarters of all available LUTs in a Xilinx Virtex-5 LX50 are required for a single masked S-box look-up table.

Nevertheless, we could show that the newly proposed constructions indeed provide a higher level of security when only considering the register update model as the leakage source. On the other hand, our results show that this leakage assumption is still not close enough to practice even when using large ROM-style tables instead of pure combinational circuits to implement the masked S-boxes. By pointing out exploitable univariate leakages of all the design profiles we showed that just stating the security of the scheme under a register update assumption (simplified by HD) is not a valid choice in any kind of masking realization, being it in combinational logic or large ROMs.

A closer collaboration of the different fields of countermeasure creation and practical evaluations would help to increase the impact of new proposals. It indeed allows a better adaption of new schemes in real-world applications. This way the industry sector would benefit not only of a theoretical proof but would appreciate the demonstration of the consistency of the theoretical claims with practice.

Acknowledgment. The authors would like to thank Elisabeth Oswald for her valuable input while shepherding this paper and the anonymous reviewers of CHES 2012 for their constructive feedbacks.

In this project Oliver Mischke has been partially funded by the European Union, Investing in your future, European Regional Development Fund.

References

1. Side-channel Attack Standard Evaluation Board (SASEBO). Further information are available via http://www.risec.aist.go.jp/project/sasebo/
2. Batina, L., Gierlichs, B., Prouff, E., Rivain, M., Standaert, F.-X., Veyrat-Charvillon, N.: Mutual Information Analysis: a Comprehensive Study. J. Cryptology 24(2), 269–291 (2011)
3. Bogdanov, A., Knudsen, L.R., Leander, G., Paar, C., Poschmann, A., Robshaw, M.J.B., Seurin, Y., Vikkelsoe, C.: PRESENT: An Ultra-Lightweight Block Cipher. In: Paillier, P., Verbauwhede, I. (eds.) CHES 2007. LNCS, vol. 4727, pp. 450–466. Springer, Heidelberg (2007)
4. Brier, E., Clavier, C., Olivier, F.: Correlation Power Analysis with a Leakage Model. In: Joye, M., Quisquater, J.-J. (eds.) CHES 2004. LNCS, vol. 3156, pp. 16–29. Springer, Heidelberg (2004)

5. Genelle, L., Prouff, E., Quisquater, M.: Secure Multiplicative Masking of Power Functions. In: Zhou, J., Yung, M. (eds.) ACNS 2010. LNCS, vol. 6123, pp. 200–217. Springer, Heidelberg (2010)
6. Genelle, L., Prouff, E., Quisquater, M.: Thwarting Higher-Order Side Channel Analysis with Additive and Multiplicative Maskings. In: Preneel, B., Takagi, T. (eds.) CHES 2011. LNCS, vol. 6917, pp. 240–255. Springer, Heidelberg (2011)
7. Gierlichs, B., Batina, L., Preneel, B., Verbauwhede, I.: Revisiting Higher-Order DPA Attacks: Multivariate Mutual Information Analysis. In: Pieprzyk, J. (ed.) CT-RSA 2010. LNCS, vol. 5985, pp. 221–234. Springer, Heidelberg (2010)
8. Guo, J., Peyrin, T., Poschmann, A., Robshaw, M.: The LED Block Cipher. In: Preneel, B., Takagi, T. (eds.) CHES 2011. LNCS, vol. 6917, pp. 326–341. Springer, Heidelberg (2011)
9. Hämäläinen, P., Alho, T., Hännikäinen, M., Hämäläinen, T.D.: Design and Implementation of Low-Area and Low-Power AES Encryption Hardware Core. In: DSD 2006, pp. 577–583. IEEE Computer Society (2006)
10. Maghrebi, H., Prouff, E., Guilley, S., Danger, J.-L.: A First-Order Leak-Free Masking Countermeasure. In: Dunkelman, O. (ed.) CT-RSA 2012. LNCS, vol. 7178, pp. 156–170. Springer, Heidelberg (2012)
11. Mangard, S., Oswald, E., Popp, T.: Power Analysis Attacks: Revealing the Secrets of Smart Cards. Springer (2007)
12. Mangard, S., Pramstaller, N., Oswald, E.: Successfully Attacking Masked AES Hardware Implementations. In: Rao, J.R., Sunar, B. (eds.) CHES 2005. LNCS, vol. 3659, pp. 157–171. Springer, Heidelberg (2005)
13. Moradi, A.: Statistical Tools Flavor Side-Channel Collision Attacks. In: Pointcheval, D., Johansson, T. (eds.) EUROCRYPT 2012. LNCS, vol. 7237, pp. 428–445. Springer, Heidelberg (2012)
14. Moradi, A., Mischke, O., Eisenbarth, T.: Correlation-Enhanced Power Analysis Collision Attack. In: Mangard, S., Standaert, F.-X. (eds.) CHES 2010. LNCS, vol. 6225, pp. 125–139. Springer, Heidelberg (2010)
15. Moradi, A., Poschmann, A., Ling, S., Paar, C., Wang, H.: Pushing the Limits: A Very Compact and a Threshold Implementation of AES. In: Paterson, K.G. (ed.) EUROCRYPT 2011. LNCS, vol. 6632, pp. 69–88. Springer, Heidelberg (2011)
16. Nikova, S., Rechberger, C., Rijmen, V.: Threshold Implementations Against Side-Channel Attacks and Glitches. In: Ning, P., Qing, S., Li, N. (eds.) ICICS 2006. LNCS, vol. 4307, pp. 529–545. Springer, Heidelberg (2006)
17. Nikova, S., Rijmen, V., Schläffer, M.: Secure Hardware Implementation of Nonlinear Functions in the Presence of Glitches. J. Cryptology 24(2), 292–321 (2011)
18. Prouff, E., Rivain, M.: A Generic Method for Secure SBox Implementation. In: Kim, S., Yung, M., Lee, H.-W. (eds.) WISA 2007. LNCS, vol. 4867, pp. 227–244. Springer, Heidelberg (2008)
19. Prouff, E., Roche, T.: Higher-Order Glitches Free Implementation of the AES Using Secure Multi-party Computation Protocols. In: Preneel, B., Takagi, T. (eds.) CHES 2011. LNCS, vol. 6917, pp. 63–78. Springer, Heidelberg (2011)
20. Rivain, M., Prouff, E.: Provably Secure Higher-Order Masking of AES. In: Mangard, S., Standaert, F.-X. (eds.) CHES 2010. LNCS, vol. 6225, pp. 413–427. Springer, Heidelberg (2010)
21. Shibutani, K., Isobe, T., Hiwatari, H., Mitsuda, A., Akishita, T., Shirai, T.: *Piccolo*: An Ultra-Lightweight Blockcipher. In: Preneel, B., Takagi, T. (eds.) CHES 2011. LNCS, vol. 6917, pp. 342–357. Springer, Heidelberg (2011)
22. Xilinx, Inc. Virtex-5 Libraries Guide for HDL Designs (2009), http://www.xilinx.com/support/documentation/sw_manuals/xilinx11/virtex5_hdl.pdf

Appendix

									y							
		0	**1**	**2**	**3**	**4**	**5**	**6**	**7**	**8**	**9**	**a**	**b**	**c**	**d**	**e**	**f**
	00	a8	f8	f0	00	d9	62	fd	39	4a	bd	af	06	a9	35	e1	df
	01	14	5b	82	0d	9b	d4	29	17	b9	02	f7	95	3e	65	79	d7
	02	7d	e4	ba	8b	cc	dc	1d	b5	87	71	07	fa	ef	d5	48	2f
	03	a7	e3	b2	6f	aa	ed	4d	a0	81	c0	8c	15	e0	19	9e	f1
	04	84	6b	4c	da	93	eb	58	2b	d3	27	33	76	b8	51	96	a3
	05	f4	c5	75	ae	d2	30	85	fb	64	38	3f	5c	9c	66	98	c1
	06	bb	63	a4	73	52	fc	9d	8d	24	25	31	cf	e2	57	9f	c3
x	**07**	8f	f3	20	7f	3b	bc	bf	1a	54	03	91	0a	67	a5	16	10
	08	c7	e8	b3	21	13	72	0f	7a	01	88	e5	d1	f5	7c	40	ee
	09	97	4e	83	94	ad	5d	04	c4	32	a1	e7	92	43	b7	1e	e9
	0a	12	70	50	1f	a6	36	05	77	f6	ea	46	28	56	7b	55	db
	0b	61	34	b4	2e	9a	a2	6d	86	4f	cb	ab	ce	8a	6c	99	42
	0c	6a	5a	3d	ca	59	11	53	3c	ac	74	b6	c8	3a	89	2d	47
	0d	2a	cd	de	0c	5f	26	23	4b	c6	b0	7e	6e	f2	c2	b1	fe
	0e	18	37	0b	d6	e6	d0	49	8e	41	c9	69	44	d8	90	be	5e
	0f	0e	f9	60	ff	1b	ec	09	78	80	1c	dd	08	45	68	22	2c

Fig. 9. The G function selected in our experiments in the **First CT-RSA** profiles, values for the input as $(x, y) \in \mathbb{F}_2^5 \times \mathbb{F}_2^4$ (in hexadecimal format)

Efficient and Provably Secure Methods for Switching from Arithmetic to Boolean Masking

Blandine Debraize

Gemalto, 6 rue de la Verrerie, 92197 Meudon Cedex, France
blandine.debraize@gemalto.com

Abstract. A large number of secret key cryptographic algorithms combine Boolean and arithmetic instructions. To protect such algorithms against first order side channel analysis, it is necessary to perform conversions between Boolean masking and arithmetic masking. Louis Goubin proposed in [5] an efficient method to convert from Boolean to arithmetic masking. However the conversion method he also proposed in [5] to switch from arithmetic to Boolean is less efficient and could be a bottleneck in some implementations. Two faster methods were proposed in [2] and [9], both using precomputed tables. We show in this paper that the algorithm in [2] is bugged, and propose an efficient correction. Then, we propose an alternative to the algorithm in [9] with a valuable timing/memory tradeoff. This new method offers better security in practice and is well adapted for 8-bit architectures in terms of time performance (3.3 times faster than Goubin's algorithm for one single conversion).

Keywords: side channel analysis, differential power analysis, Boolean masking, arithmetic masking, conversion from arithmetic to Boolean masking.

1   Introduction

In 1999, the concept of Differential Power Analysis (DPA) was introduced in [7] by Paul Kocher. It consists in retrieving information about the secret key of an algorithm by analyzing the power consumption curves generated by the device in which the algorithm is implemented, during its execution. It was extended to some other techniques like CPA (Correlation Power Analysis), and EMA (Electromagnetic Analysis), based on similar principles. All these attacks relying on physical leakage of an electronic device are more generically called side channel analysis.

Countermeasures were soon developed to thwart these attacks. The most commonly used method, initially proposed in [1] and [6], consists in splitting all key-dependant intermediate variables processed during the execution of the algorithm into several shares. The value of each share, considered independently from the other ones, is randomly distributed and independent of the value of the

E. Prouff and P. Schaumont (Eds.): CHES 2012, LNCS 7428, pp. 107–121, 2012.

secret key: thefefore, the power leakage of one share does not reveal any secret information. It is shown in [1] that the number of power curves needed to mount an attack grows exponentially with the number of shares. When only two shares are used, the method comes to masking all intermediate data with random. In this case it is said that the implementation is protected against first order DPA. For algorithms that combine Boolean and arithmetic operations, two different kinds of masking must be used: Boolean masking and arithmetic masking. A large number of algorithms have this property: all software oriented finalists of the eSTREAM stream cipher competition [4], some other stream ciphers like Snow 2.0 [3] and Snow 3G, the block cipher IDEA [8], and several hash function designs used for HMAC constructions. The security of DPA-protected implementations of such ciphers strongly depends on the security of conversions between arithmetic and Boolean masking in both directions.

Two secure conversion algorithms (one for each direction) were proposed by Goubin in [5], but the arithmetic to Boolean method of [5] is quite slow and can be a bottleneck in some implementations. Then a second arithmetic to Boolean algorithm using two precomputed tables was proposed by Jean-Sébastien Coron and Alexei Tchulkine in [2]. Finally, an extension of the method of [2] was proposed by Olaf Neiße and Jürgen Pulkus in [9], allowing to reduce memory consumption.

In this paper we first recall the mechanisms of these three methods, showing that the Coron-Tchulkine algorithm is not correct in most cases. Then we propose a modification of Coron-Tchulkine's algorithm, correcting the bug and improving time performance. We also propose a new fast and secure arithmetic to Boolean conversion technique. Finally we give some performance comparisons between all methods.

2   Definitions and Previous Work

The masking technique introduced in [1] and [6] consists in splitting each intermediate variable that appears in the cryptographic algorithm, using a secret sharing scheme. Therefore, an attacker must analyze multiple point distributions, which requires a number of power curves exponential in the number of shares. To protect implementations against first order DPA, this technique has to be applied with two shares.

For algorithms that combine Boolean and arithmetic functions, two kinds of masking are used:

1. *Boolean masking*: $x' = x \oplus r$
2. *Arithmetic masking*: $x' = x - r \bmod 2^K$.

Here $\oplus$ is the exclusive or. The variable x refers to the secret intermediate data, r to the random value used to obtain the masked data x', these three data having size K.

The conversion algoritms from one masking to another must also be secure against side channel analysis. This means that all intermediate variables must be independent of the secret data.

2.1   First Secure Method

In [5] Louis Goubin proposed an efficient method to convert a Boolean masking into an arithmetic masking, relying on the fact that the function $f_{x'}(r) = (x' \oplus r) - r$ is affine in r over the field with two elements.

An algorithm converting from arithmetic to Boolean masking was also proposed in [5], based on the following recursion formula:

$$(A + r) \oplus r = u_{K-1}, \quad \text{where:} \begin{cases} u_0 = 0, \\ \forall k \geq 0, u_{k+1} = 2[u_k \wedge (A \oplus r) \oplus (A \wedge r)]. \end{cases}$$

But this method is less efficient than from Boolean to arithmetic, as the number of operations is linear in the size of the intermediate data.

2.2   Coron-Tchulkine Method

In [2], Jean-Sébastien Coron and Alexei Tchulkine proposed a second method to convert from arithmetic to Boolean masking. This method is based on the use of two precomputed tables. Let us recall its principle: two tables G and C are generated during the precomputation phase of the algorithm. Both tables have size 2^k, where k is the size of the processed data; the value of k is typically 4 or 8. For example if $k = 4$, a 32-bit variable is divided into 8 nibbles: the algorithm works then in 8 steps, each step processing one nibble of the 32-bit variable. Table G converts a nibble from arithmetic to Boolean masking, while Table C manages carries coming from the modular addition. Indeed, let us consider a masked data x' splitted into n nibbles $x'_{n-1}||...||x'_i||...||x'_0$: each value $x_i = x'_i + r$ can be possibly more than 2^k. In this case the carry must be added to the nibble x'_{i+1} before its conversion. As the carry value is correlated to the secret data, it must be masked. Therefore, for each input x'_i, the table C outputs the carry value c masked by the addition of a random k-bit value γ. Both Tables G and C can be described as follows:

Algorithm 2.1. Precomputation of tables	
Table G generation	Table C generation
Input: a nibble size k	Input: a k-bit value r.
1. Generate a random k-bit r	1. Generate a random k-bit γ
2. For $A = 0$ to $2^k - 1$ do	2. For $A = 0$ to $2^k - 1$ do
$\quad G[A] = (A + r) \oplus r$	$\quad C[A] \leftarrow \begin{cases} \gamma, & \text{if } A + r < 2^k \\ \gamma + 1 \bmod 2^k, & \text{if } A + r \geq 2^k \end{cases}$
3. Output G and r.	3. Output C and γ.

Finally the conversion phase can be described by the following algorithm, where the symbol $||$ means concatenation:

Algorithm 2.2. Conversion of a $(n \cdot k)$-bit variable
Input: (A, R) such that $x = A + R \bmod 2^{n \cdot k}$
and r, γ generated during precomputation phase
1.   For $i = 0$ to $n - 1$ do
2.   Split A into $A_h \| A_l$ and R into $R_h \| R_l$ such that
A_l and R_l have size k
3. $A \leftarrow A - r \bmod 2^{(n-i) \cdot k}$
4. $A \leftarrow A + R_l \bmod 2^{(n-i) \cdot k}$
5.   if $i < n - 1$ do
6. $A_h \leftarrow A_h + C[A_l] \bmod 2^{(n-i-1) \cdot k}$
7. $A_h \leftarrow A_h - \gamma \bmod 2^{(n-i-1) \cdot k}$
8. $x'_i \leftarrow G[A_l] \oplus R_l$
9. $x'_i \leftarrow x'_i \oplus r$
10. $A \leftarrow A_h$ and $R \leftarrow R_h$
11.   Output $x' = x'_{n-1} \| ... \| x'_i \| ... \| x'_0$

Let us specify that the value A_h and A_l are updated at the same time as A: the value A is splitted into A_h and A_l throughout all the algorithm. This remark is true for all conversion algorithms of this paper.

But this algorithm is actually not correct in case $n > 2$. Indeed, let us suppose that the following propositions are true together:

1. $n > 2$,
2. γ takes the value $2^k - 1$,
3. The carry equals 1.

Then in the first iteration of the loop of Algorithm 2.2 (i.e. when $i = 0$), the size of A_h is greater than k. In this case the value $A_h + C[A_l] - \gamma$ does not equal $A_h + 1$. Thus the table C generation must be modified to obtain an algorithm outputting always the correct value.

In Section 3, we propose a method combining correction and time performance improvement of Coron-Tchulking method.

2.3   Neiße-Pulkus Method

A third method was proposed in 2004 by Olaf Neiße and Jürgen Pulkus in [9]. As Coron-Tchulkine algorithm, it is based on the precomputation of tables. The principle is first to store the values of each possible nibble updated in the new masking mode in a 2^k-entry table, as Table G of Section 2.2. The carry is also stored in a 2^k-entry table C, but contrary to Coron-Tchulkine method, it is here stored unmasked. Tables G and C can be possibly combined in one table to reduce RAM space requirements.

The carry is masked during conversion step by the fact that sometimes the direct value of the intermediate variable is processed and sometimes its complement is processed. A random bit z generated at the beginning of each conversion step decides if the complement is used or not.

The method is based on the fact that, for any l-bit value x and its complement $\bar{x}$, the equation $x + \bar{x} + 1 = 2^l$ holds. Thus for a bit z, if $\tilde{x}$ denotes x when $z = 0$, and $\bar{x}$ when $z = 1$, we obtain $\tilde{x} = x - z \bmod 2^l$. And for two l-bit values x_1 and x_2 it can easily proved that $\widetilde{x_1 + x_2} = \widetilde{x_1} + \widetilde{x_2} + z \bmod 2^l$.

Let us take the notations of Section 2.2 (Algorithm 2.2): a random k-bit mask r used as input and output mask of Table G, and (A, R) such that $x = A + R \bmod 2^{n \cdot k}$. Then $\tilde{x} - (r||\cdots||r) = \tilde{A} + \tilde{R} - (r||\cdots||r) + z$, where each nibble of $\tilde{x} - (r||\cdots||r)$ is taken as input of the table, the output being the corresponding nibble of $\tilde{x} \oplus (r||\cdots||r)$. Thus the bit z must be added to the intermediate variable A at the beginning of each conversion step. At the end, as $\tilde{A} \oplus \tilde{R} = A \oplus R$, the correct result is then obtained.

The main principle of the conversion algorithm of [9] can be summarized by Algorithm 2.3.

Security of the Method. The authors of [9] claim that their algorithm is resistant against DPA. From a DPA-only point of view, the value of the bit $C[A_l]$ (line 9 of Algorithm 2.4) is indeed independent of the value of the secret data, due to the fact that for a k-bit value w, the number of k-bit values r such that $w + r \geq 2^k$ is w, and the number of k-bit values r such that $\bar{w} + r \geq 2^k - 1$ is $2^k - w - 1$, inducing a constant number $2^k - 1$ of possible r contributing to a non-zero carry.

But in practice this algorithm may pose a security problem, as the value $-z \bmod 2^{n \cdot k}$ is manipulated several times during one conversion step: as z is a random bit, this value is either 0 or 0xFF...FF. It could be distinguished by the attacker in some context, using SPA techniques. With this information, the attacker could mount a DPA attack, using the fact that the carries are then unmasked. This implies that the behavior of the component in terms of power and electromagnetic leakage must be studied very carefully before choosing this conversion method.

Algorithm 2.3. Conversion of a $(n \cdot k)$-bit variable

Input: (A, R) such that $x = A + R \bmod 2^{n \cdot k}$ and r, generated
       during precomputation phase

1.   Generate a random bit z
2. $Z \leftarrow -z \bmod 2^{n \cdot k}$
3. $A \leftarrow (A \oplus Z) - (r||\cdots||r) \bmod 2^{n \cdot k}$
4. $R \leftarrow R \oplus Z$
5.   For $i = 0$ to $n - 1$ do
6.   Split A into $A_h||A_l$ and R into $R_h||R_l$ such that
        A_l and R_l have size k
7. $A \leftarrow A + R_l \bmod 2^{(n-i) \cdot k}$
8.   if $i < n - 1$ do
9. $A_h \leftarrow A_h + C[A_l] \bmod 2^{(n-i-1) \cdot k}$
10. $x'_i \leftarrow G[A_l] \oplus R_l$
11. $A \leftarrow A_h$ and $R \leftarrow R_h$
12.   Output $x' = (x'_{n-1}||...||x'_i||...||x'_0) \oplus (r||\cdots||r)$

The final method proposed in [9] is slightly different from Algorithm 2.3, as a technique is added in [9] to reduce memory consumption (it is recalled in

Section 4.2). The use of this technique implies a decrease in the speed of the algorithm. As our paper mainly focuses on time performance, we propose in Appendix C another modification of the algorithm in order to reach maximal speed. This algorithm is used for the performance tests described at section 5.

3   Correction and Improvement of Coron-Tchulkine Method

We show in Appendix A that the immediate possible corrections of Coron-Tchulkine method keeping the size of the carry's mask γ to k bits are not first order DPA resistant: the size of γ must be at least $(n-1) \cdot k$ to obtain complete independence of intermediate data from the secret key.

Let us remark that both the information provided by Table G of Coron-Tchulkine method (update of the nibble in the new masking mode) and the information of Table C (additively masked carry) can be summarized in one unique table T whose outputs have size $n \cdot k$:

Algorithm 3.1. Table T generation
1.   Generate a random k-bit r and a random $(n \cdot k)$-bit γ
2.   For $A = 0$ to $2^k - 1$ do
$\qquad T[A] = ((A + r) \oplus r) + \gamma \bmod 2^{n \cdot k}$
3.   Output T, r and γ

The number of entries of the table is 2^k, and the size of each entry is $\frac{n \cdot k}{8}$ bytes. For typical values $k = 8$ and $n = 4$, the memory consumption is doubled here compared to the adaptation of Neiße-Pulkus method proposed in Appendix C.

The resulting conversion algorithm is as follows:

Algorithm 3.2. Conversion of a $(n \cdot k)$-bit variable
Input: (A, R) such that $x = A + R \bmod 2^{n \cdot k}$
$\qquad\qquad$ and r, γ generated during precomputation phase
1.   For $i = 0$ to $n - 1$ do
2.   Split A into $A_h \| A_l$ and R into $R_h \| R_l$,
$\qquad\quad$ such that A_l and R_l have size k
3. $A \leftarrow A - r \bmod 2^{(n-i) \cdot k}$
4. $A \leftarrow A + R_l \bmod 2^{(n-i) \cdot k}$
5. $A \leftarrow A_h \| 0 + T[A_l] \bmod 2^{n \cdot k}$
6. $A \leftarrow A - \gamma \bmod 2^{n \cdot k}$
7. $x'_i \leftarrow A_l \oplus R_l$
8. $x'_i \leftarrow A_l \oplus r$
9. $A \leftarrow A_h$ and $R \leftarrow R_h$
10.   Output $x' = x'_0 \| ... \| x'_i \| ... \| x'_{n-1}$

Here, as T's outputs have the same size as the processed data, if the value $A+r$ is greater than 2^k during the precomputation of T, the $(k+1)^{\text{th}}$ least significant bit of $T[A]$ is automatically set to 1 before being masked by the addition of γ.

During the conversion algorithm, the carry is added to the current variable (line 5) at the same time as the nibble A_l is updated.

The use of one table instead of two is clearly an advantage in terms of time performance. But it is still possible to reduce the execution time of the conversion algorithm by moving some instructions out of the loop. These improvements are described in Appendix B.

4   New Method

In terms of performance, for a 16-bit or an 8-bit processor, the drawback of the method described in Section 3 is the fact that the size of the manipulated data is the same as the size of the intermediate data of the algorithm. Indeed, the typical size for intermediate data is 32 bits: the time of the conversion algorithm is then multiplied by 2 with a 16-bit processor, and by 4 with an 8-bit processor. In this section we propose a method more appropriate for processors whose registers have size smaller than the intermediate data of the algorithm.

4.1   Principle

Let us remark that using precomputed tables to keep data masked during the algorithm execution comes to treating masked information as memory address information. As a carry is a 1-bit information, our goal in this section is to apply this principle to 1-bit information.

Let us suppose that instead of being masked arithmetically as proposed in Sections 2.2 and 3, carries are protected by Boolean masks. The protection comes then to adding by exclusive or a random bit to the carry value. For example, if we call ρ such a random bit, a 2-entry table C can be generated during the precomputed step in the following way:

Algorithm 4.1. Table C generation
1.   Generate a random bit ρ and a random $(n \cdot k)$-bit value λ
2. $C[\rho] = \lambda$
3. $C[\rho \oplus 1] = \lambda + 1 \bmod 2^{n \cdot k}$
4.   Output C and λ.

Now let us suppose that a carry c, protected by the Boolean mask ρ, is manipulated during the conversion algorithm. Thus the masked value $b = c \oplus \rho$ can be used in the following way to add the carry c to the value A_h in a secure way:

Algorithm 4.2. Carry addition
Inputs: – a value A_h (masked arithmetically),
– a carry bit b (c masked in a Boolean way)
– C, λ generated during precomputation phase
1. $A_h = A_h + C[b] \bmod 2^{n \cdot k}$
2. $A_h = A_h - \lambda \bmod 2^{n \cdot k}$
3.   Output A_h

It is easy to convince oneself about:

1. *The correctness of the method*: whatever the value of ρ is, the value $C[b]$ is equal to the carry c added to λ modulo $2^{n \cdot k}$.
2. *The resistance against order 1 DPA*: all processed intermediate variables are independent of the unmasked values. The masked carry is treated as information about the address of a RAM location. This address is independent from the value of the carry, as it changes from one execution to the other.

In next Section, we propose a method combining the information of both the nibble to be updated and of the carry masked by exclusive or, using this combination as an address in a conversion table.

4.2 Algorithm

In this section a new method for switching from arithmetic to Boolean masking is proposed. As the method described in Section 3, it requires the precomputation of one table T whose outputs must contain information about both the nibble updated in the new masking mode and the next carry bit. Here the output of T is directly the value $(A + r + c) \oplus r$, where c is the carry resulting from the previous addition.

The precomputed table T has the following properties:

- The carry value is masked by exclusive or with a random bit.
- During conversion phase, the choice of the address in the table not only depends on the value of the nibble but also on the value of the maksed previous carry. This implies T has size 2^{k+1}. The needed amount of memory is then doubled compared to Neiße-Pulkus method, and is the same compared to the correction of Coron-Tchulkine method for typical values $k = 8$ and $n = 4$.

The value of the random bit used to mask carries decides during precomputation step if the values of the addresses of nibbles of type $(A + r) \oplus r$ are greater or less than the addresses of nibbles of type $(A + r + 1) \oplus r$. And during conversion step, the value of the masked carry is used to compute the address of the next nibble to be loaded from the table.

The method is outlined in Algorithms 4.3 and 4.4. Here again, all processed variable are independent of secret data, inducing resistance of the algorithm against first order DPA.

Algorithm 4.3. Table T generation

1. Generate a random k-bit r and a random bit ρ
2. For $A = 0$ to $2^k - 1$ do
   $$T[\rho || A] \quad = (A + r) \oplus (\rho || r)$$
   $$T[(\rho \oplus 1) || A] = (A + r + 1) \oplus (\rho || r)$$
3. Output T, r and ρ

If $k = 8$, the time of the conversion phase is optimized. But in this case the size of the output data of the table is $k + 1 = 9$ bits. This implies that this data needs two bytes to be stored, and the size of the table in RAM is then 1024 bytes. This amount of memory is possible today on many embedded components, but could still be too large in some cases. As in [9], the fact that the Boolean masking of a secret data $x'_b = x \oplus r$ and the arithmetic masking of the same data $x'_a = x - r$ mod 2^k have always the same least significant bit can be used to reduce the size of the precomputed table. Thus storing the least significant bit of $(A + r) \oplus r$ or of $(A + r + 1) \oplus r$ is not necessary. The place of this useless bit can be taken by the carry bit. The resulting algorithm is then slower but the needed amount of memory is reduced by half. In our method it is then 512 bytes.

Algorithm 4.4. Conversion of a $n \cdot k$-bit variable

Input: (A, R) such that $x = A + R$ mod $2^{n \cdot k}$,
          r, ρ generated during precomputation phase

1. $A \leftarrow A - (r||...||r||...||r)$ mod $2^{n \cdot k}$
2. $\beta \leftarrow \rho$
3. For $i = 0$ to $n - 1$ do
4.   Split A into $A_h||A_l$ and R into $R_h||R_l$,
          such that A_l and R_l have size k.
5. $A \leftarrow A + R_l$ mod $2^{(n-i) \cdot k}$
6. $\beta||x'_i \leftarrow T[\beta||A_l]$
7. $x'_i \leftarrow x'_i \oplus R_l$
8. $A \leftarrow A_h$ and $R \leftarrow R_h$
9. Output $x' = (x'_0||...||x'_i||...||x'_{n-1}) \oplus (r||...||r||...||r)$

---

5 Performance Tests

In this section we propose some performance comparisons between Goubin's method and the three methods based on precomputed tables. The versions chosen for the tests are the ones that are optimized in terms of time performance:

- Algorithms C.1 and C.2 (Appendix C) for the modified Neiße-Pulkus method (named Mod. N.-P. in Tables 1, 2 and 3).
- Algorithms B.1 and B.2 (Appendix B) for the correction of Coron-Tchulkine method (named Mod. C.-T. in Tables 1, 2 and 3).
- Algorithms 4.3 and 4.4 (Section 4.2) for the new method proposed in this paper.

We first chose to perform C implementations of these algorithms and test them on 8051 architectures: one 8-bit and one 16-bit microprocessors. Table 1 and Table 2 summarize the performance comparison results for both components. These results are given in clock cycles numbers, computed with the help of a simulation tool. The size of the data to be converted from arithmetic to Boolean masking is 32 bits, as it is the most common size for intermediate data of cryptographic algorithms. For the table-based algorithms, two nibble sizes were tested: $k = 4$ and $k = 8$. The size of the precomputed tables in RAM are given in number of bytes.

Table 1. Smart card 8-bit microprocessor

	Goubin's method	Mod. N.-P. $k = 4$	Mod. N.-P. $k = 8$	Mod. C.-T. $k = 4$	Mod. C.-T. $k = 8$	New method $k = 4$	New method $k = 8$
Precomputation time	10325	2562	40274	18589	109391	3166	93007
Conversion time	39213	15479	9208	13969	7060	11720	6111
Table size	0	16	512	64	1024	32	1024

Table 2. Smart card 16-bit microprocessor

	Goubin's method	Mod. N.-P. $k = 4$	Mod. N.-P. $k = 8$	Mod. C.-T. $k = 4$	Mod. C.-T. $k = 8$	New method $k = 4$	New method $k = 8$
Precomputation time	86	377	3734	921	5933	439	5174
Conversion time	934	558	308	512	274	445	257
Table size	0	16	512	64	1024	32	1024

Table 3. Smart card 32-bit microprocessor

	Goubin's method	Mod. N.-P. $k = 4$	Mod. N.-P. $k = 8$	Mod. C.-T. $k = 4$	Mod. C.-T. $k = 8$	New method $k = 4$	New method $k = 8$
Precomputation time	15.1	9.6	156.2	25.5	188.8	12.1	180.3
Conversion time	32.9	12.9	10.3	12.1	8	14.9	9.2
Table size	0	16	512	64	1024	32	1024

We also performed performance comparison tests for the same algorithms in ARM assembler on a 32-bit 26 MHz microprocessor. In Table 3 the time results of these tests are given in microseconds.

The generation of random numbers is required by all methods. For Goubin's algorithm (see [5]), the size of the random value is 32 bits, and only one such random word is necessary for each execution. For this reason, the time of this generation is set in precomputation step. Thus we remark that, depending on the chip, the time of the generation of the random values is generally not negligible for these conversion algorithms. This also explains the time difference between the precomputation steps of the Neiße-Pulkus method and of the Coron-Tchulkine method (one random byte is generated in N.-P. against four in C.-T. method).

From the figures given in Table 1 and Table 2, we remark that the new method is more efficient on these microcontrollers than the improved correction of Coron-Tchulkine algorithm. This confirms the fact that the new method is better adapted to 8-bit and 16-bit architectures. On the 8-bit architecture, the conversion step of the new method is the fastest.

Choosing $k = 4$, the improved Neiße-Pulkus and the new method are both faster than Goubin's algorithm on all architectures, even for a single conversion, with a small amount of needed memory. Neiße-Pulkus method is about twice faster on the 32-bit microcontroller, and the new method about three times faster on the 8-bit microcontroller.

6   Conclusion

In this paper we sought the fastest methods for switching from arithmetic to Boolean masking. We first analyzed the two known methods [2,9] based on pre-computed tables: we showed that the algorithm proposed in [2] is not correct and proposed an improved correction. We also proposed a new method, which is well adapted for 8-bit architectures in terms of time preformance. As the correction of [2], it offers better security against side channel analysis in practice than the algorithm in [9].

References

1. Chari, S., Jutla, C.S., Rao, J.R., Rohatgi, P.: Towards Sound Approaches to Counteract Power-Analysis Attacks. In: Wiener, M. (ed.) CRYPTO 1999. LNCS, vol. 1666, pp. 398–412. Springer, Heidelberg (1999)
2. Coron, J.-S., Tchulkine, A.: A New Algorithm for Switching from Arithmetic to Boolean Masking. In: Walter, C.D., Koç, Ç.K., Paar, C. (eds.) CHES 2003. LNCS, vol. 2779, pp. 89–97. Springer, Heidelberg (2003)
3. Ekdahl, P., Johansson, T.: A New Version of The Stream Cipher SNOW. In: Nyberg, K., Heys, H.M. (eds.) SAC 2002. LNCS, vol. 2595, pp. 47–61. Springer, Heidelberg (2003)
4. eSTREAM. ECRYPT Stream Cipher Project, IST-2002-507932, http://www.ecrypt.eu.org/stream/
5. Goubin, L.: A Sound Method for Switching between Boolean and Arithmetic Masking. In: Koç, Ç.K., Naccache, D., Paar, C. (eds.) CHES 2001. LNCS, vol. 2162, pp. 3–15. Springer, Heidelberg (2001)
6. Goubin, L., Patarin, J.: DES and Differential Power Analysis (The "Duplication" Method). In: Koç, Ç.K., Paar, C. (eds.) CHES 1999. LNCS, vol. 1717, pp. 158–172. Springer, Heidelberg (1999)
7. Kocher, P.C., Jaffe, J., Jun, B.: Differential Power Analysis. In: Wiener, M. (ed.) CRYPTO 1999. LNCS, vol. 1666, pp. 388–397. Springer, Heidelberg (1999)
8. Lai, X., Massey, J.L.: A Proposal for a New Block Encryption Standard. In: Damgård, I.B. (ed.) EUROCRYPT 1990. LNCS, vol. 473, pp. 389–404. Springer, Heidelberg (1991)
9. Neiße, O., Pulkus, J.: Switching Blindings with a View Towards IDEA. In: Joye, M., Quisquater, J.-J. (eds.) CHES 2004. LNCS, vol. 3156, pp. 230–239. Springer, Heidelberg (2004)

Appendices

A Security Weaknesses of Two Immediate Modifications of Coron-Tchulkine Method

Two immediate modifications of the method proposed in [2] seems possible. We call these two possible tables C' and C'':

Algorithm A.1. Carry table C' generation	**Algorithm A.2.** Carry table C'' generation
Input: a random r of k bits.	Input: a random r of k bits.
1. Generate a random k-bit γ such that $\gamma < 2^k - 1$ 2. For $A = 0$ to $2^k - 1$ do $C'[A] \leftarrow \begin{cases} \gamma, & \text{if } A + r < 2^k \\ \gamma + 1, & \text{if } A + r \geq 2^k \end{cases}$ 3.Output C' and γ.	1. Generate a random k-bit γ 2. For $A = 0$ to $2^k - 1$ do $C''[A] \leftarrow \begin{cases} \gamma, & \text{if } A + r < 2^k \\ \gamma + 1, & \text{if } A + r \geq 2^k \end{cases}$ 3. Output C'' and γ.

But both corrections imply that some manipulated data are not completely decorrelated from the value of the secret data. Indeed, the least significant bit of the output of Table C' is correlated to the corresponding carry bit. Let us compute the correlation coefficient.

Using the notations from Section 2.2, let us call b_0 the least significant bit of the output of Table C', and c the corresponding carry value. Let us define a set χ:

$$\chi = \{ \, (A, r, \gamma) : A \in \mathbb{N} \text{ such that } A < 2^n,$$
$$r \in \mathbb{N} \text{ such that } r < 2^n,$$
$$\gamma \in \mathbb{N} \text{ such that } \gamma < 2^n - 1\}$$

The correlation coefficient between b_0 and c conditionned on the subset χ is as follows:

$$Cor(b_0, \, c \mid \chi) = |P(b_0(x) = c(x) \mid x \in \chi) - P(b_0(x) \neq c(x) \mid x \in \chi)|$$
$$= \left| \frac{1}{\#\chi} \sum_{x \in \chi} (-1)^{b_0(x) \oplus c(x)} \right|$$

Actually the value $b_0(x) \oplus c(x)$ neither depends on the value of A nor on the value of r, but only depends on the value of the least significant bit of γ. We call this bit γ_0. As 2^{k-1} times out of $2^k - 1$, $\gamma_0 = 0$, and $2^{k-1} - 1$ times out of $2^k - 1$, $\gamma_0 = 1$, we have:

$$Cor\left(b_0, c \mid \chi\right) = \left| \frac{1}{2^k - 1} \sum_{x \in \chi} (-1)^{\gamma_0(x)} \right|$$

$$= \left| \frac{1}{2^k - 1} \left(\sum_{\gamma < 2^k - 1, \gamma_0 = 0} (-1)^0 + \sum_{\gamma < 2^k - 1, \gamma_0 = 1} (-1)^1 \right) \right|$$

$$= \left| \frac{1}{2^k - 1} \left(2^{k-1} - (2^{k-1} - 1) \right) \right| = \frac{1}{2^k - 1}$$

If $k = 4$, the correlation coefficient is then $\frac{1}{15}$.

It could be shown in a similar way that the correlation coefficient between the most significant bit of the output of Table C'' (the $(k+1)^{\text{th}}$ bit) and the carry bit has value $\frac{31}{256}$.

Many other modifications of carry Table C are possible, each one corresponding to a specific interval in which the random carry's mask γ takes its values. In each of these cases it can be shown in a way similar as above that if this interval is smaller than $[0, 2^{(n-1) \cdot k} - 1]$, a correlation exists between the output of the table and the carry bit.

In case the interval is $[0, 2^{(n-1) \cdot k} - 1]$, as the addition of Table C's output is performed modulo at most $2^{(n-1) \cdot k}$ (line 6 of Algorithm 2.2), the precomputed addition "γ+carry" can be also performed modulo $2^{(n-1) \cdot k}$ without the lack of correctness of the initial Coron-Tchulkine method.

B Time Performance Improvement of Algorithm 3.2

To reduce the execution time of Algorithm 3.2, some instructions can be set out of the loop. Three of them can be removed from the loop without weakening security:

- The arithmetic masking with the random r (line 3) can be performed before the loop.
- The subtraction of the value γ (line 6) can be performed before the loop (inducing a modification of the precomputed table T).
- The Boolean unmasking with the random r (line 8) can be performed after the loop.

Indeed, in case these instructions are moved out of the loop, all nibbles of A but one remain masked with the initial mask R during the execution of the algorithm, and the lasting nibble is masked by the random value r. All intermediate variables are then independent of secret data throughout the execution.

Some extra calculations must be performed during precomputation step, allowing to minimize the cost of the subtraction of γ during conversion step. The improved version of the method is then as follows:

Algorithm B.1. Table T generation

1. Generate a random k-bit r and a random $((n-1) \cdot k)$-bit γ
2. Compute $\Gamma = \sum_{i=1}^{k-1} 2^{i \cdot k} \cdot \gamma \bmod 2^{n \cdot k}$
3. $\gamma' \leftarrow \gamma || r$
4. For $A = 0$ to $2^k - 1$ do
   $T[A] = \left(A + \gamma' \bmod 2^{n \cdot k}\right) \oplus r$
5. Output T, r and Γ

Algorithm B.2. Conversion of a $n \cdot k$-bit variable

Input: (A, R) such that $x = A + R \bmod 2^{n \cdot k}$
  and r, E generated during precomputation phase

1. $A \leftarrow A - (r||...||r||...||r) \bmod 2^{n \cdot k}$
2. $A \leftarrow A - \Gamma \bmod 2^{n \cdot k}$
3. For $i = 0$ to $n - 1$ do
4.   Split A into $A_h||A_l$ and R into $R_h||R_l$,
     such that A_l and R_l have size k
5. $A \leftarrow A + R_l \bmod 2^{(n-i) \cdot k}$
6. $A \leftarrow A_h||0 + T[A_l] \bmod 2^{n \cdot k}$
7. $x'_i \leftarrow A_l \oplus R_l$
8. $A \leftarrow A_h$ and $R \leftarrow R_h$
9. Output $x' = (x'_0||...||x'_i||...||x'_{n-1}) \oplus (r||...||r||...||r)$

C   Time Performance Improvement of Neiße-Pulkus Method

The principle of the method described in [9] is a direct extension of the Coron-Tchulkine method. Therefore the two precomputed tables of [9] can be generated at the same time in one unique table exactly the same way as proposed in section 3 for Coron-Tchulkine method.

The adapted version of Neiße-Pulkus method is then as follows:

Algorithm C.1: Table T generation

1. Generate a random k-bit r
2. For $A = 0$ to $2^k - 1$ do
   $T[A] = (A + r) \oplus r$
3. Output T and r

Algorithm C.2. Conversion of a $n \cdot k$-bit variable

Input: (A, R) such that $x = A + R \mod 2^{n \cdot k}$

  and r generated during precomputation phase

1. Generate a random bit z
2. $Z \leftarrow -z \mod 2^{n \cdot k}$
3. $A \leftarrow (A \oplus Z) - (r || \cdots || r) \mod 2^{n \cdot k}$
4. $R \leftarrow R \oplus Z$
5. For $i = 0$ to $n - 1$ do
6.   Split A into $A_h || A_l$ and R into $R_h || R_l$,
      such that A_l and R_l have size k
7. $A \leftarrow A + R_l \mod 2^{(n-i) \cdot k}$
8. $A \leftarrow A_h || 0 + T[A_l] \mod 2^{n \cdot k}$
9. $x'_i \leftarrow A_l \oplus R_l$
10. $A \leftarrow A_h$ and $R \leftarrow R_h$
11. Output $x' = (x'_0 || ... || x'_i || ... || x'_{n-1}) \oplus (r || ... || r || ... || r)$

Let us remark that the instructions inside the loop are the same in algorithm B.2 and in algorithm C.2. Concerning memory consumption, for typical values $k = 8$ and $n = 4$, the size of T's outputs is here 9 bits: the required amount of memory is then 512 bytes. It is reduced by half compared to the improved Coron-Tchulkine algorithm of Appendix B.

A Differential Fault Attack on the Grain Family of Stream Ciphers

Subhadeep Banik, Subhamoy Maitra, and Santanu Sarkar

Applied Statistics Unit, Indian Statistical Institute,
203 B T Road, Kolkata 700 108, India
{s.banik_r,subho}@isical.ac.in, sarkar.santanu.bir@gmail.com

Abstract. In this paper we study a differential fault attack against the Grain family of stream ciphers. The attack works due to certain properties of the Boolean functions and corresponding choices of the taps from the LFSR. The existing works, by Berzati et al. (2009) and Karmakar et al. (2011), are applicable only on Grain-128 exploiting certain properties of the combining Boolean function h. That idea could not easily be extended to the corresponding Boolean function used in Grain v1. Here we show that the differential fault attack can indeed be efficiently mounted for the Boolean function used in Grain v1. In this case we exploit the idea that there exists certain suitable α such that $h(\mathbf{x}) + h(\mathbf{x} + \alpha)$ is linear. In our technique, we present methods to identify the fault locations and then construct set of linear equations to obtain the contents of the LFSR and the NFSR. As a countermeasure to such fault attack, we provide exact design criteria for Boolean functions to be used in Grain like structure.

Keywords: Fault Attacks, Countermeasures, Grain v1, Grain-128, Grain-128a, LFSR, NFSR, Stream Cipher.

1 Introduction

The Grain v1 stream cipher is in the hardware profile of the eStream portfolio [1] that has been designed by Hell, Johansson and Meier in 2005 [15]. It is a synchronous bit oriented stream cipher, although it is possible to achieve higher throughput at the expense of additional hardware. The physical structure of Grain is simple as well as elegant and it has been designed so as to require low hardware complexity. Following certain attacks on the initial design of the cipher, the modified versions Grain v1 [15], Grain-128 [16] and Grain-128a [2] were proposed after incorporating certain changes. Analysis of this cipher is an area of recent interest in as evident from numerous cryptanalytic results related to this family [3–5, 8–13, 19, 20, 22, 23, 27, 28].

Fault attacks are known to be very efficient against stream ciphers in general, and have received a lot of attention in recent cryptographic literature [6, 7, 17, 18, 21]. For differential fault attack scenario in stream ciphers, the attacker is allowed to inject faults in the internal state. Then by analyzing the difference in the faulty

E. Prouff and P. Schaumont (Eds.): CHES 2012, LNCS 7428, pp. 122–139, 2012.

and the fault-free keystreams, one should be able to deduce the complete or partial information about the internal state/secret key. The most common method of injecting faults is by using laser shots or clock glitches [24, 25]. Though the fault attacks usually rely on optimistic assumptions and study the cipher in a model that is weaker than the original version, they are not unrealistic as evident from literature. In this paper too, the model we study is a follow up of existing state-of-the-art literature [5, 19]. A detailed justification of the feasibility of such fault model is presented in [5, Section IIIB]. Before proceeding further, let us now present the fault model.

1. Similar to [5], we consider that the attacker is able to reset the system with the original Key-IV and start the cipher operations again. The work [19] requires a different assumption, where the IVs need to be modified in each initialization.
2. The attacker can inject a fault at any one random bit location of the LFSR or NFSR. As a result of the fault injection, the binary value in the bit-location (where the fault has been injected) is toggled. The attacker is not allowed to choose the location where he wants to inject the fault. However, as assumed in both [5, 19] the fault in any bit may be reproduced at any later stage of operation, once injected.
3. Similar to [5], we inject faults in the LFSR only, whereas the NFSR has been used for fault injection in [19].
4. The attacker has full control over the timing of fault injection, i.e., it is possible to inject the fault precisely at any stage of the cipher operation.

OUR CONTRIBUTION. Grain-128 has been successfully cryptanalyzed by employing fault attacks [5, 19]. However, Grain-v1 employs Boolean function of different kind, and thus such fault attacks may not immediately work against this cipher. In this paper we have tried to explore a generic fault attack on the structure of the Grain family of stream ciphers and thus in particular the idea works for Grain v1 too. The works presented in [5, 19] exploited the fact that the Boolean function g in Grain-128 is quadratic and the function h has only one cubic term other than the quadratic terms. This is not the scenario in Grain v1, where the Boolean functions are of more complicated structure in their Algebraic Normal Form. We point out that there are still problems in the choice of such functions as in Grain v1 [15] and suggest suitable choice instead of that one.

The novel idea of our fault attack is based on certain specific observations related to the output Boolean function h. For Grain v1, h is a 5-variable function with the differential property that $h(s_0, s_1, s_2, s_3, s_4) + h(1 + s_0, 1 + s_1, s_2, s_3, 1 + s_4) = s_2$. This helps us in determining one of the LFSR bits of the internal state and repeating this several times we get the complete LFSR state. Then we further note that h can be written as $s_4 \cdot u(s_0, s_1, s_2, s_3) + v(s_0, s_1, s_2, s_3)$, where $u(s_0, s_1, s_2, s_3) + u(s_0, 1 + s_1, s_2, 1 + s_3) = 1$. This helps us in determining the NFSR bits. To highlight our contribution in this paper, we like to refer to the following comment from [19]:

"The attack may be extended to Grain-like ciphers with higher degree feedback functions and output functions. However, determining fault locations can be a challenging task if linear terms are removed from output bit expression. Higher degree feedback functions and output functions will however certainly increase the attack complexity as mostly nonlinear equations will be obtained."

We show that the complexity of the attack is not exactly related to the degree of the output functions. A q-variable Boolean function (say, $h : \{0,1\}^q \to \{0,1\}$) with high degree can also be attacked in a similar manner if there exists certain suitable $\alpha \in \{0,1\}^q$ such that $h(\mathbf{x}) + h(\mathbf{x} + \alpha)$ is linear. That is, higher degree functions may not increase the attack complexity as linear equations may actually be available using clever techniques instead of nonlinear ones.

An integral part of any fault attack is to identify the register location where the fault has been injected. We outline a novel technique of identifying the fault location in the LFSR by using an optimal length Signature vector technique.

We also point out that there exists a pool of 5-variable Boolean functions which are of matching parameters as proposed in [15] and possess additional properties that help in resisting the kind of differential fault attack that we describe here. That is, we present specific countermeasures to this kind of fault attack that relies on proper choice of the nonlinear combining Boolean function.

ORGANIZATION OF THIS PAPER. In this section, we proceed with the details of the Grain family (in particular Grain v1). Next, in Section 2, we present a broad description of the actual attack. The implementation of the attack on Grain v1 along with the fault location identification routine is explained in Section 3. The countermeasure corresponding to this attack with respect to proper choice of Boolean functions is described in Section 4. Section 5 concludes the paper.

We abuse the + notation for Boolean XOR, i.e., $GF(2)$ addition as well as standard arithmetic addition. However, that will be clear from the context.

1.1   Brief Description of Grain Family

The exact structure of the Grain family is explained in Figure 1. It consists of an n-bit LFSR and an n-bit NFSR. Certain bits of both the shift registers are taken as inputs to a combining Boolean function, whence the keystream is produced. The update function of the LFSR is given by the equation $y_{t+n} = f(Y_t)$, where $Y_t = [y_t, y_{t+1}, \ldots, y_{t+n-1}]$ is an n-bit vector that denotes the LFSR state at the tth clock interval and f is a linear function on the LFSR state bits obtained from a primitive polynomial in $GF(2)$ of degree n. The NFSR state is updated as $x_{t+n} = y_t + g(X_t)$. Here, $X_t = [x_t, x_{t+1}, \ldots, x_{t+n-1}]$ is an n-bit vector that denotes the NFSR state at the tth clock interval and g is a non-linear function of the NFSR state bits.

The output keystream is produced by combining the LFSR and NFSR bits as $z_t = h'(X_t, Y_t) = \bigoplus_{a \in A} x_{t+a} + h(X_t, Y_t)$, where A is some fixed subset of $\{0, 1, 2, \ldots, n-1\}$.

Key Loading Algorithm (KLA). The Grain family uses an n-bit key K, and an m-bit initialization vector IV, with $m < n$. The key is loaded in the NFSR and the IV is loaded in the 0th to the $(m-1)^{th}$ bits of the LFSR. The remaining m^{th} to $(n-1)^{th}$ bits of the LFSR are loaded with some fixed pad $P \in \{0,1\}^{n-m}$. Hence at this stage, the $2n$ bit initial state is of the form $K\|IV\|P$.

Key Scheduling Algorithm (KSA). After the KLA, for the first $2n$ clocks, the keystream produced at the output point of the function h' is XOR-ed to both the LFSR and NFSR update functions, i.e., during the first $2n$ clock intervals, the LFSR and the NFSR bits are updated as $y_{t+n} = z_t + f(Y_t)$, $x_{t+n} = y_t + z_t + g(X_t)$.

Pseudo-Random keystream Generation Algorithm (PRGA). After the completion of the KSA, z_t is no longer XOR-ed to the LFSR and the NFSR but it is used as the Pseudo-Random keystream bit. Therefore during this phase, the LFSR and NFSR are updated as $y_{t+n} = f(Y_t)$, $x_{t+n} = y_t + g(X_t)$.

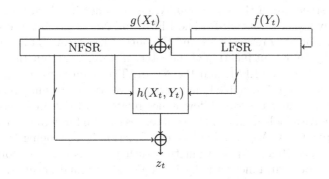

Fig. 1. Structure of Stream Cipher in Grain Family

One may note that given any arbitrary state and the information about its evolution (the number of clocks in KSA or PRGA), one can calculate the corresponding state S_0^K at the beginning of the KSA. This is because the state update functions in both the KSA and PRGA in the Grain family are one-to-one and invertible. Hence one can construct the KSA^{-1} routine that given an input $2n$ bit vector denoting the internal state of the cipher at the end of the KSA, returns the $2n$ bit vector giving internal state of the cipher at the beginning of the KSA. One can similarly describe a PRGA^{-1} routine that inverts one round of the PRGA.

As we will consider Grain v1 for our attack description, let us describe it now. In Grain v1, the size of Key is $n = 80$ bits and the IV is of size $m = 64$ bits. The pad used in the KLA is $P = \texttt{0xFFFF}$. The LFSR update rule is given by $y_{t+80} = y_{t+62} + y_{t+51} + y_{t+38} + y_{t+23} + y_{t+13} + y_t$. The NFSR state is updated as

$x_{t+80} = y_t + g(x_{t+63}, x_{t+62}, x_{t+60}, x_{t+52}, x_{t+45}, x_{t+37}, x_{t+33}, x_{t+28}, x_{t+21}, x_{t+15},$
$x_{t+14}, x_{t+9}, x_t)$, where $g(x_{t+63}, x_{t+62}, x_{t+60}, x_{t+52}, x_{t+45}, x_{t+37}, x_{t+33},$
$x_{t+28}, x_{t+21}, x_{t+15}, x_{t+14}, x_{t+9}, x_t)$
$= x_{t+62} + x_{t+60} + x_{t+52} + x_{t+45} + x_{t+37} + x_{t+33} + x_{t+28} + x_{t+21} + x_{t+14} + x_{t+9}$
$+x_t + x_{t+63}x_{t+60} + x_{t+37}x_{t+33} + x_{t+15}x_{t+9} + x_{t+60}x_{t+52}x_{t+45} + x_{t+33}x_{t+28}x_{t+21}$
$+x_{t+63}x_{t+45}x_{t+28}x_{t+9} + x_{t+60}x_{t+52}x_{t+37}x_{t+33} + x_{t+63}x_{t+60}x_{t+21}x_{t+15}$
$+x_{t+63}x_{t+60}x_{t+52}x_{t+45}x_{t+37} + x_{t+33}x_{t+28}x_{t+21}x_{t+15}x_{t+9}$
$+x_{t+52}x_{t+45}x_{t+37}x_{t+33}x_{t+28}x_{t+21}.$

The output keystream is produced by combining the LFSR and NFSR bits as
$z_t = \bigoplus_{a \in A} x_{t+a} + h(y_{t+3}, y_{t+25}, y_{t+46}, y_{t+64}, x_{t+63})$, where
$A = \{1, 2, 4, 10, 31, 43, 56\}$ and $h(s_0, s_1, s_2, s_3, s_4) = s_1 + s_4 + s_0 s_3 + s_2 s_3 + s_3 s_4 + s_0 s_1 s_2 + s_0 s_2 s_3 + s_0 s_2 s_4 + s_1 s_2 s_4 + s_2 s_3 s_4.$

2   Broad Idea of the Generic Differential Fault Attack

In this section we will describe the generic fault attack idea on any cipher with the physical structure of the Grain family, i.e. a cipher in which there is an n-bit LFSR driving an n-bit NFSR. The LFSR and NFSR are being updated by feedback functions f, g respectively and the output keystream bit at each round is generated by an output function of the internal state, i.e., a function of certain locations from both the LFSR and the NFSR. The main nonlinear part of the output function is the 5-variable function h and we study this function carefully.

For this, let us first describe a few issues related to Boolean functions. The readers may have a look at [26] and the references therein for detailed background on Boolean functions. A q-variable Boolean function is a mapping from the set $\{0,1\}^q$ to the set $\{0,1\}$. Apart from the truth table, another important way to represent a Boolean function is by its Algebraic Normal Form (ANF). A q-variable Boolean function $h(x_1, \ldots, x_q)$ can be considered to be a multivariate polynomial over $GF(2)$. This polynomial can be expressed as a sum of products representation of all distinct k-th order products ($0 \leq k \leq q$) of the variables. More precisely, $h(x_1, \ldots, x_q)$ can be written as

$$a_0 + \bigoplus_{1 \leq i \leq q} a_i x_i + \bigoplus_{1 \leq i < j \leq q} a_{ij} x_i x_j + \cdots + a_{12\ldots q} x_1 x_2 \ldots x_q,$$

where the coefficients $a_0, a_{ij}, \ldots, a_{12\ldots q} \in \{0, 1\}$. This is the ANF representation of h. The number of variables in the highest order product term with nonzero coefficient is called the *algebraic degree*, or simply the degree of h and denoted by $deg(h)$. Functions of degree at most one are called affine functions. Given the above background, let us present the following definition.

Definition 1. *Consider a q-variable Boolean function F. A non-zero vector $\alpha \in \{0,1\}^q$ is said to be an affine differential of F if $F(\mathbf{x}) + F(\mathbf{x} + \alpha)$ is an affine function. A Boolean function is said to be affine differential resistant if it does not have any affine differential.*

We propose to recover the secret key used in the cipher by observing and analyzing the difference between the fault-free and faulty keystreams. Our attack algorithm attempts to recover the internal state of the cipher after the completion of KSA (or equivalently when PRGA is about to begin). Since both the PRGA and the KSA of Grain family is completely invertible, one can then run the KSA^{-1} routine to determine the secret key K. As we have pointed out earlier, one can obtain a set of linear equations if there exist affine differentials corresponding to the function h and one such corresponding affine function should be on the variables that come from the locations of the LFSR only.

Given this background, the attack will comprise of the following steps:

1. The attacker is allowed to reset the cipher with the original Key-IV and restart cipher operations.
2. The attacker can inject a fault at any one random bit location of the LFSR. As a result of the fault injection, the binary value in the bit-location (where the fault has been injected) is toggled. The attacker is not allowed to choose the location of the LFSR where he wants to inject the fault. However, the fault in any LFSR bit may be reproduced at any later stage of operation, once injected.
3. Initially the attacker injects a fault (may be more than one in a few cases) in a randomly chosen position of the LFSR and identifies the fault location by comparing the original (fault-free) and faulty keystream.
4. The attacker has full control over the timing of fault injection, i.e., it is possible to inject the fault precisely at any stage of the cipher operation. Thus, knowing the fault location, (i) it is possible to restart the cipher operations with the original Key-IV, (ii) inject further faults in the same location (in our case either two or four faults) at specific PRGA rounds.
5. In this case, by comparing the original (fault-free) and faulty keystream in certain PRGA rounds, we obtain linear equations with respect to the LFSR state bits at the beginning of the PRGA. We run the fault attack suitable number of times so that we have several such linear equations and solving them we get the LFSR state. It is possible to obtain the linear equations (and thus to solve them efficiently) due to certain property of the Boolean function h.
6. Similarly as above, comparing the original (fault-free) and faulty keystream in certain other PRGA rounds, we obtain linear equations with respect to the NFSR bits at certain PRGA round and thus get back the NFSR state at the beginning of the PRGA. One can then run the KSA^{-1} routine to determine the secret key K.

In the next section we detail this algorithm with respect to Grain v1.

3   Differential Fault Analysis: Case Study with Grain v1

Our attack is generic and it works for any version of the Grain family. For Grain-128 our attack works in a similar broad framework as in [5], though the exact details of the signatures, construction of linear equations and the way of exploiting the Boolean functions need to follow the method we describe below. Since the existing works [5, 19] will not work on Grain v1 due to comparatively complicated output function h, we concentrate on this version as a case study to explain our novel approach. Further, we would like to point out that to the best of our knowledge there is no existing fault attack on Grain v1 available in literature. Moreover, our attack strategy works on any generic Grain like structure and points out the importance of properly choosing the Boolean functions and the LFSR, NFSR locations that will be fed into the functions.

3.1   Obtaining the Location of the Fault

Our attack model assumes that the attacker is allowed to toggle the value at exactly one random location of the LFSR. The attacker, however can not explicitly choose the location where the fault is to be injected. In order for the attack to succeed, it is very important that it will be possible to identify the location of the LFSR where the fault has been induced.

Some Definitions and Notations. Let $S_0 \in \{0, 1\}^{160}$ be the initial state of the Grain v1 PRGA, and S_{0,Δ_ϕ} be the initial state resulting after injecting fault in LFSR location $\phi \in [0, 79]$. Let $Z = [z_0, z_1, \ldots, z_l]$ and $Z^\phi = [z_0^\phi, z_1^\phi, \ldots, z_l^\phi]$ be the first l keystream bits produced by S_0 and S_{0,Δ_ϕ} respectively. The task for the fault location identification routine is to determine the fault location ϕ by analyzing the difference between Z and Z^ϕ. Initially we have taken the value of $l = 80$. After describing the fault location identification strategy in detail, we will study the value of l more critically.

We define an 80 bit vector E_ϕ over GF(2) whose i^{th} element $E_\phi(i)$ is the logical XNOR (complement of XOR) of the ith elements of Z and Z^ϕ, i.e., $E_\phi(i) = 1 + z_i + z_i^\phi$ (here + should be interpreted as $\oplus$). Since S_0 can have 2^{144} values (each arising from a different combination of the 80 bit key and 64 bit IV, rest 16 padding bits are fixed), each of these choices of S_0 may lead to different patterns of E_ϕ. The bitwise logical AND of all such vectors E_ϕ is denoted as the Signature vector Sgn_ϕ for the fault location ϕ.

The Sgn_ϕ Pattern. Note that whenever $Sgn_\phi(i)$ is 1 this implies that the ith keystream bit produced by S_0 and S_{0,Δ_ϕ} is equal for all choices of S_0. Calculating the Signature vectors by this method is a computationally infeasible task. We will describe a method to calculate them efficiently as below.

For Grain v1, two initial states of the PRGA $S_0, S_{0,\Delta_{79}} \in \{0, 1\}^{160}$ which differ only in the 79th position of the LFSR, produce identical output bits in 68 specific positions among the initial 80 keystream bits produced during the PRGA. If an input differential is introduced in the 79th LFSR position, then at all rounds numbered $k \in [0, 79] \setminus \{15, 33, 44, 51, 54, 57, 62, 69, 72, 73, 75, 76\}$, the difference exists in positions that do not provide input to the Boolean function h and hence at these clocks the keystream bit produced by the two states are essentially the same. At all other clock rounds the difference appears at positions which provide input to h. Hence the keystream produced at these clocks may be different. Following the explanation given above, we can write Sgn_{79} in hexadecimal notation, $Sgn_{79} = $ FFFE FFFF BFF7 EDBD FB27, which has $80 - 12 = 68$ many 1's and rest 0's.

Generalizing the above idea, for two PRGA initial states $S_0, S_{0,\Delta_\phi} \in \{0, 1\}^{160}$ which differ only in the ϕth LFSR location, an analysis of the differential trails shows that out of the first 80 keystream bits produced by them, the bits at a certain fixed rounds are guaranteed to be equal. Thus by performing the above analysis for all fault locations ϕ ($0 \leq \phi \leq 79$), it is possible to calculate all the Signature vectors. A table containing the vectors for each fault location ϕ is available in Table 1.

Table 1. Fault Signature Vectors Sgn_ϕ for $0 \leq \phi \leq 79$ in hexadecimal notation for Grain v1

ϕ	Sgn_ϕ	ϕ	Sgn_ϕ	ϕ	Sgn_ϕ	ϕ	Sgn_ϕ
0	FFFF 3F7F CB93 A080 0000	20	FFFF BEFF F3B7 F4A9 3808	40	FFFE DFFF B3EE ED31 3B40	60	FFFD FFBF EDEF F93A 7E52
1	FFFF 9FBF E5C9 D040 0000	21	FFFF DF7F F9DB FA54 9C04	41	FFFF 6FFF D9F7 7698 9DA0	61	FFFE FFDF F6F7 FC9D 3F29
2	FFFF CFDF F2E4 E820 0000	22	FFFF EFBF FCED FD2A 4E02	42	FFFF B7FF ECFB BB4C 4ED0	62	FFFF 7FEF DB7B F64E 9D90
3	7FFF E7EF F972 7410 0000	23	FFFF 77DF DE72 F694 2501	43	FFFF DBFF F67D DDA6 2768	63	FFFF BFF7 EDBD FB27 4EC8
4	BFFF F3F7 FCB9 3A08 0000	24	FFFF BBEF EF39 7B4A 1280	44	FFFF EDFF FB3E EED3 13B4	64	7FFF DFFB F6DE FD93 A764
5	DFFF F9FB FE5C 9D04 0000	25	7FFF DDF7 F79C BDA5 0940	45	FFFF F6FF FD9F 7769 89DA	65	BFFF EFFD FB6F 7EC9 D3B2
6	EFFF FCFD FF2E 4E82 0000	26	BFFF EEFB FBCE 5ED2 84A0	46	7FFF FB7F FECF BBB4 C4ED	66	DFFF F77E FDB7 BF64 E9D9
7	F7FF FE7E FF97 2741 0000	27	DFFF F77D FDE7 2F69 4250	47	BFFF FDBF FF67 DDDA 6276	67	EFFF FBFF 7EDB DFB2 74EC
8	FBFF FF3F 7FCB 93A0 8000	28	EFFF FBBE FEF3 97B4 A128	48	DFFF FEDF FFB3 EEED 313B	68	F7FF FDFF BF6D EFD9 3A76
9	FDFF FF9F BFE5 C9D0 4000	29	F7FF FDDF 7F79 CBDA 5094	49	EFFF FF6F FFD9 7776 989D	69	FBFF FEFF DFB6 F7EC 9D3B
10	FEFF FFCF DFF2 E4E8 2000	30	FBFF FEEF BFBC E5ED 284A	50	F7FF FFB7 FFEC FBBB 4C4E	70	FDFF FF7F EFDB 7BF6 4E9D
11	FF7F FFE7 EFF9 7274 1000	31	FDFF FF77 DFDE 72F6 9425	51	FBFF 7FDB DFF2 74FC A40B	71	FEFF FFBF F7ED BDFB 274E
12	FFBF FFF3 F7FC B93A 0800	32	FEFF FFBB EFEF 397B 4A12	52	FDFF BFED EFF9 3A7E 5205	72	FF7F FFDF FBF6 DEFD 93A7
13	FFDF 7FF9 DBFA 549C 0400	33	FF7F FFDD F7F7 9CBD A509	53	FEFF DFF6 F7FC 9D3F 2902	73	FFBF FFEF FDFB 6F7E C9D3
14	FFEF BFFC EDFD 2A4E 0200	34	FFBF FFEE FBFB CE5E D284	54	FF7F EFFB 7BF4 4E9F 9481	74	FFDF FFF7 FEFD B7BF 64E9
15	FFF7 DFFE 76FE 9527 0100	35	FFDF FFF7 7DFD E72F 6942	55	FFBF F7FD BDFF 274F CA40	75	FFEF FFFB FF7E DBDF B274
16	FFFB EFFF 3B7F 4A93 8080	36	FFEF FFFB BBEF E397 B4A1	56	FFDF FBFE BFFF 93A7 E520	76	FFF7 FFFD FF6F 6DEF D93A
17	FFFD F7FF 9DBF A549 C040	37	FFF7 FFFD DF7F 79CB DA50	57	FFEF FDFF 6F7F C9D3 F290	77	FFFB FFFE FFDF B6F7 EC9D
18	FFFE FBFF CEDF D2A4 E020	38	FFFB 7FFE CFBB B4C4 ED00	58	FFF7 FEFF 7B8F E4E9 F948	78	FFFD FFFF 7FEF DB7B F64E
19	FFFF 7DFF E76F E952 7010	39	FFFD BFFF 67DD DA62 7680	59	FFFB FF7F DBDF F274 FCA4	79	FFFE FFFF BFF7 EDBD FB27

Steps for Location Identification. As mentioned above, the task for the fault identification routine is to determine the value of ϕ given the vector E_ϕ. For any element $V \in \{0, 1\}^l$ define the set $B_V = \{i : 0 \leq i < l, V(i) = 1\}$. Now define a relation $\preceq$ in $\{0, 1\}^l$ such that for 2 elements $V_1, V_2 \in \{0, 1\}^l$, we will have $V_1 \preceq V_2$ if $B_{V_1} \subseteq B_{V_2}$.

Now we check the elements in B_{E_ϕ}. By definition, these are the PRGA rounds i during which $z_i = z_i^\phi$. By the definition of Signature vector proposed above, we know that for the correct value of ϕ, $B_{Sgn_\phi} \subseteq B_{E_\phi}$ and hence $Sgn_\phi \preceq E_\phi$. So our strategy would be to search all the Signature vectors and formulate the candidate set $\Psi_0 = \{\psi : 0 \leq \psi \leq 79, Sgn_\psi \preceq E_\phi\}$. If $|\Psi_0|$ is 1, then the single element in Ψ_0 will give us the fault location ϕ. However, this may not necessarily be the case always. If Ψ_0 has more than one element, we will be unable to decide conclusively at this stage.

In such a scenario we reset the cipher with the original Key-IV and this time apply the fault at the same location ϕ at the beginning of the 80th PRGA round and record the next 80 keystream bits $Z^\phi(80) = [z_{80}^\phi(80), z_{81}^\phi(80), \ldots, z_{159}^\phi(80)]$, where $z_i^\phi(t)$ denotes the ith keystream bit produced due to a fault on LFSR location ϕ at the beginning of PRGA round t. Let the corresponding fault-free bits be denoted by $Z(80) = [z_{80}, z_{81}, \ldots, z_{159}]$. Now reformulate and recalculate the vector E_ϕ so that its ith element is the logical XNOR of the ith elements of $Z(80)$ and $Z^\phi(80)$. We now search over the Signature vectors in the candidate set Ψ_0 and narrow down the set of possible candidates to $\Psi_1 = \{\psi : \psi \in \Psi_0, Sgn_\psi \preceq E_\phi\}$. Clearly, $|\Psi_1| \leq |\Psi_0|$, and so if $|\Psi_1| = 1$ then the fault location ϕ is the single element in Ψ_1. If not, we repeat the above process for another round, i.e. reset and apply the fault at the PRGA round 160 etc. If after k rounds of this process, $|\Psi_{k-1}| = 1$, then the single element in Ψ_{k-1} gives us the desired location ϕ.

Length of Signature Vector. In the idea given above, we have considered the length of the signature vector $l = 80$. It may be noted that that fault identification routine is also possible if we increase or decrease the length of the signature vector. So what guidelines must be followed to choose an optimal signature length. Intuitively the following considerations seem to be useful.

1. The signature vector must be long enough so as to uniquely identify the fault location applying one or more faults.
2. The length of the signature vector must be such that the average number of faults required to identify the fault location can be minimized.

We shall see how each of the above considerations affect the choice of the length l of the Signature vector. For example, by simply looking at the Signature vectors (one may refer to Table 1), one can deduce for sure that l can not be less than or equal to 16, otherwise $\phi = 0, 1, 2, 19, 20, 21, 22, 23, 24, 41, 42, 43, 44, 45, 61, 62, 63$ will have the same Signature vectors. We will give a better bound on l in the following Lemma.

Lemma 1. *The LFSR fault location can not be uniquely identified if the length of the signature vector Sgn_ϕ is less than or equal to 44.*

Proof. Take $l = 44$. Studying the Signature vectors, one can check that $Sgn_{40} =$ FFFE DFFF B3E and $Sgn_{79} =$ FFFE FFFF BFF. Note that, for all locations $i \in [0, 43]$ such that $Sgn_{40}(i) = 1$, the value of $Sgn_{79}(i)$ is also 1. This implies that $Sgn_{40} \preceq Sgn_{79}$. Now consider the case with the fault location $\phi = 79$. Then by the definition of the signature vector we have $Sgn_{79} \preceq E_\phi$. Since $\preceq$ is a partial order on $\{0,1\}^l$, this implies that $Sgn_{40} \preceq E_\phi$ and so whenever $\phi = 79$ the fault location identification routine will never be able to narrow down the set of possible candidates Ψ_k to only $\{79\}$ for any value of k. It is easy to check that the same argument holds for any $l < 44$. ☐

Whenever $l \geq 45$ a simple exhaustive search through the Signature vectors for all fault locations, will show that $Sgn_{\phi_1} \not\preceq Sgn_{\phi_2}$ for any two fault locations

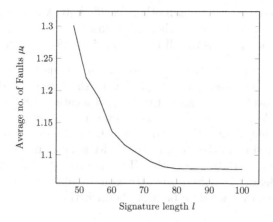

Fig. 2. Average number of faults vs Length of Signature

$0 \le \phi_1 \ne \phi_2 \le 79$. Further, we have to choose some $l \ge 45$ so that the average number of faults, for determining the fault location uniquely, can be minimized. Finding, this optimal value of l mathematically is a difficult task, and hence we choose to determine the optimal value by performing computer simulations. By taking the average over 2^{20} uniformly randomly chosen Key-IV pairs for Grain v1, for every signature length $l \ge 45$ we get the curve of Average number of faults μ_l vs Length of Signature l given in Figure 2.

We can see that after $l = 80$, $\mu_l = 1.08$ becomes almost constant for increasing values of l. So the length of the Signature vector has been chosen to be 80 bits.

A similar analysis for Grain-128 shows that the minimum Signature length must be greater than or equal to 62. For $l = 128$, the value of μ_l is around 1.001.

3.2   Determining the LFSR Internal State

Once the fault location ϕ has been identified we can proceed towards determining the LFSR internal state at the beginning of the PRGA. Depending on the value of ϕ we do one of the following.

- If $0 \le \phi \le 37$, we disregard the faulty keystream bits, and reset the cipher and look to hit another LFSR location.
- If $38 \le \phi \le 41$, we reset the cipher and apply faults at the location ϕ at the beginning of PRGA rounds $0, 20$ and record the faulty keystream bits at certain specific PRGA rounds. We then reset the cipher and look to hit another LFSR location.
- If $42 \le \phi \le 79$, we reset the cipher and apply faults at the location ϕ at the beginning of PRGA rounds $0, 20$ and record the faulty keystream bits at certain specific PRGA rounds. We reset the cipher again and apply faults at the location ϕ at the beginning of PRGA rounds $204, 224$ and record the

faulty keystream bits at certain other specific PRGA rounds. We then reset the cipher and look to hit another LFSR location

– We continue this process till all LFSR locations 38 to 79 have been hit.

We would like to point out that each double fault (injected at PRGA rounds $0, 20$ or $204, 224$) yields one linear equation in the initial LFSR state bits of the PRGA. By injecting 2 faults in the 4 LFSR locations 38 to 41 and 4 faults in the 38 LFSR locations 42 to 79, we obtain a set of 80 independent linear equations in the initial LFSR state bits, which can be solved to get the entire LFSR state at the start of the PRGA. The faulty keystream bits recorded in this phase will be again used to recover the NFSR internal state as will be explained in Section 3.3. Before describing the attack in detail let us state the following symbolic notations that we shall be using henceforth.

Some Notations

1. $S_t = [x_0^t, x_1^t, \ldots, x_{79}^t \ \ y_0^t, y_1^t, \ldots, y_{79}^t]$ is used to denote the internal state of the cipher at the beginning of round t of the PRGA. Thus x_i^t (y_i^t) denotes the ith NFSR (LFSR) bit at the start of round t of the PRGA. When $t = 0$, we use $S_0 = [x_0, x_1, \ldots, x_{79} \ \ y_0, y_1, \ldots, y_{79}]$ to denote the internal state for convenience.
2. $S_t^{\phi}(t_1, t_2)$ is used to denote the internal state of the cipher at the beginning of round t of the PRGA, when a fault has been injected in LFSR location ϕ at the beginning of the t_1^{th} and the t_2^{th} PRGA round.
3. $z_i^{\phi}(t_1, t_2)$ denotes the keystream bit produced in the ith PRGA round, after faults have been injected in LFSR location ϕ at the beginning of the t_1^{th} and the t_2^{th} PRGA round. z_i is the fault-free i^{th} keystream bit.

Beginning the Attack. We start by making the following observation about the output Boolean function h in Grain v1: $h(s_0, s_1, s_2, s_3, s_4) + h(1 + s_0, 1 + s_1, s_2, s_3, 1 + s_4) = s_2$. Hence h is not affine differential resistant. Note that s_0, s_1, s_2, s_3 correspond to LFSR locations $3, 25, 46, 64$ respectively and s_4 corresponds to the NFSR location 63. This implies that if two internal states S and S_Δ be such that they differ in LFSR locations $3, 25$ and NFSR location 63 and in no other location that contributes inputs to the output keystream bit, then the difference of the keystream bit produced by them will be equal to the value in LFSR location 46. Getting differentials at exactly these 3 locations may be difficult by injecting a single fault, but may be achieved if we faulted the same LFSR location twice, as will be explained by the following lemma.

Lemma 2. *If a fault is injected in LFSR location $38 + r$ ($0 \leq r \leq 41$), at the beginning of the PRGA rounds λ and $\lambda + 20$ ($\lambda = 0, 1, \ldots$), then in round number $55 + \lambda + r$ of the PRGA, the faulty internal state $S_{55+\lambda+r}^{38+r}(\lambda, \lambda + 20)$ and the fault-free internal state $S_{55+\lambda+r}$ will differ in LFSR locations $3, 25$ and NFSR location 63 and in none of the other 9 tap locations that contributes to the output keystream bit.*

Proof. The proof requires the analysis of the differential trail of the successive PRGA rounds. A differential Δ introduced in LFSR location $38+r$ $(0 \leq r \leq 41)$, at the beginning of rounds λ and $\lambda+20$ of the PRGA, will certainly reside on the LFSR locations 3, 25 and NFSR location 63 at the beginning of round $55+\lambda+r$ of the PRGA. The differential also does not affect any other location involved in the computation of the output keystream bit in round $55 + \lambda + r$. $\square$

The above lemma implies that if $\lambda = 0$, i.e., if faults are injected at the beginning of the PRGA and round 20 at location $38 + r$, $0 \leq r \leq 41$ of the LFSR, then in the PRGA round $55 + r$ we will have

$$z_{55+r} + z_{55+r}^{38+r}(0,20) = y_{46}^{55+r} \quad \forall r \in [0,41].$$

Now since the NFSR does not influence the LFSR during the PRGA, y_{46}^{55+r} is a linear function of the initial LFSR bits $y_0, y_1, \ldots, y_{79}$ for all $0 \leq r \leq 41$. For example, by analyzing the LFSR we have $y_{46}^{55} = y_3 + y_{16} + y_{21} + y_{26} + y_{34} + y_{41} + y_{44} + y_{54} + y_{59} + y_{65} + y_{72}$.

So in this process, we obtain 42 linear equations in the original LFSR bits $y_0, y_1, \ldots, y_{79}$. We need another 38 equations such that the resulting 80 equations are linearly independent. We have attempted to find the remaining 38 equations by resetting the cipher and then introducing faults later in the PRGA. If we let $\lambda = 204$, i.e., if double faults were introduced in LFSR locations $42 + r$ with $0 \leq r \leq 37$ at the beginning of the PRGA rounds 204 and 224, then by the previous analysis it may be deduced that

$$z_{263+r} + z_{263+r}^{42+r}(204,224) = y_{46}^{263+r} \quad \forall r \in [0,37].$$

This provides us with another 38 equations. We have observed that these equations are linearly independent. Writing these equations in matrix notation, we have $LY = W$. The rows of the matrix L is defined by the linear functions $y_{46}^{55}, y_{46}^{56}, \ldots, y_{46}^{96}, y_{46}^{263}, \ldots, y_{46}^{300}$. Further, $Y = [y_0 \ y_1 \ \ldots \ y_{79}]^T$ and W is the column vector defined as follows

$$W(r) = z_{55+r} + z_{55+r}^{38+r}(0,20) \quad 0 \leq r \leq 41,$$

$$W(42+r) = z_{263+r} + z_{263+r}^{42+r}(204,224) \quad 0 \leq r \leq 37.$$

Since the matrix L and its inverse can be pre-computed beforehand, the vector $Y = L^{-1}W$ can be calculated immediately after applying the faults and calculating W.

Note that for the second round of fault injections the choice of fault locations $42 \leq \phi \leq 79$ and PRGA rounds 204, 224 is by no means unique. By searching over various values of λ, one may be able to obtain a set of linearly independent equations for other choices of fault locations and PRGA rounds.

Remark 1. If the function h in Grain v1 had been affine differential resistant, then such linear equations could not have been formed. Instead one had to consider a set of nonlinear equations to get Y. As referred in [19], solving such nonlinear equations is more challenging task and in that case the fault attack we describe here would have been less efficient. The method works in a similar manner for Grain-128 and Grain-128a. For example, the output function in Grain-128 is of the form $h(s_0, s_1, s_2, s_3, s_4, s_5, s_6, s_7, s_8) = s_0 s_1 + s_2 s_3 + s_4 s_5 + s_6 s_7 + s_0 s_4 s_8$, where s_0 and s_4 corresponds to NFSR variables. One can check that for any $\alpha \in \{001000000, 000100000, 000000100, 000000010\}$, $h(\mathbf{x}) + h(\mathbf{x} + \alpha)$ is a linear function of LFSR variables only.

3.3   Determining the NFSR Internal State

Once the LFSR internal state of the initial PRGA round is known, one can then proceed to determine the NFSR internal state. In [4] it was shown, that this could have been done efficiently for the initial version of the cipher i.e. Grain v0. After the attack in [4] was reported, the designers made the necessary changes to Grain v1, Grain-128 and Grain-128a so that for these new ciphers, determining the NFSR state form the knowledge of the LFSR state was no longer straightforward. In order to determine the NFSR bits, we look into the decomposition of the Boolean function h in more detail. The attack we will describe in this section can be mounted due to the following observations on the Grain output function h.

A. $h(\cdot)$ can be written in the form $s_j \cdot u(\cdot) + v(\cdot)$ where s_j corresponds to a variable which takes input from an NFSR tap location;
B. There exists a differential β such that $u(\mathbf{s}) + u(\mathbf{s} + \beta) = 1$;
C. $v(\mathbf{s}) + v(\mathbf{s} + \beta) =$ a function of variables that takes input from LFSR locations only.

For Grain v1, $h(s_0, s_1, s_2, s_3, s_4) = s_4 \cdot u(s_0, s_1, s_2, s_3) + v(s_0, s_1, s_2, s_3)$, where $u(s_0, s_1, s_2, s_3) = 1 + s_3 + s_0 s_2 + s_1 s_2 + s_2 s_3$, and $v(s_0, s_1, s_2, s_3) = s_1 + s_0 s_3 + s_2 s_3 + s_0 s_1 s_2 + s_0 s_2 s_3$. Thus we note that (i) u, v are functions on the LFSR bits only, (ii) $u(s_0, s_1, s_2, s_3) + u(s_0, 1 + s_1, s_2, 1 + s_3) = 1$ and (iii) $v(s_0, s_1, s_2, s_3) + v(s_0, 1 + s_1, s_2, 1 + s_3) = 1 + s_0 + s_2$. Hence h satisfies all the properties listed above.

The fault-free keystream bit at the tth round can now be rewritten as $z_t = \bigoplus_{a \in A} x_a^t + x_{63}^t \cdot u(y_3^t, y_{25}^t, y_{46}^t, y_{64}^t) + v(y_3^t, y_{25}^t, y_{46}^t, y_{64}^t)$. Consider two internal states S_t and $S_{t,\Delta}$ which differ in the LFSR locations 25 and 64 and in no other location, that provides input to h. If z_t and $z_{t,\Delta}$ are the keystream bits produced by S_t and $S_{t,\Delta}$ in that round, then using the previous observation we can see that

$$z_t + z_{t,\Delta} = x_{63}^t + v(y_3^t, y_{25}^t, y_{46}^t, y_{64}^t) + v(y_3^t, 1 + y_{25}^t, y_{46}^t, 1 + y_{64}^t).$$

Let $c_t = \left[v(y_3^t, y_{25}^t, y_{46}^t, y_{64}^t) + v(y_3^t, 1 + y_{25}^t, y_{46}^t, 1 + y_{64}^t) \right]$. Since the LFSR internal state is already available, c_t can be computed immediately, and hence the difference of the two keystream bits plus the value of c_t gives us the value at the NFSR location 63 at round t of the PRGA. In the next Lemma, we shall investigate when this differential pattern in the internal state is obtained by employing the same fault injection strategy in the previous subsection.

Lemma 3. *Let $S_0, S_1, S_2, \ldots$ be the successive internal states of the PRGA for Grain v1. Then the faulty state $S_t^\phi(0, 20)$ will differ from S_t at LFSR locations $25, 64$ and none of the other 10 tap locations that feed the output function for the following values of ϕ, t: (i) $\phi = 51 + r$, $t = 91 + r$ for $0 \leq r \leq 28$, (ii) $\phi = 62 + r$, $t = 55 + r$ for $0 \leq r \leq 17$, (iii) $\phi = 62 + r$, $t = 75 + r$ for $0 \leq r \leq 15$.*

Proof. The proof follows from an analysis of the differential trails of Grain v1 PRGA, and is similar to the proof for Lemma 2.   □

The Lemma essentially implies that if faults are injected at the beginning of the PRGA and round 20 at location $51 + r$ of the LFSR ($0 \leq r \leq 28$), then in the PRGA round $91 + r$ we will have

$$z_{91+r} + z_{91+r}^{51+r}(0, 20) + c_{91+r} = x_{63}^{91+r} \quad \forall r \in [0, 28].$$

Also, the following equations hold:

$$z_{55+r} + z_{55+r}^{62+r}(0, 20) + c_{55+r} = x_{63}^{55+r} \quad \forall r \in [0, 17],$$

$$z_{75+r} + z_{75+r}^{62+r}(0, 20) + c_{75+r} = x_{63}^{75+r} \quad \forall r \in [0, 15].$$

Since the LHS of all the above equations are known, we can therefore calculate the value of the NFSR location 63 for all PRGA rounds 55, 56, ..., 72, 75, 76, ..., 119. Because of the shifting property of the NFSR, the equations $x_{i-1}^j = x_i^{j+1} \ \forall i \in [1, 79]$ hold. Therefore knowing $x_{63}^{55}, x_{63}^{56}, \ldots, x_{63}^{72}, x_{63}^{75}, x_{63}^{76}, \ldots, x_{63}^{119}$ is equivalent to knowing $x_{15}^{103}, x_{16}^{103}, \ldots, x_{32}^{103}, x_{35}^{103}, x_{36}^{103}, \ldots, x_{79}^{103}$, i.e., we now know 63 out of the 80 NFSR state bits of S_{103}.

Finding the Remaining Bits. Any bits of the NFSR internal state not found out in the previous subsection could be obtained by performing an exhaustive search over them. However, if h is such that both u, v are functions on the LFSR bits only then the attack can be further simplified. Since the function h in Grain v1 satisfies this property, we proceed to determine the remaining 17 NFSR bits of S_{103}. These may be found by a combination of solving equations and guesswork. Since the 80 LFSR bits of S_0 have already been found in the previous section, one can efficiently calculate the 80 LFSR bits of S_{103} by running the Grain v1 PRGA routine. This is because the LFSR evolves independently during the PRGA. Then, by observing the fault-free output keystream bits we can write the following equations:

$$z_{102+\gamma} = x_{0+\gamma}^{103} + x_{1+\gamma}^{103} + x_{3+\gamma}^{103} + x_{9+\gamma}^{103} + x_{30+\gamma}^{103} + x_{42+\gamma}^{103} + x_{55+\gamma}^{103} + u_{102+\gamma} x_{62+\gamma}^{103} + v_{102+\gamma},$$

for $\gamma = 0, 1, \ldots, 14$, where $u_i = u(y_3^i, y_{25}^i, y_{46}^i, y_{64}^i)$ and $v_i = v(y_3^i, y_{25}^i, y_{46}^i, y_{64}^i)$. Since the LFSR initial state is known, u_i, v_i are available. Consider the set of 15 equations given above. In the last equation it can be seen that x_{14}^{103} is the only unknown and hence its value may be easily calculated. Once x_{14}^{103} is known, x_{13}^{103} becomes the only unknown in the 14th equation and its value too may be immediately calculated. Backtracking in this manner one can calculate upto x_5^{103} from the 6th equation. At this point we have calculated the value of 73 NFSR bits of S_{103}. The 5th equation is

$$z_{106} = x_4^{103} + x_5^{103} + x_7^{103} + x_{13}^{103} + x_{34}^{103} + x_{46}^{103} + x_{59}^{103} + u_{106}x_{66}^{103} + v_{106}.$$

This equation has two unknowns x_4^{103} and x_{34}^{103} and so the value of either unknown can not be calculated conclusively. Similarly the 4th equation has two unknowns x_3^{103} and x_{33}^{103}. If we try out all the possibilities of $x_{34}^{103}, x_{33}^{103}$ then the value of the remaining 5 unknowns may be calculated uniquely. So we do an exhaustive search over the 2 bits (4 possible candidates) for S_{103}. The correct S_{103} may be found out by observing the keystream bits $z_{103}, z_{104}, \ldots$, as required. We eliminate any candidate S_{103} vector that does not produce the required keystream bit sequence. This routine thus gives us the entire S_{103} vector. Note that in order to recover the NFSR state one does not have to inject any additional faults other than those already injected to determine the LFSR state.

Remark 2. If the function h in Grain v1 were such that it could not be decomposed into u and v as above, then the attack would not have been as straightforward. The attack here is efficient as u and v are of certain nice structures and their inputs are from LFSR bits only. The LFSR bits are already known after the recovery of the LFSR bits and that helps in recovering the NFSR state easily. It can be checked that the output function of Grain-128 and Grain-128a also follows properties **(A)**, **(B)**, **(C)** given at the beginning of this section and thus renders them vulnerable to this attack.

3.4   Finding the Secret Key and Complexity of the Attack

It is known that the KSA and PRGA routines in the Grain family are invertible. Once we have all the bits of S_{103}, by running the inverse PRGA routine 103 times, we obtain the initial PRGA state S_0. Thereafter, by running the inverse KSA routine one can recover the secret key.

The attack complexity directly depends on the number of fault experiments to be performed such that all of locations in $[38, 79]$ of the LFSR are covered. To have this, the expected number of fault experiments is $80 \cdot \sum_{i=1}^{42} \frac{1}{i} \approx 344$. In each fault experiment, the fault identification routine requires μ_l faults and simulation results show that the expected value of μ_l is 1.08. Further depending on the LFSR location hit, during the attack phase, one needs to inject 2 or 4 extra faults for determining the internal state. Therefore, the expected number of faults that our attack needs is $344 \times (1.08) + 4 \times 38 + 2 \times 4 \approx 2^{9.05}$.

To determine the internal state, we have to perform one matrix multiplication, and solve a set of 78 linear equations and then exhaustively search over 2 variables. After that, 103 invocations of the $PRGA^{-1}$ routine and a single invocation of the KSA^{-1} routine are needed to determine the Secret Key.

Thus the dominant time/memory consuming process in our attack is the multiplication of $L^{-1}W$ which requires around 80×80 bits to store L^{-1} and

$80^2 \approx O(2^{12.6})$ bit operations to calculate the product. Further storing the Sgn_ϕ patterns also requires 80×80 bits as described in Table 1.

As stated before, this is the first reported fault attack on Grain v1. Two fault attacks [5, 19] have been reported against Grain-128 and that is the reason direct comparison is not possible. However, one may note that our resource requirements are either favorable or comparable to that of [5, 19].

4   Countermeasure: Choice of Proper Boolean Function

In [5], it was suggested that one of the methods to prevent such fault attacks was to keep two identical implementations of both the shift registers in the cipher hardware. Naturally this needs additional hardware.

One important question here is what are the reasons such that the fault attack can be efficiently implemented. We have already seen that the source of the weakness lies with the output Boolean function h. Our attack is possible as there exists the vector $\alpha = [1, 1, 0, 0, 1]$ such that $h(\mathbf{s}) + h(\mathbf{s} + \alpha)$ is an affine Boolean function. This function h, used in Grain v1, is clearly not affine differential resistant. In [15], the designers clearly specify the reasons for choosing this particular output function.

> "This filter function is chosen to be balanced, correlation immune of the first order and has algebraic degree 3. The nonlinearity is the highest possible for these functions, namely 12."

In view of the fault attack presented here, we need affine differential resistant functions with the same parameters. One may refer to [26] to have many such functions in the class of rotation symmetric Boolean functions and we describe the ANF of one of those as below:

$F(s_0, s_1, s_2, s_3, s_4) = s_0 s_1 + s_1 s_2 + s_2 s_3 + s_3 s_4 + s_4 s_0 + s_0 s_2 + s_1 s_3 + s_2 s_4 + s_3 s_0 + s_4 s_1 + s_0 s_1 s_3 + s_1 s_2 s_4 + s_2 s_3 s_0 + s_3 s_4 s_1 + s_4 s_0 s_2$. This function can be realized with a few extra logic gates as below. The gate count is presented as per the calculation of [15].

	Gate Requirement					
	NAND2	NAND3	NAND4	NAND5	NAND6	Gate Count
h [15]	8	1	9	1	0	30
F (our)	8	0	10	2	1	38

Proper cryptographic choice of h with possibly higher number of variables with efficient implementation in terms of low gate counts is an important open question. Further, we should also note that the decomposition of h in u, v that possess properties **(A)**, **(B)**, **(C)** given in Section 3.3 helps in mounting an efficient fault attack. We further note that the function F described above does not satisfy property **(B)** if s_4 is the only variable that takes input from an NFSR location. This implies that even if the initial LFSR state of the PRGA is made known to the attacker, the attacker will be unable to apply the attack given in Section 3.3 to the function F.

5   Conclusion

In this paper we have described a differential fault attack that works on all the versions of Grain. Such attacks were studied earlier on Grain-128 in [5, 19]. However, the attacks could not be mounted on Grain v1 due to the different structure of the output function $h(\cdot)$. Here we show that the function of Grain v1 too has some weakness in terms of having affine differentials. By this we mean that there exists certain suitable α such that $h(\mathbf{x}) + h(\mathbf{x} + \alpha)$ is linear. Our attack works due to this observation and corresponding choices of the taps from the LFSR. That is, from a general perspective, the differential fault attack can be mounted on Grain like structures even with Boolean functions of higher degree. We also provide examples of functions that are affine differential resistant and suggest use of such functions in Grain family as a countermeasure. Our work provides clear direction in choosing the output Boolean function and its inputs from the locations of the LFSR and the NFSR.

Acknowledgments The authors like to thank the Centre of Excellence in Cryptology, Indian Statistical Institute for relevant support towards this research.

Reference

1. The ECRYPT Stream Cipher Project. eSTREAM Portfolio of Stream Ciphers (revised on September 8, 2008)
2. Ågren, M., Hell, M., Johansson, T., Meier, W.: A New Version of Grain-128 with Authentication. In: Symmetric Key Encryption Workshop 2011. DTU, Denmark (2011)
3. Aumasson, J.P., Dinur, I., Henzen, L., Meier, W., Shamir, A.: Efficient FPGA Implementations of High-Dimensional Cube Testers on the Stream Cipher Grain-128. In: SHARCS - Special-purpose Hardware for Attacking Cryptographic Systems (2009)
4. Berbain, C., Gilbert, H., Maximov, A.: Cryptanalysis of Grain. In: Robshaw, M. (ed.) FSE 2006. LNCS, vol. 4047, pp. 15–29. Springer, Heidelberg (2006)
5. Berzati, A., Canovas, C., Castagnos, G., Debraize, B., Goubin, L., Gouget, A., Paillier, P., Salgado, S.: Fault Analysis of Grain-128. In: IEEE International Workshop on Hardware-Oriented Security and Trust, pp. 7–14 (2009)
6. Berzati, A., Canovas-Dumas, C., Goubin, L.: Fault Analysis of Rabbit: Toward a Secret Key Leakage. In: Roy, B., Sendrier, N. (eds.) INDOCRYPT 2009. LNCS, vol. 5922, pp. 72–87. Springer, Heidelberg (2009)
7. Blömer, J., Seifert, J.P.: Fault Based Cryptanalysis of the Advanced Encryption Standard (AES). In: Wright, R.N. (ed.) FC 2003. LNCS, vol. 2742, pp. 162–181. Springer, Heidelberg (2003)
8. Bjørstad, T.E.: Cryptanalysis of Grain using Time/Memory/Data tradeoffs, v1.0 (February 25, 2008), http://www.ecrypt.eu.org/stream.
9. De Cannière, C., Küçük, Ö., Preneel, B.: Analysis of Grain's Initialization Algorithm. In: Vaudenay, S. (ed.) AFRICACRYPT 2008. LNCS, vol. 5023, pp. 276–289. Springer, Heidelberg (2008)

10. Dinur, I., Güneysu, T., Paar, C., Shamir, A., Zimmermann, R.: An Experimentally Verified Attack on Full Grain-128 Using Dedicated Reconfigurable Hardware. In: Lee, D.H., Wang, X. (eds.) ASIACRYPT 2011. LNCS, vol. 7073, pp. 327–343. Springer, Heidelberg (2011)
11. Dinur, I., Shamir, A.: Breaking Grain-128 with Dynamic Cube Attacks. In: Joux, A. (ed.) FSE 2011. LNCS, vol. 6733, pp. 167–187. Springer, Heidelberg (2011)
12. Englund, H., Johansson, T., Turan, M.S.: A Framework for Chosen IV Statistical Analysis of Stream Ciphers. In: Srinathan, K., Rangan, C.P., Yung, M. (eds.) INDOCRYPT 2007. LNCS, vol. 4859, pp. 268–281. Springer, Heidelberg (2007)
13. Fischer, S., Khazaei, S., Meier, W.: Chosen IV Statistical Analysis for Key Recovery Attacks on Stream Ciphers. In: Vaudenay, S. (ed.) AFRICACRYPT 2008. LNCS, vol. 5023, pp. 236–245. Springer, Heidelberg (2008)
14. Fredricksen, H.: A survey of full length nonlinear shift register cycle algorithms. SIAM Rev. 24, 195–221 (1982)
15. Hell, M., Johansson, T., Meier, W.: Grain - A Stream Cipher for Constrained Environments. ECRYPT Stream Cipher Project Report 2005/001 (2005), http://www.ecrypt.eu.org/stream
16. Hell, M., Johansson, T., Maximov, A., Meier, W.: A Stream Cipher Proposal: Grain-128. In: IEEE International Symposium on Information Theory, ISIT 2006 (2006)
17. Hoch, J.J., Shamir, A.: Fault Analysis of Stream Ciphers. In: Joye, M., Quisquater, J.-J. (eds.) CHES 2004. LNCS, vol. 3156, pp. 240–253. Springer, Heidelberg (2004)
18. Hojsík, M., Rudolf, B.: Differential Fault Analysis of Trivium. In: Nyberg, K. (ed.) FSE 2008. LNCS, vol. 5086, pp. 158–172. Springer, Heidelberg (2008)
19. Karmakar, S., Roy Chowdhury, D.: Fault Analysis of Grain-128 by Targeting NFSR. In: Nitaj, A., Pointcheval, D. (eds.) AFRICACRYPT 2011. LNCS, vol. 6737, pp. 298–315. Springer, Heidelberg (2011)
20. Khazaei, S., Hassanzadeh, M., Kiaei, M.: Distinguishing Attack on Grain. ECRYPT Stream Cipher Project Report 2005/071 (2005), http://www.ecrypt.eu.org/stream
21. Kircanski, A., Youssef, A.M.: Differential Fault Analysis of Rabbit. In: Jacobson Jr., M.J., Rijmen, V., Safavi-Naini, R. (eds.) SAC 2009. LNCS, vol. 5867, pp. 197–214. Springer, Heidelberg (2009)
22. Knellwolf, S., Meier, W., Naya-Plasencia, M.: Conditional Differential Cryptanalysis of NLFSR-based Cryptosystems. In: Abe, M. (ed.) ASIACRYPT 2010. LNCS, vol. 6477, pp. 130–145. Springer, Heidelberg (2010)
23. Lee, Y., Jeong, K., Sung, J., Hong, S.: Related-Key Chosen IV Attacks on Grain-v1 and Grain-128. In: Mu, Y., Susilo, W., Seberry, J. (eds.) ACISP 2008. LNCS, vol. 5107, pp. 321–335. Springer, Heidelberg (2008)
24. Skorobogatov, S.P.: Optically Enhanced Position-Locked Power Analysis. In: Goubin, L., Matsui, M. (eds.) CHES 2006. LNCS, vol. 4249, pp. 61–75. Springer, Heidelberg (2006)
25. Skorobogatov, S.P., Anderson, R.J.: Optical Fault Induction Attacks. In: Kaliski Jr., B.S., Koç, Ç.K., Paar, C. (eds.) CHES 2002. LNCS, vol. 2523, pp. 2–12. Springer, Heidelberg (2003)
26. Stanica, P., Maitra, S.: Rotation symmetric Boolean functions - Count and cryptographic properties. Discrete Applied Mathematics (DAM) 156(10), 1567–1580 (2008)
27. Stankovski, P.: Greedy Distinguishers and Nonrandomness Detectors. In: Gong, G., Gupta, K.C. (eds.) INDOCRYPT 2010. LNCS, vol. 6498, pp. 210–226. Springer, Heidelberg (2010)
28. Zhang, H., Wang, X.: Cryptanalysis of Stream Cipher Grain Family. IACR Cryptology ePrint Archive 2009: 109 (2009), http://eprint.iacr.org/2009/109

Algebraic Side-Channel Attacks Beyond the Hamming Weight Leakage Model

Yossef Oren[1], Mathieu Renauld[2],
François-Xavier Standaert[2], and Avishai Wool[1]

[1] Cryptography and Network Security Lab., School of Electrical Engineering
Tel-Aviv University, Ramat Aviv 69978, Israel
{yos,yash}@eng.tau.ac.il
[2] Université catholique de Louvain, Crypto Group
Place du Levant 3, B-1348 Louvain-la-Neuve, Belgium
{mathieu.renauld,fstandae}@uclouvain.be

Abstract. Algebraic side-channel attacks (ASCA) are a method of crypt-analysis which allow performing key recoveries with very low data complexity. In an ASCA, the side-channel leaks of a device under test (DUT) are represented as a system of equations, and a machine solver is used to find a key which satisfies these equations. A primary limitation of the ASCA method is the way it tolerates errors. If the correct key is excluded from the system of equations due to noise in the measurements, the attack will fail. On the other hand, if the DUT is described in a more robust manner to better tolerate errors, the loss of information may make computation time intractable. In this paper, we first show how this robustness-information tradeoff can be simplified by using an optimizer, which exploits the probability data output by a side-channel decoder, instead of a standard SAT solver. For this purpose, we describe a way of representing the leak equations as vectors of aposteriori probabilities, enabling a natural integration of template attacks and ASCA. Next, we put forward the applicability of ASCA against devices which do not conform to simple leakage models (e.g. based on the Hamming weight of the manipulated data). We finally report on various experiments that illustrate the strengths and weaknesses of standard and optimizing solvers in various settings, hence demonstrating the versatility of ASCA.

1 Introduction

In an algebraic side-channel attack (ASCA), the attacker is provided with a **device under test** (DUT) which performs a cryptographic operation (e.g. encryption). While performing this operation the device emits a measurable **side-channel leakage** that is expected to be data dependent. A typical example of such a leakage is a power consumption or electromagnetic radiation trace. As a result of the data dependence, a certain amount of **leaks** is modulated into the trace. These leaks are functions of the internal state of the DUT, which can teach the attacker about intermediate computations at various stages of the cryptographic operation. The trace, and its embedded leaks, are subjected to some **noise** due to interference and to the limitations of the measurement setup

E. Prouff and P. Schaumont (Eds.): CHES 2012, LNCS 7428, pp. 140–154, 2012.

[15, §1.2]. In order to recover the secret key from a power trace using an ASCA, the attacker generally performs different steps as we describe next.

1. In a first offline phase, the DUT is analyzed in order to identify the position of the leaking operations in the traces, for instance by using classical side-channel attacks like CPA [5] or template attacks [7].
2. Next, in a second offline phase, the DUT is profiled and a decoding process is devised, in order to map between a single power trace and a vector of leaks. A common output of the decoder would be the Hamming weight of the processed data as in [18], but many other decoders are possible.
3. After the offline phase, the attacker is provided with a small number of power traces (typically, a single trace). The traces are accompanied by **auxiliary information** such as known plaintext and ciphertext. The decoding process is applied to the power trace, and a vector of leaks is recovered. This vector of leaks may contain some errors, e.g. due to the effect of noise.
4. The leak vector, together with a formal description of the algorithm implemented in the DUT, is represented as a system of equations. This equation set also includes any additional auxiliary information.
5. A machine solver evaluates the equation set and attempts to find a candidate key satisfying it. In the case of an optimizing solver, a goal function is also specified to define the optimality of each candidate solution. The solver may fail to terminate in a tractable time, or otherwise return a candidate key.
6. Eventually, and optionally, some post-processing can be used, e.g. in order to brute force the remaining key candidates provided by the solver.

As indicated in the above list, there are several conditions which must all hold true before such an attack succeeds. First, the correct key should not be excluded from the set of solutions to the equation system. This can happen if the traces are too noisy, or if the decoder is not adapted to the attacked device. Next, the solver should not run for an intractable time. This can happen if not enough side-channel information is provided. Finally, the returned key should be the correct key, or at least a key close enough to allow an efficient enumeration.

Related Work. ASCA were introduced by Renauld et al. in [17,18], and first applied to the block ciphers PRESENT [4] and AES [14]. These works showed how keys can be recovered from a single measurement trace of these algorithms implemented in an 8-bit microcontroller, provided that the attacker can identify the Hamming weights of several intermediate computations during the encryption process. Already in these papers, it was observed that noise was the main limiting factor for efficient ASCA. To mitigate this issue, a heuristic solution was introduced in [18], and further elaborated in [22]. The main idea was to adapt the leakage model in order to trade some loss of information for more robustness, for example by grouping hard to distinguish Hamming weight values together into sets. We will denote this approach as set-ASCA. Other improvements regarding the error tolerance of ASCA have also been discussed in [13]. In parallel, an alternative proposal was introduced at CHES 2010, and denoted as Tolerant ASCA (TASCA) [15]. Here, the idea was to include the imprecise Hamming weights in the equation set, and to deal with these imprecisions via the solver. The authors showed how leaking implementations of Keeloq [9] could be attacked in this way, and recently extended their results to the AES case [16].

Our Contribution. One primary limitation of the ASCA method lies in its intolerance to errors: if the correct key is excluded from the system of equations, the attack will fail. This can be somehow mitigated by the robustness vs. information tradeoff, but only up to a certain point, as the loss of too much information makes the computation time intractable. In this work, we first show how an optimizing solver can use probability data to retain the robustness required to be error-tolerant, while losing less information than a SAT solver, at the cost of a larger problem representation. For this purpose, we describe a novel way of describing measurement equations directly as vectors of aposteriori probability, using the objective function of the optimizing solver. Next, we discuss the generalization of ASCA from the case of Hamming weight leakages to generic (template-based) models. We show that template attacks and ASCA can be naturally integrated, both with standard solvers and optimizers. We additionally provide experimental results allowing to put forward the strengths and weaknesses of the newly proposed probabilistic TASCA and set-ASCA. Overall, the resulting attacks allow strongly reduced data complexity template attacks, when compared to standard divide-and-conquer key recovery attacks.

Document Structure. The rest of this paper is organized as follows. Section 2 describes our experimental setup. Section 3 discusses how to exploit probabilistic information in TASCA and evaluates the performances of this improved attack, compared to set-ASCA and the original TASCA. Section 4 investigates attacks against a device that does not leak according to the well-known Hamming weight leakage model. Finally, concluding remarks are given in Section 5.

2   Experimental Setup

Our analysis considers two simulated implementations of the AES Rijndael in 8-bit microcontrollers as DUT. We assumed that no leaks from the key expansion process are available to the solver and that the DUT performs round key expansion in advance. This corresponds to a more challenging scenario, as it was established in [10] that the Hamming weights leaked from an 8-bit microcontroller implementation of the AES during key expansion are sufficient for full key recovery, even without any additional state information. We also assumed that the plaintext and the ciphertext are known to the attacker. We finally exploited the information the device leaks about the 8-bit operands commuting on its data bus. In total, it corresponds to 100 values per round, as described below:

- The **AddRoundKey** operation leaks information about the 16 state bytes after the XOR with the key, as well as information about the key bytes themselves, giving a total of 32 leaks per round.

- The **SubBytes** operation is implemented as a look-up table (LUT) and leaks information about its 16 output state bytes (and not any other internal state information), for a total of 16 leaks per round.
- The **ShiftRows** operation does not leak any information.
- The **MixColumns** operation is implemented using 8-bit XTIME and XOR operations as specified in [8, §5.1], and leaks 36 additional bytes of internal state and 16 leaks for its final state, resulting in a total of 52 leaks per round.

Note that the optimizer we used to perform TASCA was not memory-efficient enough to represent the entire AES encryption in equation form. As a result, we provided it with a known plaintext and the cipher equations for the first round of encryption only. By contrast, the SAT solver was provided with a plaintext/ciphertext pair, and all the cipher equations. In order to have comparable experiments, we only exploited the 100 first round leakages, in both cases.

Regarding the leakage models, we considered two different scenarios. First, we used the templates obtained from a PIC microcontroller. As illustrated in Figure 1, this device closely follows a Hamming weight leakage model. Next, we used the templates obtained from the AES S-box implemented in a 65-nanometer CMOS technology, previously analyzed in [19]. In particular, we selected one of the S-boxes for which the leakage model is not correlated with the Hamming weight of the manipulated data, as illustrated in Figure 2. In both cases, the signal-to-noise ratio was similar and relatively high (with the variance of the signal approximately 10 times larger than the noise variance), yet leading to some decoding errors, as will be investigated next. These setups were selected in order to illustrate the efficiency of ASCA in different implementation contexts.

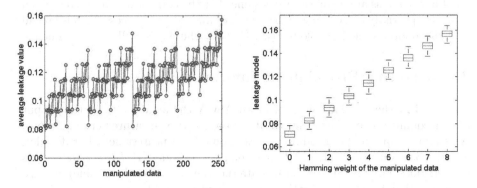

Fig. 1. PIC leakage model: average values (left) and grouped by HWs (right)

The solver used for the TASCA experiments was SCIP version 1.2.0 compiled for Windows 64-bit [3]. This solver is currently the best non-commercial solver available for non-linear optimization problems, as listed by [12]. The solver was run on a quad-core Intel Core i7 950, running at 3.06GHz with 8MB cache. For the set-ASCA experiments we used CryptoMiniSAT 2.9 [21]. This solver won several prizes in SAT competitions (SAT Race 2010 [20] and SAT competition

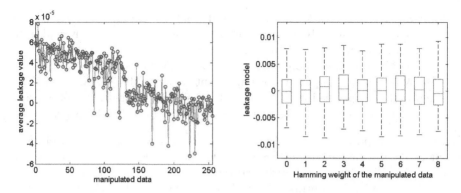

Fig. 2. 65nm S-box leakage model: average values (left) and grouped by HWs (right)

2011 [1]) and is well adapted to deal with cryptographic problems, as XOR operations (very frequent in cryptographic algorithms) are managed by the solver using specific optimized clauses. The solver was run on on a quad-core Intel Core Intel Xeon X5550 processor, running at 2.67 GHz with 8MB cache.

Finally, we note that because of the previously mentioned memory limitation issue, the success condition is defined differently for set-ASCA and TASCA. In the set-ASCA scenario, the entire encryption operation is included in the equation set, meaning that the solver either outputs the correct key or otherwise runs for an intractable time. By contrast, the TASCA solver only exploits the first round equations and, therefore, can sometimes return an incorrect key. We deal with this condition by measuring the amount of incorrect bytes in the result – if 4 bytes or less are incorrect, we assume that the correct key can be recovered from this partially correct key by brute force[1] and declare success. We also recall that the amount of leaks exploited (i.e. 100) was the same in all our experiments.

3   Exploiting Probabilistic Information

As stated in the previous section, in an ASCA the attacker takes the output of a decoding process and converts it into a series of measurement equations. An example for such a decoding process would be a nearest neighbor decoder, a naïve Bayes decoder [11, §13.2] or a template decoder [7]. In most cases this decoder does not only output "hard" data (i.e. the most likely leak value) but also some additional "soft" information, such as confidence information, a ranking of

---

[1] Assume that e of the 16 bytes are incorrect. The attacker must go over all $\binom{16}{e} \approx 2^{4e}/e!$ possible locations for those errored bytes, then try $256^e = 2^{8e}$ possible candidate assignments for these positions, resulting in an approximate total effort of $2^{4e} \cdot 2^{8e} = 2^{12e}$ AES operations. Most modern Intel CPUs have a native implementation of AES (AES-NI), which allows a sustained rate of more than 2^{31} AES operations per second [2]. Thus, an attacker can use a single machine with an AES-NI implementation to probe the neighborhood of a candidate key and find the correct key within less than 24 hours, even if 4 of the 16 bytes are incorrect.

several possible leaks by decreasing order of likelihood or, most generally, a full vector listing the aposteriori probability for each possible leak value, conditioned on the received trace. In this section, we discuss an improvement of TASCA which is capable of taking advantage of this soft (probabilistic) information.

For this purpose, let us start from the standard scenario of a Hamming weight-based ASCA. In this context, the tradeoff between robustness and information is generally achieved by choice of the **set size** k. It defines the number of acceptable values for each individual side-channel leak in the equation set, relative to the apriori selection of a leakage model (e.g. the Hamming weight of the manipulated data). The value of k can be either determined as a global constant for all equations in the set (e.g. as in [16,18]), or determined on a per-leak basis according to some heuristic (e.g. as in [13,22]). In the case of a precisely-defined equation set, in which $k = 1$, only the most likely value output by the decoder is accepted. This representation provides the most information, but it cannot tolerate any errors. As the set size k grows, so does the robustness of the equation set, but this comes at the price of a loss of information. The original work on ASCA [17] investigated only the case of $k = 1$. Thus, the single value chosen as most likely by the decoder was entered into the equation set. In the set-ASCA experiments of [13,18,22], more than one value was listed as possible to the solver, sacrificing information for robustness against errors. In this case, each leakage equation would accept the k most likely values, as output by the decoder. The TASCA attack of [16] also uses a set, and additionally uses a goal function to mark one of the value in the set as likelier than the others, without further quantification of this likelihood. This representation is more informative than in a set-ASCA, but it still does not fully take advantage of the information provided by the decoder.

We now present a more expressive way of representing the probability information provided by the decoding phase. In the most general case, the decoder outputs a full probability vector for each leak, listing the aposteriori probability of the leak having each possible value, conditioned on the specific trace being received. This output is typical for e.g. template decoders [7]. In the case of a Hamming weight-based template decoder, each potential leak will have an associated vector of 9 aposteriori probabilities corresponding to Hamming weights 0 to 8. If we further assume that individual leaks are uncorrelated, then the combined probability of all leaks in the trace is proportional to the product of the individual aposteriori probabilities. Of course, most of these combinations are impossible, since they violate the cipher equations. The goal of the solver in this case would be to find the set of leaks that maximizes the product of aposteriori probabilities while still corresponding to a valid encryption. As shown in [18], the exact values of the Hamming weight leaks provide enough information to uniquely and efficiently find the correct key of an AES encryption. If we define the exact value of leak i as x_i, we can define the objective of the attack as:

$$x_1 \cdots x_m = \arg \max_{x_1 \cdots x_m} \prod_{i=1 \cdots m} \Pr(x_i | trace) \; s.t. \text{ cipher eq'ns are satisfied.}$$

Since the goal function of the SCIP solver is expressed as a sum of integers which must be minimized, we represent the objective using this equivalent expression:

$$x_1 \cdots x_m = \arg \min_{x_1 \cdots x_m} \sum_{i=1 \cdots m} - \log \left(\Pr(x_i | trace) \right) \; s.t. \, \text{cipher eq'ns are satisfied.}$$

The representation of a probability vector $\overline{p_x}$ for a certain leaked Hamming weight x with set size k as a side-channel leak equation is thus split into two parts: the *constraint set* and the *goal term*. The constraint set is very straightforward – it considers k different events called "HW(x) is 0", "HW(x) is 1", etc., describes each event in terms of the relevant combination of bits in the leaked byte, and finally requires that one and only one of these events be true in for each leak in a satisfiable solution. The goal term matches each event with a corresponding probability. Each probability p is represented in the goal term as $- \lfloor C \log p \rfloor$, where C is an implementation parameter. The goal terms of all leaks in the system are then summed together to create the global goal function.

3.1   Experimental Validation

In order to compare set-ASCA, basic TASCA and TASCA with probabilities in the Hamming weight leakage model, we designed a first set of experiments, based on simulated leakages from the PIC device illustrated in Figure 1. For each experiment, we list the *decoding success rate* – the proportion of traces for which all 100 correct leaks are included in the 100 k-sized sets provided by the decoder – and the *key recovery success rate* – the proportion of traces for which the solver returned the correct key within a reasonable time. We also report on the (median and maximum) solving time and show the average *number of correct key bytes* in case of successful attacks. For set-ASCA, both the plaintext and ciphertext are included in the equation system, meaning that an attack can only succeed when every 16 key bytes are correct. On the other hand, in TASCA instances only the plaintext is used, meaning that when the set size increases, several keys can be valid according to the algebraic representation. In this latter case the average number of correct key bytes can be below 16. As explained in Section 2, the attack is still considered successful when at least 12 out of the 16 key bytes are correct. Our results are summarised in Table 1.

These experiments lead to a number of interesting observations. First and as expected, they clearly illustrate the information vs. robustness tradeoff. That is, the probability of decoding success grows as the set size grows (better robustness). However, this impacts the performances of the different attacks in different manners. For set-ASCA, the solving time quickly increases to the point of intractability, because of a lack of information. By contrast, the basic TASCA are more resistant to the loss of information: as the set size grows the running time increases, but the key recovery probability is much less affected. Yet, the limited information available to the solver causes parts of the key to be recovered incorrectly in some cases, which then requires an additional brute forcing step. Combining probabilistic information with a set size of 3 finally allowed the

Table 1. set-ASCA, basic TASCA and probabilistic TASCA experimental results against the PIC microcontroller simulated leakages with Hamming Weight model

attack	set size	decoding success	key rec. success	med. solving time	max. solving time	# of correct key bytes
set-ASCA	1	0%	0%	N/A	N/A	N/A
set-ASCA	2	83%	83%	2 seconds	6 seconds	16
set-ASCA	3	100%	0%	24+ hours	24+ hours	N/A
basic TASCA	1	0%	0%	N/A	N/A	N/A
basic TASCA	2	83%	75%	43.7 minutes	11.8 hours	14.48
basic TASCA	3	100%	80%	16.8 hours	66 hours	13.25
prob. TASCA	1	0%	0%	N/A	N/A	N/A
prob. TASCA	2	83%	82%	56.7 minutes	10.07 hours	15.88
prob. TASCA	3	100%	100%	8.2 hours	143 hours	16

optimizer to recover the correct key in virtually all experiments. It also reduced the running time compared to the basic TASCA. Yet, both TASCA approaches are still much slower than the set-ASCA approach when it succeeds (e.g. for set sizes $k \leq 2$), due to the more complex design of the optimizer. This is also reflected by the larger memory requirements of the TASCA solving phase.

Summarizing, the TASCA approaches allow improved flexibility as they systematically deal with the information vs. robustness tradeoff during the solving phase. By contrast, set-ASCA shift this problem to the decoder phase. In case of low-noise scenarios, or whenever the adversary can average the measurements, set-ASCA is the method of choice because of its reduced memory requirements and solving times. It also allows exploiting all the leaks (i.e. not only the first round ones). By contrast, the more the measurements are noisy and/or hard to interpret by the adversary (e.g. because of countermeasures), the more the TASCA approaches becomes interesting, thanks to its optimizing features.

4   Beyond the Hamming Weight Model

As illustrated in Section 2, Figure 2, the leakage of certain devices (e.g. in 65nm and smaller technologies) cannot always be precisely expressed with simple models. As a result, it is interesting to investigate how ASCA/TASCA can be extended towards these more challenging scenarios. In this section, we show how to move from Hamming weight-based models to more generic ones.

For this purpose, let us assume that the attacker has performed template-based profiling of the DUT [7]. Given a power trace, the he can now create a probability vector for each leak, where each entry in this vector matches a certain possible leak value, and each value in the vector is the *aposteriori* probability of this leak conditioned on the power trace being processed. Assuming the DUT has an 8-bit architecture, each such vector contains 256 entries. The decoding process will output a number of such vectors – one for every leak in the equation set. As for the Hamming weight model, we use this side-channel information to restrict the size of the solution space. In order to do so, we define a parameter called the **support size** k', which is comparable (though not identical) to the set size k in the previous section. It corresponds to the amount of possible values associated to each leak. These values are chosen according to the probability vector: the k' most probable values are considered possible, and the others are rejected. Hence, the value of k' must be carefully chosen in order to avoid rejecting the correct value from the set of possible ones, making the problem unsolvable.

Representing this generic leakage model as clauses or equations is less easy than for the Hamming weight model. For the set-ASCA, the easiest way to represent a set of k' possible values for a leaked byte x is to exclude all impossible values. For example, in order to exclude the value $x = 9 \Leftrightarrow (x_0, ..., x_7) = (0, 0, 0, 0, 1, 0, 0, 1)$, we add to the SAT problem the clause $(x_0 \cup x_1 \cup x_2 \cup x_3 \cup -x_4 \cup x_5 \cup x_6 \cup -x_7)$. Each set of k' possible values is thus translated into $256 - k'$ clauses with 8 literals per clause. In order to speed up the solving process, we additionally apply some simplification techniques, e.g. reducing the length and number of closes. For the TASCA, the representation of a probability vector $\overline{p_x}$ for a certain leaked byte x with support size k' as a side-channel leak equation is again split into two parts: the *constraint set* and the *goal term*. The constraint set describes k' different events called "x is 0", "x is 1", etc., and requires that one and only one of these events is true for each leak. The goal term matches each event with a corresponding probability. An example of such a representation can be found in Appendix A. Since this representation is especially suited for template-based profiling, we call it **template** TASCA and set-ASCA.

4.1   Impact of the Support Size and Goal Function

A first natural question in this new setting is: how small must the support size be for the attacks to succeed, and what is the impact of the probabilistic information that can be added to the optimizer? To answer it, we designed an experiment in which we compared many pairs of template-TASCA instances of single-round AES with different support sizes. In each pair, one of the instances was provided with an **unweighted** probability vector (that is, all nonzero elements in the probability vector are considered of equal probability), while the other was provided with a **weighted** vector function. The latter one was simulated (independent of any actual leakage model) such that the single correct byte value always had a higher probability than all the other ones in the support. For the rest, the instances were identical, with only plaintext provided, such that the solver could potentially output an incorrect key as in the previous section.

The results of this experiment are illustrated in Figure 3. As we can see, they can be divided into four distinct phases. In the first phase (support sizes up to 10), the performance of the weighted and unweighted instances is identical, probably because enough information is available in the support of the function, making the additional information in the goal function redundant. In the second phase (support sizes 10 − 50) both the weighted and the unweighted instances end in successful key recovery, but the weighted instances are faster by two orders of magnitude. We see that in this range there is still enough information

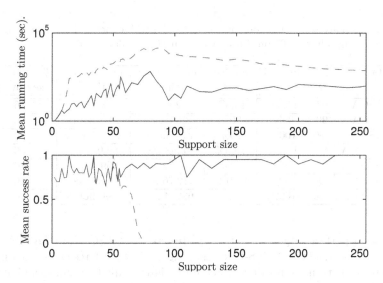

Fig. 3. Template-TASCA attacks with weighted (solid line) and unweighted (dashed line) probability vectors: experimental running time and success rate

in the unweighted instances to precisely specify the correct key, but the added information of the goal function allows the optimizer to reach the correct answer more quickly. In the third phase (support sizes 50 – 70) the success rate of the unweighted instances slowly falls to 0, probably because more and more incorrect keys can satisfy the constraint set. However, the additional information in the goal function causes the optimizer to prefer the likeliest solution, which in our case was the correct one. Finally, in the fourth phase (support sizes 70 and up) the large amount of possible keys in the support makes the success rate of the unweighted instances marginally small. Note that even with a full support ($k' = 256$) the performance was still good. This implies that all information about the instances can be encoded into the goal function and not into the constraints, and thus that the correct key will never be excluded from the equation set.

Furthermore, it could be argued that the running time of the weighted templates is faster than that of the unweighted ones because the correct guess is always the highest ranked. To investigate this scenario, we repeated the same experiment, this time setting the rank of the correct key candidate to 2. The change in rank caused the running time of the weighted case to increase, but had no effect on the success rate. We verified this behavior for ranks of up to 14.

4.2   Experimental Validation

As in the previous section, we verified the effectiveness of our attacks by performing several experiments. This time, we considered a DUT where the simulated leakages are generated according to the model of the 65nm ASIC implementing one AES S-box presented in Section 2. As illustrated in Figure 2, the leakage function of this device is very different from the Hamming weight model. Thus, it

Table 2. Template set-ASCA and probabilistic TASCA experimental results against simulated leakages from a 65nm S-box, with generic template model

attack	set size	dec. SR	key rec. SR	med. solving time	max. solving time	# of correct key bytes
set-ASCA	64	15.5%	15.5%	2 sec.	2 sec.	16
set-ASCA	90	90%	90%	265 sec.	24+ hours	16
set-ASCA	100	100%	29%	24+ hours	24+ hours	16
prob. TASCA	64	15.5%	15.5%	35.88 sec.	86.03 sec.	16
prob. TASCA	90	90%	90%	245.72 sec.	869.4 sec.	16
prob. TASCA	100	100%	100%	342.76 sec.	21271 sec.	16
prob. TASCA	256	100%	100%	62254 sec.	48+ hours	16

constitutes a perfect target for our template-based set-ASCA and TASCA. In a first step, we profiled the AES S-box, resulting in 256 templates corresponding to the 256 possible transition values. Each univariate template assumes a Gaussian noise and was characterized by a mean value μ and a noise standard deviation σ. In a second step, we used Bayesian inversion to simulate the classification probability $\Pr\left(x_i|trace\right)$ from the template output $\Pr\left(trace|x_i\right)$.

The results of the attacks are summarized in Table 2, where we selected different support sizes k'. As expected, smaller support sizes lead to more unsatisfiable/unsolvable problems, but these problems are solved faster, meaning a higher success rate for the computation phase. As soon as $k' \geq 100$, all the problems are solvable, but the solving process becomes much longer. We compared two attacks: the set-ASCA and the probabilistic TASCA. Both essentially confirmed our previous observations. Namely, the set-ASCA instances are very fast to solve for low support sizes, but suddenly increase in difficulty between $k' = 90$ and $k' = 100$. By comparison, probabilistic TASCA instances for low support sizes are much slower to solve than set-ASCA ones. Nevertheless, the difficulty of solving TASCA instances increases slower than for set-ASCA ones. In the end, probabilistic TASCA is able to solve problems with support size $k' = 256$, which is totally infeasible for set-ASCA (as $k' = 256$ means no side-channel information for set-ASCA instances). Summarizing, we again observe a tradeoff between efficiency (set-ASCA) and flexibility (probabilistic TASCA).

Besides, Table 3 presents a comparison of the Hamming weight model and the template model in terms of set size and support size. For each set size, *i.e.* for each number of possible Hamming weight values, the table details the corresponding average support size $\bar{k}'$ and the minimum and maximum support sizes k'_{min} and k'_{max}. For instance if $k = 2$, the two consecutive Hamming weight values $\mathrm{HW}(x) = \{0 \text{ or } 1\}$ correspond to $k'_{min} = 9$ possible transition values out of 256. Similarly, the two Hamming weight values $\mathrm{HW}(x) = \{3 \text{ or } 4\}$ correspond to $k'_{max} = 126$ possible transition values out of 256. On average, a leak that is represented by a set of 2 possible Hamming weight values can also be represented by a set of $\bar{k}' = 95$ possible transition values. Contrarily to the attacks using the template model where the support size is the same for every leak, the

attacks using the Hamming weight model present different support sizes. Therefore, some sets of Hamming weight values offer more information than others. In
other words, the Hamming weight information is not uniformly distributed over
the 100 considered leaks in the first AES round. This table allows us to compare
and better understand the results from Table 1 and Table 2. For example, we
observe that solving set-ASCA problems with the Hamming weight model for
set size $k = 2$ takes about 2 seconds, while solving set-ASCA problems with the
template model for a similar support size $k' = 90$ takes more than 250 seconds.
Hence, set-ASCA seems to take advantage of the non-uniform information proposed by the Hamming weight model. This confirms observations already made
in [6]: the SAT solver usually exploits small parts of the equation system where
the information is most concentrated. Interestingly, the same is not true for probabilistic TASCA: template instances with support size 90 or 100 are faster to
solve than Hamming weight instances with set size 2. Our hypothesis is that for
probabilistic TASCA, the goal function contains more information when using
the template model than the Hamming weight model, as the Hamming weight
model does not make any distinction between different transition values with
the same weight. As a consequence, the advantage offered by non-uniform information is counterbalanced by a less informative goal function.

Table 3. Comparison between set sizes (Hamming weight model) and corresponding
average $\bar{k}'$, minimum k'_{min} and maximum k'_{max} support sizes (template model)

Set size k	$\bar{k}'$	k'_{min}	k'_{max}
1	50	1	70
2	95	9	126
3	134	37	186

5   Concluding Remarks

In this paper we showed how both optimizers and solvers can be used to perform
ASCA even if the leakage function does not conform to the Hamming weight
model. The solver-based approach (set-ASCA) was shown to be faster than the
optimizer-based approach (TASCA) when a high degree of robustness is not required (for example, if the traces can be preprocessed by averaging many traces).
However, in cases when robustness is required, the optimizer approach was shown
to be both faster and with higher success rate than the solver-based approach.
This is due to the additional flexibility afforded by the optimizer goal function,
which allowed us to construct a generic representation of the measured leak as a
vector of aposteriori probabilities. The new flexible representation presented in
this paper allows TASCA and set-ASCA attacks to be used as a natural match
for template attacks. To carry out a combined Template-TASCA or Template-
set-ASCA, the attacker should not only create templates for the original key
bytes, but also for all intermediate values. The solver step will then replace any
traditional post-processing step used in template attacks such as brute-force key

enumeration. As a result, we further illustrated how an ASCA can be used effectively as a post-processing step of a template attack, dramatically reducing its data complexity. We believe that attention should be given to this capability when evaluating the security of systems using template attacks.

Future Work. Optimizers are less efficient than solvers in terms of running time, but since a solver does not have any efficient way of representing the objective function which contains the aposteriori probabilities, its running time quickly becomes intractable when high robustness is desired. It may be possible to increase the robustness of the solver-based approach by finding a better way of choosing the set size k or support size k'. For example, instead of choosing the k' most likely value, the solver can set a threshold probability and include in its support all values with a higher probability than this threshold. The solver might also be used in an adaptive manner - slowly increasing the support size while the solver returns unsatisfiability, until we reach the minimal sized support for which a solution exists. Quite naturally, the opposite approach would be interesting too. Namely, the search for improved optimizers, allowing to represent more complex problems with reduced memory efficiency would be another way to close the gap between set-ASCA and TASCA. Finally, it would be interesting to carefully investigate the connection between the offline and online phases of a template attack on the success of template-TASCA and template set-ASCA. A better model obtained through better profiling in the offline phase should intuitively allow the use of lower-quality data in the online attack phase, and vice versa.

Acknowledgements. Mathieu Renauld is a PhD student funded by the Walloon region through the SCEPTIC project. François-Xavier Standaert is an associate researcher of the Belgian Fund for Scientific Research (FNRS-F.R.S.). This work has been funded in part by the ERC project 280141 (acronym CRASH) and with support from Wallonia-Brussels International. The authors wish to thank the anonymous reviewers for their encouraging and insightful comments.

References

1. SAT 2011 Competition, `http://www.cril.univ-artois.fr/SAT11/phase2.pdf`
2. Akdemir, K., Dixon, M., Feghali, W., Fay, P., Gopal, V., Guilford, J., Ozturc, E., Worlich, G., Zohar, R.: Breakthrough AES Performance with Intel AES New Instructions. Technical report, Intel Corporation (October 2010), `http://software.intel.com/file/27067`
3. Berthold, T., Heinz, S., Pfetsch, M.E., Winkler, M.: SCIP – Solving Constraint Integer Programs. SAT 2009 competitive events booklet (2009)
4. Bogdanov, A., Knudsen, L.R., Leander, G., Paar, C., Poschmann, A., Robshaw, M.J.B., Seurin, Y., Vikkelsoe, C.: PRESENT: An Ultra-Lightweight Block Cipher. In: Paillier, P., Verbauwhede, I. (eds.) CHES 2007. LNCS, vol. 4727, pp. 450–466. Springer, Heidelberg (2007)
5. Brier, E., Clavier, C., Olivier, F.: Correlation Power Analysis with a Leakage Model. In: Joye, M., Quisquater, J.-J. (eds.) CHES 2004. LNCS, vol. 3156, pp. 16–29. Springer, Heidelberg (2004)

6. Carlet, C., Faugère, J.-C., Goyet, C., Renault, G.: Analysis of the algebraic side channel attack. J. Cryptographic Engineering 2(1), 45–62 (2012)
7. Chari, S., Rao, J.R., Rohatgi, P.: Template Attacks. In: Kaliski Jr., B.S., Koç, Ç.K., Paar, C. (eds.) CHES 2002. LNCS, vol. 2523, pp. 13–28. Springer, Heidelberg (2003)
8. Daemen, J., Rijmen, V.: AES Proposal: Rijndael (1998)
9. Dawson, S.: Code Hopping Decoder using a PIC16C56. Microchip confidential, leaked online in 2002 (1998)
10. Mangard, S.: A Simple Power-Analysis (SPA) Attack on Implementations of the AES Key Expansion. In: Lee, P.J., Lim, C.H. (eds.) ICISC 2002. LNCS, vol. 2587, pp. 343–358. Springer, Heidelberg (2003)
11. Manning, C.D., Raghavan, P., Schtze, H.: Introduction to Information Retrieval. Cambridge University Press, New York (2008)
12. Manquinho, V., Roussel, O.: Pseudo-Boolean Competition 2009 (July 2009), http://www.cril.univ-artois.fr/PB09/
13. Mohamed, M.S.E., Bulygin, S., Zohner, M., Heuser, A., Walter, M.: Improved Algebraic Side-Channel Attack on AES. Cryptology ePrint Archive, Report 2012/084 (2012), http://eprint.iacr.org/
14. Information Technology Laboratory (National Institute of Standards and Technology). Announcing the Advanced Encryption Standard (AES). Computer Security Division, Information Technology Laboratory, National Institute of Standards and Technology, Gaithersburg, MD (2001)
15. Oren, Y., Kirschbaum, M., Popp, T., Wool, A.: Algebraic Side-Channel Analysis in the Presence of Errors. In: Mangard, S., Standaert, F.-X. (eds.) CHES 2010. LNCS, vol. 6225, pp. 428–442. Springer, Heidelberg (2010), http://iss.oy.ne.ro/TASCA
16. Oren, Y., Wool, A.: Tolerant Algebraic Side-Channel Analysis of AES. Cryptology ePrint Archive, Report 2012/092 (2012), http://iss.oy.ne.ro/TASCA-eprint
17. Renauld, M., Standaert, F.-X.: Algebraic Side-Channel Attacks. In: Bao, F., Yung, M., Lin, D., Jing, J. (eds.) Inscrypt 2009. LNCS, vol. 6151, pp. 393–410. Springer, Heidelberg (2010)
18. Renauld, M., Standaert, F.-X., Veyrat-Charvillon, N.: Algebraic Side-Channel Attacks on the AES: Why Time also Matters in DPA. In: Clavier, C., Gaj, K. (eds.) CHES 2009. LNCS, vol. 5747, pp. 97–111. Springer, Heidelberg (2009)
19. Renauld, M., Standaert, F.-X., Veyrat-Charvillon, N., Kamel, D., Flandre, D.: A Formal Study of Power Variability Issues and Side-Channel Attacks for Nanoscale Devices. In: Paterson, K.G. (ed.) EUROCRYPT 2011. LNCS, vol. 6632, pp. 109–128. Springer, Heidelberg (2011)
20. Sinz, C.: SAT-Race 2010 (2010), http://baldur.iti.uka.de/sat-race-2010/results.html
21. Soos, M.: CryptoMiniSat2, http://www.msoos.org/cryptominisat2/
22. Zhao, X., Wang, T., Guo, S., Zhang, F., Shi, Z., Liu, H., Wu, K.: SAT based Error Tolerant Algebraic Side-Channel Attacks. In: 2011 Conference on Cryptographic Algorithms and Cryptographic Chips, CASC 2011 (July 2011)

A Appendix: A Sample Template-TASCA Instance

The appendix demonstrates the format of leak equations used in a Template-TASCA attack, following the notation introduced in Subsection 4. The equations are given in the OPB format supported by the SCIP solver [3].

Assume that during the cryptographic operation the DUT processes two bytes x and y. Using a template profiling step, the attacker creates a model of the leakages produced by the processing of x and y. Given a trace, the attacker can now use this information to calculate vectors of aposterioti probabilities for x and for y $(\overline{p_x}, \overline{p_y})$, conditioned on the specific trace having been received. The support size has been set to $k' = 4$. The vectors passed to the solver are $\overline{p_x} = \{\frac{1}{2}, \frac{1}{3}, \frac{1}{12}, \frac{1}{12}, 0 \cdots 0\}, \overline{p_y} = \{\frac{1}{5}, \frac{1}{5}, \frac{1}{5}, 0, 0, \frac{2}{5}, 0 \cdots 0\}$. The attacker also chooses the implementation parameter $C = 10$ to efficiently capture the probability information while limiting the ultimate size of the goal term. The attacker then uses the aposteriori probability vectors to generate the following equations:

* Leak Equations:
```
+1 ˜x_is_00 +1   ˜x_0 ˜x_1 ˜x_2 ˜x_3 ˜x_4 ˜x_5 ˜x_6 ˜x_7 = 1;
+1 ˜x_is_01 +1    x_0 ˜x_1 ˜x_2 ˜x_3 ˜x_4 ˜x_5 ˜x_6 ˜x_7 = 1;
+1 ˜x_is_02 +1   ˜x_0  x_1 ˜x_2 ˜x_3 ˜x_4 ˜x_5 ˜x_6 ˜x_7 = 1;
+1 ˜x_is_03 +1    x_0  x_1 ˜x_2 ˜x_3 ˜x_4 ˜x_5 ˜x_6 ˜x_7 = 1;
+1  x_is_00 +1 x_is_01 +1 x_is_02 +1 x_is_03 = 1;

+1 ˜y_is_00 +1   ˜y_0 ˜y_1 ˜y_2 ˜y_3 ˜y_4 ˜y_5 ˜y_6 ˜y_7 = 1;
+1 ˜y_is_01 +1    y_0 ˜y_1 ˜y_2 ˜y_3 ˜y_4 ˜y_5 ˜y_6 ˜y_7 = 1;
+1 ˜y_is_02 +1   ˜y_0  y_1 ˜y_2 ˜y_3 ˜y_4 ˜y_5 ˜y_6 ˜y_7 = 1;
+1 ˜y_is_05 +1    y_0 ˜y_1  y_2 ˜y_3 ˜y_4 ˜y_5 ˜y_6 ˜y_7 = 1;
+1  y_is_00 +1 y_is_01 +1 y_is_02 +1 y_is_05 = 1;
```

* Goal term:
```
min: +6 x_is_00 +10 x_is_01 +24 x_is_02 +24 x_is_03 ...
     +16 y_is_00 +16 y_is_01 +16 y_is_02 +9 y_is_05 ;
```

In addition to these leak equations, the instance will also contain additional equations which describe the cryptographic meaning of the variables x and y, as well as equations which capture the auxiliary information available to the attacker (such as known plaintext).

Selecting Time Samples
for Multivariate DPA Attacks

Oscar Reparaz, Benedikt Gierlichs, and Ingrid Verbauwhede

KU Leuven Dept. Electrical Engineering-ESAT/SCD-COSIC and IBBT
Kasteelpark Arenberg 10, B-3001 Leuven-Heverlee, Belgium
{oscar.reparaz,benedikt.gierlichs,ingrid.verbauwhede}@esat.kuleuven.be

Abstract. Masking on the algorithm level, i.e. concealing all sensitive intermediate values with random data, is a popular countermeasure against DPA attacks. A properly implemented masking scheme forces an attacker to apply a higher-order DPA attack. Such attacks are known to require a number of traces growing exponentially in the attack order, and computational power growing combinatorially in the number of time samples that have to be exploited jointly. We present a novel technique to identify such tuples of time samples before key recovery, in black-box conditions and using only known inputs (or outputs). Attempting key recovery only once the tuples have been identified can reduce the computational complexity of the overall attack substantially, e.g. from months to days. Experimental results based on power traces of a masked software implementation of the AES confirm the effectiveness of our method and show exemplary speed-ups.

Keywords: Time sample selection, multivariate side-channel attack, masking, reverse-engineering.

1 Introduction

Side-channel attacks are used to break implementations of cryptographic algorithms in embedded devices. Since the introduction by Kocher [11] in the late nineties, they have been refined and a series of countermeasures have been designed to thwart them. A particularly popular countermeasure against Differential Power Analysis (DPA) attacks [12] is d-order masking [4,7], since it enjoys a formal proof of security against higher-order DPA attacks [4,15] of order d or less. d-order masking is based on splitting every sensitive intermediate value in $d + 1$ shares and we consider the case that they are manipulated at distinct times, as is typical for software implementations. $d + 1$-order DPA and $d + 1$-variate Mutual Information Analysis (MIA) attacks [5,17] (from now on referred to as multivariate attacks together) allow to break d-order masked implementations by analyzing tuples of $d + 1$ time samples, corresponding to all shares of a masked sensitive variable, from each trace. However, multivariate attacks are significantly more difficult to mount than univariate attacks for two reasons. First, attacks exploiting higher-order moments are exponentially more

E. Prouff and P. Schaumont (Eds.): CHES 2012, LNCS 7428, pp. 155–174, 2012.

sensitive to noise as the masking order d increases [4,19]. As a consequence, the number of traces required to mount a successful attack grows exponentially in d. Second, multivariate attacks need to search over $d+1$-tuples of time samples. The computational complexity of the attacks therefore grows combinatorially in the attack order $d+1$. Hence, secure implementations use a masking order d in combination with a suitable noise level to ensure that an attack will require a sufficiently large number of traces and a heavy amount of computation, such that the attack becomes impractical.

Related Work. Most related works on non-profiled multivariate attacks start from the assumption that the time samples where the shares of the targeted, masked sensitive variable leak are known, and focus on the key recovery [5,10,15,17,18,20,23]. Few related works tackle the problem of identifying (tuples of) interesting time samples before key recovery, and they do so with heuristic approaches. Agrawal et al. [1] describe a method to identify tuples of time samples that requires a chosen input adversarial model and that can only exploit the leakage of single bits. Their method is tailored to Boolean masking and the measurements can not be re-used for key recovery, due to the way the inputs are chosen. Oswald et al. [16] essentially propose an exhaustive search over all $d+1$-tuples of time samples in a small time window that is selected based on an *educated guess*. The interpretation of *educated guess* is left to the practitioner. Note that the guess does not select tuples of time samples, but a window of time samples that has to be searched for a tuple exhaustively in combination with key recovery. This method can be applied with known inputs or outputs and, in principle, to any masking scheme. The approach suggested by Lemke and Paar [14] and Gierlichs et al. [5] is to examine the empirical variance of several power traces when the input data is kept constant, i.e. it requires a chosen input adversarial model. In an ideal case, the variance is then caused only by masking, and therefore time samples with high variance mostly correspond to time samples where the masks or masked variables are being processed. Note that also this method does not identify tuples of time samples but a set of samples that has to be searched for a tuple exhaustively in combination with key recovery. The measurements can not be re-used for key recovery and, in principle, the method can be applied to any masking scheme.

In summary, the *educated guess* of Oswald et al. is the only method described in the literature that can be applied in black-box conditions and with known inputs or outputs.

Contribution. We present a novel method for identifying interesting $d+1$-tuples of time samples before key recovery. It is not heuristic but systematic and ranks all possible $d+1$-tuples of time samples in a given window according to their dependency on, informally speaking, "typical attack targets". It does not provide a qualitative yes/no decision, but instead ranks tuples with respect to a meaningful metric such that there is a natural order in which to attack them. Our technique can lead to a substantial improvement in the computational efficiency of multivariate attacks compared to exhaustive search over the same window of time samples, since it retains only a small fraction of all possible $d+1$-tuples for key recovery. The relative improvement depends on the size of the subkeys that are attacked. In absolute terms, the improvement becomes more pronounced with increasing attack order $d+1$, increasing size of the time window, and increasing number of traces.

Our approach is based on mutual information and is fully generic: it applies to attacks of any order $d+1$, including univariate attacks against unmasked implementations, it applies to all possible masking schemes, it requires only a known input or output scenario, it can traverse S-boxes, locate shares of the masked S-box output, and it does not require any restrictive assumptions on the device leakage behavior. In other words, our method does not require more restrictive assumptions than a generic MIA attack [6].

Paper Organization. In Sect. 2 we introduce our notation, recall the basics of masking and discuss state-of-the-art multivariate attacks. In Sect. 3 we present our technique together with an analysis of how and why it works. We discuss its efficiency, impact, and possible refinements in Sect. 4. In Sect. 5 we present experimental results that validate our proposal and highlight some of its interesting properties. Section 6 concludes the paper.

2   Preliminaries

In this paper we consider only non-profiled, multivariate attacks. *Interesting* tuples are tuples of time samples that carry leakage of all shares of a masked variable that is a (possibly keyed) function of the plaintext.

2.1   Notation

Capital letters in bold face, e.g. $\mathbf{M}$, denote random variables. Lowercase letters, e.g. m, denote a specific value of $\mathbf{M}$, e.g. $\mathbf{M} = m$. $\mathbf{M}_i$ are mask bytes, $\mathbf{P}$ is a plaintext byte, $\mathbf{K}$ is a key byte, and S-box is a cryptographic S-box. $\mathbf{L}(t)$ is the random variable corresponding to the measured side-channel leakage at time t. $t_{\mathbf{M}}$ denotes the instant when the device is manipulating the random variable $\mathbf{M}$. $\mathbf{I}(\mathbf{A}; \mathbf{B}; \mathbf{C})$ denotes the multivariate mutual information between $\mathbf{A}$, $\mathbf{B}$ and $\mathbf{C}$ [2,5] and is computed as

$$\mathbf{I}(\mathbf{A}; \mathbf{B}; \mathbf{C}) = \mathbf{I}(\mathbf{A}; \mathbf{B}) - \mathbf{I}(\mathbf{A}; \mathbf{B}|\mathbf{C}). \tag{1}$$

Note that, if $\mathbf{A}$ and $\mathbf{B}$ are independent, $\mathbf{I}(\mathbf{A}; \mathbf{B}) = 0$ and $\mathbf{I}(\mathbf{A}; \mathbf{B}; \mathbf{C}) \leq 0$.

2.2   Masking

Masking was introduced by Goubin and Patarin [7] and by Chari et al. [4] (together with a proof of security) as a sound approach to protect implementations against first-order DPA attacks. In a d-order masked implementation, every sensitive variable $\mathbf{Z}$ is randomly split into $d + 1$ shares $\mathbf{M}_1, \ldots, \mathbf{M}_d, \mathbf{V}$ satisfying

$$\mathbf{M}_1 \star \ldots \star \mathbf{M}_d \star \mathbf{V} = \mathbf{Z}, \tag{2}$$

where $\star$ is some suitable group operator.

The security of properly implemented masking schemes relies on the fact that even if the adversary manages to know any information about up to d shares out of $d + 1$ (for example, via side-channel leakage), he cannot learn any information about the sensitive variable $\mathbf{Z}$.

Throughout the paper we assume that the shares $\mathbf{M}_1, \ldots, \mathbf{M}_d, \mathbf{V}$ are manipulated (and leak) separately at different time instants. Further, we assume that these time instants and the values of the shares are unknown to the adversary.

2.3   Multivariate Attacks

Masked implementations of order d can in theory always be broken by $d + 1$-variate attacks as originally proposed by Messerges [15] and Chari et al. [4]. They exploit the statistical dependence between the leakage of the $d+1$ shares and the sensitive variable $\mathbf{Z}$. There are essentially two different methods for performing multivariate attacks.

The first approach [4,10,15,16,18,23] consists in reducing the problem to a univariate scenario by preprocessing each trace, and then running a first-order attack on the preprocessed traces. The preprocessing generates a new trace from all possible $d + 1$-tuples of distinct time samples of the original trace, where for each tuple the $d + 1$ time samples are combined with a so-called *combination function* (typically the absolute difference [15] or the centered product [18]). The second approach, proposed by Prouff and Rivain [17] and Gierlichs et al. [5], does not rely on a preprocessing step but directly uses multivariate MIA for the attack.

A major shortcoming of both methods is that they suffer from the effect known as "combinatorial explosion" and hence combinatorial time complexity in $d+1$. Both methods aim to recover subkeys while, at the same time, searching for a suitable $d + 1$-tuple of time samples in the traces. In the first approach, the preprocessed traces are $\binom{L}{d+1}$ time samples long, where L is the trace length. These traces have to be processed for each hypothesis on the subkey. In the second approach, the distinguisher should be computed for each of the $\binom{L}{d+1}$ $d + 1$-tuples, and for each hypothesis on the subkey.

Hence, it is very important to identify the interesting tuples (or to narrow down a window of time samples as much as possible) prior to key recovery in order to keep the computational complexity of a multivariate attack at a feasible level.

3   Identifying Interesting Tuples of Time Samples

In this section we explain how to identify interesting $d+1$-tuples of time samples prior to key recovery. Note that we focus our attention on this aspect and that key recovery is not the primary focus of the paper. For clarity in the exposition, in what follows we assume a first-order Boolean masking scheme (two shares) and a noise-free scenario. The practical results presented in Sect. 5 are based on measured power traces.

3.1   Core Idea

Let us consider a scenario with fixed plaintext, fixed key, and sensitive intermediate value $\mathbf{Z} = F_k(p)$, where F_k is some key-dependent function (for example, $F_k(p) = \texttt{S-box}(p \oplus k)$). The key observation is that the mutual information between the leakages at time instants corresponding to the manipulation of the mask $\mathbf{M}_1$ and the masked intermediate value $\mathbf{V} = \mathbf{M}_1 \oplus F_k(p)$ is non-zero. That is,

$$\mathbf{I}(\mathbf{L}(t_{\mathbf{M}_1}); \mathbf{L}(t_{\mathbf{V}})) > 0 . \tag{3}$$

The interpretation is straightforward: leakage at $t_{\mathbf{V}}$ depends only on $\mathbf{V}$, which varies in function of only the mask $\mathbf{M}_1$ (since the plaintext and the key are fixed), and some information about the mask is leaked at $t_{\mathbf{M}_1}$. Hence, the information shared between leakage at $t_{\mathbf{M}_1}$ and $t_{\mathbf{V}}$ is non-zero. On the other hand, the information shared between leakage at two unrelated time samples t_0 and t_1 is zero

$$\mathbf{I}(\mathbf{L}(t_0); \mathbf{L}(t_1)) = 0 \tag{4}$$

because no relation exists between data handled at t_0 and at t_1. Thus, Eqs. (3) and (4) allow us to distinguish pairs of time samples that contain leakage of dependent variables (case of Eq. (3)) from those pairs that contain leakage of independent variables, that are irrelevant for the multivariate attack (case of Eq. (4)). Note that not all pairs of time samples that contain leakage of dependent variables carry some information about the key. For example, if the same mask is manipulated at t_0 and t_1, then $\mathbf{I}(\mathbf{L}(t_0); \mathbf{L}(t_1)) > 0$.

We stress that the value of $\mathbf{K}$ need not be known, and no hypothesis on it be made.

The General Case. In the above example we required a fixed plaintext and thus a chosen plaintext scenario. We can relax this assumption and instead work with known (varying) plaintexts. Suppose that the device is manipulating the plaintext byte $\mathbf{P}$, the mask $\mathbf{M}_1$ and the masked intermediate value $\mathbf{V}$ such that $\mathbf{V} = \mathbf{M}_1 \oplus F_k(\mathbf{P})$ at time instants $t_{\mathbf{P}}$, $t_{\mathbf{M}_1}$ and $t_{\mathbf{V}}$, respectively. The natural extension of the core observation to known varying plaintexts is that $\mathbf{L}(t_{\mathbf{P}})$, $\mathbf{L}(t_{\mathbf{M}_1})$ and $\mathbf{L}(t_{\mathbf{V}})$ are not independent, and therefore the mutual information between them is non-zero

$$\mathbf{I}(\mathbf{L}(t_{\mathbf{M}_1}); \mathbf{L}(t_{\mathbf{V}}); \mathbf{L}(t_{\mathbf{P}})) \neq 0 . \tag{5}$$

At three unrelated time samples t_0, t_1 and t_2, on the other hand, the mutual information is zero

$$\mathbf{I}(\mathbf{L}(t_0); \mathbf{L}(t_1); \mathbf{L}(t_2)) = 0 . \tag{6}$$

The interpretation follows the same lines as in the particular case. Leakage at t_V depends only on V, which now varies in function of the plaintext and the mask (since the key is fixed), and some information about the mask and the plaintext is leaked at t_{M_1} and t_P, respectively. Thus, the information shared between $L(t_{M_1})$, $L(t_V)$ and $L(t_P)$ is non-zero.

Note that it is not necessary to search for t_P, nor is it necessary for t_P to physically exist in the power traces. By assumption, the plaintext is known, so it is possible to substitute $L(t_P)$ with the leakage of the known plaintext under some hypothesized leakage model $\tilde{L}(P)$. This makes the analysis faster since one has to search only for a pair of time instants (t_{M_1} and t_V) instead of searching for a triplet. The choice of $\tilde{L}$ will be discussed in Sect. 3.3.

We can hence use Eqs. (5) and (6) to distinguish dependent triplets from independent triplets. In addition, and contrary to the particular case with fixed plaintext, all identified tuples are now interesting tuples and all relate to the *specific* plaintext byte P. Most of them carry some information about the key and can be useful for a key recovery attack. The only possible type of tuple that will be identified as interesting although it does not carry some information about the key is the one corresponding to all shares of the specific, masked plaintext byte. We discuss this in more detail in Sect. 3.3.

3.2   Suggested Workflow for Multivariate Attacks

The previous observations allow an attacker to identify interesting $d + 1$-tuples of time samples prior to key recovery. Again, we use $d = 1$ in the explanation. The proposed workflow divides an attack in three phases:

Step 1. (Window selection) The adversary uses any available mean to narrow down the time window to analyze. For example, the adversary could select a small window based on an *educated guess* [16], if possible. Obviously, care has to be taken to not discard too many time samples since the window must contain at least one interesting $d + 1$-tuple.

Step 2. (Tuple selection) The adversary estimates $I(L(t_1); L(t_2); \tilde{L}(P))$ for all (t_1, t_2) with $t_1 > t_2$ in the remaining window, and keeps a list of pairs of time samples yielding negative mutual information with large absolute value.

Step 3. (Key recovery attack) The adversary performs the preferred strategy for a bivariate attack on traces consisting only of the pairs of time samples in the list. These traces consist of a few pairs of time samples, and hence the key recovery step is much faster.

3.3   Which Tuples of Time Samples Pop Up?

The adversary has freedom to choose the hypothesized leakage model $\tilde{L}$ for the plaintext. Depending on the choice of $\tilde{L}$, different tuples of time samples will be identified. In this section we analyze two cases.

$\tilde{\mathbf{L}}$ **is the identity function.** When the adversary computes the mutual information between time samples and a plaintext byte, i.e. $\tilde{\mathbf{L}}(\mathbf{P}) = \mathbf{P}$, he will be able to identify all tuples corresponding to all shares of any (sensitive) variable of the form $\mathbf{Z} = F_k(\mathbf{P})$. In particular, the method is able to identify the shares $(\mathbf{M}_1, \mathbf{V})$ with $\mathbf{V} = \mathbf{P} \oplus \mathbf{M}_1$, $\mathbf{V} = \mathbf{P} \oplus \mathbf{K} \oplus \mathbf{M}_1$ and $\mathbf{V} = \texttt{S-box}(\mathbf{P} \oplus \mathbf{K}) \oplus \mathbf{M}_1$, since the key is fixed.

This result is useful, as it allows the attacker to locate both the masked variables *before* the S-box (masked plaintext and masked S-box input) as well as the masked variables *after* the S-box (masked S-box output). Note that it is irrelevant if the masks before and after the S-box are the same. If the mask does not change, the identified tuples of time samples will share one component.

$\tilde{\mathbf{L}}$ **is an approximation of the device leakage behavior.** If the attacker chooses $\tilde{\mathbf{L}}$ as an approximation of the leakage behavior $\mathbf{L}$, he will be able to identify all tuples of time samples corresponding to all shares of any (sensitive) variable of the form $\mathbf{Z} = F_k(\mathbf{P})$ appearing *before* the S-box (e.g. masked plaintext and masked S-box input). For a typical S-box, he will not be able to identify tuples of time samples corresponding to shares of sensitive variables *after* the S-box. The intuitive reasoning behind this is that knowledge of the distribution of the plaintext's *leakage* does not give sufficient information for guessing the distribution of the S-box output's leakage. The advantage of this choice, compared to the identity function, is the ease of estimation, see Sect. 4.1. Disadvantages are that one cannot locate shares of masked variables after the S-box and that one relies on an assumption about the device leakage behavior.

Note that we compute the mutual information according to Eq. (1), and not as

$$
\begin{aligned}
\mathbf{I}((\mathbf{L}(t_0), \mathbf{L}(t_1)); \tilde{\mathbf{L}}(\mathbf{P})) = \\
\mathbf{I}(\mathbf{L}(t_0); \tilde{\mathbf{L}}(\mathbf{P})) + \mathbf{I}(\mathbf{L}(t_1); \tilde{\mathbf{L}}(\mathbf{P})) - \mathbf{I}(\mathbf{L}(t_0); \mathbf{L}(t_1); \tilde{\mathbf{L}}(\mathbf{P})),
\end{aligned}
\tag{7}
$$

where the last of the three terms is in turn given by Eq. (1) [2]. The reasoning for this choice is straightforward. The first two terms of Eq. (7) capture first-order leakage of variables that depend on $\tilde{\mathbf{L}}(\mathbf{P})$, e.g. unmasked plaintext, unmasked S-box input and, depending on the choice of $\tilde{\mathbf{L}}$, unmasked S-box output. By assumption, the masking scheme is properly implemented and there is no first-order leakage of sensitive variables. Hence, the only first-order leakage that these terms could capture is that of the unmasked plaintext, which is of no use for our purpose. By omitting the two terms and using Eq. (1) we ensure that only interesting tuples yield non-zero mutual information.

Moreover, Eq. (1) allows us to target very specific tuples. For our interesting tuples it holds that $\mathbf{I}(\mathbf{L}(t_{\mathbf{M}_1}), \mathbf{L}(t_{\mathbf{V}})) = 0$ such that interesting tuples yield strictly negative mutual information, see (1).

4   Discussion

In this section we discuss several aspects of the proposed workflow for multivariate attacks, such as its efficiency, refinements and additional applications.

4.1   Efficiency Analysis

We evaluate the efficiency of the proposed workflow with respect to the running time and the number of traces needed, and we compare these numbers to those of a "classical" multivariate MIA attack that uses exhaustive search instead of step 2. Although the proposed method is not limited to a particular multivariate attack technique for step 3, using multivariate MIA here allows us to draw important conclusions regarding the efficiency of the proposed workflow, since the numbers can be directly compared. In both cases we focus the attacks on the (masked) S-box output. According to the previous section, this choice implies that step 2 of the proposed workflow uses the identity function $\tilde{\mathbf{L}}(\mathbf{P}) = \mathbf{P}$. We analyze two different scenarios:

(a) Unknown leakage behavior. Step 3 of the proposed workflow and the "classical" MIA both use the identity leakage model, or possibly some truncated identity leakage model in case of a bijective S-box. The point here is that both step 3 and the "classical" MIA use the same leakage model.
(b) Known leakage behavior $\mathbf{L}$ equal to Hamming weight leakage. Step 3 of the proposed workflow and the "classical" MIA both use the Hamming weight leakage model.

Running Time. We assume that after step 1 the traces are L time samples long and contain at least one tuple of time samples corresponding to all shares of the masked S-box output. We further assume that all attacks are provided with sufficiently many traces, i.e. there are no PDF estimation problems.

   In scenario (a) the running time of the "classical" MIA attack is given by $\binom{L}{d+1} \times \alpha \times |K|$, where $\binom{L}{d+1}$ is the number of $d+1$-tuples of time samples to analyze, α is the time it takes to compute the MIA distinguisher for one $d+1$-tuple of time samples and one subkey hypothesis using the identity leakage model, and $|K|$ is the number of subkey hypotheses. In scenario (b) the running time of the "classical" MIA attack is $\binom{L}{d+1} \times \beta \times |K|$, where β is the time it takes to compute the MIA distinguisher for one $d+1$-tuple of time samples and one subkey hypothesis using the Hamming weight leakage model.

   In scenario (a) the running time of step 2 of the proposed workflow is given by $\binom{L}{d+1} \times \alpha$, and the running time of step 3 is $|K| \times \alpha \times \gamma$, where γ is the number of $d+1$-tuples in the list of interesting tuples generated in step 2. We have that $\gamma \geq 1$ and typically γ is much smaller than L. The combined running time of steps 2 and 3 is $\binom{L}{d+1} \times \alpha + |K| \times \alpha \times \gamma$. In scenario (b) the running time of step 2 is again $\binom{L}{d+1} \times \alpha$ and the running time of step 3 is $|K| \times \beta \times \gamma$. The combined running time of both steps is $\binom{L}{d+1} \times \alpha + |K| \times \beta \times \gamma$. Note that in both scenarios (a) and (b), the total running time of the proposed workflow is dominated by step 2. Table 1 summarizes these numbers and shows that the proposed workflow essentially runs $|K|$ times faster.

   So far we have limited this analysis to attacks against a single subkey. For attacking multiple subkeys, it may be that only recovering the first subkey is hard and that the interesting tuples of time samples related to the other subkeys

Table 1. Running time of MIA attacks using the proposed and the "classical" workflow

	Proposed workflow	"Classical" MIA	Improvement factor
Scenario (a)	$\binom{L}{d+1} \times \alpha + |K| \times \alpha \times \gamma$ $\approx \binom{L}{d+1} \times \alpha$	$\binom{L}{d+1} \times \alpha \times |K|$	$\approx |K|$
Scenario (b)	$\binom{L}{d+1} \times \alpha + |K| \times \beta \times \gamma$ $\approx \binom{L}{d+1} \times \alpha$	$\binom{L}{d+1} \times \beta \times |K|$	$\approx |K| \times \beta/\alpha$

can be easily guessed once the tuple related to the first subkey has been found. But it may also be that recovering the other subkeys requires basically the same computation as recovering the first subkey. In either case, the improvement factor is essentially $|K|$. In the latter case, this improvement applies to recovering *each* subkey, which is not obvious since we express the improvement as a factor. Further, we note that the improvement factor is independent of the masking order d and the window size L. However, in absolute terms the running time improvement increases substantially with increasing attack order $d + 1$, L and the number of traces. Finally, we point out that the analysis holds independently of the method used to estimate the mutual information, as long as we assume that all involved estimations of mutual information use the same method.

Number of Traces Needed. It is not straightforward to make a precise but general statement about the number of traces needed for our method to successfully locate interesting tuples. Many factors play a role. We make a brief assessment and describe two of the effects that have to be considered.

First, we consider an idealized scenario where steps 2 and 3 succeed as soon as the same precision for the estimations is achieved. In this case, in scenario (b) (Hamming weight leakage model), step 2 may require more traces to pinpoint the interesting $d+1$ tuples of time samples than step 3 to recover the key. This is due to the fact that, in the attack step, the estimation of $\mathbf{I}(\mathbf{L}(t_\mathbf{V}); \mathbf{L}(t_\mathbf{M}); \mathrm{HW}(\mathbf{Z}))$ with $\mathbf{Z} = \mathtt{S\text{-}Box}(\mathbf{P} \oplus k)$ for a hypothesized k requires generally less traces than an equally precise estimation of $\mathbf{I}(\mathbf{L}(t_\mathbf{V}); \mathbf{L}(t_\mathbf{M}); \mathbf{P})$ in the tuple selection step. This is because of the different number of classes for $\mathrm{HW}(\mathbf{Z})$ and for $\mathbf{P}$. In the case of AES, there are 256 different possible values for $\mathbf{P}$, while there are only nine different possible values for $\mathrm{HW}(\mathbf{Z})$. Nevertheless, since step 2 requires a larger number of traces, these traces must be obtained and may be used in step 3. A "classical" $d+1$-variate MIA attack requires the same (smaller) number of traces as step 3.

In scenario (a) ((possibly truncated) identity leakage model) the previous effect is typically less pronounced and thus the difference in the number of traces required in each step is smaller. The same holds for the difference in the number of traces needed for step 2 and a "classical" $d + 1$-variate MIA attack.

Second, the precision of the mutual information estimates required in step 3 to distinguish the correct key hypothesis from incorrect ones may not be the same as the precision required to distinguish an interesting tuple from a non-interesting one in step 2. The relation between these precisions can be almost arbitrary. However, it should typically hold that the precision required by an attack against the S-box output in step 3 is not higher than the precision required in step 2.

Summarizing, in scenario (a) the proposed workflow offers a running time improvement factor in the order of magnitude of $|K|$, possibly at the cost of an increased number of traces. In scenario (b) the proposed workflow requires more traces than a "classical" attack but still offers an interesting running time improvement factor. It offers a trade off. Whether the trade off is attractive depends on the ratio β/α in the running time improvement factor, and on how many more traces are required.

4.2   On the S-Box

The fact that the method can distinguish all $d + 1$ tuples corresponding to all shares of any (sensitive) variable of the form $\mathbf{Z} = F_k(\mathbf{P})$ can be used to traverse bijective S-boxes without making any hypothesis on the subkey. This is because the S-box input is a keyed permutation of the plaintext, and the S-box output is a permutation of the S-box input. Both permutations are transparent to mutual information when using the identity function $\tilde{\mathbf{L}}(\mathbf{P}) = \mathbf{P}$.

It is less obvious, nevertheless true, that the method also works in the case of non-injective S-boxes, as for instance in DES. The reasoning is similar to the above. The S-box input is a keyed permutation of the plaintext. The S-box output is not a permutation of the S-box input, but a non-injective function of it. Therefore, if we use the identity function and condition on the plaintext, the S-box can be traversed just like a bijective S-box, and interesting tuples of time samples after the S-box can be identified. Note that a non-injective S-box cannot be traversed from output to input in the same way.

4.3   Additional Applications

The method described in this paper is fully generic and does not place any restrictive assumption on the specific targeted implementation. However, the method benefits from the available specificities of the implementation. For example, an adversary could mount the following strategy if he knows that the device's leakage behavior is close to the Hamming weight model. Using the Hamming weight model, the adversary first locates tuples corresponding to the S-box input to narrow down the time window. Then, using the identity function, he searches in that window for the S-box output.

The adversary could also locate tuples corresponding to the S-box input of the next S-box lookup to further narrow down the time window.

Bit-tracing [9] is a technique used to track the time instants when a predictable variable is handled in the execution flow of an unknown implementation. This is a useful technique to reverse-engineer unknown implementations. The ideas in Section 3.1 can be exploited to track masked variables during the execution of an algorithm. Note that the fact that the proposed method can traverse S-boxes (by the arguments given in Sect. 4.2) can also lead to a significant speed-up in the bit-tracing process of masked implementations.

4.4 Estimation of Mutual Information

We note that any suitable method for estimating the mutual information or the required probability distributions, e.g. histograms [6], kernel density estimation [17], B-splines [21], statistical moments [13], parametric methods [17], and any similar metric, e.g. Kullback-Leibler divergence [22], Kolmogorov-Smirnov test [22], Cramér-von-Mises test [22], can be used.

Available knowledge about the device leakage behavior, e.g. close to the Hamming weight model, can be used to speed up the estimations. Here we do not refer to the choice of $\tilde{L}$ but to the leakage variables.

4.5 Key Recovery Step

By construction, our method identifies tuples of time samples that correspond to all shares of a masked (sensitive) variable. It does so irrespective of the particular dependencies between each share and its side-channel leakage. Therefore, a generic multivariate MIA attack with (possibly truncated) identity leakage model appears to be most suited to exploit the unknown dependencies, in general. However, if standard assumptions approximate the leakage behavior good enough or the specific leakage behavior is known, the identified tuples can be exploited more efficiently with adapted multivariate MIA or higher-order DPA attacks.

The proposed method can identify interesting tuples that relate to a specific plaintext byte, but it cannot *per se* focus on interesting tuples that correspond to a *specific* function of that plaintext byte. As a consequence, the method will in general not discriminate between interesting tuples that correspond to all shares of the masked plaintext, the masked S-box input or the masked S-box output. Clearly, the latter is preferable for an attack. In our experiments we noted that enough interesting tuples corresponding to all shares of the masked S-box output appeared at the top of the ranked list.

5 Experiments

In this section we present experimental results of our method, insight on its computation and a performance evaluation. We note that all "numbers of traces" reported in this section cannot be generalized to other platforms and implementations.

5.1 Measurements

We use an 8-bit microcontroller of Atmel's AVR family in a smart card plastic body as platform for our experiments. The microcontroller runs a first-order Boolean masked implementation of AES-128 encryption that follows the lines of [8]. This concrete implementation uses six independent mask bytes for one encryption. Before the SubBytes operation, all state bytes are protected by the same mask $\mathbf{M}_0$. After the SubBytes operation, all state bytes are protected by the same mask $\mathbf{M}_1$. Before MixColumns, each column of the state is remasked with $\mathbf{M}_2, \ldots, \mathbf{M}_5$. After MixColumns, each column of the state is masked with $\mathbf{M}'_2, \ldots, \mathbf{M}'_5$ that depend on $\mathbf{M}_2, \ldots, \mathbf{M}_5$. Shiftrows does not affect the masking and after the next AddRoundKey operation, all state bytes are again protected by $\mathbf{M}_0$ due to the masked key schedule. Note that the six masks are re-used to protect all rounds. There are no additional countermeasures.

We obtained 50 000 power traces from encryptions of randomly chosen plaintexts with a fixed key and random masks. The card was clocked at 4MHz and we used a sampling frequency of 200MS/s.

5.2   Selection of a Time Window: Step 1

To reduce the computational burden, we restricted the measurements to cover only the first 1.5 rounds of the encryption. This was done based on an educated guess on the SPA features present in the power traces. Then, we compressed the traces by integration to one point per clock cycle. As a result, each compressed trace comprises 800 points. The subsequent analyses were carried out on these compressed traces.

5.3   Computation of the Method: Step 2

To show the full potential of the method, we chose $\tilde{\mathbf{L}}$ to be the identity function. In what follows, $\mathbf{P}$ refers to the third plaintext byte, an arbitrary choice. We estimate densities with histograms (using nine bins for each dimension unless otherwise stated, because we expect a leakage behavior close to the Hamming weight/distance model) and we use ˆ to indicate estimates, e.g. $\hat{\mathbf{I}}$ is an estimate of $\mathbf{I}$. The computation of step 2 is split into two terms:

$$\hat{\mathbf{I}}(\mathbf{L}(t_0); \mathbf{L}(t_1); \mathbf{P}) = \hat{\mathbf{I}}(\mathbf{L}(t_0); \mathbf{L}(t_1)) - \hat{\mathbf{I}}(\mathbf{L}(t_0); \mathbf{L}(t_1)|\mathbf{P}) . \tag{8}$$

In our experiments, we noted that a straightforward computation of this expression can result in inconvenient estimation errors. The reason lies in the different number of traces used to estimate each term on the right side of Eq. (8). $\hat{\mathbf{I}}(\mathbf{L}(t_0); \mathbf{L}(t_1))$ is computed with all available traces, say T. The second term is computed as

$$\hat{\mathbf{I}}(\mathbf{L}(t_0); \mathbf{L}(t_1)|\mathbf{P}) = \sum_{p=0}^{255} \hat{\Pr}(\mathbf{P} = p)\hat{\mathbf{I}}(\mathbf{L}(t_0); \mathbf{L}(t_1)|\mathbf{P} = p) \tag{9}$$

and for the computation of each summand about $T/256$ traces are used. This difference in the number of traces translates into different estimation accuracies for each term in Eq. (8), burying the small relevant difference between them due to the effect of $\mathbf{P}$ in the larger difference due to the different estimation accuracies.

To amend this, since we have that $\mathbf{I}(\mathbf{L}(t_0); \mathbf{L}(t_1)) = \mathbf{I}(\mathbf{L}(t_0); \mathbf{L}(t_1)|\mathbf{D})$ for a uniformly distributed *dummy* random variable $\mathbf{D}$ that is independent of the leakages and taking values in $\{0, \ldots, 255\}$, we can compute $\hat{\mathbf{I}}(\mathbf{L}(t_0); \mathbf{L}(t_1))$ in a way that resembles Eq. (9) and approximate it by

$$\hat{\mathbf{I}}(\mathbf{L}(t_0); \mathbf{L}(t_1)|\mathbf{D}) = \sum_{d=0}^{255} \hat{\Pr}(\mathbf{D} = d)\hat{\mathbf{I}}(\mathbf{L}(t_0); \mathbf{L}(t_1)|\mathbf{D} = d). \qquad (10)$$

This leads to equally (in-)accurate estimates for both terms in Eq. (8) and the difference between them is mostly due to the effect of $\mathbf{P}$.

To illustrate the effectiveness of step 2 we compute $\hat{\mathbf{I}}(\mathbf{L}(t_0); \mathbf{L}(t_1)|\mathbf{D})$ and $\hat{\mathbf{I}}(\mathbf{L}(t_0); \mathbf{L}(t_1)|\mathbf{P})$ from 50 000 measurements using the same bin distributions for both terms. We use this relatively large number of traces to present aesthetically pleasant figures. Far less traces are sufficient for the method to work.

Figure 1 (left) shows a plot of the values of the first term of Eq. (8), i.e. $\hat{\mathbf{I}}(\mathbf{L}(t_0); \mathbf{L}(t_1))$ computed as Eq. (10), for $t_0, t_1 \in \{1, \ldots, 800\}$ and $t_0 \neq t_1$. It is obviously sufficient to compute the values only for $t_0 < t_1$ or $t_0 > t_1$. The x- and y-axes both denote time. We plot a mean trace next to each of them for orientation. The values of mutual information are represented by different colors according to the color bar on the left side. We blank out most pairs of time samples, those that yield small values of mutual information, by plotting them in white. All pairs that yield mutual information values above a certain threshold are plotted in black.

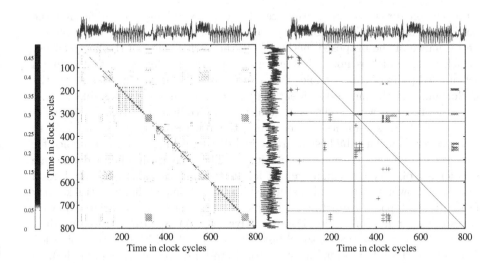

Fig. 1. Left: Matrix of $\hat{\mathbf{I}}(\mathbf{L}(t_0); \mathbf{L}(t_1))$ values. The color bar is in units of bits. A mean trace is plotted next to the axes. Right: Above diagonal, 'x': 100 pairs of time samples where a multivariate MIA attack succeeds. Below diagonal, '+': 100 top ranked pairs in the list of step 2

We can see that the locations of the pairs have a clear structure and could possibly aid reverse-engineering of the implementation. Since we know the implementation, we can easily relate parts of the figure to operations: AddRoundKey (approx. index 100 to 150), SubBytes (approx. index 200 to 300), remasking (approx. 300 to 350), four parts of MixColumn (approx. index 350 to 500), AddRoundKey (approx. index 550 to 600), followed by SubBytes and remasking in round two. These pairs are, however, not yet interesting pairs because it is not clear if they can be exploited by an attack (see the discussion of Eqs. (3) and (4)).

Next, we rank the list of pairs according to the result of Eq. (8). The 100 top ranked pairs in the list, i.e. negative mutual information and large absolute value, are depicted in the lower triangle of Figure 1 (right) with '+' symbols.

For the sake of comparison, we include in the upper triangle of the figure the 100 pairs of time samples where a multivariate MIA attack on the third key byte (using the Hamming weight leakage model on predicted S-box output values and 50 000 traces) achieves the largest nearest-rival distinguishing score [24], marked with 'x' symbols.

The partial match between the upper and the lower triangular matrix serves as a first visual evidence for the effectiveness of the method. In particular, the method is able to identify pairs corresponding to both shares of the S-box output of a specific state byte (here the third) without making any hypothesis about the key.

5.4   Performance Evaluation of Step 2

This section details the performance of the proposed method in finding the pairs that can be exploited for key recovery. Informally, we aim to decouple the performance of the proposed method from the performance of the key recovery attack itself, which is not the focus of this paper. To do so, we first define a set of *good* pairs of time samples that can be attacked and then we analyze the performance of the method in identifying *good* pairs among all possible pairs.

More precisely, we define sets of good pairs by running an attack on all pairs using 50 000 measurements and retaining the 100 resp. 290 pairs that lead to key recovery and have highest nearest-rival distinguishing score. Our choice for the size of the sets is somewhat arbitrary. The idea is simply to define one smaller set of very good pairs and a larger set that contains additional good pairs with lower nearest-rival distinguishing score. Since different attacks may favor different pairs, we define such sets for three cases: multivariate MIA on the S-box output, Correlation Power Analysis (CPA) [3] with centered product combination function [18] on the S-box output and CPA with same combination function on the S-box input (all using the Hamming weight leakage model). In total, we hence define six sets of good pairs.

Once the sets of good pairs are defined, we run step 2 parametrized by the number of traces. For each number of traces, we repeat the run of step 2 on 100 randomly chosen sets of traces and, each time, keep the position of the best ranked good pair in the list generated by step 2. In other words, we test the pairs in descending order of their ranking (rank 1 is best) and stop as soon as a pair is good. This ranking position is the minimum size of the list from step 2 required for step 3 to succeed in that particular run for a given attack technique. Recall that, by definition, an attack on a good pair succeeds with a comfortable nearest-rival distinguishing score (albeit the absolute margin for a CPA attack on the S-box input is a lot smaller). We hence evaluate only the performance of step 2.

The distributions of the ranks of the best ranked good pairs are shown as boxplots in Fig. 2 (sets of 100 good pairs) and in Fig. 3 (sets of 290 good pairs). For both figures, the used attack techniques are, from left to right: MIA S-box output, CPA S-box output and CPA S-box input.

In the boxplots, the central mark is the median (2nd quartile) and the box edges (solid) represent the 1st and the 3rd quartile. The whiskers (dashed) extend to $q_3 + 1.5(q_3 - q_1)$ and $q_1 - 1.5(q_3 - q_1)$, where q_1 and q_3 are the 1st and 3rd quartiles, respectively. Outliers are marked with '+' symbols.

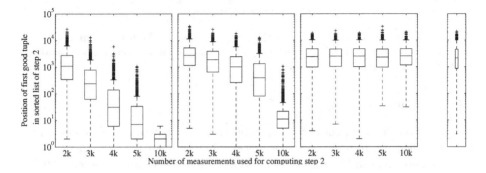

Fig. 2. Distribution of the ranking of the first *good* pair in the list of step 2. Left to right: MIA S-box output, CPA S-box output, CPA S-box input, hypothetical random method. 100 good pairs

For comparison, the rightmost boxplot in each figure shows the distribution that a hypothetical method that ranks the pairs at random, instead of step 2, would produce. These distributions are independent of an attack technique and only relate to the number of good pairs among all pairs, here 100 resp. 290 out ouf $800 \times 799/2 = 319\,600$.

One can observe that the proposed method begins to identify good pairs (i.e. to perform better than a random guess) that are exploitable by multivariate MIA or CPA attacks on the S-box output when 3 000 traces or more are available. As the number of traces increases, the medians of the distributions become smaller, i.e. good pairs move steadily toward the top of the list.

One can also observe that our method ranks good pairs for multivariate MIA slightly higher than good pairs for CPA on the S-box output. On the other hand, the method is not able to identify good pairs for a CPA attack on the S-box input better than a random guess. We note that both behaviors are not a property of our method but probably related to our test platform and the implementation.

In the case of larger lists of 290 good pairs, the previously made observations mostly hold. As expected, the medians of the distributions are smaller than in the case of 100 good pairs, simply because even a random guess becomes more likely to succeed. In addition, we can observe that the method now ranks good pairs for multivariate MIA and CPA on the S-box output almost equally well.

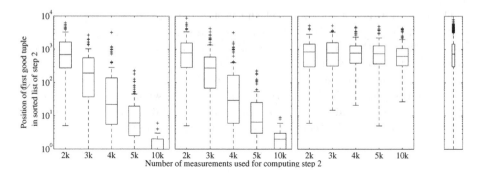

Fig. 3. Distribution of the ranking of the first *good* pair in the list of step 2. Left to right: MIA S-box output, CPA S-box output, CPA S-box input, hypothetical random method. 290 good pairs

5.5   Practical Attacks

The above results highlight important properties of our method and demonstrate that it is effective. In practice, one is however less interested in the exact rank of the first good pair in the sorted list, and more interested in the success rate of an attack end-to-end. This clearly involves the performance of our method *and* the efficiency of the attack used in step 3.

Table 2 shows success rates for steps 2 and 3 together. First we use a given number of randomly chosen traces to compute step 2. Then we attack the $\gamma = 10$ resp. 100 best ranked pairs with multivariate MIA, CPA on the S-box output and CPA on the S-box input (as described before) in step 3, using the same traces. We repeat this procedure 100 times. For the numbers in the first row of the table, we considered an attack successful if the correct key leads to the smallest correlation (or mutual information) value (negative sign and highest absolute value), over all evaluated γ pairs. For the numbers in the second row of the table, we additionally required the correct key to stand out at least by a factor of 1.5 compared to the nearest rival (left) and by a factor of at least 2 (right).

A first observation is that a CPA attack on the S-box input does not work in our concrete scenarios. CPA attacks on the S-box output converge slightly faster toward 100% success rate than multivariate MIA attacks on the S-box output. We can further see that, given enough traces, both attacks in step 3 eventually reach 100% success, even if we attack only the top ten pairs of step 2 and require the correct key to stand out by a factor of at least 1.5. These results confirm that the combination of steps 2 and 3 works in practice, and that step 2 is able

Table 2. Success rates for steps 2 and 3 together, for several parameters: number of traces, size γ of the list of step 2, key recovery attack

Number of traces		2k	3k	4k	5k	10k		2k	3k	4k	5k	10k
MIA S-box output	$\gamma = 100$	3	15	59	83	100	$\gamma = 10$	0	3	34	53	100
CPA S-box output		11	41	75	95	100		1	11	48	66	100
CPA S-box input		1	0	0	1	0		2	0	0	1	0
MIA S-box output	$\gamma = 10$	0	2	15	35	100	$\gamma = 10$	0	0	7	17	90
CPA S-box output	factor 1.5	0	3	28	52	98	factor 2	0	0	10	17	78
CPA S-box input		0	0	0	0	0		0	0	0	0	0

to identify exploitable pairs of time samples. Interestingly, one can further see that multivariate MIA attacks on the S-box output have a small advantage over CPA attacks on the S-box output, if we require the correct key to stand out by a factor of at least 2.

5.6   Computational Efficiency

In Tab. 3 we present empirical execution times for our implementations of the proposed workflow (steps 2 and 3) and the strategy that uses exhaustive search instead of step 2. Step 3 of the proposed workflow was performed with multivariate MIA on the S-box output (using the Hamming weight leakage model and list size $\gamma = 100$). For the exhaustive search strategy we evaluated two variants: multivariate MIA on the S-box output (using the Hamming weight leakage model) and CPA on the S-box output (with centered product preprocessing). All implementations were executed on the same processor on a single core. We note that the absolute execution times are heavily implementation-dependent and thus relative speed-ups are more interesting, since they are less tied to the particular implementation used.

Table 3. Empirical execution times for steps 2 and 3 ($\gamma = 100$) of the proposed workflow and several attacks using exhaustive search

Number of traces	step 2 + step 3	Exhaustive search		Improvement factor
5 000	2m30s + 2s	MIA-HW	1h48m	43
		CPA	2h 48m	68
50 000	10m24s + 19s	MIA-HW	17h18m	97
		CPA	23h32m	132

A first observation regarding Tab. 3 is the speed-up achieved by the proposed workflow, compared to exhaustive search, when multivariate MIA is used for key recovery. This is a directly interpretable result that corresponds to scenario (b)

in Sect. 4.1. The improvement factor in this case is of 43 when 5 000 and 97 when 50 000 traces are used, respectively. We observe that, for our implementations, the factor β/α depends on the number of traces.

One can further see that, for our implementations, applying the proposed workflow is even advantageous if exhaustive search is done with CPA. It achieves an improvement factor of 68 in the running time of the attack when 5 000 traces are used, and an improvement factor of 132 when 50 000 traces are used. However, we stress that this result is not universally valid. The speed-ups are heavily affected by the relative efficiency of our implementations of linear correlation and mutual information estimation.

As a final observation concerning Tab. 3, we remark the validity of the approximation we made in Tab. 1: the running time of the proposed workflow is dominated by step 2. Step 3 contributes at most 3% to the total running time if the list size is $\gamma = 100$.

6   Conclusion

Multivariate DPA attacks can suffer from the effect known as "combinatorial explosion" and hence combinatorial time complexity in the number of time samples that have to be exploited jointly. We presented a novel technique to identify such interesting tuples of time samples before key recovery. Compared to previous work on this topic, our method is not heuristic but systematic and works in black-box conditions using only known inputs (or outputs). Our technique can lead to a substantial improvement in the computational efficiency of multivariate attacks compared to exhaustive search over the same window of time samples, since it retains only a small fraction of all possible tuples for key recovery. Our approach is based on mutual information and is fully generic, i.e. it does not require more restrictive assumptions than a generic MIA attack. Experimental results based on power traces of a masked software implementation of the AES confirm the effectiveness of the technique, highlight some of its interesting properties and attest attractive running time improvements. An aspect that is not fully explored in this paper and left for future work is a thorough analysis of the number of traces needed for the technique to work.

Acknowledgments. We thank the anonymous reviewers for their thorough evaluation and insightful comments.

This work was supported in part by the Research Council of KU Leuven: GOA TENSE (GOA/11/007), by the IAP Programme P6/26 BCRYPT of the Belgian State (Belgian Science Policy), by the European Commission through the ICT programme under contract ICT-2007-216676 ECRYPT II, by the Flemish Government FWO G.0550.12N and by the Hercules Foundation AKUL/11/19. Benedikt Gierlichs is Postdoctoral Fellow of the Fund for Scientific Research - Flanders (FWO).

References

1. Agrawal, D., Archambeault, B., Rao, J.R., Rohatgi, P.: The EM Side-Channel(s). In: Kaliski Jr., B.S., Koç, Ç.K., Paar, C. (eds.) CHES 2002. LNCS, vol. 2523, pp. 29–45. Springer, Heidelberg (2003)
2. Batina, L., Gierlichs, B., Prouff, E., Rivain, M., Standaert, F.-X., Veyrat-Charvillon, N.: Mutual Information Analysis: A Comprehensive Study. Journal of Cryptology 24(2), 269–291 (2011)
3. Brier, E., Clavier, C., Olivier, F.: Correlation Power Analysis with a Leakage Model. In: Joye, M., Quisquater, J.-J. (eds.) CHES 2004. LNCS, vol. 3156, pp. 16–29. Springer, Heidelberg (2004)
4. Chari, S., Jutla, C.S., Rao, J.R., Rohatgi, P.: Towards Sound Approaches to Counteract Power-Analysis Attacks. In: Wiener, M. (ed.) CRYPTO 1999. LNCS, vol. 1666, pp. 398–412. Springer, Heidelberg (1999)
5. Gierlichs, B., Batina, L., Preneel, B., Verbauwhede, I.: Revisiting Higher-Order DPA Attacks: Multivariate Mutual Information Analysis. In: Pieprzyk, J. (ed.) CT-RSA 2010. LNCS, vol. 5985, pp. 221–234. Springer, Heidelberg (2010)
6. Gierlichs, B., Batina, L., Tuyls, P., Preneel, B.: Mutual Information Analysis. In: Oswald, E., Rohatgi, P. (eds.) CHES 2008. LNCS, vol. 5154, pp. 426–442. Springer, Heidelberg (2008)
7. Goubin, L., Patarin, J.: DES and Differential Power Analysis (The "Duplication" Method). In: Koç, Ç.K., Paar, C. (eds.) CHES 1999. LNCS, vol. 1717, pp. 158–172. Springer, Heidelberg (1999)
8. Herbst, C., Oswald, E., Mangard, S.: An AES Smart Card Implementation Resistant to Power Analysis Attacks. In: Zhou, J., Yung, M., Bao, F. (eds.) ACNS 2006. LNCS, vol. 3989, pp. 239–252. Springer, Heidelberg (2006)
9. Joye, M., Olivier, F.: Side-Channel Analysis. In: Encyclopedia of Cryptography and Security, 2nd edn., pp. 1198–1204 (2011)
10. Joye, M., Paillier, P., Schoenmakers, B.: On Second-Order Differential Power Analysis. In: Rao, J.R., Sunar, B. (eds.) CHES 2005. LNCS, vol. 3659, pp. 293–308. Springer, Heidelberg (2005)
11. Kocher, P.C.: Timing Attacks on Implementations of Diffie-Hellman, RSA, DSS, and Other Systems. In: Koblitz, N. (ed.) CRYPTO 1996. LNCS, vol. 1109, pp. 104–113. Springer, Heidelberg (1996)
12. Kocher, P.C., Jaffe, J., Jun, B.: Differential Power Analysis. In: Wiener, M. (ed.) CRYPTO 1999. LNCS, vol. 1666, pp. 388–397. Springer, Heidelberg (1999)
13. Le, T.-H., Berthier, M.: Mutual Information Analysis under the View of Higher-Order Statistics. In: Echizen, I., Kunihiro, N., Sasaki, R. (eds.) IWSEC 2010. LNCS, vol. 6434, pp. 285–300. Springer, Heidelberg (2010)
14. Lemke-Rust, K., Paar, C.: Gaussian Mixture Models for Higher-Order Side Channel Analysis. In: Paillier, P., Verbauwhede, I. (eds.) CHES 2007. LNCS, vol. 4727, pp. 14–27. Springer, Heidelberg (2007)
15. Messerges, T.S.: Using Second-Order Power Analysis to Attack DPA Resistant Software. In: Koç, Ç.K., Paar, C. (eds.) CHES 2000. LNCS, vol. 1965, pp. 238–251. Springer, Heidelberg (2000)
16. Oswald, E., Mangard, S., Herbst, C., Tillich, S.: Practical Second-Order DPA Attacks for Masked Smart Card Implementations of Block Ciphers. In: Pointcheval, D. (ed.) CT-RSA 2006. LNCS, vol. 3860, pp. 192–207. Springer, Heidelberg (2006)

17. Prouff, E., Rivain, M.: Theoretical and Practical Aspects of Mutual Information Based Side Channel Analysis. In: Abdalla, M., Pointcheval, D., Fouque, P.-A., Vergnaud, D. (eds.) ACNS 2009. LNCS, vol. 5536, pp. 499–518. Springer, Heidelberg (2009)
18. Prouff, E., Rivain, M., Bevan, R.: Statistical Analysis of Second Order Differential Power Analysis. IEEE Trans. Computers 58(6), 799–811 (2009)
19. Schramm, K., Paar, C.: Higher Order Masking of the AES. In: Pointcheval, D. (ed.) CT-RSA 2006. LNCS, vol. 3860, pp. 208–225. Springer, Heidelberg (2006)
20. Standaert, F.-X., Veyrat-Charvillon, N., Oswald, E., Gierlichs, B., Medwed, M., Kasper, M., Mangard, S.: The World Is Not Enough: Another Look on Second-Order DPA. In: Abe, M. (ed.) ASIACRYPT 2010. LNCS, vol. 6477, pp. 112–129. Springer, Heidelberg (2010)
21. Venelli, A.: Efficient Entropy Estimation for Mutual Information Analysis Using B-Splines. In: Samarati, P., Tunstall, M., Posegga, J., Markantonakis, K., Sauveron, D. (eds.) WISTP 2010. LNCS, vol. 6033, pp. 17–30. Springer, Heidelberg (2010)
22. Veyrat-Charvillon, N., Standaert, F.-X.: Mutual Information Analysis: How, When and Why? In: Clavier, C., Gaj, K. (eds.) CHES 2009. LNCS, vol. 5747, pp. 429–443. Springer, Heidelberg (2009)
23. Waddle, J., Wagner, D.: Towards Efficient Second-Order Power Analysis. In: Joye, M., Quisquater, J.-J. (eds.) CHES 2004. LNCS, vol. 3156, pp. 1–15. Springer, Heidelberg (2004)
24. Whitnall, C., Oswald, E.: A Comprehensive Evaluation of Mutual Information Analysis Using a Fair Evaluation Framework. In: Rogaway, P. (ed.) CRYPTO 2011. LNCS, vol. 6841, pp. 316–334. Springer, Heidelberg (2011)

Unified and Optimized Linear Collision Attacks and Their Application in a Non-profiled Setting

Benoît Gérard* and François-Xavier Standaert**

UCL Crypto Group, Université catholique de Louvain
Place du Levant 3, B-1348, Louvain-la-Neuve, Belgium

Abstract. Side-channel collision attacks are one of the most investigated techniques allowing the combination of mathematical and physical cryptanalysis. In this paper, we discuss their relevance in the security evaluation of leaking devices with two main contributions. On the one hand, we suggest that the exploitation of linear collisions in block ciphers can be naturally re-written as a Low Density Parity Check Code decoding problem. By combining this re-writing with a Bayesian extension of the collision detection techniques, we succeed in improving the efficiency and error tolerance of previously introduced attacks. On the other hand, we provide various experiments in order to discuss the practicality of such attacks compared to standard DPA. Our results exhibit that collision attacks are less efficient in classical implementation contexts, e.g. 8-bit microcontrollers leaking according to a linear power consumption model. We also observe that the detection of collisions in software devices may be difficult in the case of optimized implementations, because of less regular assembly codes. Interestingly, the soft decoding approach is particularly useful in these more challenging scenarios. Finally, we show that there exist (theoretical) contexts in which collision attacks succeed in exploiting leakages whereas all other non-profiled side-channel attacks fail.

1 Introduction

Most side-channel attacks published in the literature and used to evaluate leaking cryptographic devices are based on a divide-and-conquer strategy. Kocher et al.'s Differential Power Analysis (DPA) [10], Brier et al.'s Correlation Power Analysis (CPA) [5] and Chari et al.'s Template Attacks (TA) [6] are notorious examples. However, alternatives to these standard approaches have also been investigated, e.g. by trying to combine side-channel information with classical cryptanalysis. The collision attacks introduced by Schramm et al. at FSE 2003 are among the most investigated solutions for this purpose [19]. While initially dedicated to the DES, they have then been applied to the AES [18] and improved in different directions over the last years, as witnessed by the recent works of Ledig et al. [11], Bogdanov [2,3,4], Moradi et al. [13,14] and Clavier et al. [7].

---

* Postdoctoral researcher supported by Walloon region MIPSs project.
** Associate researcher of the Belgian Fund for Scientific Research (FNRS-F.R.S.). This work has been funded in part by the ERC project 280141 (acronym CRASH).

E. Prouff and P. Schaumont (Eds.): CHES 2012, LNCS 7428, pp. 175–192, 2012.

From an application point of view, collision attacks differ from standard side-channel attacks by their underlying assumptions. Informally, divide-and-conquer distinguishers essentially assume that a cryptographic device leaks information that depends on its intermediate computations, under a given leakage model. The leakage model is generally obtained either from engineering intuition, in the case of non-profiled attacks such as DPA and CPA, or through a preliminary estimation of the chip measurements probability distribution, in the case of profiled attacks such as TA. By contrast, collision attacks do not require a precise knowledge of the leakage distribution. They rather trade this need for a combination of two other assumptions: (*i*) the distribution of a couple of measurements corresponding to the intermediate computation of identical values can be distinguished from the one corresponding to different values; (*ii*) the adversary is able to divide each measurement trace corresponding to the encryption of a plaintext into sub-traces corresponding to elementary operations, e.g. the execution of block cipher S-boxes. In other words, collision attacks trade the need of precise leakage models for the need to detect identical intermediate computations, together with a sufficient knowledge of the operations scheduling in the target device. Interestingly, the knowledge of precise leakage models has recently been shown to be problematic in non-profiled attacks [22], e.g. in the case of devices with strongly non-linear leakage functions. Hence, although the existence of such devices remains an open question [16], they at least create a theoretical motivation for understanding the strengths and weaknesses of collision attacks.

This paper brings two main contributions related to this state-of-the-art.

First, we observe that many previous collision attacks do not efficiently deal with errors (i.e. when the correct value of a key-dependent variable is not the likeliest indicated by the leakages), and rely on add-hoc solutions for this purpose. In order to handle erroneous situations more systematically, we introduce two new technical ingredients. On the one hand, we propose to re-write side-channel collision attacks as a Low Density Parity Check (LDPC) decoding problem. On the other hand, we describe a (non-profiled) Bayesian extension of collision detection techniques. We show that these tools are generic and allow successful key recoveries with less measurement data than previous ones, by specializing them to two exemplary attacks introduced by Bogdanov [2,3] and Moradi et al. [13].

Second, we question the relevance of side-channel collision attacks and their underlying assumptions, based on experimental case studies. For this purpose, we start by showing practical evidence that in "simple" scenarios, the efficiency of these attacks is lower than the one of more standard attacks, e.g. the non-profiled extension of Schindler's stochastic approach [17], described in [8]. We then observe that in actual software implementations, the detection of collisions can be difficult due to code optimizations. As a typical example, we observe that the leakage behavior of different AES S-boxes in an Atmel microcontroller may be different, which prevents the detection of a collision with high confidence for these S-boxes. We conclude by exhibiting an (hypothetical) scenario were the leakage function is highly non-linear (i.e. in the pathological example from [22]), collision attacks lead to successful key recoveries whereas all non-profiled attacks fail.

2   Background

2.1   Notations

In order to simplify the understanding of the paper, we will suppose that the targeted block cipher is the AES Rijndael. Hence, the number of S-boxes considered is 16, and these S-boxes manipulate bytes. Nevertheless, all the following statements can be adapted to another key alternating cipher, by substituting the correct size and number of S-boxes. In this context, the first-round subkey and plaintexts are all 16-byte states. We respectively use letters k and x for the key and a plaintext, and use subscripts to point to a particular byte:

$$x \stackrel{\text{def}}{=} (x_1, x_2, \ldots, x_{16}) \quad , \quad k \stackrel{\text{def}}{=} (k_1, k_2, \ldots, k_{16}).$$

Next, the attackers we will consider have access to a certain number of side-channel traces, corresponding to the encryption of different plaintexts encrypted using the same key k. We denote with n_t the number of different inputs encrypted, and with $x^{(1)}, \ldots, x^{(n_t)}$ the corresponding plaintexts. Each trace obtained is composed of 16 sub-traces corresponding to the 16 S-box computations $t^{(i)} \stackrel{\text{def}}{=} (t_1^{(i)}, \ldots t_{16}^{(i)})$. Each sub-trace is again composed of a number ℓ of points (or samples). Hence, the sub-trace corresponding to the a-th S-box will be denoted as $t_a^{(i)} \stackrel{\text{def}}{=} (t_{a,1}^{(i)}, \ldots, t_{a,\ell}^{(i)})$. Furthermore, we will use the corresponding capital letters X, K and T to refer to the corresponding random variables.

2.2   Linear Collision Attacks

Linear collision attacks are based on the fact that if an attacker is able to detect a collision between two (first-round) S-box executions, then he obtains information about the key. Indeed, if a collision is detected, e.g. between the computation of S-box a for plaintext $x^{(i_a)}$ and S-box b for plaintext $x^{(i_b)}$, this attacker obtains a linear relation between the two corresponding input bytes:

$$x_a^{(i_a)} \oplus k_a = x_b^{(i_b)} \oplus k_b.$$

This relation allows him to decrease the dimension of the space of possible keys by 8, removing keys for which $k_a \oplus k_b \neq x_a^{(i_a)} \oplus x_b^{(i_b)}$. A linear system can then be built by combining several equations, and solving this system reveals (most of) the key. Naturally, the success of the attack mainly depends on the possibility to detect collisions. Two main approaches have been considered for this purpose.

In the first approach, simple statistics such as the Euclidean distance [18] or Pearson's correlation coefficient [19], are used as detection metrics. In this case, the detection of a collision can be viewed as a binary hypothesis test. It implies to define an acceptance region (i.e. a threshold on the corresponding statistic). As a result, a collision may not be detected and a false collision may be considered as a collision. This second point is the most difficult to overcome, as a false-collision implies adding a false equation in the system, which in turn implies the attack failure. Heuristic solutions based on binary and ternary vote have then been proposed in [3] to mitigate this issue. In binary vote, the idea is to observe the same supposed collision using many traces, and to take a hard decision by comparing the number of times the collision detection procedure returns `true` with some threshold. Ternary vote is based on the fact that if there is a collision between two values, then the output of the collision-detection procedure should be the same when comparing both traces with a third one.

An alternative approach is the correlation-enhanced attack introduced by Moradi et al. [13]. This approach is somehow orthogonal to the first one, since we are not in the context of binary hypothesis testing anymore. Namely, instead of only returning `true` or `false`, a comparison procedure directly returns the score obtained using the chosen statistic (e.g. Pearson's correlation coefficient). Hence, when comparing two sub-traces $t_a^{(i)}$ and $t_b^{(j)}$, we obtain a score that is an increasing function of the likelihood of $K_a \oplus K_b$ being equal to $x_a^{(i)} \oplus x_b^{(j)}$.

Besides, the authors of [13] combined their attack with a pre-processing of the traces, that consists in building "on-the-fly" templates of the form:

$$\bar{t}_a^{(x)} = \frac{\sum_{i, x_a^{(i)} = x} t_a^{(i)}}{\#\{i, x_a^{(i)} = x\}}. \tag{1}$$

Such a pre-processing is typically useful to extract first-order side-channel information (i.e. difference in the mean values of the leakage distributions).

3   General Framework for Linear Collision Attacks

In this section, we propose a general framework for describing the different linear collision attacks that have been proposed in the literature. One important contribution of this framework is to represent these attacks as a decoding problem. In particular, we argue that a natural description of collision attacks is obtained through the theory of LDPC codes, designed by Gallager in 1962 [9].

3.1   Collision Attacks as an LDPC Decoding Problem

We start with the definition of LDPC codes.

Definition 1. *LDPC codes (graph representation). Let $\mathcal{G}$ be a bipartite graph with m left nodes and r right nodes. Let us denote by $\mathcal{G}_E$ the set of edges i.e. $(i, j) \in \mathcal{G}_E$ if and only if the i-th left node and the j-th right node are adjacent. This graph defines a code $\mathcal{C}$ of length m over $\mathbb{F}_q^m$, such that for $w = (w_1, w_2, \ldots, w_m) \in \mathbb{F}_q^m$, we have:*

$$w \in \mathcal{C} \Longleftrightarrow \forall 1 \leq j \leq r, \bigoplus_{i, (i,j) \in \mathcal{G}_E} w_i = 0.$$

This code is said to be an (m, i, j) LDPC code if the maximum degree for a left nodes is i and the maximum degree for a right nodes is j.

In general, left nodes are called message nodes while right nodes are named check nodes, since they correspond to conditions for code membership. This definition can be directly related to our collision attack setting. First observe that a collision between S-boxes a and b provides information on the variable:

$$\Delta K_{a,b} \overset{\text{def}}{=} K_a \oplus K_b.$$

It follows that the vector $\Delta K \overset{\text{def}}{=} (\Delta K_{1,2}, \ldots, \Delta K_{15,16})$ determines a coset of K of size 2^8. Hence, it can be seen as a codeword of an LDPC code of dimension 15 and length 120. This LDPC code corresponding to our problem has a very particular structure: the set of check nodes only contains right nodes of degree equal to 3. These nodes correspond to the linear relationships:

$$\Delta K_{a,b} \oplus \Delta K_{a,c} = \Delta K_{b,c}, \quad \forall\, 1 \le a < b < c \le 16.$$

Therefore, finding the key in a linear collision attack consists in finding the likeliest codeword of the aforementioned LDPC code, and then exhaustively testing the keys derived from this system by setting K_1 to each of its 2^8 possible values. This LDPC formulation for the linear collision attack problem allows the use of a decoding algorithm to recover the likeliest system of equations. In general, it is well known that the performances of such a decoder can be drastically improved when soft information is available. Interestingly, soft information is naturally available in our context, e.g. through the scores obtained for each possible value of a variable $\Delta K_{a,b}$. Nevertheless, these scores do not have a direct probabilistic meaning. This observation suggests that a Bayesian extension of the statistics used for collision detection, where the scores would be replaced by actual probabilities, could be a valuable addition to collision attacks, in order to boost the decoder performances. As will be shown in Section 5, this combination of LDPC decoding and Bayesian statistics can indeed lead to very efficient attacks.

3.2   General Framework

A general description of linear collision attacks is given in Algorithm 1 and holds in five main steps. First, the traces may be prepared with a `PreProcessTraces` procedure. For example, signal processing can be applied to align traces or to remove noise. Instantiations of this procedure proposed in previous attacks [3,13] will be discussed in Section 4.1. Next, the scores $S_{a,b} \overset{\text{def}}{=} (S_{a,b}(\delta))_{\delta \in \mathbb{F}_{256}}$ corresponding to the possible values δ of the variables $\Delta K_{a,b}$ are extracted (with the `ComputeStatistics` procedure). Different techniques have again been proposed for this purpose in the literature. In order to best feed the LDPC decoder, the scores can be turned into distributions for the variables $\Delta K_{a,b}$, thanks to an `ExtractDistributions` procedure. As will be discussed in Section 4.2, this can be obtained by normalization, or by applying a Bayesian extension of the computed statistics. In particular, we will show how meaningful probabilities can be outputted for two previously introduced similarity metrics (in a non-profiled setting). Using these distributions, the `LDPCDecode` procedure then returns a list of the ℓ most likely codewords that correspond to the most likely consistent

---

Algorithm 1. General framework for linear-collision attacks

Input: n_t plaintexts $x^{(1)}, \ldots, x^{(n_t)}$ and the corresponding traces $t^{(1)}, \ldots, t^{(n_t)}$.
Output: The key k used by the targeted device.
$(\bar{t}_1, \ldots, \bar{t}_{16}) \leftarrow \texttt{PreProcessTraces}(x^{(1)}, \ldots, x^{(n_t)}, t^{(1)}, \ldots, t^{(n_t)})$;
foreach $1 \leq a < b \leq 16$ **do**
  $\quad\lfloor\ S_{a,b} \leftarrow \texttt{ComputeStatistics}(\bar{t}_a, \bar{t}_b)$;
$\Pr[\Delta K] \leftarrow \texttt{ExtractDistributions}(S_{1,2}, \ldots, S_{15,16})$;
$\{S_1, \ldots, S_\ell\} \leftarrow \texttt{LDPCDecode}(\Pr[\Delta K])$;
foreach *system* S_i *and key candidate* k *compatible with equations in* S_i **do**
  $\quad$**if** $\texttt{TestKey}(k)$ **then**
  $\qquad\lfloor$ **return** k;

return failure;

---

systems $\{S_1, \ldots, S_\ell\}$ of 120 equations (with S_i more likely than S_{i+1}). Such a decoding algorithm is detailed in Section 4.3 for the case $\ell = 1$. Finally, the 2^8 full keys fulfilling S_1 are tested in the `TestKey` procedure. The correct key is returned if found otherwise keys fulfilling S_2 are tested and so on. If the correct key does not fulfill any of the S_i's, then `failure` is returned.

4   Instantiation of the Framework Procedures

Following the previous general description, we now propose a few exemplary instantiations of its different procedures. Doing so, we show how to integrate previously introduced collision attacks in our framework.

4.1   Pre-processing

Pre-processing the traces is frequently done in side-channel analysis, and collision attacks are no exceptions. For example, Bogdanov's attacks take advantage of averaging (by measuring the power consumption of the same plaintext several times), in order to reduce the measurement noise [2,3,4]. Similarly, Moradi et al. [13] start by building the "on-the-fly" templates defined in Equation (1). This latest strategy shows good results in attacks against unprotected implementations with first-order leakages and our experiments in Section 5 will exploit it[1].

4.2   Information Extraction

The use of an LDPC soft-decoding algorithm requires to extract distributions for the variables $\Delta K_{a,b}$. As mentioned in Section 3.1, such distributions can be obtained heuristically, by normalizing scores obtained with classical detection techniques. But the optimal performances of a soft-decoder are only reached when these distributions correspond to actual probabilities $\Pr\left[\Delta K_{a,b} = \delta \big| S(a,b,\delta)\right]$.

---

[1] By contrast, averaging is detrimental in the case of masked implementations with only second-order leakages, as detailed in [7]. As an alternative, the authors of this paper concatenate sub-traces $t_a^{(i)}$ corresponding to the same plaintext x and form a vector $\bar{t}_a$ by collecting these concatenated sub-traces for different plaintext values.

While such probabilities are easily computed in profiled attacks, obtaining them in a non-profiled setting requires more efforts and some assumptions. In this section, we first introduce general tools that may be applied to any given detection technique for this purpose. For illustration, we then apply them to both the Euclidean distance (ED) and the correlation-enhanced (CE) detection techniques.

Bayesian Extensions: General Principle. The naive approach for extracting distributions from scores $S(a, b, \delta)$, obtained for a candidate $\Delta K_{a,b} = \delta$, is to apply normalization:

$$\text{Norm}(S(a, b, \delta)) \stackrel{\text{def}}{=} \frac{S(a, b, \delta)}{\sum_{\delta'} S(a, b, \delta')}.$$

As already mentioned, such normalized scores are not directly meaningful since they do not correspond to actual probabilities $\Pr\left[\Delta K_{a,b} = \delta \big| S(a, b, \delta)\right]$. Therefore, and as an alternative, we now propose a Bayesian technique for computing scores that corresponds to these probabilities and is denoted as:

$$\text{BayExt}(S(a, b, \delta)) \approx \Pr\left[\Delta K_{a,b} = \delta \big| S(a, b, \delta)\right],$$

where the $\approx$ symbol recalls that the distributions are estimated under certain (practically relevant) assumptions. For this purpose, we introduce the next model.

Model 1. *Let T (resp. T') be the sub-trace corresponding to the execution of an S-box with input X (resp. X'). Let $S(T, T')$ be a statistic extracted from the pair of traces (T, T') (typically the Euclidean distance or a correlation coefficient). Then, there exists two different distributions $\mathcal{D}_c$ and $\mathcal{D}_{nc}$ such that:*

$$\Pr\left[S(T, T') = s\right] = \begin{cases} \Pr_{\mathcal{D}_c}\left[S = s\right] & \text{if } X = X', \\ \Pr_{\mathcal{D}_{nc}}\left[S = s\right] & \text{otherwise.} \end{cases}$$

We note that in theory, the distribution of the statistic in the non-collision case should be a mixture of different distributions, corresponding to each pair of non-colliding values. However, in the context of non-profiled attacks, estimating the parameters of these distributions (mean and variance, typically) for each component of the mixture would require a large amount of measurement traces (more than required to successfully recover the key). Hence, we model this mixture as a global distribution. As will be clear from our experimental results, this heuristic allows us to perform successful attacks with small amounts of measurement traces. Model 1 directly implies that the distribution of $\Delta K_{a,b}$ can be expressed using $\Pr_{\mathcal{D}_c}[\cdot]$ and $\Pr_{\mathcal{D}_{nc}}[\cdot]$, as stated in the following Lemma.

Lemma 1. *Let $\Sigma \stackrel{\text{def}}{=} (s_i, \Delta x_i)_{1 \leq i \leq n}$ be the set of observed statistics s_i and the corresponding suggested value Δx_i for a given XOR of key bytes $\Delta K_{a,b}$. Then:*

$$\Pr\left[\Delta K_{a,b} = \delta \big| \Sigma\right] \propto \prod_{i=1}^{n} \Pr\left[S = s_i \big| \Delta x_i, \Delta K_{a,b} = \delta\right],$$

$$\propto \prod_{i, \Delta x_i = \delta} \Pr_{\mathcal{D}_c}\left[S = s_i\right] \prod_{i, \Delta x_i \neq \delta} \Pr_{\mathcal{D}_{nc}}\left[S = s_i\right],$$

where $\mathrm{Pr}_{\mathcal{D}_c}[S = s_i]$ *(resp.* $\mathrm{Pr}_{\mathcal{D}_{nc}}[S = s_i]$*) denotes the distribution of the statistic* S *when resulting from the comparison between to identical (resp. different) inputs. Moreover, if for any* i, $\mathrm{Pr}_{\mathcal{D}_{nc}}[S = s_i]$ *is non-zero, then:*

$$\mathrm{Pr}\left[\Delta K_{a,b} = \delta \middle| \Sigma\right] \propto \prod_{i, \Delta x_i = \delta} \frac{\mathrm{Pr}_{\mathcal{D}_c}[S = s_i]}{\mathrm{Pr}_{\mathcal{D}_{nc}}[S = s_i]}.$$

Proof. The first line is a direct application of Bayes' relation, the second results from Model 1, and the final formula is obtained dividing by $\prod_i \mathrm{Pr}_{\mathcal{D}_{nc}}[S = s_i]$. ◇

In order to solve our estimation problem, we have no other *a priori* information on $\mathcal{D}_c$ and $\mathcal{D}_{nc}$ than their non-equality. This problem is a typical instance of data clustering. That is, the set of observations s_i is drawn from a mixture of two distributions $\mathcal{D}_c$ and $\mathcal{D}_{nc}$, with respective weights 2^{-8} and $(1 - 2^{-8})$. For both detection metrics in this paper, we show next that it is easy to theoretically predict one out of the two distributions. We will then estimate the parameters of the other distribution based on this prediction and some additional measurements. Lemma 2 (proven in Appendix A) provides formulas to estimate the non-collision distribution parameters based on the collision ones. Moving from the collision to the non-collision distribution can be done similarly.

Lemma 2. *Let* $\mathcal{D}$ *be a mixture of two distributions* $\mathcal{D}_c$ *and* $\mathcal{D}_{nc}$ *with respective weights* 2^{-8} *and* $1 - 2^{-8}$. *Let us denote by* $\bar{\mu}$ *and* $\bar{\sigma}^2$ *estimates for the expected value and variance of* $\mathcal{D}$ *obtained from observed values. Similarly, we denote* $(\bar{\mu}_c, \bar{\sigma}^2{}_c)$ *estimates obtained for* $\mathcal{D}_c$. *Then, we can derive the following estimates for expected value and variance of* $\mathcal{D}_{nc}$:

$$\bar{\mu}_{nc} = \frac{\bar{\mu} - 2^{-8}\bar{\mu}_c}{1 - 2^{-8}}, \quad and \quad \bar{\sigma}^2{}_{nc} = \frac{\bar{\sigma}^2 - 2^{-16}\bar{\sigma}^2{}_c}{(1 - 2^{-8})^2}.$$

Specialization to the Euclidean Distance Detection. The Euclidean distance (ED) has been proposed as a detection tool in [18] and investigated in a profiled setting in [4]. The Euclidean distance between two traces T and T' equals:

$$ED(T, T') \stackrel{\text{def}}{=} \sum_{j=1}^{\ell} (T_j - T'_j)^2.$$

Let us first detail a natural non-Bayesian use of this similarity metric. Then, we will specialize the aforementioned framework in order to provide formulas to compute actual probabilities $\mathrm{Pr}\left[\Delta K_{a,b} \middle| \bar{t}_a, \bar{t}_b\right]$ from observed Euclidean distances. In general, the smaller is the Euclidean distance between traces $\bar{T}_a^{(i_a)}$ and $\bar{T}_b^{(i_b)}$, the more probable is the value $x_a^{(i_a)} \oplus x_b^{(i_b)}$ for the variable $\Delta K_{a,b}$ is. Hence, we will consider the opposite of $ED\left(\bar{T}_a^{(i_a)}, \bar{T}_b^{(i_b)}\right)$ as the score to normalize:

$$S_{ED}(a, b, \delta) \stackrel{\text{def}}{=} \mathrm{Norm}\left(d_0 - \max_{x_a^{(i_a)} \oplus x_b^{(i_b)} = \delta} ED\left(\bar{T}_a^{(i_a)}, \bar{T}_b^{(i_b)}\right)\right), \tag{2}$$

where d_0 is chosen such that all values $d_0 - \max_{x_a^{(i_a)} \oplus x_b^{(i_b)} = \delta} ED\left(\bar{T}_a^{(i_a)}, \bar{T}_b^{(i_b)}\right)$ are strictly positive. Note that if a single trace is given, only a single Euclidean distance can be computed between each pair of S-boxes a and b. By contrast, many Euclidean distances can be computed per pair of S-boxes when the number of traces increases. This justifies the use of a (heuristic) max function to select which Euclidean distance will be retained to compute the scores. Let us now consider the application of the Bayesian framework to the use of ED. We will use ED_B to refer to this Bayesian extension of ED. As mentioned earlier, efficiently deriving probabilities from scores in a non-profiled setting requires to make some assumptions. In the following, we consider the frequent case where the leakage is the sum of a deterministic part (that depends on an intermediate computation result) and a white Gaussian noise, as proposed in [17] and stated in Model 2.

Model 2. *For any input byte X_a^i and for any point j in the corresponding sub-trace T_a^i, the power consumption $T_{a,j}^i$ is the sum of a deterministic value $L_j(X_a^i)$ and some additive white Gaussian noise $N_{a,j}^i$ of variance σ_j^2:*

$$T_{a,j}^i = L_j(X_a^i) + N_{a,j}^i.$$

To lighten notation, we will omit superscripts and subscripts when a statement applies for all inputs and sub-traces. This model admittedly deviates from the distribution of actual leakage traces, since the noise in different samples can be correlated. We note again that in non-profiled attacks, it is not possible to obtain any information on the covariances of this Gaussian noise. Nevertheless, taking points in the trace that are far enough ensures that these covariances are small enough for this approximation to be respected in practice, as will be confirmed experimentally in Section 5. In such a context, the variables $(T_j - T_j')$ are drawn according to the Gaussian distribution $\mathcal{N}(L_j(X) - L_j(X'), 2\sigma_j^2)$. As a result, the normalized Euclidean distance becomes:

$$ED_B(T, T') \overset{\text{def}}{=} \sum_{j=1}^{\ell} \frac{(T_j - T_j')^2}{2\sigma_j^2}, \qquad (3)$$

and can be modeled using χ^2 distribution family (i.e. sums of squared Gaussian random variables). Indeed, each term of the sum is distributed according to a non-central χ^2 distribution with non central parameter $(T_j - T_j')^2$. In the case of a collision, this parameter vanishes and all the terms are drawn according to a central χ^2 distribution. Hence $\mathcal{D}_c$ is a χ^2 distribution with ℓ degrees of freedom. In the case of $\mathcal{D}_{nc}$, the attacker has no knowledge of $(L_j(X) - L_j(X'))^2$ and is unable to directly estimate the distribution. Yet, as previously mentioned it is possible to obtain a good approximation of this distribution from the parameters of $\mathcal{D}_c$ using Lemma 2. Experiments show that the shape of $\mathcal{D}_{nc}$ quickly tends towards a Gaussian distribution when increasing the number of points ℓ. Combining these observations, we obtain the following score:

$$\text{BayExt}(S_{ED}(a,b,\delta)) \overset{\text{def}}{=}$$
$$\text{Norm}\left(\exp\left[\frac{1}{2}\sum_{x_a^{(i_a)} \oplus x_b^{(i_b)} = \delta} \frac{\left(EDB\left(\bar{T}_a^{(i_a)}, \bar{T}_b^{(i_b)}\right) - \mu_{nc}\right)^2}{\sigma_{nc}^2} - EDB(\bar{T}_a^{(i_a)}, \bar{T}_b^{(i_b)})\right]\right). \qquad (4)$$

Remark. If averages are performed during the `PreProcessTraces` procedure, the number of traces used to compute values $\bar{T}_a^{(i_a)}$ may be different. Hence, the normalized Euclidean distance EDB has to take this into account when comparing sub-traces $\bar{T}_a^{(i_a)}$ and $\bar{T}_b^{(i_b)}$. This is done by replacing $2\sigma_j^2$ by $\sigma_j^2\left(\frac{1}{\#(a,i_a)} + \frac{1}{\#(b,i_b)}\right)$ in (3), where $\#(a,i_a)$ is the number of traces averaged to obtain $\bar{T}_a^{(i_a)}$.

Specialization to the Correlation-Enhanced Detection. We now consider the use of Pearson's correlation coefficient as detection tool. Let us recall that for two vectors U and V having the same length and mean values $\bar{U}$ and $\bar{V}$, the correlation coefficient is defined as:

$$\rho(U,V) \overset{\text{def}}{=} \frac{\sum_i (U_i - \bar{U})(V_i - \bar{V})}{\sqrt{\sum_i (U_i - \bar{U})^2}\sqrt{\sum_i (V_i - \bar{V})^2}}.$$

Many papers take advantage of this comparison metric in the side-channel literature. In the following, we focus on the correlation-enhanced solution proposed in [13], as it generally provides the best results. This attack applies to "on-the-fly" templates $\bar{t}$ such that $\bar{T}_a^{(x)}$ contains the sub-traces obtained by averaging the computations of an S-box a with plaintext byte x. The detection is based on the fact that if $\Delta K_{a,b} = \delta$, then traces $\bar{T}_a^{(x)}$ should correspond to (i.e. be similar with) traces $\bar{T}_b^{(x \oplus \delta)}$. We denote the permutation of the vector $\bar{T}_b$ that contains the $\bar{T}_b^{(x \oplus \delta)}$'s for increasing x values as $\bar{T}_b^{\oplus \delta}$. Then, the normalized score for a given δ is obtained with $S_{CE}(a,b,\delta) \overset{\text{def}}{=} \text{Norm}\left(\rho\left(\bar{T}_a, \bar{T}_b^{\oplus \delta}\right)\right)$. In practice, some values of $T_a^{(x)}$ or $T_b^{(x \oplus \delta)}$ may not be defined if few traces are used. In the case where at least one of the two traces is undefined, the coordinate will be ignored in the computation of the correlation coefficient. As for the Euclidean distance, we propose to apply the Bayesian extension to the use of the correlation-enhanced detection. The distribution of the correlation coefficient is not easy to handle, but we can approximate it with a Gaussian one using the Fisher transform of this coefficient, as proposed in [12]: $CE_B(a,b,\delta) \overset{\text{def}}{=} \text{arctanh } S_{CE}(a,b,\delta)$. Asymptotically, the random variables $CE_B(a,b,\delta)$ are normally distributed with mean equal to the expected value of $\rho\left(\bar{T}_a, \bar{T}_b^{\oplus \delta}\right)$, and variance $(N-3)^{-1}$, where N is the number of coordinates used to compute the correlation coefficient. Given these modified statistics, we now derive the corresponding Bayesian extension. For non-collisions, the correlation coefficient has an expected value of 0. Let us denote the expected value of the correlation when a collision occurs as μ_c. Since both distributions have the same variance (say σ^2), Lemma 1 translates into:

$$\Pr\left[\Delta K = \delta \middle| CE_B(a, b, \delta) = s\right] \propto \frac{e^{-\frac{(s-\mu_c)^2}{\sigma^2}}}{e^{-\frac{s^2}{\sigma^2}}} \propto \exp\left[\frac{s^2 - (s-\mu_c)^2}{\sigma^2}\right] \propto \exp\left[\frac{2s}{\sigma^2}\right].$$

In practice, it turned out that distributions are really close to Gaussian, even for a small number of traces used, but their variance did not tend towards the expected value $(N-3)^{-1}$. Hence, in our following experiments, we rather used:

$$S_{CE\_B}(a, b, \delta) \stackrel{\text{def}}{=} \text{Norm}\left(e^{2\, CE_B(a,b,\delta)}\right). \tag{5}$$

Again, results in Section 5 show that even if based on slightly incorrect models, the impact of these Bayesian extensions on the attack efficiency is positive.

4.3   LDPC Decoding

A soft-decoding algorithm for non-binary LDPC codes can be found in [1] and is presented in Algorithm 2. It consists in iterating a belief propagation stage a certain number of times. Let us recall that a code can be represented using a bipartite graph: left nodes are message nodes and correspond to positions of the codeword; right nodes are check nodes that represent redundancy constraints. The attacker receives distributions for the message nodes. The belief propagation step boils down to updating these message node distributions according to the adjacent message nodes (that is, message nodes sharing a common check node). Such a decoding algorithm actually works for any linear code, but quickly becomes intractable as the degree of check nodes increases. In the case of LDPC codes, this degree is small (by definition) which makes the algorithm run efficiently. In our particular context where check nodes have degree 3, information from adjacent nodes can further be exploited through the convolution:

$$P_{a,c} * P_{b,c}(\delta) \stackrel{\text{def}}{=} \sum_{\alpha \in \mathbb{F}_{256}} P_{a,c}(\delta) P_{b,c}(\delta \oplus \alpha).$$

In addition, the corresponding graph has small cycles and the propagation is very fast (after only two iterations, a position has been influenced by all others). Hence, the number of iterations of the while loop can be considered as a constant. Note that the convolution of two probability tables can be computed using a fast Walsh transform. Indeed, for a field of q elements, the q convolutions can be computed in $\Theta(q \ln q)$. Hence, Algorithm 2 has a complexity of $\Theta(q \ln q)$.

5   Experiments

We now present experiments obtained in three different settings. The first set of attacks targets a reference implementation of the AES and confirms the relevance of the tools we introduced. Following, a second set of attacks targeting an optimized implementation of AES (namely, the *furious* implementation from [15]) is presented. The main observation is that small code optimizations may lead

---

Algorithm 2. Proposition for `LDPCSoftDecoding` procedure

Input: The distributions $\Pr\left[\Delta K_{a,b} = \delta\right]$.
Output: The likeliest consistent system $\mathcal{S}$.
foreach $1 \leq a < b \leq 16$, $\delta \in \mathbb{F}_{256}$ **do**
$\quad\lfloor\ P_{a,b}(\delta) \leftarrow \Pr\left[\Delta K_{a,b} = \delta\right];$

while $\left(\underset{\delta}{\operatorname{argmax}}\ P_{1,2}(\delta), \ldots, \underset{\delta}{\operatorname{argmax}}\ P_{15,16}(\delta)\right)$ *is not a codeword* **do**
$\quad$**foreach** $1 \leq a < b \leq 16$ **do**
$\quad\quad$**foreach** $\delta \in \mathbb{F}_{256}$ **do**
$\quad\quad\quad\lfloor\ P_{a,b}(\delta) \leftarrow P_{a,b}(\delta) \cdot \prod_{c \notin \{a,b\}} P_{a,c} * P_{b,c}(\delta);$
$\quad\quad\ P_{a,b} \leftarrow \frac{P_{a,b}}{\|P_{a,b}\|_1};$

return $\left(\underset{\delta}{\operatorname{argmax}}\ P_{1,2}(\delta), \ldots, \underset{\delta}{\operatorname{argmax}}\ P_{15,16}(\delta)\right);$

---

to variations in S-boxes leakage functions, which in turn results in less efficient attacks. For these two first sets of attacks, we measured the power consumption of an Atmel microcontroller running the target AES implementations at 20MHz, by monitoring the voltage variations over a small resistor inserted in the supply circuit. We then conclude this paper by investigating a theoretical setting where leakage functions are not linear. This final experiment motivates the potential interest of collision attacks compared to other non-profiled distinguishers.

In the following, we compare collision attacks (Coll) using the Euclidean distance (ED) and the correlation-enhanced (CE) detection techniques, with the non-profiled variant of Schindler et al.'s stochastic approach [17], described in [8] and using a 9-element basis including the target S-boxes output bits. Collision attacks have been performed using the instantitation of `PreProcessTraces` procedure from [13], defined by (1). As mentioned in the introduction, and for all our experiments, we assumed that we were able to divide each leakage traces in 16 sub-traces corresponding to the 16 AES S-boxes. For the real measurements, the detection metrics were directly applied to these sub-traces, following the descriptions in the previous sections. As for the simulated ones in Section 5.3, we generated univariate leakages for each S-box execution, according to the hypothetical (linear and non-linear) leakage functions and a Gaussian noise.

5.1 Attacking the Reference Implementation

In this first experiment, we consider a favorable setting where each table look-up is performed with the same register (our ASM code provided in Appendix B). The goal is to emphasize the gain obtained from the LDPC formulation of the problem and the Bayesian extension. The 2^8-th order success rates (defined in [20]) obtained in this case are given in Figure 1. Original collision attacks directly extract the key from scores obtained with the ED or CE metrics. Attacks taking advantage of the LDPC decoder are marked with an (L), and the use of the

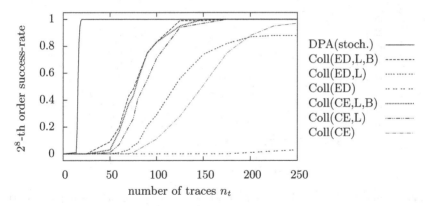

Fig. 1. Order 2^8 success rates of attacks using the homemade implementation

Bayesian extension is denoted by $(B)^2$. As expected, using the LDPC decoding algorithm greatly improves the attacks performances. Moreover, using the Bayesian extension also provides a non-negligible gain. Interestingly, when both tools are combined, ED and CE detection metrics seem to be equivalent in terms of data complexity. This may be a good empirical indication that the error correcting codes approach we propose really extracts all the available information.

5.2 Attacking the Optimized Implementation

In this next experiment, we targeted the AES *furious* implementation. This optimized implementation is a more challenging target, since the S-box layer and the ShiftRows operation are interleaved. Moreover, the table looks-up are performed from different registers. Due to these optimizations, the leakage functions of the different S-boxes are not so similar anymore (see Appendix B). Hence, the correct key is unlikely to correspond to the most likely codeword. A direct consequence of this more challenging context is that the success rate of order 2^8 is not suited to evaluate the attack performances (i.e. the correct key may be rated beyond the 2^8 first ones by the attack). As an alternative, we estimated the median rank of the correct key among the 2^{128} possible values.

In that case, one should use a decoding algorithm with $\ell > 1$ and test the $2^8 \cdot \ell$ corresponding keys afterwards. Unfortunately, the efficiency of such a list-decoding algorithm depends on the shape of distributions $\Pr[\Delta K]$, themselves being highly dependent on the similarity metrics used to compare traces, and the device running the cipher. Therefore, we left the design of such an algorithm as a scope for further research. Yet, and in order to be able to analyze the attack performances, we used an ad-hoc list-decoding algorithm that consists in enumerating key classes from a subset of 15 positions of dimension 15 (using the algorithm in [21]), for which the corresponding distributions have a small

---

[2] Since the Bayesian extension does not modify the ordering of the scores, using it only makes sense when applying the LDPC decoding algorithm.

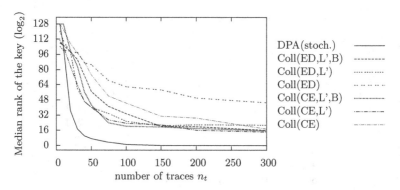

Fig. 2. Estimates of the median key rank when attacking the *furious* implementation

entropy. Note that the belief propagation part of the decoder in Algorithm 2 can be used (or not) before performing the enumeration. To avoid confusion, we will denote by (L') the use of this belief propagation step before enumerating. Computing the median rank of the key also becomes intensive as this rank becomes large. Hence, we decided to enumerate keys up to the 2^{20}-th first ones, and in the cases where the correct key was not found, estimated the key rank by multiplying correct subkey ranks. These heuristics naturally have to be taken into consideration when analyzing the results in Figure 2. However, we believe that they provide a fair understanding of the different attacks we investigated. Namely, as in Figure 1, the soft-decoding algorithm allows great performance improvements in the *furious* implementation case-study. By contrast, the Bayesian extensions were less directly useful. This observation again relates to the different leakage models observed for different S-boxes. As the parameter estimation in the Bayesian extensions requires a sufficient precision to be exploitable, they were only useful after approximately 150 traces in this more challenging scenario. Note finally that most collision attacks are stuck around rank 2^{15}. This can be explained by one of the S-boxes leaking in a drastically different way than the others in our implementation. As a result, we were only able to recover 14 bytes out of the 15 ones from this optimized implementation (even with large number of measurements). This leads to a median rank of roughly 2^7 for the correct system, an a median rank $2^{15=7+8}$ for the correct master key.

5.3   Simulated Experiments with Non-linear Leakages

In the previous experiments, the non-profiled variant of Schindler et al.'s stochastic approach consistently gave better results than the (improved) collision attacks investigated. As a result, one can naturally question the interest of such attacks in a security evaluation context. In order to discuss this issue, this final section analyzes the relevance of collision attacks in a purely theoretical setting. We used two different sets of simulated traces for this purpose: the first ones were generated using a leakage function of which the output is a linear function of the S-boxes output bits; the second ones were generated with a leakage function of

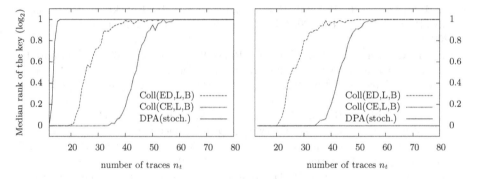

Fig. 3. 2^8-th order success rate for linear (left) and non-linear (right) leakage functions

which the output is a highly non-linear function of the S-boxes output bits (i.e. a situation emulating the worst case scenario from [22]). The results corresponding to these alternative scenarios are in Figure 3. As expected, the linear leakages in the left part of the figure are efficiently exploited by all attacks, with an improved data complexity for the stochastic approach. By contrast, in the right part of the figure, the stochastic approach is unable to exploit the non-linear leakages, and only collision attacks lead to successful key recoveries[3]. This confirms that there exist situations in which non-profiled collision attacks are able to exploit information leakage that no other non-profiled attack can. We leave the quest for such leakage functions (or protected circuits) as a scope for further research.

Acknowledgements. We would like to thank Jean-Pierre Tillich for providing useful information and advices about the decoding of LDPC codes.

References

1. Bennata, A., Burshtein, D.: Design and analysis of nonbinary LDPC codes for arbitrary discrete-memoryless channels. IEEE Transactions on Information Theory 52, 549–583 (2006)
2. Bogdanov, A.: Improved Side-Channel Collision Attacks on AES. In: Adams, C., Miri, A., Wiener, M. (eds.) SAC 2007. LNCS, vol. 4876, pp. 84–95. Springer, Heidelberg (2007)
3. Bogdanov, A.: Multiple-Differential Side-Channel Collision Attacks on AES. In: Oswald, E., Rohatgi, P. (eds.) CHES 2008. LNCS, vol. 5154, pp. 30–44. Springer, Heidelberg (2008)
4. Bogdanov, A., Kizhvatov, I.: Beyond the Limits of DPA: Combined Side-Channel Collision Attacks. IEEE Transactions on Computers 61(8), 1153–1164 (2012)
5. Brier, E., Clavier, C., Olivier, F.: Correlation Power Analysis with a Leakage Model. In: Joye, M., Quisquater, J.-J. (eds.) CHES 2004. LNCS, vol. 3156, pp. 16–29. Springer, Heidelberg (2004)

---

[3] Increasing the basis with non-linear elements would not allow solving this issue as long as only non-profiled attacks are considered. It would lead to more precise leakage models both for the correct key candidates and the wrong ones, by over-fitting.

6. Chari, S., Rao, J.R., Rohatgi, P.: Template Attacks. In: Kaliski Jr., B.S., Koç, Ç.K., Paar, C. (eds.) CHES 2002. LNCS, vol. 2523, pp. 13–28. Springer, Heidelberg (2003)

7. Clavier, C., Feix, B., Gagnerot, G., Roussellet, M., Verneuil, V.: Improved Collision-Correlation Power Analysis on First Order Protected AES. In: Preneel, B., Takagi, T. (eds.) CHES 2011. LNCS, vol. 6917, pp. 49–62. Springer, Heidelberg (2011)

8. Doget, J., Prouff, E., Rivain, M., Standaert, F.-X.: Univariate side channel attacks and leakage modeling. Journal of Cryptographic Engineering 1(2), 123–144 (2011)

9. Gallager, R.G.: Low density parity check codes. Transactions of the IRE Professional Group on Information Theory IT-8, 21–28 (1962)

10. Kocher, P.C., Jaffe, J., Jun, B.: Differential Power Analysis. In: Wiener, M. (ed.) CRYPTO 1999. LNCS, vol. 1666, pp. 388–397. Springer, Heidelberg (1999)

11. Ledig, H., Muller, F., Valette, F.: Enhancing Collision Attacks. In: Joye, M., Quisquater, J.-J. (eds.) CHES 2004. LNCS, vol. 3156, pp. 176–190. Springer, Heidelberg (2004)

12. Mangard, S.: Hardware Countermeasures against DPA – A Statistical Analysis of Their Effectiveness. In: Okamoto, T. (ed.) CT-RSA 2004. LNCS, vol. 2964, pp. 222–235. Springer, Heidelberg (2004)

13. Moradi, A., Mischke, O., Eisenbarth, T.: Correlation-Enhanced Power Analysis Collision Attack. In: Mangard, S., Standaert, F.-X. (eds.) CHES 2010. LNCS, vol. 6225, pp. 125–139. Springer, Heidelberg (2010)

14. Moradi, A.: Statistical Tools Flavor Side-Channel Collision Attacks. In: Pointcheval, D., Johansson, T. (eds.) EUROCRYPT 2012. LNCS, vol. 7237, pp. 428–445. Springer, Heidelberg (2012)

15. Poettering, B.: Fast AES implementation for Atmel's AVR microcontrollers, http://point-at-infinity.org/avraes/

16. Renauld, M., Kamel, D., Standaert, F.-X., Flandre, D.: Information Theoretic and Security Analysis of a 65-Nanometer DDSLL AES S-Box. In: Preneel, B., Takagi, T. (eds.) CHES 2011. LNCS, vol. 6917, pp. 223–239. Springer, Heidelberg (2011)

17. Schindler, W., Lemke, K., Paar, C.: A Stochastic Model for Differential Side Channel Cryptanalysis. In: Rao, J.R., Sunar, B. (eds.) CHES 2005. LNCS, vol. 3659, pp. 30–46. Springer, Heidelberg (2005)

18. Schramm, K., Leander, G., Felke, P., Paar, C.: A Collision-Attack on AES: Combining Side Channel and Differential-Attack. In: Joye, M., Quisquater, J.-J. (eds.) CHES 2004. LNCS, vol. 3156, pp. 163–175. Springer, Heidelberg (2004)

19. Schramm, K., Wollinger, T., Paar, C.: A New Class of Collision Attacks and Its Application to DES. In: Johansson, T. (ed.) FSE 2003. LNCS, vol. 2887, pp. 206–222. Springer, Heidelberg (2003)

20. Standaert, F.-X., Malkin, T., Yung, M.: A Unified Framework for the Analysis of Side-Channel Key Recovery Attacks. In: Joux, A. (ed.) EUROCRYPT 2009. LNCS, vol. 5479, pp. 443–461. Springer, Heidelberg (2009)

21. Veyrat-Charvillon, N., Gérard, B., Renauld, M., Standaert, F.-X.: An optimal key enumeration algorithm and its application to side-channel attacks. Cryptology ePrint Archive, Report 2011/610 (2011), http://eprint.iacr.org/2011/610

22. Veyrat-Charvillon, N., Standaert, F.-X.: Generic Side-Channel Distinguishers: Improvements and Limitations. In: Rogaway, P. (ed.) CRYPTO 2011. LNCS, vol. 6841, pp. 354–372. Springer, Heidelberg (2011)

A   Proof of Lemma 2

We want to express the mean and the variance of a mixture of two distributions as a function of the means and variances of these distributions. Let us recall that $\mathcal{D}$ is a mixture of two distributions $\mathcal{D}_c$ and $\mathcal{D}_{nc}$ with respective weights 2^{-8} and $1-2^{-8}$. We denote by μ and σ^2 the expected value and variance of $\mathcal{D}$, by (μ_c, σ_c^2) the expected value and variance of $\mathcal{D}_c$, and by $(\mu_{nc}, \sigma_{nc}^2)$ the expected value and variance of $\mathcal{D}_{nc}$. Let X_c and X_{nc} be respectively drawn according to $\mathcal{D}_c$ and $\mathcal{D}_{nc}$ and X be the mixture $2^{-8}X_c + (1 - 2^{-8})X_{nc}$. Then, due to the linearity of the operator, $\mathbb{E}(X) = \mathbb{E}\left(2^{-8}X_c + (1 - 2^{-8})X_{nc}\right) = 2^{-8}\mu_c + (1 - 2^{-8})\mu_{nc}$. Thus, it follows that $\mu_{nc} = \frac{\mu - 2^{-8}\mu_c}{1 - 2^{-8}}$. Concerning the variance, a slightly more difficult calculus leads to the claimed result. First, we use the relationship $\mathbb{V}(X) = \mathbb{E}\left(X^2\right) - \mathbb{E}(X)^2$ and we develop its first term in:

$$\mathbb{E}\left(X^2\right) = \mathbb{E}\left(2^{-16}X_c^2 + 2 2^{-8}(1 - 2^{-8})X_c X_{nc} + (1 - 2^{-8})^2 X_{nc}^2\right).$$

Since X_c and X_{nc} are independent variables, we have:

$$\mathbb{E}\left(X^2\right) = 2^{-16}\mathbb{E}\left(X_c^2\right) + 2^{-7}(1 - 2^{-8})\mathbb{E}(X_c)\mathbb{E}(X_{nc}) + (1 - 2^{-8})^2\mathbb{E}\left(X_{nc}^2\right).$$

We then notice that $\mathbb{E}\left(X_c^2\right) = \sigma_c^2 + \mu_c^2$ (the same holds for $\mathbb{E}\left(X_{nc}^2\right)$). Hence:

$$\mathbb{E}\left(X^2\right) = 2^{-16}(\sigma_c^2 + \mu_c^2) + 2^{-7}(1 - 2^{-8})\mu_c\mu_{nc} + (1 - 2^{-8})^2(\sigma_{nc}^2 + \mu_{nc}^2).$$

Now returning to $\mathbb{V}(X) = \mathbb{E}\left(X^2\right) - \mathbb{E}(X)^2$, we finally observe that many terms in μ vanish to yield $\mathbb{V}(X) = 2^{-16}\sigma_c^2 + (1 - 2^{-8})^2\sigma_{nc}^2$.

B   Additional Details about the Target Implementations

Our experiments are based on two different implementations of the AES: a reference one that has been designed such that S-boxes looks-ups have similar leakage functions, and the *furious* implementation. They respectively execute the AES S-box in four instructions (mov SR,ST22; mov ZL,SR; lpm SR,Z; mov ST22,SR) and three instructions (mov H1,ST21; mov ZL,ST22; lpm ST21,Z). We observed that the most leaking operation in the S-box computation was the mov operation, that stores the input of the S-box in the ZL register. Hence, in the reference implementation, we first copy intermediate values in a register SR, that is the same for all 16 S-boxes computations. Then, SR is updated and the output is copied back to the initial state register. On the contrary, we can see that in the *furious* implementation, the ZL register is directly updated from the state register (here ST22), and the answer directly goes back to the state register. In addition, the *furious* implementation combines the S-box layer with the ShiftRows operation. It explains why the output is stored in ST21 and not ST22. The optimizations in the *furious* implementation are the main reason of the poor results of the attacks performed. As a simple illustration, we plotted the templates of the leakage points used in the attacks for different S-boxes in Figure 4 (for the reference implementation) and Figure 5 (for the *furious* one).

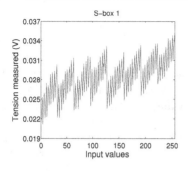

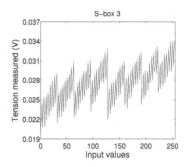

Fig. 4. Leakage functions for the reference implementation

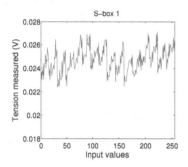

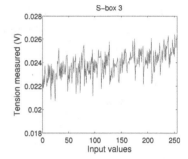

Fig. 5. Leakage functions for the *furious* implementation

C   About Time Complexity

Metrics used in Section 5 for analyzing experiments only consider the success rate of the attacks as a function of their data complexity. We consider the time complexity of the proposed collision attack in this section. When attacking n_s S-boxes processing n_b-bit words, these complexities for our different procedures are:

ComputeStatistics	$O\left(n_s^2 n_t^2 \ell\right)$
ExtractDistributions	$O\left(n_s^2 (n_t^2 + 2^{n_b})\right)$
LDPCDecode	$O\left(n_s^2 n_b 2^{n_b}\right)$

When using the pre-processing technique that has a cost $\Theta\left(n_s n_t \ell\right)$, the complexity of the procedure ComputeStatistics is decreased to $O\left(n_s^2 2^{2n_b} \ell\right)$. Hence, it turns out that, in realistic contexts, collision attacks can be performed in a negligible time compared to the on-line acquisition and the final key search phases (a similar comment applies to stochastic attacks). Furthermore, by carefully profiling the number of cycles needed to perform the different steps of the attacks, we observed that the slight time overhead induced by the use of a Bayesian extension and/or an LDPC decoding algorithm is positively balanced by the reduction of the data complexity, hence leading to globally more efficient attacks.

Towards Super-Exponential Side-Channel Security with Efficient Leakage-Resilient PRFs

Marcel Medwed[1,*], François-Xavier Standaert[1,**], and Antoine Joux[2]

[1] UCL Crypto Group, Université catholique de Louvain
Place du Levant 3, B-1348, Louvain-la-Neuve, Belgium
[2] DGA and Université de Versailles Saint-Quentin-en-Yvelines, Laboratoire PRISM
45 avenue des États-Unis, F-78035 Versailles Cedex, France

Abstract. Leakage-resilient constructions have attracted significant attention over the last couple of years. In practice, pseudorandom functions are among the most important such primitives, because they are stateless and do not require a secure initialization as, e.g. stream ciphers. However, their deployment in actual applications is still limited by security and efficiency concerns. This paper contributes to solve these issues in two directions. On the one hand, we highlight that the condition of bounded data complexity, that is guaranteed by previous leakage-resilient constructions, may not be enough to obtain practical security. We show experimentally that, if implemented in an 8-bit microcontroller, such constructions can actually be broken. On the other hand, we present tweaks for tree-based leakage-resilient PRFs that improve their efficiency and their security, by taking advantage of parallel implementations. Our security analyses are based on worst-case attacks in a noise-free setting and suggest that under reasonable assumptions, the side-channel resistance of our construction grows super-exponentially with a security parameter that corresponds to the degree of parallelism of the implementation. In addition, it exhibits that standard DPA attacks are not the most relevant tool for evaluating such leakage-resilient constructions and may lead to overestimated security. As a consequence, we investigate more sophisticated tools based on lattice reduction, which turn out to be powerful in the physical cryptanalysis of these primitives. Eventually, we put forward that the AES is not perfectly suited for integration in a leakage-resilient design. This observation raises interesting challenges for developing block ciphers with better properties regarding leakage-resilience.

1 Introduction

Physical attacks, in which adversaries take advantage of the peculiarities of the devices on which cryptographic operations are running, are an important concern for modern security applications. They typically include side-channel attacks (where the adversary monitors the leakage due to the cryptographic

---

[*] Postdoctoral researcher funded by the 7th framework European project TAMPRES.
[**] Associate researcher of the Belgian Fund for Scientific Research (FNRS-F.R.S.). This work has been funded in part by the ERC project 280141 (acronym CRASH).

E. Prouff and P. Schaumont (Eds.): CHES 2012, LNCS 7428, pp. 193–212, 2012.

computations [26]), fault attacks (where the adversary tries to force a device to perform erroneous computations [6]), tampering attacks (where the adversary probes a few wires of the implementation [4]) and memory attacks (where the adversary directly monitors parts of the memory [18]). As a result, a variety of hardware-level countermeasures have been proposed, in order to reduce the amount of information an implementation may provide to adversaries. Over the last few years, such hardware-level countermeasures have been complemented by significant efforts to extend the formal guarantees of provable security from cryptographic algorithms towards cryptographic implementations. For this purpose, various models have been introduced, trying to capture physical reality in abstract terms, with the goal of allowing meaningful reasoning about physical security. Examples of models to formalize side-channel attacks include Micali and Reyzin's physically observable cryptography [29] and Dziembowski and Pietrzak's leakage-resilient cryptography [14]. Examples of models to formalize fault attacks, tampering attacks and memory attacks can also be found, e.g. in [3,16,21,22]. Eventually, more general abstractions, such as the auxiliary input model [11], or the bounded retrieval model [10,13], have been introduced for similar purposes. Quite naturally, these different models raise numerous questions about their relevance to practice, e.g. regarding the correspondence (or lack thereof) between the basic assumptions used in proofs and what can actually be guaranteed by hardware designers. In general, there remain many open problems to answer in order to specify a fully satisfying model (see, e.g. [38,39]).

Nevertheless, and somewhat independent of the practical relevance of the models used to formalize physical security issues, it may very well be that (small variations of) ideas proposed in these theoretical works actually provide significantly enhanced security against large categories of "practical" side-channel attacks (such as surveyed in [28]), when analyzed in more restricted frameworks such as [41]. In this paper, we follow this direction and focus on the security of symmetric cryptographic primitives such as block and stream ciphers. This focus is naturally motivated by the fact that such low-cost algorithms are among the most frequently considered targets, e.g. for power and EM analysis.

Related work. The very idea to prevent side-channel attacks at the protocol level relying on key updates refers to Kocher [25]. Following, Pseudo-Random Number Generators (PRNGs) allowing security against side-channel attacks have first been proposed and analyzed in a specialized setting [33] (with noisy Hamming weight or identity leakage functions). Leakage-resilient stream ciphers based on an "alternating structure", with a proof of security in the standard model have then been described in [14,34]. Similar constructions without alternating structure, but with a proof in a model relying on a random-oracle assumption can be found in [42,43]. Finally, an attempt to prove the security of a PRG without alternating structure and in the standard model, assuming non-adaptive leakage functions, was also suggested in [43], and was later shown to require large amounts of public randomness for the proof to hold in [15]. It remains an open question to determine if the exact construction proposed in [43], using only two alternating public values, can be proven secure or attacked in a practical setting.

While theoretically interesting (and efficiently implemented), these stream ciphers and PRNGs all suffer from the limitation that they require a secure initialization mechanism. As already clear in [33], this problem of initialization is an important issue for deploying leakage-resilient constructions in real-world devices, as it typically implies much larger performance overheads. Roughly speaking, and taking AES-based designs as an example, all previous leakage-resilient PRNGs can output one 128-bit block every two AES executions. By contrast, the initialization mechanism proposed at ASIACCS 2008 [33] requires 128 AES executions per block (if the best security against side-channel attacks is privileged). Following solutions did not allow any improvement in this respect.

Interestingly, this requirement of a secure initialization process for PRNGs can actually be translated into the need of a leakage-resilient PseudoRandom Function (PRF). As a consequence, it was first observed in [42] that a tree-based construction such as the one of Goldreich, Goldwasser and Micali (GGM) [17] inherently brings improved resistance against side-channel attacks. Again taking an AES-based example, these PRFs allow ensuring that every intermediate key in the tree is only used twice. The construction in [42] was proven secure against side-channel attacks under a random-oracle assumption, together with the observation that leakage-resilience for stateless PRFs requires to limit the leakage function to be non-adaptive. Next, Dodis and Pietrzak constructed a leakage-resilient PRF and proved its security in the standard model (with non-adaptive leakages as well) [12], by applying a GGM-like construction to the stream cipher with alternating structure from [34]. How to replace the alternating structure by alternating public randomness is additionally discussed in [15].

Finally, the construction of PseudoRandom Permutations (PRPs) was first discussed at CRYPTO 2010 [12]. In this paper, Dodis and Pietrzak describe efficient attacks (with non-adaptive leakage functions) against Feistel ciphers constructed from leakage-resilient PRFs. It is shown in [15] that more positive results can be obtained in a known-plaintext (rather than chosen-plaintext) adversarial scenario. However, as for the PRF constructions listed above, the practicality of these PRP constructions is strongly limited by performance overheads.

In this work, we tackle two important questions related to leakage-resilient PRFs.

First, we study their implementation in leaking devices and the security level that they provide against standard side-channel attacks. Doing so, we put forward the difference between a side-channel attack with bounded data complexity and a side-channel attack with bounded number of measurements. As previous constructions in [12,15,42] all guarantee a bounded data complexity, but do not prevent unbounded number of measurements for the PRF executed on the same inputs, we exhibit attacks in low-cost (8-bit) microcontrollers taking advantage of these capabilities. We use these experiments to argue about the need of parallelism for leakage-resilient PRF implementations in general.

Second, we study how to exploit parallelism in the implementations of PRFs, in order to significantly improve their efficiency. For this purpose, we take advantage of a careful selection of the public values used in an AES-based construction. By enforcing that these public values are such that all the bytes corresponding to the first-round S-boxes are identical, we succeed in significantly reducing the success rate of a Differential Power Analysis (DPA) against our implementations. Doing so, we also observe that DPA is not the most suitable tool for cryptanalyzing such leakage-resilient designs, and describe advanced attacks exploiting lattice reduction, that allow us to better evaluate worst-case security levels. This analysis puts forward the need to enumerate a permutation of the AES bytes, which offers an interesting security parameter (as the number of such permutations grows as a factorial function). Furthermore, this trick allows us to reduce the 128 AES iterations per block, required in the execution of previous PRFs constructions, down to 17 (and even less if the PRF is used for encryption in counter mode). This results in an overhead factor that is much more in line with the ones of other countermeasures against side-channel attacks. While our proposal goes against the requirement of independent public randomness in [15], it is backed up by a practical security analysis, which again raises the question whether this requirement is motivated by the physics or by proof artifacts. Besides, our proposal could be integrated into the PRF of CRYPTO 2010, in which the alternative structure removes the need of public randomness in the proofs.

We finally remark that the techniques analyzed in this paper raise interesting challenges for the design of new block ciphers allowing efficient implementations when inserted in leakage-resilient PRFs, or for the direct design of ad hoc constructions. As suggested by the AES-128 instance that we study in Section 6, the Rijndael algorithm may not be the best suited block cipher for this purpose, and we suggest a few directions that could lead to improved solutions.

2   The Leakage Resilient PRF Constructions

In the first sections of this paper, we will base our discussions on the GGM construction depicted in the left part of Figure 1. Let $F_k(x)$ denote the PRF indexed by k and evaluated on x. Further, let the building blocks $E_{k^i}(p_j^i)$ denote the application of a block cipher E to a plaintext p_j^i under a key k^i (the figure takes the example of $E = $ AES-128 with $1 \leq i \leq 128$ and $0 \leq j \leq 1$). Let also $x(i)$ denote the ith bit of x. The PRF first initializes $k^0 = k$ and then iterates as follows: $k^{i+1} = E_{k^i}(p_0^i)$ if $x(i) = 0$ and $k^{i+1} = E_{k^i}(p_1^i)$ if $x(i) = 1$. Eventually, the $(n+1)^{th}$ intermediate key k^{128} is the PRF output as $F_k(x)$.

In this basic version, the execution of the PRF guarantees that any side-channel adversary will at most observe the leakage corresponding to two plaintexts p_0^i and p_1^i per intermediate key. This implies 128 executions of the AES-128 to produce a single 128-bit output. A straightforward solution to trade improved performances for additional leakage is to increase the number of observable plaintexts per intermediate key. If one has N_p such plaintexts per stage, the number of AES-128 executions to produce a 128-bit output is divided by $\log_2(N_p)$. However, as will be discussed in Section 3, such a tradeoff scales badly and very rapidly decreases the side-channel security of an implementation. The more efficient alternative that we propose in this paper is based on a slight variation of this idea, illustrated in the right part of Figure 1. It can be viewed as a GGM construction with $N_p = 256$, but where the same set of 256 carefully chosen plaintexts

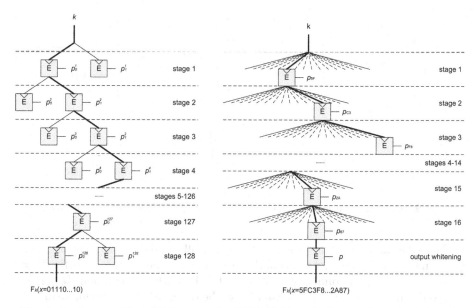

Fig. 1. Leakage-resilient PRFs: straight GGM (left) and efficient alternative (right)

is re-used in each PRF stage, excepted for the last stage where $N_p = 1$. Note that this is in contrast with the proof of leakage-resilience in [15], that requires all the p_j^i's to be public random values that are independently picked prior to encryption. In terms of efficiency, this proposal reduces the number of stages of a PRF based on the AES-128 to 17 (i.e. 16 plus one final whitening). As will be seen in Section 5, it also leads to interesting practical security guarantees.

3   Bounded Data Complexity May Not Be Enough

Let us consider the PRF construction in the left part of Figure 1 with the AES-128 as block cipher. As previously mentioned, for each intermediate key k^i, this construction prevents adversaries to mount side-channel attacks with data complexity larger than 2. By contrast, nothing prevents the repetition of large number of measurements for the same input p_j^i. In this section, we investigate whether this condition of bounded data complexity is sufficient to guarantee practical security against side-channel attacks. For this purpose, we set up experimental attacks against an implementation of the AES in an 8-bit microcontroller, with limited data complexity (i.e. small N_p values). For illustration, we used the measurement setup and statistical tools previously described in [40]. More precisely, we considered template attacks in principal subspaces, using only power consumption measurements, and Principal Component Analysis (PCA) as dimensionality reduction technique. We exploited templates for two target operations, namely the AddRoundKey and SubBytes in the first AES round.

The results of attacks targeting the first AES master key byte are given in Figure 2 for $N_p = 1, 2$. The number of measurements used in the attacks is given

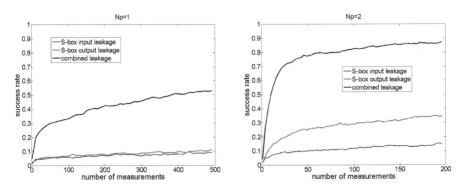

Fig. 2. Experimental attacks against the AES with bounded data complexity

on the x-axis, and their first-order success rate (following the definition in [41]) is given on the y-axis. The right part of the figure corresponds to the result for $N_p = 2$, i.e. the exact data complexity tolerated by the PRF construction. It can be observed that high success rates can already be obtained with our simple attack setting. In fact, due to the 8-bit bus of our microcontroller, even attacks with data complexity $N_p = 1$ allow reaching non-negligible success rates. As shown in Appendix C, Figure 8, this success rate dramatically increases with N_p, clearly suggesting that enhancing the PRF efficiency in this direction is not acceptable for security reasons. Admittedly, this simple scenario may not be reflective of better protected or larger, parallel devices. But it at least suggests that the security assumptions in all previous works on leakage-resilient PRFs overlook the important difference between data complexity and number of measurements.

As a result, two natural directions can be envisioned. On the one hand, one could design new (stateful) PRFs ensuring a bounded number of measurements. This would essentially correspond to the storage of all the intermediate nodes that have been computed in previous invocations of the tree-based PRF in Figure 1. Although different security vs. memory tradeoffs could be considered, this solution is hardly realistic from an implementation cost point of view. On the other hand, one could investigate the impact of large (parallel) implementations, where the bounded data complexity would be better reflected in the attacks' success rates. The following section investigates this second option.

4   Efficiently Exploiting Parallelism

In this section, we study how parallelism improves the security against DPA attacks and the efficiency of a tree-based PRF. For this purpose, we will mainly focus on one step of the constructions in Figure 1, and take the example of $k^0 = k$. In this context, there are three main parameters to consider when evaluating the side-channel security of the PRF, next denoted as N_p, N_s and σ_n^2. First, the adversary is allowed to encrypt N_p (for now, random) plaintexts p_j $(1 \geq j \geq N_p)$ under the key k. Second, we target an AES-like block cipher where N_s S-boxes are executed in parallel. Finally, the leakage measurements are affected by a noise with variance σ_n^2. Let us denote the bytes of the plaintexts as $p_j[i]$ and the bytes of the key as $k[i]$. We will consider leakages of the form:

$$l_j = \sum_{i=1}^{N_s} \mathsf{L}(\mathsf{S}(p_j[i] \oplus k[i])) + n, \tag{1}$$

with S the AES S-box, L a leakage function and n a Gaussian-distributed noise with variance σ_n^2. In such a setting, parallelism essentially depends on the number of S-boxes N_s. Increasing this parameter typically allows increasing the amount of "algorithmic noise" in the attacks, as we now detail. For illustration, we considered a Hamming weight leakage function $\mathsf{L} = \mathsf{W}_{\mathsf{H}}(\cdot)$ with $\sigma_n^2 = 0$, and a DPA adversary using Bayesian templates [9]. The left part of Figure 3 summarizes the joint effect of N_s and N_p in this random plaintext scenario, where the guessing entropy (in $\log_2$ scale) of the first master-key byte is used as evaluation metric [41]. It indicates the average position of the correct key byte in the scored list provided by the DPA attack, and thus reflects the key-search complexity of an adversary who is given such a list. One can clearly see the strong impact of increasing N_p, as the key search complexity decreases almost exponentially with it. In addition, the random plaintext scenario allows directly recovering information on all key bytes, by applying a straightforward divide-and-conquer strategy.

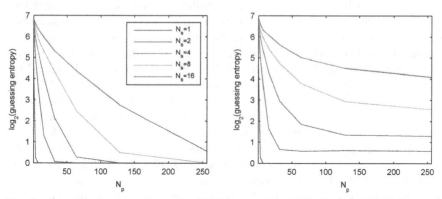

Fig. 3. Guessing entropy of the first AES master-key byte ($\log_2$ scale). Left: random (uniform) plaintext scenario. Right: carefully chosen plaintext scenario.

Careful selection of the plaintexts. The previous discussion highlighted that parallelism is not sufficient to guarantee security against side-channel attacks. In this section, we propose to tweak the PRF design with carefully chosen plaintexts, in order to prohibit the application of standard divide-and-conquer strategies. For this purpose, we define our plaintexts as the concatenation of N_s identical values j, i.e. $p_j = j\|j\|\ldots\|j$, with $1 \le j \le N_p$ and N_p limited by the S-box input space. Under conditions discussed later in this section, the effect of this measure is that in a DPA attack, the predictions corresponding to the N_s key bytes cannot be distinguished anymore. That is, all key bytes are targeted at the same time. As a result, and even when increasing N_p, not all the N_s key bytes can be highly ranked by the attack. This effect can be seen in the right part of Figure 3 and is reflected in a higher guessing entropy for the target key byte.

One important consequence of this observation is that the reduced guessing entropy of one key byte does not directly translate towards more key bytes. Indeed, the adversary now has to reconstruct a full key from a single score vector (rather than N_s ones in the random plaintext scenario). Intuitively, the task of reconstructing the full key could be divided into two steps: (1) picking a subset of N_s key bytes and (2) afterwards determining their order. Probably the most important result for this countermeasure is that even if the N_s correct key bytes are always ranked in the N_s first positions, task (2) still has a complexity of $N_s!$, a number which grows super-exponentially. We now discuss the conditions upon which this security parameter can actually be observed:

1. The leakage function L in Equation (1) has to be identical for all S-boxes.
2. Side-channel attacks exploiting unknown ciphertexts should be hard.

As far as the first condition is concerned, it is admittedly a new type of assumption. Therefore we investigated its practicality in Section 6, based on an FPGA case study. Our conclusions can be summarized as follows. (a) This assumption is indeed implementation-dependent. That is, we were able to identify both implementations with close to identical leakage models for all S-boxes, and implementations in which these models exhibit significant differences. (b) Even in the cases where significant differences occur, these differences could not be exploited. Essentially, this is because constructing the models (byte per byte, as imposed by computational constraints) has to be done for uniform plaintexts. That is, we assume parts of the bytes in the implementation to produce independent algorithmic noise (binomially, or approximately Gaussian distributed). By contrast, during an attack, the plaintexts are carefully chosen for all the bytes, hence generating a strongly key-dependent noise that was not characterized during profiling. As a result, the modeled leakage and the leakage during an attack do not match, which prohibits successful key recoveries. Summarizing, our experiments provide good indication that our assumption is sufficiently fulfilled for power measurements. As for EM measurements, it depends on the localization capabilities of the adversary. As discussed in [27], Chapter 3, distinguishing structures of a few hundred gates in complex circuits is a non-trivial task, especially for deep-submicron technologies. Hence, we believe that our countermeasure rules out an important part of low-cost EM attacks, and leave the investigation of advanced localization issues as an interesting question for future research.

As far as the second condition is concerned, first note that only the ciphertexts of the last step in the right construction of Figure 1 are given to the adversary. But for this last iteration, only one public plaintext p can be queried (i.e. the data complexity is bounded to one). For all the other steps, the ciphertexts remain internal intermediate values. In this context, we just observe that most DPA attacks against block cipher implementations are based on the knowledge of either the plaintexts or the ciphertexts. To our knowledge, the best attacks in fully unknown input conditions are algebraic ones, e.g. [35,36], which are hardly realistic in large parallel devices. Hence, it is reasonable to assume that the most critical threat against this PRF construction is taking advantage of the carefully chosen plaintexts. This scenario is investigated next.

5   Worst Case Security Analyses

The previous section argued that breaking an AES-based leakage-resilient PRF taking advantage of parallelism could be at least as hard as enumerating a permutation over the AES S-boxes. This would correspond to $16! \approx 2^{44}$ for AES-128, $24! \approx 2^{79}$ for Rijndael-192 and $32! \approx 2^{117}$ for Rijndael-256. Hence, a natural question is to determine whether one can hope for more security, i.e. independent of their order, how difficult is the task of finding the correct N_s key bytes of the PRF? For this purpose, a first strategy is to apply a standard DPA attack and to enumerate the keys from the single score vector it provides. As discussed in Appendix A, this does not lead to any efficient key recovery and suggests large security guarantees. However, it turns out that in view of the design tweaks used in our PRF construction, standard DPA attacks are not anymore the most relevant tool for their security evaluation. In the rest of this section, we discuss two alternative techniques for attacking a PRF implementation. In both cases, the attacks rely on strong assumptions. Namely, we assume that the adversary has a perfect knowledge of the leakage function and can average his measurements in order to obtain noiseless leakages. As discussed in Appendix B, producing such noiseless traces may require significant amounts of measurement data. These conditions are motivated by the goal of investigating worst-case security. In this setting, we first describe an iterative type of DPA attack that significantly improves over the one in Appendix A. Next, we analyze the impact of advanced attacks using lattice reduction. In both cases, the results underline that the PRF construction does not offer much more security than what is bounded by the time needed to enumerate the permutation, if perfect measurements are available.

5.1   An Iterative DPA-Like Attack

The aim of the iterative DPA attack is to recover the correct set of key bytes in the PRF implementation, exploiting the fact that one correct key byte is ranked at position one with high probability (see Appendix A). It works by iteratively removing the algorithmic noise corresponding to the best rated key bytes.

In the beginning of the attack, the adversary has an empty set of recovered key bytes and mounts a first DPA. As a result, he adds a first key byte to the set of recovered key bytes, corresponding to the highest rank in his (single) score vector. Next, he mounts a second DPA, this time adding the algorithmic noise corresponding to the already recovered key byte to his predictions. As a result, he adds a second key byte to the set of recovered key bytes. This procedure is repeated until a set of N_s key bytes is recovered. Simulated DPA attacks in a noise free scenario and assuming that the adversary exactly knows the leakage model (i.e. in the same worst-case conditions as in Section 4) show that this simple strategy succeeds for $N_s = 16$ and $N_p = 256$ with a probability of $\approx 59\%$. However, as soon as we either increase N_s or decrease N_p, the success rate drops, as exhibited in Table 1. Additionally, there is no obvious way to (1) tell immediately if a wrong key byte was picked and (2) efficiently recover the master key from an incorrect set of key bytes. In the next section, we show that advanced attacks based on lattice reduction provide a more robust and systematic way to exploit the side-channel leakage of our PRF implementation.

Table 1. Success rates for the iterative DPA attack

	$N_p = 4$	8	16	32	64	128	256
$N_s = 2$	0.0	30.3	81.4	98.6	99.7	99.7	99.5
4	0.0	0.0	13.0	78.3	96.8	97.3	97.8
8	0.0	0.0	0.0	10.1	69.8	89.2	91.1
16	0.0	0.0	0.0	0.0	3.8	36.7	58.8
32	0.0	0.0	0.0	0.0	0.0	0.2	2.9

5.2 Advanced Attacks Using Lattice Reduction

Let us first recall Equation (1) that describes our noiseless leakages:

$$l_j = \sum_{i=1}^{N_s} \mathsf{L}(\mathsf{S}(p_j[i] \oplus k[i])).$$

Similarly to the previous section, the goal of a lattice-reduction attack is to recover the vector of key bytes $k = \{k[1], k[2], \ldots, k[N_s]\}$ up to a permutation, from a vector of noiseless leakages $\overline{\mathbf{L}} = \{l_1, l_2, \ldots, l_{N_p}\}$. To simplify the analysis, we first assume that all key bytes in the vector k are distinct[1]. In this context, we denote byte-wise hypothetical leakage values as $l_a^b = \mathsf{L}(\mathsf{S}(b \oplus a))$, where b (resp. a) represents an hypothetical plaintext byte (resp. key byte). Next, we define a N_p-dimension vector $\overline{l}_a = \{l_a^1, l_a^2, \ldots, l_a^{N_p}\}$. Our problem can now be restated as finding a subset $\mathcal{K}$ of $[0, 1, \ldots, 255]$ containing N_s elements such that $\overline{\mathbf{L}} = \sum_{a \in \mathcal{K}} \overline{l}_a$. This turns the initial problem into a vectorial knapsack problem. To solve this knapsack problem, we can either try generic algorithms as in [5,20,37], or a lattice-based approach [24]. It is well-known that the lattice-based approach is very efficient for some knapsack problems and fails to work for other parameters. Since our context is quite specific, the parameters we are concerned with are not covered in standard textbooks. Moreover, our parameters are fixed and an asymptotic analysis does not make sense in this case. As a consequence, we decided to investigate the practical performance of a lattice reduction attack. As will be clear next, this lattice reduction approach is surprisingly efficient and the security estimates obtained by analyzing exhaustive

---

[1] With this assumption, an exhaustive search on k can be achieved by trying all choices of N_s key values among 256. Under this exhaustive search attack, the security for $N_s = 16$ is 83 bits, for $N_s = 24$ it is 111 bits and for $N_s = 32$ it is 135 bits.

search (in footnote 2) are overoptimistic. We note that the lattice based approach also outperforms the results obtained with generic algorithms. Hence, we only focus on this solution in the rest of the section. Taking the wost-case example of $N_p = 256$, we can construct the lattice spanned by the columns of the following matrix:

$$\begin{pmatrix} \kappa\, \bar{l}_0 & \kappa\, \bar{l}_1 & \cdots & \kappa\, \bar{l}_{255} & \kappa\, \overline{\mathbf{L}} \\ 1 & 0 & \cdots & 0 & 0 \\ 0 & 1 & \cdots & 0 & 0 \\ \vdots & \vdots & \ddots & \vdots & 0 \\ 0 & 0 & \cdots & 1 & 0 \end{pmatrix},$$

where κ is a large enough constant to guarantee that any short vector in this lattice has its 256 first rows equal to 0. There exists a short vector of squared-norm N_s in the lattice which contains 0 in the first 256 rows and such that the next 256 rows are the characteristic vector of the set $\mathcal{K}$, i.e. there is a 1 in row $257 + i$ iff i is in $\mathcal{K}$ (all other rows contain zeros). Hence, if we find a short vector containing exactly N_s 1s, it can be converted into a set of keys which is compatible with the observed leakages. Note that for large values of N_p (e.g. 256), we expect only one solution for $\mathcal{K}$, which is experimentally verified next.

Note finally that if there are collisions in the vector of keys, we can still apply the same method with a minor change: the expected short vector becomes an encoding of a multiset. In particular, a key byte which appears twice is encoded by a 2. As a consequence, the principle of the attack is left unchanged. However, due to the presence of squares in the computation of the norm, the expected short vector has a larger norm which lowers the probability of success.

Experimenting the Attack. As in the previous sections, we decided to consider a Hamming weight leakage function in our evaluations. In addition to the previously described case with $N_p = 256$, we again experimented with truncated versions of the vector $\overline{\mathbf{L}}$, i.e. with smaller N_p's. For each pair of parameters N_s, N_p, we performed 100 independent experiments (except for the case $N_p = 256$ where we performed 1000 experiments) and extracted the success rate and average execution time of the LLL algorithm using the FPLLL library [1] of Cadé, Pujol and Stehlé on an Intel Core i7-2820QM processor clocked at 2.30GHz. These results are given in Table 2. Note that a TBD entry means that we have

Table 2. Measured success rates and average timing for the lattice-based attack

	$N_p = 256$	254	252	251	250	249	248	247	246	245
$N_s = 16$	100	100	100	100	100	100	100	100	100	100
	1.3s	1.4s	1.4s	1.4s	1.5s	1.5s	3.1s	34.8s	73.0s	131.4s
24	99.9	100	100	100	100	100	100	100	TBD	TBD
	1.4s	1.4s	1.4s	1.4s	1.5s	1.5s	3.1s	35.5s	≈88s	≈143s
32	79.6	79	79	83	80	79	76	TBD	TBD	TBD
	1.4s	1.5s	1.5s	1.5s	1.6s	1.6s	3.3s	≈33s	≈81s	≈140s

only performed a single test in order to determine an approximate running time but no meaningful probability of success. These results clearly exhibit that the LLL-based approach outperforms the heuristic iterative DPA in the previous section. Yet, one can observe that decreasing the number of leakages in large implementations (e.g. for $N_s = 32$) leads to significant increases of the execution times. More detailed results for the $N_p = 256$ case are presented in Table 3, also reflecting the fraction of key vectors containing collisions in our experiments.

Table 3. Additional data for $N_p = 256$

N_s	Key vectors w/o collisions	Successes	Key vectors w/ collisions	Successes	Overall fraction	Timing
16	610	610	390	390	100 %	1.3s
24	328	328	672	671	99.9 %	1.4s
32	141	137	859	659	79.6 %	1.4s
40	40	23	960	479	50.2 %	1.6s

Improving the success rate. In order to improve the probability of success when N_s grows, we can also combine the lattice reduction approach with a partial exhaustive search. The idea is to guess the contribution of some fixed vector, to subtract this guessed contribution from the target vector and to re-run the attack without the guessed vector and with a smaller short vector.

6   Practical Instantiation Issues

The previous sections of this paper suggest that a leakage-resilient PRF offers interesting security arguments compared to state-of-the-art countermeasures against side-channel attacks. Motivated by the need to understand the impact of different parameters in a PRF implementation, our analysis was mostly based on idealized leakage functions. In this section, we complement this view with a first discussion of some important practical instantiation issues.

Performance Evaluation. We evaluated the hardware performance of our construction based on AES-128 (LRPRF-128) and Rijndael-192 (LRPRF-192). For this purpose, we opted for a fully parallel, encryption-only implementation of the algorithms. In addition to these block ciphers, the PRF designs also contain a register to store the x value and some control logic to operate the blocks. Using Synopsis Design Compiler 2010 and the STM 65nm CMOS standard cell library, this resulted in an area of 9.97 kGE (resp. 14.43 kGE) for LRPRF-128 (resp. LRPRF-192). An encryption takes 10 (resp. 12) cycles plus 2 cycles to load the key and the plaintext. Thus, the complete PRF evaluation with a 16 (resp. 24) byte value for x takes $17 \times 12 + 2 = 206$ (resp. $25 \times 14 + 2 = 352$) cycles, where the additional two are again for loading the key and the x value. This is in fact in line with state-of-the-art protected implementations like the one by Moradi et al. from Eurocrypt 2011 [30]. Their threshold implementation of AES-128 takes 266 cycles at an area of 11.12 kGE. In addition, we mention that given some memory overheads, the PRF construction gains particular interest when used for encryption in counter mode. It enables starting the PRF evaluation from intermediate results of previous evaluations. For instance, producing a 512-bit keystream can be done in only $(17 + 2 + 2 + 2) \times 12 + 2 = 278$ cycles (given that there is no overflow of the least significant byte in the IV).

Investigation of Leakage Models. In order to analyze the practicality of the requirement that the S-box leakage models must be identical, we performed a case study on the SASEBO evaluation platform [2] and measured the power consumption of two circuits. The first one consisted of two block-RAM based S-boxes and the second one implemented two S-boxes following Canright's approach [8]. For both circuits, we acquired one million traces and built templates from them. That is, for each S-box within a circuit, we characterized 256 Gaussian distributions corresponding to the 256 possible inputs. We then used the mean values of these distributions as the leakage model for an S-box. The extracted leakage models can be seen in Figure 4. For block-RAM based S-boxes, they show a Pearson correlation of 0.996. This means that for carefully designed implementations, the requirement of identical leakage models can indeed be fulfilled. By contrast for the Canright implementation, there was a visible layout difference between the two instances on the FPGA. Therefore, also the models differed and the correlation of the mean values decreased to 0.686. From this case study, we can conclude that for some implementations, there are leakage differences which can be extracted by an adversary with profiling capabilities. In the next paragraph, we discuss whether these differences can be exploited.

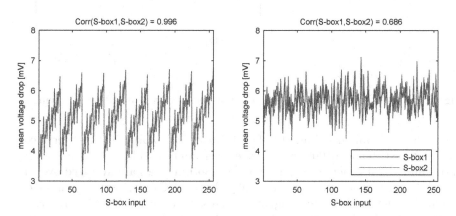

Fig. 4. Leakage models: block-RAM based S-box (left) and Canright S-box (right)

Impact of Algorithmic Noise. In traditional DPA attacks, the algorithmic noise is considered to be Gaussian and, due to uniformly distributed inputs, averages out for a sufficient number of inputs. In our case on the other hand, all inputs are determined by a single byte and a fixed key. Therefore the algorithmic noise cannot be averaged out and in addition, is fully determined by the unknown part of the key. Clearly, directly profiling such kind of noise is computationally

hard (it corresponds to performing a DPA directly on the full master key).
Therefore, and in order to analyze the effect of this key-dependent algorithmic
noise, we performed the following simulated experiment. First, we implemented
$N_s = \{2, 4, 8\}$ Canright S-boxes on the FPGA (i.e. we considered the most
different leakage models). For each S-box S_i with $i \in [1; N_s]$, we measured 400
traces for each input, while keeping the inputs to the other S-boxes at zero. This
way we could build precise leakage models L_i without acquiring any algorithmic
noise. To simulate real traces where all S-boxes operate in parallel, we then built
the overall leakage function as: $L'(p) = \sum_{i=1}^{N_s} L_i(S_i(p[i])) + n$, where p is the
N_s-byte input, $p[i]$ is the ith byte of the input, and n an Gaussian distributed
measurement noise estimated from our data set. This leakage description was
then used to simulate our N_s S-box device from which we could sample traces for
arbitrary inputs. From this point, we proceeded as usual. That is, we built N_s
templates (now including algorithmic noise) by sampling 100 million traces from
$L'(\cdot)$. Next, we launched template attacks by sampling 300 times 256 000 traces
(for 300 different keys). The results of these attacks can be seen in Figure 5.

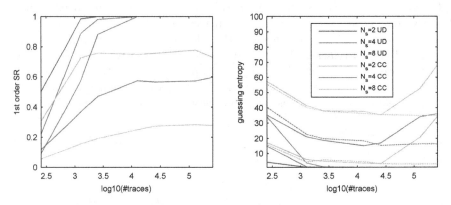

Fig. 5. Impact of algorithmic noise reflected by the success rate and guessing entropy

For the uniformly distributed (UD) plaintexts, all subkeys are recovered correctly after 3 000 and 13 000 traces respectively, indicated by a guessing entropy
of one and a first-order success rate of one. Both metrics are averaged over the
N_s S-boxes. By contrast, for the carefully chosen (CC) plaintexts, it can be observed that the success rates stagnate at the same time as when they reach one
for the uniformly distributed plaintexts. This is because for some subkeys the
models will fit. Hence, those subkeys can be recovered with good probability. But
for the remaining subkeys it is not possible to carry out a successful recovery. In
the case of a template attack (represented with plain curves in the figure), this
means that the probability of the correct key will diminish at some point, which
is the reason why the guessing entropy increases again after 26 000 traces. In the
case of a correlation attack using the templates' mean value as model (instead
of the usual Hamming weight model [7]), this effect vanishes, as represented by
the dotted lines. That is, the "hard to recover" subkeys then stagnate at fixed

ranks in the lists (corresponding to a fixed correlation coefficient value), rather than decreasing towards a probability zero, due to an incorrect model. Thus, we can conclude that even if there are actual differences in the leakage models of the different S-boxes of a PRF implementation, and strong profiling is possible for the adversary, the key-dependent algorithmic noise prevents the building of a sound leakage model. For example, already for 8 parallel S-boxes, each subkey remains with a guessing entropy of ≈ 35 in our case study. It would further increase with more parallel S-boxes (the previously described PRF implementations would have at least 16 and 24 ones, respectively). Hence, the only way to perform successful key recoveries in these cases would be to build templates for the full key, which is unrealistic for computational reasons.

Preventing DPA Attacks against MixColumns. The central result of our security analyses is that performing a side-channel attack against our parallel PRF implementation should at least require to enumerate a permutation over N_s S-boxes. However, this implicitly assumes that the only path for performing a DPA is this operation, which neglects the possibility to mount attacks against MixColumns. In general, such attacks are more computationally intensive, as they require guessing 2^{32} key candidates. Yet, this remains achievable with modern computers. Given that such attacks succeed, it would only remain to enumerate a permutation of the MixColumns operations (i.e. $4! \times 4$ for the AES-128, where the factor 4 relates to the fact that the adversary would recover the 32-bit subkeys up to a byte-wise rotation). However, this attack may not always be applicable in practice, and can be made more computationally intensive, as we now discuss. First note that in a hardware implementation, the adversary may have to target the Hamming distance between the state register values before and after the first round. But one byte of this value depends on five key bytes and four bytes of this value depend on eight key bytes, which is harder to guess. In addition, there is a simple and general trick to increase the amount of bytes to guess after the MixColumns transform. Namely, one just has to switch the order of MixColumns and AddRoundKey. This requires that the key schedule applies the inverse MixColumns operation to the round keys before outputting them. Since the only non-linear operation in the key schedule is SubWord, the key schedule can operate on accordingly recoded round keys and, instead of applying SubWord, apply the sequence MixColumns, SubWord, InvMixColumns. The costs of this accounts for one InvMixColumns and one MixColumns unit. Finally, any attack against larger subkeys would still have a bounded data complexity of 256 with key-dependent algorithmic noise. Summarizing, the protection of MixColumns against DPA can be enhanced by different architectural means. Besides, and quite interestingly, this discussion highlights that the AES is not the best suited algorithm for integration in our leakage-resilient PRF. Hence, it suggests the design of block ciphers with more convenient diffusion layers for this purpose as another interesting scope for further research.

Security against Fault Attacks. Finally, there is an additional advantage to our construction. Usually, the resources for side-channel attacks and fault protections cannot be shared. For leakage-resilient PRFs, on the other hand, we can provide a first-order fault protection based on temporal redundancy, by just repeating the last step of the construction. Taking the LRPRF-128 as an example, we would perform 18 instead of 17 encryptions. This accounts for an overhead of only 5.8%, rather than the usual 100% for block ciphers.

7   Conclusions and Consequences for Block Cipher Design

This paper describes tweaks to improve both the practical security and the efficiency of leakage-resilient PRFs. They allow quantifying physical security with a parameter that has super-exponential impact on the time complexity of a successful attack. They also open the paths towards real world applications, as their performance overheads are in line with other countermeasures against side-channel attacks. In particular, the only known countermeasure with an exponential security parameter is masking. But increasing the number of masks in a block cipher implementation is generally (much) more expensive than increasing its parallelism. Next, our results suggest interesting challenges for the design of new block ciphers, as the AES Rijndael appears not to be an ideal candidate for integration in leakage-resilient constructions. Possible tracks for investigation include modifying the number and size of S-boxes (that directly affect the security vs. efficiency tradeoff of the PRF), reducing the number of rounds in the inner steps of the construction, and improving diffusion layers in order to avoid the possible attacks after the diffusion layer described in Section 6.

References

1. http://perso.ens-lyon.fr/xavier.pujol/fplll/
2. http://staff.aist.go.jp/akashi.satoh/sasebo/en/board/sasebo.html
3. Akavia, A., Goldwasser, S., Vaikuntanathan, V.: Simultaneous Hardcore Bits and Cryptography against Memory Attacks. In: Reingold, O. (ed.) TCC 2009. LNCS, vol. 5444, pp. 474–495. Springer, Heidelberg (2009)
4. Anderson, R., Kuhn, M.: Low Cost Attacks on Tamper Resistant Devices. In: Christianson, B., Crispo, B., Lomas, M., Roe, M. (eds.) Security Protocols 1997. LNCS, vol. 1361, pp. 125–136. Springer, Heidelberg (1998)
5. Becker, A., Coron, J.-S., Joux, A.: Improved generic algorithms for hard knapsacks. In: Paterson (ed.) [32], pp. 364–385
6. Boneh, D., DeMillo, R.A., Lipton, R.J.: On the Importance of Checking Cryptographic Protocols for Faults. In: Fumy, W. (ed.) EUROCRYPT 1997. LNCS, vol. 1233, pp. 37–51. Springer, Heidelberg (1997)
7. Brier, E., Clavier, C., Olivier, F.: Correlation Power Analysis with a Leakage Model. In: Joye, M., Quisquater, J.-J. (eds.) CHES 2004. LNCS, vol. 3156, pp. 16–29. Springer, Heidelberg (2004)
8. Canright, D.: A Very Compact S-Box for AES. In: Rao, J.R., Sunar, B. (eds.) CHES 2005. LNCS, vol. 3659, pp. 441–455. Springer, Heidelberg (2005)
9. Chari, S., Rao, J.R., Rohatgi, P.: Template Attacks. In: Kaliski Jr., B.S., Koç, Ç.K., Paar, C. (eds.) CHES 2002. LNCS, vol. 2523, pp. 13–28. Springer, Heidelberg (2003)

10. Crescenzo, G.D., Lipton, R.J., Walfish, S.: Perfectly secure password protocols in the bounded retrieval model. In: Halevi, Rabin (eds.) [19], pp. 225–244
11. Dodis, Y., Kalai, Y.T., Lovett, S.: On cryptography with auxiliary input. In: Mitzenmacher, M. (ed.) STOC, pp. 621–630. ACM (2009)
12. Dodis, Y., Pietrzak, K.: Leakage-Resilient Pseudorandom Functions and Side-Channel Attacks on Feistel Networks. In: Rabin, T. (ed.) CRYPTO 2010. LNCS, vol. 6223, pp. 21–40. Springer, Heidelberg (2010)
13. Dziembowski, S.: Intrusion-resilience via the bounded-storage model. In: Halevi, Rabin (eds.) [19], pp. 207–224
14. Dziembowski, S., Pietrzak, K.: Leakage-resilient cryptography. In: FOCS, pp. 293–302. IEEE Computer Society (2008)
15. Güneysu, T., Lyubashevsky, V., Pöppelmann, T.: Practical Lattice-Based Cryptography: A Signature Scheme for Embedded Systems. In: Prouff, E., Schaumont, P. (eds.) CHES 2012. LNCS, vol. 7428, pp. 530–547. Springer, Heidelberg (2012)
16. Gennaro, R., Lysyanskaya, A., Malkin, T., Micali, S., Rabin, T.: Algorithmic tamper-proof (atp) security: Theoretical foundations for security against hardware tampering. In: Naor (ed.) [31], pp. 258–277
17. Goldreich, O., Goldwasser, S., Micali, S.: How to construct random functions. J. ACM 33(4), 792–807 (1986)
18. Halderman, J.A., Schoen, S.D., Heninger, N., Clarkson, W., Paul, W., Calandrino, J.A., Feldman, A.J., Appelbaum, J., Felten, E.W.: Lest we remember: Cold boot attacks on encryption keys. In: van Oorschot, P.C. (ed.) USENIX Security Symposium, pp. 45–60. USENIX Association (2008)
19. Halevi, S., Rabin, T. (eds.): TCC 2006. LNCS, vol. 3876. Springer, Heidelberg (2006)
20. Howgrave-Graham, N., Joux, A.: New Generic Algorithms for Hard Knapsacks. In: Gilbert, H. (ed.) EUROCRYPT 2010. LNCS, vol. 6110, pp. 235–256. Springer, Heidelberg (2010)
21. Ishai, Y., Prabhakaran, M., Sahai, A., Wagner, D.: Private Circuits II: Keeping Secrets in Tamperable Circuits. In: Vaudenay, S. (ed.) EUROCRYPT 2006. LNCS, vol. 4004, pp. 308–327. Springer, Heidelberg (2006)
22. Ishai, Y., Sahai, A., Wagner, D.: Private Circuits: Securing Hardware against Probing Attacks. In: Boneh, D. (ed.) CRYPTO 2003. LNCS, vol. 2729, pp. 463–481. Springer, Heidelberg (2003)
23. Joux, A. (ed.): EUROCRYPT 2009. LNCS, vol. 5479. Springer, Heidelberg (2009)
24. Joux, A., Stern, J.: Lattice reduction: A toolbox for the cryptanalyst. J. Cryptology 11(3), 161–185 (1998)
25. Kocher, P.C.: Leak resistant cryptographic indexed key update. US Patent
26. Kocher, P.C., Jaffe, J., Jun, B.: Differential Power Analysis. In: Wiener, M. (ed.) CRYPTO 1999. LNCS, vol. 1666, pp. 388–397. Springer, Heidelberg (1999)
27. Lomne, V.: Power and electro-magnetic side-channel attacks: Threats and countermeasures. PhD thesis, Universite de Montpellier II (2010)
28. Mangard, S., Oswald, E., Popp, T.: Power analysis attacks - revealing the secrets of smart cards. Springer (2007)
29. Micali, S., Reyzin, L.: Physically observable cryptography (extended abstract). In: Naor (ed.) [31], pp. 278–296
30. Moradi, A., Poschmann, A., Ling, S., Paar, C., Wang, H.: Pushing the limits: A very compact and a threshold implementation of AES. In: Paterson (ed.) [32], pp. 69–88
31. Naor, M. (ed.): TCC 2004. LNCS, vol. 2951. Springer, Heidelberg (2004)

32. Paterson, K.G. (ed.): EUROCRYPT 2011. LNCS, vol. 6632. Springer, Heidelberg (2011)
33. Petit, C., Standaert, F.-X., Pereira, O., Malkin, T., Yung, M.: A block cipher based pseudo random number generator secure against side-channel key recovery. In: Abe, M., Gligor, V.D. (eds.) ASIACCS, pp. 56–65. ACM (2008)
34. Pietrzak, K.: A leakage-resilient mode of operation. In: Joux (ed.) [23], pp. 462–482
35. Renauld, M., Standaert, F.-X.: Algebraic Side-Channel Attacks. In: Bao, F., Yung, M., Lin, D., Jing, J. (eds.) Inscrypt 2009. LNCS, vol. 6151, pp. 393–410. Springer, Heidelberg (2010)
36. Renauld, M., Standaert, F.-X., Veyrat-Charvillon, N.: Algebraic Side-Channel Attacks on the AES: Why Time also Matters in DPA. In: Clavier, C., Gaj, K. (eds.) CHES 2009. LNCS, vol. 5747, pp. 97–111. Springer, Heidelberg (2009)
37. Schroeppel, R., Shamir, A.: A $t=o(2^{n/2})$, $s=o(2^{n/4})$ algorithm for certain np-complete problems. SIAM J. Comput. 10(3), 456–464 (1981)
38. Standaert, F.-X.: How Leaky Is an Extractor? In: Abdalla, M., Barreto, P.S.L.M. (eds.) LATINCRYPT 2010. LNCS, vol. 6212, pp. 294–304. Springer, Heidelberg (2010)
39. Standaert, F.-X.: Leakage resilient cryptography: a practical overview. In: ECRYPT Workshop on Symmetric Encryption (SKEW 2011), Copenhagen, Denmark (2011), http://perso.uclouvain.be/fstandae/PUBLIS/96_slides.pdf
40. Standaert, F.-X., Archambeau, C.: Using Subspace-Based Template Attacks to Compare and Combine Power and Electromagnetic Information Leakages. In: Oswald, E., Rohatgi, P. (eds.) CHES 2008. LNCS, vol. 5154, pp. 411–425. Springer, Heidelberg (2008)
41. Standaert, F.-X., Malkin, T., Yung, M.: A unified framework for the analysis of side-channel key recovery attacks. In: Joux (ed.) [23], pp. 443–461
42. Standaert, F.-X., Pereira, O., Yu, Y., Quisquater, J.-J., Yung, M., Oswald, E.: Leakage resilient cryptography in practice. In: Sadeghi, A.-R., Naccache, D. (eds.) Towards Hardware-Intrinsic Security. Information Security and Cryptography, pp. 99–134. Springer, Heidelberg (2010)
43. Yu, Y., Standaert, F.-X., Pereira, O., Yung, M.: Practical leakage-resilient pseudorandom generators. In: Al-Shaer, E., Keromytis, A.D., Shmatikov, V. (eds.) ACM CCS, pp. 141–151. ACM (2010)

A   Security against Standard DPA Attacks

The result of a standard (template-based) DPA attack against our scheme is a single vector, in which all possible subkeys are ranked according to their probability. From this, a full key consisting of N_s bytes has to be reconstructed. Ideally, the set of the N_s correct subkeys would be ranked first and all other subkeys would have a low probability. However, looking at the distribution of the subkeys within such a vector after a noise-free attack shows that this is not the case

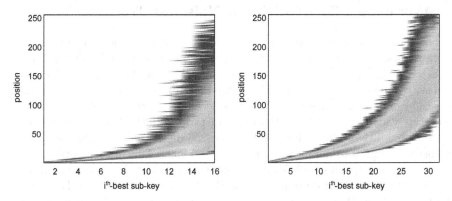

Fig. 6. Distribution of the correct key bytes within a probability vector

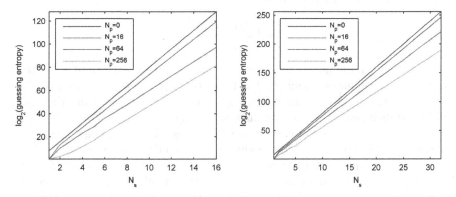

Fig. 7. Estimated and extrapolated guessing entropy of the full key

(mainly because of the algorithmic noise). Figure 7 illustrates where the correct subkeys can be found within the vector on average. Whereas the best-ranked correct key byte can be almost with certainty found at position one, some of the correct subkeys are ranked much lower. Starting from such a vector, the optimal adversarial strategy is to enumerate full keys according to their probability, where all up-to-permutation-identical keys have the same probability. Following this strategy, we estimated and extrapolated the guessing entropy. This was done by generating a probability vector for N_s values between one and seven with a constant noise variance which would correspond to an algorithmic noise of 16 or 32 parallel S-boxes. From each vector we sampled 2^{30} random full keys and checked the position of the correct full key within this set. Afterwards, we scaled this position to 256^{N_s} and added the complexity for the permutation. Finally, since we could observe a power-law for the guessing entropy, we extrapolated these values up to 16 and 32 S-boxes using the slope in log-scale. The very fact that the end points of these extrapolations suggest a security of 2^{84} and 2^{185} show that standard DPA cannot be the optimal strategy.

B   Averaging Effort to Obtain Noiseless Traces

As the security evaluations in Section 5 both consider noiseless traces, an interesting question is to determine the averaging effort that would be needed to obtain such high quality information from an actual implementation. For this purpose, we measured a fully parallel AES-128 FPGA implementation on the SASEBO evaluation platform [2]. We considered traces close to noise-free if we can correctly identify the 128-bit Hamming distance value for 256 measurements, with a probability of 0.90. Thus, each of the 256 measurements must be classified correctly with a probability of $(0.9)^{1/256} = 0.9996$. This in turn corresponds to a confidence interval of $3.54\,\sigma$, assuming a normal distribution of the noise. Thus, to allow error-free decoding, we need the mean values for the Hamming weights to be twice that value apart, meaning $7.08\,\sigma$. Given the distance between the mean values $\Delta\mu$ and the standard deviation of our measurements, we can calculate the number of traces to average as:

$$n = \left(\frac{7.08\,\sigma}{\Delta\mu}\right)^2.$$

As we get only sample means from our measurements, we calculated n for the average and the minimum value of $\Delta\mu$. The latter one is also motivated by the fact that, if due to the power model the $\Delta\mu$ values are not equidistant, then the smallest distance determines n. For the average $\Delta\mu$ we found $n = 6.9 * 10^3$ and for the minimum value we found $n = 8.8 * 10^6$. To get a close to noise-free mean trace for every plaintext, we additionally need to multiply this number by 256, thus we need to acquire a total number of $1.78 * 10^6$ and $2.27 * 10^9$ traces for the minimum and average $\Delta\mu$, respectively. The actual number of traces to acquire most likely lies somewhere between these two extreme values. Hence, it suggests that the averaging effort can be expected to be non-negligible.

C   Additional Figure

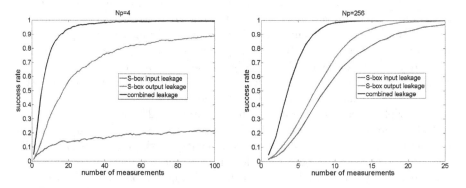

Fig. 8. Experimental attacks against the AES with bounded data complexity

Practical Leakage-Resilient Symmetric Cryptography

Sebastian Faust[1,*], Krzysztof Pietrzak[2,**], and Joachim Schipper[2,**]

[1] Århus University
[2] IST Austria

Abstract. Leakage resilient cryptography attempts to incorporate side-channel leakage into the black-box security model and designs cryptographic schemes that are provably secure within it. Informally, a scheme is *leakage-resilient* if it remains secure even if an adversary learns a bounded amount of arbitrary information about the schemes internal state. Unfortunately, most leakage resilient schemes are unnecessarily complicated in order to achieve strong provable security guarantees. As advocated by Yu et al. [CCS'10], this mostly is an artefact of the security proof and in practice much simpler construction may already suffice to protect against *realistic* side-channel attacks. In this paper, we show that indeed for simpler constructions leakage-resilience can be obtained when we aim for relaxed security notions where the leakage-functions and/or the inputs to the primitive are chosen *non-adaptively*. For example, we show that a three round Feistel network instantiated with a leakage resilient PRF yields a leakage resilient PRP if the inputs are chosen non-adaptively (This complements the result of Dodis and Pietrzak [CRYPTO'10] who show that if a adaptive queries are allowed, a super-logarithmic number of rounds is necessary.) We also show that a minor variation of the classical GGM construction gives a leakage resilient PRF if both, the leakage-function and the inputs, are chosen non-adaptively.

1 Introduction

Traditional cryptographic security notions only consider adversaries who get black-box access to the primitive at hand. That is, an adversary can only observe the input/output behavior of the cryptosystem, but gets no other information about its inner workings. Unfortunately, such black-box security notions are often insufficient to guarantee real-world security of cryptosystems. This is due to information inadvertently emitting from the physical implementation of the cryptosystem, which can be exploited by *side-channel attacks*. In the last years a

---

* Sebastian Faust acknowledges support from the Danish National Research Foundation and The National Science Foundation of China (under the grant 61061130540) for the Sino-Danish Center for the Theory of Interactive Computation, within part of this work was performed; and from the CFEM research center, supported by the Danish Strategic Research Council.
** Supported by the European Research Council/ERC Starting Grant 259668-PSPC.

E. Prouff and P. Schaumont (Eds.): CHES 2012, LNCS 7428, pp. 213–232, 2012.

large body of theoretical work attempts to incorporate these side-channel attacks into the security model and to design new cryptographic schemes that provably protect against them. Despite important progress in this area, only very few works in the theory community consider how to protect symmetric primitives against leakage attacks. That is somewhat surprising as symmetric primitives such as pseudorandom number generators and block ciphers are the "working horses" of cryptography and are by far the most frequent target of side-channel attacks. Moreover, as frequently pointed out [22,23], many of the recent theoretical constructions are rather involved and use techniques which only seem to be required to enable the security proof, and do not necessarily contribute to the real-world security of the system. In this work, we show that *simpler* and *more natural* constructions of important *symmetric* primitives such as pseudorandom functions (PRFs) and pseudorandom permutations (PRPs) are provable leakage resilience if we aim for weaker security notions.

1.1   Modeling Leakage Resilience and Weaker Security Notions

As most previous works on leakage-resilient symmetric primitives [4,20,1,23], we follow Dziembowski and Pietrzak [4] who structure the computation into *time steps* and require that the leakage given to the adversary is some bounded amount of *arbitrary* polynomial-time computable information about the data/state that is used during this step. The latter restriction that the leakage function is only applied to the state touched in an invocation was suggested by [19] under the term "only computation leaks information". As the number of invocations of a scheme is usually unbounded, also the amount of leakage can become arbitrarily large.

On Granular Leakage Resilience (gLR). Typically, a time step is one invocation of the scheme that leaks independently from the computation in the previous and next time step. This could for example be the computation of a signature [5], or the generation of a block of pseudorandom bits for stream-ciphers [4,20]. In this work, we will follow [1,22] and consider a more fine grained notion where the construction of some leakage resilient (LR) scheme CS requires several invocations of an underlying cryptographic primitive P,[1] and we require that each invocation of P leaks independently. We call this notion *granular leakage-resilience*, or gLR for short. We notice that in the literature on leakage resilience even more fine grained models have been considered [9,3]

As side-channel leakage is often a global phenomenon (e.g., in power analysis attacks the adversary measures the global power consumption of the device), the question arises whether such a locality restriction still suffices to model relevant leakages in practice. For certain important leakage classes, we can answer this question affirmatively. For instance, the prominent Hamming weight leakage function can be computed independently from the Hamming weight of the local states. A similar observation works for any *affine* leakage function.

---

[1] Concretely, we will consider the cases where CS is a LR-PRF and P a wPRF, and the case where CS is a LR-PRP and P a LR-PRF.

Formally, we model granularity as follows. Let τ_i be the state that is used by the computation (keys, inputs, randomness) in time step i. Before each such step, the adversary can adaptively choose a leakage function f_i, and after this step has been processed, she learns $f_i(\tau_i)$.

On Non-adaptive Leakage Resilience (naLR). Besides granularity, another natural relaxation of leakage resilience, which has been considered in e.g. [23,1,22], is to require that the adversary has to fix the leakage functions in advance before seeing any leakage or outputs. This notion is called non-adaptive leakage resilience, or naLR for short. In the leakage setting, a fully adaptive choice of the leakage function may be an overly powerful model to capture side-channel attacks, as in practice the leakage function is often fixed in advance by the device and the measurement equipment (for more discussion on this cf. [23,1,22]). Also, as discussed in [23], for stateless cryptographic schemes that do not allow to evolve the secret state, such as PRFs or PRPs, one simply cannot achieve security against adaptively chosen leakage functions: the adversary can just learn the state bit-by-bit by picking for each observation a different leakage function.[2]

1.2   Our Contributions

In this work, we study various new and existing constructions of leakage-resilient pseudorandom objects. In a nutshell our results can be summarized as follows:

1. We revisit the work of Yu et al. [23] and show that the proof of the proposed (more natural) construction of a non-adaptive leakage-resilient (naLR) stream cipher has a subtle flaw. We propose a simple solution to this problem which unfortunately is impractical.
2. Inspired by the work of Dodis and Pietrzak [1], we show how to construct a nagLR non-adaptive PRF which is simpler and more natural and avoids the alternating structure used in [1].
3. We prove that a Feistel network with only 3 rounds, each instantiated with a non-adaptive leakage-resilient non-adaptive PRF, yields a non-adaptive leakage-resilient non-adaptive PRP. This completes a result of [1] who showed that a leakage-resilient PRP requires a superlogarithmic number of rounds instantiated with a leakage-resilient PRF.

We elaborate on these results further below.

Section 2: Yu et al. [23] Revisited. The first leakage resilient symmetric primitive was the stream-cipher construction proposed by Dziembowski and Pietrzak in [4]. This construction has later been simplified in [20] using a weak PRF and is illustrated in Figure 1.

---

[2] In this paper we not only differentiate between adaptive/non-adaptive leakage, but also adaptive/non-adaptive PRFs. In the latter non-adaptive means the adversary fixes all inputs in advance. We use the convention that non-adaptive leakage-resilient PRF means the leakage functions are chosen non-adaptively, whereas leakage-resilient non-adaptive PRF means the inputs to the PRF are chosen non-adaptively.

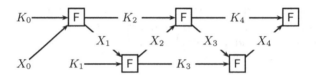

Fig. 1. Construction SC_{ALT} of a leakage resilient stream-cipher from any (weak) PRF
F [20]. The initial secret key is X_0, K_0, K_1, the output is $X_0, X_1, \ldots$.

The constructions from [4,20] use an alternating structure (cf. Figure 1) and requires the secret state to hold *two* secret keys K_i, K_{i+1} for the underlying weak PRF F (a weak PRF is only guaranteed to be random on random inputs). The alternating structure enforces independence between inputs X_i and keys K_i, but seems mostly motivated by the security proof rather than contributing much to the scheme's real-world security.

Yu et al. [23] advocate that already much simpler constructions will be secure against most practically relevant side-channel attacks. They propose a more natural construction SC_{SEQ} from any wPRF F as illustrated in Figure 2. The secret state of this scheme consists of only a *single* secret key K_i for F, and two fixed *public* random values p_0, p_1 which are used alternately as inputs to F. This scheme is not leakage resilience if the leakage functions can be chosen adaptively, which is easily seen by the so-called "precomputation" attack: as we know p_0, p_1, in the ith round we can choose a leakage function f_i which (using its input K_{i-1}) computes a future key K_t (for some $t > i$) and leaks some bits about it. As we can do this for any $i < t$, we can (for a sufficiently large t) learn the entire K_t.

It is claimed in [23] that the construction from Figure 2 is a naLR stream cipher. Note that the precomputation attack becomes infeasible if one must choose the leakage functions f_i before seeing p_0, p_1, as now $f_i(K_{i-1}, p_{i-1 \bmod 2})$ cannot compute the future key $K_{i+1} = \mathsf{F}(K_i, p_{i \bmod 2})$. Unfortunately, as we discuss in Section 2, the main technical lemma used in their proof has a subtle flaw, and thus the security proof is incorrect. Currently, we do not know if the construction is actually insecure, or if the proof can be salvaged. Our counterexample showing that their main lemma is flawed does not lead to an actual attack on the naLR security of the cipher.

The proof in [23] uses a lemma from [20] which states that the output $\mathsf{F}(K, X)$ of a weak PRF F is pseudorandom, even if K, X only have high pseudoentropy and are *independent*. The flaw in their proof roots from the fact that the input p_0 is reused every second round (cf. Figure 2), and thus already in the 3rd round, where one computes $K_3 \leftarrow \mathsf{F}(K_2, p_0)$, the key K_2 is not independent from p_0, which means one cannot apply the lemma from [20] directly.

This dependence problem disappears if one uses fresh public random inputs $p_0, p_1, p_2, \ldots$ for every round instead of alternating the two values p_0 and p_1, we will denote this construction by SC_{SEQ}^+. Of course SC_{SEQ}^+ is pretty much useless in practice as its description size (i.e. the public inputs $p_0, p_1, \ldots$) is linear in the length of the output it can generate. Nonetheless, the observation that SC_{SEQ}^+ is naLR will be useful for constructing nagLR non-adaptive PRFs as discussed below.

Section 3: Leakage-Resilient PRFs. Dodis and Pietrzak [1] construct a nagLR PRF. Their basic idea is to use the leakage resilient stream-cipher from [20] in a tree-like construction (inspired by the classical GGM construction.). Their construction is rather involved, as the alternating structure of the stream-cipher must be preserved within the tree like structure of the GGM transformation.[3]

We propose a much simpler construction illustrated in Figure 3, which we get by using the naLR stream cipher SC^+_{SEQ} (discussed in the previous section) within a GGM-like tree-structure. One may expect that starting with naLR stream-cipher like SC^+_{SEQ} and use it within GGM, we obtain a naLR PRF. Surprisingly, we show that this intuition is wrong. In fact, our construction in Figure 3 can be completely broken even using only non-adaptive leakage.

Our attack exploits the fact that, even though the leakage-functions cannot be adaptively chosen, the inputs to the PRF can be chosen adaptively. In particular, the choice of the inputs can depend on the public values p_i. Intuitively, this allows us to commit to exponentially many leakage functions (one for each input to the PRF) at the beginning, and only later, when we learn the p_i's we can choose which leakage function to choose by choosing the appropriate input to the PRF adaptively.[4] On the positive side, we show that our construction $\Gamma^{F,m}$ is a nagLR non-adaptive PRF, that is, it is secure if *not only* the leakage-function, but *also* the inputs to $\Gamma^{F,m}$ are chosen non-adaptively. This, of course, is a strong assumption, but for some important applications, like the initialization of a stream cipher [22], such a non-adaptive PRF is sufficient (in fact, here even a weak PRF is sufficient). Also, we would like to mention that in practice many side-channel attacks, such as DPA attacks, work by measuring the power consumption of the device on *random* inputs. Our security analysis incorporates such important attacks where the adversary exploits leakages from random inputs to the cryptographic scheme. We emphasize that, of course, our construction is an adaptively secure PRF in the black-box sense.

Section 4: Leakage-Resilient PRPs. A classical result by Luby and Rackoff [16] shows that a three-round Feistel (cf. Figure 4) network, where each round is instantiated with a secure PRF, is a secure PRP. Dodis and Pietrzak [1] show that three-round Feistel networks cannot be leakage resilient. More precisely, they show that every Feistel network with a constant number of rounds (using any perfectly leakage resilient round functions, e.g. a random oracle) can be broken using only very simple leakage (e.g., the Hamming-weight of the inputs to the round functions). On the positive side, they show that a Feistel network with a super-logarithmic number of rounds instantiated with $\mathcal{L}$-LR PRFs is a $\mathcal{L}$-gLR PRP for any class $\mathcal{L}$ of leakage functions. Here, $\mathcal{L}$ is some class of

---

[3] Whereas the GGM construction is just a simple tree, the construction of [1] is a graph with tree-width 3.

[4] Let us mention that for the attack we require that the leakage functions are aware (i.e. get as input) which node in the tree they are leaking from. Modeling granular leakage like this makes our positive results stronger, but the attacks more artificial. We don't know if our construction can be broken with non-adaptive leakage where the leakage-function is oblivious about the node it is leaking from.

admissible leakage functions, which in our case will usually be all polynomial-time computable functions with range $\{0,1\}^\lambda$ for some $\lambda \in \mathbb{N}$.

The aforementioned attack requires that one can query the PRF adaptively. We show that this is inherent by proving that a 3-round Feistel instantiated with $\mathcal{L}$-LR PRFs yields a $\mathcal{L}$-gLR non-adaptive PRP. This again illustrates the power of non-adaptivity in the leakage setting.

1.3   More Related Work

We notice that an alternative way to construct symmetric leakage resilient primitives is by using techniques from leakage resilient circuit compilers. Leakage-resilient circuit compilers allow to transform any circuit, e.g., an implementation of the AES, into a transformed circuit that is protected against certain classes of leakage attacks. This line of research was initiated by Ishai et al. [14] who show security against probing attacks. This result was recently generalized to a setting where leakages can be described by an AC0 circuit [6]. The works that are most relevant to ours are recent leakage-resilient circuit compilers in the "only computation leaks" setting [15,9,3,10]. While on the positive side such compilers allow to provably protect any cryptographic scheme against certain classes of leakage, they typically make strong granularity assumptions and are inefficient.[5]

An approach exploiting parallelism to achieve practically efficient leakage-resilient block-ciphers was put forward by Medwed, Standaert and Joux in these proceedings [18].

1.4   Notation and Basic Definitions

In this section, we present some basic notation and definitions that will be used throughout this paper.

Strings & Sets. Concatenation of two strings x, y is denoted $x \| y$, or, if no confusion is possible, simply xy. For $X \in \{0,1\}^n$ we denote with $X[i]$ the ith bit of X and with $X_{|i}$ the i bit prefix of X. $[a, b]$ denotes the interval $\{a, a+1, \ldots, b\}$, $[b]$ is short for $[1, b]$. For a set $\mathcal{X}$, $X \in_R \mathcal{X}$ denotes that X is assigned a value sampled uniformly at random from $\mathcal{X}$. For a distribution D, we denote $X \leftarrow D$ the random variable X sampled from the distribution D. To abbreviate notation, we often identify random variables with their distribution.

Functions. $\mathcal{R}_{m,n}$ denotes the set of all functions $\{0,1\}^m \to \{0,1\}^n$, $\mathcal{P}_n$ the set of all permutation over $\{0,1\}^n$.

Distance. With $\delta^D(X; Y)$ we denote the advantage of a circuit D in distinguishing the random variables X, Y, i.e.: $\delta^D(X; Y) \stackrel{\text{def}}{=} |\Pr[D(X) = 1] - \Pr[D(Y) = 1]|$. $\Delta(X; Y) \stackrel{\text{def}}{=} \max_D \delta^D(X; Y)$ denotes the statistical distance of X and Y. With $\delta_s(X; Y)$ we denote $\max_D \delta^D(X; Y)$ where the maximum is over all circuits D of size s.

---

[5] Circuits that make use of techniques from [15,9,3,10] grow by a factor of n^2 compared to an unprotected circuit, where n as a statistical security parameter.

Entropies. We recall some basic definitions for different types of entropy.

Definition 1. *A random variable Z has min-entropy k, denoted $H_\infty(Z) = k$, if for all z in the range of Z we have $\Pr[Z = z] \le 2^{-k}$.*

A "computational" version of min-entropy called HILL-pseudoentropy was introduced in [12].

Definition 2. *We say X has HILL pseudoentropy k, denoted by $H^{\mathrm{HILL}}_{\epsilon,s}(X) \ge k$, if there exists a distribution Y with min-entropy $H_\infty(Y) = k$ where $\delta_s(X;Y) \le \epsilon$.*

Dodis et al. [2], and Hsiao et al. [13] extended the above notions to analyze what happens to the min-entropy (resp. HILL-pseudoentropy) of a random variable X given a possibly correlated random variable Z.

Definition 3. *Let (X, Z) be a pair of random variables. The* average min-entropy *of X conditioned on Z is defined as*

$$\widetilde{H}_\infty(X|Z) = -\log \sum_{z \in Z} \Pr[Z = z] 2^{-H_\infty(X|Z=y)}$$

A computational version was given in [13] and is formally defined as follows:

Definition 4. *Let (X, Z) be a pair of random variables. X has conditional HILL pseudoentropy at least k conditioned on Z, denoted $\widetilde{H}^{\mathrm{HILL}}_{\epsilon,s}(X|Z) \ge k$ if there exists a collection of distributions Y_z for each $z \in Z$, giving rise to a joint distribution (Y, Z), such that $\widetilde{H}_\infty(Y|Z) \ge k$ and $\delta_s((X, Z); (Y, Z)) \le \epsilon$.*

Pseudorandomness. Pseudorandomness is a fundamental and extremely useful cryptographic concept. Informally, an object (such as a bit-string, function or permutation) is pseudorandom if (1) it can be efficiently implemented using a small amount of randomness and (2) it cannot be distinguished from the corresponding uniformly random object by any efficient algorithm. A basic building block to generate pseudorandomness that will be used a basic building block in our constructions is a weak pseudorandom function (weak PRF). In contrast to standard PRFs, the notion of a weak PRF is weaker, as its output only has to be pseudorandom for random inputs. We recall the definition of (weak) PRFs/PRPs below.

Definition 5. *A function $\mathsf{F} : \{0,1\}^k \times \{0,1\}^m \to \{0,1\}^n$ is an (ϵ, s, q)-pseudorandom function (PRF) if no adversary $\mathcal{A}$ of size s can distinguish $\mathsf{F}(K, \cdot)$ (instantiated with a random key K) from a random function $\mathsf{R} \leftarrow \mathcal{R}_{m,n}$. More precisely, for any $\mathcal{A}$ of size s that can make up to q queries to its oracle, we have*

$$|\Pr[K \leftarrow \{0,1\}^k : \mathcal{A}^{\mathsf{F}(K,\cdot)} \to 1] - \Pr[\mathsf{R} \leftarrow \mathcal{R}_{m,n} : \mathcal{A}^{\mathsf{R}(\cdot)} \to 1]| \le \epsilon. \quad (1)$$

A non-adaptive PRF is defined similarly, except that we only consider non-adaptive adversaries who must choose the queries $X_1, \ldots, X_q$ before seeing any outputs. A weak PRF is defined similarly, except that the inputs $X_1, \ldots, X_q$ are chosen uniformly at random and not chosen by $\mathcal{A}$.

A (non-adaptive/weak) pseudorandom permutation (PRP) is defined analogously, except that we require $\mathsf{F}(K, .)$ to be a permutation for every K.

2   Stream Ciphers

2.1   Yu et al. [23] Revisited

A stream cipher is a function $\mathsf{SC} : \{0,1\}^k \to \{0,1\}^k \times \{0,1\}^n$ that, for every key K_0, defines a sequence $X_1, X_2, \ldots$ of outputs which are recursively defined as

$$(K_{i+1}, X_{i+1}) = \mathsf{SC}(K_i)$$

The security notion for stream ciphers requires that for a random initial secret key $K_0 \in_R \{0,1\}^k$, the outputs $X_1, X_2, \ldots, X_\ell$ are pseudorandom.

A stream cipher is leakage-resilient [4] if, for any ℓ, the outputs $X_\ell, X_{\ell+1}, \ldots$ are pseudorandom given $X_0, X_1, \ldots, X_{\ell-1}$ and a bounded amount of adaptively chosen leakage $\Lambda_0, \Lambda_1, \ldots, \Lambda_{\ell-1}$. This leakage is computed as follows: for any $i = 0, 1, \ldots, \ell - 2$, before $(K_{i+1}, X_{i+1}) \leftarrow \mathsf{SC}(K_i)$ is computed, an adversary chooses a leakage function f_i with range $\{0,1\}^\lambda$ (the parameter $\lambda \in \mathbb{N}$ bounds the amount of leakage we allow per round), and then gets $\Lambda_i = f_i(K_i')$ where $K_i' \subseteq K_i$ is the part of the secret state which is accessed during the evaluation of $\mathsf{SC}(K_i)$.

Yu, Standaert, Pereira and Yung [23] propose a construction, SC_{SEQ}, illustrated in Figure 2. As outlined in the introduction this construction is vulnerable to the precomputation attack if the leakage functions can be chosen adaptively depending on the public values. In [23] it is claimed that it satisfies a relaxed notion of leakage-resilience where the leakage functions $f_1, f_2, \ldots$ are chosen non-adaptively.

The construction is initialized with a secret key $K_0 \in_R \{0,1\}^k$ for a wPRF $\mathsf{F} : \{0,1\}^k \times \{0,1\}^n \to \{0,1\}^m$ and two *public* random values $p_0, p_1 \in_R \{0,1\}^n$ (although these values are public, it will be crucial that the adversary chooses the leakage functions *before* seeing these values.) The output is recursively computed as

$$(K_{i+1}, X_{i+1}) \leftarrow \mathsf{F}(K_i, p_{i \bmod 2})$$

The proofs in [20,23] use a lemma which states that the output of a weak PRF on a random input is pseudorandom even if the key is not uniform, but only has high min-entropy.

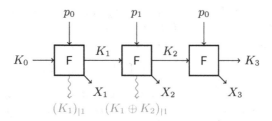

Fig. 2. The stream cipher construction SC_{SEQ} from a weak PRF F from [23]. K_0 is the secret initial key, p_0, p_1 are public random values and $X_1, X_2, \ldots$ is the output. The leakage leading to our counterexample to Lemma 3 from [23] is shown in gray.

Proposition 1 (wPRF with non-uniform keys, Lemma 2 from [20]).
Let $\mathsf{F} : \{0,1\}^k \times \{0,1\}^n \to \{0,1\}^m$ *be a* (ϵ, s, q) *secure weak PRF,* $X \in_R \{0,1\}^n$
be uniform and $K \in \{0,1\}^k$ *be any random variable which is independent of* X
and has min-entropy $H_\infty(K) \geq k - \lambda$ *for some* $\lambda \in \mathbb{N}$, *then*

$$(X, \mathsf{F}(K,X)) \text{ is pseudorandom.} \tag{2}$$

Quantitatively, $(X, \mathsf{F}(K,X))$ *cannot be distinguished by adversaries of size* $\approx$
$s\epsilon^2$ *with advantage* $\approx \epsilon 2^\lambda$, *so we have a loss of* [6] ϵ^2 *in circuit size and* 2^λ *in*
distinguishing advantage. The reduction makes $O(\lambda/\epsilon^2)$ *queries, so* q *has to be*
at least that large.

The other main ingredient of the proof is a theorem from [4],[7] which states that
a pseudorandom value $Z \in \{0,1\}^k$ has whp. HILL pseudoentropy almost $k - \lambda$
given any λ bits of auxiliary information A about Z. In our case, Z will be
$(X, \mathsf{F}(K,X))$ as in eq.(2) and A will be leakage $f(X,X) \in \{0,1\}^\lambda$. Concretely,
we get

Proposition 2. *For* F, X, K *as in Proposition 1 and* f *any leakage function*
with range $\{0,1\}^\lambda$

$$\Pr[H^{\mathrm{HILL}}_{\epsilon',s'}(X, \mathsf{F}(K,X) \mid f(K,X)) \geq n + m - 2\lambda] \geq 1 - 2^{-\lambda} \tag{3}$$

where $s' \approx s\epsilon^4 2^{4\lambda}$ *and* $\epsilon' = \epsilon 2^{2\lambda}$, *so setting, say* $\lambda = \log(\epsilon^{-1}/4)$,[8] *we get* $s' \approx$
$s\epsilon^5, \epsilon' = \sqrt{\epsilon}$.[9]

Before we turn to the problem with the security proof in [23], let us consider
a slightly different construction which we will call SC^+_{SEQ}. This construction
is defined like SC_{SEQ}, except that we use a fresh random input p_i (for $i =$
$0, \ldots, L - 1$) in *every* round, i.e. $(K_{i+1}, X_{i+1}) \leftarrow \mathsf{SC}(K_i, p_i)$. Of course this is
not a practical construction as we can output at most L blocks (where L denotes
the number of the public p_i values.) But it illustrates the proof idea, and we will
use this construction as a starting point to construct leakage-resilient PRFs in
the next section.

Theorem 1. *The construction* SC^+_{SEQ} *is a* naLR *stream cipher. The amount* λ
of leakage tolerated per round depends on F *as explained in Footnote 8.*

---

[6] Let us note that there is a typo in the conference version of [20] (the t^2 in eq.(3)
shoud be t), suggesting that the loss in circuit size is only ϵ, not ϵ^2.

[7] A more general "dense model theorem" was independently given in [21], cf. [7] for a
good overview.

[8] I.e. the leakage bound $\lambda = \log(\epsilon^{-1})/4$ is a function of the distinguishing advantage
of the best distinguisher for the weak PRF F: If F is secure against polynomial-size
distinguishers (i.e. $\epsilon = \omega(\log k)$), λ is superlogarithmic in the security parameter k.
If F is exponentially hard, λ can be linear in k.

[9] Due to the very loose reductions in [20,7], these bounds will not imply any practical
security guarantees if instantiated with a standard block cipher where k is typically
something like 128 or 256. To get practical bounds, one would have to make idealized
assumptions like assuming F is a random orcale [23].

Proof. By the definition of a leakage-resilient stream cipher, we have to consider the following random experiment: an adversary $\mathcal{A}$ chooses some $L' \in [L]$ and leakage functions $f_1, \ldots, f_{L'} : \{0,1\}^k \times \{0,1\}^n \to \{0,1\}^\lambda$. Then we sample $K_0 \in_R \{0,1\}^k$, $p_0, \ldots, p_{L-1} \in_R \{0,1\}^n$ and flip a coin $b \leftarrow \{0,1\}$.

The adversary gets the public values $p_0, \ldots, p_{L-1}$, the outputs $X_1, \ldots, X_{L'}$ and leakage $\Lambda_1, \ldots, \Lambda_{L'}$, where $\Lambda_i = f_i(K_{i-1}, p_{i-1})$.

If $b = 0$ the adversary gets a random $Z \in_R \{0,1\}^{(L-L')m'}$, if $b = 1$ she gets the remaining outputs $X_{L'+1}, \ldots, X_L$. We must prove, that she cannot guess b with probability much better than $1/2$.

We will prove that $K_{L'}$ is indistinguishable from a $\tilde{K}_{L'}$ which has $k - \lambda$ bits of average min-entropy given the view $\mathsf{view}_{L'}$ of the adversary after L' rounds, where view_i denotes the view of the adversary after the ith round, i.e.[10]

$$\mathsf{view}_i = \{p_0, \ldots, p_{i-1}, X_1, \ldots, X_i, \Lambda_1, \ldots, \Lambda_i\}$$

This will prove the theorem, as by eq.(3) and the fact that the $p_{L'+1}, \ldots, p_L$ are all chosen uniformly at random the remaining outputs $X_{L'+1}, \ldots, X_L$ will be pseudorandom. To see that $K_{L'}$ has high conditional pseudoentropy given $\mathsf{view}_{L'}$ we proceed in rounds, showing that for any $j \le L'$, if K_{j-1} has high conditional pseudoentropy given view_{j-1}, then K_j has high pseudoentropy given view_j. For $j = 1$ this follows directly from eq.(3) as $(K_1, X_1) \leftarrow \mathsf{F}(K_0, p_0)$, where K_0 and p_0 are uniform.

After the first round whp. K_1 has conditional pseudoentropy $k - 2\lambda$ given view_1. Thus, there exists a $\tilde{K}_1$ with average min-entropy $\tilde{H}_\infty(\tilde{K}_1 | \mathsf{view}_1) = k - 2\lambda$ that is indistinguishable from K_1 (given view_1). Because of this, in the above experiment we can replace K_1 with $\tilde{K}_1$ and the probability that $\mathcal{A}$ will finally guess b correctly can only change by a negligible amount (otherwise $\mathcal{A}$ would constitute a distinguisher for K_1 and $\tilde{K}_1$.) We proceed as above for L' rounds (replacing K_i with $\tilde{K}_i$ for all $i = 1, \ldots, L'$) concluding that $K_{L'}$ is indistinguishable from a $\tilde{K}_{L'}$ where $\tilde{H}_\infty(\tilde{K}_{L'} | \mathsf{view}_{L'}) = k - 2\lambda$. As the p_i are independent of K_i, we get by Proposition 1 the claimed statement. $\square$

Let us go back to the construction SC_{SEQ} from [23], where we alternate between two inputs p_0, p_1 instead of using a fresh p_i for every round. Towards proving that this construction is a naLR stream-cipher, we can proceed as in the proof of Theorem 1 for the first two rounds arguing that K_1 and K_2 are indistinguishable from $\tilde{K}_1, \tilde{K}_2$ satisfying $H_\infty(\tilde{K}_i | \mathsf{view}_i) = k - 2\lambda$, but the 3rd step becomes more difficult.

The reason is that (in our adapted experiment, where K_i got replaced with $\tilde{K}_i$ for $i = 1, 2$) we compute $K_3 \leftarrow \mathsf{F}(\tilde{K}_2, p_0)$; but p_0 is clearly *not* random (and independent) given the view of $\mathcal{A}$, as p_0 was already used in the first round. Thus we cannot just apply Proposition 1 eq. (2) to conclude that the next key K_3 to be computed has high conditional pseudoentropy.

---

[10] Note that we only include $p_0, \ldots, p_{i-1}$ into view_i, but in the actual security experiment the adversary gets to see all the $p_0, \ldots, p_L$ right away. We can do so as we only consider non-adaptive adversaries and the p_i's are chosen uniformly at random.

The authors of [23] are well aware of this problem. In order to "enforce" independence between $\tilde{K}_2$ and p_0, they put forward a lemma which claims these values become independent when given the leakage from the previous round (for clarity, we only state their lemma for the case of K_3)

Lemma 1 (Lemma 3 [23]). $\tilde{K}_2$ and $\{p_0, p_1, X_1, X_2, \Lambda_1\}$ are independent given $\{p_1, \Lambda_2\}$.

Although this approach looks promising, unfortunately, it turns out that this lemma is wrong (already for $\lambda = 1$) as can be seen by a simple counterexample illustrated in Figure 2: choose leakage functions f_1, f_2 that output the first bits $\Lambda_1 = K_1[1]$ and $\Lambda_2 = K_1[1] \oplus K_2[1]$ of K_1 and $K_1 \oplus K_2$ respectively.

First, we observe that in our adapted experiment where we replace the K_i's (having only conditional pseudoentropy) with $\tilde{K}_i$'s (having min-entropy), Λ_1, Λ_2 will be the first bits of $\tilde{K}_1$ and $\tilde{K}_1 \oplus \tilde{K}_2$. To see this, just note that, e.g., K_1 and $\tilde{K}_1$ are indistinguishable given $\Lambda_1 = K_1[1]$, this can only be the case if K_1 and $\tilde{K}_1$ agree on the first bit. To see why the lemma is flawed, we first observe that if $\{p_1, \Lambda_2 = \tilde{K}_1[1] \oplus \tilde{K}_2[1]\}$ is known, then given $\Lambda_1 = \tilde{K}_1[1]$ we can compute $\tilde{K}_2[1] = \tilde{K}_1[1] \oplus \tilde{K}_1[1] \oplus \tilde{K}_2[1]$. Now, the lemma claims $\tilde{K}_2$ is independent of $\{p_0, p_1, X_1, X_2, \tilde{K}_1[1]\}$ given $\{p_1, \tilde{K}_1[1] \oplus \tilde{K}_2[1]\}$, which by our observation means that $\tilde{K}_2[1]$ is already determined (i.e., has no entropy) given $\{p_1, \tilde{K}_1[1] \oplus \tilde{K}_2[1]\}$, but this is not true as shown by the claim below.

Claim. If $\tilde{K}_2[1]$ has no entropy given $\{p_1, \tilde{K}_1[1] \oplus \tilde{K}_2[1]\}$ then F is not a wPRF.

Proof. We will show that if F is a secure weak PRF, then $\{p_1, \tilde{K}_1[1], \tilde{K}_2[1]\}$ is close to being uniform (which implies the claim.) The value p_1 is uniform by definition. To see that $\{p_1, \tilde{K}_1[1]\}$ is uniform recall that $K_1[1] = \tilde{K}_1[1]$, and $K_1 = \mathsf{F}(K_0, p_0)$. Clearly, every individual bit of K_1 (in particular $K_1[1]$) must be close to uniform as otherwise we could distinguish K_1 from uniform (and thus break the security of F) by just outputting this bit. Similarly, $K_2 = \mathsf{F}(\tilde{K}_1, p_1)$ is pseudorandom given $\{p_1, \tilde{K}_1[1]\}$, and thus $\tilde{K}_2[1] = K_2[1]$ is close to uniform given $\{p_1, \tilde{K}_1[1]\}$.

3   Leakage-Resilient PRFs

In [1] Dodis and Pietrzak construct a nagLR PRF from any wPRF F. Informally, a PRF is leakage resilient if its outputs on all "fresh" inputs are pseudorandom, even if the adversary can query the PRF, and besides the outputs also gets the leakage from these computations.[11] In this section we propose a much simpler construction than the one from [1], which is a nagLR non-adaptive PRF, that is,

---

[11] Let us remark that this is not the only meaningful notion of leakage-resilience for PRF. Instead of requiring that only fresh outputs look pseudorandom, we could ask for a simulator that can efficiently fake leakage. A notion along this lines in a somewhat different context and for nog-continious leakage (called "seed incompressibility") has been considered in [11].

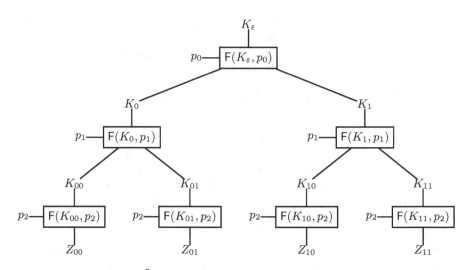

Fig. 3. Illustration of $\Gamma^{\mathsf{F},m}$ for $m = 2$. p_0, p_1 and p_2 are the random public values and $K_\varepsilon = K$ is the initial random key of the PRF. The output of the PRF for each $X \in \{0,1\}^2$ is represented by Z_X, i.e., the leaves of the tree.

it remains secure if not only the leakage function, but also the inputs are chosen non-adaptively.

We define a naLR PRF by considering an adversary $\mathcal{A}$ who has access to two oracles: the challenge and the leakage oracle. The first is as in Definition 5, i.e., either it is the pseudorandom function $\mathsf{F}(K, \cdot)$, or a random function $\mathsf{R} \leftarrow \mathcal{R}$. The latter oracle $\mathsf{F}^f(K, \cdot)$ can be queried on some input $X \in \{0,1\}^m$ and returns $\mathsf{F}(K, X)$ together with the leakage $f(K, X)$, where f is the leakage function non-adaptively chosen at the beginning of the experiment.[12] Of course, the queries to the two oracles must be disjoint.

Definition 6. *[$\mathcal{L}$-naLR (non-adaptive) PRF] A function* $\mathsf{F} : \{0,1\}^k \times \{0,1\}^m \rightarrow \{0,1\}^n$ *is a* (ϵ, s, q)*-secure* $\mathcal{L}$*-naLR PRF if for any* $\mathcal{A}$ *of size* s *that can make up to* q *disjoint queries to its two oracles, and for any leakage function* $f \in \mathcal{L}$, *we have*

$$\left| \Pr_{K \leftarrow \{0,1\}^k} [\mathcal{A}^{\mathsf{F}(K,\cdot),\mathsf{F}^f(K,\cdot)} = 1] - \Pr_{\mathsf{R} \leftarrow \mathcal{R}_{m,n} \; K \leftarrow \{0,1\}^k} [\mathcal{A}^{\mathsf{R}(\cdot),\mathsf{F}^f(K,\cdot)} = 1] \right| \leq \epsilon.$$

We will mostly omit the parameters ϵ, s *and* q *and say that* F *is a* $\mathcal{L}$*-naLR PRF if* ϵ *is some negligible function in* k *and* s, q *are superpolynomial in* k.

A $\mathcal{L}$*-naLR non-adaptive PRF is defined equivalently, except that* $\mathcal{A}$ *must choose the* q *PRF input queries non-adaptively.*

---

[12] Note that we allow the adversary to only choose one leakage function. On could also consider a stronger non-adaptive notion where the adversary can initially choose a different leakage function for every query to be made.

Recall that naLR security denotes $\mathcal{L}$-naLR security where $\mathcal{L}$ is the class of all efficiently computable functions with range $\{0,1\}^\lambda$ for some $\lambda \in \mathbb{N}$. In this section, we will only consider this special case, but we gave the general definition as it will be used in the next section. As outlined in the introduction, stateless (cf. Footnote 8) na-LR naPRFs don't exist, and thus following [22,1], we consider a "granular" nagLR-security notion, informally discussed in Section 1.1. Our construction $\Gamma^{\mathsf{F},m}$, illustrated in Figure 3, is inspired by the classical GGM construction of a PRF from a PRG [8]. On input $X \in \{0,1\}^m$ computes its output Z_X by invoking a wPRF F $m+1$ times sequentially. The inputs to the $m+1$ invocations are fixed random public values $p_0,\ldots,p_m$. The ith bit of the input $X[i]$ determines which half of the output of F in the ith invocation is used as a key for the $(i+1)$th invocation.

Let us define the PRF $\Gamma^{\mathsf{F},m} : \{0,1\}^{k+(m+1)\ell} \times \{0,1\}^m \to \{0,1\}^n$, which uses a wPRF F $: \{0,1\}^k \times \{0,1\}^\ell \to \{0,1\}^{2k}$ as main building block. The secret key is $K \leftarrow \{0,1\}^k$ and moreover we sample $m+1$ random public values $p = p_0,\ldots,p_m \leftarrow \{0,1\}^\ell$. Below we define how the output Z_X is computed by $\Gamma^{\mathsf{F},m}(K,p,X)$ in pseudocode. We explicitly state which bit of the input is read as this will determine the inputs that the leakage functions will get. With $\mathsf{F}_0(K,X)$ and $\mathsf{F}_1(K,X)$ we denote the function computing $\mathsf{F}(K,X)$ but only outputting the left and right half of the output, respectively.

PRF $\Gamma^{\mathsf{F},m}(K,(p_0,\ldots,p_m),X)$, where $X \in \{0,1\}^m$ and $K \leftarrow \{0,1\}^k$:
    Set $i := 0$ and $K_\varepsilon := K$
    Repeat:
        $i := i+1$.
        Read the input bit $X[i]$.
        Compute $K_{X_{|i}} := \mathsf{F}_{X[i]}(K_{X_{|i-1}}, p_{i-1})$.
    Until $i = m$
    Compute $Z_X := \mathsf{F}(K_X, p_m)$.
    Output Z_X.

We think of the above computation as being performed in $m+1$ time steps. Each of the m loops, and the final computation of Z_X, is a time step. Thus, the nagLR non-adaptive PRF security notion allows the adversary to initially choose a leakage function $f : \{0,1\}^\ell \times \{0,1\}^k \times \{0,1\} \to \{0,1\}^\lambda$ and inputs to the two oracles. For every input X to the $\mathsf{F}^f(K,.)$ oracle, the adversary gets $F(K,X)$ and leakage

$$f(p_0, K_\varepsilon, X[1]), f(p_1, K_{X_{|1}}, X[2]), \ldots, f(p_{m-1}, K_{X_{|m-1}}, X[m]), f(p_m, K_X, 0) \tag{4}$$

As in [1], we can actually handle somewhat stronger leakage functions which not only get the bit $X[i]$ of X touched in the ith time step, but all the bits $X_{|i}$ of X touched so far, i.e.

$$f(p_0, K_\varepsilon, X_{|1}), f(p_1, K_{X_{|1}}, X_{|2}), \ldots, f(p_{m-1}, K_{X_{|m-1}}, X_{|m-1}), f(p_m, K_X, X) \tag{5}$$

The interpretation here is, that the leakage function f knows exactly at which node of the tree it is. We are now ready to prove our main theorem in this section

Theorem 2. *If* $\mathsf{F} : \{0,1\}^k \times \{0,1\}^\ell \to \{0,1\}^{2k}$ *is a weak PRF, then* $\Gamma^{\mathsf{F},m}$ *is a* nagLR *non-adaptive PRF. The amount of leakage* λ *per time step (i.e., for each invocation of* F) *depends on the security of* F *(cf. Footnote 8)*

Proof. Let $\mathcal{A}$ be an adversary which initially chooses a leakage function f (as described above), q distinct inputs $x_1, \ldots, x_q$ and some q_0, meaning that the first q_0 queries will be leakage queries, and the last $q_1 := q - q_0$ queries are challenge queries. We sample a random key $K \leftarrow \{0,1\}^k$ and a coin $b \leftarrow \{0,1\}$ determining if we're in the real or random experiment (note that we do not yet sample the $p_0, \ldots, p_m \leftarrow \{0,1\}^\ell$.)

We now evaluate all q queries simultaneously, going down the tree as illustrated in Figure 3 layer by layer, sampling the random p_i's as we go down (this parallel evaluation is only possible as the queries are chosen non-adaptively.) It will be convenient to give the adversary the leakage for all q queries (even though the last q_1 challenge queries are not supposed to leak at all), except for the very last layer. In the last layer, we evaluate the first q_0 queries, and give the adversary this outputs together with the leakage. If $b = 0$ (which means we're in the real experiment) the adversary gets the outputs, and random values otherwise.

Below we formally describe how the leakage is computed. As just mentioned, we give the adversary more power than required for nagLR-security. Concretely, in item 2 below, she gets leakage from internal nodes on all queries, not just he leakage queries. Set $i := 0$, sample a random $K = K_\varepsilon$, and then the outputs and leakage are computed layer by layer as follows:

1. sample a random p_i and give it to the adversary.
2. compute $K_{I\|0}\|K_{I\|1} := \mathsf{F}(K_I, p_i)$ and leakage $\Lambda_I := f(K_I, p_i, I)$ for all i bit prefixes of $x_1, \ldots, x_q$. Give all the computed leakage to the adversary.
3. If $i < m - 1$ then set $i := i + 1$ and go back to step 1, otherwise go to next step (at this point we have computed K_{x_i} for all queries x_i.)
4. sample a random p_m and give it to the adversary.
5. Compute the final outputs $Z_{x_i} := \mathsf{F}(K_{x_i}, p_m) = \Gamma^{\mathsf{F},m}(K, x_i)$ and leakage $\Lambda_{x_i} := f(X_{x_1}, p_m, x_i)$ for $i = 1, \ldots, q_0$. Give this outputs and leakage to the adversary.
6. If $b = 0$, for $i = q_0 + 1, \ldots, q$, compute $Z_{x_i} := \mathsf{F}(K_{x_i}, p_m)$, otherwise, if $b = 1$, sample random $Z_{x_i} \leftarrow \{0,1\}^{2n}$. Give this values to the adversary.

We denote by view_0 the view of the adversary in the above experiment if $b = 0$, and with view_m if $b = 1$. To prove the theorem we must show that view_0 and view_m are computationally indistinguishable. We will consider hybrid views $\mathsf{view}_1, \ldots, \mathsf{view}_{m-1}$, and show that for every $i = 1, \ldots, m$, view_{i-1} and view_i indistinguishable.

Consider the computation $K_0\|K_1 = \mathsf{F}(K_\varepsilon, p_0), \Lambda_\varepsilon = f(K_\varepsilon, p_0)$ in the first layer. As K_ε has min-entropy $n - 2\lambda$ (in fact, in this first layer, this key is even uniform) and p_0 is uniform, by Proposition 1 $K_0 \| K_1$ is pseudorandom given p_0, and by Proposition 2 $K_0 \| K_1$ has (whp.) HILL pseudoentropy $2n - 2\lambda$ when additionally given Λ_ε. The first hybrid view$_1$ is dervied from the hybrid view$_0$ by replacing this $K_0 \| K_1$ with a random variable $\tilde{K}_0 \| \tilde{K}_1$ which has min-entorpy $2n - 2\lambda$ given p_0, Λ_ε. By the definition of HILL pseudoentropy, such a $\tilde{K}_0 \| \tilde{K}_1$ exists, where view$_0$ and view$_1$ are computationally indistinguishable. Thus, in view$_1$, the inputs $\tilde{K}_0$ and $\tilde{K}_1$ to the first layer (which are outptus from the zero layer) have min-entropy $n - 2\lambda$, and by Proposition 1, each outputs of this layer will have pseudoentropy $n - 2\lambda$ given the entire view of the adversary. The hybrid view$_2$ is derived from view$_1$ by replacing this outputs which have min-entropy $n - 2\lambda$, and so on, until we get the hybrid view$_{m-1}$ which is indistinguishable from view$_0$. In view$_{m-1}$, the inputs $\tilde{K}_{x_i}$ to the last layer has min-entropy $n - 2\lambda$. We choose p_m uniformly at random, and it follows by Proposition 1, that the "challange" outputs $Z_{x_i} := \mathsf{F}(\tilde{K}_{x_i}, p_m)$ are pseudorandom, and thus indistinguishable from view$_m$ which is derived from view$_{m-1}$ by replacing all challange outputs by uniformly random values (as in the case $b = 1$.)   □

3.1   An Adaptive Attack against Our Construction $\Gamma^{\mathsf{F},m}$

In Theorem 2 we showed that $\Gamma^{\mathsf{F},m}$ is a nagLR non-adaptive PRF. As discussed in Section 1.2, it trivially is not a naLR non-adaptive PRF or gLR non-adaptive PRF, i.e. the non-adaptivity and granularity for the leakage are necessary. It is a natural question whether it is a nagLR PRF like the (much more sophisticated) construction from [1]).

We answer this question negatively and show a simple attack against $\Gamma^{\mathsf{F},m}$. The attack allows the adversary to learn leakage that reveals the first λ bits of $\Gamma^{\mathsf{F},m}(K, X)$ for an input X that has not yet been queried. Clearly, this breaks the security of the PRF as required by Definition 6. Suppose $m = \ell + 1$, then the attack works as follows:

1. Define $f(p_{m-1}, K_I, I)$ to be the first λ bits of $\mathsf{F}(\mathsf{F}_0(K_I, p_{m-1}), I)$.
2. Learn the public values $p_0, \dots, p_m \in \{0, 1\}^\ell$.
3. Query the leakage oracle for $p_m \| 1$ and obtain $\Gamma^{\mathsf{F},m}(K, p_m \| 1)$ and, from the leakage, the first λ bits of

$$\mathsf{F}(\mathsf{F}_0(K_{p_m}, p_{m-1}), p_m) = \Gamma^{\mathsf{F},m}(K, p_m \| 0).$$

Thus, for a leakage query $p_m \| 1$ the attack reveals the first λ bits of $\Gamma^{\mathsf{F},m}(K, p_m \| 0)$. We emphasize that this attack is rather artificial and most likely will not affect the real-world security of our construction. However, it illustrates that any attempt to prove the security of $\Gamma^{\mathsf{F},m}$ in an adaptive setting must fail (indeed, this attack works even if we assign a different public value to every node – details are omitted in this extended abstract).

Let us emphasize that this attack requires that for each execution of the weak PRF F the corresponding leakage function is "aware" of its current position in the tree, that is, we need a leakage function as in eq.(5) and not eq.(4). Although for our positive result considering a stronger leakage model only strengthens the result, for an attack we would like the model to be as weak as possible and stick with leakage functions that only get whatever is touched, and nothing beyond that, as required for nagLR PRFs. We do not know if such an attack exists against $\Gamma^{\mathsf{F},m}$.

4   Leakage-Resilient PRPs

In the previous section we gave a simple construction of a PRF which is secure against non-adaptive leakage if queried on non-adaptively chosen inputs. In practice, one usually doesn't use pseudorandom functions, but rather pseudorandom permutations (PRPs). In particular, block ciphers, the work horses of cryptography, are assumed to be PRPs. Block ciphers are also the main targets of side-channel cryptanalysts, thus coming up with leakage-resilient PRPs is a particularly worthwhile task.

In the standard setting (i.e. without leakage), Luby and Rackoff [16] famously showed that one can construct a PRP from a PRF by using a three-round Feistel network as illustrated in Figure 4. With one round more one even gets a strong PRP, i.e. an object that is indistinguishable from a uniformly random permutation even if one can query it from both sides.

To prove that a 3-round Feistel using PRFs as round functions is a PRP one proceeds in two steps.[13] First one shows that a 3-round Feistel instantiated with *uniformly random functions* is indistinguishable from a *uniformly random permutation* (this step is completely information theoretic). In the second step one then observes that a 3-round Feistel instantiated with URFs (uniformly random functions) is indistinguishable from a 3-round Feistel using PRFs. This second step follows by a simple hybrid argument where we replace the pseudorandom round functions with uniformly random functions one by one. A restricted case of the statement claiming only non-adaptive security and using only random functions as round functions, is given by the proposition below.

Proposition 3 (3-Round Feistel is Non-Adaptively Secure PRP). *For any $n, q \in \mathbb{N}$ and $x_1, \ldots, x_q \in \{0, 1\}^{2n}$ consider the distributions:*

- *Sample $\mathsf{P} \in_R \mathbf{P}_{2n}$ and, for $i \in [q]$, set $y_i = \mathsf{P}(x_i)$.*
- *Sample $\mathsf{F}_1, \mathsf{F}_2, \mathsf{F}_3 \in_R \mathbf{R}_n$ and for $i \in [q]$ set $z_i = \Phi_{\mathsf{F}_1, \mathsf{F}_2, \mathsf{F}_3}(x_i)$ (as in Figure 4)*

then

$$\Delta([y_1, \ldots, y_q], [z_1, \ldots, z_q]) \leq \frac{q^2}{2^n}$$

Proof (sketch). Consider the values $c_i = \mathsf{F}_1(a_i) \oplus b_i$ for $i = 1, \ldots, q$ (where $x_i = a_i \| b_i$, cf. Figure 4.) As F_1 is a URF, these c_i's will contain a collision with probability at most $q(q-1)/2^{n+1}$. Assuming they are all distinct, the $u_i = \mathsf{F}_2(c_i) \oplus a_i$'s are uniformly random as F_2 is a URF. As they are uniformly random,

---

[13] This proof template follows [17]; the original proof of Luby and Rackoff [16] is "direct", but also more complicated.

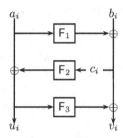

Fig. 4. 3-round Feistel Network $\Phi_{F_1,F_2,F_3} : \{0,1\}^{2n} \to \{0,1\}^{2n}$ with round functions $F_i : \{0,1\}^n \to \{0,1\}^n$

they also will contain a collision with with probability at most $q(q-1)/2^{n+1}$. This implies the values $z_i = u_i\|v_i$ are $2 \cdot q(q-1)/2^{n+1} = q(q-1)/2^n$ close to uniform over $\{0,1\}^{2n}$. The uniform distribution over q elements over $\{0,1\}^{2n}$ is $q(q-1)/2^{2n}$ close to the distribution of the $y_1, \ldots, y_q$ (which is uniform, but without repetition.) Thus, as statistical distance obeys the triangle inequality, the z_i's are $q(q-1)/2^n + q(q-1)/2^{2n} \le q^2/2^n$ close to the y_i's.   □

Proposition 3 also holds if the inputs x_i are chosen adaptively, but the proof for this case is significantly more delicate. The proof of Proposition 3 above uses the fact that the inputs $c_1, \ldots, c_q$ to the second round function (cf. Figure 4) are all distinct (with high probability). The adaptive case also goes along these lines, but here one has to argue that the c_i's are also "hidden", as an adaptive adversary who could "guess" the c_i values could compute inputs to the Feistel network where the outputs partially collide.

As shown in [1], it is already sufficient to get some simple leakage (e.g. the Hamming Weight) of the c_i values to launch such an attack. This attack can be adapted to work on Feistel networks with any number r of rounds, but its complexity (i.e. number of adaptive queries) grows exponentially in r. Still, this implies that a constant-round Feistel network, instantiated with leakage-resilient PRFs, can be broken in polynomial time, and thus is *not* a leakage-resilient PRP.

The queries to the Feistel network made in the [1] attack are adaptive, and here we show that this is indeed crucial. By Theorem 3 below, a 3-round Feistel is a *non-adaptively* secure leakage-resilient PRP if instantiated with leakage-resilient PRFs. The notion of leakage-resilience achieved by the PRP is inherited from the underlying PRF. If the round functions are $\mathcal{L}$-naLR PRFs, then we get a $\mathcal{L}$-nagLR PRP.

More formally, we initially choose a bit $b \in \{0,1\}$ and three keys k_1, k_2, k_3 for F which defines the round functions $F_i(.) = F(k_i, .)$ for $i = \{1, 2, 3\}$, and if $b = 1$ a random permutation $P \in_R \mathbf{P}_{2n}$ (using lazy sampling.) The adversary can initially choose three leakage functions $f_1, f_2, f_3 \in \mathcal{L}$, distinct inputs $a_i\|b_i$ for $i = 1, \ldots, q$ and some q_0 which specifies that the first q_0 inputs are leakage queries, and the last $q_1 := q - q_0$ are challange queries (as we consider non-adaptive queries, we can wlog. assume the queries are ordered like this.) She then gets, for every $i \le q_0$, the outputs $u_i\|v_i = \Phi_{F_1,F_2,F_3}(a_i\|b_i)$ and the leakage

$f_1(k_1, a_i), f_2(k_2, c_i)$ and $f_3(k_3, u_i)$ (so, each round of the Feistel network is considered a time-step which leaks independently.) For the queries $i > q_0$ she gets the regular output $\Phi_{F_1,F_2,F_3}(a_i\|b_i)$ if $b = 0$ and the random $P(a_i\|b_i)$ otherwise. Note that besides the evaluation of the round functions F_i, one also has to compute three XORs. It would be cheating to assume that this XORs are leakage free. We go to the other extreme, and assume the XORs leak completely by giving the adversary the entire c_i value for every leakage query. This c_i together with the known values a_i, b_i, u_i, v_i specifies all the inputs/outputs to the three XOR computations (e.g. the first XOR takes as inputs b_i and $b_i \oplus c_i$.)

Theorem 3. *Let* $F : \{0,1\}^\ell \times \{0,1\}^n \to \{0,1\}^n$ *be an* (q, ϵ, s)*-secure* $\mathcal{L}$*-naLR non-adaptive PRF. Then the three round Feistel network* Φ_{F_1,F_2,F_3}*, where each* $F_i = F(k_i, .)$ *is an independent instantiation of* F*, is a* (q, ϵ', s')*-secure* $\mathcal{L}$*-nagLR non-adaptive PRP where*

$$\epsilon' = 3\epsilon + q^2/2^n \quad s' = s - poly(q, n)$$

Proof. Let $x_1, \ldots, x_{q_0}$ and $x'_1, \ldots, x'_{q_1}$ (where $q_0 + q_1 = q$) denote the non-adaptively chosen leakage and challenge queries. Let $k_i \leftarrow \{0,1\}^\ell$ be randomly chosen keys for the round functions $F_i(.) = F(k_i, .)$. Let $f_1, f_2, f_3 \in \mathcal{L}$ denote the leakage functions chosen by the adversary. The adversary gets (with $x_i \overset{\text{def}}{=} a_i\|b_i$ and c_i, u_i, v_i as in Figure 4)

$$y_i = \Phi_{F_1,F_2,F_3}(x_i) \quad \Lambda_i \overset{\text{def}}{=} \{f_1(k_1, a_i), f_2(k_2, c_i), f_3(k_3, u_i), c_i\}$$

We must prove that the outputs $y'_1, \ldots, y'_{q_1}$, where

$$y'_i = \Phi_{F_1,F_2,F_3}(x'_i)$$

are pseudorandom given $y_1, \ldots, y_{q_0}$ and $\Lambda_1, \ldots, \Lambda_{q_0}$.

Claim. The c_i's corresponding to the q queries are distinct with probability at least $q(q-1)/2^{n+1} + \epsilon$.

Proof. To see this, let δ denote the probability that the c_i's collide; we can construct a non-adaptive q-query distinguisher for F with advantage $\delta - q(q-1)/2^{n+1}$ (note that as F is an ϵ-secure PRF this will imply that $\delta \leq q(q-1)/2^{n+1} + \epsilon$ as claimed.) This distinguisher simply queries its oracle (which is either a URF or $F(k, .)$) on inputs $a_1, \ldots, a_q$, obtaining $z_1, \ldots, z_q$; the oracle outputs 1 if and only if any of the $z_i \oplus b_i$ collide. If the outputs come from a URF, this probability is $q(q-1)/2^{n+1}$, whereas if they come from $F(k, .)$ this probability is δ by definition. This concludes the proof of the claim. □

Now assume all the c_i's are distinct. Conditioned on this, we can show by a similar argument that also all the $u_i = F_2(k_2, c_i) \oplus a_i$ will be distinct with probability $q(q-1)/2^{n+1} + \epsilon$.

Assume the c_i's and u_i's are all distinct and recall that $v_i = F_3(k_3, u_i) \oplus c_i$. Then it follows from the $\mathcal{L}$-naLR non-adaptive PRF security of F that the $y'_i = u_{q_0+i}\|v_{q_0+i}$ values for $i = 1, \ldots, q_1$ are pseudorandom given $y_1, \ldots, y_{q_0}$ and $\Lambda_1, \ldots, \Lambda_{q_0}$, as $F_2(k_2, .)$ and $F_3(k_3, .)$ are queried on distinct inputs in the first q_0 and the last q_1 queries. □

References

1. Dodis, Y., Pietrzak, K.: Leakage-Resilient Pseudorandom Functions and Side-Channel Attacks on Feistel Networks. In: Rabin, T. (ed.) CRYPTO 2010. LNCS, vol. 6223, pp. 21–40. Springer, Heidelberg (2010)
2. Dodis, Y., Reyzin, L., Smith, A.: Fuzzy Extractors: How to Generate Strong Keys from Biometrics and Other Noisy Data. In: Cachin, C., Camenisch, J. (eds.) EUROCRYPT 2004. LNCS, vol. 3027, pp. 523–540. Springer, Heidelberg (2004)
3. Dziembowski, S., Faust, S.: Leakage-Resilient Circuits without Computational Assumptions. In: Cramer, R. (ed.) TCC 2012. LNCS, vol. 7194, pp. 230–247. Springer, Heidelberg (2012)
4. Dziembowski, S., Pietrzak, K.: Leakage-resilient cryptography. In: 49th FOCS, pp. 293–302. IEEE Computer Society Press (October 2008)
5. Faust, S., Kiltz, E., Pietrzak, K., Rothblum, G.N.: Leakage-Resilient Signatures. In: Micciancio, D. (ed.) TCC 2010. LNCS, vol. 5978, pp. 343–360. Springer, Heidelberg (2010)
6. Faust, S., Rabin, T., Reyzin, L., Tromer, E., Vaikuntanathan, V.: Protecting Circuits from Leakage: the Computationally-Bounded and Noisy Cases. In: Gilbert, H. (ed.) EUROCRYPT 2010. LNCS, vol. 6110, pp. 135–156. Springer, Heidelberg (2010)
7. Fuller, B., Reyzin, L.: Computational entropy and information leakage, http://www.cs.bu.edu/~reyzin/research.html
8. Goldreich, O., Goldwasser, S., Micali, S.: How to construct random functions. Journal of the ACM 33, 792–807 (1986)
9. Goldwasser, S., Rothblum, G.N.: Securing Computation against Continuous Leakage. In: Rabin, T. (ed.) CRYPTO 2010. LNCS, vol. 6223, pp. 59–79. Springer, Heidelberg (2010)
10. Goldwasser, S., Rothblum, G.N.: How to compute in the presence of leakage. Electronic Colloquium on Computational Complexity (ECCC) 19, 10 (2012)
11. Halevi, S., Myers, S., Rackoff, C.: On Seed-Incompressible Functions. In: Canetti, R. (ed.) TCC 2008. LNCS, vol. 4948, pp. 19–36. Springer, Heidelberg (2008)
12. Håstad, J., Impagliazzo, R., Levin, L.A., Luby, M.: A pseudorandom generator from any one-way function. SIAM Journal on Computing 28(4), 1364–1396 (1999)
13. Hsiao, C.-Y., Lu, C.-J., Reyzin, L.: Conditional Computational Entropy, or Toward Separating Pseudoentropy from Compressibility. In: Naor, M. (ed.) EUROCRYPT 2007. LNCS, vol. 4515, pp. 169–186. Springer, Heidelberg (2007)
14. Ishai, Y., Sahai, A., Wagner, D.: Private Circuits: Securing Hardware against Probing Attacks. In: Boneh, D. (ed.) CRYPTO 2003. LNCS, vol. 2729, pp. 463–481. Springer, Heidelberg (2003)
15. Juma, A., Vahlis, Y.: Protecting Cryptographic Keys against Continual Leakage. In: Rabin, T. (ed.) CRYPTO 2010. LNCS, vol. 6223, pp. 41–58. Springer, Heidelberg (2010)
16. Luby, M., Rackoff, C.: How to Construct Pseudo-random Permutations from Pseudo-random Functions (Abstract). In: Williams, H.C. (ed.) CRYPTO 1985. LNCS, vol. 218, p. 447. Springer, Heidelberg (1986)
17. Maurer, U.M.: Indistinguishability of Random Systems. In: Knudsen, L.R. (ed.) EUROCRYPT 2002. LNCS, vol. 2332, pp. 110–132. Springer, Heidelberg (2002)
18. Medwed, M., Standaert, F.-X., Joux, A.: Towards Super-Exponential Side-Channel Security with Efficient Leakage-Resilient PRFs. In: Prouff, E., Schaumont, P. (eds.) CHES 2012. LNCS, vol. 7428, pp. 193–212. Springer, Heidelberg (2012)

19. Micali, S., Reyzin, L.: Physically Observable Cryptography (Extended Abstract). In: Naor, M. (ed.) TCC 2004. LNCS, vol. 2951, pp. 278–296. Springer, Heidelberg (2004)

20. Pietrzak, K.: A Leakage-Resilient Mode of Operation. In: Joux, A. (ed.) EURO-CRYPT 2009. LNCS, vol. 5479, pp. 462–482. Springer, Heidelberg (2009)

21. Reingold, O., Trevisan, L., Tulsiani, M., Vadhan, S.P.: Dense subsets of pseudo-random sets. In: 49th FOCS, pp. 76–85. IEEE Computer Society Press (October 2008)

22. Standaert, F.-X., Pereira, O., Yu, Y., Quisquater, J.-J., Yung, M., Oswald, E.: Leakage resilient cryptography in practice. In: Towards Hardware Intrinsic Security: Foundation and Practice, pp. 105–139 (2010)

23. Yu, Y., Standaert, F.-X., Pereira, O., Yung, M.: Practical leakage-resilient pseudo-random generators. In: ACM CCS 2010, pp. 141–151. ACM Press (2010)

A Statistical Model for DPA
with Novel Algorithmic Confusion Analysis

Yunsi Fei[1], Qiasi Luo[2,*], and A. Adam Ding[3]

[1] Department of Electrical and Computer Engineering
Northeastern University, Boston, MA 02115
[2] Marvell Technology Group Ltd., Santa Clara, CA 95054
[3] Department of Mathematics, Northeastern University, Boston, MA 02115

Abstract. Side-channel attacks (SCAs) exploit weakness in the physical implementation of cryptographic algorithms, and have emerged as a realistic threat to many critical embedded systems. However, no theoretical model for the widely used differential power analysis (DPA) has revealed exactly what the success rate of DPA depends on and how. This paper proposes a statistical model for DPA that takes characteristics of both the physical implementation and cryptographic algorithm into consideration. Our model establishes a quantitative relation between the success rate of DPA and a cryptographic system. The side-channel characteristic of the physical implementation is modeled as the ratio between the difference-of-means power and the standard deviation of power distribution. The side-channel property of the cryptographic algorithm is extracted by a novel algorithmic confusion analysis. Experimental results on DES and AES verify this model and demonstrate the effectiveness of algorithmic confusion analysis. We expect the model to be extendable to other SCAs, and provide valuable guidelines for truly SCA-resilient system design and implementation.

Keywords: Side-channel attack, differential power analysis.

1 Introduction

Cryptographic algorithms are widely used in various computer systems to ensure security. Despite the security strength of the algorithm, the leaked side-channel information of the cryptosystem implementation, like power consumption of smart cards and timing information of embedded processors, can be exploited to recover the secret key. Differential power analysis (DPA) is one of the early effective SCAs which analyzes the correlation between intermediate data and power consumption to reveal the secret [1]. Over the past decade, there has been many other successful power analysis attacks, including Correlation Power Attack (CPA) [2], Mutual Information Attack (MIA) [3], Partitioning Power Analysis (PPA) [4], etc. Other side-channel information, like electromagnetic emanations [5,6] and timing information [7], can also be exploited. A real secure

---

* This work was done while the author was with University of Connecticut.

E. Prouff and P. Schaumont (Eds.): CHES 2012, LNCS 7428, pp. 233–250, 2012.

system must be designed with countermeasures to be SCA-resilient. Common countermeasures include masking [8], power-balanced logic [9], and random delays [10]. To measure the SCA resilience of a system or the effectiveness of a countermeasure, several generic metrics are used, such as *number of measurements, success rate* [11,12], *guessing entropy* [13] and *information theoretic metric* [13,14]. One commonly used metric for evaluating a system's SCA resilience is the success rate, i.e., the probability that a specific SCA is successful with certain complexity constraint. For a cryptosystem, a low success rate for a SCA on it indicates its high resilience against such SCA.

Intuitively, both the *physical implementation* and *cryptographic algorithm* would affect the SCA resilience of a cryptosystem. An ideal implementation with countermeasures could reduce the side-channel leakage to minimum. Different cryptographic algorithms may have different intrinsic SCA-related properties. Accurately evaluating different implementations of the same cryptographic algorithm and comparing different cryptographic algorithms, in terms of their SCA resilience, are challenging issues. However, such analysis and theoretical modeling will reveal system-inherent parameters that affect its SCA resilience, and in practice will greatly facilitate advances in the design and implementation of real secure cryptosystems.

Related Work. There has been many related research efforts attempting to address the above issues. However, the effects of the algorithm and implementation were not clearly decoupled and better quantitative model is needed to understand their interaction. In [15], an approach is presented to model the DPA signal-to-noise ratio (SNR) of a cryptographic system, which does not further reveal how the SNR determines the ultimate SCA resilience. In [16], the relation between the difference-of-means power consumption and key hypotheses is analyzed and utilized to improve the DPA efficiency, without examining characteristics of the algorithm. [17] presented a statistical model for CPA, which illustrated well the effect of SNR on the power of CPA. However, they did not consider the interaction between the incorrect keys and thus the formula does not numerically conform to the empirical overall success rate for CPA (see Appendix A). Work in [18] exhibits DPA-related properties of SBoxes in cryptographic algorithms and introduces a new notion of *transparency order of an SBox*, without considering the implementation aspect. A framework presented in [13] unifies the theory and practice of SCA with a combination of information theory and security metrics. A quantitative analysis between the metrics and cryptographic system would be a nice complement to the general framework.

Our Contributions. In this paper, we proposes a statistical analysis model for DPA. To the best of our knowledge, this is the first analytic model for the success rate of DPA on cryptographic systems, and also the first model extracting SCA related characteristics from both the physical implementation and cryptographic algorithm. The physical implementation is represented by the power difference related to the select function and standard deviation of power waveforms. The ratio between them defines the SCA resilience of an implementation. The SCA-related property of a cryptographic algorithm is characterized by algorithmic confusion analysis. A confusion matrix is generated to measure the statistical correlation between different key candidates in DPA.

The rest of the paper is organized as follows. Section 2 introduces notions and fundamentals in cryptographic algorithms, DPA procedures, and statistical aspects in SCAs. Section 3 presents the algorithmic confusion analysis with definitions of confusion and collision coefficients. Our model for the success rate of DPA is proposed in Section 4. The model is verified with experimental results on DES and AES in Section 5. Section 6 discusses more implications of the model and its possible applications. Finally conclusions are drawn in Section 7.

2 Preliminaries

2.1 Randomness of Cryptographic Algorithm

Cryptographic algorithms are designed to be robust against cryptanalysis with two well-known statistical properties [19]. *Confusion* makes the statistical relation between the the ciphertext and key as complex as possible; *diffusion* makes the statistical relation between the ciphertext and plaintext as complex as possible. With deliberate design, an encryption algorithm is *perfectly secret* if each bit in the ciphertext C is purely random [20]:

Theorem 1. *Suppose b_C is one bit of the ciphertext C for a perfectly secret encryption algorithm, b_C has the same probability to be 0 or 1:*

$$\Pr[b_C = 1] = \Pr[b_C = 0] = \frac{1}{2}.$$

2.2 Differential Power Analysis Procedure

All SCAs have a common hypothesis test procedure. We next give an introduction on the earliest discovered and important DPA procedure.

- Side-channel *measurements* obtain physical side-channel information W, i.e., waveforms of power consumption collected from devices. Denote the *waveform population* as $\mathcal{W} = \{W_1, \ldots, W_{N_m}\}$, where W_i is a (time series) measurement with a certain input, and N_m is the total number of measurements for the cryptographic system. Each W is a time series as $W = \{w^1, \ldots, w^{N_p}\}$, where N_p is the number of points in W.
- *Key hypotheses* enumerate all possible values of the subkey k under attack, denoted as $\langle k \rangle = \{k_0, \ldots, k_{N_k-1}\}$, where N_k is the total number of key guesses, and $N_k = 2^{l_k}$ with l_k as the subkey bit-length.
- *Select function* ψ for DPA is one single bit b_d of *intermediate data* d computed from the plaintext M or ciphertext C and a key, written as $\psi = b_d$. The value of ψ is either 1 or 0.
- *Correlation* between ψ for each key hypothesis and $\mathcal{W}$ is computed for a specific attack. The correlation for DPA is the difference of means (DoM) δ,

i.e., the difference between the average power consumption of two waveform groups partitioned with $\psi = 1$ and 0, written as:

$$\delta = \frac{\sum \mathcal{W}_{\psi=1}}{N_{\psi=1}} - \frac{\sum \mathcal{W}_{\psi=0}}{N_{\psi=0}} \tag{1}$$

where $N_{\psi=1}$ and $N_{\psi=0}$ are the numbers of measurements with $\psi = 1$ and $\psi = 0$ respectively, under a particular key hypothesis. Given enough number of measurements, the DoM δ_c for the correct key k_c converges to the power difference ϵ related to the bit b_d under attack, written as $\lim\limits_{N_m \to \infty} \delta_c = \epsilon$, where $N_m = N_{\psi=1} + N_{\psi=0}$.

- *Testing* with the maximum likelihood method chooses the key hypothesis which has the maximum correlation (DoM in DPA) as the correct key.

2.3 Central Limit Theorem

The basic statistical aspect of our model is the Central Limit Theorem [21], considering the various noises in leakage measurements and the sampling process for side-channel cryptanalysis, i.e., the leakage measurement is for a set of random inputs rather than enumerating the entire input space. Consider a random distribution $\mathcal{X} = \{x_1, x_2, x_3, \ldots\}$. The mean value and standard deviation of the population are μ and σ, respectively. Randomly select a sample of size N_x from the population we get the mean value:

$$\bar{X} = \frac{1}{N_x} \sum_{i=1}^{N_x} x_i.$$

When N_x is sufficiently large, $\bar{X}$ is approximately normally distributed, $\mathcal{N}(\mu_{\bar{X}}, \sigma_{\bar{X}})$, with $\mu_{\bar{X}} = \mu$ and $\sigma_{\bar{X}} = \frac{\sigma}{\sqrt{N_x}}$.

DPA is a sampling process on the entire waveform population, which is usually regarded as normally distributed [22]. Denote the standard deviation of the waveform population as $\sigma_{\mathcal{W}}$. Thus the two mean terms for the DoM computation in Equation (1) are normal random variables with distribution $\mathcal{N}\left(\epsilon + b, \sigma_{\mathcal{W}}/\sqrt{N_{\psi=1}}\right)$ and $\mathcal{N}\left(b, \sigma_{\mathcal{W}}/\sqrt{N_{\psi=0}}\right)$, respectively. Here b denotes the mean power consumption for the waveform group $\psi = 0$. Since both $N_{\psi=0}$ and $N_{\psi=1}$ are approximately $\frac{N_m}{2}$ according to Theorem 1, δ_c is a random variable with normal distribution $\mathcal{N}(\mu_{\delta_c}, \sigma_{\delta_c})$ as $\mu_{\delta_c} = \epsilon$ and $\sigma_{\delta_c} = 2\frac{\sigma_{\mathcal{W}}}{\sqrt{N_m}}$. This statement still holds for large N_m by the Central Limit Theorem when we drop the normal distribution assumption on the waveform population.

3   Algorithmic Confusion Analysis

A chosen select function involves a certain SBox of the cryptographic algorithm (a preset computation given as a lookup table) and a subkey. In this section, we attempt to reveal properties of the algorithm that would indicate its resilience to DPA. The analysis is only algorithm and select function related, and independent on the leakage measurements.

3.1   Confusion Coefficient

Assume the select function for DPA is chosen as a bit in the last-round encryption, which is dependent on several bits of the ciphertext, the subkey, and the corresponding SBox. Two key hypotheses k_i and k_j would have two corresponding $\psi|k_i$ and $\psi|k_j$. The values of $\psi|k_i$ and $\psi|k_j$ can be different or the same. We find out that the probability that $\psi|k_i$ is different or the same with $\psi|k_j$ reveals DPA-related property of the cryptographic algorithm.

We name a *confusion coefficient* after the confusion property of cryptographic algorithms defined in [19]. The *confusion coefficient* κ over two keys (k_i, k_j) is defined as:

$$\kappa = \kappa(k_i, k_j) = \Pr\left[(\psi|k_i) \neq (\psi|k_j)\right] = \frac{N_{(\psi|k_i)\neq(\psi|k_j)}}{N_t}$$

where N_t is the total number of values for the relevant ciphertext bits, and $N_{(\psi|k_i)\neq(\psi|k_j)}$ is the number of occurrences for which different key hypotheses k_i and k_j result in different ψ values. For example, in our DPA attack on DES (Data Encryption Standard) algorithm, N_t is $2^7 = 128$.

Similarly, the *complementary confusion coefficient* or *collision coefficient* ξ over (k_i, k_j) is defined as:

$$\xi = \xi(k_i, k_j) = \Pr\left[(\psi|k_i) = (\psi|k_j)\right] = \frac{N_{(\psi|k_i)=(\psi|k_j)}}{N_t}$$

We have $\kappa + \xi = 1$ and $0 \leq \kappa < 1$ and $0 < \xi \leq 1$. For a perfectly secret cryptographic, we have:

Lemma 1. *Confusion Lemma (see Appendix B for the proof).*

$$\Pr\left[(\psi|k_i) = 0, (\psi|k_j) = 1\right] = \Pr\left[(\psi|k_i) = 1, (\psi|k_j) = 0\right] = \frac{1}{2}\kappa$$

$$\Pr\left[(\psi|k_i) = 1, (\psi|k_j) = 1\right] = \Pr\left[(\psi|k_i) = 0, (\psi|k_j) = 0\right] = \frac{1}{2}\xi.$$

For three different keys k_h, k_i and k_j, we further define a three-way confusion coefficient:

$$\tilde{\kappa} = \tilde{\kappa}(k_h, k_i, k_j) = \Pr\left[(\psi|k_i) = (\psi|k_j), \psi|k_i \neq (\psi|k_h)\right].$$

Lemma 2. $\tilde{\kappa}(k_h, k_i, k_j) = \frac{1}{2}[\kappa(k_h, k_i) + \kappa(k_h, k_j) - \kappa(k_i, k_j)]$. *(See Appendix C)*

3.2   Confusion Coefficient and DPA

The power measurements are for one key embedded in the cryptographic system, i.e., the correct key, denote as k_c. Denote k_g as one of the incorrect key guesses. Suppose the DoM for k_c and k_g are δ_c and δ_g, respectively. The difference between the two DoMs is $\Delta(k_c, k_g) = (\delta_c - \delta_g)$. We have obtained the mean and variance of $\Delta(k_c, k_g)$ (see Appendix D):

$$
\begin{aligned}
E[\Delta(k_c, k_g)] &= 2\kappa(k_c, k_g)\epsilon \\
Var[\Delta(k_c, k_g)] &= 16\kappa(k_c, k_g)\frac{\sigma_W^2}{N_m} + 4\kappa(k_c, k_g)\xi(k_c, k_g)\frac{\epsilon^2}{N_m}
\end{aligned}
\tag{2}
$$

Hence, $\lim_{N_m \to \infty} \Delta(k_c, k_g) = 2\kappa(k_c, k_g)\epsilon$.

4 Statistical Model for DPA

In DPA, to successfully distinguish the correct key k_c from other key hypotheses, it requires the DoM of k_c to be larger than that of all other keys, written as: $\delta_{k_c} > \{\delta_{\langle \overline{k_c} \rangle}\}$, where $\langle \overline{k_c} \rangle$ denotes all the incorrect keys, i.e., $\{k_0, \ldots, k_{N_k-1}\}$ excluding k_c, and $\{\delta_{\langle \overline{k_c} \rangle}\}$ denotes $\{\delta_{k_0}, \ldots, \delta_{k_{N_k-1}}\}$ excluding δ_{k_c}. The success rate to recover the correct key, SR, is defined as the probability for $\delta_{k_c} > \delta_{\langle \overline{k_c} \rangle}$:

$$\text{SR} = \text{SR}\left[k_c, \langle \overline{k_c} \rangle\right] = \Pr\left[\delta_{k_c} > \{\delta_{\langle \overline{k_c} \rangle}\}\right]$$

The overall success rate is against $(N_k - 1)$ wrong keys. We next show the derivation of the success rates starting from the simple one-key success rate.

1-key Success Rate. We first consider the 1-key success rate, i.e., the success rate of k_c over an incorrect key k_g chosen out of $\langle \overline{k_c} \rangle$, written as:

$$\text{SR}_1 = \text{SR}\left[k_c, k_g\right] = \Pr\left[\delta_{k_c} > \delta_{k_g}\right] = \Pr\left[\Delta(k_c, k_g) > 0\right].$$

From Section 2.3, $\Delta(k_c, k_g)$ is the difference of two normal random variables, therefore follows distribution $\mathcal{N}\left(\mu_{\Delta(k_c,k_g)}, \sigma_{\Delta(k_c,k_g)}\right)$. From Equation (2),

$$\mu_{\Delta(k_c,k_g)} = 2\kappa(k_c, k_g)\epsilon, \qquad \sigma_{\Delta(k_c,k_g)} = 2\sqrt{\frac{\kappa(k_c, k_g)}{N_m}}\sqrt{4\sigma_{\mathcal{W}}^2 + \xi(k_c, k_g)\epsilon^2}.$$

Let $\Phi(x) = \frac{1}{2}[1 + erf(\frac{x}{\sqrt{2}})]$ denote the cumulative distribution function (cdf) of the standard normal distribution, where erf(x) is the error function $erf(x) = \frac{2}{\sqrt{\pi}}\int_{-\infty}^{x} e^{-t^2/2}dt$. Since $\frac{\Delta(k_c,k_g)-\mu_{\Delta(k_c,k_g)}}{\sigma_{\Delta(k_c,k_g)}}$ is a standard normal random variable,

$$\text{SR}_1 = \Pr\left[\Delta(k_c, k_g) > 0\right] = 1 - \Phi(-\frac{\mu_{\Delta(k_c,k_g)}}{\sigma_{\Delta(k_c,k_g)}}) = \Phi(\frac{\mu_{\Delta(k_c,k_g)}}{\sigma_{\Delta(k_c,k_g)}})$$

$$= \frac{1}{2}\left[1 + \text{erf}\left(\frac{\mu_{\Delta(k_c,k_g)}}{\sqrt{2}\sigma_{\Delta(k_c,k_g)}}\right)\right] = \frac{1}{2}\left[1 + \text{erf}\left(\sqrt{\frac{\kappa(k_c, k_g)}{(\frac{2\sigma_{\mathcal{W}}}{\epsilon})^2 + \xi(k_c, k_g)}}\sqrt{\frac{N_m}{2}}\right)\right] \tag{3}$$

This is a function of confusion coefficients $\kappa(k_c, k_g)$, the ratio of ϵ to $\sigma_{\mathcal{W}}$, and the number of measurements, N_m. Overall, the higher $\epsilon/\sigma_{\mathcal{W}}$, $\kappa(k_c, k_g)$, and N_m are, the higher the success rate is, i.e., more susceptible to DPA.

2-keys Success Rate. Next we consider the 2-keys success rate, i.e., the success rate of k_c over two chosen incorrect keys k_{g_1} and k_{g_2}, written as:

$$\text{SR}_2 = \text{SR}\left[k_c, \{k_{g_1}, k_{g_2}\}\right] = \Pr\left[\delta_{k_c} > \delta_{k_{g_1}}, \delta_{k_c} > \delta_{k_{g_2}}\right] = \Pr\left[y_1 > 0, y_2 > 0\right]$$

where

$$y_1 = \Delta(k_c, k_{g_1}) = \delta_{k_c} - \delta_{k_{g_1}}, \qquad y_2 = \Delta(k_c, k_{g_2}) = \delta_{k_c} - \delta_{k_{g_2}}.$$

Since y_1 and y_2 are random variables with normal distribution, $Y_2 = [y_1, y_2]^T$ is a random vector with two-dimension normal distribution as $\mathcal{N}(\boldsymbol{\mu}_2, \boldsymbol{\Sigma}_2)$, where

$$\boldsymbol{\mu}_2 = \begin{bmatrix} \mu_{y_1} \\ \mu_{y_2} \end{bmatrix} = \begin{bmatrix} 2\kappa(k_c, k_{g_1})\epsilon \\ 2\kappa(k_c, k_{g_2})\epsilon \end{bmatrix}, \qquad \boldsymbol{\Sigma}_2 = \begin{bmatrix} \mathrm{Cov}(y_1, y_1) & \mathrm{Cov}(y_1, y_2) \\ \mathrm{Cov}(y_1, y_2) & \mathrm{Cov}(y_2, y_2) \end{bmatrix}.$$

The covariances in $\boldsymbol{\Sigma}_2$ are (see Appendix E for the proof):

$$\mathrm{Cov}(y_1, y_1) = 16\kappa(k_c, k_{g_1})\frac{\sigma_W^2}{N_m} + 4\kappa(k_c, k_{g_1})\xi(k_c, k_{g_1})\frac{\epsilon^2}{N_m}$$

$$\mathrm{Cov}(y_2, y_2) = 16\kappa(k_c, k_{g_2})\frac{\sigma_W^2}{N_m} + 4\kappa(k_c, k_{g_2})\xi(k_c, k_{g_2})\frac{\epsilon^2}{N_m}$$

$$\mathrm{Cov}(y_1, y_2) = 16\tilde{\kappa}(k_c, k_{g_1}, k_{g_2})\frac{\sigma_W^2}{N_m} + 4[\tilde{\kappa}(k_c, k_{g_1}, k_{g_2}) - \kappa(k_c, k_{g_1})\kappa(k_c, k_{g_2})]\frac{\epsilon^2}{N_m}.$$

Let $\Phi_2(\boldsymbol{x})$ denote the cdf of the 2-dimension standard normal distribution.

$$\mathrm{SR}_2 = \Phi_2(\sqrt{N_m}\boldsymbol{\Sigma}_2^{-1/2}\boldsymbol{\mu}_2) \tag{4}$$

which is a function of the ratio ϵ/σ_W, sample size N_m, and confusion coefficients $\kappa(k_c, k_{g_1})$, $\kappa(k_c, k_{g_2})$ and $\kappa(k_{g_1}, k_{g_2})$.

$(\mathbf{N_k - 1})$-**keys success rate.** The overall success rate is the success rate of k_c over all other $(N_k - 1)$ keys $\langle \overline{k_c} \rangle$,

$$\mathrm{SR} = \mathrm{SR}_{N_k-1} = \mathrm{SR}\left[k_c, \langle \overline{k_c} \rangle\right] = \Pr\left[\delta_{k_c} > \{\delta_{\langle \overline{k_c} \rangle}\}\right] = \Pr\left[Y > 0\right]$$

where Y is the $(N_k - 1)$-dimension vector of differences between δ_{k_c} and $\delta_{\langle \overline{k_c} \rangle}$:

$$Y = \delta_{k_c} - \delta_{\langle \overline{k_c} \rangle} = \left[\Delta(k_c, k_{g_1}), \dots, \Delta(k_c, k_{g_{N_k-1}})\right]^T = [y_1, \dots, y_{N_k-1}]^T.$$

Y is randomly distributed with $\mathcal{N}(\boldsymbol{\mu}_Y, \boldsymbol{\Sigma}_Y)$. The mean is:

$$\boldsymbol{\mu}_Y = 2\epsilon\boldsymbol{\kappa} \tag{5}$$

where $\boldsymbol{\kappa}$ denotes a $(N_k - 1)$-dimension *confusion vector* for the correct key k_c with entries $\kappa(k_c, k_{g_i})$, $i = 1, \dots, N_k - 1$. The elements in the $(N_k - 1) \times (N_k - 1)$ matrix $\boldsymbol{\Sigma}_Y$ are covariances between $y_1, \dots, y_{N_k-1}$. Thus

$$\boldsymbol{\Sigma}_Y = 16\frac{\sigma_W^2}{N_m}\boldsymbol{K} + 4\frac{\epsilon^2}{N_m}(\boldsymbol{K} - \boldsymbol{\kappa}\boldsymbol{\kappa}^T) \tag{6}$$

where $\boldsymbol{\kappa}^T$ denotes the transpose of $\boldsymbol{\kappa}$, and $\boldsymbol{K}$ is the $(N_k - 1) \times (N_k - 1)$ *confusion matrix* of the cryptographic algorithm for k_c, with elements $\{\varkappa_{ij}\}$ as:

$$\varkappa_{ij} = \begin{cases} \kappa(k_c, k_{g_i}) & \text{if } i = j \\ \tilde{\kappa}(k_c, k_{g_i}, k_{g_j}) & \text{if } i \neq j. \end{cases}$$

The confusion matrix K fully depicts the relation between all the key candidates in the algorithm, and Equation (6) shows how it affects the success rate.

Let $\Phi_{N_k-1}(x)$ denote the cdf of the $(N_k - 1)$-dimension standard normal distribution.

$$\mathrm{SR} = \mathrm{SR}_{N_k-1} = \Phi_{N_k-1}(\sqrt{N_m}\Sigma_Y^{-1/2}\mu_Y). \tag{7}$$

Our statistical model for the overall success rate (SR) results in a multivariate Gaussian distribution formula. We can see that SR is determined by parameters related to both the physical implementation, ϵ and σ_W, and the cryptographic algorithm, K. ϵ and σ_W can be computed from the side-channel measurements of the cryptographic system. K is only determined by the specific selection function and cryptographic algorithm, independent of real physical implementations. Given these parameters, SR can be calculated with numerical simulations of the $(N_k - 1)$-dimension normal distribution. Our model extracts the effect of both the implementation and algorithm on SCA resilience quantitatively.

5   Experimental Results

5.1   DPA on DES

We perform DPA on DES, with the selection function on a randomly chosen bit. In our experiments, we choose the first bit of the input for the last round to evaluate the success rate model. We take the data set from DPAcontest [23] secmatv1 and focus on a single point (the 15750th point) which has the maximum DoM for k_c. Discussions on multi-point leakage will be given in Section 6.3. We generate the empirical success rate with 1000 trials as in [11,12].

To compute the theoretical success rate, we need the physical implementation parameters $SNR = \epsilon/\sigma_W$ and the confusion coefficients κ for any two keys. Since k_c has been recovered for this data set, using *all* the power measurements at the selected leakage time point (the 15750th point), we can estimate ϵ as the DoM under k_c and estimate σ_W^2 as the variance of power measurements. For a DES subkey of 6-bit, the number of key guesses, N_k, is 64, and there are (64×63) confusion coefficients. We found that they fall into nine values.

$$\{0.25, 0.3125, 0.375, 0.4375, 0.5, 0.5625, 0.625, 0.6875, 0.75\}.$$

We define these values as *characteristic confusion values* of a DES SBox. Why they end up in these nine values and what are the implications are unknown yet. However, we believe they manifest some important DPA-related properties of the SBoxes.

Fig. 1 plots the empirical success rates (the solid curves) and theoretical success rates (the dashed curves) of our model. We show the success rates against different number of key candidates for $k_c = k_{60}$. From top down, they are: $\mathrm{SR}_1 = \mathrm{SR}(k_c, k_0)$, $\mathrm{SR}_2 = \mathrm{SR}(k_c, \{k_0, k_1\})$, $\mathrm{SR}_8 = \mathrm{SR}(k_c, \{k_0, \dots, k_7\})$, $\mathrm{SR}_{32} = \mathrm{SR}(k_c, \{k_0, \dots, k_{31}\})$, and the overall success rate $\mathrm{SR}_{63} = \mathrm{SR}(k_c, \langle \overline{k_c} \rangle)$. We can see that the two curves for SR_{63} track each other very well, showing the accuracy of our theoretical model.

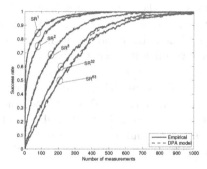 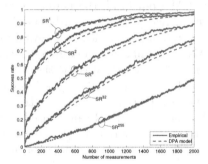

Fig. 1. Empirical and theoretical success rates of DPA on DES

Fig. 2. Empirical and theoretical success rates of DPA on AES

5.2 DPA on AES

We next perform DPA on AES. The select function is defined as the XORed value of input and output of the third bit of the sixteenth SBox in the last round of AES. We measured the power consumption data using the SASEBO GII board with AES implementation designated by DPAcontest [24]. The total number of measurements in the data set is $100,000$. The size of the AES subkey is 8, and there are (256×255) confusion coefficients, which also fall into nine characteristic confusion values of AES SBox:

$$\{0.4375, 0.453125, 0.46875, 0.484375, 0.5, 0.515625, 0.53125, 0.546875, 0.5625\}.$$

Fig. 2 shows the empirical success rates (solid curves) and theoretical success rates (dashed curves) of DPA for $k_c = k_{143}$. The two 255-keys success rate curves of empirical and theoretical track each other very well, demonstrating that the model is also very accurate for AES.

6 Discussions

Our DPA analysis builds a quantitative model for the SCA resilience of a cryptographic system over its inherent parameters, including ϵ, σ_W and $\boldsymbol{K}$. Next we present more SCA-related insights from the model about the implementation and algorithms, and how to use it to evaluate countermeasures and algorithms.

6.1 Signal and Noise of the Side Channel

Theoretically, DPA targets a portion of circuits that are related to the select function ψ, and other parts of the circuits are considered as random noise unrelated to ψ. DoM ϵ of the correct key is the power difference between $\psi = 1$ and $\psi = 0$ of the part of circuits under attack. DPA is a statistical process retrieving the DoM ϵ out of all the power consumptions. As the number of power wave-forms increases, the standard deviation of the difference between the DoMs for the correct key and incorrect keys decreases. When the standard deviations of DoMs are significantly less than ϵ, DPA has a significant success rate to recover the key. Here, ϵ indicates the signal level, and $\sigma_{\mathcal{W}}$ indicates the noise level.

We define $\epsilon/\sigma_{\mathcal{W}}$ as the *signal-to-noise ratio* (SNR) for the side channel. It is shown in Equation (3) for the 1-key success rate how the SNR determines the DPA results. The SNR can be used as a metric to measure the SCA resilience of the implementation of a cryptographic system. It is similar to the SNR defined in [15,22], however with more explicit quantitative implications in our model.

6.2   DPA-Confusion Property of Cryptographic Algorithms

Our algorithmic confusion analysis reveals the inherent property of a cryptographic algorithm, i.e., how differently the key candidates behave in DPA. Confusion coefficient is determined by both the cryptographic algorithm and selection function ψ. The eight different SBoxes in DES may have different confusion properties. Different bits in the same SBox may also have different confusion properties. Compared to DES, the confusion coefficients of an AES SBox are more concentrated near 0.5, which means the key candidates behave more randomly. For SBoxes with the same key space size, the success rates have the same dimensions, and hence the one with larger confusion coefficients leaks more information, leading to higher success rates. For two algorithms with different subkey space sizes, we need to compute the overall success rates. Comparing DES and AES, the dimension factor dominates over confusion coefficients. AES has 256 key candidates and the overall success rate is for 255-keys, making it more resilient than the 63-keys success rate of DES.

The experiments in Section 5.2 define the selection function ψ for DPA on AES as the XORed value of two intermediate data due to the characteristics of ASIC implementation. In micro-controller implementation, the select function is defined directly as one intermediate data. A good select function for attacks gives larger confusion coefficient $\kappa(k_c, k_g)$ and therefore larger success rate as shown in Equation (3). The algorithmic confusion analysis can also serve as a methodology to evaluate how good selection functions are at distinguishing the correct key.

6.3   Evaluation of DPA Countermeasure: Random Delay

Our model will be very useful for evaluating different DPA countermeasures. Here we take the method of random delay as example, which is an effective countermeasure to hide leakage [10,25,26]. We analyze the resilience of random delay under DPA to demonstrate the usage of our DPA model.

The random delay has no effect on the intermediate value. Thus it has no effect on the algorithmic confusion properties. It changes the success rate of DoM attack by affecting the signal-to-noise ratio. Random delay spreads out the original side-channel leakages along the time, and therefore lowers the signal level. We consider the simplest case of single-point leakage at time s (general cases are presented in Appendix F). Suppose the power leakage without random delay is ϵ. The distribution of the random delay is $\Pr(t) = f_{rd}(t)$, for $t = 1, 2, ..., N_{rd}$ time units. Denote the time with largest $f_{rd}(t)$ as t_{max}. Then the maximum leakage after random shifting is $\varepsilon_{rd} = \varepsilon \cdot f_{rd}(t_{max})$, and the maximum leakage time point shifts from s to $s + t_{max}$. For uniform random delay, $\varepsilon_{rd} = \varepsilon/N_{rd}$. Larger N_{rd} would decrease ε_{rd}, however, it also slows down the program and degrades the performance. This also applies to more general multiple-point leakage. Our quantitative model can therefore aid the designer to fine-tune the balance between the SCA resilience and performance. Note that our success rate model is based on the knowledge of the correct key, k_c. It is meant to be adopted by the cryptosystem designer to take SCA security as a metric in the early design stage, by evaluating the SCA resilience of their implementations and countermeasures vigorously. It does not help the attacks.

6.4 Application of the Model to Other Side-Channel Attacks

Although there have been many other effective power analysis attacks, we choose the DPA to build the success rate model for its simplicity. As a matter of fact, all the power analysis attacks can be unified and it has been shown that the most popular approaches, such as DoM test, CPA, and Bayesian attacks, are essentially equivalent on a common target device (with the same power leakage model) [27]. DPA is the simplest one in modeling, because it targets a single bit. In CPA, the select function is the Hamming weight of the SBox output rather than a single bit. In addition, the correlation is the Pearson Correlation rather than the difference-of-means. We can envision that the success rate for CPA is still dependent on the implementation-determined parameters ϵ and $\sigma_{\mathcal{W}}$, and algorithm-dependent confusion coefficients κ and matrix. However, the confusion coefficient is no longer the probability that two different keys end up with different select function values, but would be generally the mean value of squared select function difference. We can regard the DoM model as a special case of the CPA model with the number of bits as 1. In our future work, we will investigate the success rate formulas for other power analysis attacks and timing attacks and their constituent parameters.

7 Conclusions

In this paper, a theoretical model for DPA on cryptographic systems is presented. It reveals how physical implementations and cryptographic algorithms jointly affect the SCA resilience. The relation between the success rate and cryptographic systems is modeled over a multivariate Gaussian distribution. The signal-to-noise ratio between the power difference and standard deviation of the power distribution indicates how resilient an implementation is. The confusion matrix generated by algorithmic confusion analysis illustrates how the cryptographic algorithm affects the resilience. Experimental results on DES and AES verify the model. We believe that this model is innovative, provides valuable insights on side-channel characteristics of cryptosystems, and could significantly facilitate SCA-resilient design and implementations.

References

1. Kocher, P.C., Jaffe, J., Jun, B.: Differential Power Analysis. In: Wiener, M. (ed.) CRYPTO 1999. LNCS, vol. 1666, pp. 388–397. Springer, Heidelberg (1999)
2. Brier, E., Clavier, C., Olivier, F.: Correlation Power Analysis with a Leakage Model. In: Joye, M., Quisquater, J.-J. (eds.) CHES 2004. LNCS, vol. 3156, pp. 16–29. Springer, Heidelberg (2004)
3. Gierlichs, B., Batina, L., Tuyls, P., Preneel, B.: Mutual Information Analysis. In: Oswald, E., Rohatgi, P. (eds.) CHES 2008. LNCS, vol. 5154, pp. 426–442. Springer, Heidelberg (2008)
4. Le, T.-H., Clédière, J., Canovas, C., Robisson, B., Servière, C., Lacoume, J.-L.: A Proposition for Correlation Power Analysis Enhancement. In: Goubin, L., Matsui, M. (eds.) CHES 2006. LNCS, vol. 4249, pp. 174–186. Springer, Heidelberg (2006)
5. Quisquater, J.-J., Samyde, D.: ElectroMagnetic Analysis (EMA): Measures and Counter-Measures for Smart Cards. In: Attali, S., Jensen, T. (eds.) E-smart 2001. LNCS, vol. 2140, pp. 200–210. Springer, Heidelberg (2001)
6. Gandolfi, K., Mourtel, C., Olivier, F.: Electromagnetic Analysis: Concrete Results. In: Koç, Ç.K., Naccache, D., Paar, C. (eds.) CHES 2001. LNCS, vol. 2162, pp. 251–261. Springer, Heidelberg (2001)
7. Kocher, P.C.: Timing Attacks on Implementations of Diffie-Hellman, RSA, DSS, and Other Systems. In: Koblitz, N. (ed.) CRYPTO 1996. LNCS, vol. 1109, pp. 104–113. Springer, Heidelberg (1996)
8. Chari, S., Jutla, C.S., Rao, J.R., Rohatgi, P.: Towards Sound Approaches to Counteract Power-Analysis Attacks. In: Wiener, M. (ed.) CRYPTO 1999. LNCS, vol. 1666, pp. 398–412. Springer, Heidelberg (1999)
9. Tiri, K., Verbauwhede, I.: A VLSI design flow for secure side-channel attack resistant ICs. In: Proc. Design, Automation & Test in Europe, pp. 58–63 (2005)
10. Clavier, C., Coron, J.-S., Dabbous, N.: Differential Power Analysis in the Presence of Hardware Countermeasures. In: Paar, C., Koç, Ç.K. (eds.) CHES 2000. LNCS, vol. 1965, pp. 252–263. Springer, Heidelberg (2000)
11. Gierlichs, B., Lemke-Rust, K., Paar, C.: Templates vs. Stochastic Methods: A Performance Analysis for Side Channel Cryptanalysis. In: Goubin, L., Matsui, M. (eds.) CHES 2006. LNCS, vol. 4249, pp. 15–29. Springer, Heidelberg (2006)
12. Standaert, F.-X., Bulens, P., de Meulenaer, G., Veyrat-Charvillon, N.: Improving the rules of the DPA contest. Cryptology ePrint Archive, Report 2008/517 (2008), http://eprint.iacr.org/2008/517
13. Standaert, F.-X., Malkin, T.G., Yung, M.: A Unified Framework for the Analysis of Side-Channel Key Recovery Attacks. In: Joux, A. (ed.) EUROCRYPT 2009. LNCS, vol. 5479, pp. 443–461. Springer, Heidelberg (2009)
14. Veyrat-Charvillon, N., Standaert, F.-X.: Mutual Information Analysis: How, When and Why? In: Clavier, C., Gaj, K. (eds.) CHES 2009. LNCS, vol. 5747, pp. 429–443. Springer, Heidelberg (2009)

15. Messerges, T.S., Dabbish, E.A., Sloan, R.H.: Examining smart-card security under the threat of power analysis attacks. IEEE Trans. on Computers 51(5), 541–552 (2002)
16. Bevan, R., Knudsen, E.: Ways to Enhance Differential Power Analysis. In: Lee, P.J., Lim, C.H. (eds.) ICISC 2002. LNCS, vol. 2587, pp. 327–342. Springer, Heidelberg (2003)
17. Mangard, S.: Hardware Countermeasures against DPA – A Statistical Analysis of Their Effectiveness. In: Okamoto, T. (ed.) CT-RSA 2004. LNCS, vol. 2964, pp. 222–235. Springer, Heidelberg (2004)
18. Prouff, E.: DPA Attacks and S-Boxes. In: Gilbert, H., Handschuh, H. (eds.) FSE 2005. LNCS, vol. 3557, pp. 424–441. Springer, Heidelberg (2005)
19. Shannon, E.A.: Communication theory of secrecy systems. Bell System Technical Journal 28(4), 656–715 (1949)
20. Katz, J., Lindell, Y.: Introduction to Modern Cryptography. Chapman & Hall/CRC Press (2007)
21. Johnson, O.T.: Information Theory and the Central Limit Theorem. Imperial College Press (2004)
22. Mangard, S., Oswald, E., Popp, T.: Power Analysis Attacks: Revealing the Secrets of Smart Cards. Advances in Information Security. Springer-Verlag New York (2007)
23. DPA Contest, http://www.dpacontest.org/
24. Side-channel attack standard evaluation board (SASEBO). Research Center for Information Security (RCIS), http://www.rcis.aist.go.jp/special/SASEBO/index-en.html
25. Coron, J.-S., Kizhvatov, I.: An Efficient Method for Random Delay Generation in Embedded Software. In: Clavier, C., Gaj, K. (eds.) CHES 2009. LNCS, vol. 5747, pp. 156–170. Springer, Heidelberg (2009)
26. Coron, J.-S., Kizhvatov, I.: Analysis and Improvement of the Random Delay Countermeasure of CHES 2009. In: Mangard, S., Standaert, F.-X. (eds.) CHES 2010. LNCS, vol. 6225, pp. 95–109. Springer, Heidelberg (2010)
27. Mangard, S., Oswald, E., Standaert, F.-X.: One for all - all for one: unifying standard differential power analysis attacks. IET Inf. Security 5(2), 100–110 (2011)
28. Standaert, F.-X., Peeters, E., Rouvroy, G., Quisquater, J.: An overview of power analysis attacks against field programmable gate arrays. Proc. IEEE 94(2) (2006)

Appendix

A   The Related CPA Model

A model for CPA is proposed in [17] and improved in [28]. The overall success rate of CPA is given as:

$$\text{SR} = \left(\int_0^\infty \frac{1}{\frac{1}{\sqrt{N_m-3}}\sqrt{2\pi}} \exp\left\{ -\frac{(x-r)^2}{\frac{2}{N_m-3}} \right\} \mathrm{d}x \right)^{N_k-1}$$

where r is the Pearson correlation of CPA for the correct key, N_k is the number of key guesses in CPA, and N_m is the number of measurements.

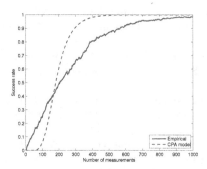 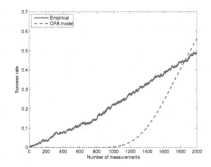

Fig. 3. Empirical and theoretical success rates of CPA on DES

Fig. 4. Empirical and theoretical success rates of CPA on AES

The CPA model assumes that the different dimensions of the overall success rate are independent, i.e., the covariances between different key guesses are 0. This is a false assumption, and therefore makes the CPA model inaccurate. We generate success rates using the CPA model for DES and AES with the same data set used in Section 5, as shown in Figs.3 and 4. The success rate curves from the CPA model do not match the empirical results.

B   Proof for the Confusion Lemma (Lemma 1)

Apply Theorem 1 to each of the key hypotheses k_i and k_j, we have:

$$\Pr\left[\psi|k_i = 1\right] = \Pr\left[\psi|k_i = 0\right] = \frac{1}{2}, \qquad \Pr\left[\psi|k_j = 1\right] = \Pr\left[\psi|k_j = 0\right] = \frac{1}{2}.$$

Because $\Pr\left[\psi|k_i = 1\right] = \Pr\left[\psi|k_i = 1, \psi|k_j = 0\right] + \Pr\left[\psi|k_i = 1, \psi|k_j = 1\right]$, and similarly for the other three probabilities above, from the definitions of the coefficients κ and ξ, we have:

$$\Pr\left[\psi|k_i = 1, \psi|k_j = 0\right] = \Pr\left[\psi|k_i = 0, \psi|k_j = 1\right] = \frac{1}{2}\kappa,$$

$$\Pr\left[\psi|k_i = 0, \psi|k_j = 0\right] = \Pr\left[\psi|k_i = 1, \psi|k_j = 1\right] = \frac{1}{2}\xi.$$

C   Proof for Lemma 2

$$
\begin{aligned}
\kappa(k_h, k_i) + \kappa(k_h, k_j) &= \Pr[(\psi|k_h) \neq (\psi|k_i)] + \Pr[(\psi|k_h) \neq (\psi|k_j)] \\
&= \Pr[(\psi|k_j) = (\psi|k_h) \neq (\psi|k_i)] + \Pr[(\psi|k_j) = (\psi|k_i) \neq (\psi|k_h)] + \\
&\quad \Pr[(\psi|k_i) = (\psi|k_h) \neq (\psi|k_j)] + \Pr[(\psi|k_i) = (\psi|k_j) \neq (\psi|k_h)] \\
&= \Pr[(\psi|k_j) \neq (\psi|k_i)] + 2\Pr[(\psi|k_j) = (\psi|k_i) \neq (\psi|k_h)] \\
&= \kappa(k_i, k_j) + 2\tilde{\kappa}(k_h, k_i, k_j).
\end{aligned}
$$

Therefore: $\tilde{\kappa}(k_h, k_i, k_j) = \frac{1}{2}[\kappa(k_h, k_i) + \kappa(k_h, k_j) - \kappa(k_i, k_j)]$.

D   Confusion Coefficient and DPA DoM

In DPA, the set of waveforms $\mathcal{W}$ is divided into two groups according to the value of ψ for one key hypothesis. Therefore, for the correct key k_c and a guessed key k_g the N_m measurements are divided into four groups $\mathcal{W}_{ij}$, $i,j = 0,1$, as shown in Table 1. For example, $\mathcal{W}_{10}$ is the group of measurements that satisfy $\psi|k_c = 1$ and $\psi|k_g = 0$, and $\mathcal{W}_{1\_}$ is the group of measurements for $\psi|k_c = 1$. We denote the number of measurements in each group by N_{ij}. We have $N_{1\_} = N_{11} + N_{10}, N_{0\_} = N_{01} + N_{00}, N_{1\_} + N_{0\_} = N_{\_1} + N_{\_0} = N_m$. Suppose the DoM for k_c and k_g are δ_c

Table 1. The four groups of waveforms $\mathcal{W}_{ij}$ and their number of measurements

	$\psi|k_c = 1$	$\psi|k_c = 0$	total
$\psi|k_g = 1$	$\mathcal{W}_{11}$ (N_{11})	$\mathcal{W}_{01}$ (N_{01})	$\mathcal{W}_{\_1}$ $(N_{\_1})$
$\psi|k_g = 0$	$\mathcal{W}_{10}$ (N_{10})	$\mathcal{W}_{00}$ (N_{00})	$\mathcal{W}_{\_0}$ $(N_{\_0})$
total	$\mathcal{W}_{1\_}$ $(N_{1\_})$	$\mathcal{W}_{0\_}$ $(N_{0\_})$	$\mathcal{W}$ (N_m)

and δ_g, respectively. The difference between the two DoMs is:

$$\Delta(k_c, k_g) = \delta_c - \delta_g = (\frac{N_{11}}{N_{1\_}} - \frac{N_{11}}{N_{\_1}})\bar{W}_{11} + (\frac{N_{10}}{N_{1\_}} + \frac{N_{10}}{N_{\_0}})\bar{W}_{10}$$
$$- (\frac{N_{01}}{N_{0\_}} + \frac{N_{01}}{N_{\_1}})\bar{W}_{01} - (\frac{N_{00}}{N_{0\_}} - \frac{N_{00}}{N_{\_0}})\bar{W}_{00} \qquad (8)$$

where $\bar{W}_{ij} = \sum W_{ij}/N_{ij}$ for $i,j = 0,1$, which are normal random variables according to the Central Limit Theorem as given in Section 2.3. Hence $\Delta(k_c, k_g)$ is also normally distributed because it is a linear combination of normal random variables. We now calculate its mean and variance.

Note that there are two sources of randomness in $\Delta(k_c, k_g)$. The first source is from the randomly selected plaintexts. Denote $\boldsymbol{\psi}_c = (\psi_1, ..., \psi_{N_m})|k_c$ and $\boldsymbol{\psi}_g = (\psi_1, ..., \psi_{N_m})|k_g$ as the values of the select function for the set of measurement plaintext under the correct key k_c and incorrect key k_g, respectively. $\boldsymbol{\psi}_c$, $\boldsymbol{\psi}_g$, and N_{ij}s are all random variables. Conditional on given $\boldsymbol{\psi}_c$ and $\boldsymbol{\psi}_g$, the partition of the measured waveforms $\mathcal{W}$ into four groups are fixed, and thus N_{ij}s are constants. There is still the second source of randomness, measurement errors in $\mathcal{W}$. Therefore $\bar{W}_{ij}$s are still random variables conditional on given $\boldsymbol{\psi}_c$ and $\boldsymbol{\psi}_g$.

Given $\boldsymbol{\psi}_c$ and $\boldsymbol{\psi}_g$, the waveforms in groups $\mathcal{W}_{11}$ and $\mathcal{W}_{10}$ have the same mean, which is larger by the amount ϵ than the mean of waveforms in $\mathcal{W}_{01}$ and $\mathcal{W}_{00}$. Without loss of generality, we assume that the theoretical means of $\bar{W}_{11}$, $\bar{W}_{10}$, $\bar{W}_{01}$ and $\bar{W}_{00}$ are ϵ, ϵ, 0 and 0, respectively. Therefore from equation (8), the conditional mean of $\Delta(k_c, k_g)$ is:

$$E[\Delta(k_c, k_g)|\boldsymbol{\psi}_c, \boldsymbol{\psi}_g] = (\frac{N_{11}}{N_{1\_}} - \frac{N_{11}}{N_{\_1}} + \frac{N_{10}}{N_{1\_}} + \frac{N_{10}}{N_{\_0}})\epsilon$$
$$= (1 - \frac{N_{11}}{N_{\_1}} + \frac{N_{10}}{N_{\_0}})\epsilon = (\frac{N_{01}}{N_{\_1}} + \frac{N_{10}}{N_{\_0}})\epsilon.$$

Now we consider the randomness in N_{ij}s, related to the algorithmic confusion analysis. According to Lemma 1, each of the waveform with $\psi|k_g = 1$ has $\kappa(k_c, k_g)$ probability to have $\psi|k_c = 0$. N_{01} and N_{10} independently follow Binomial$(N_{-1}, \kappa(k_c, k_g))$ and Binomial$(N_{-0}, \kappa(k_c, k_g))$ distributions, respectively.

$$E\left(\frac{N_{01}}{N_{-1}} + \frac{N_{10}}{N_{-0}}\right) = \frac{\kappa(k_c, k_g)N_{-1}}{N_{-1}} + \frac{\kappa(k_c, k_g)N_{-0}}{N_{-0}} = 2\kappa(k_c, k_g),$$

$$Var\left(\frac{N_{01}}{N_{-1}} + \frac{N_{10}}{N_{-0}}\right) = \frac{\kappa(k_c, k_g)\xi(k_c, k_g)}{N_{-1}} + \frac{\kappa(k_c, k_g)\xi(k_c, k_g)}{N_{-0}}$$

As $N_m \to \infty$, according to Theorem 1 and Lemma 1, $N_{1-} \simeq N_{0-} \simeq N_{-1} \simeq N_{-0} \simeq N_m/2$, $N_{10} \simeq N_{01} \simeq \kappa(k_c, k_g)N_m/2$, $N_{11} \simeq N_{00} \simeq [1 - \kappa(k_c, k_g)]N_m/2$. Thus:

$$E[\Delta(k_c, k_g)] = E\{E[\Delta(k_c, k_g)|\psi_c, \psi_g]\} = 2\kappa(k_c, k_g)\epsilon \tag{9}$$

$$Var\{E[\Delta(k_c, k_g)|\psi_c, \psi_g]\} = 4\kappa(k_c, k_g)\xi(k_c, k_g)\frac{\epsilon^2}{N_m} \tag{10}$$

The $\bar{\mathcal{W}}_{ij}$s are independent and each has the conditional variance $\sigma_{\mathcal{W}}/\sqrt{N_{ij}}$. From (8), we get:

$$E\{Var[\Delta(k_c, k_g)|\psi_c, \psi_g]\} = E\{\sigma_{\mathcal{W}}^2[(\frac{1}{N_{1-}} - \frac{1}{N_{-1}})^2 N_{11} + (\frac{1}{N_{1-}} + \frac{1}{N_{-0}})^2 N_{10}$$
$$+ (\frac{1}{N_{0-}} + \frac{1}{N_{-1}})^2 N_{01} + (\frac{1}{N_{0-}} - \frac{1}{N_{-0}})^2 N_{00}]\}$$
$$= \sigma_{\mathcal{W}}^2 \frac{16\kappa(k_c, k_g)}{N_m}.$$

Combined with (10),

$$Var[\Delta(k_c, k_g)] = E\{Var[\Delta(k_c, k_g)|\psi_c, \psi_g]\} + Var\{E[\Delta(k_c, k_g)|\psi_c, \psi_g]\}$$
$$= 16\kappa(k_c, k_g)\frac{\sigma_{\mathcal{W}}^2}{N_m} + 4\kappa(k_c, k_g)\xi(k_c, k_g)\frac{\epsilon^2}{N_m}. \tag{11}$$

E   2-keys Success Rate

For the 2-keys success rate, we have got $Y_2 = [y_1, y_2]^T$, a random vector with two-dimension normal distribution, $\mathcal{N}(\mu_2, \Sigma_2)$. Now we calculate the formula for Σ_2. From equation (8), we have (for $i = 1, 2$):

$$y_i = \Delta(k_c, k_{gi}) = (\frac{N_{11,y_i}}{N_{1-,y_i}} - \frac{N_{11,y_i}}{N_{-1,y_i}})\bar{\mathcal{W}}_{11,y_i} + (\frac{N_{10,y_i}}{N_{1-,y_i}} + \frac{N_{10,y_i}}{N_{-0,y_i}})\bar{\mathcal{W}}_{10,y_i}$$
$$- (\frac{N_{00,y_i}}{N_{0-,y_i}} - \frac{N_{00,y_i}}{N_{-0,y_i}})\bar{\mathcal{W}}_{00,y_i} - (\frac{N_{01,y_i}}{N_{-1,y_i}} + \frac{N_{01,y_i}}{N_{-0,y_i}})\bar{\mathcal{W}}_{01,y_i} \tag{12}$$

From Appendix D Equation (11), the variance of y_1 and y_2 are:

$$\text{Cov}(y_i, y_i) = 16\kappa(k_c, k_{g_i})\frac{\sigma_W^2}{N_m} + 4\kappa(k_c, k_{g_i})\xi(k_c, k_{g_i})\frac{\epsilon^2}{N_m}, \qquad i = 1, 2$$

Next we compute $\text{Cov}(y_1, y_2)$, the covariance between y_1 and y_2. We first calculate $E[\text{Cov}(y_1, y_2 | \psi_c, \psi_{g_1}, \psi_{g_2})]$. The conditional covariance between y_1 and y_2 given $(\psi_c, \psi_{g_1}, \psi_{g_2})$ is:

$$
\begin{aligned}
Cov(y_1, &y_2 | \psi_c, \psi_{g_1}, \psi_{g_2}) \\
=&4\kappa(k_c, k_{g_1})\kappa(k_c, k_{g_2})[Cov(\bar{\mathcal{W}}_{10,y_1}, \bar{\mathcal{W}}_{10,y_2}) + Cov(\bar{\mathcal{W}}_{01,y_1}, \bar{\mathcal{W}}_{01,y_2}) \\
&- Cov(\bar{\mathcal{W}}_{10,y_1}, \bar{\mathcal{W}}_{01,y_2}) - Cov(\bar{\mathcal{W}}_{01,y_1}, \bar{\mathcal{W}}_{10,y_2})]
\end{aligned}
$$

The waveforms in $\mathcal{W}_{10y_1}$ and $\mathcal{W}_{10y_2}$ are those with $\psi | k_c = 1$, different from those in $\mathcal{W}_{01y_1}$ and $\mathcal{W}_{01y_2}$, therefore: $Cov(\bar{\mathcal{W}}_{10,y_1}, \bar{\mathcal{W}}_{01,y_2}) = Cov(\bar{\mathcal{W}}_{01,y_1}, \bar{\mathcal{W}}_{10,y_2}) = 0$.

To compute $Cov(\bar{\mathcal{W}}_{10,y_1}, \bar{\mathcal{W}}_{10,y_2})$, we consider how similar they are, i.e., how many waveforms are the same between the partitions $\mathcal{W}_{10y_1}$ and $\mathcal{W}_{10y_2}$. Let $N_{10,s}$ denote the number of same waveforms between $\mathcal{W}_{10,y_1}$ and $\mathcal{W}_{10,y_2}$. Then $Cov(\sum \mathcal{W}_{10,y_1}, \sum \mathcal{W}_{10,y_1}) = N_{10,s}\sigma_W^2$. $N_{10,s} \simeq \tilde{\kappa}(k_c, k_{g_1}, k_{g_2})N_m/2$ as $N_m \to \infty$ by the definition of $\tilde{\kappa}(k_c, k_{g_1}, k_{g_2})$. Hence,

$$Cov(\bar{\mathcal{W}}_{10,y_1}, \bar{\mathcal{W}}_{10,y_2}) = \frac{Cov(\sum \mathcal{W}_{10,y_1}, \sum \mathcal{W}_{10,y_2})}{N_{10,y_1} N_{10,y_2}} = \frac{2\tilde{\kappa}(k_c, k_{g_1}, k_{g_2})}{\kappa(k_c, k_{g_1})\kappa(k_c, k_{g_2})N_m}\sigma_W^2$$

Similarly, we get the same expression for $Cov(\bar{\mathcal{W}}_{01,y_1}, \bar{\mathcal{W}}_{01,y_2})$. Thus we get

$$E[Cov(y_1, y_2 | \psi_c, \psi_{g_1}, \psi_{g_2})] = 16\tilde{\kappa}(k_c, k_{g_1}, k_{g_2})\frac{\sigma_W^2}{N_m}. \qquad (13)$$

Next we calculate $Cov[E(y_1 | \psi_c, \psi_{g_1}, \psi_{g_2}), E(y_2 | \psi_c, \psi_{g_1}, \psi_{g_2})]$. Let N_{d,y_1} denote the number of measurements where $\psi | k_c$ is different from $\psi | k_{g_1}$. From (12),

$$
\begin{aligned}
Cov[&E(y_1 | \psi_c, \psi_{g_1}, \psi_{g_2}), E(y_2 | \psi_c, \psi_{g_1}, \psi_{g_2})] \\
=&(\frac{2\epsilon}{N_m})^2 Cov(N_{d,y_1}, N_{d,y_2}) = (\frac{2\epsilon}{N_m})^2[E(N_{d,y_1}N_{d,y_2}) - E(N_{d,y_1})E(N_{d,y_2})]
\end{aligned}
$$

We re-express $N_{d,y_1} = \sum I[(\psi | k_c) \neq (\psi | k_{g_1})]$ and $N_{d,y_2} = \sum I[(\psi | k_c) \neq (\psi | k_{g_2})]$. Obviously $E(N_{d,y_1}) = N_m\kappa(k_c, k_{g_1})$ and $E(N_{d,y_2}) = N_m\kappa(k_c, k_{g_2})$.

$$N_{d,y_1}N_{d,y_2} = \sum I[(\psi | k_c) \neq (\psi | k_{g_1})] \sum I[(\psi | k_c) \neq (\psi | k_{g_2})]$$

Note that $N_{d,y_1}N_{d,y_2}$ is the sum of N_m^2 terms. Most of the terms in the sum have expectation $\kappa(k_c, k_{g_1})\kappa(k_c, k_{g_2})$ except for those N_m terms corresponding to the same waveforms, which have expectation:

$$
\begin{aligned}
E\{I[(\psi | k_c) \neq (\psi | k_{g_1})]I[(\psi | k_c) \neq (\psi | k_{g_2})]\} &= E\{I[(\psi | k_{g_1}) = (\psi | k_{g_2}) \neq (\psi | k_c)]\} \\
&= \tilde{\kappa}(k_c, k_{g_1}, k_{g_2})
\end{aligned}
$$

Hence,

$$E(N_{d,y_1}N_{d,y_2}) = N_m^2 \kappa(k_c, k_{g_1})\kappa(k_c, k_{g_2}) + N_m[\tilde{\kappa}(k_c, k_{g_1}, k_{g_2}) - \kappa(k_c, k_{g_1})\kappa(k_c, k_{g_2})]$$

This implies that

$$Cov[E(y_1|\psi_c, \psi_{g_1}, \psi_{g_2}), E(y_2|\psi_c, \psi_{g_1}, \psi_{g_2})]$$
$$= 4[\tilde{\kappa}(k_c, k_{g_1}, k_{g_2}) - \kappa(k_c, k_{g_1})\kappa(k_c, k_{g_2})]\frac{\epsilon^2}{N_m}. \tag{14}$$

Combining (13) and (14), we get:

$$Cov(y_1, y_2)$$
$$= E[Cov(y_1, y_2|\psi_c, \psi_{g_1}, \psi_{g_2})] + Cov[E(y_1|\psi_c, \psi_{g_1}, \psi_{g_2}), E(y_2|\psi_c, \psi_{g_1}, \psi_{g_2})]$$
$$= 16\tilde{\kappa}(k_c, k_{g_1}, k_{g_2})\frac{\sigma_W^2}{N_m} + 4[\tilde{\kappa}(k_c, k_{g_1}, k_{g_2}) - \kappa(k_c, k_{g_1})\kappa(k_c, k_{g_2})]\frac{\epsilon^2}{N_m}.$$

F Leakage Evaluation of Random Delay

The resilience of random delay is determined by the maximum leakage ϵ_{rd}, which is the overall leakage accumulated with random shifting. We consider two scenarios of the original leakage:

1. Single-point leakage. Only one time point s in the power consumption waveform leaks information with signal level ϵ. This is the simplified ideal case. The maximum leakage after random shifting is the original leakage distributed with the maximum probability, which is:

$$\epsilon_{rd} = \epsilon \cdot f_{rd}(t_{max}) = \epsilon \cdot \max_{0 \le t \le N_{rd}-1} \{f_{rd}(t)\}.$$

For uniform random delay, $\Pr(t) = f_{rd}(t) = 1/N_{rd}$, for $t = 0, 1, ..., N_{rd} - 1$. Hence the signal $\epsilon_{rd} = \epsilon/N_{rd}$ decreases from the original signal ϵ by a factor N_{rd}.

2. Multiple-point leakage. At time t, the leakage signal strength is $\epsilon(t)$. Then the leakage signal at time i with random delay is:

$$\epsilon_{rd}(i) = \sum_{t=0}^{N_{rd}-1} f_{rd}(t)\epsilon(i+t).$$

The maximum leakage accumulation with random delay as:

$$\epsilon_{rd} = \max_i \left\{ \sum_{t=0}^{N_{rd}-1} f_{rd}(t)\epsilon(i+t) \right\}.$$

Then the success rate of the strongest single-point DoM attack on the device with random delay can be calculated by Formula (7) using the ϵ_{rd} value.

Practical Security Analysis of PUF-Based Two-Player Protocols

Ulrich Rührmair[1] and Marten van Dijk[2]

[1] Technische Universität München, 80333 München, Germany
ruehrmair@in.tum.de
[2] RSA Laboratories, Cambridge, MA, USA
marten.vandijk@rsa.com

Abstract. In recent years, PUF-based schemes have not only been suggested for the basic tasks of tamper sensitive key storage or the identification of hardware systems, but also for more complex protocols like oblivious transfer (OT) or bit commitment (BC), both of which possess broad and diverse applications. In this paper, we continue this line of research. We first present an attack on two recent OT- and BC-protocols which have been introduced at CRYPTO 2011 by Brzuska et al. [1,2]. The attack quadratically reduces the number of CRPs which malicious players must read out in order to cheat, and fully operates within the original communication model of [1,2]. In practice, this leads to insecure protocols when electrical PUFs with a medium challenge-length are used (e.g., 64 bits), or whenever optical PUFs are employed. These two PUF types are currently among the most popular designs. Secondly, we discuss countermeasures against the attack, and show that interactive hashing is suited to enhance the security of PUF-based OT and BC, albeit at the price of an increased round complexity.

Keywords: Physical Unclonable Functions (PUFs), Cryptographic Protocols, Oblivious Transfer, Bit Commitment, Security Analysis, Interactive Hashing.

1 Introduction

Today's electronic devices are mobile, cross-linked and pervasive, which makes them a well-accessible target for adversaries. The well-known protective cryptographic techniques all rest on the concept of a secret binary key: They presuppose that devices store a piece of digital information that is, and remains, unknown to an adversary. It turns out that this requirement is difficult to realize in practice. Physical attacks such as invasive, semi-invasive or side-channel attacks carried out by adversaries with one-time physical access to the devices, as well as software attacks like application programming interface (API) attacks, viruses or Trojan horses, can lead to key exposure and security breaks. As Ron Rivest emphasized in his keynote talk at CRYPTO 2011 [22], merely calling a bit string a "secret key" does not make it secret, but rather identifies it as an interesting target for the adversary.

Indeed, one main motivation for the development of *Physical Unclonable Functions (PUFs)* was their promise to better protect secret keys. A PUF is an (at least partly) disordered physical system P that can be challenged with so-called external stimuli or

E. Prouff and P. Schaumont (Eds.): CHES 2012, LNCS 7428, pp. 251–267, 2012.

challenges c, upon which it reacts with corresponding responses r. Contrary to standard digital systems, these responses depend on the micro- or nanoscale structural disorder of the PUF. It is assumed that this disorder cannot be cloned or reproduced exactly, not even by the PUF's original manufacturer, and that it is unique to each PUF. Any PUF P thus implements a unique and individual function f_P that maps challenges c to responses $r = f_P(c)$. The tuples (c, r) are called the challenge-response pairs (CRPs) of the PUF.

Due to its complex internal structure, a PUF can avoid some of the shortcomings of classical digital keys. It is usually harder to read out, predict, or derive PUF-responses than to obtain digital keys that are stored in non-volatile memory. The PUF-responses are only generated when needed, which means that no secret keys are present permanently in the system in an easily accessible digital form. Finally, certain types of PUFs are naturally tamper sensitive: Their exact behavior depends on minuscule manufacturing irregularities, often in different layers of the IC, and removing or penetrating these layers will automatically change the PUF's read-out values. These facts have been exploited in the past for different PUF-based security protocols. Prominent examples include identification [21,9], key exchange [21], and various forms of (tamper sensitive) key storage and applications thereof, such as intellectual property protection or read-proof memory [11,14,29].

In recent years, also the use of PUFs in more advanced cryptographic protocols together with formal security proofs has been investigated. In these protocols, usually PUFs with a large challenge set and with a freely accessible challenge-response interface are employed.[1] The PUF is used similar to a "physical random oracle", which is transferred between the parties, and which can be read-out exactly by the very party who currently holds physical possession of it. Its input-output behavior is assumed to be so complex that its response to a randomly chosen challenge cannot be predicted numerically and without direct physical measurement, not even by a person who had physical access to the PUF at earlier points in time. In 2010, Rührmair [23] showed that oblivious transfer (OT) can be realized between two parties by physically transferring a PUF in this setting. He observed that via the classical reductions of Kilian [13], this implies PUF-based bit commitment and PUF-based secure multi-party computations. In the same year, the first formal security proof for a PUF-protocol was provided by Rührmair, Busch and Katzenbeisser [24]. They presented definitions and a reductionist security proof for PUF-based identification protocols. At CRYPTO 2011 Brzuska et al. [1] adapted Canetti's universal composition (UC) framework [3] to include PUFs. They gave PUF-based protocols for oblivious transfer (OT), bit commitment (BC) and key exchange (KE) and proved them to be secure in their framework.

The investigation of advanced cryptographic settings for PUF makes sense even from the perspective of a pure practitioner: Firstly, it clarifies the potential of PUFs in theory, a necessary prerequisite before this potential can be unleashed in commercial applications without risking security failures. Secondly, BC and OT protocols are

---

[1] This type of PUF sometimes has been termed *Physical Random Function* [9] or *Strong PUF* [11,26,25,24] in the literature. We emphasize that the Weak/Strong PUF terminology introduced by Guajardo et al. [11] is not to be understood in a judgemental or pejorative manner.

extremely versatile cryptographic primitives, which allow the implementation of such diverse tasks as zero-knowledge identification, the enforcement of semi-honest behavior in cryptographic protocols, secure multi-party computation (including online auctions or electronic voting), or key exchange. If these tasks shall be realized securely in practice by PUFs, a theoretical investigation of the underlying primitives — in this case BC and OT — is required first.

In this paper, we continue this line of research, and revisit the use of PUFs in OT- and BC-protocols. Particular emphasis is placed on the achievable *practical security* if well-established PUFs (like electrical PUFs with 64-bit challenge lengths or optical PUFs) are used in the protocols. We start by observing an attack on the OT- and BC-protocols of Brzuska et al. [1,2] which quadratically reduces the number of responses that a malicious player must read out in order to cheat. It works fully in the original communication model of Brzuska et al. and makes no additional assumptions. As we show, the attack makes the protocols insecure in practice if electrical PUFs with medium bitlengths around 64 bits are used, and generally if optical PUFs are employed. This has a special relevance since the use of optical PUFs for their protocols had been explicitly proposed by Brzuska et al. (see Section 8 of [2]). Secondly, we investigate countermeasures against our attack, and show that interactive hashing can be used to enhance the security of PUF-based OT and BC protocols.

Our work continues the recent trend of a formalization of PUFs, including protocol analyses, more detailed investigations of non-trivial communication settings, and formal security proofs. This trend will eventually lay the foundations for future PUF research, and seems indispensible for a healthy long-term development of the field. It also combines protocol design and practical security analyses in a novel manner.

Organization of this Paper. In Section 2 we present the protocols of Brzuska et al. in order to achieve a self-contained treatment. Section 3 gives our quadratic attack. Section 4 discusses its practical effect. Section 5 discusses countermeasures. We conclude the paper in Section 6.

2   The Protocols of Brzuska et al.

Our aim in this paper is to present a quadratic attack on two recent PUF-protocols for OT and BC by Brzuska et al. [1,2] and to discuss its practical relevance. In order to achieve a self-contained treatment, we will now present these two protocols. To keep our exposition simple, we will not use the full UC-notation of [1], and will present the schemes mostly without error correction mechanisms, since the latter play no role in the context of our attack.

The protocols use two communication channels between the communication partners: A binary channel, over which all digital communication is handled. It is assumed that this channel is non-confidential, but authenticated. And secondly an insecure physical channel, over which the PUF is sent. It is assumed that adversaries can measure adaptively selected CRPs of the PUF while it is in transition over this channel.

2.1 Oblivious Transfer

The OT protocol of [1] implements one-out-of-two string oblivious transfer. It is assumed that in each subsession the sender P_i initially holds two (fresh) bitstrings $s_0, s_1 \in \{0, 1\}^\lambda$, and that the receiver P_j holds a (fresh) choice bit b.

Brzuska et al. generally assume in their treatment that after error correction and the application of fuzzy extractors, a PUF can be modeled as a function PUF : $\{0, 1\}^\lambda \to \{0, 1\}^{rg(\lambda)}$. We use this model throughout this paper, too. In the subsequent protocol of Brzuska et al., it is furthermore assumed that $rg(\lambda) = \lambda$, i.e., that the PUF implements a function PUF : $\{0, 1\}^\lambda \to \{0, 1\}^\lambda$ (compare [1,2]).

Protocol 1: PUF-BASED OBLIVIOUS TRANSFER ([1], SLIGHTLY SIMPLIFIED DESCRIPTION)

External Parameters: The protocol has a number of external parameters, including the security parameter λ, the session identifier sid, a number N that specifies how many subsessions are allowed, and a pre-specified PUF-family $\mathcal{P}$, from which all PUFs which are used in the protocol must be drawn.

Initialization Phase: Execute once with fixed session identifier sid:

1. The receiver holds a PUF which has been drawn from the family $\mathcal{P}$.
2. The receiver measures l randomly chosen CRPs $c_1, r_1, \ldots, c_l, r_l$ from the PUF, and puts them in a list $\mathcal{L} := (c_1, r_1, \ldots, c_l, r_l)$.
3. The receiver sends the PUF to the sender.

Subsession Phase: Repeat at most N times with fresh subsession identifier ssid:

1. The sender's input are two strings $s_0, s_1 \in \{0, 1\}^\lambda$, and the receiver's input is a bit $b \in \{0, 1\}$.
2. The receiver chooses a CRP (c, r) from the list $\mathcal{L}$ at random.
3. The sender chooses two random bitstrings $x_0, x_1 \in \{0, 1\}^\lambda$ and sends x_0, x_1 to the receiver.
4. The receiver returns the value $v := c \oplus x_b$ to the sender.
5. The sender measures the responses r_0 and r_1 of the PUF that correspond to the challenges $c_0 := v \oplus x_0$ and $c_1 := v \oplus x_1$.
6. The sender sets the values $S_0 := s_0 \oplus r_0$ and $S_1 := s_1 \oplus r_1$, and sends S_0, S_1 to the receiver.
7. The receiver recovers the string s_b that depends on his choice bit b as $s_b = S_b \oplus r$. He erases the pair (c, r) from the list $\mathcal{L}$.

Comments. The protocol implicitly assumes that the sender and receiver can interrogate the PUF whenever they have access to it, i.e., that the PUF's challenge-response interface is publicly accessible and not protected. This implies that the employed PUF must possess a large number of CRPs. Using a PUF with just a few challenges does not make sense: The receiver could then create a full look-up table for all CRPs of such a PUF before sending it away in Step 3 of the Initialization Phase. This would subsequently allow him to recover both strings s_0 and s_1 in Step 6 of the protocol subsession, as he could obtain r_0 and r_1 from his look-up table. Similar observations hold for the upcoming protocol 2. Indeed, all protocols discussed in this paper require PUFs with a large number of challenges and publicly accessible challenge-response interfaces. These PUFs have sometimes been referred to as *Physical Random Functions* or also as *Strong PUFs* in the literature [11,26,25].

Furthermore, please note that no physical transfer of the PUF is envisaged during the subsessions of the protocol. According to the model of Brzuska et al., an adversary only has access to it during the initialization phase, but not between the subsessions. This protocol use has some similarities with a stand-alone usage of the PUF, in which exactly one PUF-transfer occurs between the parties.

2.2   Bit Commitment

The second protocol of [1] implements PUF-based Bit Commitment (BC) by a generic reduction to PUF-based OT. The BC-sender initially holds a bit b. When the OT-Protocol is called as a subprotocol, the roles of the sender and receiver are reversed: The BC-sender acts as the OT-receiver, and the BC-receiver as the OT-sender. The details are as follows.

Protocol 2:   PUF-BASED BIT COMMITMENT VIA PUF-BASED OBLIVIOUS TRANSFER ([1], SLIGHTLY SIMPLIFIED DESCRIPTION)

Commit Phase:

1. The BC-sender and the BC-receiver jointly run an OT-protocol (for example Protocol 1).
   (a) In this OT-protocol, the BC-sender acts as OT-receiver and uses his bit b as the choice bit of the OT-protocol.
   (b) The BC-receiver acts as OT-sender. He chooses two strings $s_0, s_1 \in \{0,1\}^\lambda$ at random, and uses them as his input s_0, s_1 to the OT-protocol.
2. When the OT-protocol is completed, The BC-sender has learned the string $v := s_b$. This closes the commit phase.

Reveal Phase:

1. In order to reveal bit b, the BC-sender sends the string (b, v) (with $v = s_b$) to the BC-receiver.

Comments.   The security of the BC-protocol is inherited from the underlying OT-protocol. Once this protocol is broken, also the security of the BC-protocol is lost. This will be relevant in the upcoming sections.

3   A Quadratic Attack on Protocols 1 and 2

We will now discuss a cheating strategy in Protocols 1 and 2. Compared to an attacker who exhaustively queries the PUF for all of its m possible challenges, we describe an attack on Protocols 1 and 2 which reduces this number to $\sqrt{m}$. As we will argue later in Section 4, this has a particularly strong effect on the protocol's security if an optical PUF is used (as has been explicitly suggested by [2]), or if electrical PUFs with medium challenge lengths of 64 bits are used.

Our attack rests on the following lemma.

Lemma 3. *Consider the vector space* $(\{0,1\}^\lambda, \oplus)$, $\lambda \geq 2$, *with basis* $\mathcal{B} = \{a_1, \ldots, a_{\lfloor \lambda/2 \rfloor}, b_1, \ldots, b_{\lceil \lambda/2 \rceil}\}$. *Let A be equal to the linear subspace generated by the vectors in* $\mathcal{B}_A = \{a_1, \ldots, a_{\lfloor \lambda/2 \rfloor}\}$, *and let B be the linear subspace generated by the vectors in* $\mathcal{B}_B = \{b_1, \ldots, b_{\lceil \lambda/2 \rceil}\}$. *Define* $S := A \cup B$. *Then it holds that:*

(i) *Any vector* $z \in \{0,1\}^\lambda$ *can be expressed as* $z = a \oplus b$ *with* $a, b \in S$, *and this expression (i.e., the vectors a and b) can be found efficiently (i.e., in at most* $\text{poly}(\lambda)$ *steps).*

(ii) *For all distinct vectors* $x_0, x_1, v \in \{0,1\}^\lambda$ *there is an equal number of combinations of linear subspaces A and B as defined above for which* $x_0 \oplus v \in A$ *and* $x_1 \oplus v \in B$.

(iii) *S has cardinality* $|S| \leq 2 \cdot 2^{\lceil \lambda/2 \rceil}$.

Proof. (i) Notice that any vector $z \in \{0,1\}^\lambda$ can be expressed as a linear combination of all basis vectors: $z = \sum u_i a_i + \sum v_j b_j$, i.e., $z = a \oplus b$ with $a \in A$ and $b \in B$. This expression is found efficiently by using Gaussian elimination.

(ii) Without loss of generality, since x_0, x_1 and v are distinct vectors, we may choose $a_1 = x_0 \oplus v \neq 0$ and $b_1 = x_1 \oplus v \neq 0$. The number of combinations of linear subspaces A and B is independent of the choice of a_1 and b_1. (Notice that if $x_0 \neq x_1$ but $v = x_0$, then the number of combinations is twice as large.)

(iii) The bound follows from the construction of S and the cardinalities of A and B, which are $|A| = 2^{\lfloor \lambda/2 \rfloor}$ and $|B| = 2^{\lceil \lambda/2 \rceil}$. $\square$

An Example. Let us give an example in order to illustrate the principle of Lemma 3. Consider the vector space $(\{0,1\}^\lambda, \oplus)$ for an even λ, and choose as subbases $\mathcal{B}_{A_0} = \{e_1, \ldots, e_{\lambda/2}\}$ and $\mathcal{B}_{B_0} = \{e_{\lambda/2+1}, \ldots, e_\lambda\}$, where e_i is the unit vector of length λ that has a one in position i and zeros in all other positions. Then the basis $\mathcal{B}_{A_0}$ spans the subspace A_0 that contains all vectors of length λ whose second half is all zero, and $\mathcal{B}_{B_0}$ spans the subspace B_0 that comprises all vectors of length λ whose first half is all zero. It then follows immediately that every vector $z \in \{0,1\}^\lambda$ can be expressed as $z = a \oplus b$ with $a \in A_0$ and $b \in B_0$, or, saying this differently, with $a, b \in S$ and $S := A_0 \cup B_0$. It is also immediate that S has cardinality $|S| \leq 2 \cdot 2^{\lambda/2}$.

Relevance for PUFs. The lemma translates into a PUF context as follows. Suppose that a malicious and an honest player play the following game. The malicious player gets access to a PUF with challenge length λ in an initialization period, in which he can query CRPs of his choice from the PUF. After that, the PUF is taken away from him. Then, the honest player chooses a vector $z \in \{0, 1\}^\lambda$ and sends it to the malicious player. The malicious player wins the game if he can present the correct PUF-responses r_0 and r_1 to two arbitrary challenges c_0 and c_1 which have the property that $c_0 \oplus c_1 = z$. Our lemma shows that in order to win the game with certainty, the malicious player does not need to read out the entire CRP space of the PUF in the initialization phase; he merely needs to know the responses to all challenges in the set S of Lemma 3, which has a quadratically reduced size compared to the entire CRP space. This observation is at the heart of the attack described below.

In order to make the attack hard to detect for the honest player, it is necessary that the attacker chooses random subspaces A and B, and does not use the above trivial choices A_0 and B_0 all the time. This fact motivates the random choice of A and B in Lemma 3. The further details are as follows.

The Attack. As in [1,2], we assume that the PUF has got a challenge set of $\{0, 1\}^\lambda$. Given Lemma 3, the OT-receiver (who initially holds the PUF) can achieve a quadratic advantage in Protocol 1 as described below.

First, he chooses uniformly random linear subspaces A and B, and constructs the set S, as described in Lemma 3. While he holds possession of the PUF before the start of the protocol, he reads out the responses to all challenges in S. Since $|S| \leq 2 \cdot 2^{\lceil \lambda/2 \rceil}$, this is a quadratic improvement over reading out all responses of the PUF.

Next, he starts the protocol as normal. When he receives the two values x_0 and x_1 in Step 3 of the protocol, he computes two challenges c_0^* and c_1^* both in set S such that

$$x_0 \oplus x_1 = c_0^* \oplus c_1^*.$$

According to Lemma 3(i), this can be done efficiently (i.e., in $poly(\lambda)$ operations). Notice that, since the receiver knows all the responses corresponding to challenges in S, he in particular knows the two responses r_0^* and r_1^* that correspond to the challenges c_0^* and c_1^*.

Next, the receiver deviates from the protocol and sends the value $v := c_0^* \oplus x_0$ in Step 4. For this choice of v, the two challenges c_0 and c_1 that the sender uses in Step 5 satisfy

$$c_0 := c_0^* \oplus x_0 \oplus x_0 = c_0^*$$

and

$$c_1 := c_0^* \oplus x_0 \oplus x_1 = c_0^* \oplus c_0^* \oplus c_1^* = c_1^*.$$

By Lemma 3(ii), Alice cannot distinguish the received value v in Step 4 from any random vector v. In other words, Alice cannot distinguish Bob's malicious behavior (i.e., fabricating a special v with suitable properties) from honest behavior. As a consequence, Alice continues with Step 6 and transmits $S_0 = s_0 \oplus r_0^*$ and $S_1 = s_1 \oplus r_1^*$. Since Bob knows both r_0^* and r_1^*, he can recover both s_0 and s_1. This breaks the security of the protocol.

Please note the presented attack is simple and effective: It fully works within the original communication model of Brzuska et al. [1,2]. Furthermore, it does not require laborious computations of many days on the side of the attacker (as certain modeling attacks on PUFs do [25]). Finally, due to the special construction we proposed, the honest players will not notice the special choice of the value v, as the latter shows no difference from a randomly chosen value.

Effect on Bit Commitment (Protocol 2). Due to the reductionist construction of Protocol 2, our attack on the oblivious transfer scheme of Protocol 1 directly carries over to the bit commitment scheme of Protocol 2 if Protocol 1 is used in it as a subprotocol. By using the attack, a malicious sender can open the commitment in both ways by reading out only $2 \cdot 2^{\lceil \lambda/2 \rceil}$ responses (instead of all 2^λ responses) of the PUF. On the other hand it can be observed easily that the hiding property of the BC-Protocol 2 is unconditional, and is not affected by our attack.

4   Practical Consequences of the Attack

What are the practical consequences of our quadratic attack, and how relevant is it in real-world applications? The situation can perhaps be illustrated via a comparison to classical cryptography. What effect would a quadratic attack have on schemes like RSA, DES and SHA-1? To start with RSA, the effect of a quadratic attack here is rather mild: The length of the modulus must be doubled. This will lead to longer computation times, but restore security without further ado. In the case of single-round DES, however, a quadratic attack would destroy its security, and the same holds for SHA-1. The actual effect of our attack on PUF-based OT and BC has some similarities with DES or SHA-1: PUFs are finite objects, which cannot be scaled in size indefinitely due to area requirements, arising costs, and stability problems. This will also become apparent in our subsequent discussion.

4.1   Electrical Integrated PUFs

We start our discussion by electrical integrated PUFs, and take the well-known Arbiter PUF as an example. It has been discussed in theory and realized in silicon mainly for challenge lengths of 64 bits up to this date [9,10,15,28]. Our attack on such a 64-bit implementation requires the read-out of $2 \cdot 2^{32} = 8.58 \cdot 10^9$ CRPs by the receiver. This read-out can be executed before the protocol (i.e., not during the protocol), and will hence not be noticed by the sender. Assuming a MHz CRP read-out rate [15] of the Arbiter PUF, the read-out takes $8.58 \cdot 10^3$ sec, or less than 144 min.

Please note that the attack is independent of the cryptographic hardness of the PUF, such as its resilience against machine learning attacks. For example, a 64-bit, 8-XOR-Arbiter PUF (i.e., an Arbiter PUF with eight parallel standard 64-bit Arbiter PUFs whose single responses are XORed at the end of the structure) is considered secure in practice against all currently known machine learning techniques [25]. Nevertheless, this type of PUF would still allow the above attack in 144 min.

Our attacks therefore enforce the use of PUFs with a challenge bitlength of 128 bits or more in Protocols 1 and 2. Since much research currently focuses on 64-bit implementations of electrical PUFs, publication and dissemination of the attack seems important to avoid their use in Protocols 1 and 2. Another aspect of our attack is that it motivates the search for OT- and BC-protocols that are immune, and which can safely be used with 64-bit implementations. The reason is that the usage of 128-bit PUFs doubles the area consumption of the PUF and negatively affects costs.

4.2 Optical PUFs

Let us now discuss the practical effect of our attack on the optical PUF introduced by Pappu [20] and Pappu et al. [21]. The authors use a cuboid-shaped plastic token of size 1 cm × 1 cm × 2.5 mm, in which thousands of light scattering small spheres are distributed randomly. They analyze the number of applicable, decorrelated challenge-response pairs in their set-up, arriving at a figure of $2.37 \cdot 10^{10}$ [21]. Brzuska et al. assume that these challenges are encoded in a set of the form $\{0,1\}^{\lambda}$, in which case $\lambda = \lceil \log_2 2.37 \cdot 10^{10} \rceil = 35$. If this number of 2^{35} is reduced quadratically by virtue of Lemma 3, we obtain on the order of $2 \cdot 2^{18} = 5.2 \cdot 10^5$ CRPs that must be read out by an adversary in order to cheat. It is clear that even dedicated measurement set-ups for optical PUFs cannot realize the MHz rates of the electrical example in the last section. But even assuming mild read-out rates of 10 CRPs or 100 CRPs per second, we still arrive at small read-out times of $5.2 \cdot 10^4$ sec or $5.2 \cdot 10^3$ sec, respectively. This is between 14.4 hours (for 10 CRPs per second) or 87 minutes (for 100 CRPs per second). If a malicious receiver holds the PUF for such a time frame before the protocol starts (which is impossible to control or prevent for the honest players), he can break the protocol's security.

Can the situation be cleared by simply scaling the optical PUF to larger sizes? Unfortunately, also an asymptotic analysis of the situation shows the same picture. All variable parameters of the optical PUF [21,20,16] are the x-y-coordinate of the incident laser beam and the spatial angle Θ under which the laser hits the token. This leads to a merely cubic complexity in the three-dimensional diameter d of the cuboid scattering token. [2] Given our attack, this implies that the adversary must only read out $O(d^{1.5})$ challenges in order to cheat in Protocols 1 and 2. If only the independent challenges are considered, the picture is yet more drastic: As shown in [31], the PUF has at most a quadratic number of *independent* challenges in d. This reduces to a merely *linear* number of CRPs which the adversary must read out in our attack. Finally, we remark that scaling up the size of the PUF also quickly reaches its limits under practical aspects: The token considered by Pappu et al. [21,20] has an area of 1 cm × 1 cm. In order to slow down the quadratic attack merely by a factor of 10, a token of area 10 cm × 10 cm would have to be used. Such a token is too large to even fit onto a smart card.

Overall, this leads to the conclusion that optical PUFs like the ones discussed in [20,21,16] cannot be used safely with the Protocols 1 and 2 in the face of our attack. Due to their low-degree polynomial CRP complexity, and due to practical size constraints, simple scaling of the PUFs constitutes no efficient countermeasure. This distinguishes the optical approach from the electrical case of the last section. This observation has a particular relevance, since Brzuska et al. had explicitly suggested optical PUFs for the implementation of their protocols (see Section 8 of [2]).

---

[2] Please note in this context that the claim of [2] that the number of CRPs of an optical PUF is super-polynomial must have been made erroneously or by mistake; our above brief analysis shows that it is at mostly cubic. The low-degree polynomial amount of challenges of the optical PUF is indeed confirmed by the entire literature on the topic, most prominently [21,20,31].

5 Potential Countermeasures

5.1 Additional PUF Transfers and Time Constraints?

Can we bind the time in which the malicious player has got access to the PUF in order to prevent our attack? The current Protocols 1 and 2 obviously are unsuited to this end; but could there be modifications of theirs which have this property? A simple approach seems the introduction of one additional PUF transfer from the sender to the receiver in the initialization phase. This assumes that the sender initially holds the PUF, transfers it to the receiver, and measures the time period within which the receiver returns the PUF. The (bounded) period in which the receiver had access to the PUF can then be used to derive a bound on the number of CRPs the receiver might know. This could be used to enforce security against a cheating receiver. Please note that a long, uncontrolled access time for the sender is no problem for the protocol's security, whence it suffices to concentrate on the receiver.

On closer inspection, however, there are significant problems with this approach. In general, each PUF-transfer in a protocol is very costly. One PUF-transfer per protocol seems acceptable, since it is often executed automatically and for free, for example by consumers carrying their bank cards to cash machines. But having two such transfers in one protocol, as suggested above, will most often ruin a protocol's practicality.

A second issue is that binding the adversarial access time *in a tight manner* by two consecutive PUF transfers is very difficult. How long will one physical transfer of the PUF take? 1 day? If the adversary can execute this transfer a few hours faster and can use the gained time for executing measurements on the PUF, our countermeasure fails. The same holds if the adversary carries out the physical transfer himself and can measures the PUF while it is in transit.

In summary, enforcing a tight time bound on the receiver's access time by two PUF transfers or also by other measures will be impossible in almost any applications. The above idea may thus be interesting as a theoretical concept for future PUF-protocol design, but cannot be considered a generally efficient and practically relevant counter-measure.

5.2 Interactive Hashing

Let us now discuss a second and more effective countermeasure: The employment of interactive hashing (IH) as a substep in OT protocols. As we will show, protocols based on IH can achieve better security properties than Protocol 1. The idea of using IH in the context of PUFs has been first been suggested by Rührmair in 2010; his OT-protocol was the first published PUF-based two-player protocol [23]. The following approach is a simplified version of his original scheme. We also give (for the first time) a security analysis of the protocol. Via the general reduction of BC to OT presented in Protocol 2, our construction for OT can also be used to implement PUF-based BC.

5.2.1 Interactive Hashing as a Security Primitive
Interactive hashing (IH) is a two-player security primitive suggested by [18,17]. It has been deployed as a protocol tool in various contexts, including zero-knowledge proofs,

bit commitment and oblivious transfer (see references in [17]. The following easily accessible and application-independent definition of IH has been given in [4]; for more a formal treatment see [27].

Definition 4 (Interactive Hashing (IH) [4]). *Interactive Hashing is a cryptographic primitive between two players, the sender and the receiver. It takes as input a string $c \in \{0,1\}^t$ from the sender, and produces as output two t-bit strings, one of which is c and the other $c' \neq c$. The output strings are available to both the sender and the receiver, and satisfy the following properties:*

1. *The receiver cannot tell which of the two output strings was the original input. Let the two output strings be c_0, c_1, labeled according to lexicographic order. Then if both strings were a priori equally likely to have been the sender's input c, then they are a posteriori equally likely as well.*
2. *When both participants are honest, the input is equally likely to be paired with any of the other strings. Let c be the sender's input and let c' be the second output of interactive hashing. Then provided that both participants follow the protocol, c' will be uniformly distributed among all $2^t - 1$ strings different from c.*
3. *The sender cannot force both outputs to have a rare property. Let $\mathcal{G}$ be a subset of $\{0,1\}^t$ representing the sender's "good set". Let G be the cardinality of $\mathcal{G}$ and let $T = 2^t$. Then if G/T is small, the probability that a dishonest sender will succeed in having both outputs c_0, c_1 be in $\mathcal{G}$ is comparably "small".*

One standard method to implement IH is by virtue of a classical technique by Naor et al. [17]. To achieve a self-contained treatment, we describe this technique in a variant introduced by Crepeau et al. [4] below. In the protocol below, let c be a t-bit string that is the input to sender in the interactive hashing. All operations take place in the binary field $\mathcal{F}_2$.

Protocol 5: INTERACTIVE HASHING [4]

1. The receiver chooses a $(t-1) \times t$ matrix $\mathbf{Q}$ uniformly at random among all binary matrices of rank $t - 1$. Let q_i be the i-th query, consisting of the i-th row of $\mathbf{Q}$.
2. For $1 \leq i \leq t - 1$ do:
   (a) The receiver sends query q_i to the sender.
   (b) The sender responds with $v_i = q_i \cdot c$.
   (c) Given $\mathbf{Q}$ and $v \in \{0,1\}^{t-1}$ (the vector of the sender's responses), both parties compute the two values of $c \in \{0,1\}^t$ consistent with the linear system $\mathbf{Q} \cdot c = v$. These solutions are labeled c_0, c_1 according to lexicographic order.

The following theorem, which is taken from [4,27], tells us about the security of the above scheme. It relates to the security definition 4.

Theorem 6 (Security of Protocol 5). *Protocol 5 satisfies all three information theoretic security properties of Definition 4. Specifically, for Property 3 of Definition 4, it ensures that a dishonest sender can succeed in causing both outputs to be in the "good set" $\mathcal{G}$ with probability at most $15.6805 \cdot G/T$, where $G = |\mathcal{G}|$ and $T = 2^t$.*

5.2.2   Oblivious Transfer

We are now presenting a PUF-based oblivious transfer protocol that uses IH as a sub-step. It bears some similarities with an earlier protocol of Rührmair [23] in the sense that it also uses interactive hashing, but is slightly simpler.

Protocol 7:   PUF-BASED 1-OUT-OF-2 OBLIVIOUS TRANSFER WITH INTERACTIVE HASHING

1. The sender's input are two strings $s_0, s_1 \in \{0,1\}^\lambda$ and the receiver's input is a bit $b \in \{0,1\}$.
2. The receiver chooses a challenge $c \in \{0,1\}^\lambda$ uniformly at random. He applies c to the PUF, which responds r. He transfers the PUF to the sender.
3. The sender and receiver execute an IH protocol, where the receiver has input c. Both get outputs c_0, c_1. Let i be the value where $c_i = c$.
4. The receiver sends $b' := b \oplus i$ to the sender.
5. The sender applies the challenges c_0 and c_1 to the PUF. Denote the corresponding responses as r_0 and r_1.
6. The sender sends $S_0 := s_0 \oplus r_{b'}$ and $S_1 := s_1 \oplus r_{1-b'}$ to receiver.
7. The receiver recovers the string s_b that depends on his choice bit b as $S_b \oplus r = s_b \oplus r_{b \oplus b'} \oplus r = s_b \oplus r_i \oplus r = s_b$.

5.2.3   Security and Practicality Analysis

We start by a security analysis of Protocol 7 in the so-called "stand alone, good PUF model", which was introduced by van Dijk and Rührmair in [6]. In this communication model, the following two assumptions are made: (i) the PUF-protocol is executed only once, and the adversary or malicious players have no access to the PUF anymore after the end of the protocol; (ii) the two players do not manipulate the used PUFs on a hardware level. We stress that whenever these two features cannot be guaranteed in practical applications, a number of unexpected attacks apply, which spoil the security of the respective protocols. Even certain impossibility results can be shown under these circumstances; see [6] for details.

In the following analysis in the stand alone, good PUF model, we assume that the adversary has the following capabilities:

1. He knows a certain number of CRPs of the PUF, and has possibly used them to build an (incomplete) predictive model of the PUF. In order to model this ability, we assume that there is a proper subset $S \subsetneq C$ of the set of all challenges C such that the adversary knows the correct responses to the challenges in S with probability one. The cardinality of S depends on the previous access times of the adversary to the PUF and the number of CRPs he has collected from other sources, for example protocol eavesdropping. It must be estimated by the honest protocol users based on the given application scenario. Usually $|S| \ll |C|$.
2. Furthermore, we assume that the adversary can correctly guess the response to a uniformly and randomly chosen challenge $c \in C \setminus S$ with probability at most ϵ, where the probability is taken over the choice of c and over the adversary's

random coins. Usually ϵ will be significantly smaller than one. To name two examples: In the case of a well-designed electrical PUF with single-bit output, ϵ will be around 0.5; in the case of a well-designed optical PUF [20,21] with multi-bit images as outputs, ϵ can be extremely small, for example smaller than 2^{-100}. Again, the honest protocol users must estimate ϵ based on the circumstances and the employed PUF.

Assuming the above capabilities and using Theorem 6, the probability that the receiver can cheat in Protocol 7 is bounded above by

$$15.6805 \cdot |S|/|C| + \epsilon,$$

a term that will usually be significantly smaller than one.

Under the presumption that this cheating probability of the receiver is indeed smaller than one, the security of Protocol 7 can be further amplified by using a well-known result by Damgard, Kilian and Savail (see Lemma 3 of [5]):

Theorem 8 (OT-Amplification [5]). *Let (p, q)-WOT be a 1-2-OT protocol where the sender with probability p learns the choice bit c and the receiver with probability q learns the other bit b_{1-c}. Assume that $p + q < 1$. Then the probabilities p and q can be reduced by running k (p, q)-WOT-protocols to obtain a $(1 - (1 - p)^k, q^k)$-WOT protocol.*

In the case of our OT-Protocol 7 it holds that $p = 0$, whence the technique of Damgard et al. leads to an efficient security amplification, and to a $(0, q^k)$-WOT protocol. The PUF does not need to be transferred k times, but one PUF-transfer suffices. We remark that the probability amplification according to Theorem 8 is not possible with Protocol 1 after our quadratic attack, since the attack leads to a cheating probability of one for the receiver, i.e., $p + q \geq 1$ in the language of Theorem 8.

Let us quantatively illustrate the security gain of Protocol 7 over Protocol 1 via a simplified back-of-the-envelope calculation: We argued earlier that via our quadratic attack, a malicious receiver who has read out $2 \cdot 2^{18}$ CRPs from an optical PUF can cheat with probability 1 (= with certainty) in Protocol 1. Let us compare this to the case that an optical PUF is used in the IH-based Protocol 7. Let us assume that the adversary has collected the same number of CRPs (= $2 \cdot 2^{18}$ CRPs) as in the quadratic attack, and that the (multi-bit) response of the optical PUF on the remaining CRPs is still hard to preduct, i.e., it cannot be predicted better than with probability $\epsilon \leq 2^{-100}$. Then by Theorem 6 and by our above analysis, the adversary's chances to break Protocol 7 are merely around $15.6805 \cdot 2^{19} \cdot 2^{-35} + 2^{-100} \approx 0.00024$. This probability can then be exponentially reduced further via Theorem 8.

On the downside, however, the IH-Protocol 5 has a round complexity that is linear (i.e., equal to $\lambda - 1$) in the security parameter λ. This is relatively significant for the optical PUF (where $\lambda = 35$) and electrical PUFs with medium bitlengths (where $\lambda = 64$). One possible way to get around this problem is to use the constant round interactive hashing scheme by Ding et al. [7]. However, this scheme has slightly worse security guarantees than the IH scheme of the last sections. Future work will analyze the exact security loss under the use of the IH scheme of Ding. A first analysis to this end can be found in van Dijk and Rührmair [6].

To summarize the discussion in this section, interactive hashing can restore the security of PUF-based OT protocols even for small sized PUFs with 64-bit challenge lengths and for optical PUFs in the stand alone, good PUF model. Via the general reduction of BC to OT given in Protocol 2, this result can be used to securely implement PUF-based BC in this model, too. However, the use of IH leads to an increased number of communication rounds that is about equal to the (binary) challenge length of the PUF, i.e., around 64 rounds for the integrated PUFs with 64 bit challenges, and around 35 rounds for optical PUFs of size 1 cm^2 [21]. It must be decided on the basis of the concrete application scenario whether such a number of rounds is acceptable.

6   Summary and Conclusions

We revisited PUF-based OT- and BC-protocols, including the recent schemes of Rührmair from Trust 2010 [23] and Brzuska et al. from Crypto 2011 [1,2]. We placed special emphasis on the security which these protocols achieve in practice, in particular when they are used in connection with widespread optical and 64-bit electrical PUF-implementations. Our analysis revealed several interesting facts.

First of all, we described a simple and efficient method by which the OT- and BC-protocol of Brzuska et al. can be attacked with probability one in practice if electrical PUFs with 64-bit challenge lengths are used, or whenever optical PUFs are employed. Since much research focuses on 64-bit implementations of electrical PUFs [9,10,15], and since Brzuska et al. had explicitly suggested optical PUFs for the implementation of their protocols (see Section 8 of [2]), the publication and dissemination of our quadratic attack seems important to avoid their use in Protocols 1 and 2. Please note that our attack is independent of the cryptographic hardness of the PUF, and is merely based on its challenge size.

Secondly, we discussed an alternative class of protocols for oblivious transfer that are based on interactive hashing techniques. They are inspired by the earlier OT-protocol of Rührmair [23]. We argued that these protocols lead to better security in practice. They can be used safely with 64-bit electrical PUFs. When used with optical PUFs, they lead to better security than the protocols of Brzuska et al., but the security margins are tighter than in the 64-bit case. In both cases, a well-known result by Damgard, Kilian and Savail [5] can be used in order to reduce the cheating probabilities exponentially.

Our discussion shows once more that PUFs are quite special cryptographic and security tools. Due to their finite nature, asymptotic constants that might usually be hidden in $O(\cdot)$- and $\Theta(\cdot)$-notations become relevant in practice and should be discussed explicitly. Furthermore, their specific nature often allows new and unexpected forms of attacks. One of the aims of our work is to bridge the gap between PUFs in theory and applications; reconciling these two fields seems a necessary prerequisite for a healthy long-term development of the field. We hope that the general methods and the approach of this paper can contribute to this goal.

Recommendations for Protocols Use and Future Work. Let us conclude the paper with a condensed recommendation for the practical implementation of PUF-based OT and BC protocols, and by a discussion of future work. Firstly, it is clear from our results that the protocol of Brzuska et al. cannot be used safely with optical PUFs a la Pappu (i.e., with non-integrated optical PUFs that have only a small or medium sized challenge set), or with electrical PUFs with challenge lengths around 64 bits.

Secondly, we showed that Protocols based on interactive hashing (IH) can achieve better security. These protocols can be employed safely with optical PUFs and with electrical PUFs of challenge length 64. Furthermore, Damgard et al.'s [5] amplification technique can be applied in order to bring the cheating probabilities arbitrarily close to zero. Nevertheless, we would like to stress once more to practical PUF-users that this analysis only applies if the protocols are employed in the stand alone, good PUF model (see Section 5.2.3 and [6]). As soon as the features of this model cannot be enforced in a given application (for example by certifying a PUF, or by erasing PUF responses at the protocol end [6]), certain new attacks apply, which spoil both the security of IH-based protocols and of the protocols of Brzuska et al. These attacks are not the topic of this publication, but have been described in all detail in [6].

If a PUF has challenge length of 128 bits or more, it seems at first sight that the protocols of Brzuska et al. could be used safely *in the stand alone, good PUF model, too,* but we stress that this recommendation is yet to be confirmed by full formal analysis. One issue is that the PUF security feature required by the protocols of Brzuska et al. is (in a nutshell) that the adversary must be unable to select two PUF-challenges with a given distance d such that he knows the two corresponding responses. This security property of a PUF is new in the literature and should yet be further investigated in future work before final recommendations are being made. In particular, it does not seem simple or straightforward to judge in practice whether a given PUF fulfills this property.

A second topic for future research is how the round complexity of the IH-based protocols can be reduced. Some steps to this end have been made by van Dijk and Rührmair in [6], where the constant-round interactive hashing scheme of Ding et al. [7] is applied to obtain contant-round PUF-based OT and BC protocols.

Acknowledgements. The authors would like to thank Stefan Wolf and Jürg Wullschleger for enjoyable discussions, and Stefan Wolf for suggesting the example in Section 3, page 256 to us. Part of this work was conducted within the physical cryptography project at the TU München.

References

1. Brzuska, C., Fischlin, M., Schröder, H., Katzenbeisser, S.: Physically Uncloneable Functions in the Universal Composition Framework. In: Rogaway, P. (ed.) CRYPTO 2011. LNCS, vol. 6841, pp. 51–70. Springer, Heidelberg (2011)
2. Brzuska, C., Fischlin, M., Schröder, H., Katzenbeisser, S.: Physical Unclonable Functions in the Universal Composition Framework. Full version of the paper. Available from Cryptology ePrint Archive (2011) (downloaded on February 28, 2012)
3. Canetti, R.: Universally Composable Security: A New Paradigm for Cryptographic Protocols. In: FOCS 2001, pp. 136–145 (2001); Full and updated version available from Cryptology ePrint Archive
4. Crépeau, C., Kilian, J., Savvides, G.: Interactive Hashing: An Information Theoretic Tool (Invited Talk). In: Safavi-Naini, R. (ed.) ICITS 2008. LNCS, vol. 5155, pp. 14–28. Springer, Heidelberg (2008)

5. Damgård, I., Kilian, J., Salvail, L.: On the (Im)possibility of Basing Oblivious Transfer and Bit Commitment on Weakened Security Assumptions. In: Stern, J. (ed.) EUROCRYPT 1999. LNCS, vol. 1592, pp. 56–73. Springer, Heidelberg (1999)

6. van Dijk, M., Rührmair, U.: Physical Unclonable Functions in Cryptographic Protocols: Security Proofs and Impossibility Results. Cryptology ePrint Archive, Report 228/2012 (2012)

7. Ding, Y.Z., Harnik, D., Rosen, A., Shaltiel, R.: Constant-round oblivious transfer in the bounded storage model. Journal of Cryptology 20(2), 165–202 (2007)

8. Gassend, B.: Physical Random Functions. MSc Thesis. MIT (2003)

9. Gassend, B., Clarke, D.E., van Dijk, M., Devadas, S.: Silicon physical random functions. In: ACM Conference on Computer and Communications Security 2002, pp. 148–160 (2002)

10. Gassend, B., Lim, D., Clarke, D., van Dijk, M., Devadas, S.: Identification and authentication of integrated circuits. Concurrency and Computation: Practice & Experience 16(11), 1077–1098 (2004)

11. Guajardo, J., Kumar, S.S., Schrijen, G.-J., Tuyls, P.: FPGA Intrinsic PUFs and Their Use for IP Protection. In: Paillier, P., Verbauwhede, I. (eds.) CHES 2007. LNCS, vol. 4727, pp. 63–80. Springer, Heidelberg (2007)

12. Impagliazzo, R., Rudich, S.: Limits on the Provable Consequences of One-Way Permutations. In: STOC 1989, pp. 44–61 (1989)

13. Kilian, J.: Founding cryptography on oblivious transfer. In: STOC (1988)

14. Kumar, S.S., Guajardo, J., Maes, R., Schrijen, G.J., Tuyls, P.: The Butterfly PUF: Protecting IP on every FPGA. In: HOST 2008, pp. 67–70 (2008)

15. Lee, J.-W., Lim, D., Gassend, B., Suh, G.E., van Dijk, M., Devadas, S.: A technique to build a secret key in integrated circuits with identification and authentication applications. In: Proceedings of the IEEE VLSI Circuits Symposium (June 2004)

16. Maes, R., Verbauwhede, I.: Physically Unclonable Functions: a Study on the State of the Art and Future Research Directions. In: Naccache, D., Sadeghi, A.-R. (eds.) Towards Hardware-Intrinsic Security, sec. 1. Springer (2010)

17. Naor, M., Ostrovsky, R., Venkatesan, R., Yung, M.: Perfect zero- knowledge arguments for NP using any one-way permutation. Journal of Cryptology (1998); Preliminary version In: Brickell, E.F. (ed.) CRYPTO 1992. LNCS, vol. 740, pp. 196–214. Springer, Heidelberg (1993)

18. Ostrovsky, R., Venkatesan, R., Yung, M.: Fair games against an all-powerful adversary. In: AMS DIMACS Series in Discrete Mathematics and Theoretical Computer Science, pp. 155–169 (1993); Preliminary version in SEQUENCES 1991

19. Majzoobi, M., Koushanfar, F., Potkonjak, M.: Lightweight Secure PUFs. In: IC-CAD 2008, pp. 607–673 (2008)

20. Pappu, R.: Physical One-Way Functions. PhD Thesis, Massachusetts Institute of Technology (2001)

21. Pappu, R., Recht, B., Taylor, J., Gershenfeld, N.: Physical One-Way Functions. Science 297, 2026–2030 (2002)

22. Rivest, R.: Illegitimi non carborundum. Invited keynote talk, CRYPTO 2011 (2011)

23. Rührmair, U.: Oblivious Transfer Based on Physical Unclonable Functions. In: Acquisti, A., Smith, S.W., Sadeghi, A.-R. (eds.) TRUST 2010. LNCS, vol. 6101, pp. 430–440. Springer, Heidelberg (2010)

24. Rührmair, U., Busch, H., Katzenbeisser, S.: Strong PUFs: Models, Constructions and Security Proofs. In: Sadeghi, A.-R., Tuyls, P. (eds.) Towards Hardware Intrinsic Security: Foundation and Practice. Springer (2010)

25. Rührmair, U., Sehnke, F., Sölter, J., Dror, G., Devadas, S., Schmidhuber, J.: Modeling Attacks on Physical Unclonable Functions. In: ACM Conference on Computer and Communications Security (2010)

26. Rührmair, U., Sölter, J., Sehnke, F.: On the Foundations of Physical Unclonable Functions. Cryptology e-Print Archive (June 2009)
27. Savvides, G.: Interactive Hashing and reductions between Oblivious Transfer variants. PhD thesis, McGill University, Montreal (2007)
28. Suh, G.E., Devadas, S.: Physical Unclonable Functions for Device Authentication and Secret Key Generation. In: DAC 2007, pp. 9–14 (2007)
29. Tuyls, P., Schrijen, G.-J., Škorić, B., van Geloven, J., Verhaegh, N., Wolters, R.: Read-Proof Hardware from Protective Coatings. In: Goubin, L., Matsui, M. (eds.) CHES 2006. LNCS, vol. 4249, pp. 369–383. Springer, Heidelberg (2006)
30. Tuyls, P., Škorić, B.: Strong Authentication with Physical Unclonable Functions. In: Petkovic, M., Jonker, W. (eds.) Security, Privacy and Trust in Modern Data Management. Springer (2007)
31. Tuyls, P., Škorić, B., Stallinga, S., Akkermans, A.H.M., Ophey, W.: Information-Theoretic Security Analysis of Physical Uncloneable Functions. In: Patrick, A.S., Yung, M. (eds.) FC 2005. LNCS, vol. 3570, pp. 141–155. Springer, Heidelberg (2005)

Soft Decision Error Correction for Compact Memory-Based PUFs Using a Single Enrollment

Vincent van der Leest[1], Bart Preneel[2], and Erik van der Sluis[1]

[1] Intrinsic-ID, Eindhoven, The Netherlands
http://www.intrinsic-id.com
[2] KU Leuven Dept. Electrical Engineering-ESAT/SCD-COSIC and IBBT, Belgium
http://www.kuleuven.be

Abstract. Secure storage of cryptographic keys in hardware is an essential building block for high security applications. It has been demonstrated that Physically Unclonable Functions (PUFs) based on uninitialized SRAM are an effective way to securely store a key based on the unique physical characteristics of an Integrated Circuit (IC). The startup state of an SRAM memory is unpredictable but not truly random as well as noisy, hence privacy amplification techniques and a Helper Data Algorithm (HDA) are required in order to recover the correct value of a full entropy secret key. At the core of an HDA are error correcting techniques. The best known method to recover a full entropy 128-bit key requires 4700 SRAM cells. Earlier work by Maes et al. has reduced the number of SRAM cells to 1536 by using soft decision decoding; however, this method requires multiple measurements (and thus also power resets) during the storage of a key, which will be shown to be an unacceptable overhead for many applications. This article demonstrates how soft decision decoding with only a single measurement during storage can reduce the required number of SRAM cells to 3900 (a 17% reduction) without increasing the size of en-/decoder. The number of SRAM cells can even be reduced to 2900 (a 38% reduction). This does increase cost of the decoder, but depending on design requirements it can be shown to be worthwhile. Therefore, it is possible to securely store a 128-bit key at a very low overhead in an IC or FPGA.

1 Introduction

Due to submicron process variations during manufacturing, every transistor in an IC has slightly different physical properties. These properties can be measured and since the process variations are uncontrollable, they result in features that cannot be copied. Therefore, it is possible to create an electronic device with a unique electronic fingerprint that offers a very strong resistance against cloning.

Physically Unclonable Functions (PUFs) are based on an electronic circuit that measures the responses of hardware to random input challenges. These responses depend on the unique and uncontrollable physical properties of the device and allow to authenticate the device. PUF responses are inherently noisy,

E. Prouff and P. Schaumont (Eds.): CHES 2012, LNCS 7428, pp. 268–282, 2012.

due to the presence of noise during the measurements. Helper Data Algorithms (HDAs) based on forward error correction have been developed to correct this noise. This paper presents an improved soft decoding algorithm for HDAs. In soft decoding, decisions are not based on the 0 or 1 value of a bit but on the probability for a bit to take the value 0 or 1.

SRAM memories have specific properties, which make them very suitable for use as PUFs: it turns out that uninitialized SRAM contains an unpredictable value because of the unbalance between two transistors; this unbalance depends on process variations and is hard to control. The unpredictable start-up value can be used for secure storage of a key. This is a convenient alternative to key storage in EEPROM, because EEPROM brings additional costs and is typically not available when a new technology node is rolled out. In addition to these benefits, SRAM PUFs are also more secure than EEPROM since the key is completely absent when the device is powered off. The SRAM values observed do contain entropy but they are not truly random. This can be resolved by using a larger SRAM in combination with privacy amplification (as shown by Guajardo et al. in [6]).

1.1 Related Work

Pappu [14] introduced the concept of PUFs in 2001 under the name Physical One-Way Functions. The proposed technology was based on obtaining a response (scattering pattern) when shining a laser on a bubble-filled transparent epoxy wafer. In 2002 this principle was translated by Gassend et al. [5] into Silicon Physical Random Functions. These functions make use of the manufacturing process variations in ICs, with identical masks, to uniquely characterize each IC. For this purpose the frequency of ring oscillators were measured. Using this method (now known as a Ring Oscillator PUF), they were able to characterize ICs. In 2004 Lee et al. [9] proposed another PUF that is based on delay measurements, the Arbiter PUF.

Besides intrinsic PUFs based on delay measurements a second type of PUF in ICs is known: the memory-based PUF. These PUFs are based on the measurement of start-up values of memory cells. This memory-based PUF type includes SRAM PUFs, which were introduced by Guajardo et al. in 2007 [6]. Furthermore, so-called Butterfly PUFs were introduced in 2008 by Kumar et al. [8], D Flip-Flop PUFs by Maes et al. [11] in 2008, and recently Buskeeper PUFs by Simons et al. [15] in 2012.

The first HDAs for generating cryptographic keys from PUFs were introduced by Linnartz et al. [10] in 2003 (as Shielding Functions) and Dodis et al. [4] in 2004 (as Fuzzy Extractors). After these introductions, secure use of Fuzzy Extractors was discussed by Boyen in [2]. A first efficient hardware implementation of an HDA was described in [1] by Bösch et al. The first HDA using soft decision error correction for memory-based PUFs was proposed by Maes et al. [12,13] in 2009.

1.2   Our Contribution

This paper introduces a new soft decision decoding scheme for HDAs used in
PUF implementations. To the best of our knowledge, this is the first ever soft decision decoder for memory-based PUFs that only requires a single PUF measurement during enrollment. This approach offers a significant increase in practical
usability over the method proposed in [12,13], as will be shown in Sect. 3.

Besides using only a single enrollment measurement, the soft decision decoding
scheme as introduced in this paper allows for an efficient hardware implementation. For that reason the construction only uses simple linear block codes such
as repetition, Reed-Muller (RM), and Golay codes. This paper will show that
these soft decision decoders offer substantial added value over their hard decision counterparts from [1] and are also efficiently implementable in hardware (in
contrast to more complex codes such as BCH and LDPC).

1.3   Paper Outline

Section 2 introduces the concept of HDAs. The state of the art of soft decision
decoding in PUF HDAs is presented in Sect. 3. The problem of the known
method for soft decision decoding is discussed together with how our proposed
method can improve this. When this has been established, Sect. 4 describes the
newly proposed method in more detail. We will compare the performance of our
new method to known implementations of hard decision decoding. Results of
this comparison can be found in Sect. 5. Conclusions are drawn in Sect. 6.

2   Helper Data Algorithms

2.1   Construction for Secure Key Storage

As stated earlier, an important application of PUFs is secure key storage [16].
Memory-based PUFs can be used for this purpose. In this paper we use SRAM
PUFs as the example of memory-based PUFs. Of all memory-based PUFs,
SRAM is the one with the best performance regarding both reproducibility and
entropy (as demonstrated in [3]). Secure key storage with PUFs makes use of
an HDA to securely store and reconstruct the key. Different constructions of
HDAs exist. The implementation used in this paper is depicted in Fig. 1. We
distinguish two phases in this HDA: enrollment and reconstruction.

Enrollment. During enrollment the key is programmed into the device, comparable to the key programming phase for other secure key storage mechanisms.
First, the response of the targeted PUF is measured. This response is called the
reference PUF response (R) and is the input of the Fuzzy Extractor [2,4,10].
This Fuzzy Extractor (FE) derives a cryptographic key from a random secret
and computes helper data W by xor-ing the encoded secret with R. In the reconstruction phase, W enables FE to reconstruct the exact same ("programmed")
cryptographic key from a new response of this specific PUF. The helper data is
stored in non-volatile memory attached to the device and is public information.

Enrollment

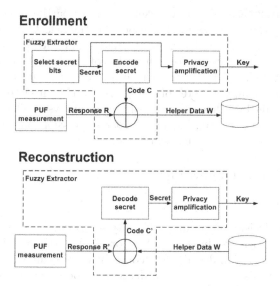

Reconstruction

Fig. 1. Enrollment and reconstruction for the HDA

Reconstruction. In the reconstruction phase the same PUF is measured again and its response (R', which is slightly different from R) is input to the FE. The FE uses W and R' to reconstruct the cryptographic key that was "programmed" during enrollment. If R' is close enough to R, the original key will be successfully reconstructed using information reconciliation.

2.2   Fuzzy Extractors

This section explains in more detail the most important building blocks of a Fuzzy Extractor.

Secret Encoding. Secret encoding is performed during enrollment and consists of selecting a random secret and encoding this secret with the chosen error correction code. In this paper we use linear error correcting codes with length n, dimension k and minimum Hamming distance d, which are listed as $[n, k, d]$ codes. The encoded secret C is xor-ed with R (this method is called code-offset technique) to create the value W, which will be used during reconstruction.

Information Reconciliation. In Fig. 1 information reconciliation can be found as "Decode secret" during reconstruction. This can be done after R' has been xor-ed with W to create C', which differs from C at the same positions that R' differs from R. Hence if R' and R are sufficiently close together (depending on how many errors can be corrected by the selected code construction), C' can be corrected into C and decoded into the secret encoded at enrollment.

Privacy Amplification. As an attacker may have partial information on the PUF (due to non-randomness in the response), the selected secret should be compressed into a cryptographic key with maximum entropy. This minimizes the knowledge of the attacker about the value of the key. According to [6] the secrecy rate for SRAM PUFs is 0.76, which indicates that for deriving a key of 128 bits with full entropy a secret of $\lceil 128/0.76 \rceil = 171$ bits is required. This is the number of secret bits that will be used for our analyses. This paper will not focus on privacy amplification, but note that compression can be achieved (for example) with a cryptographic hash function.

3   Soft Decision Decoding

3.1   State of the Art

When using soft decision decoding, reliability information about incoming bits is provided along with the bit-value of '0' or '1'. In other words, every bit at the input of a soft decision decoder is accompanied by a value that indicates the confidence level of this specific bit. Soft decision decoders can use this additional information to improve their error correcting capabilities; it is well known that on a typical Gaussian channel soft decision decoding results in an improvement of about 2 dB over hard decision. The goal is to help the decoder to output the most likely transmitted codeword and decrease the error rate at its output.

Soft decision decoding for memory-based PUFs has only been used in the literature by Maes et al. in [12,13]. Their proposal is the following:

- **During enrollment** several measurements of the (SRAM) PUF are performed. Based on these measurements an error probability for each PUF bit is derived and stored together with the helper data. The more stable the response of a specific PUF bit is during these multiple enrollment measurements, the higher the confidence level of the value of this bit will be.
- **During reconstruction** error probabilities from enrollment are used to indicate the confidence level of each individual bit. It is proven in [12,13] that using this soft decision information, less PUF bits are required to successfully reconstruct the secret bits that are used for the cryptographic key.

3.2   Motivation for Construction

The problem with the method from [12,13] is that multiple enrollment measurements are required. This has the following consequences:

- Non-volatile storage will be required in the device containing the PUF. Values of the multiple enrollment measurements need to be added in order to obtain the error probability of each bit. A key business case for PUFs is the replacement of non-volatile key storage (as described in Sect. 1). Therefore, the method from [12,13] gives up on the essential advantage of PUFs for key storage and introduces additional process steps (introducing extra delay), costs, footprint while decreasing security (because of possible attacks on non-volatile memory).

- The size of the required storage grows with the number of measurements performed. For example, when 3 enrollment measurements are performed, the sum value of each PUF bit can take on any integer value between 0 and 3. Therefore, this requires an additional 2 bits of storage per PUF bit. With 7 measurements 3 bits are required (and so on).
- Multiple measurements (and additional processing) leads to a longer time required for enrolling each PUF. This could lead to problems when enrolling millions of devices in production lines.

To solve these practical problems, we propose a new method for soft decision decoding. The requirements for this new method are the following:

- It should only use one measurement during enrollment (and reconstruction).
- It should be efficiently implementable in hardware.

The methods proposed in this paper will only focus on HDAs using the code-offset technique with linear block codes. Other codes, such as LDPC and convolutional codes, are more complex to decode and not well suited to deal with the limited amount of data available in PUF implementations. Therefore they will not be considered in this paper.

3.3 Our Proposal

The previous section has motivated why we propose a new low footprint HDA construction. This HDA should require as few PUF bits as possible in combination with low algorithmic complexity, while avoiding the implementation issues from the previous soft decision decoding method.

The method for soft decision coding proposed here is based on the concatenated codes from [1]. Figure 2 shows the flowcharts of encoding and decoding in the proposed HDA. Encoding is performed in a similar manner as for the hard decision construction (and is thus only based on a single PUF measurement). During decoding however, there are two differences with the hard decision construction:

- The repetition decoder is replaced by a quantizer, which derives probabilistic information from a single reconstruction measurement.
- The second decoder is a soft decision decoder (using the probabilistic information from the quantizer).

When using the repetition decoder as a quantizer, it "weighs" the amount of ones and zeros at its (non-probabilistic) input and outputs a (probabilistic) value between 0 and 1 that corresponds to this input. An input string consisting of s bits with i ones and $s - i$ zeros will be converted by the quantizer into an output value of i/s. These strings of length s are the sum (modulo 2) of the repetitive output of the encoder with noise of the PUF measurement. Hence, without noise the value i would either be 0 or s. So the closer i is to one of these values, the more confident the soft output value of the quantizer will be.

The soft values at the output of the quantizer are used as input for the soft decision decoder. Candidate soft decision decoders are described in Sect. 4.

Encoding

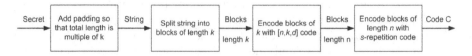

Decoding

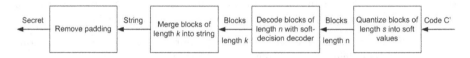

Fig. 2. Encoding and decoding as used in the proposed HDA

4   Soft Decision Decoders

We propose two methods for performing soft decision decoding, which are based on well-known error correction codes and more in particular on concatenated codes as described in [1]. Furthermore, it is important that both methods are implementable in hardware without too much overhead on resources. This rules out the more complex (BCH) codes from [1], since it is not possible to convert them into a hardware efficient soft decision code (and even the resource efficiency of some hard decision BCH implementations is questionable). To illustrate cost effectiveness of both solutions, a comparison between the hard- and soft decision implementations of these codes will be given in Sect. 5.4.

4.1   Brute-Force RM Decoder

The first proposed method is brute-force, which can be used for codes with a limited set of codewords (that is, with a small dimension k). In this method the soft input of the decoder is compared to all possible codewords. Based on Euclidean Distance, the most likely codeword from the list is selected to be decoded. In our analysis we use this method for evaluating soft decision decoding with two concatenated codes. The two constructions are repetition in combination with the Reed-Muller[16,5,8] code and with Reed-Muller[8,4,4]. It is clear that both

---

Algorithm 1. Brute-Force Soft Decision Reed-Muller Decoder

Input: String of size n consisting of soft values between 0 and 1.
Actions:
1. Calculate Euclidean Distance of input string to all possible codewords of RM code.
2. Select codeword of length n with lowest Euclidean Distance to input.
3. Decode codeword to corresponding encoded secret bits.
Output: Binary string of size k.

codes only have a limited number of codewords (32 and 16 respectively). Algorithm 1 describes how the proposed brute-force soft decision decoder works. In both constructions the repetition decoders are used as quantizers to create the soft input for the RM decoders, as described in Sect. 3.3.

4.2 Hackett Decoder

The second method is a concatenated code using repetition and Golay[24,12,8], where the Golay decoder is used for soft decision decoding as described in [7]. Again the repetition code is used as a quantizer, which produces soft values that are used as input for soft decision Golay decoder. The algorithm of this Golay decoder is described in Algorithm 2 and is visualized in Fig. 3.

Algorithm 2. Hackett Soft decision Golay Decoder

Input: String of size n consisting of soft values between 0 and 1.
Actions:
1. Convert input to corresponding hard (binary) values.
2. Based on soft input values, select 4 bits from input with least confidence.
3. Calculate overall parity of hard values.
    if parity is even -> flip least confident bit.
4. Initialize values required for loop: ED_min $= \infty$, $y =$ "hard values" and $k = 0$.
5. Error correct y using hard decision Golay[24,12,8].
6. Calculate Euclidean Distance of resulting string to soft input.
    if Euclidean Distance < ED_min -> Replace ED_min and $z = y$.
    if k < 7 -> flip two bits (as described in Table 1) to get new y, go back to step 5.
    else -> Decode codeword z to corresponding encoded secret bits.
Output: Binary string of size k.

Furthermore, Table 1 provides an overview of how the 8 different patterns are created from the original value of y by flipping bits (all possible patterns with even weight consisting of 4 bits). In this table b_0 denotes the least confident bit, b_1 the second least confident, etc. According to [7] only these 8 patterns are required (and not all 16 possibilities when flipping 4 bits), because it is already assured that the parity of all values of y are odd, which will lead to an odd number of errors. It is also claimed that hard decision decoding of an even number of errors rarely yields a codeword closer to the soft input than decoding with an odd number of errors. Therefore, patterns with an even parity are not used in this decoder.

5   Soft vs. Hard Decision Comparison

This section is dedicated to demonstrating the added value of soft decision decoding for PUF implementations. For that purpose the soft decision implementations from the previous section are compared to their hard decision counterparts based

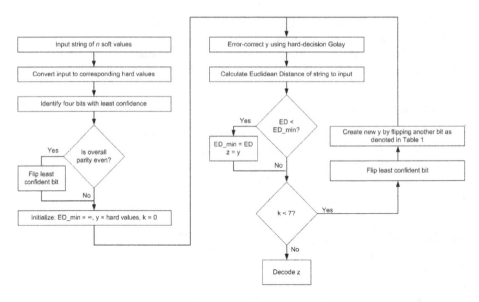

Fig. 3. Flowchart of Hackett decoder

Table 1. Bits flipped in comparison to initial value of y for different values of k

k	b_0	b_1	b_2	b_3	Bits flipped compared to $k-1$
0	0	0	0	0	-
1	1	0	0	1	b_0 and b_3
2	0	0	1	1	b_0 and b_2
3	1	0	1	0	b_0 and b_3
4	0	1	1	0	b_0 and b_1
5	1	1	0	0	b_0 and b_2
6	0	1	0	1	b_0 and b_3
7	1	1	1	1	b_0 and b_2

on error correcting performance and required resources. This will show that the number of PUF bits required for successfully reconstructing keys is much smaller when using soft decision decoding and that the additional resources required are limited.

The performance of the proposed soft decision decoders will not be compared to those from [12,13]. Even though the performance of the decoders from [12,13] is better than those presented here, this is an unfair comparison. Those decoders require multiple enrollment measurements, which leads to the problems listed in Sect. 3.2. The decoders proposed in this paper and their hard decision counterparts do not have these problems and can therefore be compared fairly.

The system used for context in this section derives 171 secret bits from an SRAM PUF. This example has been taken from [6] and is also referred to in [12,13] and [1]. The bit error rate of the PUF data (noise of PUF measurement) is called

ϵ and will be 15%, which is similar to these same references and is a good representation of noise on SRAM used at operating temperatures ranging from -40°C to +85°C (industrial temperature standard). Furthermore, the False Rejection Rate (FRR) of the 171 secret bits should be lower than 10^{-6} (i.e. the probability of incorrectly decoding a secret key with $\epsilon = 0.15 <$ one in a million).

5.1 Hard Decision Concatenated Codes

The FRR of the hard decision decoders can be calculated using a set of formulas. The first step is to calculate the error probability of the repetition decoder. This probability depends on the length of the repetition code as well as the value of ϵ and is defined as follows: When the number of bit errors at the input of the repetition decoder is higher than half of the length of the code (repetition code can only be of odd length, to avoid equal number of zeros and ones), the output of the decoder will be incorrect. This leads to the following formula:

$$Pe_{rep} = \sum_{i=\lceil s/2 \rceil}^{s} \binom{s}{i} \epsilon^i (1-\epsilon)^{s-i} = 1 - \sum_{i=0}^{\lfloor s/2 \rfloor} \binom{s}{i} \epsilon^i (1-\epsilon)^{s-i} .$$

Using the error probability of the repetition decoder in combination with the parameters of the hard decision code, the error probability of the concatenated code can be derived. When the number of errors from the repetition decoder is too high for the hard decision decoder to correct ($> \lfloor (d-1)/2 \rfloor$), the outcome of the hard decision decoder will be incorrect. The corresponding formula, where $t = \lfloor (d-1)/2 \rfloor$, is:

$$Pe_{code} = \sum_{i=t+1}^{n} \binom{n}{i} Pe_{rep}^i (1-Pe_{rep})^{n-i} = 1 - \sum_{i=0}^{t} \binom{n}{i} Pe_{rep}^i (1-Pe_{rep})^{n-i} .$$

Finally when these error probabilities are known, the total FRR of the key can be calculated (this step has been omitted in [1], which may give the false impression that the results from this paper differ from those in [1]). A key can only be reconstructed successfully when all required output blocks from the concatenated decoder (and thus all secret bits) are correct. Dividing the length of the secret by k (number of secret bits per decoding) leads to the number of output blocks that need to be decoded correctly to reconstruct the key successfully. In other words the FRR of the hard decision decoders is defined as:

$$FRR_{key} = 1 - (1 - Pe_{code})^{\lceil secret\ length/k \rceil} = 1 - (1 - Pe_{code})^{\lceil 171/k \rceil} .$$

5.2 Soft Decision Simulation Results

Unfortunately, it is not straightforward to define formulas for calculating the FRR of the soft decision decoders. In order to be able to evaluate the performance of these systems, the decoders have been simulated. These simulations have been performed by encoding random secrets with the concatenated encoders, adding random noise with $\epsilon = 0.15$, and decoding the resulting string with the soft decision decoders as described in earlier sections. The results of this simulation can be found in Fig. 4. This figure displays the FRRs of the three different concatenated soft decision decoders as a function of the length of the repetition code. Note that the repetition code can be of even length in this system, since it is used as a quantizer when decoding (and not as a hard decision repetition decoder). A threshold is set in the figure at an FRR of 10^{-6}, which allows us to derive the shortest repetition length required to achieve an FRR (for a key based on 171 secret bits) below this threshold.

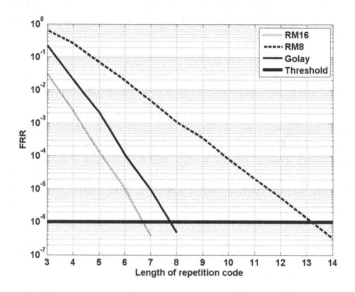

Fig. 4. Simulation results for soft decision decoders of RM[16,5,8], RM[8,4,4], and Golay[24,12,8] codes

5.3   Comparison

Based on the formulas from Sect. 5.1 and the simulation results from Sect. 5.2 the performance of the hard- and soft decision decoders can be compared. Table 2 shows the amount of SRAM required to reconstruct a key based on 171 secret bits with a total FRR below 10^{-6}. We conclude that the amount of SRAM required for soft decision decoding is significantly lower than that of hard decision decoding (RM[16,5,8]: 47% decrease, RM[8,4,4]: 44%, Golay: 38%). This shows that the soft decision decoders are very suitable when implementing an SRAM PUF HDA, which is optimized on the amount of SRAM required.

Table 2. Results hard and soft decision codes (deriving 171 secret bits, FRR $< 10^{-6}$)

Code	Type	Rep. Length	FRR	Amount of SRAM (Bytes)[a]
RM[16,5,8]	Hard	13	$1.6 \cdot 10^{-7}$	$\lceil 171/5 \rceil * 16 * 13/8 = \mathbf{910}$
RM[16,5,8]	Soft	7	$3.7 \cdot 10^{-7}$	$\lceil 171/5 \rceil * 16 * 7/8 = \mathbf{490}$
RM[8,4,4]	Hard	25	$3.4 \cdot 10^{-7}$	$\lceil 171/4 \rceil * 8 * 25/8 = \mathbf{1075}$
RM[8,4,4]	Soft	14	$3.3 \cdot 10^{-7}$	$\lceil 171/4 \rceil * 8 * 14/8 = \mathbf{602}$
Golay[24,12,8]	Hard	13	$4.0 \cdot 10^{-7}$	$\lceil 171/12 \rceil * 24 * 13/8 = \mathbf{585}$
Golay[24,12,8]	Soft	8	$4.8 \cdot 10^{-7}$	$\lceil 171/12 \rceil * 24 * 8/8 = \mathbf{360}$

[a] Calculation method for the amount of SRAM:
$\lceil$ required secret bits (171) / secret bits per codeword $\rceil$ = # required codewords
# of codewords * length codewords * repetition length = # of bits / 8 = # of Bytes

5.4 Resource Estimates

Besides a comparison based on the amount of SRAM required for each code construction, one can also compare the codes based on the total amount of resources required. For this purpose the total footprint of each construction will be estimated. One could also compare codes based on either timing or power consumption. However, we believe that these comparison are much less important when comparing hard and soft decision coders. When PUFs are used for key storage, silicon area is the main cost factor (note that this silicon is mostly inactive). Power and delay overhead play only a very minor role, since the coders are only active when generating the key (at start-up of a device). This one-time operation can be done within 1 millisecond at any realistic clock and will therefore not consume much time or power. Therefore, area will be the only factor in this comparison. Results on timing have been added to Table 3 for informational purposes only. Power consumption by the soft decision decoders has been found to be negligible in comparison to regular IC operation (since there is only consumption during a short time at power-up) and is therefore omitted.

The total footprint consists of the following components: the encoder, quantizer/repetition coder, decoder and SRAM. The required resources for each component are estimated based on synthesis and the results can be found in Table 3. In this table all estimates are based on area-optimized IC implementations. These constructions have also been implemented on FPGA, so they can be used on FPGAs with uninitialized SRAM. Since these FPGAs are rare however, the main focus of this paper is on IC implementations. Furthermore, all estimates are denoted in GE[1] and for SRAM a size of 1 GE per bit has been used as a reasonable estimate[2].

---

[1] GE – Gate Equivalent is a measure of area in any technology. 1 GE is the area of a NAND2 (standard drive strength).

[2] For TSMC 65nm standard cell library raw gate density $\approx$ 854 Kgate/mm^2, while SRAM cells are $0.499\mu m^2$ (6T) [17]. Hence, one SRAM cell is < 0.5 gates (1mm^2 / 854000 = $1.17\mu m^2$). This is without read-out circuitry, which provides significant overhead for small SRAMs. Considering a factor 2 overhead, an SRAM cell $\approx$ 1 GE.

Table 3. Resource estimates for different code constructions

Code	Type	Dec. clks	Encoder	Quant./Rep.	Decoder	SRAM	Total
RM[16,5,8]	Hard	±200	0.12 kGE	0.14 kGE	0.75 kGE	7.3 kGE	**8.3 kGE**
RM[16,5,8]	Soft	±400	0.12 kGE	0.10 kGE	1.1 kGE	3.9 kGE	**5.2 kGE**
RM[8,4,4]	Hard	±100	55 GE	0.19 kGE	0.5 kGE	8.6 kGE	**9.3 kGE**
RM[8,4,4]	Soft	±200	55 GE	0.14 kGE	0.6 kGE	4.8 kGE	**5.6 kGE**
Golay[24,12,8]	Hard	±15	0.30 kGE	0.14 kGE	1.0 kGE	4.7 kGE	**6.1 kGE**
Golay[24,12,8]	Soft	±150	0.30 kGE	0.13 kGE	3.0 kGE	2.9 kGE	**6.3 kGE**

Table 3 shows that soft decision decoding results in a substantial decrease in the SRAM size and in most cases additional overhead for the soft decoder is small. Soft decision decoders require more registers than their hard decision counterparts, since all codeword bits are now represented by a multi-bit soft value. This alone adds 24 to 72 Flip-Flops to the implementations. These are needed independent of speed/area trade-offs. Another unavoidable footprint increase is the calculation and comparison of distances. The speed/area trade-off is mainly determined by the amount of parallelism used here. Finally, the Hackett soft decision Golay decoder introduces additional logic for (among others) the selection of weak bits.

The choice of an HDA implementation depends on the parameter that should be optimized. In this example, when optimizing on the amount of SRAM used by the code construction, the soft decision Golay code should be implemented. If the total footprint of the implementation needs to be minimized, the soft decision RM[16,5,8] is the preferred choice. What is most important to notice is that the results clearly show the benefit of soft decision decoding.

The added value of soft decision decoding will increase even further when an HDA requires error correction with either a lower FRR, a higher ϵ, a larger number of secret bits or multiple keys. In those cases the SRAM will become an even more dominant factor in the total footprint of the implementation. Therefore, it will be more important to decrease the amount of SRAM required. An example in which 5 128-bits keys need to be stored with these code constructions can be found in Table 4. Here we conclude that the soft decision Golay code is favourable when SRAM is the dominant component of the footprint.

Table 4. Estimation of total footprint for different code constructions storing 5 keys

Code	Type	Encoder	Quant./Rep.	Decoder	SRAM	Total
RM[16,5,8]	Hard	0.12 kGE	0.14 kGE	0.75 kGE	36.4 kGE	**37.4 kGE**
RM[16,5,8]	Soft	0.12 kGE	0.10 kGE	1.1 kGE	19.6 kGE	**20.9 kGE**
RM[8,4,4]	Hard	55 GE	0.19 kGE	0.5 kGE	43.0 kGE	**43.7 kGE**
RM[8,4,4]	Soft	55 GE	0.14 kGE	0.6 kGE	24.1 kGE	**24.9 kGE**
Golay[24,12,8]	Hard	0.30 kGE	0.14 kGE	1.0 kGE	23.4 kGE	**24.8 kGE**
Golay[24,12,8]	Soft	0.30 kGE	0.13 kGE	3.0 kGE	14.4 kGE	**17.8 kGE**

Note: In this section the amount of non-volatile memory required to store helper data (outside of the chip) has not been taken into account. The helper data size (in bytes) is equal to that of the SRAM, hence it is clear that this size in-/decreases linearly with the SRAM size.

6   Conclusions

This paper presents a new and efficient method of soft decision error correction decoding that can be used in HDAs for memory-based PUFs. This new method is based on hard decision decoding using concatenated codes as proposed in [1], where the repetition decoder is replaced by a quantizer that creates the input for a soft decision decoder. It results in a code construction for HDAs that requires less PUF bits for error correction. Furthermore, the proposed method of soft decision decoding can be implemented efficiently in hardware and does not suffer from the same practical problems as the soft decision construction from [12,13].

Using several (hardware) implementations of soft decision decoders, the added value of soft decision decoding has been demonstrated for an HDA that derives 171 secret bits with an FRR below 10^{-6} while $\epsilon = 0.15$. The soft decision decoders decrease the number of PUF bits that are required to derive the secret bits in comparison to their hard decision counterparts by 38% to 47%. This decrease of PUF bits comes at only a limited cost in hardware resources of the decoder, which becomes even less significant when the size of the PUF becomes more dominant in the total footprint of the HDA. The optimal HDA implementation can be chosen based on the parameter that should be kept as small as possible (number of PUF bits, total footprint of HDA, etc.) in combination with the values of FRR, ϵ and secret size.

Acknowledgements. This work has been supported in part by the European Commission through the FP7 programme under contracts 238811 UNIQUE and 216676 ECRYPT II, by the IAP program P6/26 BCRYPT of the Belgian state, and by the Research Council KU Leuven through GOA TENSE (GOA/11/007). The authors would like to thank the anonymous referees for constructive comments.

References

1. Bösch, C., Guajardo, J., Sadeghi, A.-R., Shokrollahi, J., Tuyls, P.: Efficient Helper Data Key Extractor on FPGAs. In: Oswald, E., Rohatgi, P. (eds.) CHES 2008. LNCS, vol. 5154, pp. 181–197. Springer, Heidelberg (2008)
2. Boyen, X.: Reusable Cryptographic Fuzzy Extractors. In: CCS 2004, pp. 82–91. ACM, New York (2004), http://doi.acm.org/10.1145/1030083.1030096
3. Claes, M., van der Leest, V., Braeken, A.: Comparison of SRAM and FF PUF in 65nm Technology. In: Laud, P. (ed.) NordSec 2011. LNCS, vol. 7161, pp. 47–64. Springer, Heidelberg (2012)

4. Dodis, Y., Reyzin, L., Smith, A.: Fuzzy Extractors: How to Generate Strong Keys from Biometrics and Other Noisy Data. In: Cachin, C., Camenisch, J. (eds.) EUROCRYPT 2004. LNCS, vol. 3027, pp. 523–540. Springer, Heidelberg (2004)
5. Gassend, B., Clarke, D., van Dijk, M., Devadas, S.: Silicon Physical Random Functions. In: CCS 2002, pp. 148–160. ACM, New York (2002), http://doi.acm.org/10.1145/586110.586132
6. Guajardo, J., Kumar, S.S., Schrijen, G.-J., Tuyls, P.: FPGA Intrinsic PUFs and Their Use for IP Protection. In: Paillier, P., Verbauwhede, I. (eds.) CHES 2007. LNCS, vol. 4727, pp. 63–80. Springer, Heidelberg (2007)
7. Hackett, C.: An Efficient Algorithm for Soft-Decision Decoding of the (24, 12) Extended Golay Code. IEEE Transactions on Communications 29(6), 909–911 (1981)
8. Kumar, S., Guajardo, J., Maes, R., Schrijen, G.J., Tuyls, P.: The Butterfly PUF Protecting IP on Every FPGA. In: Tehranipoor, M., Plusquellic, J. (eds.) IEEE International Workshop on Hardware-Oriented Security and Trust (HOST 2008), pp. 67–70. IEEE Computer Society (2008)
9. Lee, J., Lim, D., Gassend, B., Suh, G., van Dijk, M., Devadas, S.: A Technique to Build a Secret Key in Integrated Circuits for Identification and Authentication Applications. In: IEEE Symposium on VLSI Circuits 2004, pp. 176–179. IEEE (2004)
10. Linnartz, J.-P., Tuyls, P.: New Shielding Functions to Enhance Privacy and Prevent Misuse of Biometric Templates. In: Kittler, J., Nixon, M.S. (eds.) AVBPA 2003. LNCS, vol. 2688, pp. 393–402. Springer, Heidelberg (2003)
11. Maes, R., Tuyls, P., Verbauwhede, I.: Intrinsic PUFs from Flip-Flops on Reconfigurable Devices. In: Workshop on Information and System Security (WISSec 2008), Eindhoven, NL, p. 17 (2008)
12. Maes, R., Tuyls, P., Verbauwhede, I.: Low-Overhead Implementation of a Soft Decision Helper Data Algorithm for SRAM PUFs. In: Clavier, C., Gaj, K. (eds.) CHES 2009. LNCS, vol. 5747, pp. 332–347. Springer, Heidelberg (2009)
13. Maes, R., Tuyls, P., Verbauwhede, I.: Soft Decision Helper Data Algorithm for SRAM PUFs. In: IEEE International Symposium on Information Theory (ISIT 2009), pp. 2101–2105. IEEE Press, Piscataway (2009)
14. Ravikanth, P.S.: Physical One-Way Functions. Ph.D. thesis (2001), aAI0803255
15. Simons, P., van der Sluis, E., van der Leest, V.: Buskeeper PUFs, a Promising Alternative to D Flip-Flop PUFs. In: IEEE International Symposium on Hardware-Oriented Security and Trust (HOST 2012), pp. 7–12, June 3-4 (2012), http://ieeexplore.ieee.org/stamp/ stamp.jsp?tp=&arnumber=6224311&isnumber=6224308, doi:10.1109/HST.2012.6224311
16. Škorić, B., Tuyls, P., Ophey, W.: Robust Key Extraction from Physical Uncloneable Functions. In: Ioannidis, J., Keromytis, A., Yung, M. (eds.) ACNS 2005. LNCS, vol. 3531, pp. 407–422. Springer, Heidelberg (2005)
17. Taiwan Semiconductor Manufacturing Company Limited (TSMC): 65nm technology overview, http://www.tsmc.com/english/dedicatedFoundry/technology/65nm.html

PUFs: Myth, Fact or Busted?
A Security Evaluation of Physically Unclonable Functions (PUFs) Cast in Silicon

Stefan Katzenbeisser[1], Ünal Kocabaş[1], Vladimir Rožić[3],
Ahmad-Reza Sadeghi[2], Ingrid Verbauwhede[3], and Christian Wachsmann[1]

[1] Technische Universität Darmstadt (CASED), Germany
katzenbeisser@seceng.informatik.tu-darmstadt.de,
{unal.kocabas,christian.wachsmann}@trust.cased.de
[2] Technische Universität Darmstadt and Fraunhofer SIT Darmstadt, Germany
ahmad.sadeghi@trust.cased.de
[3] KU Leuven, ESAT/COSIC, Leuven, Belgium
{vladimir.rozic,ingrid.verbauwhede}@esat.kuleuven.be

Abstract. Physically Unclonable Functions (PUFs) are an emerging technology and have been proposed as central building blocks in a variety of cryptographic protocols and security architectures. However, the security features of PUFs are still under investigation: Evaluation results in the literature are difficult to compare due to varying test conditions, different analysis methods and the fact that representative data sets are publicly unavailable.

In this paper, we present the first large-scale security analysis of ASIC implementations of the five most popular intrinsic electronic PUF types, including arbiter, ring oscillator, SRAM, flip-flop and latch PUFs. Our analysis is based on PUF data obtained at different operating conditions from 96 ASICs housing multiple PUF instances, which have been manufactured in TSMC 65 nm CMOS technology. In this context, we present an evaluation methodology and quantify the robustness and unpredictability properties of PUFs. Since all PUFs have been implemented in the same ASIC and analyzed with the same evaluation methodology, our results allow for the first time a fair comparison of their properties.

Keywords: Physically Unclonable Functions (PUFs), ASIC implementation, evaluation framework, unpredictability, robustness.

1 Introduction

Physically Unclonable Functions (PUFs) are increasingly proposed as central building blocks in cryptographic protocols and security architectures. Among other uses, PUFs enable device identification and authentication [33,28], binding software to hardware platforms [7,14,4] and secure storage of cryptographic secrets [37,17]. Furthermore, PUFs can be integrated into cryptographic algorithms [2] and remote attestation protocols [29]. Today, PUF-based security

E. Prouff and P. Schaumont (Eds.): CHES 2012, LNCS 7428, pp. 283–301, 2012.

products are already announced for the market, mainly targeting IP-protection, anti-counterfeiting and RFID applications [36,11].

PUFs typically exhibit a challenge/response behavior: When queried with a *challenge*, the PUF generates a random *response* that depends on the physical properties of the underlying PUF hardware. Since these properties are sensitive to typically varying operating conditions, such as ambient temperature and supply voltage, the PUF will always return a slightly different response each time it is stimulated. The most vital PUF properties for PUF-based security solutions are robustness and unpredictability [1]. Robustness requires that, when queried with the same challenge multiple times, the PUF should generate similar responses that differ only by a small error that can be corrected by an appropriate error correction mechanism. This is an essential requirement in PUF-based applications that must rely on the availability of data generated by or bound to the PUF and should be fulfilled under different operating conditions. Unpredictability guarantees that the adversary cannot efficiently compute the response of a PUF to an unknown challenge, even if he can adaptively obtain a certain number of other challenge/response pairs from the same and other PUF instances. With a PUF instance we denote one particular hardware implementation of a PUF design. Unpredictability is important in most PUF-based applications, such as authentication protocols, where the adversary could forge the authentication if he could predict the PUF response. Existing PUF-based security solutions typically rely on assumptions that have not been confirmed for all PUF types. For instance, most delay-based PUFs have been shown to be emulatable in software [26], which contradicts the unpredictability and unclonability properties. Hence, a systematic analysis of the security properties of real PUF implementations in hardware is fundamental for PUF-based security solutions.

In contrast to most cryptographic primitives, whose security can be related to well established (albeit unproven) assumptions, the security of PUFs relies on assumptions on physical properties and is still under investigation. The security properties of PUFs can either be evaluated theoretically, based on mathematical models of the underling physics [35,30], or experimentally by analyzing PUF implementations [10,34,8,9,16]. However, mathematical models never capture physical reality in its full extent, which means that the conclusions on PUF security drawn by this approach are naturally debatable. The main drawback of the experimental approach is its limited reproducibility and openness: Even though experimental results have been reported in literature for some PUF implementations, it is difficult to compare them due to varying test conditions and different analysis methods. Furthermore, raw PUF data is rarely available for subsequent research, which greatly hinders a fair comparison.

Our Goal and Contribution. We present the first large-scale security analysis of ASIC implementations of the five most popular electronic PUF types, including different delay-based PUFs (arbiter and ring oscillator PUFs) and different memory-based PUFs (SRAM, flip-flop and latch PUFs). Hereby, we focus on robustness and unpredictability, which are the most vital PUF properties in many security-critical applications. The ASICs have been manufactured in TSMC 65 nm CMOS technology within a multi-project wafer run and contain multiple implementations of the same PUF design. Our analysis is based on PUF data obtained from 96 ASICs at different temperatures, supply voltages and noise levels that correspond to the corner values typically tested for consumer-grade IT products. In this context, we developed an evaluation methodology for the empirical assessment of the robustness and unpredictability properties of PUFs. Since all PUFs have been implemented in the same ASIC and analyzed with the same methodology, our results allow for the first time a fair comparison of the robustness and unpredictability of these PUFs.

Our evaluation results show that all PUFs in the ASIC are sufficiently robust for practical applications. However, not all of them achieve the unpredictability property. In particular, the responses of arbiter PUFs have very low entropy, while the entropy of flip-flop and latch PUF responses are affected by temperature variations. In contrast, the ring oscillator and SRAM PUFs seem to achieve all desired properties of a PUF: Their challenge/response behavior hardly changes under different operating conditions and the entropy of their responses is quite high. Furthermore, the responses generated by different ring oscillator and SRAM PUF instances seem to be independent, which means that the adversary cannot predict the response of a PUF based the challenge/responses pairs of another PUF. However, the min-entropy, i.e., the minimum number of random bits observed in a response of the ring oscillator PUF is low, which means that some responses can be guessed with high probability.

Outline. We provide background information on PUFs in Section 2 and give an overview of the ASIC implementation of the analyzed PUFs in Section 3. We present our evaluation methodology in Section 4 and our analysis results in Section 5. Finally, we conclude in Section 6.

2   Background on PUFs

A Physically Unclonable Function (PUF) is a function that is embedded into a physical object, such as an integrated circuit [25,20]. When queried with a *challenge* x, the PUF generates a *response* y that depends on both x and the unique device-specific physical properties of the object containing the PUF. Since PUFs are subject to noise induced by environmental variations, they return slightly different responses when queried with the same challenge multiple times.

PUFs are typically assumed to be *robust*, *physically unclonable*, *unpredictable* and *tamper-evident*, and several approaches to quantify and formally define their properties have been proposed (see [1] for an overview). Informally, robustness means that, when queried with the same challenge multiple times, the PUF returns similar responses with high probability. Physical unclonability demands that it is infeasible to produce two PUFs that are indistinguishable based on their challenge/response behavior. Unpredictability requires that it is infeasible to predict the PUF response to an unknown challenge, even if the PUF can be adaptively queried for a certain number of times. Finally, a PUF is tamper-evident if any attempt to physically access the PUF changes its challenge/response behavior. The properties required from a PUF strongly depend on the application. For instance, a PUF with small challenge/response space can be easily emulated by reading out all its challenge/response pairs and creating a look-up table. While such a PUF cannot be used directly in authentication schemes (such as in [32]), it could still be used in a key storage scenario (such as in [17]), where the adversary is typically assumed not being able to interact with the PUF.

There is a variety of PUF implementations (see [20] for an overview). The most appealing ones for the integration into electronic circuits are *electronic PUFs*, which come in different flavors. *Delay-based PUFs* are based on race conditions in integrated circuits and include arbiter PUFs [15,24,18] and ring oscillator PUFs [6,32,21]. *Memory-based PUFs* exploit the instability of volatile memory cells, such as SRAM [7,9], flip-flops [19,16] and latches [31,14].

Note that memory-based PUFs can be emulated in software since the limited number of memory cells allows creating a look-up table. Further, most delay-based PUFs are subject to model building attacks that allow emulating the PUF in software [15,24,18,26]. To counter this problem, additional primitives must be used: Controlled PUFs [5] and Feed-Forward PUFs [22] use cryptographic functions or XOR-networks in hardware, respectively, to hide the responses of the underlying PUF. Furthermore, PUFs are inherently noisy and must be combined with error correction mechanisms, such as *fuzzy extractors* [3] that remove the effects of noise before the PUF response can be processed in a cryptographic algorithm. Typically, the cryptographic and error correcting components as well as the link between them and the PUF must be protected against invasive and side channel attacks.

3   The PUF ASIC

Our analysis is based on data obtained from 96 ASICs that have been manufactured in TSMC 65 nm CMOS technology within a Europractice multi-project wafer run. The ASIC has been designed within the UNIQUE[1] research project. Each ASIC implements multiple instances of three different memory-based PUFs (SRAM, flip-flop and latch PUFs) and two different delay-based PUFs (ring oscillator and arbiter PUFs). The main characteristics and number of PUF instances in the ASICs are shown in Table 1. Furthermore, the ASIC is equipped with an active core that emulates the noisy working environment of a microprocessor. When enabled, this core performs AES encryption during the PUF evaluation.

The implementation of the arbiter PUF follows the basic approach presented by Lee et al. [15] and consists of 64 delay elements and an arbiter. The delay elements are connected in a line, forming two delay paths with an arbiter placed at the end. Each challenge corresponds to a different configuration of the delay paths. More detailed, each delay element has two inputs and two outputs and

---

[1] http://www.unique-project.eu/

Table 1. Physically Unclonable Functions (PUFs) implemented in the 96 ASICs

PUF class	PUF type	Number of instances per ASIC	Total number of instances	Challenge space size	Response space size
Delay-based	Arbiter	256	24,576	2^{64}	2
	Ring oscillator	16	1,536	$32,640 \approx 2^{15}$	2
Memory-based	SRAM	4 (8 kB)	384	2^{11}	2^{32}
	Flip-flop	4 (1 kB)	384	2^8	2^{32}
	Latch	4 (1 kB)	384	2^8	2^{32}

can be configured to map inputs to outputs directly (challenge bit 0) or to switch them (challenge bit 1). During the read-out of the PUF response, the input signal propagates along both paths and, depending on which of the paths is faster, a single response bit is generated. To ensure that the delay difference results from the manufacturing process variations rather than the routing of the metal lines, a symmetric layout for the delay elements and full-custom layout blocks were used. Further, to reduce any bias the capacitive loads of the connecting metal wires was balanced and a symmetric NAND-latch was used as arbiter.

The ring oscillator PUF uses the design by Suh et al. [32]. Each ring oscillator PUF consists of 256 ring oscillators and a control logic, which compares the frequency of two different oscillators selected by the PUF challenge. Depending on which of the oscillators is faster, a single response bit is generated. The individual ring oscillators are implemented using layout macros to ensure that all oscillators have exactly the same design, which is fundamental for the correct operation of the ring oscillator PUF.

The memory-based PUFs are implemented as arrays of memory elements (SRAM cells, latches, flip-flops). All these memory elements are bi-stable circuits with two stable states corresponding to a logical 0 and 1. After power-up, each memory element enters either of the two states. The resulting state depends on the manufacturing process variations and the noise in the circuit. When challenged with a memory address, the PUF returns the 32 bit data word at that address. The implementations of the memory-based PUFs follow the SRAM PUF design by Holcomb et al. [9], the flip-flop PUF design by Maes et al. [19] and the latch PUF design by Su et al. [31]. Latch and flip-flop PUFs are implemented using the standard cells from TSMC's 65 nm low-power library. The placement and implementation of the SRAM cells of the SRAM PUF has been done by TSMC's memory compiler. The latch and flip-flop PUFs are based on standard cells using a clustered strategy, where all latches or flip-flops of the same PUF instance are grouped together in single block.

The test setup consists of an ASIC evaluation board, a Xilinx Virtex 5 FPGA and a PC (Figure 1). Each evaluation board can take five ASICs and allows controlling the ASIC supply voltage with an external power supply. The interaction with the evaluation board and the ASICs is performed by the FPGA, which is connected to a PC that controls the PUF evaluation process and stores the raw

Fig. 1. Test setup with Xilinx Virtex 5 FPGA (left) and ASIC evaluation board with five PUF ASICs (right)

PUF responses obtained from the ASICs. The tests at different temperatures have been performed in a climate chamber.

4   Our Evaluation Methodology

Many PUF-based applications require PUF responses to be reliably reproducible while at the same time being unpredictable (see, e.g., [20,1]). Hence, our empirical evaluation focuses on robustness and unpredictability.

Notation. With $|x|$ we denote the length of some bitstring x. Let E be some event, then $\Pr[E]$ denotes the probability that E occurs. We denote with $\mathsf{HW}(x)$ the Hamming weight of a bitstring x, i.e., the number of non-zero bits of x. With $\mathsf{dist}(x, y)$ we denote the Hamming distance between two bit strings x and y, i.e., the number of bits that are different in x and y.

4.1   Robustness Analysis

Robustness is the property that a PUF always generates responses that are similar to the responses generated during the enrolment of the PUF. Note that PUFs should fulfil this property under different operating conditions, such as different temperatures, supply voltages and noise levels. The robustness of PUFs can be quantified by the bit error rate $\mathsf{BER} := \frac{\mathsf{dist}(y_{E_i}, y_{E_5})}{|y_{E_5}|}$, which indicates the number of bits of a PUF response y_{E_i} that are different from the response y_{E_5} observed during enrolment. We determine the maximum BER of all PUF instances in all ASICs based on challenge/response pairs collected at different ambient temperatures ($-40\,^{\circ}\mathrm{C}$ to $+85\,^{\circ}\mathrm{C}$), supply voltages ($\pm 10\%$ of the nominal $1.2\,\mathrm{V}$) and noise levels (active core enabled and disabled), which correspond to the corner values that are typically tested for consumer grade IT products. This shows the impact of the most common environmental factors on the BER of each PUF type. We did not test different noise levels at different temperatures and supply voltages since most PUFs (except the arbiter PUF) turned out to be hardly affected by even the maximum amount of noise the active core can generate. An overview of all test cases considered for robustness is given in Table 2. We estimate the BER of all PUFs in all ASICs using the following procedure:

Table 2. Robustness test cases

Test Case	Active Core Off	On	Ambient Temperature $-40\,°C$	$+25\,°C$	$+85\,°C$	Supply Voltage 1.08 V	1.2 V	1.32 V	Iter. k
E_1	×		×			×			20
E_2	×		×				×		40
E_3	×		×					×	20
E_4	×			×		×			30
E_5	×			×			×		60
E_6	×			×				×	30
E_7	×				×	×			20
E_8	×				×		×		40
E_9	×				×			×	20
E_{11}		×		×			×		60

Step 1: Sample challenge set generation. A sample challenge set $\mathcal{X}'$ is generated for each PUF type (arbiter, ring oscillator, SRAM, flip-flop and latch PUF) and used in all subsequent steps. For all but the arbiter PUF the complete challenge space is used as a sample set. Since the arbiter PUF has an exponential challenge space, we tested it for $13,000$ randomly chosen challenges, which is a statistically significant subset and representative for the whole challenge space.

Step 2: Enrolment. For each PUF instance, the response y_i to each challenge $x_i \in \mathcal{X}'$ is obtained under nominal operating conditions (test case E_5) and stored in a database $\mathsf{DB_0}$.

Step 3: Data acquisition. For all test cases E_p in Table 2, each PUF instance is evaluated k times on each $x_i \in \mathcal{X}'$ and its responses are stored in a database DB_p for $p = 1, \dots, 11$.

Step 4: Analysis. For each PUF instance, the maximum BER between its responses in $\mathsf{DB_0}$ and its responses in $\mathsf{DB_1}, \dots, \mathsf{DB_{11}}$ over all $x_i \in \mathcal{X}'$ is computed.

4.2 Unpredictability Analysis

Unpredictability ensures that the adversary cannot efficiently compute the response of a PUF to an unknown challenge, even if he can adaptively obtain a certain number of other challenge/response pairs from the same and other PUF instances [1]. This is important in most PUF-based applications, such as authentication protocols, where the adversary can forge the authentication when he can predict a PUF response. Note that unpredictability should be independent of the operating conditions of the PUF, which could be exploited by an adversary.

The unpredictability of a PUF implementation can be estimated empirically by applying statistical tests to its responses and/or based on the complexity of the best known attack against the PUF [20,1]. Statistical tests, such as the DIEHARD [23] or NIST [27] test suite, can in principle be used to assess the

Table 3. Unpredictability test cases

Test Case	Active Core		Ambient Temperature			Supply Voltage		
	Off	On	$-40\,^\circ$C	$+25\,^\circ$C	$+85\,^\circ$C	1.08 V	1.2 V	1.32 V
E_{13}	×		×				×	
E_{14}	×			×			×	
E_{15}	×				×		×	
E_{16}	×			×		×		
E_{17}	×			×				×

unpredictability of PUF responses. However, since these test suites are typically based on a series of stochastic tests, they can only indicate whether the PUF responses are random or not. Moreover, they require more input data than the memory-based PUFs and ring oscillator PUFs in the ASIC provide. Similar as in symmetric cryptography, the unpredictability of a PUF can be estimated based on the complexity of the best known attack. There are attacks [26] against delay-based PUFs that emulate the PUF in software and allow predicting PUF responses to arbitrary challenges. These attacks are based on machine learning techniques that exploit statistical deviations and/or dependencies of PUF responses. However, emulation attacks have been shown only for simulated PUF data and it is currently unknown how these attacks perform against real PUFs [26]. Another approach is estimating the entropy of the PUF responses based on experimental data. In particular, *min-entropy* indicates how many bits of a PUF response are uniformly random. The entropy of PUFs can be approximated using the context-tree weighting (CTW) method [39], which is a data compression algorithm that allows assessing the redundancy of bitstrings [10,34,8,16].

We assess the unpredictability of PUFs using Shannon entropy, which is a common metric in cryptography and allows establishing relations to other publications that quantify the unpredictability of PUFs using entropy (such as [35,32,9,1]). We estimate the entropy and min-entropy of the responses of all available PUFs. Specifically, we first check whether PUF responses are biased by computing their Hamming weight and estimate an upper bound of the entropy of PUF responses using a compression test. Eventually, we approximate the entropy and min-entropy of the responses of all available PUFs. Our entropy estimation is more precise than previous approaches since it considers dependencies between the individual bits of the PUF responses. Furthermore, to get an indication of whether responses of *different* PUF instances are independent, we compute the Hamming distance between responses of different PUF instances.

We assess the unpredictability of all available PUFs at different temperatures and supply voltage levels (Table 3) to determine the effects of environmental variations on the unpredictability using the following procedure: We assess the unpredictability of all PUFs in the ASICs using the following procedure:

Step 1: Sample challenge set generation. For each PUF type, a sample challenge set $\mathcal{X}'$ is generated that is used in all subsequent steps. For all but the arbiter PUF, the complete challenge space is used as a sample challenge set. Since the arbiter PUF has an exponential challenge space, we again test it only for $13,000$ challenges. The subsequent analysis steps require $\mathcal{X}' := \{x' \in \mathcal{X}'' | \text{dist}(x, x') \leq k\}$, which includes a set $\mathcal{X}''$ of randomly chosen challenges and all challenges that differ in at most k bits from the challenges in $\mathcal{X}''$ (that may be known to the adversary).

Step 2: Data acquisition. For all test cases E_q in Table 3, each PUF instance is evaluated on each $x_i \in \mathcal{X}'$ and the responses y are stored in a database DB_q.

Step 3: Analysis. For each test case E_q, the responses in DB_q are analyzed as detailed in the following items:

Step 3a: Hamming weight. For each PUF instance, the average Hamming weight of all its responses y_i in DB_q is computed, which indicates whether the responses are biased towards 0 or 1.

Step 3b: CTW Compression. For each PUF instance, a binary file containing all its responses in DB_q is generated and compressed using the context-tree weighting (CTW) algorithm [38]. The resulting compression rate is an estimate of the upper bound of the entropy of the PUF responses.

Step 3c: Entropy estimation. For each PUF instance, the entropy and min-entropy of all its responses in DB_q is estimated as detailed in the next paragraph.

Step 3d: Hamming distance. For each PUF type, the Hamming distance $\text{dist}(y, y')$ of all pairs of responses (y, y') in DB_q generated by pairwise different PUF instances for the same challenge x is computed. While all previous steps consider only responses of the *same* PUF instance, the Hamming distances indicate whether responses of *different* PUF instances are independent. This is important to prevent the adversary from predicting the responses of one PUF implementation based on the challenge/response pairs of another (e.g., his own) PUF implementation, which would contradict the unpredictability property.

Entropy Estimation. Let x be the PUF challenge for which the adversary should predict the response y. Further, let $Y(x)$ be the random variable representing y. Moreover, let $W(x)$ be the random variable representing the set of all responses of the PUF except y, i.e., $W(x) = \{y' | y' \leftarrow \text{PUF}(x'); \ x' \in \mathcal{X} \setminus \{x\}\}$. We are interested in the conditional entropy

$$\mathbf{H}(Y|W) = - \sum_{x \in \mathcal{X}} \Pr\left[Y(x), W(x)\right] \cdot \log_2 \Pr\left[Y(x)|W(x)\right] \tag{1}$$

and the conditional min-entropy

$$\mathbf{H}_\infty(Y|W) = - \log_2\left(\max_{x \in \mathcal{X}}\left\{\Pr\left[Y(x)|W(x)\right]\right\}\right), \tag{2}$$

which quantify the average and minimal number of bits of y, respectively, that cannot be predicted by the adversary, even in case all other responses in $W(x)$ are known.[2] Hence, $2^{-\mathbf{H}_\infty(Y|W)}$ is an information-theoretic upper bound for the probability that an adversary guesses the PUF response y to challenge x.

However, computing Equations 1 and 2 for $W(x)$ is difficult since (1) the sizes of the underlying probability distributions are exponential in the response space size, and (2) the complexity of computing $\mathbf{H}(Y|W)$ grows exponentially with the challenge space size of the PUF to be analyzed. Hence, Equations 1 and 2 can at most be estimated by making assumptions on the physical properties of the PUFs that reduce the size of $W(x)$. In the following, we explain how we estimated these entropies for each PUF type and discuss the underlying assumptions.

Memory-based PUFs. A common assumption on memory-based PUFs is that spatially distant memory cells are independent [20,1]. A similar assumption has been used by Holcomb et al. [9], who estimate the entropy of SRAM PUF responses based on the assumption that individual bytes of SRAM are independent. However, physically neighboring memory cells can strongly influence each other, in particular when they are physically connected.[3] Hence, our entropy estimation considers dependencies between neighboring memory cells (which could be exploited by an adversary) while assuming that spatially distant memory cells are independent. More specifically, we compute the entropy of the PUF response bit $Y_{i,j}$ of the memory cell at row i and column j of the underlying memory under the worst case assumption that the values of all neighboring memory cells $W'(x) = (Y_{i-1,j}, Y_{i,j+1}, Y_{i+1,j}, Y_{i,j-1})$ are known, i.e., we compute Equations 1 and 2 for $W'(x)$.

Ring Oscillator PUFs. The ring oscillator PUFs in the ASICs compare the oscillation frequency of two ring oscillators O_i and O_j selected by the PUF challenge $x = (i, j)$ and returns a response $Y(i, j)$, depending on which of the two oscillators was faster. Since neighboring ring oscillators may affect each other (e.g., by electromagnetic induction), we consider the potential dependency between the frequencies of neighboring oscillators and assume that the frequency of spatially distant oscillators is independent. Thus, we compute Equations 1 and 2 for $W'(i, j) = (Y_{i-2,j}, Y_{i-1,j}, Y_{i+1,j}, Y_{i+2})$.

Arbiter PUFs. Arbiter PUFs measure the delay difference of two delay lines that are configured by the PUF challenge. The individual delays caused by the switches and their connections are additive, which implies that the PUF response y to a challenge x can be computed if a sufficient number of responses to challenges that are close to x are known. Hence, we compute Equations 1 and 2 for $W'(x) = \{y' \leftarrow \mathsf{PUF}(x')|x' \in \mathcal{X}', \ \mathrm{dist}(x, x') \leq k\}$, which corresponds to the

---

[2] Note that this corresponds to the game-based security definition of unpredictability by Armknecht et al. [1], which formalizes the difficulty of predicting Y in case the PUF responses in W are known.

[3] SRAM cells are typically arranged in a matrix, where all cells in a row are connected by a word line and all cells in a column are connected by a bit line.

worst case where the adversary knows responses to challenges that differ in at most k bits from the challenge whose response he must guess. Specifically, we use $\mathcal{X}$ consisting of 200 randomly chosen challenges and $k = 1$.

Computing the Entropy. To compute the entropy and min-entropy (Equations 1 and 2) for each test case E_q, we first estimate $\Pr\left[x = Y(x), w = W(x)\right]$ for each $x \in \mathcal{X}'$ by dividing the number of observations of each tuple (x, w) in database DB_q by the size of the sample challenge set $\mathcal{X}'$. Further, to compute $\Pr\left[x = Y(x)\middle| w = W(x)\right] = \Pr\left[x = Y(x), w = W(x)\right] / \Pr\left[w = W(x)\right]$, we estimated $\Pr\left[w = W(x)\right]$ by dividing the number of observations of each tuple $\left(Y(x), w = W(x)\right)$ in database DB_i by the size of $\mathcal{X}'$. Eventually, we computed Equations 1 and 2.

5 Evaluation and Results

We applied the evaluation methodology in Section 4 to all PUF instances in all ASICs. Most of our results are illustrated using *bean plots* [12] that allow an intuitive visualization of empirical probability distributions (Figures 2 to 5). Each bean shows two distributions, smoothed by a Gaussian kernel to give the impression of a continuous distribution, together with their means indicated by black bars. The distribution in black on the left side typically corresponds to data collected under normal PUF operating conditions, while the one in gray on the right side corresponds to some other test case in Table 2 and 3.

Due to space restrictions, we illustrate only the most important results and provide a detailed discussion in the full version of this paper [13].

5.1 Robustness Results

We computed the bit error rate (BER) under varying environmental conditions (Table 3). Our results show that all arbiter, ring oscillator and SRAM PUF instances have a very similar BER, while there is a big variability in the BERs of the flip-flop and latch PUF instances (Figure 2). Further, the BER of the arbiter, ring oscillator and SRAM PUF instances is below 10% for all test cases, which can be handled by common error correction schemes, such as fuzzy extractors [3]. The BER of most PUFs depends on the operating temperature (Figure 2a): Compared to $+25\,^{\circ}\text{C}$ (test case E_5), at $-40\,^{\circ}\text{C}$ (test case E_2) the BER of the flip-flop and latch PUF increases significantly, while the BER of the ring oscillator and SRAM PUF increases only slightly and the BER of the arbiter PUF hardly changes. A similar behavior of the BERs can be observed at $+85\,^{\circ}\text{C}$ (test case E_8). All PUFs in all ASICs turned out to be robust against variations of their supply voltages. Compared to nominal operating conditions (test case E_5), the distributions of the BERs only slightly increase when varying the supply voltage by 10% (test case E_4 and E_6). The arbiter PUF exhibits a significantly increased BER when operated in a noisy working environment (test case E_{11}; Figure 2b) while there is no significant change of the BER of all other PUFs. Hereby, we

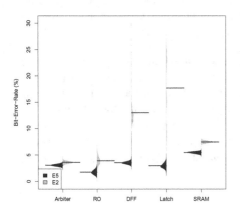 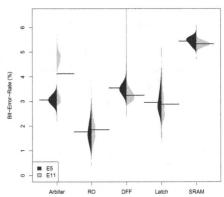

(a) Bit error rates at $+25\,°C$ (test case E_5, black) and at $-40\,°C$ (test case E_2, gray)

(b) Bit error rates with active core off (test case E_5, black) and active core on (test case E_{11}, gray)

Fig. 2. Distribution of the bit error rate (BER) in percent over all PUF instances at different ambient temperatures and noise levels. The two peaks of the BER distribution of the arbiter PUF show that those arbiter PUFs that are spatially close to the active core are more affected than those farther away.

observed that the BER of arbiter PUF instances that are spatially close to the active core significantly changes, while those that are farther away are not directly affected.

5.2   Unpredictability Results

In this section, we present the results of our unpredictability analysis. Due to the time-limited access to the climate chamber, the data required to analyze the unpredictability of the arbiter PUF at $-40\,°C$ and at $+85\,°C$ is not available. However, we show the results for normal operating conditions and different supply voltages.

Hamming Weights. To get a first indication of randomness in the PUFs, we computed the Hamming weight of their responses as described in Section 4.2. Our results show that ring oscillator and SRAM PUF responses are close to the ideal Hamming weight of 0.5, independent of the operating conditions (Figure 3), which indicates that their responses may be random. The Hamming weight of the flip-flop PUF and latch PUF responses strongly depends on the ambient temperature (Figure 3a) and is clearly biased. Supply voltage variations (test cases E_{16} and E_{17}) have no significant impact on the Hamming weight of the responses of any of the PUF instances in the ASIC (Figure 3b).

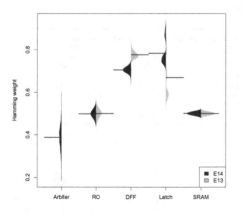 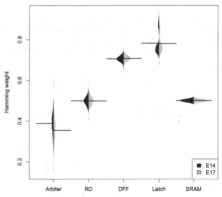

(a) Hamming weight at $+25\,^{\circ}$C (test case E_{14}, black) and $-40\,^{\circ}$C (test case E_{13}, gray)

(b) Hamming weight at 1.20 V (nominal voltage, test case E_{14}, black) and 1.32 V ($+10\%$ overvoltage, test case E_{17}, gray)

Fig. 3. Distribution of the Hamming weight over all PUF instances at different ambient temperatures and supply voltage levels. The two peaks of the Hamming weight distribution of the latch PUF may come from the fact that one of the four latch PUF instances on each ASIC is implemented in a separate power domain.

CTW Compression. The context-tree weighting (CTW) compression test gives a good indication of the upper bound of the entropy of PUF responses. The higher the compression rate, the lower the entropy of the PUF. The results of this test (Table 4) confirm the Hamming weight test results:

The compression rate of the ring oscillator and SRAM PUF responses is invariant for all test cases; the compression rates of the flip-flop and latch PUF responses do not change for different supply voltages (test case E_{16} and E_{17}), but vary with the ambient temperature (test cases E_{13}, E_{14} and E_{15}). The compression rate of the SRAM PUF responses strongly indicates that these responses are uniformly random, while there seem to be some dependencies in the responses generated by all other PUFs.

Entropy Estimation. The results of the entropy estimation described in Section 4.2 confirm the results of all previous tests and provide more insights into the entropy and min-entropy of the PUF responses (Figure 4).

The entropy of responses corresponding to neighboring arbiter PUF challenges is remarkably low, which confirms the high prediction rate of emulation attacks against arbiter PUFs reported in literature [26]. The entropy and min-entropy of the ring oscillator and SRAM PUF responses is invariant to temperature (test cases E_{13}, E_{14} and E_{15}) and supply voltage (test case E_{16} and E_{17}) variations. Moreover, the entropy and min-entropy of flip-flop and latch PUFs vary with

Table 4. CTW compression results

| Test Case | Size of PUF response after CTW compression in percent | | | | |
	Arbiter	Ring-Oscillator	Flip-Flop	Latch	SRAM
E_{13}	—	0.77	0.77	0.84	1.00
E_{14}	0.51	0.77	0.87	0.70	1.00
E_{15}	—	0.77	0.98	0.53	1.00
E_{16}	0.53	0.77	0.88	0.69	1.00
E_{17}	0.49	0.77	0.87	0.71	1.00

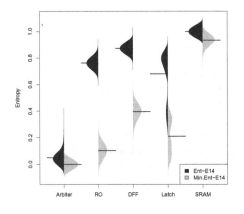
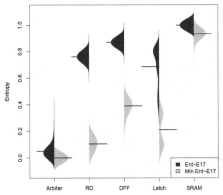

(a) Entropy (black) and min-entropy (gray) at $+25\,^\circ$C (test case E_{14})

(b) Entropy (black) and min-entropy (gray) at 1.32 V ($+10\%$, test case E_{17})

Fig. 4. Distribution of the entropy (black) and min-entropy (gray) over all PUF instances at different ambient temperatures and supply voltage levels

the operating temperature (test cases E_{13}, E_{14} and E_{15}) and are constant for different supply voltages (test case E_{16} and E_{17}).

Hamming Distances. The Hamming distance test (Section 4.2) gives an indication of whether the responses generated by different PUF instances to the same challenge are independent. Our results show that, independent of the ambient temperature (test cases E_{13}, E_{14} and E_{15}) and supply voltage (test cases E_{16} and E_{17}), the responses of different ring oscillator and SRAM PUF instances have the ideal Hamming distance of 0.5, while there seem to be dependencies between the responses generated by different arbiter PUF instances to the same challenge (Figure 5). The Hamming distance of the responses of the flip-flop PUFs changes for different temperatures and supply voltages. At $+85\,^\circ$C (test case E_{15}) the Hamming distance of the flip-flop PUF is ideal, while it is biased towards zero at $-40\,^\circ$C (test case E_{13}). Moreover, at 1.08 V (-10% undervoltage, test case E_{16}) we observed a bias of the Hamming distance towards one, while

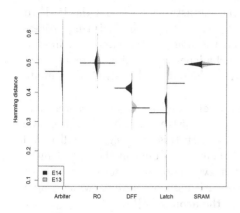 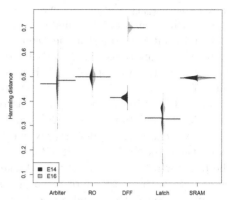

(a) Hamming distance at $+25\,^\circ$C (test case E_{14}, black) and at $-40\,^\circ$C (test case E_{13}, gray)

(b) Hamming distance at $1.20\,$V (nominal voltage, test case E_{14}, black) and at $1.08\,$V (-10% undervoltage, test case E_{16}, gray)

Fig. 5. Distribution of the Hamming distance over all PUF instances at different ambient temperatures and supply voltage levels

the Hamming distance at $1.32\,$V ($+10\%$ overvoltage, test case E_{17}) is similar to the distribution at nominal operating conditions (test case E_{14}). The Hamming distance of the responses of the latch PUFs are biased towards zero and invariant for different supply voltages.

5.3 Discussion

Our results show that arbiter, ring oscillator and SRAM PUFs are more robust to temperature variations than the latch and flip-flop PUFs. This could be due to the dual nature of these PUFs, i.e., the two delay paths, two ring oscillators, and the symmetrical structure of the SRAM cells, respectively. As discussed in Section 3, we do not have access to the internal circuit diagrams and layout of the standard cells provided by TSMC and thus can only speculate about the transistor schematics of the flip-flops and latches. Standard cell libraries typically use implementations based on transmission gates, which are more compact than static latches or flip-flops with a dual structure and there is no duality or symmetry in these transistor schematics. Further, the results of the Hamming weight and Hamming distance tests indicate that the unpredictability of PUFs with a dual structure are less affected by temperature variations.

The entropy of the arbiter PUF is remarkably low, which can be explained by the linear structure of this PUF. Note that in the arbiter PUF implementation, two signals travel along two delay paths and finally arrive at an arbiter (Section 3). In case the delay difference δ_t of the two paths is greater than the setup-time t_{setup} plus the hold time t_{hold} of the arbiter, the PUF response will be correctly generated according to which signal arrives first. However, in case $\delta_t < t_{\text{setup}} + t_{\text{hold}}$, the arbiter will be in the metastable state and the PUF response will depend on the bias of the arbiter caused by manufacturing process and/or layout variations of the arbiter and the noise in the circuit. A limited number of simulations (with 20 PUFs for 3 challenges) including extracted post layout parasitics were performed before the tape-out of the ASIC to estimate this effect.

Since the arbiter PUF design is based on delay accumulation, it is very susceptible to emulation attacks [26]. An example illustrating this fact is the case where two challenges differ in only the last bit. In this case, signals will travel along the same paths through 63 delay elements, and only in the last element the paths will be different. If the attacker knows the outcome for one challenge, he can guess the outcome of the other one with high probability, which might explain the low entropy and min-entropy of the arbiter PUFs.

5.4 Summary

The arbiter PUF responses have a very low entropy and their use in applications with strict unclonability and unpredictability requirements should be carefully considered. Further, the arbiter PUFs are susceptible to changes of their supply voltage and to environmental noise, which significantly increases the bit error rate of the PUF. However, the bit error rate stays within acceptable bounds and can be compensated by existing error correction mechanisms.

The flip-flop and latch PUFs are susceptible to temperature variations, which have a significant effect on the bit error rate and the unpredictability of the PUF responses. Hence, flip-flop and latch PUFs should not be used in an environment, where the adversary can lower the ambient temperature of the PUF, reducing the entropy of the PUF responses.

The SRAM and ring oscillator PUFs achieve almost all desired properties of a PUF: The bit error rate does not change significantly under different operating conditions, the entropy of the PUF responses is high and the responses generated by different PUF instances seem to be independent. However, the ring oscillator exhibits a low min-entropy, which might be problematic in some applications.

6   Conclusion

We performed the first large-scale analysis of the five most popular PUF types (arbiter, ring oscillator, SRAM, flip-flop and latch PUFs) implemented in ASIC. Our analysis is based on PUF data obtained from 96 ASICs, each housing several PUF instances. Our results allow for the first time a fair comparison of these PUFs. In this context, we presented an evaluation methodology for the empirical assessment of the robustness and unpredictability properties of PUFs that are fundamental in most applications of PUFs.

Our results show that the SRAM and ring oscillator PUFs seem to achieve all desired properties of a PUF. However, the arbiter PUFs have a very low entropy and the entropy of the flip-flop and latch PUFs is susceptible to temperature variations. Hence, the suitability of these PUFs for security-critical applications, such as authentication or key generation must be carefully considered.

Future work includes the analysis of stronger PUF constructions and the development of entropy estimation methodologies that also include potential dependencies between different PUF instances.

Acknowledgement. We thank all our partners Intel, Intrinsic ID, KU Leuven, and Sirrix AG, who developed the ASIC and evaluation board. Further, we thank Intrinsic ID for providing us the raw PUF data for the test cases at $-40\,°C$ and $+85\,°C$. Moreover, we thank Vincent van der Leest (Intrinsic ID), Roel Maes (KU Leuven) and our anonymous reviewers for their helpful comments. This work has been supported by the European Commission under grant agreement ICT-2007-238811 UNIQUE.

References

1. Armknecht, F., Maes, R., Sadeghi, A.R., Standaert, F.X., Wachsmann, C.: A formal foundation for the security features of physical functions. In: IEEE Symposium on Security and Privacy (SSP), pp. 397–412. IEEE Computer Society (May 2011)
2. Armknecht, F., Maes, R., Sadeghi, A.-R., Sunar, B., Tuyls, P.: Memory Leakage-Resilient Encryption Based on Physically Unclonable Functions. In: Matsui, M. (ed.) ASIACRYPT 2009. LNCS, vol. 5912, pp. 685–702. Springer, Heidelberg (2009)
3. Dodis, Y., Reyzin, L., Smith, A.: Fuzzy Extractors: How to Generate Strong Keys from Biometrics and Other Noisy Data. In: Cachin, C., Camenisch, J. (eds.) EUROCRYPT 2004. LNCS, vol. 3027, pp. 523–540. Springer, Heidelberg (2004)
4. Eichhorn, I., Koeberl, P., van der Leest, V.: Logically reconfigurable PUFs: Memory-based secure key storage. In: ACM Workshop on Scalable Trusted Computing (ACM STC), pp. 59–64. ACM, New York (2011)
5. Gassend, B., Clarke, D., van Dijk, M., Devadas, S.: Controlled physical random functions. In: Computer Security Applications Conference (ACSAC), pp. 149–160. IEEE (2002)
6. Gassend, B., Clarke, D., van Dijk, M., Devadas, S.: Silicon physical random functions. In: ACM Conference on Computer and Communications Security (ACM CCS), pp. 148–160. ACM, New York (2002)
7. Guajardo, J., Kumar, S., Schrijen, G.J., Tuyls, P.: FPGA Intrinsic PUFs and Their Use for IP Protection. In: Paillier, P., Verbauwhede, I. (eds.) CHES 2007. LNCS, vol. 4727, pp. 63–80. Springer, Heidelberg (2007)
8. Hammouri, G., Dana, A., Sunar, B.: CDs Have Fingerprints Too. In: Clavier, C., Gaj, K. (eds.) CHES 2009. LNCS, vol. 5747, pp. 348–362. Springer, Heidelberg (2009)
9. Holcomb, D.E., Burleson, W.P., Fu, K.: Power-Up SRAM state as an identifying fingerprint and source of true random numbers. IEEE Transactions on Computers 58(9), 1198–1210 (2009)
10. Ignatenko, T., Schrijen, G.J., Škorić, B., Tuyls, P., Willems, F.: Estimating the Secrecy-Rate of physical unclonable functions with the Context-Tree weighting method. In: IEEE International Symposium on Information Theory (ISIT), pp. 499–503. IEEE (July 2006)

11. Intrinsic ID: Product webpage (November 2011), http://www.intrinsic-id.com/products.html
12. Kampstra, P.: Beanplot: A boxplot alternative for visual comparison of distributions. Journal of Statistical Software 28(1), 1–9 (2008)
13. Katzenbeisser, S., Kocabaş, Ü., Rožić, V., Sadeghi, A.R., Verbauwhede, I., Wachsmann, C.: PUFs: Myth, fact or busted? A security evaluation of physically unclonable functions (PUFs) cast in silicon. Cryptology ePrint Archive (2012)
14. Kumar, S.S., Guajardo, J., Maes, R., Schrijen, G.J., Tuyls, P.: Extended abstract: The butterfly PUF protecting IP on every FPGA. In: Workshop on Hardware-Oriented Security (HOST), pp. 67–70. IEEE (June 2008)
15. Lee, J.W., Lim, D., Gassend, B., Suh, E.G., van Dijk, M., Devadas, S.: A technique to build a secret key in integrated circuits for identification and authentication applications. In: Symposium on VLSI Circuits, pp. 176–179. IEEE (June 2004)
16. van der Leest, V., Schrijen, G.J., Handschuh, H., Tuyls, P.: Hardware intrinsic security from D flip-flops. In: ACM Workshop on Scalable Trusted Computing (ACM STC), pp. 53–62. ACM, New York (2010)
17. Lim, D., Lee, J.W., Gassend, B., Suh, E.G., van Dijk, M., Devadas, S.: Extracting secret keys from integrated circuits. IEEE Transactions on Very Large Scale Integration (VLSI) Systems 13(10), 1200–1205 (2005)
18. Lin, L., Holcomb, D., Krishnappa, D.K., Shabadi, P., Burleson, W.: Low-power sub-threshold design of secure physical unclonable functions. In: International Symposium on Low-Power Electronics and Design (ISLPED), pp. 43–48. IEEE (August 2010)
19. Maes, R., Tuyls, P., Verbauwhede, I.: Intrinsic PUFs from flip-flops on reconfigurable devices (November 2008)
20. Maes, R., Verbauwhede, I.: Physically unclonable functions: A study on the state of the art and future research directions. In: Towards Hardware-Intrinsic Security. Information Security and Cryptography, pp. 3–37. Springer, Heidelberg (2010)
21. Maiti, A., Casarona, J., McHale, L., Schaumont, P.: A large scale characterization of RO-PUF. In: International Symposium on Hardware-Oriented Security and Trust (HOST), pp. 94–99. IEEE (June 2010)
22. Majzoobi, M., Koushanfar, F., Potkonjak, M.: Testing techniques for hardware security. In: International Test Conference (ITC), pp. 1–10. IEEE (October 2008)
23. Marsaglia, G.: The Marsaglia random number CDROM including the diehard battery of tests of randomness, http://www.stat.fsu.edu/pub/diehard/
24. Öztürk, E., Hammouri, G., Sunar, B.: Towards robust low cost authentication for pervasive devices. In: International Conference on Pervasive Computing and Communications (PerCom), pp. 170–178. IEEE, Washington, DC (2008)
25. Pappu, R., Recht, B., Taylor, J., Gershenfeld, N.: Physical One-Way functions. Science 297(5589), 2026–2030 (2002)
26. Rührmair, U., Sehnke, F., Sölter, J., Dror, G., Devadas, S., Schmidhuber, J.: Modeling attacks on physical unclonable functions. In: ACM Conference on Computer and Communications Security (ACM CCS), pp. 237–249. ACM, New York (2010)
27. Rukhin, A., Soto, J., Nechvatal, J., Smid, M., Barker, E., Leigh, S., Levenson, M., Vangel, M., Banks, D., Heckert, A., Dray, J., Vo, S.: A statistical test suite for random and pseudorandom number generators for cryptographic applications. Special Publication 800-22 Revision 1a, NIST (April 2010)
28. Sadeghi, A.R., Visconti, I., Wachsmann, C.: Enhancing RFID security and privacy by physically unclonable functions. In: Towards Hardware-Intrinsic Security. Information Security and Cryptography, pp. 281–305. Springer, Heidelberg (2010)

29. Schulz, S., Sadeghi, A.R., Wachsmann, C.: Short paper: Lightweight remote attestation using physical functions. In: Proceedings of the Fourth ACM Conference on Wireless Network Security (ACM WiSec), pp. 109–114. ACM, New York (2011)
30. Škorić, B., Maubach, S., Kevenaar, T., Tuyls, P.: Information-theoretic analysis of capacitive physical unclonable functions. Journal of Applied Physics 100(2) (July 2006)
31. Su, Y., Holleman, J., Otis, B.P.: A digital 1.6 pJ/bit chip identification circuit using process variations. IEEE Journal of Solid-State Circuits 43(1), 69–77 (2008)
32. Suh, E.G., Devadas, S.: Physical unclonable functions for device authentication and secret key generation. In: ACM/IEEE Design Automation Conference (DAC), pp. 9–14. IEEE (June 2007)
33. Tuyls, P., Batina, L.: RFID-Tags for Anti-counterfeiting. In: Pointcheval, D. (ed.) CT-RSA 2006. LNCS, vol. 3860, pp. 115–131. Springer, Heidelberg (2006)
34. Tuyls, P., Škorić, B., Ignatenko, T., Willems, F., Schrijen, G.J.: Entropy estimation for optical PUFs based on Context-Tree weighting methods security with noisy data. In: Security with Noisy Data, pp. 217–233. Springer, London (2007)
35. Tuyls, P., Škorić, B., Stallinga, S., Akkermans, A.H.M., Ophey, W.: Information-Theoretic Security Analysis of Physical Uncloneable Functions. In: Patrick, A.S., Yung, M. (eds.) FC 2005. LNCS, vol. 3570, pp. 141–155. Springer, Heidelberg (2005)
36. Verayo, Inc.: Product webpage (November 2011), http://www.verayo.com/product/products.html
37. Škorić, B., Tuyls, P., Ophey, W.: Robust Key Extraction from Physical Uncloneable Functions. In: Ioannidis, J., Keromytis, A.D., Yung, M. (eds.) ACNS 2005. LNCS, vol. 3531, pp. 407–422. Springer, Heidelberg (2005)
38. Willems, F.M.J.: CTW website, http://www.ele.tue.nl/ctw/
39. Willems, F.M.J., Shtarkov, Y.M., Tjalkens, T.J.: The context-tree weighting method: basic properties. IEEE Transactions on Information Theory 41(3), 653–664 (1995)

PUFKY: A Fully Functional PUF-Based Cryptographic Key Generator

Roel Maes, Anthony Van Herrewege, and Ingrid Verbauwhede

KU Leuven Dept. Electrical Engineering-ESAT/SCD-COSIC and IBBT
Kasteelpark Arenberg 10, B-3001 Leuven-Heverlee, Belgium
{roel.maes,anthony.vanherrewege,ingrid.verbauwhede}@esat.kuleuven.be

Abstract. We present PUFKY: a practical and modular design for a cryptographic key generator based on a Physically Unclonable Function (PUF). A fully functional reference implementation is developed and successfully evaluated on a substantial set of FPGA devices. It uses a highly optimized ring oscillator PUF (ROPUF) design, producing responses with up to 99% entropy. A very high key reliability is guaranteed by a syndrome construction secure sketch using an efficient and extremely low-overhead BCH decoder. This first complete implementation of a PUF-based key generator, including a PUF, a BCH decoder and a cryptographic entropy accumulator, utilizes merely 17% (1162 slices) of the available resources on a low-end FPGA, of which 82% are occupied by the ROPUF and only 18% by the key generation logic. PUFKY is able to produce a cryptographically secure 128-bit key with a failure rate $< 10^{-9}$ in 5.62 ms. The design's modularity allows for rapid and scalable adaptations for other PUF implementations or for alternative key requirements. The presented PUFKY core is immediately deployable in an embedded system, e.g. by connecting it to an embedded microcontroller through a convenient bus interface.

Keywords: Physically Unclonable Functions (PUFs), Cryptographic Key Generation, Fuzzy Extractors.

1 Introduction

An indispensable premise for the majority of cryptographic implementations is the ability to securely generate, store and retrieve keys. The required effort to meet these conditions is often underestimated in the algorithmic description of cryptographic primitives. The minimal common requirements for a secure key generation and storage are *i)* a source of true randomness that ensures unpredictable and unique fresh keys, and *ii)* a protected memory which reliably stores the key's information while shielding it completely from unauthorized parties. From an implementation perspective, both requisites are non-trivial to achieve. The need for unpredictable randomness is typically filled by applying a seeded pseudo-random bit generator (PRNG). However, the fact that such generators are difficult to implement properly was just recently made clear again by the observation [13] that a large collection of "random" public RSA keys contains many

E. Prouff and P. Schaumont (Eds.): CHES 2012, LNCS 7428, pp. 302–319, 2012.

pairs which share a prime factor, which is immediately exploitable. Implementing a protected memory is also a considerable design challenge, often leading to increased implementation overhead and restricted application possibilities, to enforce the physical security of the stored key. Countless examples can be provided of broken cryptosystems due to poorly designed or implemented key storages, or bad handling of keys. Moreover, even high-level physical protection mechanisms are often not sufficient to prevent well-equipped and motivated adversaries from discovering stored secrets [24, 25].

PUF-based key generators try to tackle both requirements at once by harvesting static, device-unique randomness and processing it into a cryptographic key. This avoids the need for both a PRNG, since the randomness is already intrinsically present in the device, and the need for a protected non-volatile memory, since the used randomness is static over the lifetime of the device and can be measured again and again to regenerate the same key from otherwise illegible random features. Since PUF responses are generally noisy and of low-entropy, a PUF-based key generator faces two main challenges: increasing the reliability to a practically acceptable level and compressing sufficient entropy in a fixed length key. Fuzzy extractors [7] perform exactly these two functions and can be immediately applied for this purpose, as suggested in a number of earlier PUF key generator proposals. In [10], Guajardo et al. propose to use an SRAM PUF for generating keys, using a fuzzy extractor configuration based on linear block codes. This idea was extended and optimized by Bösch et al. [4] who propose a concatenated block code configuration, and Maes et al. [14] who propose to use a soft-decision decoder. Yu et al. [28] propose a configuration based on ring oscillator PUFs and apply an alternative error-correction method.

Contribution. Our main contribution is a highly practical PUF-based cryptographic key generator design (PUFKY), and an efficient yet fully functional FPGA reference implementation thereof. The proposed design comprises a number of major contributions based on new insights: *i)* we propose a novel variant of a ring oscillator PUF based on very efficient Lehmer-Gray order encoding; *ii)* we abandon the requirement of information-theoretical security in favor of a much more practical yet still cryptographically strong key generation; *iii)* we counter the widespread belief that code-based error-correction, BCH decoding in particular, is too complex for efficient PUF-based key generation, by designing a highly resource-optimized BCH decoder; and *iv)* we present a global optimization strategy for PUF-based key generators based on well-defined design constraints.

Structure. In Section 2 we provide necessary background information on the individual elements of the proposed key generator. Section 3 describes the design stage, putting all these elements together in the PUFKY architecture and Section 4 provides concrete results on an optimized reference implementation of the proposed PUF and the full PUFKY design. In Section 5, we discuss some interesting details of our design and hint at possible future improvements and applications. Finally, we conclude in Section 6.

2 Background

2.1 Notation

We briefly introduce the notational conventions used throughout this work. A random variable is denoted by a capital letter X and a particular outcome thereof by a lower case letter x. A vector of length n is written as $X^n = (X_1, \ldots, X_n)$ and $\mathrm{HW}(X^n)$ is the Hamming weight of X^n. A matrix is represented by a bold faced symbol $\mathbf{A}$. $H(X)$ is the Shannon entropy of the random variable X and $H_\infty(X)$ is its min-entropy. For a random binary vector $X^n \in \{0,1\}^n$, we respectively define $R(X^n) \equiv \frac{H(X^n)}{n}$ and $R_\infty(X^n) \equiv \frac{H_\infty(X^n)}{n}$. By $B_{n,p}(t)$ we denote the binomial cumulative distribution function with parameters n and p evaluated in t, and $B_{n,p}^{-1}(q)$ is its inverse. By $\mathcal{C}(n, k, t)$ we denote a binary block code of length n, dimension k and minimal distance $2t + 1$ which is hence able to correct up to t bit errors. When $\mathcal{C}(n, k, t)$ is linear it is defined by a generator and a parity-check matrix, respectively denoted by $\mathbf{G}^{k \times n}$ and $\mathbf{H}^{n-k \times n}$, satisfying the property $\mathbf{GH}^T = \mathbf{0}$.

2.2 Physically Unclonable Functions (PUFs)

PUFs are hardware primitives which produce unpredictable and instantiation-dependent outcomes. A silicon PUF is implemented on a silicon chip and uses the intrinsic device randomness caused by chip manufacturing process variations to generate a device-unique response. Due to their physical nature, PUF responses are generally not perfectly reproducible (noisy) and not perfectly random. If we consider the response of a particular PUF instance as a binary vector X^n, the unreliability is expressed by the expected bit error rate between two evaluations x^n and x'^n of the same response: $\Pr(x_i \neq x_i')$. The entropy density $R(X^n)$ of a response expresses its relative amount of randomness. We will refer to a PUF with a maximal bit error rate p_e and an entropy density of at least ρ as a (p_e, ρ)-PUF.

A Ring Oscillator PUF (ROPUF) is a silicon PUF which generates a response based on the frequencies of on-chip digital ring oscillators. Since the exact frequency of a such oscillators is noticeably affected by process variations, an accurate measurement thereof will contain unpredictable and device-unique information. The first concept of a ROPUF was proposed by Gassend et al. [9], based on a single configurable oscillator. Concerns about predictability and robustness led to the proposal of an improved ROPUF structure by Suh and Devadas [23], which uses a number of fixed oscillators and considers the relative frequencies of oscillator pairs instead of their absolute values. Yin and Qu [27] further explored this technique by considering the frequency ordering of larger groups of oscillators which is able to produce longer bit responses. Maiti et al. [15] performed an extensive characterization of ROPUFs on a large FPGA population, justifying their qualities as silicon PUFs.

2.3 Secure Sketching

The notion of a secure sketch was proposed by Dodis et al. [7] and provides a method to reliably reconstruct the outcome of a noisy variable in such a way that the entropy of the outcome remains high. A number of possible constructions based on error-correcting codes was also proposed in [7]. In this work, we will focus on the *syndrome construction* for binary vectors.

We describe the operation of a syndrome construction secure sketch which uses a binary linear block code $\mathcal{C}(n, k, t)$ with parity-check matrix $\mathbf{H}$. The *sketch procedure* takes as input an outcome of $X^n \to x^n$ and produces a *sketch* $h^{n-k} = x^n \mathbf{H}^T$. The *recovery procedure* takes as input a different (possibly noisy) outcome of $X^n \to x'^n (= x^n \oplus e^n$ with e^n a bit error vector) and the previously generated sketch h^{n-k}, and calculates the *syndrome* $s^{n-k} = x'^n \mathbf{H}^T \oplus h^{n-k}$. Because of the linearity of the code, it is easy to show that $s^{n-k} \equiv e^n \mathbf{H}^T$. If $\mathrm{HW}(e^n) \le t$ then e^n can be decoded from s^{n-k}, which is equivalent to a decoding operation for $\mathcal{C}(n, k, t)$, and x^n can be recovered as $x^n = x'^n \oplus e^n$.

The sketch h^{n-k} needs to be stored in between sketching and recovering. The key point is that knowledge of h^{n-k} does not fully disclose the entropy of X^n, but at most $n-k$ bits thereof. This means that h^{n-k} can be stored and communicated publicly and there will still be at least $H(X^n) - (n - k)$ bits of entropy left in X^n. In the setting of cryptographic key generation, the term *helper data* is used to refer to such public information which is produced by the initial key extraction and used by subsequent key regenerations.

The design parameters of the syndrome construction are mainly determined by the selection of an appropriate linear block code $\mathcal{C}(n, k, t)$. In order to yield a meaningful secure sketch, $\mathcal{C}(n, k, t)$ needs to meet some constraints determined by the available (p_e, ρ)-PUF and by the required remaining entropy m and reliability $1 - p_{fail}$ of the output of the secure sketch. These constraints are listed in the first column of Table 1. The *practicality* constraint restricts the possible codes to ones for which a practical decoding algorithm exists. The *rate* and *correction* constraints further bound the possible code parameters as a function of the available input (p_e, ρ) and the required output (m, p_{fail}). They respectively express the requirement of not disclosing the full entropy of the PUF through the helper data, and the minimally needed bit error correction capacity in order to meet the required reliability. Bösch et al. [4] demonstrated that code concatenation offers considerable advantages when used in secure sketch constructions. Notably the use of a simple repetition code as an inner code significantly relaxes the design constraints. The parameter constraints for a syndrome construction based on the concatenation of a repetition code $\mathcal{C}_1(n_1, 1, t_1 = \frac{n_1 - 1}{2})$ as an inner code and a second linear block code $\mathcal{C}_2(n_2, k_2, t_2)$ as an outer code, are given in the second column of Table 1.

2.4 BCH Decoding

BCH codes are particularly performant cyclical linear block codes for which efficient error-decoding algorithms exist. A binary BCH code $\mathcal{C}_{BCH}(n_{BCH}, k_{BCH}, t_{BCH})$

Table 1. Parameter constraints for the syndrome construction of secure sketches, depending on the type of code construction used

	$\mathcal{C}(n,k,t)$	$\mathcal{C}_2(n_2,k_2,t_2) \circ \mathcal{C}_1(n_1,1,t_1 = \frac{n_1-1}{2})$
Practicality	$\mathcal{C}(n,k,t)$ is efficiently decodable	$\mathcal{C}_2(n_2,k_2,t_2)$ is efficiently decodable
Rate	$\frac{k}{n} > 1 - \rho$	$\frac{k_2}{n_1 n_2} > 1 - \rho$
Correction	$t \geq B_{n,p_e}^{-1}\left((1-p_{fail})^{\frac{1}{r}}\right),$ with $r = \lceil \frac{m}{k-n(1-\rho)} \rceil$	$t_2 \geq B_{n_2,1-p_e'}^{-1}\left((1-p_{fail})^{\frac{1}{r}}\right),$ with $p_e' = 1 - B_{n_1,p_e}(t_1)$ $r = \lceil \frac{m}{k_2 - n_1 n_2(1-\rho)} \rceil$

is defined for $n_{BCH} = 2^u - 1$, but BCH codes of any code length can be constructed by also considering shortened versions: $\mathcal{C}_{BCH}(n_{BCH} - v, k_{BCH} - v, t_{BCH})$.

Decoding a BCH syndrome into the most-likely bit error vector is typically performed in three steps. First, so called *syndrome evaluations* z_i are calculated by evaluating the *syndrome* s^{n-k} as a polynomial for $\alpha, \ldots, \alpha^{2t_{BCH}}$, with α a generator for $\mathbb{F}_{2^u}$. The next step is using these z_i to generate an error location polynomial Λ. This is generally accomplished with the Berlekamp-Massey (BM) algorithm. First published by Berlekamp [2] and later optimized by Massey [16], this algorithm requires the inversion of an element in $\mathbb{F}_{2^u}$ in each of its $2t_{BCH}$ iterations. In order not to have to do this costly calculation, many authors have come up with modified versions of the algorithm, e.g. [20–22]. However, these are all time-memory tradeoffs of the original inversionless BM algorithm by Burton [5], which we prefer due to its lower storage requirements. Finally, by calculating the roots of Λ, one can find the error vector e^n. This is done with the Chien search algorithm [6] by evaluating Λ for $\alpha, \ldots, \alpha^{t_{BCH}}$. If Λ evaluates to zero for α^i then the corresponding error bit $e_{n_{BCH}-i} = 1$.

2.5   Cryptographic Key Generation

To ensure their unpredictability, cryptographic keys should be generated from a random source. Recommendations for appropriate sources and best practice extraction methods can be found, e.g. in [1, 8, 12], and are used heavily in practical implementations. In addition to these best practice methods, *strong extractors* [18] have been proposed as unconditionally secure extractors of uniform randomness. However they generally induce a large entropy loss, i.e. the output length is much smaller than the entropy of the input, which is undesirable since high-entropy randomness is scarce in most implementations. To generate reliable keys from noisy non-uniform sources like PUFs, Dodis et al. [7] introduced the concept of a *fuzzy extractor*. This is basically a concatenation of a secure sketch, as described in Sect. 2.3, with a strong extractor and is able to generate information-theoretically secure keys. To obtain this very high security level, one still has to make a strong assumption about the min-entropy of the randomness source, which is often impossible. Moreover, due to the use of a strong extractor, large entropy losses need to be taken into account here, which often makes the overall key generation very impractical[1].

Another approach is considered in key generation based on PRNGs seeded from an entropic source, as described in [1, 8, 12]. Such generators obtain their initial internal state by accumulating entropy from a, usually low-quality, entropic source using an entropy accumulation function. In [1, Sect. 10.4], constructions for entropy accumulators based on a generic cryptographic hash function or a block cipher are provided. Kelsey et al. [12] also strongly recommend a cryptographic hash function for this purpose. Following this motivation, we opt for a hash function to accumulate entropy in our design. The amount of data to be accumulated to reach a sufficient entropy level, depends on the (estimated) entropy rate of the considered source. For PRNGs which produce large quantities of output data, the source entropy estimates are usually very conservative. For PUFs, entropy comes at a high implementation cost and being too conservative leads to an excessively large overhead. For this reason we are forced to consider relatively tight estimates on the remaining entropy in a PUF response after secure sketching. On the other hand, the output length of a PUF-based key generator is very limited (a single key) compared to PRNGs. In any case, the total amount of entropy which needs to be accumulated should at least match the length of the generated key.

3   Design

3.1   PUFKY Architecture

The top-level architecture of our PUFKY PUF-based key generator is shown in Fig. 1. As a PUF, we use an ROPUF which produces high-entropy outputs based on the frequency ordering of a selection of ring oscillators, as described in Section 3.2. To account for the bit errors present in the PUF response, we use a secure sketch construction based on the concatenation of two linear block codes, a repetition code $\mathcal{C}_{REP}(n_{REP}, 1, \frac{n_{REP}-1}{2})$ with n_{REP} odd and a BCH code $\mathcal{C}_{BCH}(n_{BCH}, k_{BCH}, t_{BCH})$. The design of the syndrome generation and error decoder blocks used in the secure sketching is described in Section 3.3. To accumulate the remaining entropy after secure sketching, we apply the recently proposed light-weight cryptographic hash function SPONGENT [3].

3.2   ROPUF Design

Our ROPUF design is inspired by the design from Yin and Qu [27] which generates a response based on the frequency ordering of a set of oscillators. A measure of the frequency of an oscillator is obtained by counting the number of oscillations in a

---

[1] In earlier work on PUF-based key generation with fuzzy extractors, e.g. [4, 10, 14], the additional entropy loss by the strong extractor is ignored and the resulting keys can not be considered information-theoretically secure.

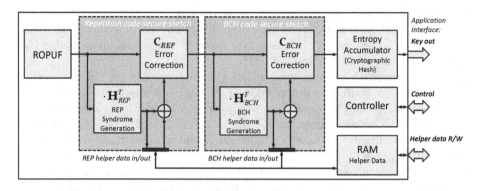

Fig. 1. PUFKY: PUF-based cryptographic key generator architecture

fixed time interval. To amortize the overhead of the frequency counters, oscillators are ordered in b batches of a oscillators sharing a counter. In total, our ROPUF design contains $b \times a$ oscillators of which sets of b can be measured in parallel. The measurement time is determined as a fixed number of cycles of an independent on-chip ring oscillator and is fixed at 87 μs. After some post-processing, an ℓ-bit response is generated based on the relative ordering of b simultaneously measured frequencies. A total of $a \times \ell$-bit responses can be produced by the ROPUF in this manner. Note that, to ensure the independence of different responses, each oscillator is only used for a single response generation. The architecture of our ROPUF design is shown in Fig. 2.

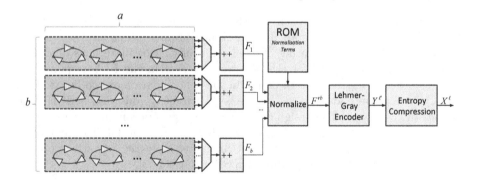

Fig. 2. ROPUF architecture

Encoding the ordering of b frequency measurements $F^b = (F_1, \ldots, F_b)$ in an ℓ-bit response $X^\ell = (X_1, \ldots, X_\ell)$, turns out to be the main design challenge for this type of ROPUF. As discussed in Section 2.3, the *quality* of the PUF responses, expressed by (p_e, ρ), will be decisive for the design constraints of the secure sketch, and by consequence for the key generator as a whole. The details of the post-processing will largely determine the final values for (p_e, ρ). We propose a three-step encoding for $F^b \rightarrow X^\ell$:

1. *Frequency Normalization:* remove structural bias from the measurements.
2. *Order Encoding:* encode the normalized frequency ordering to a stable bit vector in such a way that all ordering entropy is preserved.
3. *Entropy Compression:* compress the order encoding to maximize the entropy density without significantly increasing the bit error probability.

Frequency Normalization. Only a portion of a measured frequency F_i will be random, and only a portion of that randomness will be caused by the effects of process variations on the considered oscillator. The analysis from [15] demonstrates that F_i is subject to both device-dependent and oscillator-dependent *structural* bias. Device-dependent bias does not affect the ordering of oscillators on a single device, so we will not consider it further. Oscillator-dependent structural bias on the other hand is of concern to us since it has a potentially severe impact on the randomness of the frequency ordering. From a probabilistic viewpoint, it is reasonable to assume the frequencies F_i to be independent, but due to the oscillator-dependent structural bias we can not consider them to be *identically* distributed since each F_i has a different expected value μ_{F_i}. The ordering of F_i will be largely determined by the deterministic ordering of μ_{F_i} and not by the effect of random process variations on F_i. Fortunately, we are able to obtain an accurate estimate $\tilde{\mu}_{F_i}$ of μ_{F_i} by averaging F_i over many measurements on many devices. Subtracting this estimate from the measured frequency gives us a *normalized frequency* $F'_i = F_i - \tilde{\mu}_{F_i}$. Assuming $\tilde{\mu}_{F_i} \approx \mu_{F_i}$, the resulting normalized frequencies F'_i will be independent *and* identically distributed (i.i.d.). Calculating $\tilde{\mu}_{F_i}$ needs to be performed only once for a single design after the oscillator implementations are fixed, preferably over an initial test batch of ROPUF instances. When these normalization terms are known with high accuracy, they are included in the design, e.g. using a ROM.

Order Encoding. Sorting a vector F'^b of normalized frequencies, e.g. in ascending order, amounts to rearranging its elements in one of $b!$ possible ways. The goal of the order encoding step is to produce an ℓ'-bit vector $Y^{\ell'}$ which uniquely encodes the ascending order of F'^b. Since the elements of F'^b are i.i.d., each of the $b!$ possible orderings is equally likely to occur [26], leading to $H(Y^{\ell'}) = \log_2 b! = \sum_{i=2}^{b} \log_2 i$. An optimal order encoding has a high entropy density but a minimal sensitivity to noise on the F'_i values. We propose a Lehmer encoding of the frequency ordering, followed by a Gray encoding of the Lehmer coefficients. A Lehmer code is a unique numerical representation of an ordering which is moreover efficient to obtain since it does not require explicit value sorting. It represents the sorted ordering of F'^b as a coefficient vector $L^{b-1} = (L_1, \ldots, L_{b-1})$ with $L_i \in \{0, 1, \ldots, i\}$. It is clear that L^{b-1} can take $2 \times 3 \times \ldots \times b = b!$ possible values which is exactly the number of possible orderings. The Lehmer coefficients are calculated from F'^b as $L_j = \sum_{i=1}^{j} gt(F'_{j+1}, F'_i)$, with $gt(x, y) = 1$ if $x > y$ and

0 otherwise. The Lehmer encoding has the nice property that a minimal change in the sorted ordering caused by two neighboring values swapping places only changes a single Lehmer coefficient by ± 1. Using a binary Gray encoding for the Lehmer coefficients, this translates to only a single bit difference as preferred. The length of the binary representation becomes $\ell' = \sum_{i=2}^{b} \lceil \log_2 i \rceil$ yielding $R\left(Y^{\ell'}\right) = \frac{\sum_{i=2}^{b} \log_2 i}{\sum_{i=2}^{b} \lceil \log_2 i \rceil}$ which is close to optimal.

Entropy Compression. $R\left(Y^{\ell'}\right)$ is already quite high, but can be increased further by compressing it to X^{ℓ} with $\ell \leq \ell'$. Note that $Y^{\ell'}$ is not quite uniform over $\{0,1\}^{\ell'}$ since some bits of $Y^{\ell'}$ are biased and/or dependent. This results from the fact that most of the Lehmer coefficients, although uniform by themselves, can take a range of values which is not an integer power of two, leading to a suboptimal binary encoding. We propose a simple compression by selectively XOR-ing bits from $Y^{\ell'}$ which suffer the most from bias and/or dependencies, leading to an overall increase of the entropy density. Note that XOR-compression potentially also increases the bit error probability, but at most by a factor $\frac{\ell'}{\ell}$.

3.3   Syndrome Generation and Error Decoding for $\mathcal{C}_{REP}$ and $\mathcal{C}_{BCH}$

Repetition Code $\mathcal{C}_{REP}$. The syndrome generation of $x^{n_{REP}}$ consists of pairwise XOR-ing x_1 with each remaining bit of $x^{n_{REP}}$, or $h_i = x_1 \oplus x_{i+1}$. Error decoding is based on a Hamming weight check of the syndrome $s^{n_{REP}-1}$, which immediately yields the value for the first error bit e_1. The remaining error bits are again obtained by a pairwise XOR of e_1 with each of the syndrome bits, but this step is discarded in the syndrome construction. In our design, both syndrome generation and error decoding of a repetition code are fully combinatorial.

BCH Code $\mathcal{C}_{BCH}$. Since BCH codes are cyclical codes, their syndrome generation is a finite field division by the code's generator polynomial. This is efficiently implemented in hardware as an LFSR evaluation of length $(n_{BCH} - k_{BCH})$.

The error decoding step of a BCH code is more complex and requires the largest design effort of all elements in our secure sketch. Most BCH decoders are designed with a focus on throughput and use systolic array designs, e.g. [19, 20, 22]. Aiming for a size-optimized implementation, we propose a serialized, minimalistic coprocessor design with a 10-bit application-specific instruction set and limited conditional execution support. Although highly optimized towards BCH decoding, the architecture is generic in the sense that it can decode any BCH code, including shortened versions, requiring only a slight change of firmware and memory size. The datapath consists of two blocks: an address and a data block. To optimize array indexing, all addressing is done indirectly using a five element address RAM, which is efficiently updated by a dedicated address ALU. The output of the address RAM is directly connected to the data RAM.

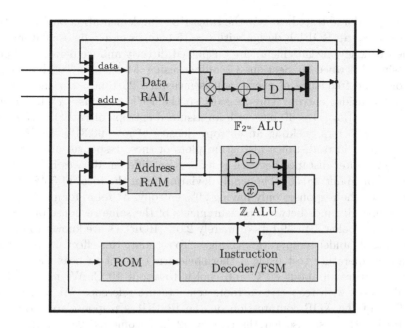

Fig. 3. BCH decoder architecture

The data block consists of data RAM and an ALU which is used mainly for multiply-accumulate operations over $\mathbb{F}_{2^u}$. To minimize the size, this ALU contains only a single register. All other necessary operands come directly from the data RAM. A high-level overview of the coprocessor architecture is shown in Fig. 3.

BCH error decoding is done in the three steps elaborated in Section 2.4. A listing of each used algorithm and their approximate runtimes can be found in Appendix A. The performance of the algorithm execution is heavily optimized using branch removal and loop unrolling. The coprocessor's instruction set can be found in Appendix B.

4 Implementation

We now present the implementation results of our PUFKY design as described in Section 3. The implementation was synthesized, configured and tested on a Xilinx® Spartan®-6 FPGA (XC6SLX45) which is a low-end FPGA in 45 nm technology, specifically targeted for embedded system solutions.

4.1 PUF Implementation and Characterization

We first test our ROPUF implementation separately to obtain its quality parameters (p_e, ρ). This characterization also produces the $\tilde{\mu}_{F_i}$ normalization terms required in the final key generator implementation as detailed in Section 3.2.

We configured and tested exactly the same PUF implementation on 10 identical FPGAs, using an ROPUF design with $b = 16$ batches of $a = 64$ oscillators each.

The frequency measurements are outputted directly and we perform all post-processing described in Section 3.2 offline, using MATLAB[2]. To characterize the noise, the frequency of every loop is measured 25 times. For the moment we don't consider entropy compression, so the PUF response X^ℓ has length $\ell = \ell' = \sum_{i=2}^{b} \lceil \log_2 i \rceil = 49$ bits with an assumed entropy of $H(X^\ell) = H(Y^{\ell'}) = \log_2 b! = 44.25$ bits, yielding an entropy density of $\rho = 90.31\%$. In Fig. 4(a), the inter- and intra-distance histogram plots of these responses are presented. The average inter-distance between responses on different devices is about 23.7 in 49 bits or about 48.4%. The small deviation from the ideal of 50% is representative for the responses only having 90% entropy. At room temperature, the average intra-distance between measurements of the same response on a single device is just below 1 in 49 bits or merely 2.0%. ROPUFs are known to become more unstable under temperature changes. To estimate this effect, we performed a rough temperature test using a thermoelectric element to heat the FPGA's die temperature to about 80°C and cool it to about 10°C. We measured the intra-distances with respect to a room temperature reference. We also studied the effect of the XOR-compression on the ROPUF's response robustness, by compressing the response lengths to $\ell = 42$ (ρ becomes 97.95%) and $\ell = 40$ (ρ becomes 98.78%). Fig. 4(b) shows the effect of both temperature and XOR-compression on the average bit error probability. Heating the FPGA die has the most severe impact on the stability of the ROPUF's responses. As expected, XOR-compression also slightly increases the bit error probability, approximately by a factor $\frac{\ell'}{\ell}$. Taking into account a 2% safety margin on the observed bit error rates, our ROPUF implementation yields a ($p_e = 12\%, \rho = 90.31\%$)-PUF for $\ell = 49$, or a ($13\%, 97.95\%$)-PUF for $\ell = 42$, or a ($14\%, 98.78\%$)-PUF for $\ell = 40$.

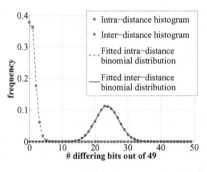

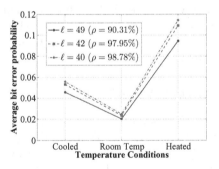

(a) Inter- and intra-distance histogram plots at room temperature, for $\ell = 49$.

(b) Scaling of p_e under heating/cooling and entropy compression to $\ell = 42$ and $\ell = 40$.

Fig. 4. Characterization of our ROPUF implementation

---

[2] In the final PUFKY implementation, all post-processing is done on the device.

4.2 Full Key Generator Implementation

Now we can start optimizing the full PUFKY design according to the constraints as expressed in Section 2.3. The main cost variable for implementation size is the number of required oscillators ($a \times b$), and for performance the number of errors the BCH decoder needs to correct (t_{BCH}). Since we target embedded systems, we aim for an as small as possible implementation at a practically acceptable performance. The optimization parameters depend on our ROPUF, expressed by the triplet (ℓ, p_e, ρ) for which concrete values are provided at the end of Section 4.1, and on the requirements for the generated key, expressed by (m, p_{fail}). For our reference implementation, we aim for a key length $m = 128$ with failure rate $p_{fail} \leq 10^{-9}$. After a thorough exploration of the design space with these parameters, we converge on the following PUFKY reference implementation:

- We select the ($p_e = 13\%, \rho = 97.95\%$)-ROPUF variant with $\ell = 42$, implementing $b = 16$ batches of $a = 53$ oscillators each.
- A secure sketch applying a concatenation of $\mathcal{C}_{REP}(7, 1, 3)$ and $\mathcal{C}_{BCH}(318, 174, 17)$. The repetition block generates 36 bits of helper data for every 42-bit PUF response and outputs 6 bits to the BCH block. The BCH block generates 144 bits of helper data once and feeds 318 bits to the entropy accumulator.
- The ROPUF generates in total $a \times \ell = 2226$ bits containing $a \times \ell \times \rho = 2180.4$ bits of entropy. The total helper data length is $53 \times 36 + 144 = 2052$. The remaining entropy after secure sketching is at least $2180.4 - 2052 = 128.4$ bits which are accumulated in an $m = 128$-bit key by a SPONGENT-128 hash function implementation.

The total size of our PUFKY reference implementation for the considered FPGA platform is 1162 slices, of which 82% is taken up by the ROPUF block. Table 2(a) lists the size of each submodule used in the design. The total time spend to extract the 128-bit key is approximately 5.62 ms (at 54 MHz). Table 2(b) lists the number of cycles spend in each step of the key extraction.

Table 2. Area consumption and runtime of our reference PUFKY implementation on a Xilinx Spartan-6 FPGA. Due to slice compression and glue logic the sum of module sizes is not equal to total size. The PUF runtime is independent of clock speed.

(a) Area consumption		(b) Runtimes	
Module	**Size [slices]**	**Step of extraction**	**Time [cycles]**
ROPUF	952	PUF output	4.59 ms
REP decoder	37	REP decoding	0
BCH syndrome calc.	72	BCH syndrome calc.	511
BCH decoder	112	BCH decoding	50320
SPONGENT-128	22	SPONGENT hashing	3990
helper data RAM	38	control overhead	489
Total	1162	*Total @ 54 MHz*	5.62 ms

5   Discussion

5.1   Some Notes on Security

Our reference PUFKY implementation uses a best-practice entropy accumulation function based on a cryptographically secure hash to generate a key from an amount of entropic data, instead of an information-theoretically secure fuzzy extractor. The large majority of currently existing key generators based on PRNGs also use the best-practice cryptographic approach. We note that, due to the modularity of the PUFKY design, it is possible to obtain an information-theoretically secure extraction with minor replacements: *i)* one needs to consider min-entropy instead of Shannon entropy in all design constraints, *ii)* one needs to replace the entropy accumulation function by a strong extractor, and *iii)* one needs to collect more (min-)entropy than the key length to account for the additional losses induced by the strong extractor. Note that all three changes do come at a rather large implementation overhead, which is the cost one pays for obtaining information-theoretical security.

From a physical security perspective, PUFs and PUF-based key generators can be assumed, like any implementation of a cryptographic primitive, to be vulnerable to side-channel attacks when no appropriate countermeasures are taken, see e.g. [11, 17]. Since our PUFKY reference implementation is a fully functional PUF-based key generator, it is the ideal test subject for side-channel analysis to identify and protect against possible side-channel leakages in a next version. Such analysis is a logical future work which we are considering. In this light, we do want to mention the inherent side-channel resistance of the error decoding blocks in syndrome-construction secure sketches. This results from the fact that no data processed by these blocks contains any information about the PUF output nor about the extracted key, but only about the public syndrome and the error on the PUF output.

5.2   Application Possibilities

The key generated by our PUFKY key generator can basically be used in any conceivable key-based security application. In its current form, the reference implementation produces cryptographically strong 128-bit keys with a failure rate $< 10^{-9}$, but similar implementations for other key parameters (or alternative PUF designs) can be produced rapidly based on our modular PUFKY architecture. Using a PUF-based key offers a number of advantages over traditional key generation, the most noteworthy being: *i)* one does not need protected non-volatile memory to permanently store the key since it can be regenerated at any time, and *ii)* the key is intrinsically bound to a particular platform instantiation which is very useful, e.g. in anticounterfeiting or HW/SW binding applications. We note that both advantages are of particular interest in the context of an FPGA-based embedded system. To demonstrate the ease of integrating a PUFKY implementation in an embedded design, we developed a bus wrapper and a software driver for connecting it to a Xilinx® MicroBlaze® embedded processor. The PUFKY interface then becomes as simple as calling the driver's getKey() function from one's embedded software application.

6 Conclusion

Developing a PUF-based cryptographic key generator is a process involving many parameters, constraints and trade-offs. In this work, we identified and formalized the generic design constraints and integrated them in a practical key generator design. We propose a complete implementation of this design based on a ring-oscillator PUF, a specialized error-correcting BCH decoder and a cryptographic entropy accumulator. Our ring-oscillator PUF produces high-entropy responses (up to 99%) based on actual physical randomness. The proposed BCH decoder design is very efficient and scalable, yet occuppies only a minimal amount of resources. As our implementation results demonstrate, the induced overhead of this BCH decoder in a PUF based key generator is certainly justifiable. Finally, the choice for a cryptographic entropy accumulator, motivated by their wide-spread use in PRNG based key generators, offers a considerable efficiency gain compared to the much more stringent design constraints for information-theoretically secure key extraction. Due to its completeness and efficieny, our PUFKY reference implementation is the first PUF-based key generator to be immediately deployable in an embedded system.

Acknowledgements. This work was supported in part by the Research Council KU Leuven: GOA TENSE (GOA/11/007), by the IAP Programme P6/26 BCRYPT of the Belgian State (Belgian Science Policy) and by the European Commission through the ICT programme under contract ICT-2007-216676 ECRYPT II. In addition, this work was supported by the Flemish Government, FWO G.0550.12N and by the European Commission through the ICT programme under contract FP7-ICT-2011-284833 PUFFIN and FP7-ICT-2007-238811 UNIQUE. Roel Maes is funded by a research grant (073369) of the Institute for the Promotion of Innovation through Science and Technology in Flanders (IWT-Vlaanderen).

References

[1] Barker, E., Kelsey, J.: Recommendation for Random Number Generation Using Deterministic Random Bit Generators. NIST Special Publication 800-90A (January 2012),
http://csrc.nist.gov/publications/nistpubs/800-90A/SP800-90A.pdf

[2] Berlekamp, E.: On Decoding Binary Bose-Chadhuri-Hocquenghem Codes. IEEE Transactions on Information Theory 11(4), 577–579 (1965)

[3] Bogdanov, A., Knežević, M., Leander, G., Toz, D., Varıcı, K., Verbauwhede, I.: SPONGENT: A Lightweight Hash Function. In: Preneel, B., Takagi, T. (eds.) CHES 2011. LNCS, vol. 6917, pp. 312–325. Springer, Heidelberg (2011)

[4] Bösch, C., Guajardo, J., Sadeghi, A.-R., Shokrollahi, J., Tuyls, P.: Efficient Helper Data Key Extractor on FPGAs. In: Oswald, E., Rohatgi, P. (eds.) CHES 2008. LNCS, vol. 5154, pp. 181–197. Springer, Heidelberg (2008)

[5] Burton, H.: Inversionless Decoding of Binary BCH codes. IEEE Transactions on Information Theory 17(4), 464–466 (1971)

[6] Chien, R.: Cyclic Decoding Procedures for Bose-Chaudhuri-Hocquenghem Codes. IEEE Transactions on Information Theory 10(4), 357–363 (1964)

[7] Dodis, Y., Ostrovsky, R., Reyzin, L., Smith, A.: Fuzzy Extractors: How to Generate Strong Keys from Biometrics and Other Noisy Data. SIAM Journal on Computing 38(1), 97–139 (2008)

[8] Eastlake, D., Schiller, J., Crocker, S.: Randomness Requirements for Security. RFC 4086 (Best Current Practice) (June 2005), http://www.ietf.org/rfc/rfc4086.txt

[9] Gassend, B., Clarke, D., van Dijk, M., Devadas, S.: Silicon Physical Random Functions. In: ACM Conference on Computer and Communications Security, pp. 148–160. ACM Press (2002)

[10] Guajardo, J., Kumar, S.S., Schrijen, G.-J., Tuyls, P.: FPGA Intrinsic PUFs and Their Use for IP Protection. In: Paillier, P., Verbauwhede, I. (eds.) CHES 2007. LNCS, vol. 4727, pp. 63–80. Springer, Heidelberg (2007)

[11] Karakoyunlu, D., Sunar, B.: Differential Template Attacks on PUF Enabled Cryptographic Devices. In: 2010 IEEE International Workshop on Information Forensics and Security (WIFS), pp. 1–6 (December 2010)

[12] Kelsey, J., Schneier, B., Ferguson, N.: Yarrow-160: Notes on the Design and Analysis of the Yarrow Cryptographic Pseudorandom Number Generator. In: Heys, H.M., Adams, C.M. (eds.) SAC 1999. LNCS, vol. 1758, pp. 13–33. Springer, Heidelberg (2000)

[13] Lenstra, A.K., Hughes, J.P., Augier, M., Bos, J.W., Kleinjung, T., Wachter, C.: Ron was wrong, Whit is right. Cryptology ePrint Archive, Report 2012/064 (2012), http://eprint.iacr.org/

[14] Maes, R., Tuyls, P., Verbauwhede, I.: Low-Overhead Implementation of a Soft Decision Helper Data Algorithm for SRAM PUFs. In: Clavier, C., Gaj, K. (eds.) CHES 2009. LNCS, vol. 5747, pp. 332–347. Springer, Heidelberg (2009)

[15] Maiti, A., Casarona, J., McHale, L., Schaumont, P.: A Large Scale Characterization of RO-PUF. In: IEEE International Symposium on Hardware-Oriented Security and Trust (HOST), pp. 94–99 (June 2010)

[16] Massey, J.: Shift-Register Synthesis and BCH Decoding. IEEE Transactions on Information Theory 15(1), 122–127 (1969)

[17] Merli, D., Schuster, D., Stumpf, F., Sigl, G.: Side-Channel Analysis of PUFs and Fuzzy Extractors. In: McCune, J.M., Balacheff, B., Perrig, A., Sadeghi, A.-R., Sasse, A., Beres, Y. (eds.) TRUST 2011. LNCS, vol. 6740, pp. 33–47. Springer, Heidelberg (2011)

[18] Nisan, N., Zuckerman, D.: Randomness is Linear in Space. Journal of Computer and System Sciences 52, 43–52 (1996)

[19] Park, J.I., Lee, H., Lee, S.: An Area-Efficient Truncated Inversionless Berlekamp-Massey Architecture for Reed-Solomon Decoders. In: IEEE International Symposium on Circuits and Systems (ISCAS), pp. 2693–2696 (May 2011)

[20] Park, J.I., Lee, K., Choi, C.S., Lee, H.: High-Speed Low-Complexity Reed-Solomon Decoder using Pipelined Berlekamp-Massey Algorithm. In: International SoC Design Conference (ISOCC), pp. 452–455 (November 2009)

[21] Reed, I., Shih, M.: VLSI Design of Inverse-Free Berlekamp-Massey Algorithm. IEEE Proceedings on Computers and Digital Techniques 138(5), 295–298 (1991)

[22] Sarwate, D., Shanbhag, N.: High-Speed Architectures for Reed-Solomon Decoders. IEEE Transactions on Very Large Scale Integration (VLSI) Systems 9(5), 641–655 (2001)

[23] Suh, G.E., Devadas, S.: Physical Unclonable Functions for Device Authentication and Secret Key Generation. In: Design Automation Conference (DAC), pp. 9–14. ACM Press (2007)

[24] Tarnovsky, C.: Deconstructing a 'Secure' Processor. In: Black Hat Federal 2010 (2010)

[25] Torrance, R., James, D.: The State-of-the-Art in IC Reverse Engineering. In: Clavier, C., Gaj, K. (eds.) CHES 2009. LNCS, vol. 5747, pp. 363–381. Springer, Heidelberg (2009)

[26] Wong, K., Chen, S.: The Entropy of Ordered Sequences and Order Statistics. IEEE Transactions on Information Theory 36(2), 276–284 (1990)

[27] Yin, C.E.D., Qu, G.: LISA: Maximizing RO PUF's Secret Extraction. In: IEEE International Symposium on Hardware-Oriented Security and Trust (HOST), pp. 100–105 (June 2010)

[28] Yu, M.-D(M.), M'Raihi, D., Sowell, R., Devadas, S.: Lightweight and Secure PUF Key Storage Using Limits of Machine Learning. In: Preneel, B., Takagi, T. (eds.) CHES 2011. LNCS, vol. 6917, pp. 358–373. Springer, Heidelberg (2011)

A   BCH Decoding Algorithms

Listed below are the three algorithms that we use for BCH decoding. More information on these algorithms and how they are used can be found in Section 2.4. We denote an array A of b elements, with each element in $\mathbb{N}$ as $A[b] \in \mathbb{N}$. Array indices start at 0, unless specifically mentioned otherwise. Both Algorithm 1 and 2 amount to polynomial evaluation, however, in the former, heavy optimization is possible since we know that every coefficient must be either 0 or 1. Algorithm 3 is the same as the one presented in [5], with a few modifications to better fit the architecture of our coprocessor.

Algorithm 1: Syndrome calculation

Input: $s^{n-k}[n-k] \in \mathbb{F}_2$
Output: $z[2t] \in \mathbb{F}_{2^u}$
Data: curArg, evalArg $\in \mathbb{F}_{2^u}$
      $i, j \in \mathbb{N}$
curArg $\leftarrow \alpha$
for $i \leftarrow 0$ **to** $2t - 1$ **do**
   $z[i] \leftarrow 0$
   evalArg $\leftarrow 1$
   for $j \leftarrow 0$ **to** $n - k - 1$ **do**
      if $s^{n-k}[j] = 1$ **then**
         $z[i] \leftarrow z[i] \oplus$ evalArg
      evalArg $\leftarrow$ evalArg $\otimes$ curArg
   curArg $\leftarrow$ curArg $\otimes \alpha$

Algorithm 2: Chien search

Input: $\Lambda[t+1] \in \mathbb{F}_{2^u}$
Output: errorLoc$[n] \in \mathbb{F}_2$
Data: curAlpha, curEval $\in \mathbb{F}_{2^u}$
      $i, j \in \mathbb{N}$
for $i \leftarrow n - 1$ **to** 0 **do**
   curEval $\leftarrow \Lambda[0]$
   curAlpha $\leftarrow \alpha$
   for $j \leftarrow 1$ **to** t **do**
      $\Lambda[j] \leftarrow \Lambda[j] \otimes$ curAlpha
      curEval $\leftarrow$ curEval $\oplus \Lambda[j]$
      curAlpha $\leftarrow$ curAlpha $\otimes \alpha$
   if curEval $= 0$ **then**
      errorLoc$[i] \leftarrow 1$
   else
      errorLoc$[i] \leftarrow 0$

Algorithm 3: Inversionless Berlekamp-Massey

Input: $z[2t] \in \mathbb{F}_{2^u}$
Output: $\Lambda[t+1] \in \mathbb{F}_{2^u}$
Data: b$[t+2], \delta, \gamma \in \mathbb{F}_{2^u}$; flag $\in \mathbb{F}_2$; k $\in \mathbb{Z}$; $i, j \in \mathbb{N}$
b$[-1] \leftarrow 0$
b$[0] \leftarrow 1$
$\Lambda[0] \leftarrow 1$
for $i \leftarrow 1$ **to** t **do**
   b$[i] \leftarrow 0$
   $\Lambda[i] \leftarrow 0$
$\gamma \leftarrow 1$
k $\leftarrow 0$
for $i \leftarrow 0$ **to** $2t - 1$ **do**
   $\delta \leftarrow 0$
   for $j \leftarrow 0$ **to** $\min(i, t)$ **do**
      $\delta \leftarrow \delta \oplus (z[i-j] \otimes \Lambda[j])$
   flag $\leftarrow (\delta \neq 0)$ & $(k \geq 0)$
   if flag $= 1$ **then**
      for $j \leftarrow t$ **to** 0 **do**
         b$[j] \leftarrow \Lambda[j]$
         $\Lambda[j] \leftarrow (\Lambda[j] \otimes \gamma) \oplus (b[j-1] \otimes \delta)$
      $\gamma \leftarrow \delta$
      k $\leftarrow -$k$ - 1$
   else
      for $j \leftarrow t$ **to** 0 **do**
         b$[j] \leftarrow$ b$[j-1]$
         $\Lambda[j] \leftarrow (\Lambda[j] \otimes \gamma) \oplus (b[j-1] \otimes \delta)$
      k $\leftarrow$ k$ + 1$

Table 3 lists formulas for the ideal and actual runtime of each algorithm. We define the ideal algorithm runtime as the total number of (not unrolled) loop iterations. Note that runtime is mainly determined by t_{BCH} in all three algorithms.

Our obtained runtimes of the syndrome and error-location calculation are particularly efficient requiring only 3–5 cycles per loop iteration with well chosen parameters for $\mathcal{C}_{BCH}$.

Table 3. The ideal and actual runtimes for the BCH decoding algorithms. The formulas for actual runtime are highest order approximations.

Algorithm	Runtime [cycles]	
	Ideal	Actual (approx.)
Syndrome calculation	$2t_{BCH} \cdot (n_{BCH} - k_{BCH})$	$40t_{BCH} \cdot \lceil \frac{n_{BCH} - k_{BCH}}{u} \rceil$
Berlekamp-Massey	$3.5 \cdot (t_{BCH}^2 + t_{BCH})$	$36t_{BCH}^2$
Error loc. calculation	$n_{BCH} \cdot t_{BCH}$	$3.6n_{BCH} \cdot t_{BCH}$

B BCH Decoder Instruction Set

Table 4 gives an overview of the instructions implemented on the BCH decoding coprocessor, their result and the number of cycles needed to execute each instruction.

Table 4. Instruction set of the BCH decoding coprocessor

Opcode	Result	Cycles
jump	$PC \leftarrow value$	2
cmp_jump	$PC \leftarrow value$ if $(comp = \mathbf{true})$	3
stop	$PC \leftarrow PC$	1
comp	$cond_i \leftarrow (comp = \mathbf{true})$	2
set_cond	$cond_i \leftarrow value$	1
load_reg	$reg \leftarrow data[addr_i]$	1
load_fixed_reg	$reg \leftarrow value$	2
load_fixed_addr	$addr_i \leftarrow value$	2
mod_addr	$addr_i \leftarrow f(addr_i)$	1
copy_addr	$addr_i \leftarrow addr_j$	1
store_reg	$data[addr_i] \leftarrow reg$	1
store_fixed	$data[addr_i] \leftarrow value$	2
rotr	$data[addr_i] \leftarrow data[addr_i] \circlearrowright 1$	1
shiftl_clr	$data[addr_i] \leftarrow data[addr_i] \ll 1$	1
shiftl_set	$data[addr_i] \leftarrow (data[addr_i] \ll 1) \mid 1$	1
gf2_add_mult	$data[addr_i] \leftarrow data[addr_i] \otimes data[addr_j]$	1
	$reg \leftarrow reg \oplus (data[addr_i] \otimes data[addr_j])$	

NEON Crypto

Daniel J. Bernstein[1] and Peter Schwabe[2]

[1] Department of Computer Science
University of Illinois at Chicago, Chicago, IL 60607–7045, USA
djb@cr.yp.to
[2] Research Center for Information Technology Innovation
and Institute of Information Science
Academia Sinica, 128 Section 2 Academia Road, Taipei 115-29, Taiwan
peter@cryptojedi.org

Abstract. NEON is a vector instruction set included in a large fraction of new ARM-based tablets and smartphones. This paper shows that NEON supports high-security cryptography at surprisingly high speeds; normally data arrives at lower speeds, giving the CPU time to handle tasks other than cryptography. In particular, this paper explains how to use a single 800MHz Cortex A8 core to compute the existing NaCl suite of high-security cryptographic primitives at the following speeds: 5.60 cycles per byte (1.14 Gbps) to encrypt using a shared secret key, 2.30 cycles per byte (2.78 Gbps) to authenticate using a shared secret key, 527102 cycles (1517/second) to compute a shared secret key for a new public key, 624846 cycles (1280/second) to verify a signature, and 244655 cycles (3269/second) to sign a message. These speeds make no use of secret branches and no use of secret memory addresses.

Keywords: vectorization-friendly cryptographic primitives, efficient software implementations, smartphones, tablets, there be dragons.

1 Introduction

The Apple A4 CPU used in the iPad 1 (2010, 1GHz) and iPhone 4 (2010, 1GHz) contains a single Cortex A8 CPU core. The same CPU core also appears in many other tablets and smartphones. The point of this paper is that the Cortex A8 achieves impressive speeds for high-security cryptography:

- 5.60 cycles per byte to encrypt a message using a shared secret key;
- 2.30 cycles per byte to authenticate a message using a shared secret key;
- 527102 cycles to compute a shared secret key for a new public key;
- 624846 cycles to verify a signature on a short message; and
- 244655 cycles to sign a short message.

This work was supported by the National Science Foundation under grant 1018836; and by the European Commission under Contract ICT-2007-216676 ECRYPT II. Permanent ID of this document: 9b53e3cd38944dcc8baf4753eeb1c5e7. Date: 2012.06.19.

E. Prouff and P. Schaumont (Eds.): CHES 2012, LNCS 7428, pp. 320–339, 2012.

We do not claim that *all* high-security cryptographic primitives run well on the Cortex A8. Quite the opposite: we rely critically on a synergy between

- the capabilities of the "NEON" vector unit in the Cortex A8 and
- the parallelizability of some carefully selected cryptographic primitives.

The primitives we use are Salsa20 [9], a member of the final portfolio from the ECRYPT Stream Cipher Project; Poly1305 [5], a polynomial-evaluation message-authentication code similar to the message-authentication code in GCM; Curve25519 [6], an elliptic-curve Diffie–Hellman system; and Ed25519 [10], an elliptic-curve signature system that was introduced at CHES 2011. The rest of this paper explains how we use NEON to obtain such high speeds for these primitives.

It is not a coincidence that our selection matches the default primitives in NaCl, the existing "Networking and Cryptography library" [13] used in applications such as DNSCrypt [45]; vectorizability was one of the design criteria for NaCl. It is nevertheless surprising that a rather small vector unit, carrying out just one arithmetic instruction per cycle, can run these primitives at the speeds listed above. A high-power Intel Core 2 CPU core (at 45nm, like the Apple A4), with a 64-bit instruction set and three full 128-bit vector units, has cycle counts of 3.98/byte, 3.32/byte, 307053, 365742, and 106542 for the same five tasks with the best reported assembly-language implementations of the same primitives in the SUPERCOP benchmarking suite [12]; the Cortex A8 ends up much more competitive than one might expect. We also do better than the 697080 Cell cycles for Curve25519 achieved in [17], even though the Cell has more powerful permutation instructions and many more registers.

Side Channels. All memory addresses and branch conditions in our software are public, depending only on message lengths. There is no data flow from secret data (keys, plaintext, etc.) to cache timing, branch timing, etc. We do not claim that our software is immune to *hardware* side-channel attacks such as power analysis, but we do claim that it is immune to *software* side-channel attacks such as [44], [2], and [47].

Benchmarking Platform. The speeds reported above were measured on a low-cost Hercules eCAFE netbook (released and purchased in 2011) containing a Freescale i.MX515 CPU. This CPU has a single 800MHz Cortex A8 core. The same machine is also visible in the SUPERCOP benchmarks as h4mx515e. Occasionally we make comparisons to benchmarks that use OpenSSL or a C compiler; the netbook is shipped with Ubuntu 10.04, and in particular OpenSSL 0.9.8k and gcc 4.4.3, neither of which we claim is optimal.

All of our software has been checked against standard test suites. We are placing our software online to maximize verifiability of our results, and are placing it into the public domain to maximize reusability. Some of our preliminary results are already online and included in various public benchmark reports, but this paper is our first formal announcement and achieves even better speeds.

More CPUs with NEON. The Cortex A8 is not the only hardware design supporting the NEON instruction set. The Apple A5 CPU used in the iPad 2

(2011, 1GHz) and iPhone 4S (2011, 800MHz) contains two Cortex A9 CPU cores with NEON units. The NVIDIA Tegra 3 CPU used in the 2011 Asus Eee Pad Transformer Prime tablet (2011, 1.3GHz) and HTC One X smartphone (2012, 1.5GHz) contains four Cortex A9 CPU cores with NEON units. Qualcomm's "Snapdragon" series of CPUs reportedly includes a different NEON microarchitecture for the older "Scorpion" cores and a faster NEON microarchitecture for the newer "Krait" cores.

We have very recently benchmarked our software on a Scorpion, obtaining cycle counts of 5.40/byte, 1.89/byte, 606824, 756795, and 511123 for the five tasks listed above. We expect that further optimization for Cortex A9 and Snapdragon will produce even better results. The rest of this paper focuses on the original Cortex A8 NEON microarchitecture.

One should not think that *all* tablets and smartphones support NEON instructions. For example, NVIDIA omitted NEON from the Cortex A9 cores in the Tegra 2; lower-cost ARM11 processors do not support NEON and continue to appear in new devices; and some devices use Intel processors with a quite different instruction set. However, Apple alone has sold more than 50 million tablets with NEON and many more smartphones with NEON, and our sampling suggests that NEON also appears in the majority of new tablets and smartphones from other manufacturers. This paper turns all of these devices into powerful cryptographic engines, capable of protecting large volumes of data while leaving the CPU with enough time to actually do something useful with that data.

2   NEON Instructions and Speeds

This section reviews NEON's capabilities. This is not a comprehensive review: it focuses on the most important instructions for our software, and the main bottlenecks in those instructions. All comments about speed refer to the NEON unit in a single Cortex A8 core.

Registers. The NEON architecture has 16 128-bit vector registers (2048 bits overall), q0 through q15. It also has 32 64-bit vector registers, d0 through d31, but these registers share physical space with the 128-bit vector registers: q0 is the concatenation of d0 and d1, q1 is the concatenation of d2 and d3, etc.

For comparison, the basic ARM architecture has only 16 32-bit registers, r0 through r15. Register r13 is the stack pointer and register r15 is the program counter, leaving only 14 32-bit registers (448 bits overall) for general use. One of the most obvious benefits of NEON for cryptography is that it provides much more space in registers, reducing the number of loads and stores that we need.

Syntax. We rarely look at NEON register names, even though we write code in assembly: we use a higher-level assembly syntax that allows any number of names for 128-bit vector registers. For example, we write

    diag3 ^= b0

and then an automatic translator produces traditional assembly language

    veor q6,q6,q14

for assembly by the standard GNU assembler gas; here the translator has selected q6 for diag3 and q14 for b0. We nevertheless pay close attention to the number of "live" 128-bit registers at each moment, reorganizing our computations to fit reasonably large amounts of work into registers.

The syntax is our own design. To build the translator we reused the existing qhasm toolkit [7] and wrote a short ARM+NEON machine-description file for qhasm. This file contains, for example, the line

```
4x r=s+t:>r=reg128:<s=reg128:<t=reg128:asm/vadd.i32 >r,<s,<t:
```

stating our syntax and the gas assembly-language syntax for a 4-way vectorized 32-bit addition, and also identifying the inputs and outputs of the instruction for the qhasm register allocator. The code examples in the rest of this paper use our syntax for the sake of readability; we do not assume that readers are already familiar with NEON.

We have also experimented extensively with writing NEON code in C, using compiler extensions for NEON instructions. However, we have found that assembly language gives us far better tradeoffs between software speed and programming effort. Assembly language has a reputation for being hard to read and write, but typical code such as

```
4x a0 = diag1 + diag0
4x b0 = a0 << 7
```

in our assembly-language syntax is as straightforward as

```
a0 = diag1 + diag0;
b0 = vshlq_n_u32(a0,7);
```

in C. The critical advantage of assembly language is that it provides more control. We frequently find that every available C compiler produces poorly scheduled code, leaving the NEON unit mostly idle; changing the C code to produce better assembly-language scheduling is a hit-and-miss affair, and it is also not clear how the compiler could be modified to do better, since the C language provides no way to express instruction priorities. Writing directly in assembly language eliminates this difficulty, allowing us to focus on higher-level questions of how to decompose larger computations (such as multiplications modulo $2^{255} - 19$) into pieces suitable for vectorization.

Arithmetic Instructions. The Cortex A8 NEON microarchitecture has one 128-bit arithmetic unit. A typical arithmetic instruction such as

```
4x a = b + c
```

occupies the NEON arithmetic unit for one cycle. This instruction partitions the 128-bit output register a into four 32-bit quantities a[0], a[1], a[2], a[3], similarly partitions b and c, and then has the same effect as

```
a[0] = b[0] + c[0]
a[1] = b[1] + c[1]
a[2] = b[2] + c[2]
a[3] = b[3] + c[3]
```

where as usual + means addition modulo 2^{32}. Readers accustomed to two-operand architectures should note that there is no requirement to split this instruction into a copy a = b followed by 4x a += c.

This instruction passes through several single-cycle NEON pipeline stages N1, N2, etc. It reads its input when it is in stage N2; if the input will not be ready then it already predicts the problem at the beginning of the pipeline and stalls there, also stalling subsequent NEON instructions. It makes its output available in stage N4, two cycles after reading the input, so another addition instruction that begins two cycles later (reaching N2 when the first instruction reaches N4) can read the output without stalling.

We comment that "addition has 2-cycle latency" would be an oversimplification, for reasons that will be clear in the next paragraph. We also warn readers that ARM's Cortex A8 manual [3] reports stage N3 for the output, even though an addition that begins the next cycle will in fact stall. This is not an isolated error in the manual, but rather an unusual convention for reporting output availability: ARM consistently lists the stage just *before* the output is ready. An online Cortex A8 cycle counter by Sobole [40] correctly displays this latency, although we encountered some other cases where it was too pessimistic.

A logical instruction such as

```
a = b ^ c
```

has the same performance as an addition. A subtraction instruction

```
4x a = b - c
```

occupies the arithmetic unit for one cycle, just like addition, but needs the c input one cycle earlier, in stage N1. Addition and subtraction thus each have latency 2 as input to an addition or to the positive part of a subtraction, but latency 3 as input to the negative part of a subtraction.

Shifting by a fixed distance is like subtraction in that it needs input in stage N1 and generates output in stage N4. NEON can combine three instructions for rotation into two instructions—

```
4x a = b << 7
4x a insert= b >> 25
```

—but the second instruction occupies the arithmetic unit for two cycles and generally causes larger latency problems than a separate shift and xor.

A pair of 32-bit multiplications, each producing a 64-bit result, uses one instruction:

```
c[0,1] = a[0] signed* b[0]; c[2,3] = a[1] signed* b[1]
```

This instruction occupies the arithmetic unit for two cycles, for a total throughput of one $32 \times 32 \to 64$-bit multiplication per cycle. This instruction reads b in stage N1, reads a in stage N2, and makes c available in stage N8. This instruction has a multiply-accumulate variant, carrying out additions for free:

```
c[0,1] += a[0] signed* b[0]; c[2,3] += a[1] signed* b[1]
```

The accumulator is normally read in stage N3, but is read much later *if* it is the result of a similar multiplication instruction. A typical sequence such as

```
c[0,1] = a[0] unsigned* b[0]; c[2,3] = a[1] unsigned* b[1]
c[0,1] += e[2] unsigned* f[2]; c[2,3] += e[3] unsigned* f[3]
c[0,1] += g[0] unsigned* h[2]; c[2,3] += g[1] unsigned* h[3]
```

takes six cycles without any stalls.

Loads, Stores, and Permutations. There is a 128-bit NEON load/store unit that runs in parallel with the NEON arithmetic unit. An aligned 128-bit or aligned 64-bit load or store consumes the load/store unit for one cycle and makes its result available in N2. Alignment is static (encoded explicitly in the instruction), not dynamic:

```
x01 aligned= mem128[input_1]; input_1 += 16
```

The load/store instruction does not allow an offset from the index register but does allow subsequent increment of the index register by the load amount or by another register. There are separate instructions for an unaligned 128-bit or unaligned 64-bit load or store, for an unaligned 64-bit load or store with an offset, and various other possibilities, each consuming the load/store unit for at least two cycles.

NEON includes a few permutation instructions that consume the load/store unit for one cycle: for example,

```
r = s[1] t[2] r[2,3]
```

takes a single cycle to replace r[0] and r[1] with s[1] and t[2] respectively, leaving r[2] and r[3] unchanged. This instruction reads s and t in stage N1 and writes r in stage N3. There are more permutation instructions that consume the load/store unit for two cycles.

Each NEON cycle dispatches at best one instruction to the arithmetic unit and one instruction to the load/store unit. These two dispatches can occur in either order. For example, a sequence of 6 single-cycle instructions of the form A LS A LS A LS will take 3 NEON cycles (A LS, A LS, A LS); a sequence LS A A LS LS A will take 3 NEON cycles (LS A, A LS, LS A); but a sequence LS LS LS A A A will take 5 NEON cycles (LS, LS, LS A, A, A).

A c-cycle instruction is dispatched in the same way as c adjacent single-cycle instructions. For example, the permutation instruction in

```
4x a2 = diag3 + diag2
   diag3 = diag3[3] diag3[0,1,2]
4x next_a2 = next_diag3 + next_diag2
```

takes two LS cycles, so overall this sequence takes two cycles (A LS, LS A). Occasional permutations thus do not cost any cycles. As another example, one can interleave two-cycle permutations with two-cycle multiplications.

3   Encrypt Using a Shared Secret Key: 5.60 Cycles/Byte for Salsa20

This section explains how to encrypt data with the Salsa20 stream cipher [9] at 5.60 Cortex A8 cycles/byte: e.g., 1.14 Gbps on an 800MHz core. The inner loop uses 4.58 cycles/byte and scales linearly with the number of cipher rounds; for example, Salsa20/12 uses 2.75 cycles/byte for the inner loop and 3.77 cycles/byte for the entire cipher. (These are long-message figures, but the per-message overhead is reasonably small: for example, a 1536-byte message with full Salsa20 uses 5.75 cycles/byte.)

For comparison, [29] reports that a new AES-128-CTR assembly-language implementation, contributed to OpenSSL by Polyakov, runs at 25.4 Cortex A8 cycles per byte (0.25 Gbps at 800MHz). There is no indication that this speed includes protection against software side-channel attacks; in fact, the recent paper [47] by Weiß, Heinz, and Stumpf demonstrated Cortex A8 cache-timing leakage of at least half the AES key bits from OpenSSL and several other AES implementations. We have written our own NEON AES-128-CTR implementation using the bitslicing approach by Käsper and Schwabe [23], protecting against side-channel attacks and at the same time setting a new Cortex A8 speed record of 19.12 cycles/byte (0.33 Gbps at 800MHz), but obviously Salsa20 is much faster.

The eBASC stream-cipher benchmarks [12] report, for Cortex A8, two other ciphers providing comparable long-message speeds: 5.77 cycles/byte for NLS v2 and 7.18 cycles/byte for TPy. NLS v2 is certainly fast, but it is limited to a 128-bit key and 2^{64} bits of output, it relies on S-box lookups that would incur extra cost to protect against cache-timing attacks, and in general it does not appear to have as large a security margin as Salsa20. We see our results as showing that the same speeds can be achieved with higher security. TPy is less competitive: it relies on random access to a large *secret* array, requiring an expensive setup for each nonce (not visible in the long-message timings) and incurring vastly higher costs for protection against cache-timing attacks.

Review of Salsa20; Non-NEON Bottlenecks. Salsa20 expands a 256-bit key and a 64-bit nonce into a long output stream, and xors this stream with the plaintext to produce ciphertext. The stream is generated in 64-byte blocks. The main bottleneck in generating each block is a series of 20 rounds, each consisting of 16 32-bit add-rotate-xor sequences such as the following:

```
s4 = x0 + x12
x4 ^= (s4 >>> 25)
```

This might already seem to be a perfect fit for the basic 32-bit ARM instruction set, without help from NEON. The Cortex A8 has two 32-bit execution units;

addition occupies one unit for one cycle, and rotate-xor occupies one unit for one cycle. One would thus expect 320 add-rotate-xor sequences to occupy both integer execution units for 320 cycles, i.e., 5 cycles per byte.

However, there is a latency of 2 cycles between the two instructions shown above, and an overall latency of 3 cycles between the availability of x0 and the availability of x4. Furthermore, the ARM architecture provides only 14 registers, but Salsa20 needs at least 17 active values: x0 through x15 together with a sum such as s4. (One can overwrite x0 with s4, but only at the expense of extra arithmetic to restore x0 afterwards.) Loads and stores occupy the execution units, taking time away from arithmetic operations. (ARM can merge two loads of adjacent registers into a single instruction, but this instruction consumes both execution units for one cycle and the first execution unit for another cycle.) There are also various overheads outside the 20-round inner loop. Compiling several different C implementations of Salsa20 with many different compiler options did not beat 15 cycles per byte.

Internal Parallelization; Vectorization; NEON Bottlenecks. Each Salsa20 round has 4-way parallelism, with 4 independent add-rotate-xor sequences to carry out at each moment. Two parallel computations hide some latencies but require 8 loads and stores per round with our best instruction schedule; three or four parallel computations would hide all latencies but would require even more loads and stores per round.

NEON has far more space in registers, and its 128-bit arithmetic unit can perform 4 32-bit operations in each cycle. The 4 operations to carry out at each moment in Salsa20 naturally form a 4-way vector operation, at the cost of three 128-bit permutations per round. Salsa20 thus seems to be a natural fit for NEON.

However, NEON rotation consumes 3 operations as discussed in Section 2, so add-rotate-xor consumes 5 operations, at least 1.25 cycles; 5 add-rotate-xor operations per output byte consume at least 6.25 cycles per byte. Furthermore, NEON latencies are even higher than basic ARM latencies. The lowest-latency sequence of instructions for add-rotate-xor is

```
4x a0 = diag1 + diag0
4x b0 = a0 << 7
4x a0 unsigned>>= 25
   diag3 ^= b0
   diag3 ^= a0
```

with total latency 9 to the next addition: the individual latencies are 3 (N4 addition output a0 to N1 shift input), 0 (but carried out the next cycle since the arithmetic unit is busy), 2 (N4 shift output b0 to N2 xor input), 2 (N4 xor output diag3 to N2 xor input), and 2 (N4 xor output diag3 to N2 addition input). A straightforward NEON implementation cannot do better than 11.25 cycles per byte.

External Parallelization. We do better by taking advantage of another level of parallelizability in Salsa20: Salsa20, like AES-CTR, generates output blocks independently as functions of a simple counter. Computing two output blocks in parallel with the following pattern of add-rotate-xor operations—

0	1	2	3	4	5	6	7	8	9	10	11	12	13	14	15	16	17	18	19	20	21
+			<<		>>			^			^		+			<<		>>		^		^
	+			<<		>>			^			^		+			<<		>>		^		^

—hides almost all NEON latencies, reducing our inner loop to 44 cycles per round for both blocks, i.e., 880 cycles for 20 rounds producing 128 bytes, i.e., 6.875 cycles per byte. Computing three output blocks in parallel still fits into NEON registers (with a slightly trickier pattern of operations—the most obvious patterns would need 18 registers), further reducing our inner loop to 6.25 cycles per byte, and alleviates latency issues enough to allow two-instruction rotations, but as far as we can tell this is outweighed by somewhat lower effectiveness of the speedup discussed in the next subsection.

Previous work on Salsa20 for other 128-bit vector architectures had vectorized *across* four output blocks. However, this needs at least 17 active vectors (and more to hide latencies), requiring extra instructions for loads and stores, more than the number of permutation instructions saved. This would also add overhead outside the inner loop and would interfere with the speedup described in the next subsection.

Interleaving ARM with NEON. We do better than 6.25 cycles per byte by using the basic ARM execution units to generate one block while NEON generates two blocks. Each round involves 23 NEON instructions for one block (20 instructions for four add-rotate-xor sequences, plus 3 permutation instructions), 23 NEON instructions for a second block, and 40 ARM instructions for a third block. The extra ARM instructions reduce the inner loop to $(2/3)6.875 \approx 4.58$ cycles per byte: the cycles for the loop are exactly the same but the loop produces $1.5\times$ as much output.

We are pushing this technique extremely close to an important Cortex A8 limit. The limit is that the entire core decodes at most two instructions per cycle, whether the instructions are ARM instructions or NEON instructions. The 880 cycles that we spend for 128 NEON output bytes have 1760 instruction slots, while we use only 920 NEON instructions, leaving 840 free slots; we use 800 of these slots for ARM instructions that generate 64 additional output bytes, and an additional 35 slots for loop control to avoid excessive code size. (Register pressure forced us to spill the loop counter, and each branch instruction has a hidden cost of 3 slots; we ended up unrolling 4 rounds.) Putting even marginally more work on the ARM unit would slow down the NEON processing, and an easy quantitative analysis shows that this would slow down the cipher as a whole.

The same limit makes ARM instructions far less effective for, e.g., the computations modulo $2^{255} - 19$ discussed later in this paper. These computations are large enough that they require many NEON loads and stores alongside arithmetic, often consuming both of the instruction slots available in a cycle. There are still some slots for ARM instructions, but these computations require an even larger number of ARM loads and stores, leaving very few slots for ARM arithmetic instructions. Furthermore, these computations are dominated by multiplications rather than rotations, and even full-speed ARM multiplications have only a fraction of the power of NEON multiplications.

Minimizing Overhead. The above discussion concentrates on the performance of the Salsa20 inner loop, but there are also overheads for initializing and finalizing each block, reading plaintext, and generating ciphertext.

The 64-byte Salsa20 output block consists of four vectors x0 x1 x2 x3, x4 x5 x6 x7, x8 x9 x10 x11, and x12 x13 x14 x15 that must be xor'ed with plaintext to produce ciphertext. NEON uses 0.125 cycles/byte to read potentially unaligned plaintext, and 0.125 cycles/byte to write potentially unaligned ciphertext, for an overhead of 0.25 cycles/byte; ARM is slower. It should be possible to reduce this overhead, at some cost in code size, by overlapping memory access with computation, but we have not yet done this.

The Salsa20 inner loop naturally uses and produces "diagonal" vectors x0 x5 x10 x15, x4 x9 x13 x3, etc. Converting these diagonal vectors to the output vectors x0 x1 x2 x3 etc. poses an interesting challenge for NEON's permutation instructions. We use the following short sequence of instructions (and gratefully acknowledge optimization assistance from Tanja Lange):

```
r0 = ...                          # x0 x5 x10 x15
r4 = ...                          # x4 x9 x14 x3
r12 = ...                         # x12 x1 x6 x11
r8 = ...                          # x8 x13 x2 x7
t4 = r0[1] r4[0] t4[2,3]          # x5 x4 - -
t12 = t12[0,1] r0[3] r4[2]        # - - x15 x14
r0 = (abab & r0) | (~abab & r12)  # x0 x1 x10 x11
t4 = t4[0,1] r8[3] r12[2]         # x5 x4 x7 x6
t12 = r8[1] r12[0] t12[2,3]       # x13 x12 x15 x14
r8 = (abab & r8) | (~abab & r4)   # x8 x9 x2 x3
r4 = t4[1]t4[0]t4[3]t4[2]         # x4 x5 x6 x7
r12 = t12[1]t12[0]t12[3]t12[2]    # x12 x13 x14 x15
r0 r8 = r0[0] r8[1] r8[0] r0[1]   # x0 x1 x2 x3 x8 x9 x10 x11
```

There are 7 single-cycle permutations here, consuming 0.11 cycles/byte, and 2 two-cycle arithmetic instructions (using abab) interleaved with the permutations. Similar comments apply to block initialization. These and other overheads increase the overall encryption costs to 5.60 cycles/byte.

4   Authenticate Using a Shared Secret Key: 2.30 Cycles/Byte for Poly1305

This section explains how to compute the Poly1305 message-authentication code [5] at 2.30 Cortex A8 cycles/byte: e.g., 2.78 Gbps on an 800MHz core. Authenticated encryption with Salsa20 and Poly1305 takes just 7.90 cycles/byte.

For comparison, [29] reports 50 Cortex A8 cycles/byte for AES-GCM and 28.9 cycles/byte for its proposed AES-OCB3; compared to the 25.4 cycles/byte of AES-CTR encryption, authentication adds 25 or 3.5 cycles/byte respectively. GCM, OCB3, and Poly1305 guarantee that attacks are as difficult as breaking the underlying cipher, with similar quantitative security bounds. Another approach, without this guarantee, is HMAC using a hash function; the Cortex A8 speed leaders in the eBASH hash-function benchmarks [12] are MD5 at 6.04 cycles/byte, Edon-R at 9.76 cycles/byte, Shabal at 12.94 cycles/byte, BMW at 13.55 cycles/byte, and Skein at 15.26 cycles/byte.

One of these authentication speeds, the "free" 3.5-cycle/byte authentication in OCB3, is within a factor of 2 of our Poly1305 speed. However, OCB3 also has two important disadvantages. First, OCB3 cannot be combined with a fast stream cipher such as Salsa20—it requires a block cipher, as discussed in [29]. Second, rejecting an OCB3 forgery requires taking the time to decrypt the forgery, a full 28.9 cycles/byte; Poly1305 rejects forgeries an order of magnitude more quickly.

Review of Poly1305. Poly1305 reads a *one-time* 32-byte secret key and a message of any length. It chops the message into 128-bit little-endian integers (and a final b-bit integer with $b \leq 128$), adds 2^{128} to each integer (and 2^b to the final integer) to obtain components $m[0], m[1], \ldots, m[\ell - 1]$, and produces the 16-byte authenticator

$$(((m[0]r^\ell + m[1]r^{\ell-1} + \cdots + m[\ell - 1]r) \bmod 2^{130} - 5) + s) \bmod 2^{128}$$

where r and s are components of the secret key. "One time" has the same meaning as for a one-time pad: each message has a new key. If these one-time keys are truly random then the attacker is reduced to blind guessing; see [5] for quantitative bounds on the attacker's forgery chance. If these keys are instead produced as cipher outputs from a long-term key then security relies on the presumed difficulty of distinguishing the cipher outputs from random.

Readers familiar with the GCM authenticated-encryption mode [32] will recognize that Poly1305 shares the polynomial-evaluation structure of the GMAC authenticator inside GCM. The general structure was introduced by den Boer [18], Johansson, Kabatianskii, and Smeets [24], and independently Taylor [43]; concrete examples include [39], [34], [4], [28], and [27]. But these proposals differ in many details, notably the choice of finite field: a field of size 2^{128} for GCM, for example, and integers modulo $2^{130} - 5$ for Poly1305.

Efficient authentication in software relies primarily on fast multiplication in this field, and secondarily on fast conversion of message bytes into elements of the field. Efficient authentication under a *one-time* key (addressing the security issues discussed in [4, Section 8, Notes], [8, Sections 2.4–2.5], [21], [14], etc.) means that one cannot afford to precompute large tables of multiples of r; we count the costs of all precomputation. Avoiding the possibility of cache-timing attacks means that one cannot use variable-index table lookups; see, e.g., the discussion of GCM security in [23, Section 2.3].

Multiplication mod $2^{130} - 5$ on NEON. We represent an integer f modulo $2^{130} - 5$ in radix 2^{26} as $f_0 + 2^{26}f_1 + 2^{52}f_2 + 2^{78}f_3 + 2^{104}f_4$. At the end of the computation we reduce each f_i below 2^{26}, and reduce f to the interval $\{0, 1, \ldots, 2^{130} - 6\}$, but earlier in the computation we use standard lazy-reduction techniques, allowing wider ranges of f and of f_i.

The most attractive NEON multipliers are the paired 32-bit multipliers, which as discussed in Section 2 produce two 64-bit products every two cycles, including free additions. The product of $f_0 + 2^{26}f_1 + \cdots$ and $g_0 + 2^{26}g_1 + \cdots$ is $h_0 + 2^{26}h_1 + \cdots$ modulo $2^{130} - 5$ where

$$h_0 = f_0 g_0 + 5 f_1 g_4 + 5 f_2 g_3 + 5 f_3 g_2 + 5 f_4 g_1,$$
$$h_1 = f_0 g_1 + \; f_1 g_0 + 5 f_2 g_4 + 5 f_3 g_3 + 5 f_4 g_2,$$
$$h_2 = f_0 g_2 + \; f_1 g_1 + \; f_2 g_0 + 5 f_3 g_4 + 5 f_4 g_3,$$
$$h_3 = f_0 g_3 + \; f_1 g_2 + \; f_2 g_1 + \; f_3 g_0 + 5 f_4 g_4,$$
$$h_4 = f_0 g_4 + \; f_1 g_3 + \; f_2 g_2 + \; f_3 g_1 + \; f_4 g_0,$$

all of which are smaller than $2^{64}/195$ if each f_i and g_i is bounded by 2^{26}. Evidently somewhat larger inputs f_i and g_i, products of sums of inputs, sums of several outputs, etc. do not pose any risk of 64-bit overflow. This computation (performed from right to left to absorb all sums into products) involves 25 generic multiplications and 4 multiplications by 5, but it is better to eliminate the multiplications by 5 in favor of precomputing $5g_1, 5g_2, 5g_3, 5g_4$, in part because those are 32-bit multiplications and in part because a multiplication input is often reused.

Rather than vectorizing within a message block, and having to search for 12 convenient pairs of 32-bit multiplications in the pattern of 25 multiplications shown above, we simply vectorize across two message blocks, using a well-known parallelization of Horner's rule. For example, for $\ell = 10$, we compute

$$((((m[0]r^2 + m[2])r^2 + m[4])r^2 + m[6])r^2 + m[8])r^2$$
$$+ ((((m[1]r^2 + m[3])r^2 + m[5])r^2 + m[7])r^2 + m[9])r$$

by starting with the vector $(m[0], m[1])$, multiplying by the vector (r^2, r^2), adding $(m[2], m[3])$, multiplying by (r^2, r^2), etc. The integer $m[0]$ is actually represented as five 32-bit words, so the vector $(m[0], m[1])$ is actually represented as five vectors of 32-bit words. The 25 multiplications shown above, times two blocks, then trivially use 25 NEON multiplication instructions costing 50 cycles, i.e., 1.5625 cycles per byte. There are, however, also overheads for reading the message and reducing the product, as discussed below.

Reduction. The product obtained above can be safely added to a new message block but must be reduced before it can be used as input to another multiplication. To reduce a large coefficient h_0, we carry $h_0 \to h_1$; this means replacing (h_0, h_1) with $(h_0 \bmod 2^{26}, h_1 + \lfloor h_0/2^{26} \rfloor)$. Similar comments apply to the other coefficients. Carrying $h_4 \to h_0$ means replacing (h_4, h_0) with $(h_4 \bmod 2^{26}, h_0 + 5\lfloor h_4/2^{26} \rfloor)$, again taking advantage of the sparsity of $2^{130} - 5$.

NEON uses 1 cycle for a pair of 64-bit shifts, 1 cycle for a pair of 64-bit masks, and 1 cycle for a pair of 64-bit additions, for a total of 3 cycles for a pair of carries (plus 2 cycles for $h_4 \to h_0$). A chain of six carries $h_0 \to h_1 \to h_2 \to h_3 \to h_4 \to$

$h_0 \to h_1$ is adequate for subsequent multiplications: it leaves h_1 below $2^{26} + 2^{13}$ and each other h_i below 2^{26}. However, each step in this chain has latency at least 5, and even aggressive interleaving of carries into the computations of h_i would eliminate only a few of the resulting idle cycles. We instead carry $h_0 \to h_1$ and $h_3 \to h_4$, then $h_1 \to h_2$ and $h_4 \to h_0$, then $h_2 \to h_3$ and $h_0 \to h_1$, then $h_3 \to h_4$, spending 3 cycles to eliminate latency problems. The selection of initial indices $(0, 3)$ here allows the longer carry $h_4 \to h_0$ to overlap two independent carries $h_1 \to h_2 \to h_3$; we actually interleave $h_0 \to h_1 \to h_2 \to h_3 \to h_4$ with $h_3 \to h_4 \to h_0 \to h_1$, being careful to keep the separate uses of h_i away from each other.

This approach consumes 23 cycles for two blocks, i.e., 0.71875 cycles per byte. As message lengths grow it becomes better to retreat from Horner's method, for example computing

$$((m[0]r^4 + m[2]r^2 + m[4])r^4 + m[6]r^2 + m[8])r^2$$
$$+ ((m[1]r^4 + m[3]r^2 + m[5])r^4 + m[7]r^2 + m[9])r$$

by starting with $(m[0], m[1])$ and $(m[2], m[3])$, multiplying by (r^4, r^4) and (r^2, r^2) respectively, adding, adding $(m[4], m[5])$, *then* reducing, etc. This eliminates half of the reductions at the expense of extending the precomputation from (r^2, r^2) to (r^4, r^4). One can easily eliminate more reductions with more precomputation, but one pays for precomputation linearly in both time and space, while the benefit becomes smaller and smaller.

For comparison, [4, Section 6] precomputed 97 powers of r for a polynomial evaluation in another field. The number 97 was chosen to just barely avoid overflow of sums of 97 intermediate values; [4] did not count the cost of precomputation. Of course, when we report long-message performance figures we blind ourselves to any constant amount of precomputation, but beyond those figures we are also careful to avoid excessive precomputation (and, for similar reasons, excessive code size). We thus settled on eliminating half of the reductions.

Reading the Message. The inner loop in our computation, with half reductions as described above, computes $fr^4 + m[i]r^2 + m[i+2]$. One input is an accumulator f; the output is written on top of f for the next pass through the loop. Two more inputs are r^2 and r^4, both precomputed. The last two inputs are message blocks $m[i]$ and $m[i+2]$; the inner loop loads these blocks and converts them to radix 2^{26}. The following paragraphs discuss the costs of this conversion.

The same computations are carried out in parallel on $m[i+1]$ and $m[i+3]$, using another accumulator. We suppress further mention of this straightforward vectorization: for example, when we say below that NEON takes 0.5 cycles for a 64-bit shift involved in $m[i]$, what we actually mean is that NEON takes 1 cycle for a pair of 64-bit shifts, where the first shift is used for $m[i]$ and the second is used for $m[i+1]$.

Loading $m[i]$ produces a vector (m_0, m_1, m_2, m_3) representing the integer $m_0 + 2^{32}m_1 + 2^{64}m_2 + 2^{96}m_3$. Our goal here is to represent the same integer (plus 2^{128}) in radix 2^{26} as $c_0 + 2^{26}c_1 + 2^{52}c_2 + 2^{78}c_3 + 2^{104}c_4$. A shift of the 64 bits

(m_2, m_3) down by 40 bits produces exactly c_4. A shift of (m_2, m_3) down by 14 bits does not produce exactly c_3, and a shift of (m_1, m_2) down by 20 bits does not produce exactly c_2, but a single 64-bit mask then produces (c_2, c_3). Similar comments apply to (c_0, c_1), except that c_0 does not require a shift.

Overall there are seven 64-bit arithmetic instructions here (four shifts, two masks, and one addition to c_4 to handle the 2^{128}), consuming 3.5 cycles for each 16-byte block. There is also a two-cycle (potentially unaligned) load, along with just six single-cycle permutation instructions; NEON has an arithmetic instruction that combines a 64-bit right shift (by up to 32 bits) with an extraction of the bottom 32 bits of the result, eliminating some 64-bit-to-32-bit shuffling.

The second message block $m[i+2]$ has a different role in $fr^4 + m[i]r^2 + m[i+2]$: it is added to the output rather than the input. We take advantage of this by loading $m[i+2]$ into a vector (m_0, m_1, m_2, m_3) and adding $m_0 + 2^{32}m_1 + 2^{64}m_2 + 2^{96}m_3$ into a multiplication result $h_0 + 2^{26}h_1 + 2^{52}h_2 + 2^{78}h_3 + 2^{104}h_4$ before carrying the result. This means simply adding m_0 into h_0, adding $2^6 m_1$ into h_1, etc. We absorb the additions into multiplications by scheduling $m[i+2]$ before the computation of h. The only remaining costs for $m[i+2]$ are a few shifts such as $2^6 m_1$, one operation to add 2^{128}, and various permutations.

The conversion of $m[i]$ and $m[i+2]$ costs, on average, 0.171875 cycles/byte for arithmetic instructions. Our total cost for NEON arithmetic in Poly1305 is 2.09375 cycles/byte: 1.5625 cycles/byte for one multiplication per block, 0.359375 cycles/byte for half a reduction per block, and 0.171875 cycles/byte for input conversion. We have not yet managed to perfectly schedule the inner loop: right now it takes 147 cycles for 64 bytes, slightly above the 134 cycles of arithmetic, so our software computes Poly1305 at 2.30 cycles/byte.

5   Compute a Shared Secret Key for a New Public Key: 527102 Cycles for Curve25519; Sign and Verify: 244655 and 624846 Cycles for Ed25519

This section explains how to compute the Curve25519 Diffie–Hellman function [6], obtaining a 32-byte shared secret from Alice's 32-byte secret key and Bob's 32-byte public key, in 527102 Cortex A8 cycles: e.g., 1517/second on an 800MHz core. This section also explains how to sign and verify messages in the Ed25519 public-key signature system [10] in, respectively, 244655 and 624846 Cortex A8 cycles: e.g., 3269/second and 1280/second on an 800MHz core. Ed25519 public keys are 32 bytes, and signatures are 64 bytes.

For comparison, openssl speed on the same machine reports

- 424.2 RSA-2048 verifications per second (1.9 million cycles),
- 11.1 RSA-2048 signatures per second (72 million cycles),
- 88.6 NIST P-256 Diffie–Hellman operations per second (9.0 million cycles),
- 388.8 NIST P-256 signatures per second (2.1 million cycles), and
- 74.5 NIST P-256 verifications per second (10.7 million cycles).

Morozov, Tergino, and Schaumont [33] report two speeds for "secp224r1" Diffie–Hellman: 15609 microseconds on a 500MHz Cortex A8 (7.8 million cycles), and 6043 microseconds on a 360MHz DSP (2 million DSP cycles) included in the same CPU, a TI OMAP 3530. Curve25519 and Ed25519 have a higher security level than secp224r1 and 2048-bit RSA; it is also not clear which of the previous speeds include protection against side-channel attacks.

Review of Curve25519 and Ed25519. Curve25519 and Ed25519 are elliptic-curve systems. Key generation is fixed-base-point single-scalar multiplication: Bob's public key is a multiple $B = bP$ of a standard base point P on a standard curve. Bob's secret key is the integer b.

Curve25519's Diffie–Hellman function is variable-base-point single-scalar multiplication: Alice, given Bob's public key B, computes aB where a is Alice's secret key. The secret shared by Alice and Bob is simply a hash of aB; this secret is used, for example, as a long-term key for Salsa20, which in turn is used to generate encryption pads and Poly1305 authentication keys.

Signing in Ed25519 consists primarily of fixed-base-point single-scalar multiplication. (We make the standard assumption that messages are short; hashing time is the bottleneck for very long messages. Our measurements use 59-byte messages, as in [12].) Signing is much faster than Diffie–Hellman: it exploits precomputed multiples of P in various standard ways. Verification in Ed25519 is slower than Diffie–Hellman: it consists primarily of double-scalar multiplication.

The Curve25519 elliptic curve is the Montgomery curve $y^2 = x^3 + 486662x^2 + x$ modulo $2^{255} - 19$, with a unique point of order 2. The Ed25519 elliptic curve is the twisted Edwards curve $-x^2 + y^2 = 1 - (121665/121666)x^2y^2$ modulo $2^{255} - 19$, also with a unique point of order 2. These two curves have an "efficient birational equivalence" and therefore have the same security.

Montgomery curves are well known to allow efficient variable-base-point single-scalar multiplication. Edwards curves are well known to allow a wider variety of efficient elliptic-curve operations, including double-scalar multiplication. These fast scalar-multiplication methods are "complete": they are sequences of additions, multiplications, etc. that always produce the right answer, with no need for comparisons, branches, etc. Completeness was proven by Bernstein [6] for single-scalar multiplication on any Montgomery curve having a unique point of order 2, and by Bernstein and Lange [11] for arbitrary group operations on any Edwards curve having a unique point of order 2.

The main loop in Curve25519, executed 255 times, has four additions of integers modulo $2^{255} - 19$, four subtractions, two conditional swaps (which must be computed with arithmetic rather than branches or variable array lookups), four squarings, one multiplication by the constant 121666, and five generic multiplications. There is also a smaller final loop (a field inversion), consisting of 254 squarings and 11 multiplications. Similar comments apply to Ed25519 signing and Ed25519 verification.

Multiplication mod $2^{255} - 19$ on NEON. We use radix $2^{25.5}$, imitating the floating-point representation in [6, Section 4] but with unscaled integers rather than scaled floating-point numbers: we represent an integer f modulo $2^{255} - 19$ as

$$f_0 + 2^{26} f_1 + 2^{51} f_2 + 2^{77} f_3 + 2^{102} f_4 + 2^{128} f_5 + 2^{153} f_6 + 2^{179} f_7 + 2^{204} f_8 + 2^{230} f_9$$

where, as in Section 4, the allowable ranges of f_i vary through the computation.

We use signed integers f_i rather than unsigned integers: for example, when we carry $f_0 \to f_1$ we reduce f_0 to the range $[-2^{25}, 2^{25}]$ rather than $[0, 2^{26}]$. This complicates carries, replacing a mask with a shift and subtraction, but saves one bit in products of reduced coefficients, allowing us to safely compute various products of sums without carrying the sums. This was unnecessary in the previous section, in part because the 5 in $2^{130} - 5$ is smaller than the 19 in $2^{255} - 19$, in part because 130 is smaller than 255, and in part because the sums of inputs and outputs naturally appearing in the previous section have fewer terms than the sums that appear in these elliptic-curve computations.

The product of $f_0 + 2^{26} f_1 + 2^{51} f_2 + \cdots$ and $g_0 + 2^{26} g_1 + 2^{51} g_2 + \cdots$ is $h_0 + 2^{26} h_1 + 2^{51} h_2 + \cdots$ modulo $2^{255} - 19$ where

$$
\begin{aligned}
h_0 &= f_0 g_0 + 38 f_1 g_9 + 19 f_2 g_8 + 38 f_3 g_7 + 19 f_4 g_6 + 38 f_5 g_5 + 19 f_6 g_4 + 38 f_7 g_3 + 19 f_8 g_2 + 38 f_9 g_1 \\
h_1 &= f_0 g_1 + f_1 g_0 + 19 f_2 g_9 + 19 f_3 g_8 + 19 f_4 g_7 + 19 f_5 g_6 + 19 f_6 g_5 + 19 f_7 g_4 + 19 f_8 g_3 + 19 f_9 g_2 \\
h_2 &= f_0 g_2 + 2 f_1 g_1 + f_2 g_0 + 38 f_3 g_9 + 19 f_4 g_8 + 38 f_5 g_7 + 19 f_6 g_6 + 38 f_7 g_5 + 19 f_8 g_4 + 38 f_9 g_3 \\
h_3 &= f_0 g_3 + f_1 g_2 + f_2 g_1 + f_3 g_0 + 19 f_4 g_9 + 19 f_5 g_8 + 19 f_6 g_7 + 19 f_7 g_6 + 19 f_8 g_5 + 19 f_9 g_4 \\
h_4 &= f_0 g_4 + 2 f_1 g_3 + f_2 g_2 + 2 f_3 g_1 + f_4 g_0 + 38 f_5 g_9 + 19 f_6 g_8 + 38 f_7 g_7 + 19 f_8 g_6 + 38 f_9 g_5 \\
h_5 &= f_0 g_5 + f_1 g_4 + f_2 g_3 + f_3 g_2 + f_4 g_1 + f_5 g_0 + 19 f_6 g_9 + 19 f_7 g_8 + 19 f_8 g_7 + 19 f_9 g_6 \\
h_6 &= f_0 g_6 + 2 f_1 g_5 + f_2 g_4 + 2 f_3 g_3 + f_4 g_2 + 2 f_5 g_1 + f_6 g_0 + 38 f_7 g_9 + 19 f_8 g_8 + 38 f_9 g_7 \\
h_7 &= f_0 g_7 + f_1 g_6 + f_2 g_5 + f_3 g_4 + f_4 g_3 + f_5 g_2 + f_6 g_1 + f_7 g_0 + 19 f_8 g_9 + 19 f_9 g_8 \\
h_8 &= f_0 g_8 + 2 f_1 g_7 + f_2 g_6 + 2 f_3 g_5 + f_4 g_4 + 2 f_5 g_3 + f_6 g_2 + 2 f_7 g_1 + f_8 g_0 + 38 f_9 g_9 \\
h_9 &= f_0 g_9 + f_1 g_8 + f_2 g_7 + f_3 g_6 + f_4 g_5 + f_5 g_4 + f_6 g_3 + f_7 g_2 + f_8 g_1 + f_9 g_0.
\end{aligned}
$$

The extra factors of 2 appear because $2^{25.5}$ is not an integer. We precompute $2f_1, 2f_3, 2f_5, 2f_7, 2f_9$ and $19g_1, 19g_2, \ldots, 19g_9$; each h_i is then a sum of ten products of precomputed quantities.

Most multiplications appear as independent pairs, computing fg and $f'g'$ in parallel, in the elliptic-curve formulas we use. We vectorize across these multiplications: we start from 20 64-bit vectors such as (f_0, f_0') and (g_0, g_0'), precompute 14 64-bit vectors such as $(2f_1, 2f_1')$ and $(19g_1, 19g_1')$, and then accumulate 10 128-bit vectors such as (h_0, h_0'). By scheduling operations carefully we fit these 54 64-bit quantities into the 32 available 64-bit registers with a moderate number of loads and stores.

Some multiplications do not appear as pairs. For those cases we vectorize *within* one multiplication by the following strategy. Accumulate the vectors $(f_0 g_0, 2f_1 g_1)$ and $(19f_2 g_8, 38f_3 g_9)$ and $(19f_4 g_6, 38f_5 g_7)$ and $(19f_6 g_4, 38f_7 g_5)$ and $(19f_8 g_2, 38f_9 g_3)$ into (h_0, h_2); accumulate $(f_0 g_2, 2f_1 g_3)$ etc. into (h_2, h_4); and so on through (h_8, h_0). Also accumulate $(f_1 g_2, 19f_8 g_3)$, $(f_3 g_0, f_0 g_1)$, etc. into (h_3, h_1); accumulate $(f_1 g_4, 19f_8 g_5)$ etc. into (h_5, h_3); and so on through (h_1, h_9). Each vector added here is a product of two of the following 27 precomputed vectors:

- $(f_0, 2f_1), (f_2, 2f_3), (f_4, 2f_5), (f_6, 2f_7), (f_8, 2f_9)$;
- $(f_1, f_8), (f_3, f_0), (f_5, f_2), (f_7, f_4), (f_9, f_6)$;
- $(g_0, g_1), (g_2, g_3), (g_4, g_5), (g_6, g_7)$;
- $(g_0, 19g_1), (g_2, 19g_3), (g_4, 19g_5), (g_6, 19g_7), (g_8, 19g_9)$;
- $(19g_2, 19g_3), (19g_4, 19g_5), (19g_6, 19g_7), (19g_8, 19g_9)$;
- $(19g_2, g_3), (19g_4, g_5), (19g_6, g_7), (19g_8, g_9)$.

We tried several other strategies, pairing inputs and outputs in various ways, before settling on this strategy. All of the other strategies used more precomputed vectors, requiring more loads and stores.

Reduction, Squaring, etc. Reduction follows an analogous strategy to Section 4. One complication is that each carry has an extra operation, as mentioned above. Another complication for vectorizing a single multiplication is that the shift distances are sometimes 26 bits and sometimes 25 bits; we vectorize carrying $(h_0, h_4) \to (h_1, h_5)$, for example, but would not have been able to vectorize carrying $(h_0, h_5) \to (h_1, h_6)$.

For squaring, like multiplication, we vectorize across two independent operations when possible, and otherwise vectorize within one operation. Squarings are serialized in square-root computations (for decompressing short signatures) and in inversions (for converting scalar-multiplication results to affine coordinates), but the critical bottlenecks are elliptic-curve operations, and squarings come in convenient pairs in all of the elliptic-curve formulas that we use.

In the end arithmetic consumes 150 cycles in generic multiplication (called 1286 times in Curve25519), 105 cycles in squaring (called 1274 times), 67 cycles in multiplication by 121666 (called 255 times), 3 cycles in addition (called 1020 times), 3 cycles in subtraction (called 1020 times), and 12 cycles in conditional swaps (called 512 times), explaining fewer than 400000 cycles. The most important source of overhead in our current Curve25519 performance, 527102 cycles, is non-arithmetic instructions at the beginning and end of each function. We are working on addressing this by inlining all functions into the main loop and scheduling the main loop as a whole, and we anticipate then coming much closer to the lower bound, as in Salsa20 and Poly1305.

Similar comments apply to Ed25519. When we submitted this paper, many Ed25519 cycles (about 50000 cycles in signing and 25000 in verification) were consumed by the SHA-512 implementation selected by SUPERCOP [12]; but a subsequent OpenSSL revision drastically improved SHA-512 performance on the Cortex A8. We have not bothered investigating SHA-512 performance in more detail: the Ed25519 paper [10] recommends switching to Ed25519-SHA-3.

References

[1] — (no editor): 9th IEEE symposium on application specific processors. Institute of Electrical and Electronics Engineers (2011). See [33]
[2] Aciiçmez, O., Brumley, B.B., Grabher, P.: New results on instruction cache attacks. In: CHES 2010 [31], pp. 110–124 (2010) Citations in this document: §1

[3] ARM Limited: Cortex-A8 technical reference manual, revision r3p2 (2010), http://infocenter.arm.com/help/index.jsp?topic=/com.arm.doc.ddi0344k/index.html. Citations in this document: §2

[4] Bernstein, D.J.: Floating-point arithmetic and message authentication (1999), http://cr.yp.to/papers.html#hash127. Citations in this document: §4, §4, §4, §4

[5] Bernstein, D.J.: The Poly1305-AES message-authentication code. In: FSE 2005 [20], pp. 32–49 (2005), http://cr.yp.to/papers.html#poly1305. Citations in this document: §1, §4, §4

[6] Bernstein, D.J.: Curve25519: new Diffie-Hellman speed records. In: PKC 2006 [49], pp. 207–228 (2006), http://cr.yp.to/papers.html#curve25519. Citations in this document: §1, §5, §5, §5

[7] Bernstein, D.J.: qhasm software package (2007), http://cr.yp.to/qhasm.html. Citations in this document: §2

[8] Bernstein, D.J.: Polynomial evaluation and message authentication (2007), http://cr.yp.to/papers.html#pema. Citations in this document: §4

[9] Bernstein, D.J.: The Salsa20 family of stream ciphers. In: [37], pp. 84–97 (2008), http://cr.yp.to/papers.html#salsafamily. Citations in this document: §1, §3

[10] Bernstein, D.J., Duif, N., Lange, T., Schwabe, P., Yang, B.-Y.: High-speed high-security signatures. In: CHES 2011 [36] (2011), http://eprint.iacr.org/2011/368. Citations in this document: §1, §5, §5

[11] Bernstein, D.J., Lange, T.: Faster addition and doubling on elliptic curves. In: Asiacrypt 2007 [30], pp. 29–50 (2007), http://eprint.iacr.org/2007/286. Citations in this document: §5

[12] Bernstein, D.J., Lange, T. (eds.): eBACS: ECRYPT Benchmarking of Cryptographic Systems, accessed 5 March 2012 (2012), http://bench.cr.yp.to. Citations in this document: §1, §3, §4, §5, §5

[13] Bernstein, D.J., Lange, T., Schwabe, P.: The security impact of a new cryptographic library (2011), http://eprint.iacr.org/2011/646. Citations in this document: §1

[14] Black, J., Cochran, M.: MAC reforgeability. In: FSE 2009 [19], pp. 345–362 (2009), http://eprint.iacr.org/2006/095. Citations in this document: §4

[15] Canteaut, A., Viswanathan, K. (eds.): Progress in cryptology—INDOCRYPT 2004, 5th international conference on cryptology in India, Chennai, India, December 20–22, 2004, proceedings. LNCS, vol. 3348. Springer, Heidelberg (2004) ISBN 3-540-24130-2. See [32]

[16] Clavier, C., Gaj, K. (eds.): Cryptographic hardware and embedded systems—CHES 2009, 11th international workshop, Lausanne, Switzerland, September 6–9, 2009, proceedings. LNCS, vol. 5747. Springer, Heidelberg (2009) ISBN 978-3-642-04137-2. See [23]

[17] Costigan, N., Schwabe, P.: Fast elliptic-curve cryptography on the Cell Broadband Engine. In: Africacrypt 2009 [35], pp. 368–385 (2009), http://cryptojedi.org/users/peter/#celldh. Citations in this document: §1

[18] den Boer, B.: A simple and key-economical unconditional authentication scheme. Journal of Computer Security 2, 65–71 (1993) ISSN 0926–227X. Citations in this document: §4

[19] Dunkelman, O. (ed.): Fast software encryption, 16th international workshop, FSE 2009, Leuven, Belgium, February 22–25, 2009, revised selected papers. LNCS, vol. 5665. Springer, Heidelberg (2009) ISBN 978-3-642-03316-2. See [14]

[20] Gilbert, H., Handschuh, H. (eds.): Fast software encryption: 12th international workshop, FSE 2005, Paris, France, February 21–23, 2005, revised selected papers. LNCS, vol. 3557. Springer, Heidelberg (2005) ISBN 3-540-26541-4. See [5]

[21] Handschuh, H., Preneel, B.: Key-recovery attacks on universal hash function based MAC algorithms. In: CRYPTO 2008 [46], pp. 144–161 (2008), https://www.iacr.org/archive/crypto2008/51570145/51570145.pdf. Citations in this document: §4

[22] Helleseth, T. (ed.): Advances in cryptology—EUROCRYPT '93, workshop on the theory and application of cryptographic techniques, Lofthus, Norway, May 23–27, 1993, proceedings. LNCS, vol. 765. Springer, Heidelberg (1994) ISBN 3-540-57600-2. See [24]

[23] Käsper, E., Schwabe, P.: Faster and timing-attack resistant AES-GCM. In: CHES 2009 [16], pp. 1–17 (2009), http://eprint.iacr.org/2009/129. Citations in this document: §3, §4

[24] Johansson, T., Kabatianskii, G., Smeets, B.J.M.: On the relation between A-codes and codes correcting independent errors. In: EUROCRYPT '93 [22], pp. 1–11 (1994) Citations in this document: §4

[25] Joux, A. (ed.): Fast software encryption—18th international workshop, FSE 2011, Lyngby, Denmark, February 13–16, 2011, revised selected papers. LNCS, vol. 6733. Springer, Heidelberg (2011) ISBN 978-3-642-21701-2. See [29]

[26] Koblitz, N. (ed.): Advances in cryptology—CRYPTO '96. LNCS, vol. 1109. Springer, Heidelberg (1996). See [39]

[27] Kohno, T., Viega, J., Whiting, D.: CWC: a high-performance conventional authenticated encryption mode. In: FSE 2004 [38], pp. 408–426 (2004), http://eprint.iacr.org/2003/106. Citations in this document: §4

[28] Krovetz, T., Rogaway, P.: Fast universal hashing with small keys and no preprocessing: the PolyR construction. In: ICISC 2000 [48], pp. 73–89 (2001), http://www.cs.ucdavis.edu/~rogaway/papers/poly.htm. Citations in this document: §4

[29] Krovetz, T., Rogaway, P.: The software performance of authenticated-encryption modes. In: FSE 2011 [25], pp. 306–327 (2011), http://www.cs.ucdavis.edu/~rogaway/papers/ae.pdf. Citations in this document: §3, §4, §4

[30] Kurosawa, K. (ed.): Advances in cryptology—ASIACRYPT 2007, 13th international conference on the theory and application of cryptology and information security, Kuching, Malaysia, December 2–6, 2007, proceedings. LNCS, vol. 4833. Springer, Heidelberg (2007) ISBN 978-3-540-76899-9. See [11]

[31] Mangard, S., Standaert, F.-X. (eds.): Cryptographic hardware and embedded systems, CHES 2010, 12th international workshop, Santa Barbara, CA, USA, August 17–20, 2010, proceedings. LNCS, vol. 6225. Springer, Heidelberg (2010) ISBN 978-3-642-15030-2. See [2]

[32] McGrew, D.A., Viega, J.: The security and performance of the Galois/Counter mode (GCM) of operation. In: INDOCRYPT 2004 [15], pp. 343–355 (2004), http://eprint.iacr.org/2004/193. Citations in this document: §4

[33] Morozov, S., Tergino, C., Schaumont, P.: System integration of elliptic curve cryptography on an OMAP Platform. In: SASP 2011 [1], pp. 52–57 (2011), http://rijndael.ece.vt.edu/schaum/papers/2011sasp.pdf. Citations in this document: §5

[34] Nevelsteen, W., Preneel, B.: Software performance of universal hash functions. In: EUROCRYPT '99 [41], pp. 24–41 (1999) Citations in this document: §4

[35] Preneel, B. (ed.): Progress in cryptology—AFRICACRYPT 2009, second international conference on cryptology in Africa, Gammarth, Tunisia, June 21–25, 2009, proceedings. LNCS, vol. 5580. Springer, Heidelberg (2009). See [17]

[36] Preneel, B., Takagi, T. (eds.): Cryptographic hardware and embedded systems—CHES 2011, 13th international workshop, Nara, Japan, September 28–October 1, 2011, proceedings. LNCS. Springer, Heidelberg (2011) ISBN 978-3-642-23950-2. See [10]

[37] Robshaw, M., Billet, O. (eds.): New stream cipher designs. LNCS, vol. 4986. Springer, Heidelberg (2008) ISBN 978-3-540-68350-6. See [9]

[38] Roy, B.K., Meier, W. (eds.): Fast software encryption, 11th international workshop, FSE 2004, Delhi, India, February 5–7, 2004, revised papers. LNCS, vol. 3017. Springer, Heidelberg (2004) ISBN 3-540-22171-9. See [27]

[39] Shoup, V.: On fast and provably secure message authentication based on universal hashing. In: CRYPTO '96 [26], pp. 313–328 (1996), http://www.shoup.net/papers. Citations in this document: §4

[40] Sobole, É.: Calculateur de cycle pour le Cortex A8 (2012), http://pulsar.webshaker.net/ccc/index.php. Citations in this document: §2

[41] Stern, J. (ed.): Advances in cryptology—EUROCRYPT '99. LNCS, vol. 1592. Springer, Heidelberg (1999) ISBN 3-540-65889-0. MR 2000i:94001. See [34]

[42] Stinson, D.R. (ed.): Advances in cryptology—CRYPTO '93: 13th annual international cryptology conference, Santa Barbara, California, USA, August 22–26, 1993, proceedings. LNCS, vol. 773. Springer, Heidelberg (1994) ISBN 3-540-57766-1, 0-387-57766-1. See [43]

[43] Taylor, R.: An integrity check value algorithm for stream ciphers. In: CRYPTO '93 [42], pp. 40–48 (1994) Citations in this document: §4

[44] Tromer, E., Osvik, D.A., Shamir, A.: Efficient cache attacks on AES, and countermeasures. Journal of Cryptology 23, 37–71 (2010), http://people.csail.mit.edu/tromer/papers/cache-joc-official.pdf. Citations in this document: §1

[45] Ulevitch, D.: DNSCrypt—critical, fundamental, and about time (2011), http://blog.opendns.com/2011/12/06/dnscrypt-%E2%80%93-critical-fundamental-and-about-time/ Citations in this document: §1

[46] Wagner, D. (ed.): Advances in cryptology—CRYPTO 2008, 28th annual international cryptology conference, Santa Barbara, CA, USA, August 17–21, 2008, proceedings. LNCS, vol. 5157. Springer, Heidelberg (2008) ISBN 978-3-540-85173-8. See [21]

[47] Weiß, M., Heinz, B., Stumpf, F.: A cache timing attack on AES in virtualization environments. In: Proceedings of Financial Cryptography 2012, to appear (2012), http://fc12.ifca.ai/pre-proceedings/paper_70.pdf. Citations in this document: §1, §3

[48] Won, D. (ed.): Information security and cryptology—ICISC 2000, third international conference, Seoul, Korea, December 8–9, 2000, proceedings. LNCS, vol. 2015. Springer, Heidelberg (2001) ISBN 3-540-41782-6. See [28]

[49] Yung, M., Dodis, Y., Kiayias, A., Malkin, T. (eds.): Public key cryptography—9th international conference on theory and practice in public-key cryptography, New York, NY, USA, April 24–26, 2006, proceedings. LNCS, vol. 3958. Springer, Heidelberg (2006) ISBN 978-3-540-33851-2. See [6]

Towards One Cycle per Bit Asymmetric Encryption: Code-Based Cryptography on Reconfigurable Hardware

Stefan Heyse and Tim Güneysu

Horst Görtz Institute for IT-Security
Ruhr-Universität Bochum
44780 Bochum, Germany
{heyse,gueneysu}@crypto.rub.de

Abstract. Most advanced security systems rely on public-key schemes based either on the factorization or the discrete logarithm problem. Since both problems are known to be closely related, a major breakthrough in cryptanalysis tackling one of those problems could render a large set of cryptosystems completely useless.

Code-based public-key schemes are based on the alternative security assumption that decoding generic linear binary codes is NP-complete. In the past, most researchers focused on the McEliece cryptosystem, neglecting the fact that the scheme by Niederreiter has some important advantages. Smaller keys, more practical plain and ciphertext sizes and less computations. In this work we describe a novel FPGA implementation of the Niederreiter scheme, showing that its advantages can result a very efficient design for an asymmetric cryptosystem that can encrypt more than 1.5 million plaintexts per seconds on a Xilinx Virtex-6 FPGA, outperforming all other popular public key cryptosystems by far.

1 Introduction

Public-key cryptosystems build the foundation for virtually all advanced cryptographic requirements, such as asymmetric encryption, key exchange and digital signatures. However, up to now most cryptosystems rely on two classes of fundamental problems to establish security, namely the factoring problem and the (elliptic curve) discrete logarithm problem. Since both are related, a significant cryptanalytical improvement will turn out a large number of currently employed security systems to be insecure overnight. This threat is further nourished by upcoming generations of powerful quantum computers that have been shown to be very effective in computing solutions to the problems mentioned above [36]. Recently, IBM announced two further breakthrough in quantum computing [9] so that such practical systems might already become available in the next 15 years. Evidently, a larger *diversification* of cryptographic primitives that are resistant against such attacks is absolutely essential for the future of public-key cryptosystems.

E. Prouff and P. Schaumont (Eds.): CHES 2012, LNCS 7428, pp. 340–355, 2012.

Addressing this, cryptosystems that settle their security on alternative hard problems have gathered much attention in the last years, such as multivariate-quadratic ($\mathcal{MQ}$-), lattice-based and code-based schemes. A drawback of these constructions have been their low efficiency, large key sizes or complex computation with respect to RSA and ECC, what typically makes it difficult to employ them on small and embedded systems. First approaches to tackle these issues on such small systems has been presented for $\mathcal{MQ}$ and McEliece cryptosystems on the last years' workshops of the CHES series [8,14].

In this work, we focus on another code-based scheme. The cryptosystem presented by Niederreiter [29] is dual to McEliece's proposal [25] but enables higher efficiency while still maintaining the same security argument. In particular, it has been shown that even after more than 30 years of thorough analysis the code-based schemes remain unbroken when security parameters and fundamental codes are appropriately chosen [5]. Furthermore, a recent result indicated that McEliece and Niederreiter cryptosystems also resist quantum computing [12].

Our Contribution: In this work, we present the first implementation of the Niederreiter scheme on reconfigurable hardware. Our implementation for Xilinx' Virtex-6 FPGAs provides 80-bit of equivalent symmetric security and can run more than 1.5 million encryption and 17000 decryption operations per second, respectively. By using only a moderate amount of memory and logic resources, our implementation even outperforms many other implementation of classical cryptosystems, such as ECC-160 and RSA-1024. This impressive throughput of our implementation has become possible due to our highly optimized constant weight encoding algorithm. Due to these optimizations and the inherent advantages of Niederreiter over McEliece, we achieve a performance that are even orders of magnitudes faster than any other McEliece implementation reported so far.

Outline: This paper is structured as follows: we start with a brief introduction to Niederreiter encryption, shortly explain necessary operations on Goppa codes and introduce constant weight encoding. Section 4 describe our actual implementations for an Xilinx Virtex6LX240 FPGA. Finally, we present our results for these platforms in Section 5.

2   Previous Work

Although proposed already more than 30 years ago, the code-based encryption such as the McEliece and Niederreiter scheme has never gained much attention due to their large secret and public keys involved. With their large and costly memory requirements they have been hardly integrated in any real-world products – yet. The first FPGA-based implementation of McEliece was proposed in [14] for a Xilinx Spartan-3AN and encrypts and decrypts data in 1.07 ms and 2.88 ms, using security parameters achieving an equivalence of 80-bit symmetric security. The authors of [38] presented another accelerator for binary McEliece encryption on a more powerful Virtex5-LX110T that encrypts and decrypts in 0.5 ms and 1.4 ms providing a similar level of security. For x86 personal computers, the most recent implementation of the binary McEliece scheme is due to Biswas and Sendrier [7] that achieves about 83-bit security according to [5]. Comparing their implementation with other public key schemes, it turns out that McEliece encryption is even faster than RSA and NTRU [4] - at the cost of larger keys. But researchers addressed the issue of large keys by replacing the original used binary Goppa codes with codes that allow more compact representations, e.g, [26,32]. However, for these schemes only PC and microcontroller implementations exist so far [2,20]. Even worse, most of the attempts to reduce the key size have been broken[15]. Note that all previous works also exclusively target the McEliece cryptosystems. To the best of our knowledge, the only published Niederreiter implementation for embedded systems is an implementation for small 8-bit AVR microcontrollers enabling encryption and decryption in 1.6 ms and 179 ms [19]. Additionally, there are some Java based implementations or Niederreiter based signatures available [33].

2.1   The Niederreiter Public Key Scheme

The Niederreiter scheme [29] is a public key cryptosystem based on linear error-correcting codes, similar to the popular McEliece cryptosystem. The secret key is an efficient decoding algorithm for an error-correcting code with dimension k, length n and error correcting capability t. To create a public key, Niederreiter defined a random $n \times n$-dimensional permutation matrix P disguising the structure of the code by computing the product $\hat{H} = M \times H \times P$. Here, M can be chosen as the $(n-k) \times (n-k)$ matrix that transforms $\hat{H}$ into systematic form. Using the public key $K_{pub} = (\hat{H}, t)$ and private key $K_{sec} = (P^{-1}, H, M^{-1})$, encryption and decryption can be defined as in Algorithm 1 and Algorithm 2, respectively.

Algorithm 1. Classical Niederreiter Message Encryption

Input: Message $m, K_{pub} = (\hat{H}, t)$
Output: Ciphertext c
 1: Encode the message m as a binary string of length n and weight t called e
 2: $c = \hat{H}e^T$
 3: **return** c

Note that Algorithm 1 employs only a simple matrix multiplication operation on the input message that was transformed into a *constant weight word* before. The necessary algorithm for constant weight encoding (Bin2CW) is given in detail in Section 2.4.

Algorithm 2. Classical Niederreiter Message Decryption

Input: Ciphertext $c, K_{sec} = (P, M, g(z), \mathcal{L})$
Output: Message m
 1: $c' \leftarrow M^{-1}c$
 2: decode c' to error vector $e' = Pe^T$
 3: $e \leftarrow P^{-1}e'$
 4: Decode the error vector e to the binary message m
 5: **return** m

Decryption is the most time-consuming process and requires several complex operations on the linear code defined over binary extension fields. In Section 2.5 we briefly introduce the required steps for decoding codewords. Note that in a modern Niederreiter description, applying the permutation P can be omitted completely. This is possible by computing a permuted support and merging the mapping of roots to bit positions via the support list with the permutation. See Section 4.2 for details.

2.2   Niederreiter vs. McEliece

The main difference between Niederreiter and McEliece is the public key. While an $(n \times k)$ generator matrix serves as public key in McEliece, Niederreiter uses a $(n \times (n - k))$ parity check matrix for this purpose. Both matrices can be used in their systematic form, leading to $((n - k) * k)$ bits storage requirement in both cases. Using this method in the McEliece case demands a CCA2 secure conversion [30,24] to stay secure, whereas Niederreiter can be used without this overhead. In McEliece, an n-bit code word with errors is used as ciphertext, whereas Niederreiter uses the $(n - k)$-bit syndrome as plaintext. This shifts the syndrome computation from the receiver to the sender of the message and therefore speeds up decryption, still maintaining high encryption performance. At the same time, the parity check matrix and related information is no longer part of the secret key, thus reducing the secret key size. However, the Niederreiter scheme requires the scrambling matrix M in any case which can be omitted when using McEliece encryption. Finally, Niederreiter encryption imposes less restrictions on the plaintext size, i.e, depending on the parameter sets and constant weight encoding algorithm, Niederreiter enables plaintext blocks with a size of only hundreds of bits instead of several thousands bits as in the case of McEliece encryption. In particular, for key transportation protocols with symmetric key sizes of 128 to 256 bits, the transfer of thousands of bits as required in the case of McEliece can be an expensive overhead.

2.3   Security Parameters

All security parameters for cryptosystems are chosen in a way to provide sufficient protection against the best known attack (whereas the notion of "sufficient" is determined by the requirements of an application). On the attempt to employ an alternative cryptosystem, it is of utmost important for a security engineer to being able to safely assess if this best attack has already been found. In this context, the work by Bernstein et al. [5] currently proposes the best attack on McEliece and Niederreiter cryptosystems so far reducing the work factor to break the McEliece scheme based on a $(1024, 524)$ Goppa code and $t = 50$ to $2^{60.55}$ bit operations. According to their findings, we summarize the security parameters for specific security levels in Table 1. The public key size column gives the size

Table 1. Security Parameters for Niederreiter Cryptosystems

Security Level	Parameters (n, k, t), errors added	Size K_{pub} in KBits	Size K_{sec} $(g(z) \mid \mathcal{L} \mid M^{-1})$ KBits
Short-term (60 bit)	$(1024, 644, 38), 38$	239	$(0.37 \mid 10 \mid 141)$
Mid-term I (80 bit)	$(2048, 1751, 27), 27$	507	$(0.29 \mid 22 \mid 86)$
Mid-term II (128 bit)	$(2690, 2280, 56), 57$	913	$(0.38 \mid 18 \mid 164)$
Long-term (256 bit)	$(6624, 5129, 115), 117$	$7,488$	$(1.45 \mid 84 \mid 2,183)$

of a systematic parity check matrix and the secret key column the size of the Goppa polynomial $g(z)$, the support $\mathcal{L}$ and the inverse scrambling matrix M^{-1}.

As can be clearly seen, the main caveat against coding based cryptosystems is the significant size of the public and private keys. For 80 bit security, for example, the parameters $m = 11, n = 2048, t = 27, k \geq 1751$ already lead to an M^{-1} of 11 KBytes and a public key size of 63 KBytes. Note that we can reduce the size of the public key from originally 74 KBytes by choosing M in such a way that it brings $\hat{H}$ to systematic form $\hat{H} = (ID_{n-k} \mid Q)$, where only the redundant part Q has to be stored.

2.4 Constant Weight Encoding and Decoding

Before encrypting a message with the Niederreiter cryptosystem, the message has to be encoded into an error vector. More precisely, the message needs to be transformed into a bit vector of length n and constant weight t. There exist quite a few encoding algorithms (e.g., those in [11,34,16]), however they are not directly applicable to the restricted execution environment of embedded systems and hardware. In this work we unfolded the recursive algorithm proposed in [35] so that it can run by an iterative state machine. During the encoding operation, one has to compute a value $d \approx \frac{ln(2)}{t} \cdot (n - \frac{t-1}{t})$ to determine how many bits of the message are encoded into the distance to the next one-bit on the error vector. But many embedded (hardware) systems do not have a dedicated floating-point and division unit so these operations should be replaced. We therefore substituted the floating point operation and division by a simple and fast table lookup. Since we still preserve all properties from [35], the algorithm will still terminate with a minor negligible loss in efficiency.

The encoding algorithm suitable for embedded systems is given in Algorithm 3.

The constant weight decoding algorithm was adapted in a similar way, and is presented in Algorithm 4.

2.5 Encoding and Decoding of Goppa Codes

In this section we briefly introduce the underlying Goppa codes that provide the fundamental arithmetic for the Niederreiter cryptosystem.

Theorem 1. *Let $g(z)$ be an irreducible polynomial of degree t over $GF(2^m)$. Then the set*

$$\Gamma(G(z), GF(2^m)) = \{(c_\alpha)_{\alpha \in GF(2^m)} \in \{0,1\}^n \mid \sum_{\alpha \in GF(2^m)} \frac{c_\alpha}{z - \alpha} \equiv 0 \bmod G(z)\} \tag{1}$$

defines a binary Goppa code C of length $n = 2^m$, dimension $k \geq n - mt$ and minimum distance $d \geq 2t + 1$. The set of the α_i is called the support $\mathcal{L}$ of the code.[42]

Algorithm 3. Encode Binary String in Constant Weight Word (Bin2CW)

Input: n, t, binary stream B
Output: $\Delta[0, \ldots, t-1]$
1:   $\delta = 0, index = 0$
2: **while** $t \neq 0$ **do**
3:    **if** $n \leq t$ **then**
4:      $\Delta[index++] = \delta$
5:      $n- = 1, t- = 1, \delta = 0$
6:    **end if**
7:    $u \leftarrow uTable[n, t]$
8:    $d \leftarrow (1 << u)$
9:    **if** $read(B, 1) = 1$ **then**
10:     $n- = d, \delta+ = d$
11:    **else**
12:     $i \leftarrow read(B, u)$
13:     $\Delta[index++] = \delta + i$
14:     $\delta = 0, t- = 1, n- = (i + 1)$
15:    **end if**
16: **end while**

Algorithm 4. Decode Constant Weight Word to Binary String (CW2Bin)

Input: $n, t, \Delta[0, \ldots, t-1]$
Output: binary stream B
1:   $\delta = 0, index = 0$
2: **while** $t \neq 0$ AND $n > t$ **do**
3:    $u \leftarrow uTable[n, t]$
4:    $d \leftarrow (1 << u)$
5:    **if** $\Delta[index] \geq d$ **then**
6:     $Write(1, B)$
7:     $\Delta[index]- = d$
8:     $n- = d$
9:    **else**
10:     $\delta = \Delta[index++]$
11:     $Write(0|\delta, B)$
12:     $n- = (\delta + 1), t- = 1$
13:    **end if**
14: **end while**

There exist fast decoding algorithms with runtime $O(n \cdot t)$ operations (e.g., [31,41]). For each irreducible polynomial $g(z)$ over $GF(2^m)$ of degree t there exists a binary Goppa code of length $n = 2^m$ and dimension $k = n - mt$. This code is capable to correct up to t errors [3] and can be represented by a $k \times n$ generator matrix G such that $C = \{mG : m \in F_2^k\}$.

Since $r = c + e \equiv e \mod g(z)$ holds, the syndrome $Syn(z)$ of a received codeword can be obtained from Equation (1) by

$$Syn(z) = \sum_{\alpha \in GF(2^m)} \frac{r_\alpha}{z - \alpha} \equiv \sum_{\alpha \in GF(2^m)} \frac{e_\alpha}{z - \alpha} \mod G(z) \qquad (2)$$

To finally recover $\underline{e}$, we need to solve the key equation $\sigma(z) \cdot Syn(z) \equiv \omega(z)$ mod $g(z)$, where $\sigma(z)$ denotes a corresponding error-locator polynomial and $\omega(z)$ denotes an error-weight polynomial. Note that it can be shown that $\omega(z) = \sigma(z)'$ is the formal derivative of the error-locator and by splitting $\sigma(z)$ into even and odd polynomial parts $\sigma(z) = a(z)^2 + z \cdot b(z)^2$, we finally determine the following equation to determine error positions:

$$Syn(z)(a(z)^2 + z \cdot b(z)^2) \equiv b(z)^2 \mod G(z) \qquad (3)$$

To solve Equation (3) for a given syndrome $Syn(z)$, the following steps have to be performed:

1. Compute an inverse polynomial $T(z)$ with $T(z) \cdot Syn(z) \equiv 1 \mod g(z)$ (or provide a corresponding table). It follows that $(T(z) + z)b(z)^2 \equiv a(z)^2$ mod $g(z)$.
2. There is a simple case if $T(z) = z \Rightarrow a(z) = 0$ s.t. $b(z)^2 \equiv z \cdot b(z)^2 \cdot Syn(z)$ mod $G(z) \Rightarrow 1 \equiv z \cdot Syn(z) \mod G(z)$ which directly leads to $\sigma(z) = z$. Contrary, if $T(z) \neq z$, compute a square root $R(z)$ for the given polynomial $R(z)^2 \equiv T(z) + z \mod G(z)$. Based on an observation by Huber [23] this can be done by a simple polynomial multiplication. We can then determine solutions $a(z), b(z)$ satisfying

$$a(z) = b(z) \cdot R(z) \mod G(z). \qquad (4)$$

   using the extended euclidean algorithm. The computation is stopped, when $a(z)$ reaches degree $\lfloor \frac{t}{2} \rfloor$. Finally, we use the identified $a(z), b(z)$ to construct the error-locator polynomial $\sigma(z) = a(z)^2 + z \cdot b(z)^2$.
3. The roots of $\sigma(z)$ denote the positions of error bits.
   Searching the roots of $\sigma(z)$ with degree t over $GF(2^m)$ is time-consuming. Besides the plain evaluation of all support elements, for this two most commonly used methods are the Chien search [10] and Horner's scheme [22]. A third method as proposed in [6] for PC platforms can not be easiliy parallelized and requires a lot of greatest common divisor and trace computations. In our work we therefore use a parallelized version of the Chien search, which concurrently evaluates all t coefficients (see Section 4.2 for details). The decoding process, as required in Step 2 of Algorithm 2 for message decryption, is finally summarized in Algorithm 5.

Algorithm 5. Decoding Goppa Codes

Input: Received syndrome $Syn(z)$ of a codeword r with up to t errors
Output: Recovered message $\hat{m}$
1: $T(z) \leftarrow Syn(z)^{-1} \mod G(z)$
2: **if** $T(z) = z$ **then**
3:    $\sigma(z) \leftarrow z$
4: **else**
5:    $R(z) \leftarrow \sqrt{T(z) + z}$
6:    Compute $a(z)$ and $b(z)$ with $a(z) \equiv b(z) \cdot R(z) \mod G(z)$
7:    $\sigma(z) \leftarrow a(z)^2 + z \cdot b(z)^2$
8: **end if**
9: Determine roots of $\sigma(z)$ and compute e
10: **return** e

3 Design Decisions

In this section we discuss the design and parameter decisions for our Niederreiter implementation on reconfigurable hardware. A primary goal of our design is high-performance, a secondary reasonable hardware costs.

3.1 Parameter Selection

With the implementation of our Niederreiter cryptosystem, we aim to provide 80-bit of equivalent symmetric security, i.e., protection that is comparable to the security of ECC and RSA with approximately 160-bit and 1024-bit, respectively. This level of security is still considered sufficient for mid-term security applications providing a reasonable cost-performance ratio and thus suitable for most embedded systems. To achieve this level of security, we selected the parameters $m = 11, n = 2048, t = 27, k \geq 1751$ resulting in a private and public key size of 13.5 and 63 KBytes to be stored on the device. This amount of memory is available in each Xilinx FPGA larger than the low-cost Spartan-3 XC3S1000, Virtex-5 XC5VLX30 and Virtex-6 XC6VLX75T, respectively [44].

The above security level was originally proposed to minimize public key size for a given security level and not to maximize performance. However, we stay with this parameters to be comparable with the existing code-based implementations, which all selected this parameter set.

For practical purposes, we fixed the size of a message block to 192 bits which can be encoded into an appropriate error vector in any case. Note that the constant weight encoding algorithm requires an input of variable length to produce a constant weight output. Experiments showed that on average 210 bits are required to construct a valid constant weight word. Fixing the input message to 192 bits and adding random bits as required, makes the algorithm practicable without leaking any security-relevant information.

The secret key consisting of the Goppa polynomial $g(z)$, the support $\mathcal{L}$ and the inverse scrambling matrix M^{-1} is stored as part of the bitstream file which configures the FPGA. Because only the Spartan3-AN class from Xilinx offers internal Flash memory to store the bitstream internally in a (somewhat) protected way, appropriate actions have to be taken to protect the bitstream when storing it in external memory. In this case it is mandatory to enable bitstream encryption using AES-256 which is available for larger Xilinx Spartan-6 and all Xilinx Virtex-FPGAs from Virtex-4. Note, however, that also the Xilinx specific bitstream encryption [43] was successfully attacked by side-channel analysis in [28].

The (larger) public key can be stored either in internal or external memory since it does not require special protection. For our implementation we opted to store the public key in internal BRAMs to allow immediate access for high-performance encryption.

3.2   Inherent Side Channel Resistants

Some research had been done regarding side channels in code-based cryptography, however, all solely focused on implementations of McEliece encryption [40,39,37,21,27]. The advantage of Niederreiter in contrast to McEliece is that the ciphertext not consists only out of a pure codeword with randomly added errors. Fault attacks cannot be easily performed by flipping random bits of the ciphertext assuming that the decoder either corrects one of the the intentionally injected errors or fails to do so. For Niederreiter encryption, the ciphertext is a syndrome polynomial. Flipping random bits will result most likely in a decoding error without leaking any information. This renders all the attacks from [40,39,37] useless. Only power analysis attacks, like the one described in [21], which directly attacks the Goppa polynomial used in the Patterson algorithm, are still possible. It also requires further investigation, if adoptions of attacks targeting the root search are possible.

4   Implementation

This section describes our implementation primarily targeting a recent Virtex-6 LX240 FPGA. Note that this device is certainly too large for our implementation but was chosen due to its availability on the Xilinx Virtex-6 FPGA ML605 Evaluation Kit for practical testing. Furthermore, we provide implementations for a Xilinx Spartan-3 and Xilinx Virtex-5 to allow fair comparisons with other work (cf. Table 4).

4.1   Encryption

The public key $\hat{H}$ is stored in an internal BRAM memory block and row-wise addressed by the output of the constant weight encoder. Multiplying a binary vector with a binary matrix is equivalent to a XOR operation of each row with input vector bit equal to one. Since this operation is trivial, we now focus on the implementation of the constant weight encoding algorithm. Input data to our cryptosystem is passed using a FIFO with a non-symmetric 8-to-1 bit aspect ratio. Hence, after a word with 8-bit length is written to the FIFO, it can be read out bit by bit. This is the equivalent to the binary stream reader presented in Algorithm 3. Its main part is implemented as a small finite state machine. Every time a valid $\Delta[i]$ has been computed, it is directly transferred to the vector-matrix-multiplier summing up the selected rows. By interleaving operations we are able to process one bit from the FIFO at *every* clock cycle. After the last $\Delta[t]$ has been computed, only the last indexed row of $\hat{H}$ has to be added to the sum. Directly afterwards the encryption operation has finished and the ciphertext becomes available. Due to the very regular structure of the vector-matrix-multiplier and the small operands of the constant weight encoder, we were able to achieve a high clock frequency of 300 MHz. Nevertheless, the logic in the constant weight encoder is still the bottleneck.

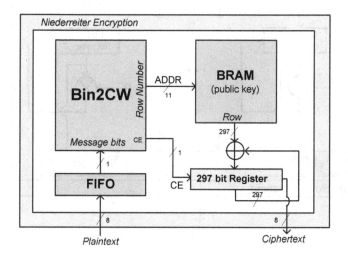

Fig. 1. Block diagram of the encryption process

4.2   Decryption

The first step in the decryption process is the multiplication by the inverse matrix M^{-1}. This 11 KByte large matrix is stored in an internal BRAM and addressed by an incrementing counter. Using this BRAM, the rows of the matrix are XORed into an intermediate register if the corresponding input bit of the ciphertext equals to one. After $(n - k) = 297$ clock cycles, this register contains the value $c' = M^{-1} * c$ as shown in Algorithm 2. Now c' is passed on to the Goppa decoder. The Goppa decoder implements standard Patterson decoding (see Algorithm 5) and returns the erroneous bit positions in the order they are found during the decoding process.

Searching roots is a time-consuming process that is highlighted by Figure 2 showing our Chien search core. Decryption performance can be boosted by instantiating two or more of these cores in parallel and let them evaluate different support elements in parallel. Beside the additional management overhead in the controlling state machine, each of this cores requires additional 620 registers and 106 LUTs. We therefore only use one core which evaluates one support element in two clock cycles and finishes the entire process after 4098 clock cycles.

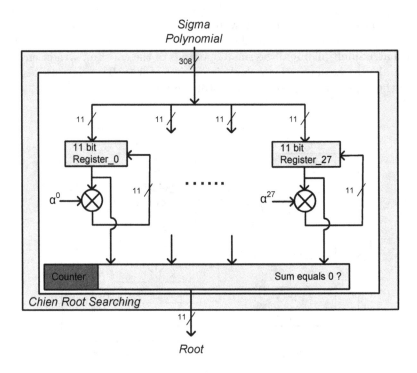

Fig. 2. Design of the Core for Chien Search

Next each root needs to be mapped to these bit positions. In this context, we constructed a table containing entries for $\mathcal{L} * P$, i.e., we merged this mapping and the reverse permutation from line 3 in Algorithm 2. As a side effect, the permutation P completely disappears from the scheme. Because the subsequent constant weight decoding algorithm expects the distance between the error bits in ascending order, we implemented a sorting circuit which sorts the error positions using a systolic implementation of bubble sort. At the same time it further computes the distance between two successive error positions. Finally, the error distances are translated into the binary message by a straightforward implementation of Algorithm 4 as developed in Section 4.1.

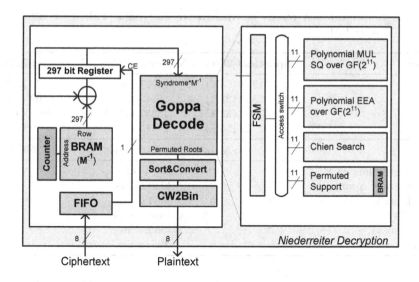

Fig. 3. Block diagram of the decryption process

5   Results

We now present the results for our implementation on three different platforms to enable a fair comparison with other work. Note that most of the differences in the number of used resources are due to architecture differences in the FPGA types, i.e., 4-input LUTs vs. 6-input LUTs and 18 KB BRAMs vs. 36 KB BRAMs in Spartan-3 and Virtex-5/6 FPGAs, respectively.

Encryption takes approximately 200 cycles and is around a factor of 72 faster than decryption that requires around 14,500 cycles. An open problem is how to transfer 1.5 million keys per second to the device, when many different public key are required, which is a typical application. Here an interface capable of transferring $1,5 * 10^6 * 63Kbyte \approx 774 \frac{Gbyte}{sec}$ would be necessary. As mentioned

Table 2. Implementation results of Niederreiter encryption with $n = 2048, k = 1751, t = 27$ after PAR

Aspect	Spartan3-2000	Virtex5-LX50	Virtex6-LX240
Slices	854 (2%)	291 (4%)	315 (1 %)
LUTs	1252 (3%)	888 (3%)	926 (1 %)
FFs	869 (2%)	930 (3%)	875 (1 %)
BRAMs	36 (90%)	18 (30%)	17 (4 %)
Frequency	150 MHz	250 MHz	300 MHz
CW Encode $e = encode(m)$		≈ 200 cycles	
Encrypt $c = e \cdot \hat{H}$		concurrently with CW Encoding	

Table 3. Implementation results of Niederreiter decryption with $n = 2048, k = 1751, t = 27$ after PAR

Aspect	Spartan3-2000	Virtex5-LX50	Virtex6-LX240
Slices	11253 (54%)	4077 (56%)	3887 (10 %)
LUTs	15559 (37%)	9743 (33%)	9409 (6 %)
FFs	13608 (33%)	13537 (47%)	12861 (4 %)
BRAMs	22 (55%)	13 (21%)	9 (2 %)
Frequency	95 MHz	180 MHz	250 MHz
Undo Scrambling $c \cdot M^{-1}$		297 cycles	
Compute $T = Syn(z)^{-1}$		4310 cycles	
Solve Equation (4)		4854 cycles	
Search Roots		4098 cycles	
Sort&Convert		85 cycles	
CW Decode		198 cycles	

Table 4. Comparison of our Niederreiter designs with single-core ECC and RSA implementations for 80 bit security

Scheme	Platform	Freq	Time/Op	Cycles/byte
This work [enc]	Virtex6-LX240T	300 MHz	0.66 µs	8.3
This work [dec]	Virtex6-LX240T	250 MHz	58.78 µs	612
McEliece [enc] [14]	Spartan3-AN1400	150 MHz	1070 µs	768
McEliece [dec] [14]	Spartan3-AN1400	85 MHz	21,610 µ	8788
This work enc	Spartan3-2000	150 MHz	1.32 µs	8.3
This work dec	Spartan3-2000	95 MHz	154 µs	612
McEliece [enc] [38]	Virtex5-LX110T	163 MHz	500 µs	389
McEliece [dec] [38]	Virtex5-LX110T	163 MHz	1400 µs	1091
This work [enc]	Virtex5-LX50T	250 MHz	0.793 µs	8.2
This work [dec]	Virtex5-LX50T	180 MHz	81 µs	612
ECC-P160 [17]	Spartan-3 1000-4	40 MHz	5.1 ms	10,200
ECC-K163 [17]	Virtex-II	128 MHz	35.75 µs	224.6
RSA-1024 random [18]	Spartan-3A	133 MHz	48.54 ms	50,436
RSA-1024 random [18]	Spartan-6	187 MHz	34.48 ms	50,373
RSA-1024 random [18]	Virtex-6	339 MHz	19.01 ms	59,258
NTRU encryption [1]	Virtex 1000EFG860	50 MHz	5 µs	8.3

above, the public-key cryptosystems RSA-1024 and ECC-P160 are assumed[1] to roughly achieve an simiar level of 80 bit symmetric security [13]. We finally compare our results to published implementations of these systems that target

---

[1] According to [13], RSA-1248 actually corresponds to 80 bit symmetric security. However, no implementation results for embedded systems are available for this key size.

similar platforms (i.e.,[14,38,17,18,1]). For a fair comparison with other existing implementations of code-based systems we also implemented our code for Spartan-3 and Virtex-5 FPGAs.

6 Conclusions

In this paper, we clearly demonstrate the very high performance that can be achieved with an efficient FPGA-based implementation of the Niederreiter scheme. Besides the more practical plaintext size and smaller public keys, the extremely high performance with more than 1.5 million encryption and 17000 decryption operations per second, respectively, makes Niederreiter an interesting candidate for applications where high throughput and many public key encryptions per second are required (e.g., upcoming car-2-car communication infrastructures).

Acknowledgements. The work described in this paper has been supported in part by the European Commission through the ICT programme under contract ICT-2007-216676 ECRYPT II. This work has been also been supported in part by the Ministry of Economic Affairs and Energy òf the State of North Rhine-Westphalia (Grant 315-43-02/2-005-WFBO-009).

References

1. Bailey, D.V., Coffin, D., Elbirt, A., Silverman, J.H., Woodbury, A.D.: NTRU in Constrained Devices. In: Koç, Ç.K., Naccache, D., Paar, C. (eds.) CHES 2001. LNCS, vol. 2162, pp. 262–272. Springer, Heidelberg (2001)
2. Berger, T.P., Cayrel, P.-L., Gaborit, P., Otmani, A.: Reducing Key Length of the McEliece Cryptosystem. In: Preneel, B. (ed.) AFRICACRYPT 2009. LNCS, vol. 5580, pp. 77–97. Springer, Heidelberg (2009)
3. Berlekamp, E.R.: Goppa codes. IEEE Trans. Information Theory IT-19(3), 590–592 (1973)
4. Bernstein, D.J., Lange, T.: ebacs: Ecrypt benchmarking of cryptographic systems (February 17, 2009), http://bench.cr.yp.to
5. Bernstein, D.J., Lange, T., Peters, C.: Attacking and Defending the McEliece Cryptosystem. In: Buchmann, J., Ding, J. (eds.) PQCrypto 2008. LNCS, vol. 5299, pp. 31–46. Springer, Heidelberg (2008)
6. Biswas, B., Herbert, V.: Efficient root finding of polynomials over fields of characteristic 2. In: WEWoRC 2009, July 7-9 (2009)
7. Biswas, B., Sendrier, N.: McEliece crypto-system: A reference implementation, http://www-rocq.inria.fr/secret/CBCrypto/index.php?pg=hymes
8. Bogdanov, A., Eisenbarth, T., Rupp, A., Wolf, C.: Time-Area Optimized Public-Key Engines: $\mathcal{MQ}$-Cryptosystems as Replacement for Elliptic Curves? In: Oswald, E., Rohatgi, P. (eds.) CHES 2008. LNCS, vol. 5154, pp. 45–61. Springer, Heidelberg (2008)
9. Chang, K.: I.B.M. Researchers Inch Toward Quantum Computer. New York Times Article (February 28, 2012), http://www.nytimes.com/2012/02/28/technology/ibm-inch-closer-on-quantum-computer.html?_r=1&hpw
10. Chien, R.: Cyclic decoding procedure for the bose-chaudhuri-hocquenghem codes. IEEE Trans. Information Theory IT-10(10), 357–363 (1964)

11. Cover, T.: Enumerative source encoding 19(1), 73–77 (1973)
12. Dinh, H., Moore, C., Russell, A.: McEliece and Niederreiter Cryptosystems That Resist Quantum Fourier Sampling Attacks. In: Rogaway, P. (ed.) CRYPTO 2011. LNCS, vol. 6841, pp. 761–779. Springer, Heidelberg (2011)
13. ECRYPT. Yearly report on algorithms and keysizes (2007-2008). Technical report, D.SPA.28 Rev. 1.1 (July 2008),
    http://www.ecrypt.eu.org/documents/D.SPA.10-1.1.pdf
14. Eisenbarth, T., Güneysu, T., Heyse, S., Paar, C.: MicroEliece: McEliece for Embedded Devices. In: Clavier, C., Gaj, K. (eds.) CHES 2009. LNCS, vol. 5747, pp. 49–64. Springer, Heidelberg (2009)
15. Faugere, J.-C., Otmani, A., Perret, L., Tillich, J.-P.: Algebraic cryptanalysis of mceliece variants with compact keys (2009)
16. Fischer, J.-B., Stern, J.: An Efficient Pseudo-random Generator Provably as Secure as Syndrome Decoding. In: Maurer, U.M. (ed.) EUROCRYPT 1996. LNCS, vol. 1070, pp. 245–255. Springer, Heidelberg (1996)
17. Güneysu, T., Paar, C., Pelzl, J.: Special-purpose hardware for solving the elliptic curve discrete logarithm problem. ACM Transactions on Reconfigurable Technology and Systems (TRETS) 1(2), 1–21 (2008)
18. Helion Technology Inc. Modular Exponentiation Core Family for Xilinx FPGA. Data Sheet (October 2008),
    http://www.heliontech.com/downloads/modexp_xilinx_datasheet.pdf
19. Heyse, S.: Low-Reiter: Niederreiter Encryption Scheme for Embedded Microcontrollers. In: Sendrier, N. (ed.) PQCrypto 2010. LNCS, vol. 6061, pp. 165–181. Springer, Heidelberg (2010)
20. Heyse, S.: Implementation of McEliece Based on Quasi-dyadic Goppa Codes for Embedded Devices. In: Yang, B.-Y. (ed.) PQCrypto 2011. LNCS, vol. 7071, pp. 143–162. Springer, Heidelberg (2011)
21. Heyse, S., Moradi, A., Paar, C.: Practical Power Analysis Attacks on Software Implementations of McEliece. In: Sendrier, N. (ed.) PQCrypto 2010. LNCS, vol. 6061, pp. 108–125. Springer, Heidelberg (2010)
22. Horner, W.G.: A new method of solving numerical equations of all orders, by continuous approximation. Philosophical Transactions of the Royal Society of London 109, 308–335 (1819)
23. Huber, K.: Note on decoding binary Goppa codes. Electronics Letters 32, 102–103 (1996)
24. Kobara, K., Imai, H.: Semantically Secure McEliece Public-Key Cryptosystems-Conversions for McEliece PKC. In: Kim, K.-C. (ed.) PKC 2001. LNCS, vol. 1992, pp. 19–35. Springer, Heidelberg (2001)
25. McEliece, R.J.: A Public-Key Cryptosystem Based On Algebraic Coding Theory. Deep Space Network Progress Report 44, 114–116 (1978)
26. Misoczki, R., Barreto, P.S.L.M.: Compact McEliece Keys from Goppa Codes. In: Jacobson Jr., M.J., Rijmen, V., Safavi-Naini, R. (eds.) SAC 2009. LNCS, vol. 5867, pp. 376–392. Springer, Heidelberg (2009)
27. Molter, H., Stöttinger, M., Shoufan, A., Strenzke, F.: A simple power analysis attack on a McEliece cryptoprocessor. Journal of Cryptographic Engineering 1, 29–36 (2011), doi:10.1007/s13389-011-0001-3
28. Moradi, A., Kasper, M., Paar, C.: Black-Box Side-Channel Attacks Highlight the Importance of Countermeasures - An Analysis of the Xilinx Virtex-4 and Virtex-5 Bitstream Encryption Mechanism. In: Dunkelman, O. (ed.) CT-RSA 2012. LNCS, vol. 7178, pp. 1–18. Springer, Heidelberg (2012)
29. Niederreiter, H.: Knapsack-Type Cryptosystems and Algebraic Coding Theory. Problems of Control and Information Theory 15, 159–166 (1986)

30. Nojima, R., Imai, H., Kobara, K., Morozov, K.: Semantic security for the McEliece cryptosystem without random oracles. Des. Codes Cryptography 49(1-3), 289–305 (2008)
31. Patterson, N.: The algebraic decoding of Goppa codes. IEEE Transactions on Information Theory 21, 203–207 (1975)
32. Persichetti, E.: Compact McEliece keys based on quasi-dyadic srivastava codes. Cryptology ePrint Archive, Report 2011/179 (2011), http://eprint.iacr.org/
33. Cayrel, P.-L.: Code-based cryptosystems: implementations, http://www.cayrel.net/research/code-based-cryptography/code-based-cryptosystems/
34. Sendrier, N.: Efficient Generation of Binary Words of Given Weight. In: Boyd, C. (ed.) Cryptography and Coding 1995. LNCS, vol. 1025, pp. 184–187. Springer, Heidelberg (1995)
35. Sendrier, N.: Encoding information into constant weight words. In: Proc. International Symposium on Information Theory, ISIT 2005, September 4-9, pp. 435–438 (2005)
36. Shor, P.W.: Polynomial-time algorithms for prime factorization and discrete logarithms on a quantum computer. SIAM J. Comput. 26(5), 1484–1509 (1997)
37. Shoufan, A., Strenzke, F., Gregor Molter, H., Stöttinger, M.: A Timing Attack against Patterson Algorithm in the McEliece PKC. In: Lee, D., Hong, S. (eds.) ICISC 2009. LNCS, vol. 5984, pp. 161–175. Springer, Heidelberg (2010)
38. Shoufan, A., Wink, T., Molter, H.G., Huss, S.A., Strenzke, F.: A Novel Processor Architecture for McEliece Cryptosystem and FPGA Platforms. In: 20th IEEE International Conference on Application-specific Systems, Architectures and Processors (July 2009)
39. Strenzke, F.: A Timing Attack against the Secret Permutation in the McEliece PKC. In: Sendrier, N. (ed.) PQCrypto 2010. LNCS, vol. 6061, pp. 95–107. Springer, Heidelberg (2010)
40. Strenzke, F., Tews, E., Gregor Molter, H., Overbeck, R., Shoufan, A.: Side Channels in the McEliece PKC. In: Buchmann, J., Ding, J. (eds.) PQCrypto 2008. LNCS, vol. 5299, pp. 216–229. Springer, Heidelberg (2008)
41. Sugiyama, Y., Kasahara, M., Hirasawa, S., Namekawa, T.: An erasures-and-errors decoding algorithm for goppa codes (corresp.). IEEE Transactions on Information Theory 22, 238–241 (1976)
42. van Tilborg, H.C.: Fundamentals of Cryptology. Kluwer Academic Publishers (2000)
43. Xilinx Inc. Advanced Security Schemes for Spartan-3A/3AN/3A DSP FPGAs, http://www.xilinx.com/support/documentation/white_papers/wp267.pdf
44. Xilinx Inc. Data Sheets and Product Information for Xilinx Spartan and Virtex FPGAs, http://www.xilinx.com/support/

Solving Quadratic Equations
with XL on Parallel Architectures

Chen-Mou Cheng[1], Tung Chou[2], Ruben Niederhagen[2], and Bo-Yin Yang[2]

[1] Intel-NTU Connected Context Computing Center,
National Taiwan University, Taipei, Taiwan
[2] Institute of Information Science, Academia Sinica, Taipei, Taiwan
{doug,blueprint,by}@crypto.tw, ruben@polycephaly.org

Abstract. Solving a system of multivariate quadratic equations (MQ) is an NP-complete problem whose complexity estimates are relevant to many cryptographic scenarios. In some cases it is required in the best known attack; sometimes it is a generic attack (such as for the multivariate PKCs), and sometimes it determines a provable level of security (such as for the QUAD stream ciphers).

Under reasonable assumptions, the best way to solve generic MQ systems is the XL algorithm implemented with a sparse matrix solver such as Wiedemann's algorithm. Knowing how much time an implementation of this attack requires gives us a good idea of how future cryptosystems related to MQ can be broken, similar to how implementations of the General Number Field Sieve that factors smaller RSA numbers give us more insight into the security of actual RSA-based cryptosystems.

This paper describes such an implementation of XL using the block Wiedemann algorithm. In 5 days we are able to solve a system with 32 variables and 64 equations over $\mathbb{F}_{16}$ (a computation of about $2^{60.3}$ bit operations) on a small cluster of 8 nodes, with 8 CPU cores and 36 GB of RAM in each node. We do not expect system solvers of the $\mathrm{F}_4/\mathrm{F}_5$ family to accomplish this due to their much higher memory demand. Our software also offers implementations for $\mathbb{F}_2$ and $\mathbb{F}_{31}$ and can be easily adapted to other small fields. More importantly, it scales nicely for small clusters, NUMA machines, and a combination of both.

Keywords: XL, Gröbner basis, block Wiedemann, sparse solver, multivariate quadratic systems.

1 Introduction

Some cryptographic systems can be attacked by solving a system of multivariate quadratic equations. For example the symmetric block cipher AES can be attacked by solving a system of 8000 quadratic equations with 1600 variables over $\mathbb{F}_2$ as shown by Courtois and Pieprzyk in [5] or by solving a system of 840 sparse quadratic equations and 1408 linear equations over 3968 variables of $\mathbb{F}_{256}$ as shown by Murphy and Robshaw in [17]. Multivariate cryptographic systems can be attacked naturally by solving their multivariate quadratic system; see for

E. Prouff and P. Schaumont (Eds.): CHES 2012, LNCS 7428, pp. 356–373, 2012.

example the analysis of the QUAD stream cipher by Yang, Chen, Bernstein, and Chen in [21].

We describe a parallel implementation of an algorithm for solving quadratic systems that was first suggested by Lazard in [11]. Later it was reinvented by Courtois, Klimov, Patarin, and Shamir and published in [4]; they call the algorithm *XL* as an acronym for *extended linearization*: XL *extends* a quadratic system by multiplying all equations with appropriate monomials and *linearizes* it by treating each monomial as an independent variable. Due to this extended linearization, the problem of solving a quadratic system turns into a problem of linear algebra.

XL is a special case of Gröbner basis algorithms (shown by Ars, Faugère, Imai, Kawazoe, and Sugita in [1]) and can be used as an alternative to other Gröbner basis solvers like Faugère's F_4 and F_5 algorithms (introduced in [7] and [8]). An enhanced version of F_4 is implemented for example in the computer algebra system Magma, and is often used as standard benchmark by cryptographers.

There is an ongoing discussion on whether XL-based algorithms or algorithms of the F_4/F_5-family are more efficient in terms of runtime complexity and memory complexity. To achieve a better understanding of the practical behaviour of XL for generic systems, we describe a parallel implementation of the XL algorithm for shared-memory systems, for small computer clusters, and for a combination of both. Measurements of the efficiency of the parallelization have been taken at small clusters of up to 8 nodes and shared-memory systems of up to 64 cores. A previous implementation of XL is PWXL, a parallel implementation of XL with block Wiedemann described in [15]. PWXL supports onl $\mathbb{F}_2$, while our implementation supports $\mathbb{F}_2$, $\mathbb{F}_{16}$, and $\mathbb{F}_{31}$. Furthermore, our implementation is modular and can be extended to other fields. Comparisons on performance of PWXL and our work will be shown in Sec. 4.3. We are planning to make our implementation available to the public.

This paper is structured as follows: The XL algorithm is introduced in Sec. 2. The parallel implementation of XL using the block Wiedemann algorithm is described in Sec. 3. Section 4 gives runtime measurements and performance values that are achieved by our implementation for a set of parameters on several parallel systems as well as comparisons to PWXL and to the implementation of F_4 in Magma.

2   The XL Algorithm

The original description of XL for multivariate quadratic systems can be found in [4]; a more general definition of XL for systems of higher degree is given in [3]. The following gives an introduction of the XL algorithm for quadratic systems; the notation is adapted from [23]:

Consider a finite field $K = \mathbb{F}_q$ and a system $\mathcal{A}$ of m multivariate quadratic equations $\ell_1 = \ell_2 = \cdots = \ell_m = 0$ for $\ell_i \in K[x_1, x_2, \ldots, x_n]$. For $b \in \mathbb{N}^n$ denote by x^b the monomial $x_1^{b_1} x_2^{b_2} \ldots x_n^{b_n}$ and by $|b| = b_1 + b_2 + \cdots + b_n$ the total degree of x^b.

XL first chooses a $D \in \mathbb{N}$ as $D := \min\{d : ((1 - \lambda)^{m-n-1}(1 + \lambda)^m)[d] \leq 0\}$ (see [22, Eq. (7)], [13,6]), where $f[i]$ denotes the coefficient of the degree-i term in the expansion of a polynomial $f(\lambda)$ e.g., $(\lambda+2)^3[2] = (\lambda^3 + 6\lambda^2 + 12\lambda + 8)[2] = 6$. XL extends the quadratic system $\mathcal{A}$ to the system $\mathcal{R}^{(D)} = \{x^b \ell_i = 0 : |b| \leq D - 2, \ell_i \in \mathcal{A}\}$ of maximum degree D by multiplying each equation of $\mathcal{A}$ by all monomials of degree less than or equal to $D-2$. Now, each monomial $x^d, |d| \leq D$ is considered a new variable to obtain a linear system $\mathcal{M}$. Note that the system matrix of $\mathcal{M}$ is sparse since each equation has the same number of non-zero coefficients as the corresponding equation of the quadratic system $\mathcal{A}$. Finally the linear system $\mathcal{M}$ is solved, giving solutions for all monomials and particularly for $x_1, x_2, \ldots, x_n$. Note that the matrix corresponding to the linear system $\mathcal{M}$ is the Macaulay matrix of degree D for the polynomial system $\mathcal{A}$ (see [12], e.g., defined in [9]).

2.1   The Block Wiedemann Algorithm

The computationally most expensive task in XL is to find a solution for the sparse linear system $\mathcal{M}$ of equations over a finite field. There are two popular algorithms for that task, the block Lanczos algorithm [16] and the block Wiedemann algorithm [2]. The block Wiedemann algorithm was proposed by Coppersmith in 1994 and is a generalization of the original Wiedemann algorithm [20]. It has several features that make it powerful for computation in XL: From the original Wiedemann algorithm it inherits the property that the runtime is directly proportional to the weight of the input matrix. Therefore, this algorithm is suitable for solving sparse matrices, which is exactly the case for XL. Furthermore, big parts of the block Wiedemann algorithm can be parallelized on several types of parallel architectures. The following paragraphs give a brief introduction to the block Wiedemann algorithm. For more details please refer to [18, Sec. 4.2] and [2].

The basic idea of Coppersmith's block Wiedemann algorithm for finding a solution $\bar{x} \neq 0$ of $B\bar{x} = 0$ for $B \in K^{N \times N}$, $\bar{x} \in K^N$ (where B corresponds to the system matrix of $\mathcal{M}$ when computing XL) is the same as in the original Wiedemann algorithm: Assume that the characteristic polynomial $f(\lambda) = \sum_{0 \leq i} f[i]\lambda^i$ of B is known. Since B is singular, it has an eigenvalue 0, thus $f(\bar{B}) = 0$ and $f[0] = 0$. We have:

$$f(B)\bar{z} = \sum_{i>0} f[i]B^i \bar{z} = B \sum_{i>0} f[i]B^{i-1}\bar{z} = 0,$$

for any vector $\bar{z} \in K^N$. Therefore, $\bar{x} = \sum_{i>0} f[i]B^{i-1}\bar{z}$, $\bar{z} \neq 0$ is a (hopefully non-zero) kernel vector and thus a solution of the linear equation system. In fact it is possible to use any *annihilating* polynomial $f(\lambda)$ of B, i.e., a polynomial $f(\lambda) \neq 0$ such that $f(B) = 0$.

Wiedemann suggests to use the Berlekamp–Massey algorithm for the computation of $f(\lambda)$. Given a linearly recurrent sequence $\{a^{(i)}\}_{i=0}^{\infty}$, the algorithm computes $c_1, \ldots, c_d$ for some d such that $c_1 a^{(d-1)} + c_2 a^{(d-2)} + \cdots + c_d a^{(0)} = 0$.

Choosing $a^{(i)} = \bar{x}^T B B^i \bar{z}$ with random vectors $\bar{x}$ and $\bar{z}$ (as delegates for BB^i) as input and $f[i] = c_{d-i}$, $0 \le i < d$ as output returns $f(\lambda)$ as an annihilating polynomial of B with high probability.

Coppersmith [2] proposed a modification of the Wiedemann algorithm that makes it more suitable for modern computer architectures by operating in parallel on a *block* of $\tilde{n}$ column vectors $\bar{z}_i$, $0 \le i < \tilde{n}$, of a matrix $z \in K^{N \times \tilde{n}}$. His block Wiedemann algorithm computes kernel vectors in three steps which are called BW1, BW2, and BW3 for the remainder of this paper. The *block sizes* of the block Wiedemann algorithm are the integers $\tilde{m}$ and $\tilde{n}$. They can be chosen freely for the implementation such that they give the best performance on the target architecture for matrix and vector operations, e.g., depending on the size of cache lines or vector registers. Step BW1 computes the first $N/\tilde{m} + N/\tilde{n} + O(1)$ elements of a sequence $\{a^{(i)}\}_{i=0}^{\infty}$, $a_i = \left(x \cdot (B \cdot B^i z) \right)^T \in K^{\tilde{n} \times \tilde{m}}$ using random matrices $x \in K^{\tilde{m} \times N}$ and $z \in K^{N \times \tilde{n}}$. This sequence is the input for the second step BW2, a block variant of the Berlekamp–Massey algorithm. It returns a matrix polynomial $f(\lambda)$ with coefficients $f[j] \in K^{\tilde{n} \times \tilde{n}}$, that is used by step BW3 to compute up to $\tilde{n}$ solution vectors in a blocked fashion similar as described above for the original Wiedemann algorithm.

3   Implementation of XL

Stage BW1 of the block Wiedemann algorithm computes $a^{(i)} = \left(x \cdot (B \cdot B^i z) \right)^T$, $0 \le i \le N/\tilde{m} + N/\tilde{n} + O(1)$. We do this efficiently using two sparse-matrix multiplications by making the random matrices x and z deliberately sparse. We compute a sequence $\{t^{(i)}\}_{i=0}^{\infty}$ of matrices $t^{(i)} \in K^{N \times n}$ defined as

$$t^{(i)} = \begin{cases} Bz & \text{for } i = 0 \\ Bt^{(i-1)} & \text{for } i > 0. \end{cases}$$

Thus, $a^{(i)}$ can be computed as $a^{(i)} = (xt^{(i)})^T$. In step BW3 we evaluate the annihilating polynomial $f(\lambda)$ by applying Horner's scheme, again using two sparse-matrix multiplications by computing

$$W^{(j)} = \begin{cases} z \cdot (f[\deg(f)]) & \text{for } j = 0, \\ z \cdot (f[\deg(f) - j]) + B \cdot W^{(j-1)} & \text{for } 0 < j \le \deg(f). \end{cases}$$

For details on the steps BW1, BW2, and BW3 please refer to [18, Sec. 4.2].

Assuming that $\tilde{m} = c \cdot \tilde{n}$ for some constant $c \ge 1$, the asymptotic time complexity of step BW1 and BW2 can be written as $O(N^2 \cdot w_B)$, where w_B is the average number of nonzero entries per row of B. Note that BW3 actually requires about half of the time of BW1 since it requires only about halve as many iterations. The asymptotic time complexity of Coppersmith's version of the Berlekamp–Massey algorithm in step BW2 is $O(N^2 \cdot \tilde{n})$. Thomé presents an improved version of Coppersmith's block Berlekamp–Massey algorithm in [19]. Thomé's version is asymptotically faster: It reduces the complexity of BW2 from $O(N^2 \cdot \tilde{n})$ to $O(N \cdot \log^2(N) \cdot \tilde{n})$. The subquadratic complexity is achieved by converting the block Berlekamp–Massey algorithm into a recursive divide-and-conquer process.

Since steps BW1 and BW3 have a higher asymptotic time complexity than Thomé's version of step BW2, we do not describe our implementation, optimization, and parallelization of Coppersmith's and Thomé's versions of step BW2 in detail in this paper for the sake of brevity. The interested reader is referred to [18, Chap. 4] for details. However, we discuss the performance of our implementations in Sec. 4.

Since the system matrix $\mathcal{M}$ has more rows than columns, some rows must be dropped randomly to obtain a square matrix B. Observe that due to the extension step of XL the entries of the original quadratic system $\mathcal{A}$ appear repeatedly in the matrix B at well-defined positions based on the enumeration scheme. Therefore, it is possible to generate the entries of B on demand spending a negligible amount of memory. However, the computation of the entry positions requires additional time; to avoid this computational overhead, we store the Macaulay matrix B in a compact memory format (see [18, Sec. 4.5.3]). This gives a significant speedup in the computation time—given that the matrix B fits into available memory.

3.1   Efficient Matrix Multiplication

All matrix multiplications of the shape $D = EF$ that we perform during XL are either multiplications of a sparse matrix by a dense matrix, or multiplications of a dense matrix by a dense matrix with matrices of a small size. For these cases, schoolbook multiplication is more efficient than the *asymptotically* more efficient Strassen algorithm or the Coppersmith–Winograd algorithm.

However, when computing in finite fields, the cost of matrix multiplications can be significantly reduced by trading expensive multiplications for cheap additions—if the field size is significantly larger than the row weight of E. This is the case for small fields like, for example, $\mathbb{F}_{16}$ or $\mathbb{F}_{31}$. We reduce the number of actual multiplications for a row r of E by summing up all row vectors of F which are to be multiplied by the same field element and performing the multiplication on all of them together. A temporary buffer $b_\alpha \in K^n, \alpha \in K$ of vectors of length n is used to collect the sum of row vectors that ought to be multiplied by α. For all entries $E_{r,c}$, row c of F is added to $b_{E_{r,c}}$. Finally, b can be reduced by computing $\sum \alpha \cdot b_\alpha, \alpha \neq 0, \alpha \in K$, which gives the result for row r of the matrix D.

With the strategy explained so far, computing the result for one row of E takes $w_E + |K| - 2$ additions and $|K| - 2$ scalar multiplications (there is no need for the multiplication by 0 and 1, and for the addition of 0). The number of actual multiplications can be further reduced by exploiting the distributivity of the scalar multiplication of vectors: Assume in the following that $K = \mathbb{F}_{p^k} = \mathbb{F}_p[x]/(f(x))$, with p prime and $f(x)$ an irreducible polynomial with $\deg(f) = k$. When $k = 1$, the natural mapping from K to $\{0, 1, \ldots, p-1\} \subset \mathbb{N}$ induces an order of the elements. The order can be extended for $k > 1$ by $\forall \beta, \gamma \in K$:

$\beta > \gamma \iff \beta[i] > \gamma[i], i = \max(\{j : \beta[j] \neq \gamma[j]\})$. We decompose each scalar factor $\alpha \in K \setminus \{0, 1, x^1, \ldots, x^{k-1}\}$ of a multiplication $\alpha \cdot b_\alpha$ into two components $\beta, \gamma \in K$ such that $\beta, \gamma < \alpha$ and $\beta + \gamma = \alpha$. Starting with the largest α, iteratively add b_α to b_β and b_γ and drop buffer b_α. The algorithm terminates when all buffers b_α, $\alpha \in K \setminus \{0, 1, x^1, \ldots, x^{k-1}\}$ have been dropped. Finally, the remaining buffers b_α, $\alpha \in \{1, x^1, \ldots, x^{k-1}\}$ are multiplied by their respective scalar factor (except b_1) and summed up to the final result. This reduces the number of multiplications to $k - 1$. All in all the computation on one row of E (with row weight w_E) costs $w_E + 2(|K| - k - 1) + k - 1$ additions and $k - 1$ scalar multiplications. For example the computations in $\mathbb{F}_{16}$ require $w_E + 25$ additions and 3 multiplications per row of a matrix E.

3.2   Parallel Macaulay Matrix Multiplication

The most expensive part in the computation of steps BW1 and BW3 of XL is a repetitive multiplication of the shape $t_{new} = B \cdot t_{old}$, where $t_{new}, t_{old} \in K^{N \times \tilde{n}}$ are dense matrices and $B \in K^{N \times N}$ is a sparse Macaulay matrix with an average row weight w_B.

For generic systems, the Macaulay matrix B has an expected number of non-zero entries per row of $(|K| - 1)/|K| \cdot \binom{n+2}{2}$. However, in our memory efficient data format for the Macaulay matrix we also store the zero entries from the original system. This results in a fixed row weight $w_B = |K| \cdot \binom{n+2}{2}$. This is highly efficient in terms of memory consumption and computation time for $\mathbb{F}_{16}$, $\mathbb{F}_{31}$, and larger fields (see [18, Chap. 4]). Since there is a guaranteed number of entries per row (i.e. the row weight w_B) we compute the Macaulay matrix multiplication in row order in a big loop over all row indices as described in the previous section.

The parallelization of the Macaulay matrix multiplication of steps BW1 and BW3 is implemented in two ways: On multi-core architectures OpenMP is used to keep all cores busy; on cluster architectures the Message Passing Interface (MPI) and InfiniBand verbs are used to communicate between the cluster nodes. Both approaches can be combined for clusters of multi-core nodes.

The strategy of the workload distribution is similar on both multi-core systems and cluster systems. Figure 1 shows an example of a Macaulay matrix. Our approach for efficient matrix multiplications (described in the previous section) trades multiplications for additions. The approach is most efficient, if the original number of scalar multiplications per row is much higher than the order of the field. Since the row weight of the Macaulay matrix is quite small, splitting the rows between computing nodes reduces the efficiency of our approach. Therefore, the workload is distributed by assigning blocks of rows of the Macaulay matrix to the computing units.

Parallelization for Shared-Memory Systems: We parallelize the data-independent loop over the rows of the Macaulay matrix using OpenMP with the directive "`#pragma omp parallel for`". The OpenMP parallelization on UMA systems encounters no additional communication cost although the pressure on

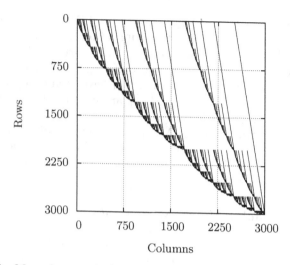

Fig. 1. Plot of a Macaulay matrix for a system with 8 variables, 10 equations, using graded reverse lexicographical (grevlex) monomial order

shared caches may be increased. On NUMA systems the best performance is achieved if the data is distributed over the NUMA nodes in a way that takes the higher cost of remote memory access into account. However, the access pattern to t_{old} is very irregular due to the structure of the Macaulay matrix: In particular, the access pattern of each core does not necessarily fully cover memory pages. Furthermore, the same memory page is usually touched by several cores. The same is true for t_{new}, since after each iteration t_{new} and t_{old} are swapped by switching their respective memory regions. Therefore, we obtained the shortest runtime by distributing the memory pages interleaved (in a round-robin fashion) over the nodes.

Parallelization for Cluster Systems: The computation on one row of the Macaulay matrix depends on many rows of the matrix t_{old}. A straightforward approach is to make the full matrix t_{old} available on all cluster nodes. This can be achieved by an all-to-all communication step after each iteration of BW1 and BW3. If B were a dense matrix, such communication would take only a small portion of the overall runtime. But since B is a sparse Macaulay matrix which has a very low row weight, the computation time for one single row of B takes only a small amount of time. In fact this time is in the order of magnitude of the time that is necessary to send one row of t_{new} to all other nodes during the communication phase. Therefore, this simple workload-distribution pattern gives a large communication overhead.

This overhead is hidden when communication is performed in parallel to computation. Today's high-performance network interconnects are able to transfer data via direct memory access (DMA) without interaction with the CPU, allowing the CPU to continue computations alongside communication. It is possible to split the computation of t_{new} into two column blocks; during computation on one block, previously computed results are distributed to the other nodes and therefore are available at the next iteration step. Under the condition that computation takes more time than communication, the communication overhead can almost entirely be hidden. Otherwise speedup and therefore efficiency of cluster parallelization is bounded by communication cost.

Apart from hiding the communication overhead it is also possible to totally avoid all communication by splitting t_{old} and t_{new} into independent column blocks for each cluster node. However, splitting t_{old} and t_{new} has an impact either on the runtime of BW1 and BW3 (if the block size becomes too small for efficient computation) or on the runtime of BW2 (since the block size has a strong impact on its runtime and memory demand).

We implemented both approaches since they can be combined to give best performance on a target system architecture. The following paragraphs explain the two approaches in detail:

a) *Operating on Two Shared Column Blocks of t_{old} and t_{new}:* For this approach, the matrices t_{old} and t_{new} are split into two column blocks $t_{old,0}$ and $t_{old,1}$ as well as $t_{new,0}$ and $t_{new,1}$. The workload is distributed over the nodes row-wise as mentioned before. First each node computes the results of its row range for column block $t_{new,0}$ using rows from block $t_{old,0}$. Then a non-blocking all-to-all communication is initiated which distributes the results of block $t_{new,0}$ over all nodes. While the communication is going on, the nodes compute the results of block $t_{new,1}$ using data from block $t_{old,1}$. After computation on $t_{new,1}$ is finished, the nodes wait until the data transfer of block $t_{new,0}$ has been accomplished. Ideally communication of block $t_{new,0}$ is finished earlier than the computation of block $t_{new,1}$ so that the results of block $t_{new,1}$ can be distributed without waiting time while the computation on block $t_{new,0}$ goes on with the next iteration step.

However, looking at the structure of the Macaulay matrix (an example is shown in Fig. 1) one can observe that this communication scheme performs much more communication than necessary. For example on a cluster of four computing nodes, node 0 computes the top quarter of the rows of matrices $t_{new,0}$ and $t_{new,1}$. Node 1 computes the second quarter, node 2 the third quarter, and node 3 the bottom quarter. Node 3 does not require any row that has been computed by node 0 since the Macaulay matrix does not have entries in the first quarter of the columns for these rows. The obvious solution is that a node i sends only these rows to a node j that are actually required by node j in the next iteration step.

This communication pattern requires to send several data blocks to individual cluster nodes in parallel to ongoing computation. This can not be done efficiently using MPI. Therefore, we circumvent the MPI API and program the network hardware directly. Our implementation uses an InfiniBand network; the same approach can be used for other high-performance networks. We access the InfiniBand hardware using the InfiniBand verbs API. Programming the InfiniBand cards directly has several benefits: All data structures that are required for communication can be prepared offline; initiating communication requires only one call to the InfiniBand API. The hardware is able to perform all operations for sending and receiving data autonomously after this API call; there is no need for calling further functions to ensure communication progress as it is necessary when using MPI. Finally, complex communication patterns using scatter-gather lists for incoming and outgoing data do not have a large overhead. This implementation reduces communication to the smallest amount possible for the cost of only a negligibly small initialization overhead.

This approach of splitting t_{old} and t_{new} into two shared column blocks has the disadvantage that the entries of the Macaulay matrix need to be loaded twice per iteration, once for each block. This gives a higher memory contention and more cache misses than when working on a single column block. However, these memory accesses are sequential. It is therefore likely that the access pattern can be detected by the memory logic and that the data is prefetched into the caches.

b) *Operating on Independent Column Blocks of t_{old} and t_{new}*: Any communication during steps BW1 and BW3 can be avoided by splitting the matrices t_{old} and t_{new} into independent column blocks for each cluster node. The nodes compute over the whole Macaulay matrix B on a column stripe of t_{old} and t_{new}. All computation can be accomplished locally; the results are collected at the end of the computation of these steps.

Although this is the most efficient parallelization approach when looking at communication cost, the per-node efficiency drops drastically with higher node count: For a high node count, the impact of the width of the column stripes of t_{old} and t_{new} becomes even stronger than for the previous approach. Therefore, this approach only scales well for small clusters. For a large number of nodes, the efficiency of the parallelization declines significantly. Another disadvantage of this approach is that since the nodes compute on the whole Macaulay matrix, all nodes must store the whole matrix in their memory. For large systems this is may not be feasible.

Both approaches for parallelization have advantages and disadvantages; the ideal approach can only be found by testing each approach on the target hardware. For small clusters approach b) might be the most efficient one although it loses efficiency due to the effect of the width of t_{old} and t_{new}. The performance of approach a) depends heavily on the network configuration and the ratio between computation time and communication time. Both approaches can be combined by splitting the cluster into independent partitions; the workload is distributed over the partitions using approach b) and over the nodes within one partition using approach a).

4   Experimental Results

This section gives an overview of the performance and the scalability of our XL implementation for generic systems. Experiments have been carried out on

Table 1. Computer architectures used for the experiments

	NUMA	Cluster
CPU		
Name	AMD Opteron 6276	Intel Xeon E5620
Microarchitecture	Bulldozer Interlagos	Nehalem
Frequency	2300 MHz	2400 MHz
Number of CPUs per socket	2	1
Number of cores per socket	16 (2 x 8)	4
Level 1 data-cache size	16 × 48 KB	4 × 32 KB
Level 2 data-cache size	8 × 2 MB	4 × 256 KB
Level 3 data-cache size	2 × 8 MB	8 MB
Cache-line size	64 byte	64 byte
System Architecture		
Number of NUMA nodes	4 sockets × 2 CPUs	2 sockets × 1 CPU
Number of cluster nodes	—	8
Total number of cores	64	64
Network interconnect	—	InfiniBand MT26428
		2 ports of 4×QDR, 32 Gbit/s
Memory		
Memory per CPU	32 GB	18 GB
Memory per cluster node	—	36 GB
Total memory	256 GB	288 GB

two computer systems: a 64-core NUMA system and an eight node InfiniBand cluster. Table 1 lists the key features of these systems.

4.1   Impact of the Block Size

We measured the impact of the block size of the block Wiedemann algorithm on the performance of the implementation on a single cluster node (without cluster communication). We used a quadratic system with 16 equations and 14 variables over $\mathbb{F}_{16}$. In this case, the degree D for the linearization is 9. The input for the algorithm is a Macaulay matrix B with $N = 817190$ rows (and columns) and row weight $w_B = 120$. To reduce the parameter space, we fix $\tilde{m}$ to $\tilde{m} = \tilde{n}$.

Figure 2 shows the runtime for block sizes 32, 64, 128, 256, 512, and 1024. Given the fixed size of the Macaulay matrix and $\tilde{m} = \tilde{n}$, the number of field operations for BW1 and BW2 is roughly the same for different choices of the block size $\tilde{n}$ since the number of iterations is proportional to $1/\tilde{n}$ and number of field operations per iteration is roughly proportional to $\tilde{n}$. However, the runtime of the computation varies depending on $\tilde{n}$.

During the i-th iteration step of BW1 and BW3, the Macaulay matrix is multiplied with a matrix $t^{(i-1)} \in \mathbb{F}_{16}^{N \times \tilde{n}}$. For $\mathbb{F}_{16}$ each row of $t^{(i-1)}$ requires $\tilde{n}/2$ bytes of memory. In the cases $\tilde{m} = \tilde{n} = 32$ and $\tilde{m} = \tilde{n} = 64$ each row thus occupies less than one cache line of 64 bytes. This explains why the best performance in BW1 and BW3 is achieved for larger values of $\tilde{n}$. The runtime of BW1 and BW3 is minimal for block sizes $\tilde{m} = \tilde{n} = 256$. In this case one row of $t^{(i-1)}$ occupies two cache

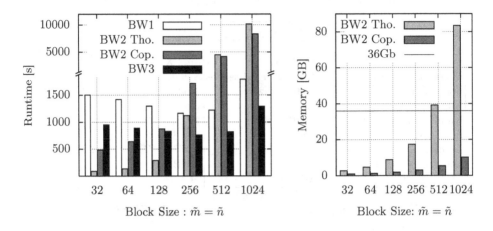

Fig. 2. Runtime and memory consumption of XL 16-14 over different block sizes on a single cluster node with two CPUs (8 cores in total) and 36 GB RAM

lines. The reason why this case gives a better performance than $\tilde{m} = \tilde{n} = 128$ might be that the memory controller is able to prefetch the second cache line. For larger values of $\tilde{m}$ and $\tilde{n}$ the performance declines probably due to cache saturation.

According to the asymptotic time complexity of Coppersmith's and Thomé's versions of the Berlekamp–Massey algorithm, the runtime of BW2 should be proportional to $\tilde{n}$. However, this turns out to be the case only for moderate sizes of $\tilde{n}$; note the different scale of the graph in Fig. 2 for a runtime of more than 2000 seconds. For $\tilde{m} = \tilde{n} = 256$ the runtime of Coppersmith's version of BW2 is already larger than that of BW1 and BW3, for $\tilde{m} = \tilde{n} = 512$ and $\tilde{m} = \tilde{n} = 1024$ both versions of BW2 dominate the total runtime of the computation. Thomé's version is faster than Coppersmith's version for small and moderate block sizes. However, by doubling the block size, the memory demand of BW2 roughly doubles as well; Figure 2 shows the memory demand of both variants for this experiment. Due to the memory–time trade-off of Thomé's BW2, the memory demand exceeds the available RAM for a block size of $\tilde{m} = \tilde{n} = 512$ and more. Therefore, memory pages are swapped out of RAM onto hard disk which makes the runtime of Thomé's BW2 longer than that of Coppersmith's version of BW2.

4.2   Scalability Experiments

The scalability was measured using a quadratic system with 18 equations and 16 variables over $\mathbb{F}_{16}$. The degree D for this system is 10. The Macaulay matrix B has a size of $N = 5\,311\,735$ rows and columns; the row weight is $w_B = 153$. Since this experiment is not concerned about peak performance but about scalability, a block size of $\tilde{m} = \tilde{n} = 256$ is used. For this experiment, the implementation of the block Wiedemann algorithm ran on 1, 2, 4, and 8 nodes of the cluster

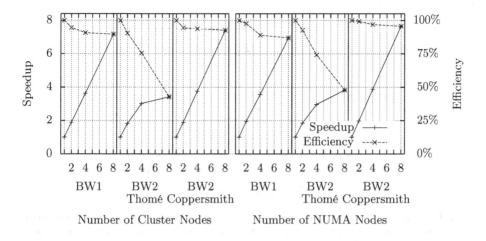

Fig. 3. Speedup and efficiency of BW1 and BW2

and on 1 to 8 CPUs of the NUMA system. The approach a) (two shared column blocks) was used on the cluster system for all node counts.

Given the runtime T_1 for one computing node and T_p for p computing nodes, the parallel efficiency E_p on the p nodes is defined as $E_p = T_1/pT_p$. Figure 3 shows the parallel speedup and the parallel efficiency of BW1 and BW2; the performance of BW3 behaves very similarly to BW1 and thus is not depicted in detail. These figures show that BW1 and Coppersmith's BW2 have a nice speedup and an efficiency of at least 90% on 2, 4, and 8 cluster nodes. The efficiency of Thomé's BW2 is only around 75% on 4 nodes and drops to under 50% on 8 nodes. In particular the polynomial multiplications require a more efficient parallelization approach. However, Thomé's BW2 takes only a small part of the total runtime for this system size; for larger systems it is even smaller due to its smaller asymptotic time complexity compared to steps BW1 and BW3. Thus, a lower scalability than BW1 and BW3 can be tolerated for BW2.

For this problem size, our parallel implementation of BW1 and BW3 scales very well for up to eight nodes. However, at some point the communication time is going to catch up with computation time: The computation time roughly halves with every doubling of the number of cluster nodes, while the communication demand per node shrinks with a smaller slope. Therefore, at a certain number of nodes communication time and computation time are about the same and the parallel efficiency declines for any larger number of nodes. We do not have access to a cluster with a fast network interconnect and a sufficient amount of nodes to measure when this point is reached, thus we can only give an estimation: Figure 4 shows the expected time of computation and communication for larger cluster sizes. We computed the amount of data that an individual node sends and receives depending on the number of computing nodes. We use the maximum of the outgoing data for the estimation of the communication time. For this particular problem size, we expect that for a cluster of around 16 nodes

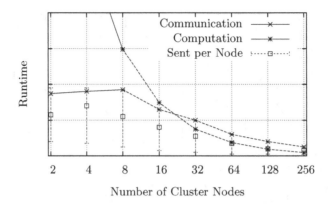

Fig. 4. Estimation of computation time vs. communication time on a cluster system. The numbers for 2, 4, and 8 nodes are measurements, the numbers for larger cluster sizes are estimations. The amount of data sent per node varies; we show the maximum, minimum, and average.

communication time is about as long as computation time and that the parallel efficiency is going to decline for larger clusters.

On the NUMA system, the scalability is similar to the cluster system. BW1 achieves an efficiency of over 85% on up to 8 NUMA nodes. The workload was distributed such that each CPU socket was filled up with OpenMP threads as much as possible. Therefore, in the case of two NUMA nodes (16 threads) the implementation achieves a high efficiency of over 95% since a memory controller on the same socket is used for remote memory access and the remote memory access has only moderate cost. When using more than one NUMA node, the efficiency declines to around 85% due to the higher cost of remote memory access between different sockets. Also on the NUMA system the parallelization of Thomé's BW2 achieves only a moderate efficiency of around 50% for 8 NUMA nodes. The parallelization scheme used for OpenMP does not scale well for a large number of threads. The parallelization of Coppersmith's version of BW2 scales almost perfectly on the NUMA system. The experiment with this version of BW2 is performed using hybrid parallelization by running one MPI process per NUMA node and one OpenMP thread per core. The overhead for communication is sufficiently small that it does not have much impact on the parallel efficiency of up to 8 NUMA nodes.

Our experiments show that the shape of the Macaulay matrix has a large impact on the performance and the scalability of XL. Currently, we are using graded reverse lexicographical order for the Macaulay matrix. However, as opposed to Gröbner basis solvers like F_4 and F_5, for XL there is no algorithmic or mathematical requirement for any particular ordering. In our upcoming research, we are going to examine if another monomial order or a redistribution of columns and rows of the Macaulay matrix has a positive impact on the performance of our implementation.

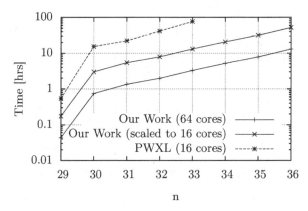

Fig. 5. Comparison of the runtime of our work and PWXL, $m = n$, $\mathbb{F}_2$

4.3 Comparison with PWXL and Magma F_4

To put our numbers into context, we compare our work with two other Gröbner basis solvers in this section: with PWXL, a parallel implementation of XL with block Wiedemann for $\mathbb{F}_2$ described in [15], and with the implementation of Faugère's F_4 algorithm [7] in the computational algebra system Magma.

Comparison with PWXL: Figure 5 compares the runtime of PWXL and our implementation for systems in $\mathbb{F}_2$ with $m = n$. We ran our XL implementation on our cluster system (see Table 1) while PWXL was running on a machine with four six-core AMD Opteron 8435 CPUs, running at 2.6 GHz.

Our implementation outperforms PWXL for the largest cases given in the paper, e.g., for $n = 33$ our implementation is 24 times faster running on 8 cluster nodes (64 CPU cores) and still 6 times faster when scaling to 16 CPU cores. This significant speedup may be explained by the fact that PWXL is a modification of the block-Wiedemann solver for factoring RSA-768 used in [10]. Therefore, the code may not be well optimized for the structure of Macaulay matrices. However, these numbers show that our implementation achieves high performance for computations in $\mathbb{F}_2$.

Comparison with F_4: Figure 6 compares time and memory consumption of the F_4 implementation in Magma V2.17-12 and our implementation of XL for systems in $\mathbb{F}_{16}$ with $m = 2n$. When solving the systems in Magma we coerce the systems into $\mathbb{F}_{256}$, because for $\mathbb{F}_{256}$ Magma performs faster than when using $\mathbb{F}_{16}$ directly. The computer used to run F_4 has an 8 core Xeon X7550 CPU running at 2.0 GHz; however, F_4 uses only one core of it. We ran XL on our NUMA system using all 64 CPU cores. For this comparison we use Coppersmith's version of BW2 since it is more memory efficient than Thomé's version.

Note that there is a jump in the graph when going from $n = 21$ to $n = 22$ for XL our implementation, similarly when going from $n = 23$ to $n = 24$

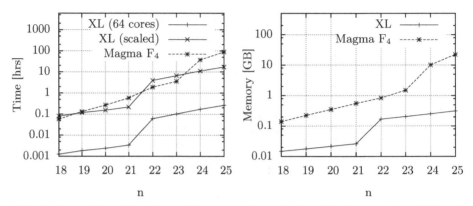

Fig. 6. Comparison of runtime and memory demand of our implementation of XL and Magma's implementation of F_4, $m = 2n$

for F_4. This is due to an increment of the degree D from 5 to 6, which happens earlier for XL. Therefore, F_4 takes advantage of a lower degree in cases such as $n = 22, 23$. Other XL-based algorithms like Mutant-XL [14] may be able to fill this gap. In this paper we omit a discussion of the difference between the degrees of XL and F_4/F_5. However, in cases where the degrees are the same for both algorithms, our implementation of XL is better in terms of runtime and memory consumption.

For $n = 25$, the memory consumption of XL is less than 2% of that of F_4. In this case, XL runs 338 times faster on 64 cores than F_4 on one single core, which means XL is still faster when the runtime is normalized to single-core performance by multiplying the runtime by 64.

4.4   Performance for Computation on Large Systems

Table 2 presents detailed statistics of some of the largest systems we are able to solve in a moderate amount of time (within at most one week). In the tables the time (BW1, BW2, BW3, and total) is measured in seconds, and the memory is measured in GB. Note that for the cluster we give the memory usage for a single cluster node. While all the fields that we have implemented so far are presented in the table, we point out that the most optimization has been done for $\mathbb{F}_{16}$.

The system with $n = 32$ variables and $m = 64$ equations over $\mathbb{F}_{16}$ listed in Table 2 is the largest case we have tested. The system was solved in 5 days on the cluster using block sizes $\tilde{m} = 256$ and $\tilde{n} = 128$. With $n = 32$ and $D = 7$ we have $N = \binom{n+D}{D} = \binom{32+7}{7} = 15\,380\,937$ and $w_B = \binom{n+2}{2} = \binom{32+2}{2} = 561$. There are roughly $N/\tilde{n} + N/\tilde{m}$ iterations in BW1 and $N/\tilde{n}$ iterations in BW3. This leads to $2N/\tilde{n} + N/\tilde{m}$ Macaulay matrix multiplications, each takes about $N \cdot (w_B + 25) \cdot \tilde{n}$ additions and $N \cdot 3 \cdot \tilde{n}$ multiplications in $\mathbb{F}_{16}$ (see Sec. 3.2). Operations performed in BW2 are not taken into account, because BW2 requires only a negligible amount of time. Therefore, solving the system using XL corresponds to computing about $(2 \cdot 15\,380\,937/128 + 15\,380\,937/256) \cdot 15\,380\,937 \cdot$

Table 2. Statistics of XL with block Wiedemann for $\mathbb{F}_2$ and $\mathbb{F}_{16}$ using Thomé's BW2, and $\mathbb{F}_{31}$ using Coppersmith's BW2

Field	Machine	m	n	D	Time in [sec] BW1	BW2	BW3	total	Memory in [GB]	Block Size $\tilde{m}, \tilde{n}$
$\mathbb{F}_2$	Cluster	32	32	7	3830	1259	2008	7116	2.4	512, 512
	Cluster	33	33	7	6315	2135	3303	11778	3.0	512, 512
	Cluster	34	34	7	10301	2742	5439	18515	3.8	512, 512
	Cluster	35	35	7	16546	3142	8609	28387	4.6	512, 512
	Cluster	36	36	7	26235	5244	15357	46944	5.6	512, 512
$\mathbb{F}_{16}$	NUMA	56	28	6	1866	330	984	3183	3.9	128,128
	Cluster				1004	238	548	1795	1.3	256,256
	NUMA	58	29	6	2836	373	1506	4719	4.6	128,128
	Cluster				1541	316	842	2707	1.6	256,256
	NUMA	60	30	7	91228	5346	64688	161287	68.8	256,128
	Cluster				53706	3023	38052	94831	10.2	256,128
	NUMA	62	31	7	145693	7640	105084	258518	76.7	256,128
	Cluster				89059	3505	67864	160489	12.1	256,128
	NUMA	64	32	7	232865	8558	163091	404551	100.3	256,128
	Cluster				141619	3672	97924	244338	15.3	256,128
$\mathbb{F}_{31}$	NUMA	50	25	6	1729	610	935	3277	0.3	64,64
	Cluster				1170	443	648	2265	0.7	128,128
	NUMA	52	26	6	2756	888	1483	5129	0.4	64,64
	Cluster				1839	656	1013	3513	0.9	128,128
	NUMA	54	27	6	4348	1321	2340	8013	0.5	64,64
	Cluster				2896	962	1590	5453	1.0	128,128
	NUMA	56	28	6	6775	1923	3610	12313	0.6	64,64
	Cluster				4497	1397	2458	8358	1.2	128,128
	NUMA	58	29	6	10377	2737	5521	18640	0.7	64,64
	Cluster				6931	2011	3764	12713	1.5	128,128

$(561 + 25) \cdot 128 \approx 2^{58.3}$ additions and about $2^{50.7}$ multiplications in $\mathbb{F}_{16}$. Since one addition in $\mathbb{F}_{16}$ requires 4 bit operations, this roughly corresponds to the computation of $4 \cdot 2^{58.3} \approx 2^{60.3}$ bit operations.

Acknowledgements. This work was in part supported by National Science Council (NSC), National Taiwan University, and Intel Corporation under Grants NSC 100-2911-I-002-001 and 101R7501. The authors would also like to thank partial sponsorship from NSC under Grants 100-2218-E-001-002 and 100-2628-E-001-004-MY3, as well as from Academia Sinica including a Career Award to the fourth author. Furthermore, this work was supported by the European Commission under Contract ICT-2007-216676 ECRYPT II. Part of this work was done when the third author was employed at Eindhoven University of Technology in the Netherlands.

References

1. Ars, G., Faugère, J.-C., Imai, H., Kawazoe, M., Sugita, M.: Comparison Between XL and Gröbner Basis Algorithms. In: Lee, P.J. (ed.) ASIACRYPT 2004. LNCS, vol. 3329, pp. 338–353. Springer, Heidelberg (2004)
2. Coppersmith, D.: Solving Homogeneous Linear Equations Over GF(2) via Block Wiedemann Algorithm. Mathematics of Computation 62(205), 333–350 (1994)
3. Courtois, N.T.: Higher Order Correlation Attacks, XL Algorithm and Cryptanalysis of Toyocrypt. In: Lee, P.J., Lim, C.H. (eds.) ICISC 2002. LNCS, vol. 2587, pp. 182–199. Springer, Heidelberg (2003)
4. Courtois, N.T., Klimov, A., Patarin, J., Shamir, A.: Efficient Algorithms for Solving Overdefined Systems of Multivariate Polynomial Equations. In: Preneel, B. (ed.) EUROCRYPT 2000. LNCS, vol. 1807, pp. 392–407. Springer, Heidelberg (2000)
5. Courtois, N.T., Pieprzyk, J.: Cryptanalysis of Block Ciphers with Overdefined Systems of Equations. In: Zheng, Y. (ed.) ASIACRYPT 2002. LNCS, vol. 2501, pp. 267–287. Springer, Heidelberg (2002)
6. Diem, C.: The XL-Algorithm and a Conjecture from Commutative Algebra. In: Lee, P.J. (ed.) ASIACRYPT 2004. LNCS, vol. 3329, pp. 323–337. Springer, Heidelberg (2004)
7. Faugère, J.-C.: A New Efficient Algorithm for Computing Gröbner Bases (F_4). Journal of Pure and Applied Algebra 139(1-3), 61–88 (1999)
8. Faugère, J.-C.: A New Efficient Algorithm for Computing Gröbner Bases without Reduction to Zero (F_5). In: ISSAC 2002, pp. 75–83. ACM (2002)
9. Faugère, J.-C., Perret, L., Petit, C., Renault, G.: Improving the Complexity of Index Calculus Algorithms in Elliptic Curves over Binary Fields. In: Pointcheval, D., Johansson, T. (eds.) EUROCRYPT 2012. LNCS, vol. 7237, pp. 27–44. Springer, Heidelberg (2012)
10. Kleinjung, T., Aoki, K., Franke, J., Lenstra, A.K., Thomé, E., Bos, J.W., Gaudry, P., Kruppa, A., Montgomery, P.L., Osvik, D.A., te Riele, H., Timofeev, A., Zimmermann, P.: Factorization of a 768-Bit RSA Modulus. In: Rabin, T. (ed.) CRYPTO 2010. LNCS, vol. 6223, pp. 333–350. Springer, Heidelberg (2010)
11. Lazard, D.: Gröbner-Bases, Gaussian Elimination and Resolution of Systems of Algebraic Equations. In: van Hulzen, J.A. (ed.) ISSAC 1983 and EUROCAL 1983. LNCS, vol. 162, pp. 146–156. Springer, Heidelberg (1983)
12. Macaulay, F.S.: The Algebraic Theory of Modular Systems. Cambridge Tracts in Mathematics and Mathematical Physics, vol. 19. Cambridge University Press (1916)
13. Moh, T.-T.: On the Method of XL and Its Inefficiency to TTM. Cryptology ePrint Archive, Report 2001/047 (2001), http://eprint.iacr.org/2001/047
14. Mohamed, M.S.E., Mohamed, W.S.A.E., Ding, J., Buchmann, J.: MXL2: Solving Polynomial Equations over GF(2) Using an Improved Mutant Strategy. In: Buchmann, J., Ding, J. (eds.) PQCrypto 2008. LNCS, vol. 5299, pp. 203–215. Springer, Heidelberg (2008)
15. Mohamed, W.S.A.E., Ding, J., Kleinjung, T., Bulygin, S., Buchmann, J.: PWXL: A Parallel Wiedemann-XL Algorithm for Solving Polynomial Equations Over GF(2). In: Cid, C., Faugère, J.-C. (eds.) SCC 2010, pp. 89–100 (2010)
16. Montgomery, P.L.: A Block Lanczos Algorithm for Finding Dependencies over GF(2). In: Guillou, L.C., Quisquater, J.-J. (eds.) EUROCRYPT 1995. LNCS, vol. 921, pp. 106–120. Springer, Heidelberg (1995)

17. Murphy, S., Robshaw, M.J.B.: Essential Algebraic Structure within the AES. In: Yung, M. (ed.) CRYPTO 2002. LNCS, vol. 2442, pp. 1–16. Springer, Heidelberg (2002)
18. Niederhagen, R.: Parallel Cryptanalysis. Ph.D. thesis, Eindhoven University of Technology (2012), http://polycephaly.org/thesis/index.shtml
19. Thomé, E.: Subquadratic Computation of Vector Generating Polynomials and Improvement of the Block Wiedemann Algorithm. Journal of Symbolic Computation 33(5), 757–775 (2002)
20. Wiedemann, D.H.: Solving Sparse Linear Equations Over Finite Fields. IEEE Transactions on Information Theory 32(1), 54–62 (1986)
21. Yang, B.-Y., Chen, C.-H., Bernstein, D.J., Chen, J.-M.: Analysis of QUAD. In: Biryukov, A. (ed.) FSE 2007. LNCS, vol. 4593, pp. 290–308. Springer, Heidelberg (2007)
22. Yang, B.-Y., Chen, J.-M.: All in the XL Family: Theory and Practice. In: Park, C., Chee, S. (eds.) ICISC 2004. LNCS, vol. 3506, pp. 67–86. Springer, Heidelberg (2005)
23. Yang, B.-Y., Chen, J.-M., Courtois, N.T.: On Asymptotic Security Estimates in XL and Gröbner Bases-Related Algebraic Cryptanalysis. In: López, J., Qing, S., Okamoto, E. (eds.) ICICS 2004. LNCS, vol. 3269, pp. 401–413. Springer, Heidelberg (2004)

Efficient Implementations of MQPKS
on Constrained Devices

Peter Czypek, Stefan Heyse, and Enrico Thomae

Horst Görtz Institute for IT Security
Ruhr University Bochum
44780 Bochum, Germany
{peter.czypek,stefan.heyse,enrico.thomae}@rub.de

Abstract. Multivariate Quadratic Public Key Schemes (MQPKS) attracted the attention of researchers in the last decades for two reasons. First they are thought to resist attacks by quantum computers and second, most of the schemes were broken. The latter may be the reason why implementations are rare. This work investigates one of the most promising member of MQPKS and its variants, namely UOV, Rainbow and enTTS. UOV resisted all kinds of attacks for 13 years and can be considered one of the best examined MQPKS. We describe implementations of UOV, Rainbow and enTTS on an 8-bit microcontroller. To address the problem of large keys, we used several optimizations and also implemented the 0/1-UOV scheme introduced at CHES 2011. To achieve a practically usable security level on the selected device, all recent attacks are summarized and parameters for standard security levels are given. To allow judgement of scaling, the schemes are implemented for the most common security levels in embedded systems 2^{64}, 2^{80} and 2^{128} bits symmetric security. This allows for the first time a direct comparison of the four schemes because they are implemented for exactly the same security levels on the same platform and also by the same developer.

Keywords: Multivariate Quadratic Signatures, MQ, Unbalanced Oil and Vinegar, UOV, Rainbow, enTTS, AVR, Embedded Device.

1 Introduction

Since Peter Shor published efficient quantum algorithms [20] to solve the problem of factorization and discrete logarithm in 1995, there is a increasing demand in investigating possible alternatives. One such class of so-called post-quantum cryptosystems is based on multivariate quadratic ($\mathcal{MQ}$) polynomials. We know that solving systems of $\mathcal{MQ}$-polynomials is hard in the worst case, as the corresponding $\mathcal{MQ}$-problem is proven to be $\mathcal{NP}$-complete [11]. Unfortunately all schemes proposed so far also need the Isomorphism of Polynomials (IP) problem to hide the trapdoor. It is not known how hard this problem is and indeed most $\mathcal{MQ}$-schemes are broken this way. So for example, the balanced Oil and Vinegar scheme [15], Sflash [4] and much more [17,16,12,7,8,23]. To encapsulate, nearly all $\mathcal{MQ}$-encryption schemes and most of the $\mathcal{MQ}$-signature schemes are

E. Prouff and P. Schaumont (Eds.): CHES 2012, LNCS 7428, pp. 374–389, 2012.

broken up to this point. There are only very few exceptions like the signature schemes HFE$^-$, Unbalanced Oil and Vinegar (UOV) and its layer based variants Rainbow and enTTS. Well, breaking the first seems to be a matter of time as some ideas of the attack against Sflash from Asiacrypt 2011 [4] might also be applicable. On the other hand, UOV resisted all kinds of attacks for 13 years. It is thought to be the most promising member of the class of $\mathcal{MQ}$-schemes.

Previous Work and Contribution. Rainbow type hardware implementations got some attention during the last years. An 0.35µm ASIC, which signs in 0.012 ms, is reported in [2]. Further [21] presents an ASIC implementation, taking only 198 clock cycles for a sign operation. An ASIC implementation of enTTS(20,28) enabling sign in 0.044 seconds running at a slow clock of 100KHz, is reported in [25]. The authors also report a MSP430 implementation signing in 71 ms and verifying in 726 ms and a 8051-compatible µC implementation signing in 198ms. At CHES 2004, Yang et al. describe an implementation of TTS targeting 8051-compatible µCs [1]. Their implementation of TTS(20,28) signs in 144ms, 170ms, 60ms and for TTS(24,32) they achieve 191ms, 227 ms, 85 ms for an i8032AH, i8051AH and W77E59, respectively. We are not aware of any implementation of UOV or Rainbow targeting small microcontrollers.

   This work describes implementations of the $\mathcal{MQ}$-signature schemes, UOV, Rainbow and enTTS, on an 8-bit microcontroller. Additionally, methods to reduce the key size are evaluated and a version of UOV published at CHES 2011 (0/1-UOV [19]) is introduced and also evaluated. To achieve a practically usable security level on the selected device, recent attacks are summarized and parameters for standard security levels are given. The actual implementations were all done by the same developer. This ensures, that we really compare different schemes and not just different skills of different developers.

Organization. Section 2 introduces $\mathcal{MQ}$-schemes in general and UOV, Rainbow and enTTS in special. Section 3 summaries recent attacks and derives parameter sets to achieve $2^{64}, 2^{80}$ and 2^{128} bit security. Afterwards, Section 4 describes our implementations before we present our results in Section 5. Finally, we conclude in Section 6 and point out some details for future improvements.

2   Multivariate Quadratic Public Key Cryptosystems

This section provides a brief introduction to UOV [14], 0/1 UOV [19], Rainbow [9] and enTTS [24]. The general idea of all these $\mathcal{MQ}$-signature schemes is to use a public multivariate quadratic map $\mathcal{P} : \mathbb{F}_q^n \to \mathbb{F}_q^m$ with

$$\mathcal{P} = \begin{pmatrix} p^{(1)}(x_1, \ldots, x_n) \\ \vdots \\ p^{(m)}(x_1, \ldots, x_n) \end{pmatrix}$$

and

$$p^{(k)}(x_1, \ldots, x_n) := \sum_{1 \leq i \leq j \leq n} \alpha_{ij}^{(k)} x_i x_j = x^\mathsf{T} \mathfrak{P}^{(k)} x,$$

where $\mathfrak{P}^{(k)}$ is the $(n \times n)$ matrix describing the quadratic form of $p^{(k)}$ and $x = (x_1, \ldots, x_n)^\mathsf{T}$. Note that we can neglect linear and constant terms as they never mix with quadratic terms and thus do not increase the security [5].

The trapdoor is given by a structured central map $\mathcal{F} : \mathbb{F}_q^n \to \mathbb{F}_q^m$ with

$$\mathcal{F} = \begin{pmatrix} f^{(1)}(u_1, \ldots, u_n) \\ \vdots \\ f^{(m)}(u_1, \ldots, u_n) \end{pmatrix}$$

and

$$f^{(k)}(u_1, \ldots, u_n) := \sum_{1 \le i \le j \le n} \gamma_{ij}^{(k)} u_i u_j = u^\mathsf{T} \mathfrak{F}^{(k)} u.$$

In order to hide this trapdoor we choose two secret linear transformations S, T and define $\mathcal{P} := T \circ \mathcal{F} \circ S$. See Figure 1 for an illustration.

Fig. 1. $\mathcal{MQ}$-Scheme in general

Unbalanced Oil and Vinegar. For the UOV signature scheme the variables u_i, $i \in V := \{1, \ldots, v\}$ are called *vinegar variables* and the remaining variables u_i, $i \in O := \{v+1, \ldots, n\}$ are called *oil variables*. The central map $\mathcal{F}$ is given by

$$f^{(k)}(u_1, \ldots, u_n) := \sum_{i \in V, j \in V} \gamma_{ij}^{(k)} u_i u_j + \sum_{i \in V, j \in O} \gamma_{ij}^{(k)} u_i u_j.$$

The corresponding matrix $\mathfrak{F}^{(k)}$ is depicted in Figure 2.

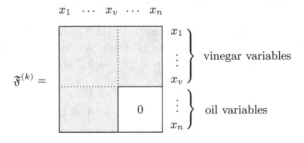

Fig. 2. Central map $\mathfrak{F}$ of UOV. White parts denote zero entries while gray parts denote arbitrary entries.

As we have m equations in $m + v$ variables, fixing v variables will yield a solution with high probability. Due to the structure of $\mathfrak{F}^{(k)}$, *i.e.* there are no quadratic terms of two oil variables, we can fix the vinegar variables at random to obtain a system of linear equations in the oil variables, which is easy to solve. This procedure is not possible for the public key, as the transformation S of variables fully mixes the variables (like oil and vinegar in a salad). Note that for UOV we can discard the transformation T of equations, as the trapdoor is invariant under this linear transformation.

Rainbow. Rainbow uses the same idea as UOV but in different layers. Current choices of parameters (q, v_1, o_1, o_2) use two layers, as it turned out to be the best choice in order to prevent MinRank attacks and preserve short signatures at the same time. We will use $q = 2^8$ throughout the paper. The central map $\mathcal{F}$ of Rainbow is divided into two layers $\mathfrak{F}^{(1)}, \dots, \mathfrak{F}^{(o_1)}$ and $\mathfrak{F}^{(o_1+1)}, \dots, \mathfrak{F}^{(o_1+o_2)}$ of form given in Figure 3.

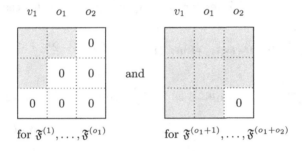

Fig. 3. Central map of Rainbow (q, v_1, o_1, o_2). White parts denote zero entries while gray parts denote arbitrary entries.

To use the trapdoor we first solve the small UOV system $\mathfrak{F}^{(1)}, \dots, \mathfrak{F}^{(o_1)}$ by randomly fixing the v_1 vinegar variables. The solution $u_1, \dots, u_{v_1+o_1}$ is now used as vinegar variables of the second layer. Solving the obtained linear system yields $u_{v_1+o_1+1}, \dots, u_{v_1+o_1+o_2}$. A formal description of Rainbow is given by the following formula.

$$f^{(k)}(u_1, \dots, u_n) := \sum_{i \in V_1, j \in V_1} \gamma_{ij}^{(k)} u_i u_j + \sum_{i \in V_1, j \in O_1} \gamma_{ij}^{(k)} u_i u_j$$
$$\text{for } k = 1, \dots, o_1$$
$$f^{(k)}(u_1, \dots, u_n) := \sum_{i \in V_1 \cup O_1, j \in V_1 \cup O_1} \gamma_{ij}^{(k)} u_i u_j + \sum_{i \in V_1 \cup O_1, j \in O_2} \gamma_{ij}^{(k)} u_i u_j$$
$$\text{for } k = o_1 + 1, \dots, o_1 + o_2$$

0/1-Unbalanced Oil and Vinegar. At CHES 2011 Petzold *et al.* [19] showed that large parts of the public key are redundant in order to prevent key recovery attacks. More precisely, S can be chosen of a special structure due to equivalent keys and thus large parts of the public and secret map are equal. Choosing this parts of $\mathcal{P}$ of a special structure, such that direct attacks on the public key do not become easier, they were able to reduce the key size and running time of the verification algorithm.

Enhanced TTS. Enhanced TTS was proposed by Yang and Chen in 2005 [24]. The general idea is the same as for Rainbow, but as TTS was designed for high speed implementation it uses as few monomials as possible. For the purpose of evaluating the security we generalize the scheme by adding more monomials. As soon as a monomial $x_i x_j$ with $x_i \in U$ and $x_j \in V$ occur in the original TTS polynomial, we just assume that all monomials $x_i x_j$ with $x_i \in U$ and $x_j \in V$ occur. This way we easily see that TTS is a very special case of the Rainbow signature scheme. There are two different scalable central maps given in [24], one is called *even* sequence and the other *odd* sequence. The following equations show the odd sequence. We restrict our implementation to this case.

$$f^{(i)} = u_i + \sum_{j=1}^{2\ell-3} \gamma_{ij} u_j u_{2\ell-2+(i+j+1 \bmod 2\ell-1)} \qquad \text{for } 2\ell-2 \le i \le 4\ell-4,$$

$$f^{(i)} = u_i + \sum_{j=1}^{\ell-2} \gamma_{ij} u_{i+j-(4\ell-3)} u_{i-j-2\ell} + \sum_{j=\ell-1}^{2\ell-3} \gamma_{ij} u_{i+j-3\ell+3} u_{i-j+\ell-2}$$

$$\text{for } i = 4\ell-3, 4\ell-2,$$

$$f^{(i)} = u_i + \gamma_{i0} u_{i-2\ell+1} u_{i-2\ell-1} + \sum_{j=4\ell-1}^{i-1} \gamma_{i,j-(4\ell-2)} u_{2(i-j)-(i \bmod 2)} u_j + \gamma_{i,i-4\ell+2} u_0 u_i$$

$$+ \sum_{j=i+1}^{6\ell-3} \gamma_{i,j-(4\ell-2)} u_{4\ell-1+i-j} u_j \qquad \text{for } 4\ell-1 \le i \le 6\ell-3.$$

If we generalize these equations to the Rainbow signature scheme, the central map is given by Figure 4.

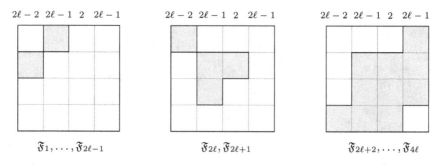

$$\begin{array}{cccc} 2\ell-2 & 2\ell-1 & 2 & 2\ell-1 \end{array} \qquad \begin{array}{cccc} 2\ell-2 & 2\ell-1 & 2 & 2\ell-1 \end{array} \qquad \begin{array}{cccc} 2\ell-2 & 2\ell-1 & 2 & 2\ell-1 \end{array}$$

$$\mathfrak{F}_1, \ldots, \mathfrak{F}_{2\ell-1} \qquad\qquad \mathfrak{F}_{2\ell}, \mathfrak{F}_{2\ell+1} \qquad\qquad \mathfrak{F}_{2\ell+2}, \ldots, \mathfrak{F}_{4\ell}$$

Fig. 4. Secret map $\mathcal{F}$ of odd sequence Enhanced TTS generalized

3   Security in a Nutshell

To provide a fair comparison between UOV, Rainbow and enTTS regarding memory consumption and running time, we first have to choose parameters of the same level of security. Therefore we briefly revisit the latest attacks and choices of parameters of all three schemes.

3.1   Security and Parameters of UOV and 0/1-UOV

Direct Attack. To forge a single signature an attacker would have to solve a system of o quadratic equations in v variables over $\mathbb{F}_q$. The usual way of finding one solution is first guessing v variables at random. This preserves one solution with high probability. The best way of solving the remaining $\mathcal{MQ}$-system of o equations and variables is to guess a few further variables and then apply some Gröbner Basis algorithm like F_4 (see Hybrid Approach of Bettale et $al.$ [3]). Recently Thomae et $al.$ showed that we can do better than guessing v variables at random [22]. Calculating these v variables through linear systems of equations allows to solve a system of $o - \lfloor \frac{v}{o} \rfloor$ quadratic equations and variables afterwards. To determine the complexity of solving a $\mathcal{MQ}$-system using a Groebner basis algorithm like F_4 we refer to [3]. In a nutshell, we first have to calculate the degree of regularity d_{reg}. For semi-regular sequences, which generic systems are assumed to be, the degree of regularity is the index of the first non-positive coefficient in the Hilbert series $S_{m,n}$ with

$$S_{m,n} = \frac{\prod_{i=1}^{m}(1 - z^{d_i})}{(1 - z)^n},$$

where d_i is the degree of the i-th equation. Then the complexity of solving a zero-dimensional (semi-regular) system using F_4 [3, Prop. 2.2] is

$$\mathcal{O}\left(\left(m\binom{n + d_{reg} - 1}{d_{reg}}\right)^{\alpha}\right),$$

with $2 \leq \alpha \leq 3$ the linear algebra constant. We used $\alpha = 2$ throughout the paper.

Key Recovery Attacks. There are two key recovery attacks known so far. The first is a purely algebraic attack called *Reconciliation* attack [6]. In order to obtain the secret key S we have to solve $\binom{k+1}{2}o$ quadratic equations in kv variables for an optimal parameter $k \in \mathbb{N}$. The second attack is a variant of the *Kipnis-Shamir* attack on the balanced Oil and Vinegar scheme [15]. The overall complexity of this attack is $\mathcal{O}(q^{v-o-1}o^4)$. Note that $v = 2o$ is very conservative in order to prevent this attack and thus v can be chosen much smaller for o large enough. As $k \geq 2$ even the Reconciliation attack will not badly benefit of choosing v smaller and direct attacks even suffer of such a choice.

Table 1. Minimal 0/1-UOV parameters achieving certain levels of security. Thereby g is the optimal number of variables to guess in the hybrid approach and k is the optimal parameter selectable for the Reconciliation attack.

security	parameter (o, v)	direct attack	Reconciliation	Kipnis-Shamir
2^{64}	$(21, 28)$	2^{67} $(g = 1)$	2^{131} $(k = 2)$	2^{66}
2^{80}	$(28, 37)$	2^{85} $(g = 1)$	2^{166} $(k = 2)$	2^{83}
2^{128}	$(44, 59)$	2^{130} $(g = 1)$	2^{256} $(k = 2)$	2^{134}

3.2 Security and Parameters of Rainbow

All attacks against UOV also apply to Rainbow. Additionally the security of Rainbow relies on the MinRank-problem. Thus we also have to take MinRank and HighRank attacks, as well as the Rainbow Band Separation attack into account. See Petzold et $al.$ [18] for an overview of the attacks and the parameters to choose.

Table 2. Minimal Rainbow parameters achieving certain levels of security. Thereby g is the optimal number of variables to guess for the hybrid approach.

security	(v_1, o_1, o_2)	direct attack	Band	MinRank	HighRank	Kipnis	Reconciliation
2^{64}	$(15, 10, 10)$	2^{67} $(g = 1)$	2^{70}	2^{141}	2^{93}	2^{125}	2^{242} $(k = 6)$
2^{80}	$(18, 13, 14)$	2^{85} $(g = 1)$	2^{81}	2^{167}	2^{126}	2^{143}	2^{254} $(k = 5)$
2^{128}	$(36, 21, 22)$	2^{131} $(g = 2)$	2^{131}	2^{313}	2^{192}	2^{290}	2^{523} $(k = 7)$

3.3 Security and Parameters of Enhanced TTS

All attacks against Rainbow also apply to enTTS. The only attack that seriously benefit from the changes made between Rainbow and enTTS is the Reconciliation attack with large k. But as the complexities of this attacks are out of reach anyway this do not affect the security. Actually the complexity is higher than the ones of all the other attacks, so we omit it. More important is the slight benefit of the Band Separation attack. For the odd sequence enTTS we derive $m + n - 1$ quadratic equations in $n - 2$ instead of n variables.

4   Implementation on AVR Microprocessors

The goal of these implementations is a fair comparison between some of the most promising $\mathcal{MQ}$-based post quantum public key schemes. All schemes were analysed in the previous section and sets of parameters with equivalent security were defined under considerations of most recent attacks. A problem when comparing such schemes is that every implementation has its own philosophy of what

Table 3. Minimal odd sequence enTTS parameters achieving certain levels of security. Thereby g is the optimal number of variables to guess for the hybrid approach.

security	(ℓ, m, n)	direct attack	Band	MinRank	HighRank	Kipnis-Shamir
2^{64}	$(7, 28, 40)$	2^{89} $(g = 1)$	2^{68}	2^{126}	2^{117}	2^{127}
2^{80}	$(9, 36, 52)$	2^{110} $(g = 2)$	2^{85}	2^{159}	2^{151}	2^{160}
2^{128}	$(15, 60, 88)$	2^{176} $(g = 3)$	2^{131}	2^{258}	2^{249}	2^{259}

is most worthy of optimization. Therefore we aim for a comparison with equal conditions for all schemes such as the same platform and implementation by the same person, also with nearly the same possible optimizations. Additionally practical figures are given in a real world scenario for signature verification and generation time. All the schemes were implemented with runtime optimization in mind.

4.1 Target Platform and Tools

An ATxMega128a1 on an xplain board was used as target device. This micro processor has a clock frequency of 32 MHz, 128KB flash program memory and 8KB SRAM. The code was written in C and optimized for embedded use. As compiler avr-gcc in version 4.5.1 and at some places assembler gcc-as 2.20.1 was used.

Polynomial Representation / Key Storage. When implementing MQPKS on microprocessors it is important to construct an efficient way of storing and reading the keys out of memory. All polynomials of an $\mathcal{MQ}$-scheme are represented by their coefficients. It is important to decide how this coefficients are processed during runtime. The coefficients of UOV and Rainbow can be easily mapped to some readout loops. This is not that easy with enTTS as only a minimal count of coefficients are used and this few coefficients are spread over three layers and six different cyclic structures. As random access on the flash memory produces a lot of addressing overhead while calculating the address each time a serial approach was chosen. All coefficients are stored in memory in the same exact order in which they are read out. There are no gaps or zeros in memory which is also memory efficient. This memory architecture allows us to read out the keys directly and simply increment the address to reach the next coefficient. The AVR instruction set allows a memory readout with a post increment in one clock cycles from SRAM or two clock cycles from Flash memory. Therefore no additional address calculation is needed. The number of coefficients to store and thus the memory consumptions in bytes is $o\left(ov + \frac{v(v+1)}{2}\right)$ for UOV, $o_1\left(o_1 v + \frac{v(v+1)}{2}\right) + o_2\left(o_2(v + o_1) + \frac{(v+o_1)(v+o_1+1)}{2}\right)$ for Rainbow and $8l^2 - 6l - 3$ for enTTS. The resulting memory requirements for specific security parameters are given in Table 5.

4.2   Arithmetic and Field

As the used microprocessor is based on an 8 bit architecture, working in $\mathbb{F}_{2^8}$ is optimal. Multiplication is done by a table look up, each element is brought to its exponential representation, processed and then transformed back to the normal polynomial representation. Every transformation from the exponential to the basis representation costs one memory access, therefore in all implementations the exponential representation is kept as long as no $\mathbb{F}_{2^8}$ addition takes place, which is a bitwise exclusive OR operation of two coefficients in the basis representation. As the coefficients of the keys are first read in by a multiplication, all keys are already stored in the exponential form. Random numbers are generated by the rand() gcc pseudo random number generator. This function is seeded with a value derived from uninitialized SRAM blocks which are arbitrary on every start up.

Inverting the Layers. All schemes require the inversion of multivariate systems of equations. As only linear systems of equation can be solved efficiently, we have to fix variables until the system gets linear and then perform a simple Gaussian elimination using LU decomposition. Here the exponential representation is also used where possible. For example the lower matrix and all variables were saved in exponential form. In enTTS the middle layer consists only of polynomials depending on already known variables. Therefore these polynomials can be inverted directly.

4.3   Key Size and Signature Runtime Reduction

The main problem of $\mathcal{MQ}$-schemes are large keys, as storage space is limited on embedded devices. Large private keys come also together with long signature time, due to the processing of more data. As the signature for a fixed message is not unique, there is a lot of redundancy that can be used to reduce the secret key S (cf. theory of equivalent keys). We used such minimal keys for UOV as well as for Rainbow. Note that there are no equivalent keys known for enTTS and thus the whole matrix S has to be stored. The special form of S has two additional side effects in addition to less space. First, also the signature time is reduced. The multiplication with the identity matrix corresponds to a copy of the signature so that only the multiplication with the remaining coefficients has to be done. For UOV this saves us $\frac{(v-1) \cdot v}{2} + \frac{(o-1) \cdot o}{2}$ equations and for Rainbow $\frac{(v-1) \cdot v}{2} + \frac{(o_1-1) \cdot o_1}{2} + \frac{(o_2-1) \cdot o_2}{2}$. The second observation is that due to the identity matrix in the vinegar × vinegar part, large parts of $\mathcal{P}$ and $\mathcal{F}$ are equal. They do not increase security an can be seen as a system parameter (cf. [19]). As required by the authors of [19] for 0/1-UOV, also a different monomial ordering was chosen according to a minimal Turán graph. This reordering prevent easier attacks on the public key. The same procedure is probably possible for Rainbow. But as no publication exists which investigated this case, it was not implemented. For enTTS this is not possible as the Tame equations in the middle layer cause to blur the variable structure and no equivalent keys are known.

4.4   Verify Runtime Reduction

In the case of 0/1-UOV, choosing the coefficient from $\mathbb{F}_2$ has another advantage besides of less memory consumption. The verification and signature generation time can be reduced. As we know that the majority of coefficients are from $\mathbb{F}_2$, we can check for a one or a zero, which leads to a copy instruction in the case of one or a skip instruction in case of zero. Only otherwise we have to perform a costly multiplication in $\mathbb{F}_{2^8}$. The effect is in our implementation not marginally visible, because the used table look up method is fast compared to a schoolbook multiplication method.

4.5   RAM Requirements

$\mathcal{MQ}$-schemes do not need a lot of RAM, in contrast to the persistent flash memory requirements. In Table 4 the requirements are listed. Besides RAM needed for persistent, counting or temporary variables, only the Gaussian elimination algorithm needs a noticeable amount of RAM. As the inversion is computed in place, only one quadratic systems at time has to be stored in RAM. In case of multiple layers the maximal requirements are defined by the largest system of equations to be solved.

Table 4. Minimal Ram Requirements for LES Solving in Bytes

security	2^{64}	2^{80}	2^{128}	general
UOV	441	784	1936	m^2
Rainbow	400	729	1849	$(o_1 + o_2)^2$
enTTS	169	289	841	$(2l - 1)^2$

4.6   Key Generation

The keys for all schemes are generated on a standard PC using a C program. Basically $T \circ \mathcal{F} \circ S = P$ has to be computed. Using the quadratic form, the composition can be written as in (1). An overview of the key generation process of 0/1 UOV with small parameters can be found in the appendix.

$$\mathfrak{P}^{(i)} = \sum_{j=1}^{m} t_{ij} S^\mathsf{T} \mathfrak{F}^{(j)} S \tag{1}$$

Another way to generate an UOV key is described in [19]. It can be done by transforming the matrix S into a matrix A_{uov} and write all coefficients of $f^{(i)}$ ordered lexicographically to the rows of Q. Then the following equation holds: $A_{uov} \cdot Q = S^T \mathfrak{F}^{(i)} S$. With this relation inverting A_{uov} is possible and therefore a inverse approach, choosing first $\mathcal{P}$ and then applying A_{uov} to get $\mathcal{F}$. For the runtime optimization the reordering of monomials can take place in A_{uov} instead of reorder the monomials in $\mathcal{P}$ and $\mathcal{F}$.

5   Results

Table 5 shows our achieved results. They are easy to compare because schemes are grouped by security level. For all schemes key size, runtime and code size are given. Where applicable the system parameter size is also included. The public and secret key sizes can be easily calculated. One element responds to one byte and no other overhead needs to be saved so the keys consists only of the coefficients of the public or secret maps and the linear transformations. In the case of 0/1-UOV a large part is fixed and declared as a system parameter, but it must be anyway saved or be easy to generate in a real world scenario, therefore thus size is also listed.

Clock cycles were count internally with two concatenated 16 bit counters which are enabled to count on every clock cycle. As the count of verify operations scales with $(\frac{n \cdot (n+1)}{2} \cdot m)$ the measured times do not surprise. As enTTS uses the largest numbers of n and m it has the lowest verify performance and the largest public keys. Rainbow is the fastest as the parameters can be chosen relatively low. The big advantage of enTTS is the small private key. Large parts of the central map are zero and have not to be saved. In terms of theoretical public key size 0/1-UOV performs the best. If the possibility to generate the system parameter on the device would exist, it would ensure the smallest public key. The gain of verification and signature time in comparison to the standard UOV is only minimal as the multiplication by table look up has no significant runtime difference in comparison to a multiplication with 0 or 1 as the 0 case is a special case and is checked anyway every time in a normal multiplication in $\mathbb{F}_{2^8}$. When measuring scalability for secret/public key size at the step from 2^{64} to 2^{128}, UOV has a increase factor of 9/9, 0/1-UOV of 9/9, Rainbow 10/11 and enTTS of 4/10. UOV scales the best in public key size, enTTS the best in private key size. Regarding the signature size, UOV has the highest expansion factor, with a message to signature ratio of approximately 2.3, followed by Rainbow with 1.7 and enTTS with 1.4.

As a comparison of an µC with an ASIC or PC implementation is meaningless, the only MQ implementation we can compare with is the one from [25]. The authors implemented enTTS(5, 20, 28) on a MSP430 running at 8 MHz. Signing requires 17.75 ms and verifying 181.5 ms, when scaled up to our clock frequency. Although, the MSP430 is a 16 bit CPU, our implementation is a factor of 3.7 faster in signing and 5.1 times faster in verifying.

Also when comparing our work with implementations of the classical signature schemes RSA and ECDSA, all four schemes perform well. E.g. for 2^{80} bit security [13] reports 203ms for a ECC sign operation, where our implementations are two to ten times faster. For the verifying operation our work is up to three times faster. Due to the short exponent in RSA-verify, [13] verifies in the same order of magnitude. But the RSA-sign operation is at least a factor of 25 slower than our work. Table 6 summarizes other implementations on comparable 8 bit platforms.

Table 5. Results

Scheme	n	m	Key Size [Byte]		System Parameter [Byte]		Clockcyles x 1000		Time[ms]@32MHz		Code Size [Byte]	
			private	public	private	public	sign	verify	sign	verify	sign	verify
enTTS(5, 20, 28)	28 20		1351	8120	*	*	153	1,126	4.79	35.22	12890	827
enTTS(5, 20, 28)[25]	28 20		1417	8680	*	*	568[1]	5,808[1]	17.75[2]	181.5[2]	-	-
uov(21, 28)	49 21		21462	25725	*	*	1,615	1,690	50.49	52.83	2188	466
0/1 uov(21, 28)	49 21		12936	4851	8526	20874	1,577	1,395	49.29	43.60	2258	578
rainbow(15, 10, 10)	35 20		9250	12600	*	*	848	1,010	26.51	31.58	4162	466
enTTS(7, 28, 40)	40 28		2731	22960	*	*	332	2,558	10.37	79.95	24898	827
uov(28, 37)	65 28		49728	60060	*	*	3,637	3,911	113.66	122.23	2188	466
0/1 uov(28, 37)	65 28		30044	11368	19684	48692	3,526	3,211	110.20	100.37	2258	578
rainbow(18, 13, 14)	45 27		19682	27945	*	*	1,740	2,214	54.38	69.19	4162	466
enTTS(9, 36, 52)	52 36		4591	49608	*	*	609	6,658	19.03	208.07	41232	827
uov(44, 59)	103 44		194700	235664	*	*	13,314	14,134	416.07	441.70	2188	466
0/1 uov(44, 59)	103 44		116820	43560	77880	192104	12,782	13,569	399.43	424.04	2258	578
rainbow(36, 21, 22)	79 43		97675	135880	*	*	8,227	9,216	257.11	288.01	4162	466
enTTS(15, 60, 88)	88 60		13051	234960	*	*	2,142	3,0789	66.94	962.17	116698	827

(Row groups labelled on the left margin: 2^{64}, 2^{80}, 2^{128})

* Not applicable
[1] Derived from values in original work
[2] Scaled to the same clock frequency

Table 6. Overview of other implemenatations on comparable platforms

Method	Time[ms]@32MHz	
	sign	verify
enTTS(5, 20, 28)[25]	17.75[1]	181.5[1]
ECC-P160 (SECG) [13]	203[1]	203[1]
ECC-P192 (SECG) [13]	310[1]	310[1]
ECC-P224 (SECG) [13]	548[1]	548[1]
RSA-1024 [13]	2,748[1]	108[1]
RSA-2048 [13]	20,815[1]	485[1]
NTRU-251-127-31 sign [10]	143[1]	-

[1] For a fair comparison with our implementation running at 32MHz, timings at lower frequencies were scaled accordingly.

6   Conclusion

In this work we present the first µC implementations of the three most common MQPKS since nearly 10 years. Additionally, we implemented for the first time 0/1-UOV on a constrained device. All recent attacks were summarized and we proposed current security parameters for 2^{64}, 2^{80} and 2^{128} bit symmetric security. Additionally, we showed that choosing $v = 2o$ for UOV is outdated. When comparing with existing MQ implementations, ours are a factor of three and five times faster in signing and verifying, respectively. We hope our implementations will inspire follow up work, to improve acceptance of MQPKS in constrained environments.

6.1   Further Improvements

There is still space for improvements and the upper limit is not reached yet. A few ideas were not implemented in this work. Saving the system parameters is not optimal. Here a replacement by a pseudo random number generator or an other generator function would reduce the public key drastically, even if verification time would be increased. In our implementation all elements of $\mathbb{F}_2$ are saved as a byte value. It would be possible to achieve smaller keys when saving 8 elements in one byte, combined with a verification function which utilizes assembler instructions maybe even a faster verification could be possible. An overall time vs. code size trade-off is still a topic to investigate. $\mathcal{MQ}$-schemes are very well scalable in regard to this trade-off.

References

1. Yang, B.-Y., Chen, J.-M., Chen, Y.-H.: TTS: High-Speed Signatures on a Low-Cost Smart Card. In: Joye, M., Quisquater, J.-J. (eds.) CHES 2004. LNCS, vol. 3156, pp. 371–385. Springer, Heidelberg (2004)
2. Balasubramanian, S., Carter, H., Bogdanov, A., Rupp, A., Ding, J.: Fast Multivariate Signature Generation in Hardware: The Case of Rainbow. In: International Conference on Application-Specific Systems, Architectures and Processors, ASAP 2008, pp. 25–30 (July 2008)
3. Bettale, L., Faugère, J.-C., Perret, L.: Hybrid Approach for Solving Multivariate Systems over Finite Fields. Journal of Mathematical Cryptology 3(3), 177–197 (2009)
4. Bouillaguet, C., Fouque, P.-A., Macario-Rat, G.: Practical key-recovery for all possible parameters of SFLASH. In: Lee, D.H., Wang, X. (eds.) ASIACRYPT 2011. LNCS, vol. 7073, pp. 667–685. Springer, Heidelberg (2011)
5. Braeken, A., Wolf, C., Preneel, B.: A Study of the Security of Unbalanced Oil and Vinegar Signature Schemes. In: Menezes, A. (ed.) CT-RSA 2005. LNCS, vol. 3376, pp. 29–43. Springer, Heidelberg (2005), http://eprint.iacr.org/2004/222/
6. Buchmann, J., Ding, J. (eds.): PQCrypto 2008. LNCS, vol. 5299. Springer, Heidelberg (2008)
7. Faugère, J.-C., Joux, A.: Algebraic Cryptanalysis of Hidden Field Equation (HFE) Cryptosystems Using Gröbner Bases. In: Boneh, D. (ed.) CRYPTO 2003. LNCS, vol. 2729, pp. 44–60. Springer, Heidelberg (2003)
8. Courtois, N.T., Daum, M., Felke, P.: On the Security of HFE, HFEv- and Quartz. In: Desmedt, Y.G. (ed.) PKC 2003. LNCS, vol. 2567, pp. 337–350. Springer, Heidelberg (2002)
9. Ding, J., Schmidt, D.: Rainbow, a New Multivariable Polynomial Signature Scheme. In: Ioannidis, J., Keromytis, A.D., Yung, M. (eds.) ACNS 2005. LNCS, vol. 3531, pp. 164–175. Springer, Heidelberg (2005)
10. Driessen, B., Poschmann, A., Paar, C.: Comparison of Innovative Signature Algorithms for WSNs. In: Proceedings of ACM WiSec 2008. ACM (2008)
11. Garey, M.R., Johnson, D.S.: Computers and Intractability — A Guide to the Theory of NP-Completeness. W.H. Freeman and Company (1979) ISBN 0-7167-1044-7 or 0-7167-1045-5
12. Goubin, L., Courtois, N.T.: Cryptanalysis of the TTM Cryptosystem. In: Okamoto, T. (ed.) ASIACRYPT 2000. LNCS, vol. 1976, pp. 44–57. Springer, Heidelberg (2000)
13. Gura, N., Patel, A., Wander, A., Eberle, H., Shantz, S.C.: Comparing Elliptic Curve Cryptography and RSA on 8-bit CPUs, pp. 119–132 (2004)
14. Kipnis, A., Patarin, J., Goubin, L.: Unbalanced Oil and Vinegar Signature Schemes. In: Stern, J. (ed.) EUROCRYPT 1999. LNCS, vol. 1592, pp. 206–222. Springer, Heidelberg (1999)
15. Kipnis, A., Shamir, A.: Cryptanalysis of the Oil and Vinegar Signature Scheme. In: Krawczyk, H. (ed.) CRYPTO 1998. LNCS, vol. 1462, pp. 257–266. Springer, Heidelberg (1998)
16. Kipnis, A., Shamir, A.: Cryptanalysis of the HFE Public Key Cryptosystem by Relinearization. In: Wiener, M. (ed.) CRYPTO 1999. LNCS, vol. 1666, pp. 19–30. Springer, Heidelberg (1999)
17. Patarin, J.: Cryptanalysis of the Matsumoto and Imai Public Key Scheme of Eurocrypt '88. In: Coppersmith, D. (ed.) CRYPTO 1995. LNCS, vol. 963, pp. 248–261. Springer, Heidelberg (1995)

18. Petzoldt, A., Bulygin, S., Buchmann, J.: Selecting Parameters for the Rainbow Signature Scheme. In: Sendrier, N. (ed.) PQCrypto 2010. LNCS, vol. 6061, pp. 218–240. Springer, Heidelberg (2010)

19. Petzoldt, A., Thomae, E., Bulygin, S., Wolf, C.: Small Public Keys and Fast Verification for Multivariate Quadratic Public Key Systems. In: Preneel, B., Takagi, T. (eds.) CHES 2011. LNCS, vol. 6917, pp. 475–490. Springer, Heidelberg (2011)

20. Shor, P.W.: Polynomial-time Algorithms for Prime Factorization and Discrete Logarithms on a Quantum Computer. SIAM J. on Computing, 1484–1509 (1997)

21. Tang, S., Yi, H., Ding, J., Chen, H., Chen, G.: High-Speed Hardware Implementation of Rainbow Signature on FPGAs. In: Yang, B.-Y. (ed.) PQCrypto 2011. LNCS, vol. 7071, pp. 228–243. Springer, Heidelberg (2011)

22. Thomae, E., Wolf, C.: Solving Underdetermined Systems of Multivariate Quadratic Equations Revisited. In: Fischlin, M., Buchmann, J., Manulis, M. (eds.) PKC 2012. LNCS, vol. 7293, pp. 156–171. Springer, Heidelberg (2012)

23. Wolf, C., Braeken, A., Preneel, B.: Efficient Cryptanalysis of RSE(2)PKC and RSSE(2)PKC (2004)

24. Yang, B.-Y., Chen, J.-M.: Building Secure Tame-like Multivariate Public-Key Cryptosystems: The new TTS. In: Boyd, C., González Nieto, J.M. (eds.) ACISP 2005. LNCS, vol. 3574, pp. 518–531. Springer, Heidelberg (2005)

25. Yang, B.-Y., Cheng, C.-M., Chen, B.-R., Chen, J.-M.: Implementing Minimized Multivariate PKC on Low-Resource Embedded Systems. In: Clark, J.A., Paige, R.F., Polack, F.A.C., Brooke, P.J. (eds.) SPC 2006. LNCS, vol. 3934, pp. 73–88. Springer, Heidelberg (2006)

A   Toy Example of 0/1 UOV Key Generation

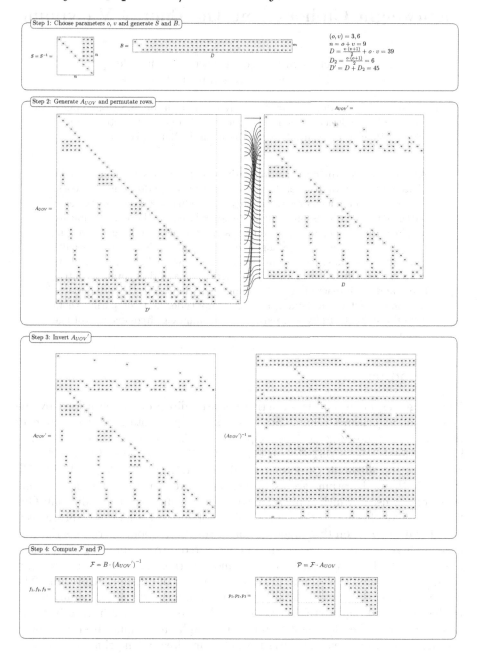

Fig. 5. 0/1 UOV Key Generation. For details see [19].

Towards Green Cryptography: A Comparison of Lightweight Ciphers from the Energy Viewpoint

Stéphanie Kerckhof, François Durvaux, Cédric Hocquet,
David Bol, and François-Xavier Standaert

Université catholique de Louvain, Institute of Information and Communication
Technologies, Electronics and Applied Mathematics, Crypto Group.
Place du Levant, 3, B-1348 Louvain-la-Neuve, Belgium

Abstract. We provide a comprehensive evaluation of several lightweight block ciphers with respect to various hardware performance metrics, with a particular focus on the energy cost. This case study serves as a background for discussing general issues related to the relative nature of hardware implementations comparisons. We also use it to extract intuitive observations for new algorithm designs. Implementation results show that the most significant differences between lightweight ciphers are observed when considering both encryption and decryption architectures, and the impact of key scheduling algorithms. Yet, these differences are moderated when looking at their amplitude, and comparing them with the impact of physical parameters tuning, e.g. frequency / voltage scaling.

1 Introduction

Lightweight cryptography is an active research direction, as witnessed by the number of algorithms aiming at "low-cost" implementations designed over the last years. Looking at block ciphers, the list includes (but is not limited to) DESXL [15], HIGHT [13], ICEBERG [22], KATAN [2], KLEIN [10], LED [11], mCrypton [16], NOEKEON [3], Piccolo [20], PRESENT [1], SEA [21] and TEA [24]. Although these algorithms are useful and inventive in many ways, determining which one to use in which application with good confidence can be difficult. One first reason for this is that the very definition of low-cost is hard to capture, as it is highly dependent on the target platform. For illustration, operations that are cheap in hardware (e.g. wire crossings) may turn out to be annoyingly expensive in software. In fact, even for a given technology, there are various criteria that could be considered to evaluate the low-cost nature of different algorithms. The implementation size (measured in gates, program memory, ...) generally comes in the first place, but power or energy can be more reflective in certain application scenarios. Besides, lightweight cryptography has mainly been developed through several independent initiatives, over an already long time period. This is in contrast with the design of standard algorithms for which the selection was/will be the result of an open competition. One outcome of the Advanced Encryption Standard (AES) and SHA3 competitions is the publication of well motivated comparative studies. Taking the example of hardware (ASIC and FPGA)

E. Prouff and P. Schaumont (Eds.): CHES 2012, LNCS 7428, pp. 390–407, 2012.

implementations, several works can be mentioned both for the AES, e.g. [7,8,23], and SHA3 candidates [9,12,14]. By contrast, only a few evaluations of lightweight algorithms are available in the literature. For example, the companion paper of the KATAN algorithm includes gate counts and throughput estimations for several ciphers [2], but they consider different technologies. A recent initiative can also be mentioned for software implementations [6]. But to the best of the authors' knowledge, there exist no systematic evaluations for hardware implementations to date.

In this paper, we compare the hardware performances of 6 block ciphers, with different block and key sizes. Namely, we considered the AES [4] and NOEKEON for 128-bit blocks and keys, HIGHT and ICEBERG for 64-bit blocks and 128-bit keys, and KATAN and PRESENT for 64-bit blocks and 80-bit keys. This choice of algorithms was motivated by having different block and key sizes, together with different styles of key scheduling and decryption. After a brief discussion underlying the relative nature of evaluation metrics for hardware implementations, we evaluate different figures of merits for these 6 candidates, with a particular focus on the energy efficiency (which explains the word "green" of our title). For this purpose, we first analyze hardware design choices and describe different architectures for encryption, decryption and encryption/decryption, with and without round unrolling and parallelization. This allows us to quantify the combinatorial cost and delays of the different ciphers, and to analyze their respective implementations. Next, we study the tuning of physical parameters, and evaluate the impact of frequency / voltage scaling on our comparisons. Doing so, we investigate the relevance of the energy per bit as a comparison criteria for lightweight block ciphers, i.e. its independence with respect to hardware design choices and frequency / voltage scaling. In other words, we question the extent to which such a metric reflects algorithmic design choices and discuss its possible biases. We answer positively and argue that it nicely summarizes the "energy efficiency" of an algorithm. We also show that the informativeness of this metric is further improved if correlated with the "performance efficiency" (usually measured with a throughput over area ratio). As a conclusion of our experiments, we finally try to extract useful suggestions for new lightweight cryptographic algorithms.

2   Evaluation Metrics for Hardware Implementations

Evaluating hardware implementations is a challenging task. In this section, and as a background to our following case study, we introduce different metrics that can be used for this purpose, together with possible shortcomings with respect to their relevance for comparing algorithms. Namely, we will consider the area, power consumption, throughput and energy cost, as summarized in Figure 1. This selection was motivated by the fact that these metrics are generally reflective of the application constraints that may be encountered in practice. In general, the most revealing units for these metrics are physical (i.e. μm^2, Watts, bit/sec and Joules). However, as these physical units can only be obtained at the very end of an implementation process, convenient first-order estimates are

application constraints	physical units	relative to	pre-layout units	HW design goals	algorithmic design goals	relevance w.r.t. algorithms
AREA	um²	time or energy constraints	#gates	share resources	reduce components cost & versatility	weakly discriminant ✳✳✳
INST. POWER (dynamic)	W (J/sec)	time or energy constraints	switching activity	reduce datapath	reduce components cost & versatility	somewhat arbitrary ✳
THROUGHPUT	bit/sec	area or power constraints	#cycles (& block size)	unroll, parallelize & pipeline	minimize the total combinatorial cost	very arbitrary ✳
ENERGY	J/enc, J/bit	area or power constraints	#cycles X POWER	unroll	minimize the total combinatorial cost	somewhat discriminant ✳✳✳

Fig. 1. Summary of evaluation metrics for hardware implementations

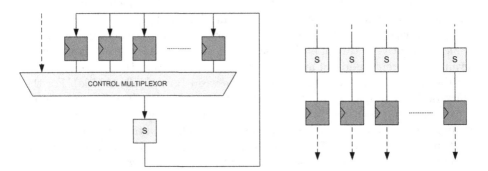

Fig. 2. Left: serial design with resource sharing. Right: unrolled & parallelized design.

obtained with the gate count, switching activity (i.e. number of bit transitions per clock cycle), number of cycles per algorithm execution and block size.

The first important observation regarding these metrics is that they are always *relative*, meaning that it is usually possible to optimize a single metric quite arbitrarily, if the other ones can be degraded. Hence, in order to make any comparison relevant, it is necessary to agree on some application objectives. For example, the area and power can be relative to time or energy constrains, while the throughput and energy can be relative to area or power constraints. As a result, optimization goals can be roughly separated as "design for low area or power" and "design for high throughput or low energy". In the first case, designers will typically share the resources (to decrease the area cost) and reduce the datapath (to limit the switching activity). Such optimizations are illustrated for the case of a block cipher S-box layer in the left part of Figure 2. Quite naturally, re-using the same component also implies that the relative cost of the S-box compared to the control logic and memory decreases, which implies a lower efficiency. Therefore, in the second case, the designer will rather unroll the S-boxes (i.e. implement them all on chip) and parallelize their computation (i.e. perform them in a single clock cycle), as illustrated in the right part of the figure. Besides, the performances of these implementations will also depend on their clock frequency, for which the maximum value f_{max} is inversely proportional to the longest combinatorial path (aka critical path) between two registers. If the clock frequency is not sufficient, inner pipelining can be used in order to cut the critical path by the addition of registers, as illustrated in Appendix, Figure 6.

Having roughly described these optimization techniques allows us to come back on the relativity of the metrics when comparing different algorithms. For example, the throughput is very arbitrary, as it can be straightforwardly improved by multiplying the circuit size. The same observation holds (to a smaller extent) for the instantaneous power consumption, as a designer could theoretically reduce his datapath to a single bit, independently of the algorithm to implement. In general, a lack of instantaneous power can also be overcome by decreasing the clock frequency and relying on decoupling capacitances. Hence, applications where this metric really matter are quite limited (RFID being the most frequent example). The area cost becomes slightly more discriminant, since increasing the sharing of resources generally implies a cost penalty in the control part. Finally, the energy per encryption is more discriminant, as it corresponds to an integral over time and is not compressible beyond what is allowed by the total combinatorial cost of an algorithm. Quite naturally, many combined metrics can also be derived, e.g. the "throughput over area ratio" is one of the most popular tool to express the performance efficiency of a given hardware implementation.

The main consequence of this relativity is that the fair comparison of hardware implementations is always specialized to a set of constraints. In the following sections, we will define our methodology for this purpose, and investigate the energy cost of different algorithms, for various hardware architectures. Beforehand, a few more comments about this evaluation are worth being mentioned.

(1) Present hardware design flows make intensive use of automated tools, of which the options highly influence the final performance. For example, imposing stronger constraints on the clock frequency can be automated in this way, at the cost of area increases. In such cases, it is useful to agree on the maximum tolerated penalty (compared to the area obtained without frequency constraints).

(2) Once all design choices have been taken, it is always possible to further tune the performances of an implementation, e.g. by taking advantage of frequency / voltage scaling. This issue will be investigated in Section 5.

(3) As technologies are shrinking to the nanometer scale, a part of their power consumption may become static (i.e. happen independent of the switching activity)[1]. As leakage currents are essentially dependent on the circuit size, it implies that the optimization goals for area and power become closer in this case. Significant leakage currents also have an impact on the energy performances.

(4) In general, comparisons are only meaningful for algorithms with the same block and key size. Yet, different block sizes can sometimes be reflected in the metrics (e.g. by computing the energy per bit rather than per block).

---

[1] Note that this effect can be mitigated by exploiting low-leakage libraries.

3   The Case of 6 Block Ciphers: Methodology

In order to make our performance evaluations as relevant as possible, we defined a strict methodology for all our implementations. It defines requirements on the target architectures, their interface and the implementation flow.

Regarding architectures, and for all the investigated ciphers, we considered encryption, decryption and encryption/decryption designs. The reference point of our evaluations is a standard loop implementation of the AES Rijndael, performing one encryption in 12 cycles, taking advantage of the efficient S-box representation of Mentens et al. [18]. This choice was mainly motivated by our low-energy consumption goal. Further reduction of the area (e.g. with 8-bit or 32-bit architectures) would lead to less energy-efficient designs. We also thought that the throughputs of these AES implementations (of a few Gbps) were large enough for a wide range of applications. Next, for all the investigated lightweight ciphers, we analyzed the generic unrolled architecture depicted in Figure 3, where N_r rounds are executed per clock cycle. Having at least one full round implemented was again motivated by our low-energy consumption objective. We used this generic architecture in order to determine the number of lightweight cipher rounds that are needed to consume the same area, or that require the same delay as an AES round. Besides, they are also interesting architectures for very low-latency implementations. As unrolling without adding pipeline is generally a suboptimal choice regarding the critical path, we further considered two implementation scenarios. In the first one, we assumed a clock frequency of 100MHz (determined by the system): it corresponds to a context where such an unrolling is indeed motivated by external constraints. Next, we estimated the maximum clock frequency. In this second case, we further investigated the impact of parallel (or pipelined) architectures. In order to exhaustively analyze our large design space (various N_r values, encryption vs. decryption vs. encryption/decryption, $f = 100$MHz vs. f_{max}), we heavily relied on generic VHDL/Verilog programming. We additionally used a common (generic as well) interface for all the ciphers, with plaintext/ciphertext (resp. key) port width corresponding to the block (resp. key) size, and a simple handshaking mechanism to control the flow.

Regarding the synthesis environment, we operated in two steps. In the first place, and in order to study the impact of architectural choices, we investigated the previously defined operating frequencies (i.e. $f = 100$MHz and f_{max}) at

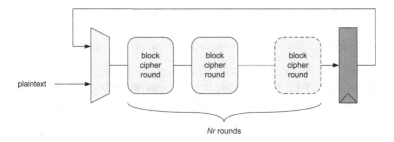

Fig. 3. Unrolled architectures with various number of rounds

1.2V, i.e. the nominal supply voltage for the technology used (see Section 4). As previously mentioned, the maximum frequency depends on the synthesis and place-and-route, since the CAD tool can further optimize a design to reach a target timing constraint at the cost of an area increase. Thus, we precisely defined the maximum frequency as the frequency obtained when the area of the design has increased by 10% compared to the unconstrained design. This step allowed us to identify the most efficient architectures for each lightweight cipher. Next, for this reduced set of architectures, we analyzed the possibilities of frequency / voltage scaling by carefully tuning the supply voltage (in Section 5).

All the ciphers were implemented following a classical ASIC flow. We used a commercial 65-nanometer CMOS low-power technology. Synthesis was performed using the Synopsys tools suite. We used the switching activity annotation, by means of behavioral simulation, in order to extract realistic power and energy figures. In addition, the supply voltage exploration was performed using the standard cells library that was re-characterized at different Vdd's. Finally, and because of space constraints, we reproduced the most informative metrics provided by our implementations in appendix, and additionally extracted some of them for illustrating our claims in the core of the paper. The remainder of our syntheses data is available online, in the full version of the paper.

4   Implementation Results at Fixed Vdd=1.2V

Using the previously defined methodology, we first reported the selected performance metrics of our different syntheses at 1.2V supply voltage in Appendix, Figures 8 to 13, where each point in the curves corresponds to a different unrolling parameter N_r. These figures allow us to evaluate the efficiency of the different ciphers implemented. In this section, we report on a number of useful observations regarding both hardware design and algorithmic design issues.

As a starting point, we looked at the area curves (given for $f = 100$MHz in Figure 8). In general, one would expect the circuit size to increase linearly with the number of rounds unrolled. However, in the case of lightweight ciphers, we observed that a number of rounds may be needed before such a linear dependency appears. This fact is in direct relation with the limited combinatorial cost of the rounds in certain ciphers (most visibly, KATAN). That is, if the cost of a round is small in comparison with the state registers and control logic, doubling the number of rounds unrolled will not double the consumed area. A similar behavior is observed for the critical path in Figure 9. If the rounds are simple enough for this critical path to be in the control logic, then doubling the amount of rounds unrolled will not result in cutting the maximum frequency by two. Again, this effect is amplified for KATAN, as its round computations only affect a few bits, the other ones being routed from the state register to itself. Overall, these figures recall that the definition of a round is arbitrary: several rounds of a lightweight ciphers are generally needed to reach the cost and delay of an AES round.

Looking at the throughput curves first confirms that in general, unrolling an implementation without pipelining it mainly makes sense if the clock frequency if fixed below the maximum one, e.g. because of system constraints. Yet, we also remark that for some ciphers, unrolling a few rounds without pipeline improves the throughput at maximum frequency too (see Figure 10). This is a consequence of "simple rounds" and the previously mentioned non-linear increase of the critical path for low N_r values. Note that even for KATAN, the throughput starts to decrease beyond $N_r = 2^6$ (this data is not included in the figures, for visibility reasons). Besides, increasing N_r at maximum frequency does not lead to the expected constant curves. This is explained by a detrimental side-effect related to the overhead cycles required to charge/discharge the plaintext and master key in their registers. Namely, this overhead becomes more significant with the number of implemented rounds (i.e. when the number of clock cycles per encryption becomes small). Note that the impact of this interfacing drawback is stronger for NOEKEON and HIGHT, as they respectively account for 2 and 3 cycles for these ciphers (one for loading the data, one per initial/final transformation).

The average power implementation results also exhibit different conclusions for $f = 100$MHz and f_{max}. In the first case, they are dominated by the switching activity in the circuit, that increases with N_r. Hence, the power is correlated with the circuit size in this case. By contrast at maximum frequency, unrolling is either neutral or implies a reduction of the average power, when the maximum frequency decreases with N_r faster than the area (e.g. for HIGHT).

Interestingly, the energy per bit (given for f_{max} in Figure 11) is remarkably similar in our two frequency contexts, because it is dominated by the switching activity in the selected low-leakage technology. This confirms that it is a reasonably discriminant metric for algorithmic comparisons. It is also quite correlated with the throughput over area metric at 100MHz in Figure 12, i.e. when unrolling the algorithms affects the throughput and area in opposite directions, with close to equivalent impact. Quite naturally, this correlation vanishes at maximum frequency, due to the inefficiency of unrolling without pipeline. Finally, the previously mentioned side-effects (such as overhead cycles and unbalanced use of logic and memory) are naturally reflected in these curves as well.

A summary of our implementation results regarding algorithm efficiency (both in terms of performances and energy) is depicted in Figure 4, where the energy per bit is represented in function of the throughput over area ratio, for our different architectures. Such figures naturally require a few cautionary remarks. First, they have to be interpreted with care, as they only provide a big picture of the implementation efficiency. Practical case studies may focus specifically on different combinations of metrics. Second, even if assuming that efficiency is indeed the design goal, comparing algorithms is difficult as their performances are sometimes close, and depend on the architectures (e.g. encryption-only, decryption-only and encryption/decryption designs lead to different ratings). Yet, we believe that a few interesting conclusions can be extracted that we now detail.

Starting with encryption designs, the comparison roughly suggests NOEKEON $\geq$ PRESENT $\approx$ KATAN $\geq$ HIGHT $\approx$ AES $\geq$ ICEBERG. This ordering is explained by different factors. Maybe the most important one is the significant differences in the key scheduling. At the extremes, NOEKEON does not have any, while for

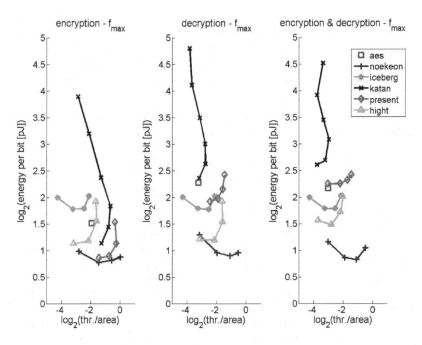

Fig. 4. Vdd = 1.2V: throughput over area ratio vs. energy per bit

ICEBERG, the key scheduling is as complex as the encryption rounds. Next, the respective block and key sizes strongly matter as well. Having larger key sizes naturally implies lower efficiency in general (but theoretically provides improved security). Less obviously (and less significantly), smaller block sizes are also negative for efficiency. For example, working on 128-bit blocks instead of 64-bit ones doubles the datapath size, but it rarely implies doubling the overall cost (thanks to the strong diffusion layers in modern ciphers). Considering the decryption architectures allows putting forward one more design issue, namely the need to perform the key scheduling "on-the-fly" in forward direction before doing it in backward direction[2] for ciphers such as AES, KATAN and PRESENT. It implies that the comparison is modified into NOEKEON $\geq$ HIGHT $\geq$ ICEBERG $\approx$ PRESENT $\geq$ AES $\approx$ KATAN. Finally, the encryption/decryption designs further indicate the possibility to efficiently share the resources between the cipher and its inverse. Here, involutional ciphers such as ICEBERG gain a particular advantage, leading to a rating: NOEKEON $\geq$ HIGHT $\approx$ ICEBERG $\geq$ AES $\approx$ PRESENT $\geq$ KATAN.

An alternative view of the performance and energy efficiency of the different algorithms is given in Appendix, Figure 13 (for $f = 100$MHz), where we plot the ratio between the throughput and the product of the area and the energy per bit. Again, such a figure requires a careful interpretation as they only provide one "global efficiency" metric. Yet, it is interesting to note that for all ciphers, the

---

[2] This choice is natural in hardware implementations as storing a fully precomputed expanded key in registers would generally require too large memory oveheads.

Table 1. Implementation results for most "globally efficient" architectures

Cipher	Mode E,D,ED	Area [μm^2]	f_{max} [MHz]	Latency [cycles]	Throughput [Mbps]	Power [mW]	Energy [pJ per bit]
AES	E	17921	444	12	4740	13,5	2,9
$N_r = 1$	D	20292	377	22	2195	10,6	4,8
	ED	24272	363	≈17	≈2997	≈12,6	≈4,4
NOEKEON	E	8011	1149	18	8173	15,0	1,8
$N_r = 1$	D	10431	1075	19	7243	14,1	1,9
	ED	10483	1075	≈18,5	≈7445	≈15,35	≈2,1
HIGHT	E	6524	641	19	2159	6,3	2,9
$N_r = 2$	D	6524	645	19	2173	6,3	2,9
	ED	8217	540	19	1820	6,1	3,3
ICEBERG	E	11377	699	17	2632	10,7	4,0
$N_r = 1$	D	11359	699	17	2632	10,7	4,0
	ED	11408	689	17	2596	10,6	4,0
KATAN	E	6231	952	17	3585	8,1	2,7
$N_r = 16$	D	8616	666	33	1292	9,8	6,1
	ED	12609	473	≈25	≈1347	≈12,7	6,4
PRESENT	E	5024	1123	17	4230	9,3	2,2
$N_r = 2$	D	6060	1041	33	2020	8,9	4,4
	ED	8213	884	≈25	≈2523	≈12,6	4,7

architecture providing the best such global efficiency has approximately the same latency. Intuitively, this suggest that the computational security of cryptographic algorithms imposes to iterate Boolean functions with a minimum complexity that is somewhat comparable for all ciphers. For illustration, we provide the complete synthesis results for these most efficient architectures for all ciphers in Table 1, where the approximate symbol means that we provide an average for ED figures.

Impact of Parallelism/Pipeline. As mentioned in the previous section, unrolling our architectures without parallelizing or pipelining becomes suboptimal at maximum frequency. In cases where the (already high) throughputs obtained in Table 1 are not sufficient for a given application, it is possible to further increase them with parallelization and pipelining. For this purpose, it is natural to start from the efficient architectures with N_r determined as in Table 1. In the first case, we just multiply several circuits as depicted in Figure 7. Since only the control part can be shared between the multiple instances, we essentially double the throughput at the cost of a doubled area (in particular, the control part is small in our implementations and this "doubling rule" was precise up to a few percents). In the case of outer pipelining, it is additionally possible to spare a few multiplexors. Yet, the trends observed for all metrics and all ciphers are essentially the same as well. In particular for the throughput over area ratio and the energy per encrypted bit (i.e. the two metrics we mainly focus on in this work), we observed similar conclusions for all the investigated ciphers. This behavior is again due to the limited cost of the control part compared to the state registers and the datapath in our block cipher implementations. As doubling the parallelism or pipeline essentially comes at the cost of a doubled area, the throughput over area ratio remains close to constant. Since the same comment applies to the energy per bit (i.e. doubling the throughput doubles the power consumption), we conclude that parallelization and pipeline do not increase the efficiency, nor do they notably affect our comparisons of algorithms.

5   Frequency / Voltage Scaling

The previous sections investigated the impact of architectural choices on the efficiency of different implementations. We used them to compare different lightweight ciphers. A natural extension of this work is to investigate the impact of physical parameters, e.g. in terms of frequency / voltage scaling. Two main questions can be investigated in this setting. First, do the algorithm comparisons remain unchanged with variable supply voltage? Second, are the efficiency differences between different lightweight ciphers significant in front of the differences when tuning a physical parameter. In order to answer these questions, we re-synthesized the "most efficient" implementations of Table 1 at different Vdd's. The library we used for this purpose is composed of cells that tolerate supply voltages from 1.2V to 0.4V. Note that, beyond the previously listed remarks about the relative nature of hardware performance comparisons, synthesis results at low supply voltage are particularly sensitive to synthesis options. This

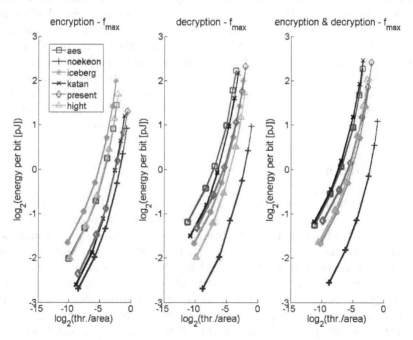

Fig. 5. Voltage scaling: throughput over area ratio vs. energy per bit

confirms the general discussion found in Saar Drimer's PhD dissertation in the context of FPGAs [5]. As a consequence, while we would expect the comparison of algorithms to be fully independent of the frequency / voltage scaling, we observed some curve overlaps in our performance evaluations. Our conclusions are as follows.

From the hardware design point of view and as expected, the critical path increases faster, as the supply voltage decreases (in Figure 14). Hence, the maximum frequency and throughput both decrease non-linearly as well, resulting in a reduction of the throughput over area ratio for low Vdd's. More positively, the reason why we lower the supply voltages is to reach lower power consuming points. This is what we observed in our experiments: the power decreases non-linearly with the supply voltage. However, due to the first observation, this power reduction is also moderated by the critical path increase. As a result, the energy per bit (represented in Figure 15) only decreases close to quadratically with the supply voltage, as expected from the energy required to switch the internal capacitances. Note that for all our syntheses, the leakage currents remained negligible (they would become significant below 0.4V in our target technology).

Regarding algorithms, we again plotted a global view of the power and performance efficiency metrics in Figure 5. This final picture allows us to answer the two previously listed questions. First, the comparisons of the different algorithms and architectures is essentially similar as the ones for Vdd = 1.2V. Yet, and as previously mentioned, some curve overlaps are noticed, due to the increased variability of our synthesis results at low supply voltage. Second, the difference between different algorithms in terms of efficiency is quite limited when compared with the impact of frequency / voltage scaling. For example, the energy per bit can be decreased by an order of magnitude when reducing Vdd (at the cost of a throughput decrease). By contrast, the difference between the various block ciphers investigated roughly corresponds to a factor 2 for this metric. Yet, the gains obtained when looking at combined metrics is non-negligible (i.e. at least it is larger than the variability due to synthesis options), in particular when looking at all architectures (i.e. not only the encryption one).

6   Conclusion

This paper provided a first comprehensive comparison of lightweight block ciphers in terms of energy (and performance) efficiency. It confirms that such ciphers do provide interesting figures compared to standard solutions such as the AES. However, the gains observed are sometimes limited and may not be sufficient to motivate the use of non standard algorithms in actual applications. Note that our conclusions are naturally restricted to an energy-oriented case-study. For example, minimizing the area would lead to totally different optimization tweaks (e.g. taking advantage of resource sharing rather than unrolling).

Synthesis results performed for different architectures and supply voltages suggest that using the smallest rounds (e.g. those of KATAN) is not the best strategy to reach energy-efficient implementations (because of the too large number of iterations required to complete each encryption). Besides, we noticed that the strong similarity in the block cipher rounds design principles does lead to remarkably comparable implementation figures. In fact, the most meaningful differences between the investigated ciphers relate to key scheduling algorithms and the efficient combination of encryption and decryption designs. Overall, we believe that these results and the general discussion about hardware performance evaluation raise interesting problems for the design of new block ciphers. Namely, finding how to make the algorithmic choices more discriminant with respect to hardware implementations is an interesting research direction.

Acknowledgements. Stéphanie Kerckhof is a PhD student funded by a FRIA grant, Belgium. François Durvaux is a PhD student funded by the Walloon region MIPSs project. Cédric Hocquet is a PhD student funded by the Walloon region MIPSs project. David Bol is a Postdoctoral Researcher of the Belgian Fund for Scientific Research (FNRS-F.R.S.). François-Xavier Standaert is an Associate Researcher of the Belgian Fund for Scientific Research (FNRS-F.R.S.). This work has been funded in part by the ERC project 280141 (acronym CRASH).

References

1. Bogdanov, A., Knudsen, L.R., Leander, G., Paar, C., Poschmann, A., Robshaw, M.J.B., Seurin, Y., Vikkelsoe, C.: PRESENT: An Ultra-Lightweight Block Cipher. In: Paillier, P., Verbauwhede, I. (eds.) CHES 2007. LNCS, vol. 4727, pp. 450–466. Springer, Heidelberg (2007)
2. De Cannière, C., Dunkelman, O., Knežević, M.: KATAN and KTANTAN — A Family of Small and Efficient Hardware-Oriented Block Ciphers. In: Clavier, C., Gaj, K. (eds.) CHES 2009. LNCS, vol. 5747, pp. 272–288. Springer, Heidelberg (2009)
3. Daemen, J., Peeters, M., Van Assche, G., Rijmen, V.: Nessie proposal: NOEKEON, http://gro.noekeon.org/
4. Daemen, J., Rijmen, V.: The Design of Rijndael: AES - The Advanced Encryption Standard. Springer (2002)
5. Drimer, S.: Security for volatile FPGAs. Technical Report UCAM-CL-TR-763, University of Cambridge, Computer Laboratory (November 2009)
6. Eisenbarth, T., Gong, Z., Güneysu, T., Heyse, S., Kerckhof, S., Indesteege, S., Koeune, F., Nad, T., Plos, T., Regazzoni, F., Standaert, F.-X., Van Oldeneel, L.: Compact implementation and performance evaluation of block ciphers in ATtiny devices (2011)
7. Elbirt, A.J., Yip, W., Chetwynd, B., Paar, C.: An FPGA implementation and performance evaluation of the AES block cipher candidate algorithm finalists. In: AES Candidate Conference, pp. 13–27 (2000)
8. Gaj, K., Chodowiec, P.: Comparison of the hardware performance of the AES candidates using reconfigurable hardware. In: AES Candidate Conference, pp. 40–54 (2000)
9. Gaj, K., Homsirikamol, E., Rogawski, M.: Fair and comprehensive methodology for comparing hardware performance of fourteen round two sha-3 candidates using FPGAS. In: Mangard, Standaert (eds.) [17], pp. 264–278

10. Gong, Z., Nikova, S., Law, Y.W.: KLEIN: A New Family of Lightweight Block Ciphers. In: Juels, A., Paar, C. (eds.) RFIDSec 2011. LNCS, vol. 7055, pp. 1–18. Springer, Heidelberg (2012)

11. Guo, J., Peyrin, T., Poschmann, A., Robshaw, M.J.B.: The led block cipher. In: Preneel, Takagi (eds.) [19], pp. 326–341

12. Henzen, L., Gendotti, P., Guillet, P., Pargaetzi, E., Zoller, M., Gürkaynak, F.K.: Developing a hardware evaluation method for SHA-3 candidates. In: Mangard, Standaert (eds.) [17], pp. 248–263

13. Hong, D., Sung, J., Hong, S., Lim, J., Lee, S., Koo, B.-S., Lee, C., Chang, D., Lee, J., Jeong, K., Kim, H., Kim, J., Chee, S.: HIGHT: A New Block Cipher Suitable for Low-Resource Device. In: Goubin, L., Matsui, M. (eds.) CHES 2006. LNCS, vol. 4249, pp. 46–59. Springer, Heidelberg (2006)

14. Kerckhof, S., Durvaux, F., Veyrat-Charvillon, N., Regazzoni, F., de Dormale, G.M., Standaert, F.-X.: Compact FPGA Implementations of the Five SHA-3 Finalists. In: Prouff, E. (ed.) CARDIS 2011. LNCS, vol. 7079, pp. 217–233. Springer, Heidelberg (2011)

15. Leander, G., Paar, C., Poschmann, A., Schramm, K.: New Lightweight DES Variants. In: Biryukov, A. (ed.) FSE 2007. LNCS, vol. 4593, pp. 196–210. Springer, Heidelberg (2007)

16. Lim, C.H., Korkishko, T.: mCrypton – A Lightweight Block Cipher for Security of Low-Cost RFID Tags and Sensors. In: Song, J.-S., Kwon, T., Yung, M. (eds.) WISA 2005. LNCS, vol. 3786, pp. 243–258. Springer, Heidelberg (2006)

17. Mangard, S., Standaert, F.-X. (eds.): CHES 2010. LNCS, vol. 6225. Springer, Heidelberg (2010)

18. Mentens, N., Batina, L., Preneel, B., Verbauwhede, I.: A Systematic Evaluation of Compact Hardware Implementations for the Rijndael S-Box. In: Menezes, A. (ed.) CT-RSA 2005. LNCS, vol. 3376, pp. 323–333. Springer, Heidelberg (2005)

19. Preneel, B., Takagi, T. (eds.): CHES 2011. LNCS, vol. 6917. Springer, Heidelberg (2011)

20. Shibutani, K., Isobe, T., Hiwatari, H., Mitsuda, A., Akishita, T., Shirai, T.: Piccolo: An ultra-lightweight blockcipher. In: Preneel, Takagi (eds.) [19], pp. 342–357

21. Standaert, F.-X., Piret, G., Gershenfeld, N., Quisquater, J.-J.: SEA: A Scalable Encryption Algorithm for Small Embedded Applications. In: Domingo-Ferrer, J., Posegga, J., Schreckling, D. (eds.) CARDIS 2006. LNCS, vol. 3928, pp. 222–236. Springer, Heidelberg (2006)

22. Standaert, F.-X., Piret, G., Rouvroy, G., Quisquater, J.-J., Legat, J.-D.: ICEBERG: An Involutional Cipher Efficient for Block Encryption in Reconfigurable Hardware. In: Roy, B., Meier, W. (eds.) FSE 2004. LNCS, vol. 3017, pp. 279–299. Springer, Heidelberg (2004)

23. Weaver, N., Wawrzynek, J.: A comparison of the AES candidates amenability to FPGA implementation. In: AES Candidate Conference, pp. 28–39 (2000)

24. Wheeler, D.J., Needham, R.M.: Tea, a Tiny Encryption Algorithm. In: Preneel, B. (ed.) FSE 1994. LNCS, vol. 1008, pp. 363–366. Springer, Heidelberg (1995)

Appendix

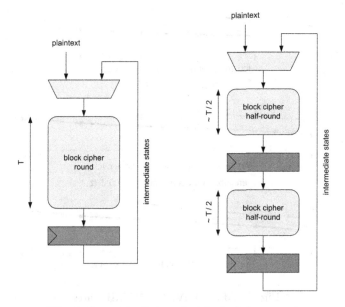

Fig. 6. Inner pipelining of a block cipher round

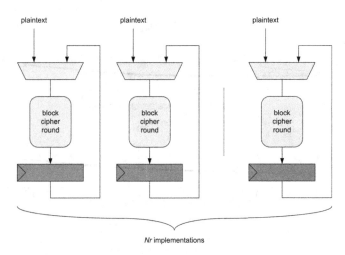

Fig. 7. Parallel architectures with various number of rounds

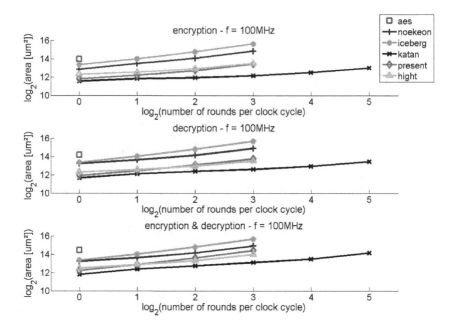

Fig. 8. Vdd = 1.2V, f = 100MHz: area

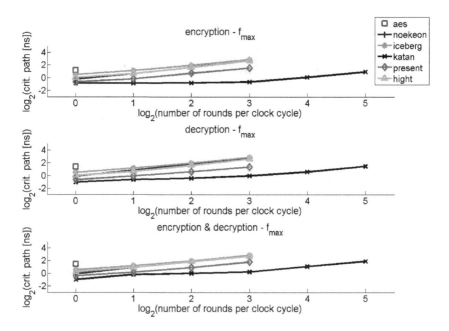

Fig. 9. Vdd = 1.2V, f_{max}: critical path

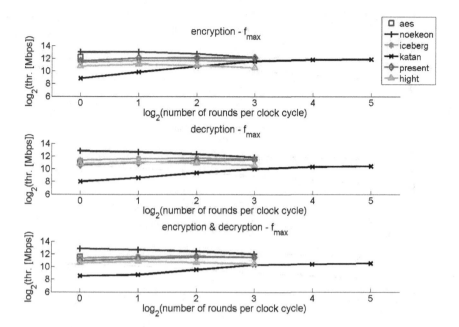

Fig. 10. Vdd = 1.2V, f_{max}: throughput

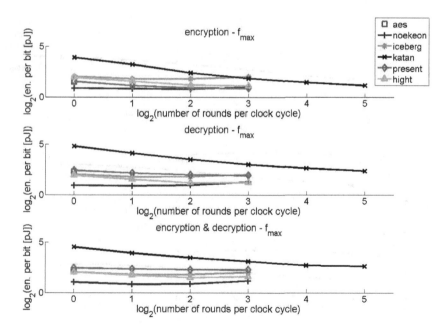

Fig. 11. Vdd = 1.2V, f_{max}: energy per bit

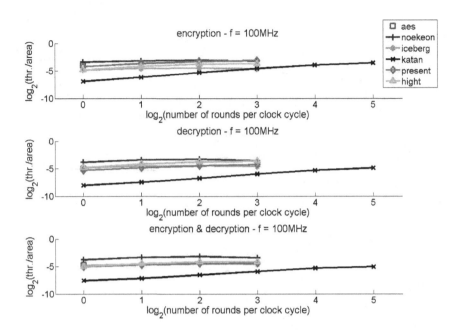

Fig. 12. Vdd = 1.2V, f = 100MHz: throughput over area ratio

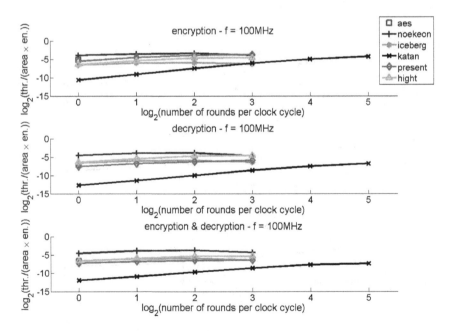

Fig. 13. Vdd = 1.2V, f = 100MHz: throughput over (area × energy per bit) ratio

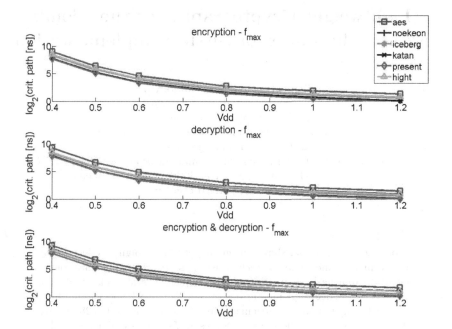

Fig. 14. Voltage scaling: critical path

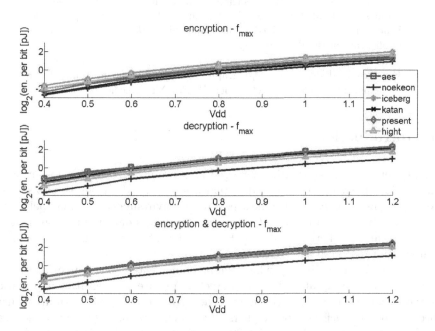

Fig. 15. Voltage scaling: energy per bit

Lightweight Cryptography for the Cloud: Exploit the Power of Bitslice Implementation

Seiichi Matsuda[1] and Shiho Moriai[2]

[1] Sony Corporation
1-7-1 Konan, Minato-ku, Tokyo 108-0075, Japan
SeiichiA.Matsuda@jp.sony.com
[2] National Institute of Information and Communications Technology (NICT)
4-2-1 Nukui-Kitamachi, Koganei, Tokyo 184-8795, Japan
shiho.moriai@nict.go.jp

Abstract. This paper shows the great potential of lightweight cryptography in fast and timing-attack resistant software implementations in cloud computing by exploiting bitslice implementation. This is demonstrated by bitslice implementations of the PRESENT and Piccolo lightweight block ciphers. In particular, bitsliced PRESENT-80/128 achieves 4.73 cycles/byte and Piccolo-80 achieves 4.57 cycles/byte including data conversion on an Intel Xeon E3-1280 processor (Sandy Bridge microarchitecture). It is also expected that bitslice implementation offers resistance to side channel attacks such as cache timing attacks and cross-VM attacks in a multi-tenant cloud environment. Lightweight cryptography is not limited to constrained devices, and this work opens the way to its application in cloud computing.

Keywords: lightweight cryptography, software implementation, bitslice implementation, cloud, block cipher, PRESENT, Piccolo.

1 Introduction

The cyber physical system has emerged as a promising direction for enriching interactions between physical and virtual worlds [14]. Many wireless sensor networks, for instance, monitor some aspect of the environment or human behaviors, and relay the data to the *cloud* for processes such as data mining, business intelligence and predictive analytics. Preservation of security and privacy in the sensed information in this system is essential.

Lightweight cryptography, which can be implemented on resource-constrained devices, is attracting attention for protecting private and sensitive information gathered on *sensors*. Recently many lightweight cryptographic primitives have been proposed, such as block ciphers, stream ciphers, hash functions, message-authentication codes [3,19,5,1,7,8,20]. Moreover, an international standard of lightweight cryptography (ISO/IEC 29192) has been developed in ISO/IEC JTC 1/SC 27.

E. Prouff and P. Schaumont (Eds.): CHES 2012, LNCS 7428, pp. 408–425, 2012.

Most lightweight cryptographic algorithms are designed to minimize the resource consumption of a *hardware implementation* such as area, power, and energy consumption, while some are software-oriented with design criteria such as low memory requirements, small code size, and limited instruction sets for low-end (e.g., 8-bit) platforms. As a result of design trade-off, some of lightweight cryptographic algorithms do not show good throughput in software implementation on mid-range to high-end microprocessors (e.g., Intel Core i7 processors) typically used for cloud computing.

This paper shows the great potential of lightweight cryptography in *fast* software implementations in cloud environments by exploiting bitslice implementation demonstrated through bitslice implementations of PRESENT and Piccolo. PRESENT and Piccolo are lightweight 64-bit block ciphers: the former was presented by Bogdanov et al. at CHES 2007 [3] and is specified in ISO/IEC 29192-2 [11], and the latter was presented by Shibutani et al. at CHES 2011 [20]. In particular, PRESENT-80/128 achieves 4.73 cycles/byte and Piccolo-80 achieves 4.57 cycles/byte on an Intel Xeon E3-1280 processor (Sandy Bridge). PRESENT-80/128 achieves 5.79 cycles/byte. Piccolo-80 achieves 5.69 cycles/byte on an Intel Core i7 870 (Nehalem), which are faster than the bitsliced AES's fastest record on the same microarchitecture (6.92 cycles/byte on an Intel Core i7 920) [12].

Only a few *software* performance data of lightweight cryptography for comparison exist in public literature. As for PRESENT, in [17] there are some software implementation results on 4-bit, 8-bit and 16-bit microcontrollers, but on a 32-bit processor, the encryption speed available is 16.2 cycles/byte on a Pentium III. In [9], it is written that optimized table-based implementations run 57 and 86 cycles/byte on a Core i7 Q720 for LED-64 and LED-128, respectively, and that they are faster than PRESENT. There is also previous work on bitsliced PRESENT by Grabher et al. [6], but their implementation results are not competitive.

It has been known that bitslice implementation is also resistant to cache timing attacks because it has no table lookups. In a multi-tenant cloud environment, cross-virtual machine (VM) attacks become new threats [18]. Bitslice implementation mitigates these risks.

The remainder of this paper is structured as follows. Section 2 shows a brief history of bitslice implementation, a use case of lightweight block ciphers in the cloud, and our target. Sections 3 and 4 respectively show bitslice implementations of lightweight block ciphers PRESENT and Piccolo, including optimizing techniques. Section 5 shows performance data and comparison with previous results, and Section 6 gives our conclusion.

2   Bitslice Implementation

Biham in 1997 introduced bitslicing as a technique for implementing cryptographic algorithms to improve the software performance of DES [2]. It was implemented on several processors and used for brute force key search of DES in the DES Challenges project in the late-1990s. The basic concept of bitslicing is

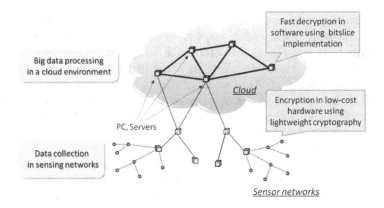

Fig. 1. Use case of decryption in bitslice implementation in the cloud

to simulate a hardware implementation in software. The entire algorithm is represented as a sequence of logical operations. On a processor with n-bit registers, a logical instruction corresponds to simultaneous execution of n hardware logical gates. In the bitslice implementation, S-boxes are computed using bit-logical instructions rather than table lookups. Since the execution time of these instructions is independent of the input and key values, the bitslice implementation is generally resistant to timing attacks.

Bitslice implementation techniques have progressed. The bitslice implementations of block ciphers presented by Biham were to encrypt/decrypt independent n blocks on a processor with n-bit registers. Matsui and Nakajima [15] demonstrated remarkable performance gain on Intel's Core 2 processor by fully utilizing its enhanced SIMD architecture. They showed a bitsliced AES running at the speed of 9.2 cycles/byte on a Core 2, which was faster than any previous standard table-based implementations. A hurdle in this implementation was that as many as n independent blocks needed to be processed simultaneously. Könighofer [13] presented an alternative implementation for 64-bit platforms that processes only four input blocks in parallel. Käsper and Schwabe [12] extended this approach and achieved a bitsliced AES in counter mode running at 7.59 cycles/byte on a Core 2.

A Use Case of Lightweight Block Ciphers in the Cloud.   In cyber physical systems, analyzing large data sets – so-called *big data* – will become a key basis of competition, underpinning new waves of productivity growth, innovation, and consumer surplus. Cloud computing will play an important role in analyzing big data, where scale-out software systems running on low-cost "commodity" platforms are expected. When sensor data need to be encrypted for privacy protection, encryption on a low-cost embedded hardware module using a lightweight block cipher will be the most cost competitive solution on the sensor side. Encrypted sensor data are collected from many sensors and decrypted on servers in the cloud when needed.

Bitslice implementation provides leverage in this use case. In most cases it can be implemented so that the sensor data size per transmission fits the block size. Encrypted sensor data from each sensor can be decrypted independently. One of the drawbacks of bitslice implementation has been the low number of applications where the encryption/decryption unit size is large, e.g., 2048-byte chunks. However, in this use case, one can simply collect encrypted sensor data from many sensors until the decryption unit size with no concern about the order, and then decrypt them by using bitslice implementation. The decryption key can be set block-by-block independently.

Our Target. We choose PRESENT and Piccolo as each representative of lightweight block ciphers based on Substitution Permutation Networks and Feistel networks, respectively. Our implementations of PRESENT and Piccolo are run on three different Intel microarchitectures: Core (45-nm), Nehalem, and Sandy Bridge. Core and Nehalem support up to Streaming SIMD Extensions (SSE) 4.1 with 16 128-bit XMM registers, and Sandy Bridge newly supports Advanced Vector Extensions (AVX) as an extension of SSE. Major enhancements of AVX are supports for 256-bit YMM registers, 256-bit floating point instruction set, and 3-operand syntax, which is also used for legacy 128-bit SSE instructions (we call this 128-bit AVX). For example, 2-operand syntax instruction `pxor xmm1, xmm2` (xmm1^=xmm2) can be expressed in 3-operand syntax as `vpxor xmm1, xmm2, xmm3` (xmm1=xmm2^xmm3). Since a source operand of an instruction is not overwritten by the result, 3-operand syntax can reduce the cost of temporary data copy to another register and reduce code size. Unfortunately, 256-bit AVX does not support integer instructions operated on 256-bit YMM registers, so we use 128-bit AVX with 3-operand using XMM registers on Sandy Bridge. Legacy SSE instructions used in our implementation, such as logical (`pand, pandn, por, pxor`), data transfer (`movdqa`), shuffle (`pshufb, pshufd`), and unpack instructions (`punpckhbw` and its variants) are supported by the three architectures. 128-bit AVX instructions used on Sandy Bridge are `vpand, vpandn, vpor, vpxor, vpshufb, vpunpckhbw` and its variants. The latency of the register-to-register operations above is one cycle. The register-to-memory operations require more cycles depending on the data dependency, memory/cache mechanism, and characteristics of each microarchitecture.

Our Implementation Approach. Our implementation handles the number of parallel blocks smaller than the original bitslice implementation. This approach enables processing operations on only 16 XMM registers without frequent loading and storing of data between XMM registers and memory, and improves convenience as a cryptographic library tool. To explore the possibility of bitslice implementation of PRESENT and Piccolo, we study several cases for the number of blocks processed in parallel: 8-, 16-, and 32-parallelism for PRESENT and 16-parallelism for Piccolo. In Section 3 and 4, at the beginning we introduce some specific optimizations for each algorithm with legacy SSE instructions, and then optimize our implementations to reduce the number of instructions of the codes by using 128-bit AVX instructions on Sandy Bridge.

3 PRESENT

PRESENT [3] is a 64-bit block cipher supporting 80- and 128-bit keys. The S/P-network of PRESENT consists of addRoundKey, sBoxLayer and pLayer with 31 rounds as shown in Fig. 2. sBoxLayer consists of 16 parallel 4-bit S-boxes and pLayer permutes bit positions of the 64-bit data state. After the final round, the state is XORed with the round key for post-whitening and output as ciphertext. We denote the 64-bit block of PRESENT by 16 4-bit data $n_0, \cdots, n_{15}$. Let $n_i = n_{i,0}||n_{i,1}||n_{i,2}||n_{i,3}$ for $0 \leq i \leq 15$, where $n_{i,j}$ is the j-th bit of n_i.

3.1 Bitsliced Representation

Our bitsliced representations for 8-, 16-, and 32-block parallel implementations are shown in Fig. 3, Fig. 4, and Figs. 5 and 6, respectively. In this paper, we denote 16 128-bit XMM registers by $r[i], 0 \leq i \leq 15$. For l-block parallel implementation, l-bit data $\mathbf{n}_{i,j}$ in the figures means the bit collection of $n_{i,j}$ gathered from the same position of each l-block. The 4-bit slicing enables us to compute 4-bit S-box using bit-logical instructions in the same way as the original 1-bit slicing [2] and the 8-bit slicing for AES [12].

We use four XMM registers for 8-block parallel implementation, eight XMM registers for 16-block parallel implementation to store the 4-bit slicing of input data, and the remaining XMM registers as temporary registers for processing sBoxLayer and pLayer.

For 32-block parallel implementation, we handle two bitsliced representations with 16 XMM registers and switch the representations of intermediate data alternately in rounds to reduce the cost of pLayer processing, i.e., the processing can be skipped every other round. Figure 5 gives the initial bitsliced representation after performing a conversion algorithm. Since there are no XMM register for temporary use, we need to move data in a XMM register to memory in the process of sBoxLayer and pLayer.

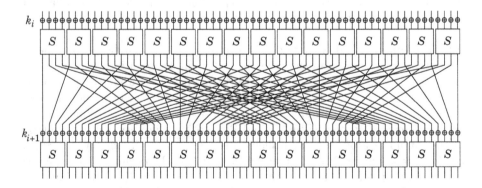

Fig. 2. The S/P network for PRESENT

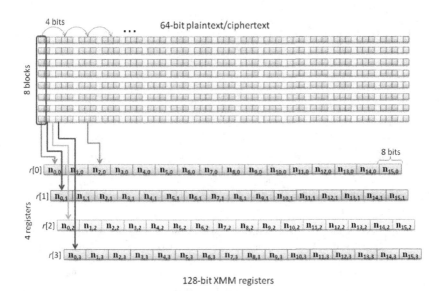

Fig. 3. Bitsliced representation of PRESENT in 8-block parallel implementation

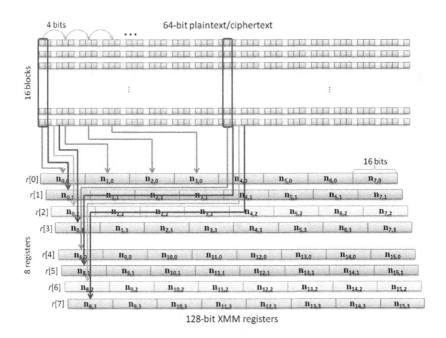

Fig. 4. Bitsliced representation of PRESENT in 16-block parallel implementation

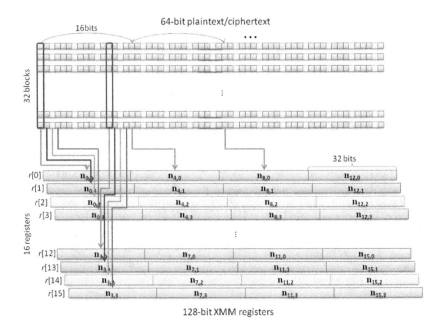

Fig. 5. First bitsliced representation of PRESENT in 32-block parallel implementation

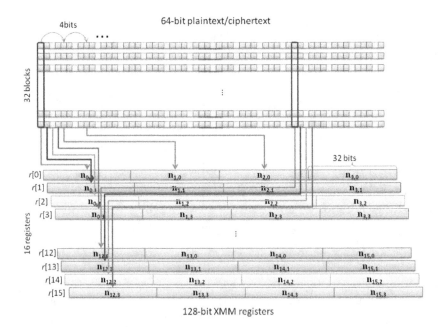

Fig. 6. Second bitsliced representation of PRESENT in 32-block parallel implementation

3.2 sBoxLayer

A smaller logical representation of S-box maximizes the advantage of bitslice implementation. One previous work reported that the logical representation of PRESENT S-box requires only 14 gates [4], in which four temporary registers were used and 3-operand logical instructions were assumed. We therefore use their logical representation for 8- and 16-block parallel implementations with 128-bit AVX on Sandy Bridge, and search for another logical representation using 2-operand instructions for the other implementations.

We took the same approach as Osvik [16] to search a software-oriented logical representation that consists of five operations and, or, xor, not, mov with only five registers (four registers for input and one register for temporary use). The instruction sequence found by our algorithm requires 20 instructions as below.

```
// Input:  r3, r2, r1, r0, tmp
// Output: r3, r2, r1, r0
 1. r2   ^= r1;    r3   ^= r1;
 2. tmp =  r2;    r2  &= r3;
 3. r1   ^= r2;    tmp ^= r0;
 4. r2  =  r1;    r1  &= tmp;
 5. r1   ^= r3;    tmp ^= r0;
 6. tmp |= r2;    r2   ^= r0;
 7. r2   ^= r1;    tmp ^= r3;
 8. r2  =  ~r2;    r0   ^= tmp;
 9. r3  =  r2;    r2  &= r1;
10. r2  |= tmp;
11. r2  =  ~r2;
```

Note that four registers r3, r2, r1, r0 of input registers contain four input bits (r3 contains the most significant bit).

3.3 pLayer

The original 1-bit slicing can compute bit-by-bit permutation like pLayer of PRESENT by only changing the order of registers with no cost. However our 4-bit slicing causes additional operations for processing pLayer in compensation for the decrease in the parallelism of bitslice implementations from 128 (size of XMM register) to 8, 16, and 32.

A combination of the shuffle byte instruction pshufb firstly introduced in Intel Supplemental SSE3 (SSSE3) and the unpack instructions for double-word punpck(h/l)dq and quad-word punpck(h/l)qdq realizes the pLayer processing. The notation h and l of h/l means high-order and low-order of 64-bit data in a 128-bit XMM register, respectively.

As the bitsliced representations for 8- and 16-block parallel implementation are almost same format, the implementation of pLayer for the 16-block requires the operations for the 8-block twice. We explain the case for the 8-block and then progress to the case for 32-block parallel implementation.

8-block Parallel Implementation. First of all, we perform `pshufb` on XMM register $r[0]$ containing $\mathbf{n}_{i,0}$ for $0 \leq i \leq 15$ in Fig. 3 as the following pattern.

$$r[0] : \mathbf{n}_{0,0}||\mathbf{n}_{4,0}||\mathbf{n}_{8,0}||\mathbf{n}_{12,0}||\mathbf{n}_{1,0}||\mathbf{n}_{5,0}||\cdots||\mathbf{n}_{10,0}||\mathbf{n}_{14,0}||\mathbf{n}_{3,0}||\mathbf{n}_{7,0}||\mathbf{n}_{11,0}||\mathbf{n}_{15,0}$$

Applying for the other registers $r[1], r[2]$, and $r[3]$ similarly, we perform the `punpckhdq` instruction on $r[0]$ and $r[1]$, which unpacks and interleaves the high-order double-word from $r[0]$ and $r[1]$ into $r[0]$. The subsequent `punpckhqdq` for $r[0]$ and $r[2]$, where $r[2]$ contains the result of `punpckhdq` for $r[2]$ and $r[3]$, can produce desired 128-bit data in register $r[0]$ as follows.

$$r[0] : \mathbf{n}_{0,0}||\mathbf{n}_{4,0}||\mathbf{n}_{8,0}||\mathbf{n}_{12,0}||\mathbf{n}_{0,1}||\mathbf{n}_{4,1}||\cdots||\mathbf{n}_{8,2}||\mathbf{n}_{12,2}||\mathbf{n}_{0,3}||\mathbf{n}_{4,3}||\mathbf{n}_{8,3}||\mathbf{n}_{12,3}$$

In the pLayer processing with legacy SSE instructions, we require 16 instructions, i.e., four `pshufb`, four `punpck(h/l)dq`, four `punpck(h/l)qdq`, and four `movdqa` for storing intermediate results. With an optimization using 128-bit AVX instructions `vpunpck(h/l)dq` and `vpunpck(h/l)qdq`, four `movdqa` become redundant, i.e., requiring 12 instructions in total.

32-block Parallel Implementation. As mentioned before, we manage two bitsliced representations for 32-block parallel implementation. These representations are constructed in such a way that the bit permutation of pLayer for the initial bitsliced representation as shown in Fig. 5 produces the other representation with only register renaming. Using the notation of the intial bitsliced representation, we can represent the updated bitsliced representation as the result of the pLayer process for the initial bitsliced representation as follows.

$$r[0] : \mathbf{n}_{0,0}||\mathbf{n}_{4,0}||\mathbf{n}_{8,0}||\mathbf{n}_{12,0}$$
$$r[4] : \mathbf{n}_{1,0}||\mathbf{n}_{5,0}||\mathbf{n}_{9,0}||\mathbf{n}_{13,0}$$
$$r[8] : \mathbf{n}_{2,0}||\mathbf{n}_{6,0}||\mathbf{n}_{10,0}||\mathbf{n}_{14,0}$$
$$r[12] : \mathbf{n}_{3,0}||\mathbf{n}_{7,0}||\mathbf{n}_{11,0}||\mathbf{n}_{15,0}$$

$$\vdots$$

$$r[3] : \mathbf{n}_{0,3}||\mathbf{n}_{4,3}||\mathbf{n}_{8,3}||\mathbf{n}_{12,3}$$
$$r[7] : \mathbf{n}_{1,3}||\mathbf{n}_{5,3}||\mathbf{n}_{9,3}||\mathbf{n}_{13,3}$$
$$r[11] : \mathbf{n}_{2,3}||\mathbf{n}_{6,3}||\mathbf{n}_{10,3}||\mathbf{n}_{14,3}$$
$$r[15] : \mathbf{n}_{3,3}||\mathbf{n}_{7,3}||\mathbf{n}_{11,3}||\mathbf{n}_{15,3}$$

The above corresponds to the second bitsliced representation as shown in Fig. 6. The pLayer processing in the next round for this representation requires an instruction sequence consisting of 16 `punpck(h/l)dq`, 16 `punpck(h/l)qdq` and 20 `movdqa` including four memory accesses for temporarily storing data twice and produces the initial bitsliced representation. Therefore the pLayer process can be computed every other round and requires 26 instructions on average. The 128-bit AVX can reduce the number of instructions from 52 to 36.

An additional operation for this trick to adjust the alignment of round keys is needed, unpacking round keys every other round in the key schedule.

4   Piccolo

Piccolo [20] is a lightweight 64-bit block cipher supporting 80-bit and 128-bit keys. Piccolo has a structure of a variant of 4-line generalized Feistel network (GFN) as shown in Fig. 7, and iterates 25 and 31 rounds for 80- and 128-bit keys, respectively. We denote a 64-bit block for Piccolo by four 16-bit words: W_0, W_1, W_2, W_3. Let $W_i = n_{4*i}||n_{4*i+1}||n_{4*i+2}||n_{4*i+3}$ for $0 \leq i \leq 3$, and let $n_j = n_{j,0}||n_{j,1}||n_{j,2}||n_{j,3}$ for $0 \leq j \leq 15$, where n_j is 4-bit data and $n_{j,k}$ is the k-th bit of n_j.

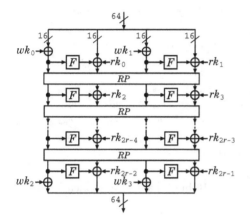

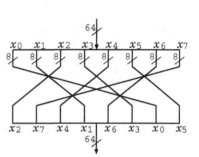

| **Fig. 7.** The structure of Piccolo | **Fig. 8.** Round permutation RP |

4.1   Bitsliced Representation

Figure 9 shows our bitsliced representation for 16-block parallel implementation of Piccolo. 16-bit data $\mathbf{n}_{i,j}$ in the figure means the bit collection of $n_{i,j}$ gathered from a same position of each 16-block for $0 \leq i \leq 15$ and $0 \leq j \leq 3$.

Since the 2-line out of the 4-line GFN is processed and the other lines pass through in one round, the data for the former and latter should be stored separately, assigning each for different registers. Then, two F-functions used in the GFN are the same, so we can pack eight 16-bit data $\mathbf{n}_{i,j}$ corresponding to the 2-line data (e.g., $W_0 \& W_2$ or $W_1 \& W_3$) on same XMM registers with the 4-bit slicing as our implementation of PRESENT.

The number of XMM registers for storing 4-bit slicing of input data is eight. Four XMM registers of the renaming eight XMM registers can be used for storing the data passed through during processing F-functions and the other four XMM registers can be used as temporary registers for processing F-functions.

If we assign the data for a line of the 4-line GFN on a XMM register for 32-block parallel implementation of Piccolo, we would need full 16 XMM registers to store 4-bit slicing of input data. It leads to more memory access than the case of PRESENT for storing the data temporarily. Therefore we think 16-block parallel implementation of Piccolo using 128-bit XMM registers is optimal parallelism.

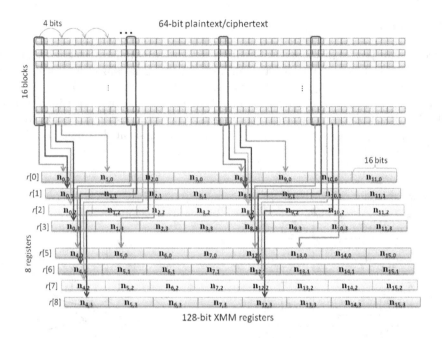

Fig. 9. Bitsliced representation of Piccolo in 16-block parallel implementation

4.2   F-Function

The F-function consists of two S-box layers and a diffusion matrix (see Fig. 10).

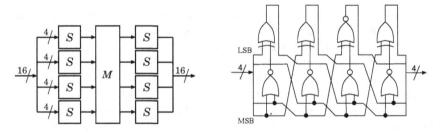

Fig. 10. F-function   **Fig. 11.** S-box S

S-box Layer. The S-box layer consists of four 4-bit bijective S-boxes S represented by the logic circuit shown in Fig. 11. A software instruction sequence of the S-box can be manually obtained from the logic circuit, which requires 15 instructions with a temporary register. We searched for a smaller instruction sequence in the similar way to the PRESENT S-box and found the following one with 13 instructions in six cycles, assuming that up to three independent instructions are issued per cycle.

```
// Input:  r3, r2, r1, r0, tmp
// Output: r0, r1, r2, r3
1. tmp =  r1;    r1  |= r2;    r3  = ~r3;
2. r0  ^= r2;    r1  ^= r3;    r3  |= r2;
3. r0  ^= r3;    r3  =  r1;
4. r3  |= r0;
5. r3  ^= tmp;   tmp |= r0;
6. r2  ^= tmp;   r3  = ~r3;
```

The notation is the same as the instruction sequence of the PRESENT S-box.

Diffusion Matrix. The following multiplication between the constant 4×4 diffusion matrix M and four 4-bit data x_0, x_1, x_2, x_3 over $\mathrm{GF}(2^4)$ defined by an irreducible polynomial $x^4 + x + 1$ outputs four 4-bit data y_0, y_1, y_2, y_3.

$$
\begin{pmatrix} y_0 \\ y_1 \\ y_2 \\ y_3 \end{pmatrix} = \begin{pmatrix} 2 & 3 & 1 & 1 \\ 1 & 2 & 3 & 1 \\ 1 & 1 & 2 & 3 \\ 3 & 1 & 1 & 2 \end{pmatrix} \cdot \begin{pmatrix} x_0 \\ x_1 \\ x_2 \\ x_3 \end{pmatrix}
$$

Let $x_i = x_{i,0}||x_{i,1}||x_{i,2}||x_{i,3}$ for $0 \le i \le 3$. Since diffusion matrix M is cyclic, 4-bit data y_i can be expressed as $y_i = 2 \cdot x_i \oplus 3 \cdot x_{i+1} \oplus x_{i+2} \oplus x_{i+3}$ where index is calculated by modulo 4. Representing each bit of the products $2 \cdot x_i$ and $3 \cdot x_i$ given in Table 1, all bits of $y_i = y_{i,0}||y_{i,1}||y_{i,2}||y_{i,3}$ are obtained as follows.

$$y_{i,0} = x_{i,1} \oplus x_{i+1,0} \oplus x_{i+1,1} \oplus x_{i+2,0} \oplus x_{i+3,0}$$
$$y_{i,1} = x_{i,2} \oplus x_{i+1,1} \oplus x_{i+1,2} \oplus x_{i+2,1} \oplus x_{i+3,1}$$
$$y_{i,2} = x_{i,0} \oplus x_{i,3} \oplus x_{i+1,0} \oplus x_{i+1,2} \oplus x_{i+1,3} \oplus x_{i+2,2} \oplus x_{i+3,2}$$
$$y_{i,3} = x_{i,0} \oplus x_{i+1,0} \oplus x_{i+1,3} \oplus x_{i+2,3} \oplus x_{i+3,3}$$

As each 128-bit XMM register contains eight 16-bit data corresponding the 2-line data, it is possible to perform two matrix calculations simultaneously, utilizing rotation of the data on the upper and lower 64-bit data in a 128-bit XMM register. The 16-bit left rotation rot_{16} on a XMM register with `pshufb` instruction is defined as follows.

$$rot_{16} : [w_0||w_1||w_2||w_3||w_4||w_5||w_6||w_7] \mapsto [w_1||w_2||w_3||w_0||w_5||w_6||w_7||w_4]$$

Note that w_i is a 16-bit data and 32-bit left rotation rot_{32} and 48-bit left rotation rot_{48} can be defined in the same way.

Table 1. Multiplication 4-bit data x_i by 2 and 3 over $\mathrm{GF}(2^4)$ defined by $x^4 + x + 1$

x_i	$x_{i,0}$	$x_{i,1}$	$x_{i,2}$	$x_{i,3}$
$2 \cdot x_i$	$x_{i,1}$	$x_{i,2}$	$x_{i,0} \oplus x_{i,3}$	$x_{i,0}$
$3 \cdot x_i$	$x_{i,0} \oplus x_{i,1}$	$x_{i,1} \oplus x_{i,2}$	$x_{i,0} \oplus x_{i,2} \oplus x_{i,3}$	$x_{i,0} \oplus x_{i,3}$

We can compute the following updated four XMM registers $r[i]$ for $0 \le i \le 3$ with 25 instructions including eight `pshufb`, 13 `pxor` and four `movdqa` instructions, using four temporary registers.

$$r[0] \leftarrow r[1] \oplus rot_{16}(r[1]) \oplus rot_{16}(r[0]) \oplus rot_{32}(r[0] \oplus rot_{16}(r[0]))$$
$$r[1] \leftarrow r[2] \oplus rot_{16}(r[2]) \oplus rot_{16}(r[1]) \oplus rot_{32}(r[1] \oplus rot_{16}(r[1]))$$
$$r[2] \leftarrow r[0] \oplus rot_{16}(r[0]) \oplus r[3] \oplus rot_{16}(r[3]) \oplus rot_{16}(r[2]) \oplus rot_{32}(r[2] \oplus rot_{16}(r[2]))$$
$$r[3] \leftarrow r[0] \oplus rot_{16}(r[0]) \oplus rot_{16}(r[3]) \oplus rot_{32}(r[3] \oplus rot_{16}(r[3]))$$

Note that translating $rot_{48} = rot_{32} \circ rot_{16}$ saves the number of `pshufb`.

4.3   Round Permutation

The round permutation (RP) permutes eight 8-bit data over 64-bit data as shown in Fig. 8. A simple implementation of RP for a XMM register holding j-th bit of 4-bit data n_i permutes four 32-bit data (e.g., $\mathbf{n}_{i,j}, \mathbf{n}_{i+1,j}$) in a 128-bit register by using double-word shuffle instruction `pshufd` as follows.

$$rp_0 : [w_0||w_1||w_2||w_3||w_4||w_5||w_6||w_7] \mapsto [w_4||w_5||w_2||w_3||w_0||w_1||w_6||w_7]$$
$$rp_1 : [w_0||w_1||w_2||w_3||w_4||w_5||w_6||w_7] \mapsto [w_0||w_1||w_6||w_7||w_4||w_5||w_2||w_3]$$

Note that we perform rp_0 and rp_1 on four XMM regsiters holding the data for $W_0 \& W_2$ and $W_1 \& W_3$, respectively. It requires only two `pshufd` instructions per bit, or eight in total per one round. Before proceeding to the next round, we need renaming four XMM regsiters holding the data for $W_0 \& W_2$ and $W_1 \& W_3$.

Remove Round Permutation.   We describe the implementation to remove RP changing the calculation of diffusion matrix M and the position of round keys with no cost in the data processing. This modification can reduce 8*(the number of rounds) instructions compared to the above implementation of RP.

Since register renaming can only switch the positions of the 2-line data, the removing RP causes the misalignment in byte position on XMM registers to effect input-output of diffusion matrix and subsequent xor with round keys and data in the previous round. The byte positions from Round 1 to 5 in normal 64-bit block with/without RP before the round process are as follows.

	byte position with RP	byte position without RP
Round 1:	$[b_0, b_1, b_2, b_3, b_4, b_5, b_6, b_7]$	$[b_0, b_1, b_2, b_3, b_4, b_5, b_6, b_7]$
Round 2:	$[b_2, b_7, b_4, b_1, b_6, b_3, b_0, b_5]$	$[b_2, \mathbf{b_3}, \mathbf{b_0}, b_1, b_6, \mathbf{b_7}, \mathbf{b_4}, b_5]$
Round 3:	$[b_4, b_5, b_6, b_7, b_0, b_1, b_2, b_3]$	$[b_0, b_1, b_2, b_3, b_4, b_5, b_6, b_7]$
Round 4:	$[b_6, b_3, b_0, b_5, b_2, b_7, b_4, b_1]$	$[b_2, b_3, b_0, \mathbf{b_1}, \mathbf{b_6}, b_7, b_4, \mathbf{b_5}]$
Round 5:	$[b_0, b_1, b_2, b_3, b_4, b_5, b_6, b_7]$	$[b_0, b_1, b_2, b_3, b_4, b_5, b_6, b_7]$

Note that b_i is 8-bit data corresponding to a pair of 4-bit data n_{2*i} and n_{2*i+1} for $0 \le i \le 7$, and the misalignment of byte position is emphasized by bold phase. The above shows that the misalignment disappears in four rounds.

In Round 2, two 8-bit data b_3, b_0 switch positions with two 8-bit data b_7, b_4 for the input data (b_2, b_3, b_6, b_7) and output data (b_0, b_1, b_4, b_5) of two F-functions. Utilizing the shuffle operations rp_1, rp_0 to cancel the effect of each misalignment, we introduce $shf_0 = rp_0 \circ rp_1$ and replace rot_{16}, rot_{32} in the original diffusion matrix with $shf_{16} = rp_0 \circ rot_{16} \circ rp_1$, $shf_{32} = rp_0 \circ rot_{32} \circ rp_0$ as below.

$$shf_0 : [x_0||x_1||x_2||x_3||x_4||x_5||x_6||x_7] \mapsto [x_4||x_5||x_6||x_7||x_0||x_1||x_2||x_3]$$
$$shf_{16} : [x_0||x_1||x_2||x_3||x_4||x_5||x_6||x_7] \mapsto [x_5||x_2||x_7||x_0||x_1||x_6||x_3||x_4]$$
$$shf_{32} : [x_0||x_1||x_2||x_3||x_4||x_5||x_6||x_7] \mapsto [x_6||x_7||x_4||x_5||x_2||x_3||x_0||x_1]$$

The new representation for calculating diffusion matrix can be expressed with 25 instructions including eight `pshufb`, four `pshufd`, and 13 `pxor` as follows.

$$r[0] \leftarrow shf_0(r[1]) \oplus shf_{16}(r[1]) \oplus shf_{16}(r[0]) \oplus shf_{32}(shf_0(r[0]) \oplus shf_{16}(r[0]))$$
$$r[1] \leftarrow shf_0(r[2]) \oplus shf_{16}(r[2]) \oplus shf_{16}(r[1]) \oplus shf_{32}(shf_0(r[1]) \oplus shf_{16}(r[1]))$$
$$r[2] \leftarrow shf_0(r[0]) \oplus shf_{16}(r[0]) \oplus shf_{16}(r[2]) \oplus shf_0(r[3]) \oplus shf_{16}(r[3])$$
$$\oplus shf_{32}(shf_0(r[2]) \oplus shf_{16}(r[2]))$$
$$r[3] \leftarrow shf_0(r[0]) \oplus shf_{16}(r[0]) \oplus shf_{16}(r[3]) \oplus shf_{32}(shf_0(r[3]) \oplus shf_{16}(r[3]))$$

We omit `movdqa` in the original diffusion matrix, utilizing `pshufd` natively supporting 3-operand. In Round 3 for the input data (b_0, b_1, b_4, b_5) of two F-functions, b_0, b_1 switches positions with b_4, b_5, but we can use the original representation owing to the calculation of two diffusion matrices independently. Since the misalignment of output data (b_2, b_3, b_6, b_7) is the same for the input data, no operations are needed. Round 4 can use the same representation in Round 2.

Therefore, we alternately call the original diffusion matrix and modified one, and adjust the data alignment for the round keys in the key schedule. With 128-bit AVX, the modified representation of diffusion matrix in Round 2 requires four more instructions compared to the original one, so the performance improvement remains about three fourths of the case with legacy SSE instructions.

5   Performance

This section summarizes the instruction counts for PRESENT and Piccolo, and shows the evaluation results of our implementations on three different computers given in Table 2.

Table 2. Computers used for benchmarking

Processor	Intel Xeon E5410	Intel Core i7 870	Intel Xeon E3-1280
Microarchitecture	Core	Nehalem	Sandy Bridge
Clock Speed	2.33 GHz	2.93 GHz	3.5 GHz
RAM	8 GB	16 GB	16 GB
OS	Linux 2.6.16.60 x86_64	Linux 3.1.10 x86_64	Linux 2.6.37.6 x86_64

Table 3. Instruction count for PRESENT and Piccolo with Legacy SSE instructions

	logical instr.	mov	shuffle	unpack	mov (mem)	xor (mem)	per round	TOTAL 80-bit	128-bit
PRESENT (8-block parallel)							40	**1444**	
addRoundKey	-	-	-	-	-	4	4	128	
sBoxLayer	17	3	-	-	-	-	20	620	
pLayer	-	4	4	8	-	-	16	496	
conversion	154	28	12	8	8	-	-	200	
PRESENT (16-block parallel)							80	**2720**	
addRoundKey	-	-	-	-	-	8	8	256	
sBoxLayer	34	6	-	-	-	-	40	1240	
pLayer	-	8	8	16	-	-	32	992	
conversion	154	32	24	16	16	-	-	232	
PRESENT (32-block parallel)							126*	**4446**	
addRoundKey	-	-	-	-	-	16	16	512	
sBoxLayer	68	12	-	-	4	-	84	2604	
pLayer	-	0/16	-	0/32	0/4	-	0/52	780	
conversion	288	78	82	64	38	-	-	550	
Piccolo (16-block parallel)							63	**1815**	**2193**
diffusion matrix	13	4/0	8/12	-	-	-	25	625	775
S-box	22	4	-	-	-	-	26	650	806
addRoundKey	4	4	-	-	-	4	12	300	372
addWhiteningKey	-	-	-	-	-	8	-	8	
conversion	154	32	24	16	16	-	-	232	

Table 3 presents the total number of instructions for PRESENT and Piccolo with the legacy SSE instruction set. The notations "logical instr." and "(mem)" in the table mean logical instructions including shift operation and instructions with memory, respectively. For the 32-block parallel implementation of PRESENT, "*" means the number of instructions per round on average. The "diffusion matrix" in Piccolo shows both the number of instructions for calculating the original diffusion matrix (left) and that for modified one (right). The "conversion" includes not just conversion process that converts input data to the bitsliced representation and reverses it to output data, but also loading input data and storing output data. Our conversion algorithm utilizes a part of the assembly code published by Käsper and Schwabe [12], which includes 84 instructions to convert eight 128-bit blocks on eight XMM registers to the bitsliced representation of 8-bit slicing on eight XMM registers with one temporary XMM register. We added a few shuffle and unpack instructions in this code to obtain desired bitsliced format.

We optimized our implementation with 128-bit AVX instructions. Owing to 3-operand syntax, the number of mov instructions in the table is zero except for register-to-memory operations. Furthermore we can use smaller instruction sequence of PRESENT S-box with 14 instructions in 8- and 16-block parallelism. The numbers of instructions for 8-, 16-, and 32-block parallel implementation

Table 4. Performance of PRESENT and Piccolo with 80-bit and 128-bit keys

Algorithm	PRESENT-80/128			Piccolo-80	Piccolo-128
Number of parallel blocks	8	16	32	16	
Xeon E3-1280 (Sandy Bridge)					
Cycles/byte	8.46	6.52	4.73	4.57	5.52
Instructions/cycle	2.04	2.48	3.10	2.61	2.61
Core i7 870 (Nehalem)					
Cycles/byte	10.88	7.26	5.79	5.69	6.80
Instructions/cycle	2.07	2.93	3.00	2.49	2.52
Xeon E5410 (Core)					
Cycles/byte	13.55	10.98	7.55	6.85	8.23
Instructions/cycle	1.67	1.93	2.30	2.07	2.08

of PRESENT with 128-bit AVX are **1106**, **2068**, and **3752**, respectively. The numbers of instructions for 16-block parallel implementation of Piccolo with 128-bit AVX are **1531** and **1849** for 80- and 128-bit keys, respectively.

Table 4 gives evaluation results. We measured the average cycles of encryptions for 1024KB random data and did not include the cost of the key schedule, which was regarded as negligible cost in our evaluation. Since the number of rounds of PRESENT is 31 for both 80- and 128-bit keys, the results of PRESENT-80 and -128 are exactly the same. The result on Xeon E5410 shows the performance of optimized code with 128-bit AVX.

For comparison, only a few software implementation results of ultra-lightweight block ciphers on general-purpose processors have been reported. A table-based implementation of LED [8] with 64- and 128-bit keys needs 57 and 86 cycles on Core i7 Q720 (1.60 GHz). Suzaki et al. [21] showed that TWINE encryption achieved 11.0 cycles/byte on Core i7 2600S (2.8 GHz, Sandy Bridge), so our implementations of PRESENT and Piccolo deliver superior performance compared with previous results and indicate an attractive option for software implementation for lightweight block ciphers on general-purpose processors.

As far as we know, besides hardware efficiency, Piccolo-80 achieves the fastest software implementation among existing 64-bit block ciphers in our implementation. Moreover, since Piccolo adopts a permutation based key schedule, which is lighter than the S-box based key schedule of PRESENT, Piccolo may have some advantage even for short message encryption. On the other hand, there are some stream ciphers with small hardware and fast software performance. For example, the public eBASC benchmarks report that TRIVIUM achieves 3.69 cycles/byte and SNOW 2.0 achieves 4.03 cycles/byte on a Nehalem CPU (dragon).

For further optimization, 256-bit AVX accelerates the performance of our bitslice implementation with low parallelism using AVX2 instruction set introduced in Haswell microarchitecture which will be released in 2013. An optimization for instruction sequences of S-box assuming both 3-operand instructions and issuing three independent instructions remains the matter of research.

6   Conclusion

This paper showed the great potential of lightweight cryptography in fast and timing-attack resistant software implementations in cloud computing by exploiting bitslice implementation. This was demonstrated by bitslice implementations of the PRESENT and Piccolo lightweight block ciphers. In particular, PRESENT-80/128 achieved 5.79 cycles/byte and Piccolo-80 achieved 5.69 cycles/byte on an Intel Core i7 processor, which is faster than the AES speed record in bitslice implementation on the same microarchitecture. We demonstrated bitslice implementation of only two lightweight block ciphers, but other lightweight block ciphers as well as other lightweight cryptographic primitives such as hash functions are worth implementing. We hope that lightweight cryptography will be used not only for constrained devices, but also for cloud computing.

Acknowledgments. The authors appreciate Kazuya Kamio and Kyoji Shibutani for useful comments and suggestions.

References

1. Aumasson, J.-P., Henzen, L., Meier, W., Naya-Plasencia, M.: QUARK: A Lightweight Hash. In: Mangard, S., Standaert, F.-X. (eds.) CHES 2010. LNCS, vol. 6225, pp. 1–15. Springer, Heidelberg (2010)
2. Biham, E.: A Fast New DES Implementation in Software. In: Biham, E. (ed.) FSE 1997. LNCS, vol. 1267, pp. 260–272. Springer, Heidelberg (1997)
3. Bogdanov, A., Knudsen, L., Leander, G., Paar, C., Poschmann, A., Robshaw, M., Seurin, Y., Vikkelsoe, C.: PRESENT: An Ultra-Lightweight Block Cipher. In: Paillier, P., Verbauwhede, I. (eds.) CHES 2007. LNCS, vol. 4727, pp. 450–466. Springer, Heidelberg (2007)
4. Courtois, N.T., Hulme, D., Mourouzis, T.: Solving Circuit Optimization Problems in Cryptography and Cryptanalysis. Cryptology ePrint Archive, Report 2011/475 (2011), http://eprint.iacr.org/2011/475
5. De Cannière, C., Dunkelman, O., Knežević, M.: KATAN and KTANTAN — A Family of Small and Efficient Hardware-Oriented Block Ciphers. In: Clavier, C., Gaj, K. (eds.) CHES 2009. LNCS, vol. 5747, pp. 272–288. Springer, Heidelberg (2009)
6. Grabher, P., Großschädl, J., Page, D.: Light-Weight Instruction Set Extensions for Bit-Sliced Cryptography. In: Oswald, E., Rohatgi, P. (eds.) CHES 2008. LNCS, vol. 5154, pp. 331–345. Springer, Heidelberg (2008)
7. Guo, J., Peyrin, T., Poschmann, A.: The PHOTON Family of Lightweight Hash Functions. In: Rogaway, P. (ed.) CRYPTO 2011. LNCS, vol. 6841, pp. 222–239. Springer, Heidelberg (2011)
8. Guo, J., Peyrin, T., Poschmann, A., Robshaw, M.: The LED Block Cipher. In: Preneel, B., Takagi, T. (eds.) CHES 2011. LNCS, vol. 6917, pp. 326–341. Springer, Heidelberg (2011)
9. Guo, J., Peyrin, T., Poschmann, A., Robshaw, M.: The LED Block Cipher. In: Preneel, B., Takagi, T. (eds.) CHES 2011. LNCS, vol. 6917, pp. 326–341. Springer, Heidelberg (2011), http://www.iacr.org/workshops/ches/ches2011/presentations/Session%207/CHES2011_Session7_2.pdf

10. Intel 64 and IA-32 Architectures Optimization Reference Manual, http://www.intel.com/
11. ISO/IEC 29192-2:2012, Information technology – Security techniques – Lightweight cryptography – Part 2: Block ciphers (2012)
12. Käsper, E., Schwabe, P.: Faster and Timing-Attack Resistant AES-GCM. In: Clavier, C., Gaj, K. (eds.) CHES 2009. LNCS, vol. 5747, pp. 1–17. Springer, Heidelberg (2009)
13. Könighofer, R.: A Fast and Cache-Timing Resistant Implementation of the AES. In: Malkin, T. (ed.) CT-RSA 2008. LNCS, vol. 4964, pp. 187–202. Springer, Heidelberg (2008)
14. Lee, E.: Cyber Physical Systems: Design Challenges. EECS Department, University of California, Berkeley (2008)
15. Matsui, M., Nakajima, J.: On the Power of Bitslice Implementation on Intel Core2 Processor. In: Paillier, P., Verbauwhede, I. (eds.) CHES 2007. LNCS, vol. 4727, pp. 121–134. Springer, Heidelberg (2007)
16. Osvik, D.A.: Speeding up Serpent. In: AES Candidate Conference, pp. 317–329 (2000)
17. Poschmann, A.: Lightweight Cryptography – Cryptographic Engineering for a Pervasive World. Cryptology ePrint Archive, Report 2009/516 (2009), http://eprint.iacr.org/2009/516
18. Ristenpart, T., Tromer, E., Shacham, H., Savage, S.: Hey, you, get off of my cloud: exploring information leakage in third-party compute clouds. In: Al-Shaer, E., Jha, S., Keromytis, A. (eds.) ACM Conference on Computer and Communications Security, pp. 199–212. ACM (2009)
19. Shamir, A.: SQUASH – A New MAC with Provable Security Properties for Highly Constrained Devices Such as RFID Tags. In: Nyberg, K. (ed.) FSE 2008. LNCS, vol. 5086, pp. 144–157. Springer, Heidelberg (2008)
20. Shibutani, K., Isobe, T., Hiwatari, H., Mitsuda, A., Akishita, T., Shirai, T.: Piccolo: An Ultra-Lightweight Blockcipher. In: Preneel, B., Takagi, T. (eds.) CHES 2011. LNCS, vol. 6917, pp. 342–357. Springer, Heidelberg (2011)
21. Suzaki, T., Minematsu, K., Morioka, S., Kobayashi, E.: TWINE: A Lightweight, Versatile Block Cipher. In: Leander, G., Standaert, F. (eds.) ECRYPT Workshop on Lightweight Cryptography 2011, pp. 146–169 (2011)

Low-Latency Encryption –
Is "Lightweight = Light + Wait"?*

Miroslav Knežević, Ventzislav Nikov, and Peter Rombouts

NXP Semiconductors, Leuven, Belgium

Abstract. The processing time required by a cryptographic primitive implemented in hardware is an important metric for its performance but it has not received much attention in recent publications on lightweight cryptography. Nevertheless, there are important applications for cost effective low-latency encryption. As the first step in the field, this paper explores the low-latency behavior of hardware implementations of a set of block ciphers. The latency of the implementations is investigated as well as the trade-offs with other metrics such as circuit area, time-area product, power, and energy consumption. The obtained results are related back to the properties of the underlying cipher algorithm and, as it turns out, the number of rounds, their complexity, and the similarity of encryption and decryption procedures have a strong impact on the results. We provide a qualitative description and conclude with a set of recommendations for aspiring low-latency block cipher designers.

1   Introduction

As cryptography is becoming ever more pervasive in modern technology, new applications regularly emerge. Some of these new applications also introduce new requirements on the implementation such as ultra fast response times. Applications such as Car2X communication (e.g. automotive road tolling, intelligent transport systems), high speed networking (optical links), and secure storage devices (e.g. memories, solid-state disks, super-speed USB 3.0), just to name a few, all require an instant response. Besides these there are also applications that require moderately high throughput but have limited maximum clock frequencies, e.g. FPGA, or strict area requirements that preclude the use of highly pipelined architectures.

Cryptographic primitive design is a balancing act between several aspects such as cryptographic strength, implementation cost, execution speed, power consumption, etc. Which trade-offs are the right ones to make is determined by

* The authors would like to thank Bruce Murray for his valuable comments and suggestions to improve the work. The authors also thank Hans De Kuyper for his valuable inputs. The work has been supported by the European Commission through the ICT program under contract ICT-2007-216646 (European Network of Excellence in Cryptology – ECRYPT II), through the Tamper Resistant Sensor Node (TAMPRES) project with contract number 258754 and through the Internet of Things - Architecture (IoT-A) project with contract number 257521.

E. Prouff and P. Schaumont (Eds.): CHES 2012, LNCS 7428, pp. 426–446, 2012.

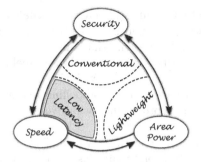

Fig. 1. Typical trade-offs in cryptography

the application. In the past, different applications have led to different corners of the design space to be explored. The most important of these are depicted in Fig. 1.

Government applications have typically favored cryptographic strength over aspects such as cost and speed, although these aspects usually do play an important role in selection processes like the former AES competition [1] and currently the SHA-3 competition [31]. The use of these algorithms in applications such as mainframe systems has resulted in the development of high throughput implementations, both in hardware and software.

More recently the advent of RFID and other wireless technologies sparked an interest in a new field: low-power and low-cost cryptography. The first primitives to be explored were stream ciphers, for example in the eSTREAM project [15], followed by a whole range of block ciphers such as TEA [37], NOEKEON [13], MINI-AES [12], MCRYPTON [29], SEA [36], HIGHT [23], DESXL [27], CLEFIA [35], PRESENT [9], MIBS [24], KATAN/KTANTAN [10], PRINTcipher [26], KLEIN [18], LED [20], PICCOLO [34], and others. The field has recently been expanded by the introduction of several new low-cost hash functions such as DM-PRESENT [20], KECCAK-f[400]/-f[200] [7,25], QUARK [6], PHOTON [19], and SPONGENT [8].

We have identified a new range of applications; those that require very fast response times and for which there is no established research field yet. Note that although most of the high-speed implementations available in literature do achieve tremendous throughput, their response time is generally not that fast. This is due to their extensive use of pipelining which enables them to process multiple messages at the same time, but in order to encrypt a single message block, this type of implementation still needs multiple clock cycles, i.e. typically more than 20. An example of this is a recent work from Mathew et al. [30], presenting a reconfigurable AES encrypt/decrypt hardware accelerator targeted for content-protection in high-performance microprocessors which, manufactured in 45 nm CMOS technology, achieves 53 Gb/s throughput. Another example comes from Hodjat and Verbauwhede [22] where area-throughput trade-offs of a fully pipelined AES implementation are described and a throughput of 30 Gb/s to 70 Gb/s is achieved.

In other words, a high throughput is usually achieved by common signal processing techniques such as pipelining and parallel processing, while achieving a low latency, on the other hand, still remains a challenge. As a consequence one could ask the following questions: What is the minimum achievable latency for a given security level? Do designs that inherently have lower latency also achieve higher throughput when implemented in a pipelined fashion? And does "lightweight" necessarily mean "light + wait?" These are all interesting questions and as it seems there are a lot of compelling reasons to take a closer look at the latency behavior of cryptographic primitives.

Our Contribution. We introduce the new field of low-latency encryption; highlight the differences with lightweight and classical cryptography, and by bringing several important applications to light we try to motivate further research in this field.

We identify several well-known lightweight block ciphers as possible candidates to yield low-latency implementations. By examining this set of ciphers in the context of low-latency encryption, our work provides the first results in the field. We therefore develop a framework that examines the low-latency behavior of cryptographic primitives on the following aspects:

- Minimal achievable latency.
- Its impact on the circuit size.
- Its impact on the power and energy consumption.

We link the collected data to the cipher design decisions and show that results are strongly influenced by their properties. More specifically, the number of rounds, the round's complexity (e.g. the S-box size, MDS (Maximum Distance Separable) matrices defined over different fields versus binary matrix), and the similarity of encryption and decryption procedures have a significant influence on the algorithm's performance. Our work concludes with a set of recommendations for aspiring low-latency block cipher designers.

Organization of the Paper. The remainder of this paper is organized as follows. In Section 2, we provide a short description of the block ciphers we have chosen to investigate. Our contributions – the implementation results, comparisons, and discussion – are presented in Sections 3 and 4. We first investigate the minimum achievable latency in Section 3.1 and then evaluate the impact optimization for low latency has on area in Section 3.2. Our study continues by combining the two previously described metrics in section Section 3.3 where the results for the time-area product are presented and in Section 3.4, we have a closer look at the impact low-latency implementations have on the power and energy consumption. We elaborate more on our results and conclude in Section 4.

2   Preliminaries

There are many algorithms to choose from for a comparative study of low-latency behavior, but in order to draw meaningful conclusions about hardware

performance a set of candidate algorithms should be chosen with similar properties. We therefore focus on algorithms that are expected to result in low-latency
implementations. Since hardware implementations of hash functions generally
require more area to implement [16] and stream ciphers usually need a large
number of initialization rounds [11, 21] we chose to focus on block ciphers only.
Furthermore, it is expected that lightweight block ciphers yield good results in
terms of implementation cost, even in a fully-unrolled implementation. Besides
latency as our primary goal, we consider silicon area as a very important factor
in practical implementations of encryption algorithms and, therefore, we restrict
our candidates to lightweight block ciphers but include AES as the reference cipher. In order to reduce the number of candidates to a manageable number, we
further restrict the set to ciphers with the well-studied SPN structure.

This results in the following list of seven lightweight SPN block ciphers:
AES [14, 32], KLEIN [18], LED [20], MCRYPTON [29], MINI-AES [12], NOEKEON [13],
PRESENT [9]. We provide a brief description of each cipher and refer for more
details to their original descriptions in the literature.

AES [14, 32], designed by Daemen and Rijmen in 1997, has become not only
a NIST standard but also the most used block cipher nowadays. The cipher
has not been considered lightweight until the work of Feldhofer et al. [17] who
provided the smallest implementation at the time, requiring only 3400 GE.[1] AES
is an iterated block cipher with a block-size of 128 bits and three possible key
lengths of 128, 192, and 256 bits. In this work, we consider only the 128-bit key
version which consists of 10 rounds. The word size is 8 bits, i.e. the data elements
are considered as elements of the field $GF(2^8)$. Each round of AES consists of
the following operations: *SubBytes*, *ShiftRows*, *MixColumns*, and *AddRoundKey*.
The operation *SubBytes* (S-layer) is defined as the simultaneous application of
the S-Box (inversion in $GF(2^8)$) to each element of the state. The permutation
layer (P-layer) consists of *ShiftRows* and *MixColumns* operations. The *ShiftRows*
operation is defined as the simultaneous left rotation of the row i of the state by
i positions. The *MixColumns* operation pre-multiplies each column of the state
by an MDS matrix defined over $GF(2^8)$. The *KeySchedule* derives the round
key from the secret key, by applying once the S-Box and some simple linear
operations. Finally, *AddRoundKey* XORs the round key to the current state.

NOEKEON [13] is a 128-bit block cipher with a 128-bit key, proposed by
Daemen, Peeters, Van Assche, and Rijmen in 2000. NOEKEON is a self-inverse,
bit-sliced cipher and can be considered as the predecessor of modern lightweight
block ciphers. It has 16 rounds and each of them consists of the following operations: *Theta*, *Pi1*, *Gamma*, and *Pi2*. The operation *Gamma* is an involutive
non-linear mapping (S-layer), in which S-boxes operate independently on 32 4-
bit tuples. *Pi1* and *Pi2* perform simple cyclic shifts. *Theta* is a linear mapping
that first XORs the working key to the state and then performs a simple linear
transformation of the state. Therefore, *Theta* acts partially as *AddRoundKey*

---

[1] The current smallest implementation of AES comes from Poschmann et al. [33] and
consumes only 2400 GE, which is comparable to the size of some of the first proposed
lightweight block ciphers.

and, together with *Pi1* and *Pi2*, forms the P-layer of the cipher. The *KeySchedule* is very simple – a so-called working key is derived from the secret key and then XORed to the state at each round. For the encryption procedure, the working key is simply equal to the secret key. Note that the self-inverse property of the cipher has big advantages when both encryption and decryption need to be implemented on the same circuit.

MINI-AES [12], or a small scale variant of AES, has been described by Cid, Murphy, and Robshaw in 2005 in order to provide a suitable framework for comparing different cryptanalytic methods. In this paper, we consider a 10-round MINI-AES with a block-size of 64 bits, a key length of 64 bits, and a word size of 4 bits. The main difference between AES and the version of MINI-AES we chose to examine is that the S-box and the MDS matrix are defined over the field $GF(2^4)$. Therefore, the selected instance of MINI-AES can be considered as a lightweight version of the AES cipher.

MCRYPTON [29] is a 64-bit block cipher supporting three different key length (64, 96, and 128 bits), designed by Lim and Korkishko in 2006 and is one of the first lightweight SPN block ciphers. Each round of MCRYPTON consists of the following operations: *NonLinear Substitution* γ, *Column-wise bit Permutation* π, *Column-to-row Transposition* τ, and *Key Addition* σ. The operation γ (S-layer) consists of 16 nibble-wise substitutions using four 4-bit S-boxes (S_0, S_1, S_2, S_3, all affine equivalents to the inversion in $GF(2^4)$ and such that $S_2 = S_0^{-1}$ and $S_3 = S_1^{-1}$). The P-layer consists of π and τ operations. The π operation is an involutional bit-wise matrix multiplication. The τ operation simply transposes the state and is thus an involution. The *KeySchedule* is simple and consists of two stages: a round key generation through a nonlinear S-box transformation and a key variable update through a simple rotation. Finally, the σ operation XORs the round key to the state. Independent of the key length, MCRYPTON always uses 12 rounds with a slightly different *KeySchedule*. Note that decryption and encryption can share most of the round operations and that the *KeySchedule* allows a direct derivation of the last round key.

PRESENT [9], designed by Bogdanov et al., was proposed in 2007 and established itself as one of the most prominent lightweight block ciphers. It has recently been adopted as a standard in ISO/IEC 29192-2. The 31-round cipher has a block-size of 64 bits and comes with an 80-bit or 128-bit key. Each round of PRESENT consists of the following operations: *sBoxLayer*, *pLayer* and *AddRoundKey*. The *sBoxLayer* is defined as the simultaneous application of a very light 4-bit S-Box to each nibble of the state. The *pLayer* is a simple bitwise permutation. The *KeySchedule* rotates the key variable, XORs a constant and applies the S-box to the key variable. *AddRoundKey* XORs the 64 most significant bits of the key variable to the state. Note that the *pLayer* provides a rather slow diffusion of the cipher, which results in the considerably high number of rounds.

KLEIN [18] is a rather young lightweight cipher proposed by Gong, Nikova, and Law in 2010. It is a block cipher with a fixed 64-bit block-size and a variable key length of 64, 80 or 96 bits. Each round of the cipher consists of the following operations: *SubNibbles, RotateNibbles, MixNibbles,* and *AddRoundKey.* The operation *SubNibbles* (S-layer) is defined as the simultaneous application of an involutive 4-bit S-Box to each element of the state. The P-layer consists of *RotateNibbles* and *MixNibbles* operations. The *RotateNibbles* operation rotates the state two bytes to the left. The *MixNibbles* coincides with the AES *MixColumns* operation, i.e. pre-multiplies each column of the state by an MDS matrix defined over $GF(2^8)$. The *KeySchedule* derives the round key from the secret key, by applying two S-Boxes and some simple linear operations. Finally, the *AddRoundKey* XORs the round key to the state. KLEIN-64/80/96 uses 12/16/20 rounds respectively.

LED [20], designed by Guo, Peyrin, Poschmann, and Robshaw in 2011, is one of the most recent lightweight ciphers. It is a nibble-based 64-bit block cipher with two variants taking 64-bit and 128-bit keys. Each round of LED consists of the following operations: *AddConstants, SubCells, ShiftRows,* and *MixColumnsSerial.* Once every 4 rounds the *AddRoundKey* operation is applied. The *SubCells* (S-layer) reuses the PRESENT S-box and applies it to each 4-bit element of the state. *MixColumnsSerial* uses an MDS matrix defined over $GF(2^4)$ for linear diffusion that is suitable for compact serial implementation since it can be represented as a power of a very simple binary matrix. *AddConstants* XORs a constant to the state at each round. *ShiftRows* operates by rotating row i of the array state by i cell positions to the left. *AddConstants, ShiftRows,* and *MixColumnsSerial* form the P-layer of the cipher. The 64-bit key variant consists of 32 rounds while the 128-bit key variant consists of 48 rounds. The cipher has no *KeySchedule*, meaning the same key is XORed to the state using *AddRoundKey*, once every 4 rounds.

The resulting set of block ciphers represents a wide spectrum of building blocks for the S-layer, P-layer, and the key schedule. In summary, AES is (the only) byte-oriented block cipher (i.e. byte-based S- and P-layers) with an MDS P-layer; NOEKEON has a nibble-based S-layer, a bit-based P-layer and it is a self-inverse bit-sliced cipher; MINI-AES is a nibble-oriented cipher (i.e. nibble-based S- and P-layers) with an MDS P-layer; MCRYPTON has a nibble-based S-layer, a bit-wise matrix for the P-layer with a specific key schedule; PRESENT has a nibble-based S-layer and a very simple bit permutation for the P-layer; KLEIN has a nibble-based S-layer and a byte-based MDS P-layer (equivalent to AES); finally, LED is a nibble-oriented block cipher (i.e. nibble-based S- and P-layers) with an MDS P-layer and no key schedule.

Note that KLEIN and AES share the same MDS matrix; LED and PRESENT share the same S-layer; MINI-AES and LED have different nibble oriented MDS matrices; and the S-layer of MINI-AES and MCRYPTON are close (affine equivalents) to each other. Therefore, we have a variety of building blocks: bit-, nibble- and byte-oriented blocks; different complexity of S-boxes; either simple matrices, the MDS ones, or just a simple permutation as the P-layer. All this allows us to investigate how different elements influence the overall performance when low-latency encryption is the ultimate goal.

3   Hardware Evaluation

In this section, we provide an extensive hardware evaluation of the seven block ciphers which we identified in the previous section. Besides the cryptographic properties of a cipher, the chosen architecture has a significant influence on the overall performance. As our goal is to evaluate designs with the lowest achievable latency, we mainly focus on 1-cycle and 2-cycle based architectures. More specifically, a 1-cycle based architecture represents a fully-unrolled architecture which requires a single clock cycle for its execution. Similarly, a 2-cycle based architecture needs two clock cycles in order to execute its computation. Since the term low-latency implies a low number of clock cycles for the algorithm execution (recall the systems with a limited clock frequency), we do not evaluate architectures that require three or more clock cycles.

We then distinguish between encryption (ENC) only and encryption/decryption (ENC/DEC) architectures. Moreover, as will become apparent later, some of the implemented ciphers benefit from the inherent similarities between encryption and decryption datapaths. In these cases, we also provide figures for a more compact but still slightly slower implementation that shares the datapath. Figure 2 depicts all the evaluated architectures, however for readability we only report results for (ENC/DEC) architectures. The results for the architectures supporting encryption only are provided in Appendix B.

The presented results are obtained in 90 nm CMOS technology, synthesized with the Cadence RTL compiler version 10.10-p104. In order to have a better overview on the hardware performance, we always provide figures for both time-constrained and unconstrained designs. By time-constrained, we mean a design that achieves the minimum possible critical path at the expense of a large area overhead. An unconstrained design consumes the minimum possible area with the drawback of being a slower circuit. In both cases, this only refers to the synthesis tool constraints and not to the actual RTL code, which in fact remains

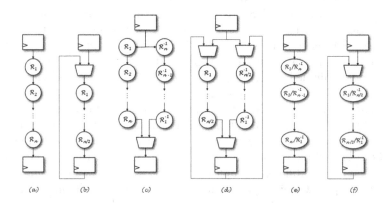

Fig. 2. Six evaluated architectures: (a) 1-cycle based, ENC-only. (b) 2-cycle based, ENC-only. (c) 1-cycle based, ENC/DEC. (d) 2-cycle based, ENC/DEC. (e) 1-cycle based, ENC/DEC, shared datapath. (f) 2-cycle based, ENC/DEC, shared datapath.

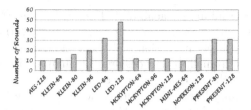

Fig. 3. Number of rounds of the tested ciphers

the same. The code of all designs is written in Verilog and tested against the available test vectors. The data showing the implementation results is provided in Table 1, Appendix A.

Although we rank the ciphers according to their hardware performance, we do not attempt to define *the most efficient* one with respect to *all* evaluated criteria. We believe that depending on the application requirements, the selection of the most efficient design could be based on any of the following criteria: area, latency, time-area product, power, or energy. Moreover, if more than one criterion influences the final decision, we believe that it is rather trivial to combine the presented data and obtain a unique benchmark. As the evaluated ciphers provide different security levels, there is no easy way to fairly compare them against each other. While it is rather obvious that a cipher with a block-size of 64 bits will perform better in terms of area than one with 128 bits, the influence of the key length remains rather vague. With this evaluation, we bring to light the influence of similar and other design decisions on the final hardware performance.

Recall that all the evaluated ciphers have a block-size of 64 bits, except AES and NOEKEON which have a 128-bit block-size. Some ciphers support different key lengths and therefore we evaluate 128-bit key AES; 64-bit, 80-bit, and 96-bit key KLEIN; 64-bit and 128-bit LED; 64-bit, 96-bit, and 128-bit MCRYPTON; 64-bit key MINI-AES; 128-bit key NOEKEON; and 80-bit and 128-bit key PRESENT. Finally, as most of the obtained results will be highly correlated with the number of cipher rounds, we provide Fig. 3, which visualizes this metric.

3.1   Latency

We define latency as a measure of time needed for a certain design to complete a defined (computational) task. In our context, the computational task is defined as an encryption of a single message block and the latency is calculated as:

$$\text{Latency} = N \cdot t_{cp} \ ,$$

where N is the number of clock cycles needed for the encryption of a single message block and t_{cp} is the critical path of the circuit. In order to highlight the difference between latency and throughput, we outline that the latency truly depends on the inherent properties of a cryptographic algorithm, while the throughput does not – it can be simply increased using the common signal processing techniques such as pipelining and parallel computations.

Figure 4 shows the minimum achievable latency for the ENC/DEC module of all the evaluated ciphers. NOEKEONs-128 denotes a NOEKEON implementation with a shared datapath for encryption and decryption, and it is clearly marked in gray to set it apart from the other designs. The figure further reveals that, in general, there is only a slight advantage of 1-cycle based architectures over 2-cycle based ones, but minimal latency is obtained with a 1-cycle based architecture as expected. The designs that show the highest performance are certainly MINI-AES and MCRYPTON (all key lengths). Being around 30 % slower, KLEIN-64 is the third best candidate. The lowest performance comes from LED-128, which is more than 5 times slower than MINI-AES. AES, for example, achieves 70 % slower critical path than MINI-AES.

What is interesting to observe is that the latency of certain designs, i.e. KLEIN and LED, depends on the key length, while for others, i.e. MCRYPTON and PRESENT, this is not the case. This links directly to the number of rounds, which in case of KLEIN and LED increases for larger key lengths, while it remains constant for MCRYPTON and PRESENT (recall Fig. 3).

In Appendix B, we provide the results for ENC-only architectures (see Fig. 12) The results show that the performance of certain designs, e.g. KLEIN-64 and MINI-AES, certainly degrades when the decryption path is embedded into the design. In order to explain this in more detail, we provide Fig. 5 where we depict the average latency per round of each cipher for both ENC-only and ENC/DEC architectures. It is easy to see that the decryption datapath of AES, KLEIN, and MINI-AES is considerably slower than that of the encryption. To a lesser extent this also holds for LED, MCRYPTON, and PRESENT. NOEKEON is the only cipher that does not suffer from this property. We also observe a clear correlation between the average latency per round and the complexity of the round.

An unconstrained design of PRESENT, on the other hand, shows a somewhat unexpected result. Its unconstrained ENC-only architecture (see Fig. 12, Appendix B) seems to be slower than its ENC/DEC architecture. This result is explained by the fact that, when unconstrained, the synthesis tool optimizes designs for area while the timing is less important. When time-constrained, however, the synthesis tool makes a significant effort to optimize for timing and therefore the ENC/DEC architecture of PRESENT becomes slower than its ENC-only architecture.

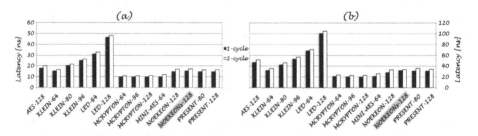

Fig. 4. Minimum latency [ns] for ENC/DEC module: (a) Time-constrained. (b) Unconstrained.

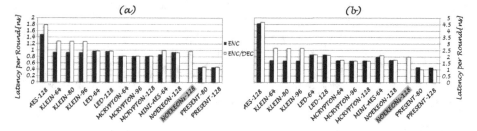

Fig. 5. Average latency [ns] per round: (a) Time-constrained. (b) Unconstrained.

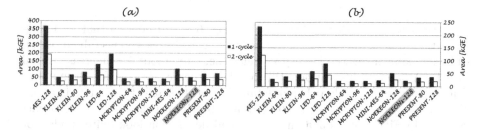

Fig. 6. Minimum area [kGE] for ENC/DEC module: (a) Time-constrained. (b) Unconstrained.

Finally, we note that the ratio between the latency of the unconstrained and time-constrained designs ranges from 2.63 for AES (ENC/DEC) to only 1.30 for NOEKEON (ENC-only, Fig. 12, Appendix B), which illustrates the *elasticity* of the design's latency.

3.2 Area

Similar to the previous subsection, we first provide results for the circuit size of all the evaluated cipher variants. Secondly, we elaborate on the area per round distribution, where we observe several interesting results. Note that the area is expressed in gate equivalence (GE) units, representing the relative size of the circuit compared to a simple 2-input NAND gate.

Figure 6 illustrates the area for ENC/DEC architectures. In contrast to the latency figures, the advantage for 2-cycle based architectures is clear: 2-cycle based architectures consume approximately half of the area of the 1-cycle based architectures. We also observe a significant correlation between the number of cipher rounds and the circuit size. MINI-AES and MCRYPTON again show the best result, followed by the approximately 25 % larger KLEIN-64 implementation. PRESENT comes as the next one with about 60 % overhead. Not surprisingly, the largest circuit size is shown by AES, which is more than 9 times larger than MINI-AES. From the lightweight ciphers, LED-128 consumes the biggest area and it is more than 4 times larger than MINI-AES.

An interesting property can be observed in NOEKEON where due to their inherent similarity the datapaths for encryption and decryption can be shared.

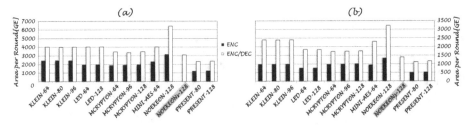

Fig. 7. Average area [GE] per round: (a) Time-constrained. (b) Unconstrained.

Denoted with NOEKEONs-128 (grayed) in Fig. 6 it can be seen that an implementation with shared datapath results in significant area savings (about 50 %), while not influencing the latency as much, i.e. only about 5 % increase (recall Fig. 4). A similar observation, still to a lesser degree, is true for MCRYPTON. When implemented with a shared datapath (not depicted) this results in about 30 % area savings with about 20 % timing overhead compared to the results depicted in Fig. 4. Although encryption and decryption look quite similar for MCRYPTON, two layers of multiplexors per round are needed in the shared datapath in order to choose the correct S-boxes. This extra logic multiplied by the unrolling factor results in quite a significant latency and area overhead for the total design.

Figure 7, which illustrates the average area per round for each cipher (except AES, since its round size goes well beyond the other values – 23 kGE for unconstrained and 37 kGE for time-constrained), shows that PRESENT has the smallest round amongst all ciphers, which is not surprising, as its round consists of an S-layer and a very light P-layer (wiring only). The P-layers of other ciphers involve more complex operations such as multiplication with an MDS matrix for MINI-AES, for example, or variations thereof for other ciphers. Note also that the average area per round of NOEKEON is relatively large. This is due to its block size of 128 bits; twice that of the other ciphers. This only confirms our initial assumption that both the number of cipher rounds and their complexity have a significant influence on hardware performance.

There are a number of observations about the area per round distribution that we illustrate here using KLEIN-80 as an example (see Fig. 8); although the same observation holds to a higher or lesser extent for most of the evaluated ciphers. The first is that due to the higher complexity of decryption, the critical path passes through the decryption datapath, which therefore becomes considerably larger than the encryption datapath when time-constrained. NOEKEON is the only cipher exempt from this effect, while the effect is barely noticeable in the case of LED. When constraints are relaxed, this effect naturally fades away, although remaining slightly noticeable even in unconstrained implementations.

Another observation that can be made for both time-constrained and unconstrained implementations, and holds over all the evaluated ciphers, is the considerably smaller area taken by the last few rounds of an unrolled design. For example in the time-constrained implementation of KLEIN-80, the last round is more than

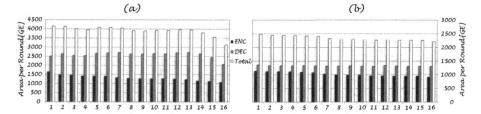

Fig. 8. Area distribution [GE] per round of KLEIN-80: (a) Time-constrained. (b) Unconstrained.

25 % smaller in size than the largest (in this case the second) round. For all other ciphers this difference always remains above 20 %. This phenomenon is explained by the fact that the logic gates used in the last rounds require considerably lower driving strength since they drive less logic than the middle rounds and can therefore be smaller. We further address this observation in Section 4.

The third observation that could be drawn from Fig. 8 is a noticeable swing in area in the first 13 rounds of the time-constrained KLEIN-80 implementation (similar observation holds for all other ciphers as well). This is however an effect introduced by the synthesis tool and is caused by insertion of a significant number of buffer cells in order to strengthen (and thus speed up) the signal propagation throughout the combinational network of the circuit which happens periodically, several rounds after each other.

The ratio between the size of time-constrained and unconstrained designs spans the range from 1.66 for KLEIN-80 to 2.22 for NOEKEONs. This ratio defines the *elasticity* of the design's area and is an indication of the overhead in area needed to achieve the smallest possible critical path of the design.

3.3   Time-Area Product

Although it is a simple combination of the two previously described metrics, we still provide graphs for the time-area product as this is an often used criterion for selecting the final implementation. Figure 9 illustrates the time-area product for the ENC/DEC architecture.

Again, the highest performance with respect to this criterion is shown by MINI-AES and all the versions of MCRYPTON. With more than 60 % overhead, KLEIN-64 takes the third place, while the lowest performance is again shown by LED-128. For all the tested ciphers it holds that the 2-cycle based architecture provides between 40 % and 45 % more efficiency with respect to this metric.

When moving from unconstrained to time-constrained designs the highest gain is shown by AES with 40 % decrease of the time-area product, while NOEKEONs achieves even a negative gain with 8 % increase of the time-area product. In general this ratio (time-area of unconstrained versus time-area of time-constrained designs) ranges between 0.85 and 1.00 which reflects in a rather small overall improvement.

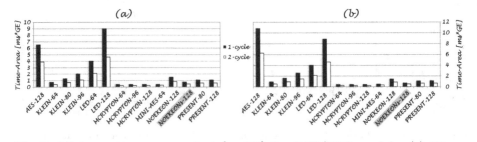

Fig. 9. Minimum time-area product [ms·GE] for ENC/DEC module: (a) Time-constrained. (b) Unconstrained.

3.4   Power and Energy

The results for the average power consumption are obtained by taking into account the switching activity of the circuit and are based on synthesis results. While accurate power measurement is only possible once the circuit is manufactured, we believe that our estimates are still reliable when it comes to comparing the power consumption between different designs. We note here that the term *average* is relative, since we consider designs with very low latency. Therefore, when considering a fully unrolled design (1-cycle), the average power is measured, and hence averaged, over a single clock cycle which in fact reflects the instantaneous power consumption. For the 2-cycle based designs, the power is averaged over two clock cycles. In order to eliminate the data dependency, we average the power consumption over 100 random vector inputs for each measurement.

Since the power consumption is linearly related to the operating frequency, this metric directly influences the value of the measured power. Our strategy of setting the operating frequency is simple in this case – we set the frequency as the reciprocal of the critical path. Therefore, the power consumption of each design is measured during its shortest possible execution time. The energy consumption is normalized over the number of processed bits, i.e. the message block-size, and calculated as:

$$E = \frac{P \cdot \text{Latency}}{B} = \frac{P \cdot N \cdot t_{cp}}{B} \ ,$$

where P is the average power, N is the number of clock cycles needed for the encryption of a single message block, t_{cp} is the critical path of the circuit, and B is the message block-size.

Figures 10 – 11 illustrate the power and energy consumption, respectively. The most power and energy efficient designs are again MINI-AES, MCRYPTON, and KLEIN-64, while LED consumes the most. Surprisingly, a large design such as AES consumes much less energy than most of the lightweight ciphers. This in fact relates to the number of rounds, which in case of AES is only 10, as well as to its block size of 128 bits (energy is normalized over the block size).

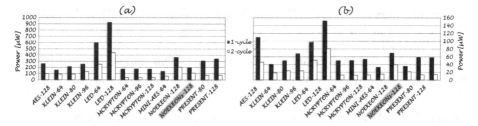

Fig. 10. Power consumption [μW] for ENC/DEC module: (a) Time-constrained. (b) Unconstrained.

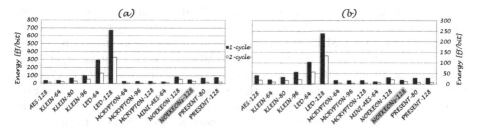

Fig. 11. Energy consumption [fJ/bit] for ENC/DEC module: (a) Time-onstrained. (b) Unconstrained.

4   Discussion and Conclusions

The ciphers we have evaluated within our framework are mainly designed for lightweight applications. They were not designed to satisfy the low-latency requirement imposed by new applications. Therefore, some of the ciphers which provide very good lightweight properties, e.g. LED and PRESENT, demonstrate quite a low hardware performance when it comes to the low-latency behavior. Still, we believe that by looking at the solutions offered by lightweight cryptography and understanding how their inherent properties influence the low-latency behavior one makes the very first step towards building an efficient low-latency cryptographic primitive. We summarize our results and give some guidelines for designing low-latency algorithms. In this context, we mainly address hardware properties of the algorithms.

S-Box. AES is the only cipher with an 8-bit S-box which is significantly larger than the 4-bit S-boxes used by the other ciphers. In theory, a cryptographically strong 8-bit S-box is on average 32 times larger than a cryptographically strong 4-bit S-box. In practice, due to the characteristics of standard cell libraries, this ratio is smaller but remains around 20. This fact strongly encourages the use of cryptographically strong 4-bit (or even 3-bit) S-boxes where possible. We stress here that even among the 4-bit (or 3-bit) S-boxes there are significant differences in circuit size [28].

Number of Rounds. Although both LED and PRESENT use 4-bit S-boxes, thus having a relatively lightweight round, the number of rounds they consist of is considerably large (see Fig. 3). When a design is (partially) unrolled, the number of rounds becomes a significant factor in the algorithm's performance. While this is obvious in the context of the circuit's latency, once we target low-latency design, also the area overhead becomes significant. This implies a higher power and energy consumption as well. We therefore suggest to minimize the number of rounds of the cryptographic algorithm.

Round Complexity. An interesting conclusion comes from comparing for example the MINI-AES and PRESENT algorithms. While the PRESENT round is very lightweight (it consist of the S-layer and the P-layer, which is in fact only wiring in hardware), the algorithm still needs a relatively large number of rounds in order to achieve good cryptographic properties. MINI-AES, on the other hand, has only 10 rounds and achieves good cryptographic properties by having a heavier P-layer, i.e. an MDS matrix, which efficiently increases the number of active S-boxes at low-cost. To illustrate, the P-layer of MINI-AES is about 30 % larger than its S-layer and therefore 10 rounds of MINI-AES versus 31 rounds of PRESENT seem to be a very good design choice. We, therefore, suggest to reduce the number of rounds at the cost of (slightly) heavier round. Finding a lightweight P-layer with good cryptographic properties is of a high importance here. Similar to MINI-AES, MCRYPTON demonstrates a very good selection for the P-layer (a bitwise matrix multiplication) while KLEIN's P-layer (a byte oriented MDS) seems to be rather heavy.

Key Schedule. When comparing KLEIN and LED on one side with MCRYPTON and PRESENT on the other, we observe that the number of rounds of KLEIN and LED increases with the key length, which is certainly an undesired property. This is not the case with MCRYPTON and PRESENT where the number of rounds remains constant even if the key length changes. Additionally, LED and NOEKEON ciphers come without key schedule, i.e. the same round key is used in all rounds. Although the key schedule is not within the critical path, this feature reduces the complexity of the circuit and it is, therefore, beneficial for the implementation cost of low-latency designs.

Heterogenous Constructions. As we already observed in Fig. 8, the last few rounds of the unrolled implementations are smaller in area than the middle ones. This leads to an interesting conclusion: we suggest to design cryptographic primitives with heterogenous rounds. Namely, designing the algorithm such that the last few rounds are more complex, and thus larger in area, would reduce the number of rounds and reduce the complexity of the whole design. This would, obviously, have consequences for lightweight (round-based) implementation of the algorithm, but here we only consider the low-latency requirements. To further illustrate this observation, we provide Fig. 17 in Appendix C, where the area per round distribution is given for the PRESENT-80 block cipher assuming several different timing constraints.

Encryption and Decryption Procedures. Although Fig. 8 shows only the results for KLEIN, it illustrates a trend common to all ciphers (except NOE-KEON). The figure clearly shows that there is a noticeable imbalance between the encryption and decryption datapaths for most of the tested ciphers. The explanation of this phenomenon is rather simple. Most of the ciphers are designed with the efficiency of the encryption procedure in mind. Therefore, the S-box and the P-layer are often chosen such that their complexity is smaller than that of their inverses. This fact indeed favors the approach of NOEKEON, where the same hardware resources can be reused for both encryption and decryption. This approach not only saves a significant amount of area, but also reduces the latency of the implementation. We also observe that although MCRYPTON has (nearly) involutional layers there is a non-negligible cost to reuse them for both encryption an decryption (due to the required insertion of multiplexors).

Conclusion. We have introduced the domain of low-latency encryption, clearly distinguishing it from the domains of lightweight and conventional encryption. Six well-known lightweight SPN block ciphers, including AES, were selected based on their properties and identified as possible candidates to yield good low-latency behavior. We evaluated their hardware performance within the context of low-latency encryption, thereby providing the first results in the field. It has been shown that the obtained results (i.e. latency, area, power, and energy consumption) are strongly influenced by the design properties such as the number of rounds, the round's complexity, and the similarity between encryption and decryption procedures. We hope that our results will inspire others to design new and efficient low-latency cryptographic primitives.

References

1. FIPS Pub. 197: Specification for the AES (November 2001), http://csrc.nist.gov/pub-lications/fips/fips197/fips-197.pdf
2. Biryukov, A. (ed.): FSE 2007. LNCS, vol. 4593. Springer, Heidelberg (2007)
3. Robshaw, M., Billet, O. (eds.): New Stream Cipher Designs. LNCS, vol. 4986. Springer, Heidelberg (2008)
4. Mangard, S., Standaert, F.-X. (eds.): CHES 2010. LNCS, vol. 6225. Springer, Heidelberg (2010)
5. Preneel, B., Takagi, T. (eds.): CHES 2011. LNCS, vol. 6917. Springer, Heidelberg (2011)
6. Aumasson, J.-P., Henzenz, L., Meier, W., Naya-Plasencia, M.: Quark: a lightweight hash. In: Cryptographic Hardware and Embedded Systems — CHES 2010 [4], pp. 1–15
7. Bertoni, G., Daemen, J., Peeters, M., Assche, G.V.: Keccak sponge function family main document (version 2.1). Submission to NIST (2010), http://keccak.noekeon.org/Keccak-main-2.1.pdf
8. Bogdanov, A., Knežević, M., Leander, G., Toz, D., Varici, K., Verbauwhede, I.: SPONGENT: A lightweight hash function. In: Cryptographic Hardware and Embedded Systems — CHES 2011 [5], pp. 312–325
9. Bogdanov, A., Knudsen, L.R., Leander, G., Paar, C., Poschmann, A., Robshaw, M.J.B., Seurin, Y., Vikkelsoe, C.: PRESENT: An Ultra-Lightweight Block Cipher. In: Paillier, P., Verbauwhede, I. (eds.) CHES 2007. LNCS, vol. 4727, pp. 450–466. Springer, Heidelberg (2007)

10. De Cannière, C., Dunkelman, O., Knežević, M.: KATAN and KTANTAN — A Family of Small and Efficient Hardware-Oriented Block Ciphers. In: Clavier, C., Gaj, K. (eds.) CHES 2009. LNCS, vol. 5747, pp. 272–288. Springer, Heidelberg (2009)

11. Cannière, C.D., Preneel, B.: Trivium. In: The eSTREAM Finalists [3], pp. 244–266

12. Cid, C., Murphy, S., Robshaw, M.J.B.: Small Scale Variants of the AES. In: Gilbert, H., Handschuh, H. (eds.) FSE 2005. LNCS, vol. 3557, pp. 145–162. Springer, Heidelberg (2005)

13. Daemen, J., Peeters, M., Rijmen, V., Assehe, G.V.: Nessie Proposal: Noekeon (2000), http://gro.noekeon.org/

14. Daemen, J., Rijmen, V.: The Design of Rijndael: AES - The Advanced Encryption Standard. Springer (2002)

15. European Network of Excellence in Cryptology – ECRYPT. The eSTREAM Project (2004), http://www.ecrypt.eu.org/stream/

16. Feldhofer, M., Rechberger, C.: A Case Against Currently Used Hash Functions in RFID Protocols. In: Meersman, R., Tari, Z., Herrero, P. (eds.) OTM 2006 Workshops, Part I. LNCS, vol. 4277, pp. 372–381. Springer, Heidelberg (2006)

17. Feldhofer, M., Wolkerstorfer, J., Rijmen, V.: AES Implementation on a Grain of Sand. IEE Proceedings Information Security 152(1), 13–20 (2005)

18. Gong, Z., Nikova, S., Law, Y.W.: KLEIN: A New Family of Lightweight Block Ciphers. In: Juels, A., Paar, C. (eds.) RFIDSec 2011. LNCS, vol. 7055, pp. 1–18. Springer, Heidelberg (2012)

19. Guo, J., Peyrin, T., Poschmann, A.: The PHOTON Family of Lightweight Hash Functions. In: Rogaway, P. (ed.) CRYPTO 2011. LNCS, vol. 6841, pp. 222–239. Springer, Heidelberg (2011)

20. Guo, J., Peyrin, T., Poschmann, A., Robshaw, M.: The LED Block Cipher. In: Cryptographic Hardware and Embedded Systems — CHES 2011 [5], pp. 326–341

21. Hell, M., Johansson, T., Maximov, A., Meier, W.: The Grain Family of Stream Ciphers. In: The eSTREAM Finalists [3], pp. 179–190

22. Hodjat, A., Verbauwhede, I.: Area-Throughput Trade-offs for Fully Pipelined 30 to 70 Gbits/s AES Processors. IEEE Transactions on Computers 55(4), 366–372 (2006)

23. Hong, D., Sung, J., Hong, S., Lim, J., Lee, S., Koo, B., Lee, C., Chang, D., Lee, J., Jeong, K., Kim, H., Kim, J., Chee, S.: HIGHT: A New Block Cipher Suitable for Low-Resource Device. In: Goubin, L., Matsui, M. (eds.) CHES 2006. LNCS, vol. 4249, pp. 46–59. Springer, Heidelberg (2006)

24. Izadi, M., Sadeghiyan, B., Sadeghian, S., Khanooki, H.: MIBS: A New Lightweight Block Cipher. In: Garay, J.A., Miyaji, A., Otsuka, A. (eds.) CANS 2009. LNCS, vol. 5888, pp. 334–348. Springer, Heidelberg (2009)

25. Kavun, E., Yalcin, T.: A Lightweight Implementation of Keccak Hash Function for Radio-Frequency Identification Applications. In: Ors Yalcin, S.B. (ed.) RFIDSec 2010. LNCS, vol. 6370, pp. 258–269. Springer, Heidelberg (2010)

26. Knudsen, L., Leander, G., Poschmann, A., Robshaw, M.: PRINTcipher: A Block Cipher for IC-Printing. In: Cryptographic Hardware and Embedded Systems — CHES 2010 [4], pp. 16–32

27. Leander, G., Paar, C., Poschmann, A., Schramm, K.: New Lightweight DES Variants. In: 14th International Workshop on Fast Software Encryption — FSE 2007 [2], pp. 196–210

28. Leander, G., Poschmann, A.: On the Classification of 4 Bit S-Boxes. In: Carlet, C., Sunar, B. (eds.) WAIFI 2007. LNCS, vol. 4547, pp. 159–176. Springer, Heidelberg (2007)

29. Lim, C., Korkishko, T.: mCrypton – A Lightweight Block Cipher for Security of Low-Cost RFID Tags and Sensors. In: Song, J.-S., Kwon, T., Yung, M. (eds.) WISA 2005. LNCS, vol. 3786, pp. 243–258. Springer, Heidelberg (2006)

30. Mathew, S., Sheikh, F., Kounavis, M., Gueron, S., Agarwal, A., Hsu, S., Kaul, H., Anders, M., Krishnamurthy, R.: 53 Gbps Native $GF(2^4)^2$ Composite-Field AES-Encrypt/Decrypt Accelerator for Content-Protection in 45 nm High-Performance Microprocessors. IEEE Journal of Solid-State Circuits 46(4), 767–776 (2011)

31. National Institute of Standards and Technology (NIST). Cryptographic Hash Algorithm Competition, http://csrc.nist.gov/groups/ST/hash/sha-3/index.html

32. National Institute of Standards and Technology (NIST). FIPS 197: Advanced Encryption Standard (November 2001)

33. Poschmann, A., Moradi, A., Khoo, K., Lim, C.-W., Wang, H., Ling, S.: Side-Channel Resistant Crypto for Less than 2,300 GE. Journal of Cryptology 24, 322–345 (2011)

34. Shibutani, K., Isobe, T., Hiwatari, H., Mitsuda, A., Akishita, T., Shirai, T.: Piccolo: An Ultra-Lightweight Blockcipher. In: Cryptographic Hardware and Embedded Systems — CHES 2011 [5], pp. 342–357

35. Shirai, T., Shibutani, K., Akishita, T., Moriai, S., Iwata, T.: The 128-bit blockcipher CLEFIA. In: 14th International Workshop on Fast Software Encryption — FSE 2007 [2], pp. 181–195

36. Standaert, F.-X., Piret, G., Gershenfeld, N., Quisquater, J.-J.: SEA: A Scalable Encryption Algorithm for Small Embedded Applications. In: Domingo-Ferrer, J., Posegga, J., Schreckling, D. (eds.) CARDIS 2006. LNCS, vol. 3928, pp. 222–236. Springer, Heidelberg (2006)

37. Wheeler, D., Needham, R.: TEA, a Tiny Encryption Algorithm. In: Preneel, B. (ed.) FSE 1994. LNCS, vol. 1008, pp. 363–366. Springer, Heidelberg (1995)

A Hardware Performance (Data)

In Table 1, we summarize hardware figures for all the tested block ciphers. The best (smallest) values in each column are marked in bold. Since all the values are obtained based on synthesis results, we believe that the metrics including area, latency, and time-area product are estimated with a good accuracy. On the other hand, we believe that accurate power and energy estimation can only be done after place and route is performed and, therefore, we do not provide a detailed report on these two metrics.

Table 1. Hardware performance of all the tested ciphers (90 nm CMOS, synthesis results)

						Time-constrained						
		1-cycle						2-cycle				
		ENC			ENC/DEC			ENC			ENC/DEC	
	L	A	T-A	L	A	T-A	L	A	T-A	L	A	T-A
AES-128	14.8	218.1	3.227	17.8	366.6	6.525	16.6	118.1	1.961	20.2	191.8	3.874
KLEIN-64	11.2	29.0	0.325	15.3	48.2	0.737	12.2	14.6	0.179	16.4	24.9	0.409
KLEIN-80	14.8	39.0	0.577	20.3	63.7	1.293	15.8	19.6	0.310	21.4	32.6	0.697
KLEIN-96	18.4	48.6	0.893	25.3	79.9	2.021	19.6	24.5	0.481	26.4	41.3	1.089
LED-64	30.9	62.0	1.917	31.2	128.7	4.014	32.2	32.2	1.038	32.8	63.5	2.081
LED-128	46.0	93.4	4.296	46.6	193.1	8.999	47.4	47.9	2.269	48.2	96.0	4.625
MCRYPTON-64	9.7	**22.5**	0.218	**9.8**	41.3	0.405	**10.4**	**11.7**	**0.124**	**10.8**	**20.9**	**0.225**
MCRYPTON-96	9.7	22.7	0.221	**9.8**	40.4	**0.396**	**10.4**	12.1	0.126	**10.8**	21.1	0.228
MCRYPTON-128	9.7	23.2	0.225	**9.8**	41.4	0.406	**10.4**	12.1	0.125	11.0	21.0	0.231
MINI-AES-64	**8.6**	23.0	**0.198**	9.9	**40.0**	**0.396**	**10.4**	12.5	0.130	12.0	22.0	0.265
NOEKEON-128	14.9	50.0	0.745	14.8	102.5	1.517	16.6	26.1	0.433	17.0	49.6	0.844
NOEKEONs-128	-	-	-	15.5	49.5	0.768	-	-	-	17.4	27.1	0.471
PRESENT-80	14.3	36.9	0.528	14.8	72.3	1.070	16	19.2	0.308	16.4	37.6	0.616
PRESENT-128	14.3	38.1	0.544	14.7	73.8	1.084	16	19.6	0.313	16.6	37.1	0.615
						Unconstrained						
		1-cycle						2-cycle				
		ENC			ENC/DEC			ENC			ENC/DEC	
	L	A	T-A	L	A	T-A	L	A	T-A	L	A	T-A
AES-128	45.5	103.6	4.715	46.6	232.2	10.820	43	62.3	2.677	51.6	122.0	6.293
KLEIN-64	20.4	11.8	0.240	31.9	28.8	0.918	25.2	7.7	0.194	35.2	15.7	0.553
KLEIN-80	26.9	15.7	0.422	42.1	38.2	1.610	32.2	10.1	0.325	46.0	20.7	0.951
KLEIN-96	33.5	19.7	0.659	53.1	47.9	2.544	39.6	12.6	0.500	57.0	25.8	1.470
LED-64	68.8	24.5	1.688	68.5	58.9	4.038	71	14.8	1.053	71.0	29.7	2.109
LED-128	102.5	36.6	3.754	100.6	88.1	8.858	103.2	21.9	2.258	105.0	44.1	4.629
MCRYPTON-64	20.2	11.7	0.235	20.7	**20.6**	0.427	22	**6.6**	0.146	23.4	**11.3**	0.264
MCRYPTON-96	19.9	11.8	0.235	20.1	20.8	**0.418**	21	6.8	**0.143**	**22.6**	11.5	**0.259**
MCRYPTON-128	20.2	12.0	0.242	**20.0**	21.0	0.419	21.2	7.0	0.148	22.8	11.6	0.265
MINI-AES-64	**19.6**	**9.4**	**0.184**	20.9	23.0	0.481	21.6	6.7	0.145	25.8	13.0	0.335
NOEKEON-128	27.6	21.3	0.587	27.9	51.6	1.438	32.4	13.8	0.446	33.0	26.6	0.878
NOEKEONs-128	-	-	-	31.8	22.3	0.710	-	-	-	33.6	15.1	0.507
PRESENT-80	36.6	15.0	0.548	31.0	34.8	1.078	33.6	9.2	0.308	36.0	18.9	0.682
PRESENT-128	35.9	15.7	0.564	30.8	36.3	1.117	33.6	9.7	0.327	34.2	20.0	0.685

L – Latency [ns]
A – Area [kGE]
T-A – Time-Area product [ms×GE]

B Hardware Performance for ENC-Only Modules

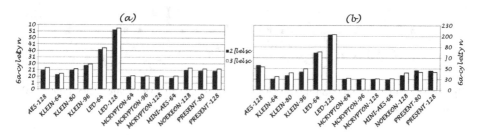

Fig. 12. Minimum latency [ns] for ENC-only module: (a) Time-constrained. (b) Unconstrained.

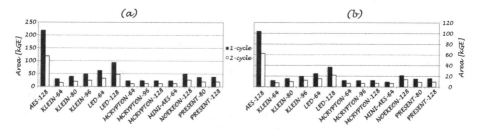

Fig. 13. Minimum area [kGE] for ENC-only module: (a) Time-constrained. (b) Unconstrained.

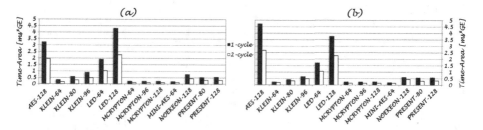

Fig. 14. Minimum time-area product [ms·GE] for ENC-only module: (a) Time-constrained. (b) Unconstrained.

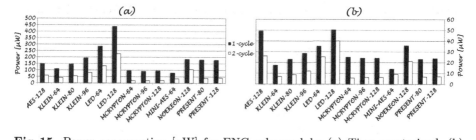

Fig. 15. Power consumption [μW] for ENC-only module: (a) Time-constrained. (b) Unconstrained.

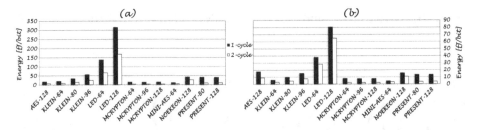

Fig. 16. Energy consumption [fJ/bit] for ENC-only module: (a) Time-constrained. (b) Unconstrained.

C Area per Round Distribution of PRESENT-80 ENC-Only

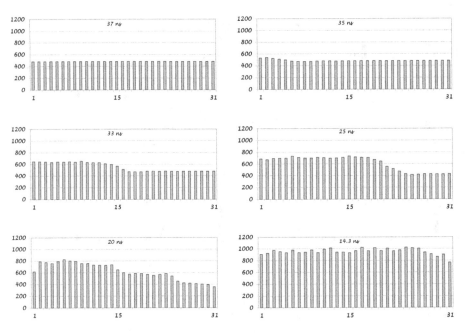

Fig. 17. Area [GE] per round distribution of the PRESENT-80 ENC-only architecture

Attacking RSA–CRT Signatures
with Faults on Montgomery Multiplication

Pierre-Alain Fouque[1,2], Nicolas Guillermin[3], Delphine Leresteux[3],
Mehdi Tibouchi[4], and Jean-Christophe Zapalowicz[2]

[1] École normale supérieure
pierre-alain.fouque@ens.fr
[2] INRIA Rennes
jean-christophe.zapalowicz@inria.fr
[3] DGA IS
nicolas.guillermin@m4x.org,
delphine.leresteux@dga.defense.gouv.fr
[4] NTT Secure Platform Laboratories
tibouchi.mehdi@lab.ntt.co.jp

Abstract. In this paper, we present several efficient fault attacks against
implementations of RSA–CRT signatures that use modular exponentiation algorithms based on Montgomery multiplication. They apply to any
padding function, including randomized paddings, and as such are the
first fault attacks effective against RSA–PSS.

The new attacks work provided that a small register can be forced to
either zero, or a constant value, or a value with zero high-order bits. We
show that these models are quite realistic, as such faults can be achieved
against many proposed hardware designs for RSA signatures.

Keywords: Fault Attacks, Montgomery Multiplication, RSA–CRT, PSS.

1 Introduction

The RSA signature scheme is one of the most used schemes nowadays. An RSA
signature is computed by applying some encoding function to the message, and
raising the result to d-th power modulo N, where d and N are the private exponent and the public modulus respectively. This modular exponentiation is the
costlier part of signature generation, so it is important to implement it efficiently.
A very commonly used speed-up is the RSA–CRT signature generation, where
the exponentiation is carried out separately modulo the two factors of N, and
the results are then recombined using the Chinese Remainder Theorem. However, when unprotected, RSA–CRT signatures are vulnerable to the so-called
Bellcore attack first introduced by Boneh et al. in [3], and later refined in multiple publications such as [31]: an attacker who knows the padded message and is
able to inject a fault in one of the two half-exponentiations can factor the public
modulus using a faulty signature with a simple GCD computation.

Many workarounds have been proposed to patch this vulnerability, including
extra computations and sanity checks of intermediate and final results. A recent

E. Prouff and P. Schaumont (Eds.): CHES 2012, LNCS 7428, pp. 447–462, 2012.

taxonomy of these countermeasures is given in [24]. The simplest countermeasure may be to verify the signature before releasing it. This is reasonably cheap if the public exponent e is small and available in the signing device. In some cases, however, e is not small, or even not given—e.g. the JavaCard API does not provide it [22]. Another approach is to use an extended modulus. Shamir's trick [25] was the first such technique to be proposed; later refinements were suggested that also protect CRT recombination when it is computed using Garner's formula [2, 7, 30, 9]. Finally, yet another way to protect RSA–CRT signatures against faults is to use redundant exponentiation algorithms, such as the Montgomery Ladder. Papers including [14, 24] propose such countermeasures. Regardless of the approach, RSA–CRT fault countermeasures tend to be rather costly: for example, Rivain's countermeasure [24] has a stated overhead of 10% compared to an unprotected implementation, and is purportedly more efficient than previous works including [14, 30].

Relatedly, while Boneh et al.'s original fault attack does not apply to RSA signatures with probabilistic encoding functions, some extensions of it were proposed to attack randomized ad hoc padding schemes such as ISO 9796-2 and EMV [10, 12]. However, Coron and Mandal [11] were able to prove that Bellare and Rogaway's padding scheme RSA–PSS [1] is secure against *random* faults in the random oracle model. In other words, if injecting a fault on the half-exponentiation modulo the second factor q of N produces a result that can be modeled as uniformly distributed modulo q, then the result of such a fault cannot be used to break RSA–PSS signatures. It is tempting to conclude that using RSA–PSS should enable signers to dispense with costly RSA–CRT countermeasures. In this paper, we argue that this is not necessarily the case.

Our Contributions. The RSA–CRT implementations targeted in this paper use the state-of-the-art modular multiplication algorithm due to Montgomery [20], which avoids the need to compute actual divisions on large integers, replacing them with only multiplications and bit shifts. A typical implementation of the Montgomery multiplication algorithm will use small registers to store precomputed values or short integer variables throughout the computation. The size of these registers varies with the architecture, from a single bit in certain hardware implementations to 16 bits, 32 bits or more in software. This paper presents several fault attacks on these small registers during Montgomery multiplication, that cause the result of one of the half-exponentiations to be unusually small. The factorization of N can then be recovered using a GCD, or an approximate common divisor algorithm such as [15, 5, 8].

We consider three models of faults on the small registers. In the first model, one register can be forced to zero. In that case, we show that causing such a fault in the inverse Montgomery transformation of the result of a half-exponentiation, or a few earlier consecutive Montgomery multiplications, yields a faulty signature which is a multiple of the corresponding factor q of N. Hence, we can factor N by taking a simple GCD. In the second model, another register can be forced to some (possibly unknown) constant value throughout the inverse Montgomery transformation of the result of a half-exponentiation, or a few earlier consecutive

Montgomery multiplications. A faulty signature in this model is a close multiple of the corresponding factor q of N, and we can thus factor N using an approximate common divisor algorithm. Finally, the third model makes it possible to force some of the higher-order bits of one register to zero. We show that, while injecting one such fault at the end of the inverse Montgomery transformation results in a faulty signature that isn't usually close enough to a multiple of q to reveal the factorization of N on its own, a moderate number of faulty signatures (a dozen or so) obtained using that process are enough to factor N.

The RSA padding scheme used for signing, whether deterministic or probabilistic, is irrelevant in our attacks. In particular, RSA–PSS implementations are also vulnerable. Of course, this does not contradict the security result due to Coron and Mandal [11], as the faults we consider are strongly non-random. Our results do suggest, however, that exponentiation algorithms based on Montgomery multiplication are quite sensitive to a very realistic type of fault attacks and that using RSA–CRT countermeasures is advisable even for RSA–PSS.

Organization of the Paper. In §2, we recall some background material on the Montgomery multiplication algorithm, on modular exponentiation techniques, and on RSA–CRT signatures. Our new attacks are then described in §§3–5, corresponding to three different fault models: null faults, constant faults, and zero high-order bits faults. Finally, in §6, we discuss the applicability of our fault models to concrete hardware implementations of RSA–CRT signatures, and find that many proposed designs are vulnerable.

2 Preliminaries

2.1 Montgomery Multiplication

Proposed by Montgomery in [20], the Montgomery multiplication algorithm provides a fast way method for computing modular multiplications and squarings. Indeed, the Montgomery multiplication algorithm only uses multiplications, additions and shifts, and its cost is about twice that of a simple multiplication (compared to 2.5 times for a multiplication and a Barett reduction), without imposing any constraint on the modulus.

Usually, one of two different techniques is used to compute Montgomery multiplication: either Separate Operand Scanning (SOS), or Coarsely Integrated Operand Scanning (CIOS). Consider a device whose processor or coprocessor architecture has r-bit registers (typically $r = 1, 8, 16, 32$ or 64 bits). Let $b = 2^r$, q be the (odd) modulus with respect to which multiplications are carried out, k the number of r-bit registers used to store q, and $R = b^k$, so that $q < R$ and $\gcd(q, R) = 1$. The SOS variant consists in using the Montgomery reduction after the multiplication: for an input A such that $A < Rq$, it computes $\mathrm{Mgt}(A) \equiv AR^{-1} \pmod{q}$, with $0 \le \mathrm{Mgt}(A) < q$. The CIOS mixes the reduction algorithm with the previous multiplication step: considering x and y with $xy < Rq$, it computes $\mathrm{CIOS}(x, y) = xyR^{-1} \bmod q$ with $\mathrm{CIOS}(x, y) < q$.

```
1: function SIGN_RSA-CRT(m)
2:     M ← μ(m) ∈ ℤ_N        ▷ message
       encoding
3:     M_p ← M mod p
4:     M_q ← M mod q
5:     S_p ← M_p^{d_p} mod p
6:     S_q ← M_q^{d_q} mod q
7:     t ← S_p - S_q
8:     if t < 0 then t ← t + p
9:     S ← S_q + ((t · π) mod p) · q
10:    return S
```

Fig. 1. RSA–CRT signature generation with Garner's recombination. The reductions d_p, d_q modulo $p-1, q-1$ of the private exponent are precomputed, as is $\pi = q^{-1} \bmod p$.

```
1: function CIOS(x, y)
2:     a ← 0
3:     y_0 ← y mod b
4:     for j = 0 to k - 1 do
5:         a_0 ← a mod b
6:         u_j ← (a_0 + x_j · y_0) · q' mod b
7:         a ← ⌊ (a + x_j · y + u_j · q) / b ⌋
8:     if a ≥ q then a ← a - q
9:     return a
```

Fig. 2. The Montgomery multiplication algorithm. The x_i's and y_i's are the digits of x and y in base b; $q' = -q^{-1} \bmod b$ is precomputed. The returned value is $(xy \cdot b^{-k} \bmod q)$. Since $b = 2^r$, the division is a bit shift.

Figure 2 presents the main steps of the CIOS variant, which will be used thereafter. However, replacing the CIOS by the SOS or any other variant proposed in [17] does not protect against any of our attacks.

2.2 Exponentiation Algorithms Using Montgomery Multiplication

Montgomery reduction is especially interesting when used as part of a modular exponentiation algorithm. A large number of such exponentiation algorithms are known, including the Square-and-Multiply algorithm from either the least or the most significant bit of the exponent, the Montgomery Ladder (used as a side-channel countermeasure against cache analysis, branch analysis, timing analysis and power analysis), the Square-and-Multiply k-ary algorithm (which boasts greater efficiency thanks to fewer multiplications), etc. The first three exponentiation algorithms will be considered in this paper, and two of those are detailed in Figure 3.

Note that using the Montgomery multiplications inside any exponentiation algorithm requires all variables to be in Montgomery representation ($\bar{x} = xR \bmod q$ is the Montgomery representation of x) before applying the exponentiation process. In line 2 of each algorithm from Figure 3, the message is transformed into Montgomery representation by computing $CIOS(x, R^2) = xR^2R^{-1} \bmod q = \bar{x}$. At the end, the very last CIOS call allows to revert to the classical representation by performing a Montgomery reduction: $CIOS(\bar{A}, 1) = (\bar{A} \cdot 1)R^{-1} \bmod q = ARR^{-1} \bmod q = A$. Finally the other CIOS steps compute the product in Montgomery representation: $CIOS(\bar{A}, \bar{B}) = (AR)(BR)R^{-1} \bmod q = \overline{AB}$.

2.3 RSA–CRT Signature Generation

Let $N = pq$ be a n-bit RSA modulus. The public key is denoted by (N, e) and the associated private key by (p, q, d). For a message M to be signed, we note

Square-and-Multiply LSB	Montgomery Ladder
	1: **function** $\text{EXP}_{\text{LADDER}}(x, e, q)$
	2: $\bar{x} \leftarrow \text{CIOS}(x, R^2 \bmod q)$
1: **function** $\text{EXP}_{\text{LSB}}(x, e, q)$	3: $A \leftarrow R \bmod q$
2: $\bar{x} \leftarrow \text{CIOS}(x, R^2 \bmod q)$	4: **for** $i = t$ down to 0 **do**
3: $A \leftarrow R \bmod q$	5: **if** $e_i = 0$ **then**
4: **for** $i = 0$ to t **do**	6: $\bar{x} \leftarrow \text{CIOS}(A, \bar{x})$
5: **if** $e_i = 1$ **then**	7: $A \leftarrow \text{CIOS}(A, A)$
6: $A \leftarrow \text{CIOS}(A, \bar{x})$	8: **else if** $e_i = 1$ **then**
7: $\bar{x} \leftarrow \text{CIOS}(\bar{x}, \bar{x})$	9: $A \leftarrow \text{CIOS}(A, \bar{x})$
8: $A \leftarrow \text{CIOS}(A, 1)$	10: $\bar{x} \leftarrow \text{CIOS}(\bar{x}, \bar{x})$
9: **return** A	11: $A \leftarrow \text{CIOS}(A, 1)$
	12: **return** A

Fig. 3. Two of the exponentiation algorithms considered in this paper. In each case, $e_0, \ldots, e_t$ are the bits of the exponent e (from the least to the most significant), b is the base in which computations are carried out ($\gcd(b, q) = 1$) and $R = b^k$.

$S = m^d \bmod N$ the corresponding signature, where m is deduced from M by an encoding function, possibly randomized. A well-known optimization of this operation is the RSA–CRT which takes advantage of the decomposition of N into prime factors. By replacing a full exponentiation of size n by two $n/2$, it divides the computational cost by a factor of around 4. Therefore RSA–CRT is almost always employed: for example, OpenSSL as well as the JavaCard API [22] use it.

Recovering S from its reductions S_p and S_q modulo p and q can be done either by the usual CRT reconstruction formula (1) below, or using the recombination technique (2) due to Garner:

$$S = (S_q \cdot p^{-1} \bmod q) \cdot p + (S_p \cdot q^{-1} \bmod p) \cdot q \bmod N. \tag{1}$$

$$S = S_q + q \cdot (q^{-1} \cdot (S_p - S_q) \bmod p). \tag{2}$$

Garner's formula (2) does not require a reduction modulo N, which is interesting for efficiency reasons and also because it prevents certain fault attacks [4]. On the other hand, it does require an inverse Montgomery transformation $S_q = CIOS(\bar{S}_q, 1)$, whereas that step is not necessary for formula (1), as it can be mixed with the multiplication with $q^{-1} \bmod p$. This is an important point, as some of our attacks specifically target the inverse Montgomery transformation. The main steps of the RSA–CRT signature generation with Garner's recombination are recalled in Figure 1.

3   Null Faults

We first consider a fault model in which the attacker can force the register containing the precomputed value $q' = (-q \bmod b)$ to zero in certain calls to the CIOS algorithm during the computation of S_q.

Under suitable conditions, we will see that such faults can cause the q-part of the signature to be erroneously evaluated as $\widetilde{S}_q = 0$, which makes it possible to retrieve the factor q of N from one such faulty signature $\widetilde{S}$, as $q = \gcd(\widetilde{S}, N)$.

3.1   Attacking CIOS($A, 1$)

Suppose first that the fault attacker can force q' to zero in the very last CIOS computation during the evaluation of S_q, namely the computation of CIOS($A, 1$). In that case, the situation is quite simple.

Theorem 1. *A faulty signature $\widetilde{S}$ generated in this fault model is a multiple of q (for any of the exponentiation algorithms considered herein and regardless of the encoding function involved, probabilistic or not).*

Proof. The faulty value $\widetilde{q'} = 0$ causes all of the variables u in the CIOS loop to vanish; indeed, for $j = 0, \ldots, k - 1$, they evaluate to:

$$\widetilde{u}_j = (a_0 + A_j \cdot 1) \cdot \widetilde{q'} \bmod 2^r = 0.$$

As a result, the value $\widetilde{S}_q$ computed by this CIOS loop can be written as:

$$\widetilde{S}_q = \left\lfloor \left(\left\lfloor \cdots \left\lfloor \left(\lfloor A_0 \cdot 2^{-r} \rfloor + A_1 \right) \cdot 2^{-r} \right\rfloor + \cdots \right\rfloor + A_{k-1} \right) \cdot 2^{-r} \right\rfloor.$$

Now, the values A_j are r-words, i.e. $0 \leq A_j \leq 2^r - 1$. It follows that each of the integer divisions by 2^r evaluate to zero, and hence $\widetilde{S}_q = 0$. As a result, the faulty signature $\widetilde{S}$ is a multiple of q as stated.   □

It is thus easy to factor N with a single faulty signature $\widetilde{S}$, by computing $\gcd(\widetilde{S}, N)$. Note also that if this last CIOS step is computed as CIOS($1, A$) instead of CIOS($A, 1$), the formulas are slightly different but the result still holds.

3.2   Attacking Consecutive CIOS Steps

If Garner recombination is not used or the computation of CIOS($A, 1$) is somehow protected against faults, a similar result can be achieved by forcing q' to zero in earlier calls to CIOS, provided that a certain number of successive CIOS executions are faulty.

Assuming that the values $\bar{x}$ and A in Montgomery representation are uniformly distributed modulo q before the first faulty CIOS, we show in the full version of this paper [13]that faults across $\ell = \lceil \log_2 \lceil \log_2 q \rceil \rceil$ iterations in the loop of the exponentiation algorithm are enough to ensure that $\widetilde{S}_q$ will evaluate to zero with probability at least $1/2$. For example, if q is a 512-bit prime, we have $\ell = 9$. This means that forcing q' to zero in 9 iterations (from 9 to 18 calls to CIOS depending on the exponentiation algorithm under consideration and on the input bits) is enough to factor the modulus at least 50% of the time—and more faulty iterations translate to higher success rates.

Table 1. Success rate of the null fault attack on consecutive CIOS steps, for a 512-bit prime q and $r = 16$. 100 faulty signatures were computed for each parameter set. For the Square-and-Multiply MSB and Montgomery Ladder algorithms, we compare success rates when faults start at the beginning of the loop vs. at a random iteration.

	S&M LSB	S&M MSB		Montgomery Ladder	
Faulty iterations	(%)	Start (%)	Anywhere (%)	Start (%)	Anywhere (%)
8	31	93	62	45	30
9	65	100	93	87	76
10	89	100	100	99	93

Simulation results. We have carried out a simulation of null faults on consecutive CIOS steps for each of the three exponentiation process algorithms, with varying numbers of faulty iterations; for the Square-and-Multiply MSB and the Montgomery Ladder algorithms, two sets of experiments have been conducted for each parameter set: one with faults starting from the first iteration, and another one with faults starting from a random iteration somewhere in the exponentiation loop. Results are collected in Table 1.

4   Constant Faults

In this section, we consider a different fault model, in which the fault attacker can force the variables u_j in the CIOS algorithm to some (possibly unknown) constant value $\widetilde{u}$.

Just as with null faults, we consider two scenarios: one in which the last CIOS computation is attacked, and another in which several inner consecutive CIOS computations in the exponentiation algorithm are targeted.

4.1   Attacking CIOS($A, 1$)

Faults on all iterations. Consider first the case when faults are injected in all iterations of the very last CIOS computation. In other words, the device computes CIOS($A, 1$), except that the variables u_j, $j = 0, \ldots, k - 1$, are replaced by a fixed, possibly unknown value $\widetilde{u}$. In that case, we show that a single faulty signature is enough to factor N and recover the secret key. The key result is as follows (the proof can be found in the full version [13]).

Theorem 2. *Let $\widetilde{S}$ be a faulty signature obtained in the fault model described above. Then, $(2^r - 1) \cdot \widetilde{S}$ is a close multiple of q with error size at most 2^{r+1}, i.e. there exists an integer T such that:*

$$\left| (2^r - 1) \cdot (\widetilde{S} + 1) - qT \right| \leq 2^{r+1}.$$

Thus, a single faulty signature yields a value $V = (2^r - 1) \cdot (\widetilde{S} + 1) \bmod N$ which is very close to a multiple of q. It is easy to use this value to recover q itself. Several methods are available:

- If r is small (say 8 or 16), it may be easiest to just use exhaustive search: q is found among the values $\gcd(V + X, N)$ for $|X| \le 2^{r+1}$, and hence can be retrieved using around 2^{r+2} GCD computations.
- A more sophisticated option, which may be interesting for $r = 32$, is the baby step, giant step-like algorithm by Chen and Nguyen [5], which runs in time $\widetilde{O}(2^{r/2})$.
- Alternatively, for any r up to half of the size of q, one can use Howgrave-Graham's algorithm [15] based on Coppersmith techniques. It is the fastest option unless r is very small (a simple implementation in Sageruns in about 1.5 ms on our standard desktop PC with a 512-bit prime q for a any r up to ≈ 160 bits, whereas exhaustive search already takes over one second for $r = 16$).

Faults on most iterations. Howgrave-Graham's algorithm is especially relevant if the constant faults do not start at the very first iteration in the CIOS loop. More precisely, suppose that the fault attacker can force the variables u_j to a constant value $\widetilde{u}$ not for all j but for $j = j_0, j_0 + 1, \ldots, k - 1$ for some j_0.

Then, the same computation as in the proof of Theorem 2 yields the following bound on $\widetilde{S}_q$:

$$\frac{\widetilde{u} \cdot q}{2^r - 1} - 2^{r j_0} - 2 < \widetilde{S}_q \le \frac{\widetilde{u} \cdot q}{2^r - 1} + 2^{r j_0} + 1.$$

It follows that $(2^r - 1) \cdot \widetilde{S}$ is a close multiple of q with error size $\lesssim 2^{r(j_0+1)}$.

Now note that Howgrave-Graham's algorithm [15] will recover q given N and a close multiple with error size at most $q^{1/2-\varepsilon}$. This means that one faulty signature $\widetilde{S}$ is enough to factor N as long as $j_0 + 1 < k/2$, i.e. the constant faults start in the first half of the CIOS loop.

4.2   Attacking Other CIOS Steps

As in §3.2, if Garner recombination is not used or $\text{CIOS}(A, 1)$ is protected against faults, we can adapt the previous attack to target earlier calls to CIOS and still reveal the factorization of N. However, the attack requires two faulty signatures with the same constant fault $\widetilde{u}$. Details are given in the full version [13].

In short, depending on the ratios $q/2^{\lceil \log_2 q \rceil}$ and $\widetilde{u}/(2^r - 1)$, two faulty signatures $\widetilde{S}, \widetilde{S}'$ with the same faulty value $\widetilde{u}$ have a certain probability of being equal modulo q. Thus, we recover q as $\gcd(N, \widetilde{S} - \widetilde{S}')$. This attack works with the Square-and-Multiply LSB and Montgomery Ladder algorithms, but not with Square-and-Multiply MSB exponentiation.

Simulation results are presented in Table 2. For various 512-bit primes q, the attack has been carried out for 1000 pairs of random messages, with a random constant fault $\widetilde{u}$ for each pair. It is successful if the two resulting faulty signatures $\widetilde{S}, \widetilde{S}'$ satisfy $\gcd(N, \widetilde{S} - \widetilde{S}') = q$.

Table 2. Success rate of the constant fault attack on successive CIOS steps, when using Square-and-Multiply LSB exponentiation with random 512-bit primes q and $r = 16$

$q/2^{\lceil \log_2 q \rceil}$	0.666	0.696	0.846	0.957
Success rate (%)	36	34.4	26.7	20.4

5   Zero High-Order Bits Faults

In this section, we consider yet another fault model, in which the fault attacker targets the very last iteration in the evaluation of $\text{CIOS}(A, 1)$ during the computation of S_q. We assume that the attacker is able to force a certain number h of the highest-order bits of u_{k-1} to zero, possibly but not necessarily all of them (i.e. $1 \leq h \leq r$). Then, while a single faulty signature is typically not sufficient to factor the modulus, multiple such signatures will be enough if h is not too small. More precisely, we prove the following theorem in the full version of this paper [13]:

Theorem 3. *Let $\widetilde{S}$ be a faulty signature obtained in this fault model. Then, $\widetilde{S}$ is a close multiple of q with error size at most $2^{-h} \cdot q + 1$, i.e. there exists an integer T such that $|\widetilde{S} - qT| \leq 2^{-h} \cdot q + 1$.*

Now, recovering q from faulty signatures of the form $\widetilde{S}$ is a partial approximate common divisor (PACD) problem, as we know one exact multiple of q, namely N, and several close multiples, namely the faulty signatures. Since the error size $\approx q/2^h$ is rather large relative to q, the state-of-the-art algorithm to recover q in that case is the one proposed by Cohn and Heninger [8] using multivariate Coppersmith techniques.

The algorithm by Cohn and Heninger is likely to recover the common divisor $q \approx N^{1/2}$ given ℓ close multiples $\widetilde{S}^{(1)}, \ldots, \widetilde{S}^{(\ell)}$ provided that the error size is significantly less than $N^{(1/2)^{1+1/\ell}}$. Hence, if the faults cancel the top h bits of u_{k-1}, we need ℓ of them to factor the modulus, where:

$$\ell \gtrsim -\frac{1}{\log_2 \left(1 - \frac{h}{\log_2 q}\right)}. \tag{3}$$

In practice, if a few more faults can be collected, it is probably preferable to simply use the linear case of the Cohn-Heninger attack (the case $t = k = 1$ in their paper [8]), since it is much easier to implement (as it requires only linear algebra rather than Gröbner bases) and involves lattice reduction in a lattice of small dimension that is straightforward to construct. We examine this method in more details in the full version of this paper [13], and find that it makes it possible to factor N provided that:

$$\ell \gtrsim \frac{\log_2 q}{h} \tag{4}$$

Table 3. Theoretical minimum number ℓ of zero higher-order h-bit faulty signatures required to factor a balanced 1024-bit RSA modulus N using the general Cohn-Heninger attack or the simplified linear one

Number h of zero top bits	48	40	32	24	16
Minimum ℓ with the general attack	8	9	11	15	22
Minimum ℓ with the linear attack	11	13	16	22	32

Table 4. Experimental success rate of the simplified (linear) Cohn-Heninger attack with ℓ faulty signatures when N is a balanced 1024-bit RSA modulus. Timings are given for our Sage implementation on a single core of a Core 2 CPU at 3 GHz.

Number ℓ of faulty signatures	11	12	13	14	15	16	17	18
Success rate with $h = 48$ (%)	23	100	100	100	100	100	100	100
Success rate with $h = 40$ (%)	0	0	2	100	100	100	100	100
Success rate with $h = 32$ (%)	0	0	0	0	0	0	99	100
Average CPU time (ms)	33	35	38	41	45	49	54	59

which is always a worse bound than (3) but usually not by a very large margin. Table 3 gives the theoretical number of faulty signatures required to factor N for various values of h, both in the general attack by Cohn and Heninger and in the simplified linear case.

We carried out a simulation of the linear version of the attack on a 1024-bit modulus N with various values of h, and found that it works very well in practice with a number of faulty signatures consistent with the theoretical minimum. The results are collected in Table 4. The attack is also quite fast: a naive implementation in Sage runs in a fraction of a second on a standard PC.

6   Fault Models

In this section we discuss how realistic the setup of the attacks described above can be. In principle, all the RSA–CRT implementations using Montgomery multiplication may be vulnerable, but we have to note that the fault setup (and how realistic it is) depends heavily on implementation choices, since many variations around the algorithm from Figure 2 have been proposed in recent literature.

After a discussion about the tools needed to get the desired effects, we focus on several implementation proposals [29, 18, 16, 21, 28, 19, 6], chosen for their relevance.

6.1   Characteristics of the Perturbation Tool

First all the perturbations needed to carry out our attacks need to be controlled and local to some gates of the chip. Therefore, the attacker needs to identify the localization of the vulnerable gates and registers. The null fault attacks described in §3 need either a q' value set to 0, or multiple consecutive faults in line 6 of the main loop of $CIOS(A, 1)$ or during multiple consecutive $CIOS$. The attacks described in §4 also need these multiple consecutive faults. Considering that state-of-art secure micro-controllers embed desynchronization countermeasures such as clock jitters and idle cycles, if the target of the perturbation is some shared logic with other treatments (like in the ALU of a CPU), the fault must be accurately space and time controlled, and the effects must be repeatable as well. Identification of the good cycles to inject the perturbation may be a very difficult task, and our attacks seem to be irrelevant. The only exception may be the null fault of §3, if the fault is injected when the q' register is loaded.

Nevertheless, many secure microcontrollers embed an isolated modular arithmetic acceleration coprocessor. A large proportion of them specifically use the Montgomery multiplication CIOS algorithm (or one of its described variants [17]). Therefore, if the q' or the u_j value is isolated in a specific small size register, a unique long duration perturbation can be sufficient for our attack to succeed. The duration of the perturbation varies with the implementation choices and can vary from one cycle to $\log_2 q$, which does not exceed a hundred microseconds on actual chips. To get this kind of effect, laser diodes are the best-suited tool, since the duration of the spot is completely controlled by the attacker [26].

6.2   Analysis of Classical Implementations of the Montgomery Multiplication

The public Montgomery architectures can be divided in 3 different categories :

- the first one [29, 18, 16] contains variations on the Tenca and Koç *Multiple Word Radix-2 Montgomery Multiplication* algorithm (MWR2MM) [29], which can be seen as a CIOS algorithm with $r = 1$. The characteristic of these implementations is that they use no multiplier. They are then suited for constrained area.
- the second category [28, 19] is an intermediate where r is a classical size for embedded architecture, such as 8,16 or 32 bits. They can be used for intermediate area/latency trade-offs.
- the last category [21, 6] propose a version of CIOS/SOS with only one loop, implying that $r \geq \lceil \log_2 q \rceil$. The main difficulty of these implementation techniques is to deal with the very large multiplications they require . For that purpose they use interpolation techniques, like Karatsuba in [6] or RNS in [21]. These implementations are designed to achieve the shortest latency.

Architectures Based on MWR2MM ($r = 1$). In this kind of architecture, q' cannot be manipulated, since it is always equal to 1, so no wire or register carries its value. On the other hand, the value of u_j is computed at every loop of the CIOS, and since it is only one bit, a simple shot on the logic driving the register during the final multiplication $CIOS(A, 1)$ is sufficient to get an exploitable result ($u_j = 0$ corresponds to the null fault of §3, and $u_j = 1$ to the constant fault of §4).

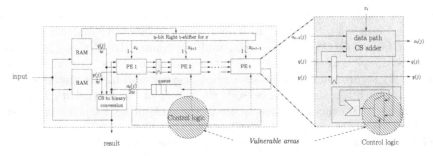

Fig. 4. Systolic Montgomery Multiplier of [29] and potential target of the fault

The first proposal [29] is a fully systolic [1] array of processing elements (PE) executing consecutively line 6 of the CIOS algorithm in one cycle, and line 7 in k cycles from LSB to MSB. Figure 4 proposes an overview of the architecture. Each PE consists of a w-word carry save adder, able to compute a w word addition and to keep the carry for the next cycle. In the figure, $T(j)$ stands for the j-th least significant w word of T.

At each clock cycle, the PE presents the computed result $a_i(j)$ to the next one, and the value u_i is kept in the PE for the computation of the next word $a_i(j + 1)$. The value of u_i is computed before the word $a_i(0)$ is presented, and then is kept in each PE during the whole computation of a_i in a register. This architecture has the great advantage of being completely scalable (whatever the number of PEs and the size of M, this architecture can compute the expected result as long as the RAM are correctly dimensioned).

To achieve our attack, the register keeping u_i can be the targeted, but every PE must be targeted simultaneously in order to get the correct result. Therefore it is more interesting to target the control logic responsible for the sequencing of the register loading, since all the PEs are connected.

In [18], the authors manage to get rid of the CS to binary converter by re-designing the CS adder of every PE. The vulnerability to our attack is therefore the same, since the redesign does not affect the targeted area.

Huang et al. [16] proposed a new version of the data dependency in the MWR2MM algorithm and rearranged the architecture of [29], in a semi systolic form. Figure 5 gives an overview of the architecture. In this architecture, the intermediate value a_i is manipulated in carry save format A specific PE, PE_0 is specialized in generating the u_i values at each cycle. while the j-th PE is in charge of computing the sequence $a_i(j)$.

This architecture is very vulnerable to our attacks, since a simple n-cycle long shot on the right logic in the PE_0 (see Figure 5) is sufficient to get the expected result.

According to the authors, the design works at 100 MHz on their target platform (a Xilinx Virtex II FPGA), therefore the duration of the perturbation is at least $10\,\mu s$ for a 1024 bits multiplication (2048 bits RSA) if the Garner recombination is

---

[1] Meaning that all the PEs are the same.

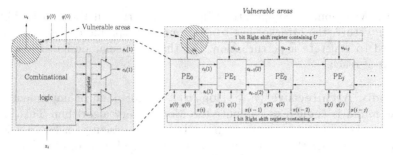

Fig. 5. Overview of the [16] architecture and potential target of the fault

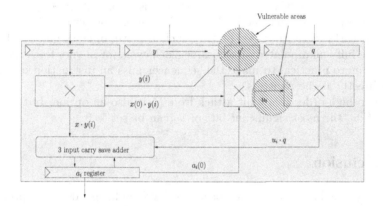

Fig. 6. Overview of the [19] architecture and potential target of the fault

used (using the attack from §3.1 or §4.1). If classical CRT reconstruction is used, according to Table 1, 200 μs will be enough for a null fault.

As a conclusion we can see that this kind of implementation is very vulnerable, since the setup of the attack is quite simple.

High Radix Architecture ($1 < r < \lceil \log_2 q \rceil$). In this type of implementation the value $q' = -q^{-1} \bmod 2^r$ is computed in a r-bit register, unless the quotient pipelining approach [23] is used.

For example, the implementation of [19] is described in Figure 6. It relies on the coordinated usage of multiplier blocks of the Xilinx Virtex II together with specifically designed carry save adders. The values u_j can be the target of any fault described in this paper, but it may be easier to put once for all the q' register to 0, with a 100% success rate for the attack if properly carried out. Another implementation is mentioned in [19] with a four-deep pipeline, but it suffers from the same vulnerability.

The attack may be more difficult to achieve on the architecture of [28, Figure 4]. First, it uses quotient determination [23], and therefore does not need to store q' anywhere. Second, the multiplier in charge of computing u_j is shared for all the Montgomery computation. In order to carry out the attack of §4 on this architecture, the attacker has to determine the specific cycles where u_j is computed to generate a perturbation. For that particular design, the attacks seem out of reach.

Full Radix Architecture ($r \geq \lceil \log_2 q \rceil$). In this kind of implementation, a single round is enough to compute the Montgomery algorithm. This implementation choice reports all the complexity on the design of a $\log_2 q \times \log_2 q$ multiplier. To reduce the full complexity of the big multiplication, interpolation techniques are used. In [6], a classical nested Karatsuba multiplication is used, whereas [21] proposes RNS.

In these architectures, a specific laser shot must swap all the u_0 or q' at the same time to produce a null fault. To have a chance, a better solution is to use non invasive attacks (in the sense of [27]), such as power or clock glitches. Indeed u_0 or q' are fully manipulated on the same clock cycle (or in very few), therefore it may be more practical to make the sequencer miss an instruction instead of aiming directly at the registers.

The zero high-order bits fault attack from §5 is also an option. In the architecture of [6], the most significant bits of u_0 can be set to 0.

7   Conclusion

In this paper, we have shown that specific realistic faults can defeat unprotected RSA–CRT signatures with any padding scheme, probabilistic or not. While it is not difficult to devise suitable countermeasures (for example, checking that S_q is not too small before outputting a signature is enough to thwart all of our attacks), this underscores the fact that relying on probabilistic signature schemes does not, in itself, protect against faults.

References

1. Bellare, M., Rogaway, P.: Probabilistic signature scheme. US Patent 6266771 (2001)
2. Blömer, J., Otto, M., Seifert, J.-P.: A new CRT-RSA algorithm secure against Bellcore attacks. In: Jajodia, S., Atluri, V., Jaeger, T. (eds.) ACM Conference on Computer and Communications Security, pp. 311–320. ACM (2003)
3. Boneh, D., DeMillo, R.A., Lipton, R.J.: On the Importance of Checking Cryptographic Protocols for Faults. In: Fumy, W. (ed.) EUROCRYPT 1997. LNCS, vol. 1233, pp. 37–51. Springer, Heidelberg (1997)
4. Brier, É., Naccache, D., Nguyen, P.Q., Tibouchi, M.: Modulus Fault Attacks against RSA-CRT Signatures. In: Preneel, B., Takagi, T. (eds.) CHES 2011. LNCS, vol. 6917, pp. 192–206. Springer, Heidelberg (2011)
5. Chen, Y., Nguyen, P.Q.: Faster Algorithms for Approximate Common Divisors: Breaking Fully-Homomorphic-Encryption Challenges over the Integers. In: Pointcheval, D., Johansson, T. (eds.) EUROCRYPT 2012. LNCS, vol. 7237, pp. 502–519. Springer, Heidelberg (2012)

6. Chow, G.C.T., Eguro, K., Luk, W., Leong, P.: A Karatsuba-based Montgomery multiplier. In: FPL 2010, pp. 434–437 (2010)

7. Ciet, M., Joye, M.: Practical fault countermeasures for Chinese remaindering based cryptosystems. In: Breveglieri, L., Koren, I. (eds.) FDTC, pp. 124–131 (2005)

8. Cohn, H., Heninger, N.: Approximate common divisors via lattices. Cryptology ePrint Archive, Report 2011/437, (2011), http://eprint.iacr.org/ (to appear at ANTS-X)

9. Coron, J.-S., Giraud, C., Morin, N., Piret, G., Vigilant, D.: Fault attacks and countermeasures on Vigilant's RSA-CRT algorithm. In: Breveglieri et al. [4], pp. 89–96

10. Coron, J.-S., Joux, A., Kizhvatov, I., Naccache, D., Paillier, P.: Fault Attacks on RSA Signatures with Partially Unknown Messages. In: Clavier, C., Gaj, K. (eds.) CHES 2009. LNCS, vol. 5747, pp. 444–456. Springer, Heidelberg (2009)

11. Coron, J.-S., Mandal, A.: PSS Is Secure against Random Fault Attacks. In: Matsui, M. (ed.) ASIACRYPT 2009. LNCS, vol. 5912, pp. 653–666. Springer, Heidelberg (2009)

12. Coron, J.-S., Naccache, D., Tibouchi, M.: Fault Attacks Against EMV Signatures. In: Pieprzyk, J. (ed.) CT-RSA 2010. LNCS, vol. 5985, pp. 208–220. Springer, Heidelberg (2010)

13. Fouque, P.-A., Guillermin, N., Leresteux, D., Tibouchi, M., Zapalowicz, J.-C.: Attacking RSA–CRT signatures with faults on Montgomery multiplication. Cryptology ePrint Archive, Report 2012/172 (2012), http://eprint.iacr.org/ (Full version of this paper)

14. Giraud, C.: An RSA implementation resistant to fault attacks and to simple power analysis. IEEE Trans. Computers 55(9), 1116–1120 (2006)

15. Howgrave-Graham, N.: Approximate Integer Common Divisors. In: Silverman, J.H. (ed.) CaLC 2001. LNCS, vol. 2146, pp. 51–66. Springer, Heidelberg (2001)

16. Huang, M., Gaj, K., Kwon, S., El-Ghazawi, T.: An Optimized Hardware Architecture for the Montgomery Multiplication Algorithm. In: Cramer, R. (ed.) PKC 2008. LNCS, vol. 4939, pp. 214–228. Springer, Heidelberg (2008)

17. Koç, Ç.K., Acar, T.: Analyzing and comparing Montgomery multiplication algorithms. IEEE Micro 16(3), 26–33 (1996)

18. McIvor, C., McLoone, M., McCanny, J.: Modified Montgomery modular multiplication and RSA exponentiation techniques. IEE Proceedings - Computers and Digital Techniques 151(6), 402–408 (2004)

19. Mentens, N., Sakiyama, K., Preneel, B., Verbauwhede, I.: Efficient pipelining for modular multiplication architectures in prime fields. In: Proceedings of the 17th ACM Great Lakes Symposium on VLSI, GLSVLSI 2007, pp. 534–539. ACM, New York (2007)

20. Montgomery, P.L.: Modular multiplication without trial division. Mathematics of Computation 44, 519–521 (1985)

21. Nozaki, H., Motoyama, M., Shimbo, A., Kawamura, S.-I.: Implementation of RSA Algorithm Based on RNS Montgomery Multiplication. In: Koç, Ç.K., Naccache, D., Paar, C. (eds.) CHES 2001. LNCS, vol. 2162, pp. 364–376. Springer, Heidelberg (2001)

22. Oracle. JavaCard 3.0.1 Platform Specification, http://www.oracle.com/technetwork/java/javacard/overview/

23. Orup, H.: Simplifying quotient determination in high-radix modular multiplication. In: IEEE Symposium on Computer Arithmetic 1995, pp. 193–193 (1995)

24. Rivain, M.: Securing RSA against Fault Analysis by Double Addition Chain Exponentiation. In: Fischlin, M. (ed.) CT-RSA 2009. LNCS, vol. 5473, pp. 459–480. Springer, Heidelberg (2009)
25. Shamir, A.: Improved method and apparatus for protecting public key schemes from timing and fault attacks. Patent Application, WO 1998/052319 A1 (1998)
26. Skorobogatov, S.: Optical fault masking attacks. In: Breveglieri et al. [4], pp. 23–29
27. Skorobogatov, S.P., Anderson, R.J.: Optical Fault Induction Attacks. In: Kaliski Jr., B.S., Koç, Ç.K., Paar, C. (eds.) CHES 2002. LNCS, vol. 2523, pp. 2–12. Springer, Heidelberg (2003)
28. Suzuki, D.: How to Maximize the Potential of FPGA Resources for Modular Exponentiation. In: Paillier, P., Verbauwhede, I. (eds.) CHES 2007. LNCS, vol. 4727, pp. 272–288. Springer, Heidelberg (2007)
29. Tenca, A.F., Koç, Ç.K.: A Scalable Architecture for Montgomery Multiplication. In: Koç, Ç.K., Paar, C. (eds.) CHES 1999. LNCS, vol. 1717, pp. 94–108. Springer, Heidelberg (1999)
30. Vigilant, D.: RSA with CRT: A New Cost-Effective Solution to Thwart Fault Attacks. In: Oswald, E., Rohatgi, P. (eds.) CHES 2008. LNCS, vol. 5154, pp. 130–145. Springer, Heidelberg (2008)
31. Yen, S.-M., Moon, S.-J., Ha, J.C.: Hardware Fault Attackon RSA with CRT Revisited. In: Lee, P.J., Lim, C.H. (eds.) ICISC 2002. LNCS, vol. 2587, pp. 374–388. Springer, Heidelberg (2003)

Reduce-by-Feedback: Timing Resistant and DPA-Aware Modular Multiplication Plus: How to Break RSA by DPA

Michael Vielhaber

Hochschule Bremerhaven, FB2, An der Karlstadt 8, D–27568 Bremerhaven, Germany
Universidad Austral de Chile, Instituto de Matemáticas, Casilla 567, Valdivia, Chile
vielhaber@gmail.com

Abstract. We (re-) introduce the Reduce-By-Feedback scheme given by Vielhaber (1987), Benaloh and Dai (1995), and Jeong and Burleson (1997).

We show, how to break RSA, when implemented with the standard version of Reduce-by-Feedback or Montgomery multiplication, by Differential Power Analysis. We then modify Reduce-by-Feedback to avoid this attack. The modification is not possible for Montgomery multiplication.

We show that both the original and the modified Reduce-by-Feedback algorithm resist timing attacks.

Furthermore, some VLSI-specific implementation details (delayed carry adder, re-use of MUX tree and logic) are provided.

Keywords: Reduce-by-Feedback, modular multiplication, Montgomery multiplication, timing analysis, differential power analysis.

1 Introduction

RSA, Diffie-Hellman (over $\mathbb{F}_p$), and elliptic curve schemes (over $\mathbb{F}_p$) use modular multiplication as their computational kernel. This is usually implemented as Montgomery multiplication [12] (1985), which is fast and has timing independent of the values. Montgomery treats the bits of the first factor to be multiplied from the LSB towards the left, and works with the residue classes $[x \cdot (2^L)^{-1}] \mod N$, where $[x]$ are the standard residue classes, and L is the length (in bits) of the operands, e.g. $L = \lceil \log_2(N) \rceil$.

There exists, however, an algorithm that avoids the mapping from $[x]$ to $[x \cdot (2^L)^{-1}] \mod N$, by working the bits of the first factor from MSB downwards to the right: Reduce-by-Feedback [15,16,20] (1987) (Sections 3 and 4).

The Reduce-by-Feedback algorithm preserves the immunity against timing attacks (Section 5), the constant shift amount of 1,2,3, or 4 bits per clock cycle, depending on the implementation effort, and all other advantages of Montgomery multiplication.

E. Prouff and P. Schaumont (Eds.): CHES 2012, LNCS 7428, pp. 463–475, 2012.

Additionally, a DPA attack against RSA implemented by Montgomery multiplication or Reduce-by-Feedback (Section 6), can be avoided by a modification of Reduce-by-Feedback (Section 7). This modification can not be applied to Montgomery multiplication, as far as we can see.

An overview about implementations of modular multiplication is given in [6].

2 Multiplication by Shift-and-Add

It is worthwhile to recall the Shift-and-Add algorithm, since Reduce-by-Feedback is constructed completely analogously, retaining its properties:

Algorithm 1. *Shift-and-Add*
Parameters:
operand length l [e.g. $= 1024$]
shift length per clock cycle z [e.g. $= 3$], with $Z := 2^z$ [e.g. $= 8$]
IN $A, B < 2^l$ *// factors, where $A = \sum_{k=0}^{l-1} a_k 2^k = \sum_{k=0}^{\lceil l/z \rceil - 1} \alpha_k Z^{\lceil l/z \rceil - 1 - k}$*
OUT M *// product $M = A \cdot B$*
Algorithm:
$M := 0$
FOR $k := 0$ TO $\lceil l/z \rceil - 1$
    $M := (M << z) + \alpha_k \cdot B$
ENDFOR

Some trivial, but remarkable properties of Shift-and-Add are:

  (i) The coefficient α_k lies in the range $\{0, 1, \ldots, Z - 1\}$, thus Z possible multiples of B are to be taken into account. Note that α_0 is the MSB part.
 (ii) We have exactly $\lceil l/z \rceil$ cycles to go in the loop, a fixed timing.
(iii) It is sufficient to store the multiples for $\alpha \geq Z/2$, and $\alpha = 0$, by supplying shifted copies for the smaller cases, e.g. cases $3 \cdot B, 6 \cdot B$ (for $\alpha = 3$ and 6) from $12 \cdot B$, $\alpha = 12$ for $z = 4, Z = 16$.
 (iv) The "1-off trick" [15,16,20,7]: A further saving is possible by replacing the odd multiples by the next higher even ones, and subtract $Z \cdot B$ in the next clock cycle:
$$((\alpha_k \cdot B) << z) + \alpha_{k+1} \cdot B = (((\alpha_k + 1) \cdot B) << z) + (\alpha_{k+1} - Z) \cdot B.$$
Putting $C_{\alpha,k} := 1$, iff α_k is odd, 0 otherwise, we then set

$$\overline{\alpha}_k := \alpha_k + C_{\alpha,k} - Z \cdot C_{\alpha,k-1} \text{ and } M := (M << z) + \overline{\alpha}_k \cdot B.$$

Hence, (iii) and (iv) combined leave us with the necessary multiples $\pm(Z/2 + 2), \pm(Z/2 + 4), \ldots, \pm Z, 0$, where we first applied (iv), then (iii).

While these are still $Z/2$ choices, and including shifts we again have Z multiples, as are necessary by using base Z, the $\pm$ comes for free in hardware as two's complement, taking the inverse outputs $\overline{Q}$ of the register latches. Only the $Z/4$ multiples $Z/2 + 2, Z/2 + 4, \ldots, Z$ have to be stored in hardware.

3 Reduce-by-Feedback

History:
This algorithm was first introduced in 1987 by Vielhaber [15],
also [16], *and in 1990 the German patent* [20] *was granted.*
Beth and Gollmann describe the algorithm in [2], *in 1989.*
Benaloh and Dai rediscovered the algorithm and gave a talk at
the Rump Session of CRYPTO'95 [1], *patenting it in the United*
States in 1998 as [19].
Finally, Jeong and Burleson re-re-discovered the algorithm in
1997, when it appeared in the journal article [7].

3.1 The Algorithm

The original idea stems from the analogy with LFSR's: The z bits running off
in front for each Shift-and-Add step are fed back into the accumulator:

Let $K \equiv 2^{l+2z+1} \mod N, 0 \le K < N$.

Also, partition M into its lower $l + z + 1$ bits and the higher part,
$M_H = \lfloor M/2^{l+z+1} \rfloor, M_L = M \mod 2^{l+z+1}, M = (M_H|M_L)$. Then

$$(M_H|M_L) << z = M_H \cdot 2^{l+2z+1} + M_L \cdot 2^z \equiv M_H \cdot K + M_L \cdot 2^z \mod N$$

The *Shift-and-Add-with-Reduce-by-Feedback* algorithm now runs as follows (note
that μ_k is M_H):

Algorithm 2. *Shift-and-Add-with-Reduce-by-Feedback*
$M := 0, C_{\alpha,-1} := 0, C_{\mu,-1} := 0$
FOR $k := 0$ TO $\lceil l/z \rceil - 1$
$\quad C_{\alpha,k} := \alpha_k \mod 2, \overline{\alpha}_k := \alpha_k + C_{\alpha,k} - Z \cdot C_{\alpha,k-1}$
$\quad \mu_k := \lfloor M/2^{l+z+1} \rfloor$
$\quad C_{\mu,k} := \mu_k \mod 2, \overline{\mu}_k := \mu_k + C_{\mu,k} - Z \cdot C_{\mu,k-1}$
$\quad M := ((M \mod 2^{l+z+1}) << z) + \overline{\alpha}_k \cdot B + \overline{\mu}_k \cdot K$
ENDFOR
// $M = A \cdot B \mod N, 0 \le M < 2^l$ (*not necessarily* $M < N$)

Reduce-by-Feedback preserves the 4 properties of Shift-and-Add:

(*i*) The standard range for the multiples of K is $\mu_k \in \{-1, 0, 1, \ldots, 2^z\}$.
(*ii*) The FOR loop excutes exactly $\lceil l/z \rceil$ times, each run comprising a shift and
2 additions. This amount is independent of the values.
(*iii*) The multiples of K required according to (*i*) can be restricted to $\mu_k \in$
$\{0\} \cup \{Z/2 + 1, \ldots, Z\}$, supplying the others by shifting.
(*iv*) The odd multiples can be traded for negative ones, applying the "1-off
trick". Hence in total we need $\alpha_k, \mu_k \in \{0, \pm(Z/2+2), \pm(Z/2+4), \ldots, \pm Z\}$,
with 0 and $\pm$ for free in hardware.

Reduce-by-Feedback is thus *completely analogous* to Shift-and-Add.

3.2   Overflow Avoidance

We check overflow avoidance by proving the inequality

$$-Z \leq \overline{\mu}_k = M_H \leq Z, \forall k$$

by induction.

We have $0 \leq B, K < 2^l$ and $0 \leq M_L < 2^{l+z+1}$. Including the "1-off trick", we require $-Z \leq \overline{\alpha}_k, \overline{\mu}_k \leq Z$, and $\overline{\alpha}_k, \overline{\mu}_k$ being even. This is true for $\overline{\alpha}_k, \forall k$ and can be assumed for $\overline{\mu}_0 = 0$ at the start.

Then

$$-1 \cdot 2^{l+z+1} \leq (M_L << z) + \overline{\alpha}_k \cdot B + \overline{\mu}_k \cdot K < 2^{l+z+1} \cdot (2^z + 1/2 + 1/2)$$

i.e. $-1 \leq M_H^+ \leq 2^z$. As with C_α, we put $C_{\mu,k} = 1$, if μ_k is odd and has to be increased by the "1-off trick", $C_{\mu,k} = 0$ otherwise, and then have

$$\overline{\mu}_{k+1} = M_H^+ + C_{\mu,k+1} - C_{\mu,k} \cdot Z \in \{-Z, -Z+2, \ldots, Z-2, Z\},$$

which proves the induction step.

Therefore, the accumulator M never exceeds the range $-1 \cdot 2^{l+z+1} \leq M < (Z+1) \cdot 2^{l+z+1}$ and the even multiples of B up to $\pm Z \cdot B$ are sufficient.

4   Implementation Issues

4.1   Re-use of MUX Tree

Since the choice of the correct multiples, $\overline{\alpha}_k \cdot B + \overline{\mu}_k \cdot K$, is completely analogous for B and for K, we may use the same logic (calculation of decision variables, MUX tree, shifter) first for the part $\overline{\alpha}_k \cdot B$ (in one half clock cycle), and then for $\overline{\mu}_k \cdot K$ (in the other half clock cycle), as described in [15,16,20].

This 1:1 analogy between Shift-and-Add and Reduce-by-Feedback was the central idea of the algorithm and leads to very compact VLSI designs:

Mapping the implementation in [16] to current 65 nm rules, and naïvely assuming a shrinking factor $(65/1000)^2$, this would roughly lead to $13 \cdot (1000/65)^2 \approx$ 3000 bits/mm^2, or a full 4096 bit RSA with control unit on about 1.5 square millimeters.

4.2   Delayed-Carry-Adder

Brickell [3] introduced the Delayed-Carry-Adder, a chain of halfadders instead of full adders, and where the resulting double register has the property $c_{i+1} \wedge s_i = 0$.

The advantage of the Delayed-Carry-Adder is the locality of carries. We do not have to wait for carry propagation and thus addition is fast. At the end of a multiplication, however, the final Delayed-Carry result has eventually to be added into the standard form, which may lead to a timing attack (see Section 5).

Nevertheless, without carry-save techniques, this carry propagation problem would arise at each addition intead of just once at the end.

Table 1. Boolean logic for Delayed-Carry adder

Standard Boolean function	Using NAND-2 gates
$d_i := s_i \wedge b_i,\quad t_i := s_i \oplus b_i$	$\overline{d}_i := \overline{s_i \wedge b_i},\quad t_i := s_i \oplus b_i$
$e_i := t_i \wedge k_i,\quad u_i := t_i \oplus k_i$	$\overline{e}_i := \overline{t_i \wedge k_i},\quad u_i := t_i \oplus k_i$
$f_i := c_i \vee d_{i-1}$ (which are not both 1, due to $c_{i+1} \wedge s_i = 0$)	$f_i := \overline{c}_i \wedge \overline{d}_{i-1}$
$g_{i+1} := u_i \wedge f_i,\ v_i := u_i \oplus f_i$	$\overline{g}_{i+1} := \overline{u_i \wedge f_i},\ v_i := u_i \oplus f_i$
$h_{i+1} := e_i \vee g_i$ (not both 1: $e_i = 1 \Rightarrow u_i = 0$)	$h_{i+1} := \overline{\overline{e}_i \wedge \overline{g}_i}$
$c_{i+1}^+ := v_i \wedge h_i,\ s_i^+ := v_i \oplus h_i$	$\overline{c}_{i+1}^+ := \overline{v_i \wedge h_i},\ s_i^+ := v_i \oplus h_i$

Also, we have to take extra care when dealing with the upper part M_H (μ_k) of the accumulator, see next subsection.

The addition $(c,s)^+ := (c,s) + b + k$ usually requires two full adders in carry-save technique. With Brickell's delayed-carry scheme, we add as follows, where (c,s) is the delayed-carry register, (b) and (k) are the terms $\alpha \cdot B$ and $\mu \cdot K$, respectively. t, u, v are intermediate sum terms, d, e, f, g, h are intermediate carries. In NAND-logic, the variable c will only be used invertedly.

This leaves us with 4 halfadders plus two OR's, the equivalent of two full adders. We thus need the same number of gate equivalents, but the result now has the Delayed-Carry Property $c_{i+1} \wedge s_i = 0$, which is crucial, when calculating μ_k (see next paragraph).

4.3 How to Keep the Invariant When Using the Delayed-Carry Representation

We feed back the z leading MSB bits, which have to be in the range $-1, 0, \ldots, Z$ (assumption for overflow avoidance).

With delayed-carry, we have $c_{i+1} \wedge d_i = 0$, hence the following patterns are the highest values possible (shown for the case $z = 3, Z = 8$), Table 1.

As can be seen in Table 2, cases 4 and 5 would lead to an overflow ($M_H > Z = 8$) due to the Delayed-Carry representation. We avoid this by looking further to the right and (cases 1 and 2) detect and avoid a subsequent overflow already in the previous cycle.

4.4 Fast Computation of MUX Control Variables

It is crucial that the clock frequency depends only on the data propagation within the bit slices, and not on the control module.

In each clock cycle, we add $\overline{\alpha} \cdot B$ and $\overline{\mu} \cdot K$ to the delayed-carry register (c,s). In the two previous half cycles, we choose these multiples by the same hardware (MUX, shifter, logic), which is not time-critical for $\overline{\alpha} \cdot B$, since in principle, all values α are known. On the other hand, $\overline{\mu}$ depends on the addition just performing in the half cycle $(k+1, H)$, while the next multiple $\overline{\mu} \cdot K$ must be selected in $(k+1, L)$. We proceed as follows (see [15][16]):

Having calculated $(M_H)_{k+1}$ in half cycle $(k+1, H)$, immediately afterwards we need $\overline{\mu}_{k+1}$ in half cycle $(k+1, L)$. We therefore have to precompute as much as

Table 2. MSB sum of Delayed-Carry-Adder

1	$c_{2^{l+z+1}+2,1,0;-1,-2}$	0 0 0 | 0 1	sum is 8 with carry, OK, avoids case 4
	$s_{2^{l+z+1}+2,1,0;-1,-2}$	1 1 1 | 1 1	
	$M_{H,2^{l+z+1}+3,2,1,0;-1,-2}$	1 0 0 0 | 0 0	
2	$c_{2^{l+z+1}+2,1,0;-1,-2}$	0 0 0 | 1 1	sum is 8 with carry, OK, avoids case 5
	$s_{2^{l+z+1}+2,1,0;-1,-2}$	1 1 1 | 1 0	
	$M_{H,2^{l+z+1}+3,2,1,0;-1,-2}$	1 0 0 0 | 0 1	
3	$c_{2^{l+z+1}+2,1,0;-1,-2}$	0 0 1 | 1 1	sum is 8, OK
	$s_{2^{l+z+1}+2,1,0;-1,-2}$	1 1 1 | 0 0	
	$M_{H,2^{l+z+1}+3,2,1,0;-1,-2}$	1 0 0 0 | 1 1	
4	$c_{2^{l+z+1}+2,1,0;-1,-2}$	0 1 1 | 1 1	sum is 9, to be avoided by case 1
	$s_{2^{l+z+1}+2,1,0;-1,-2}$	1 1 0 | 0 0	
	$M_{H,2^{l+z+1}+3,2,1,0;-1,-2}$	1 0 0 1 | 1 1	
5	$c_{2^{l+z+1}+2,1,0;-1,-2}$	1 1 1 | 1 1	sum is 11, to be avoided by case 2
	$s_{2^{l+z+1}+2,1,0;-1,-2}$	1 0 0 | 0 0	
	$M_{H,2^{l+z+1}+3,2,1,0;-1,-2}$	1 0 1 1 | 1 1	

Table 3. Precomputation of control variables

Clock cycle	Half cycle	Selection	Computation
k	H	$\overline{\alpha}_k \cdot B$	$(M_H\|M_L)_k := \ldots$
k	L	$\overline{\mu}_k \cdot K$	
$k+1$	H	$\overline{\alpha}_{k+1} \cdot B$	$(M_H\|M_L)_{k+1} := ((M_L)_k << z) + \overline{\alpha}_k \cdot B + \overline{\mu}_k \cdot K$
$k+1$	L	$\overline{\mu}_{k+1} \cdot K$	

we can: In (k, H), we already compute a partial sum $(M_H)_k \cdot Z + \overline{\alpha}_k \cdot B$ for the bit positions of M_H, including 2 more bits to the right, as described in the previous paragraph, to avoid possible overflow in the future. We then add the part $\overline{\mu}_k \cdot K$ in (k, L), for these bit positions. We also add 0,1,2,3 to obtain the four possible final values for $\overline{\mu}$, and for all four possibilities, we precompute the MUX control variables for the next choice of $\overline{\mu} \cdot K$. The only missing part are up to 3 carries from the lower part, M_L, of the sum. In this way, terminating $(k + 1, H)$, we obtain the new sum $(M_H)_{k+1}$, and immediately select the MUX-control values to fetch $\overline{\mu}_{k+1} \cdot K$ in $(k + 1, L)$ from the 4 precomputed sets.

The full-custom implementation in [16] achieves a control unit faster than the bit slices. We have this design goal also for the FPGA implementation. It remains to be verified though, whether this will apply or whether the FPGA architecture (6-input LUTs instead of a chain of half-adders) will actually make the bit slices even faster.

5   Timing Attacks

We may trivially find the Hamming weight of the exponent by just counting multiplications and squarings. To prevent this, we would have to either do both in parallel, wasting space, or introduce dummy multiplications, wasting time.

In any case, this issue is independent of the implementation of modular multiplication.

As Kocher [9] points out, however, apart from the Hamming weight, we can indeed recover the full exponent — provided that multiplication time is sensitive on the values, some lead to faster calculation than others.

The attack by Schindler [13] on Montgomery multiplication can easily be overcome by introducing a dummy subtraction, costing a single clock cycle. There is no analogue of this attack against Reduce-by-Feedback.

Therefore, with Reduce-by-Feedback as well as with Montgomery multiplication (+dummy), timing attacks are ruled out during the modular multiplication, taking in any case exactly $\lceil l/z \rceil$ cycles. The result is then in a delayed-carry- or carry-save-register.

The final carry however, may introduce timing information. Either

(i) we use carry-look-ahead logic, space-intensive, or
(ii) we keep the result in delayed-carry-form, space-intensive, or
(iii) we wait until the longest carry chain ($l + z$ bits) will have passed, time-intensive, or
(iv) we use interrupt techniques, efficient, but time-variant.

The variation due to carries in case (iv) is the only potential information leak for a timing attack. This is though independent of Reduce-by-Feedback (or Montgomery multiplication), but a consequence of using carry-save or delayed-carry techniques.

Up to here, this concerned the modular multiplication as building block. As to the exterior loop, exponentiation, Square-and-Multiply, there must of course be the same number of clock cycles between any two multiplications and/or squarings to avoid a timing/DPA mix just concentrating on the transition between two of them. Otherwise, use the double-add scheme by Joye [8] in the multiplicative version "square-multiply", wasting time though. However, this does not concern modular multiplication proper, but exponentiation.

6   How to Break RSA with Differential Power Analysis

Both Reduce-by-Feedback and Montgomery multiplication make RSA susceptible to the following DPA [10] attack. For other attacks against RSA see the power attack by Yen et al. [18], and the timing attack by Miyamoto et al. [11].

Now to our DPA attack: Every multiplication (in this section this includes squarings) starts with an empty accumulator $M = 0$, and also a zero adjustment value μ_0 (both for Reduce-by-Feedback and Montgomery multiplication).

The first factor, A, will on average start with z zeroes every Z'th multiplication. In this case, $\alpha_0 \cdot B = 0$, while the term will be nonzero otherwise.

For $\mu_0 = \alpha_0 = 0$ (in terms of Reduce-by-Feedback), we compute

$$M^+ = (M << z) + \alpha_0 \cdot B + \mu_0 \cdot K = (0 << z) + 0 + 0 = 0,$$

hence the register M was empty before the step and is overwritten again with zeroes.

If, on the other hand, $\alpha_0 \neq 0$,

$$M^+ = (M << z) + \alpha_0 \cdot B + \mu_0 \cdot K = 0 + \alpha_0 \cdot B + 0 \neq 0,$$

and roughly half of the flip-flops of register M will change state from 0 to 1. This gives a strong difference in power consumption during this first cycle of the multiplication, compared to $M^+ = M = 0$, a "point-of-interest" in terms of the template attack [4].

We focus only on this information (about half a bit for $z = 3$) and will assume that we can distinguish between $A < \frac{1}{Z} \cdot 2^l$, case $\alpha_0 = 0$, and $A \geq \frac{1}{Z} \cdot 2^l$, case $\alpha_0 \neq 0$, for every multiplication step.

We assume that we have access to the public RSA modulus N and to several known ciphertexts $\chi_1, \chi_2, \ldots$. We observe the decryptions $\chi_i^d \mod N$ for a fixed unknown exponent d (unblinded case). We compute the multiplication chains for all 2^L possible initial segments of d of a certain length L. These segments will consist of L squarings and furthermore L', $0 \leq L' \leq L$, multiplications, depending on the number of 1's in the segment. For each hypothetical segment, we do the corresponding calculations (multiplications and squarings) and memorize the sequence of initial coefficients α_0 of length $L + L'$.

We now observe the actual H/W decryption and obtain a sequence $\{= 0, \neq 0\}^{2L}$, whose first $L+L'$ components we check against all possible initial segments.

The per-symbol information is $-\left(\log_2\left(\frac{1}{Z}\right) \cdot \frac{1}{Z} + \log_2\left(\frac{Z-1}{Z}\right) \cdot \frac{Z-1}{Z}\right) = 1.0$, 0.811, 0.544, and 0.337 bits for $z = 1, 2, 3$, and 4, respectively. Hence, 1,2,2,3 decryptions χ_i should be sufficient.

The crucial case is, however, the large set of initial segments leading to the sequence $(\neq 0)^{L'}$, in the case that this is the actual observation. We expect this to happen with probability $\left(\frac{Z-1}{Z}\right)^{L'}$, thus leading to $\left(\frac{Z-1}{Z}\right)^{L'} \times 2^L$ cases. We set $L' := 0.5L$ from now on and consider C decryptions $\chi_1, \ldots, \chi_C$, whose outcomes ($\alpha = 0$ or $\alpha \neq 0$) we assume independent.

The expected number of segments which always lead to $(\neq 0)^{L+L'}$, in all C decryptions, is then

$$\left(\frac{Z-1}{Z}\right)^{1.5L \cdot C} \times 2^L.$$

To have uniqueness, we want this size down to 1, hence $\left(\frac{Z-1}{Z}\right)^{1.5L \cdot C} \times 2^L = 1$ or $C = -1/(1.5 \log_2(7/8))$, which gives $C = 0.67, 1.61, 3.47$, and 7.16 for $z = 1, 2, 3$, and 4, respectively. Therefore, $C \geq Z/2$ samples (asymptotically $Z \cdot \ln(2)/1.5$ samples) are necessary.

We now compare the $C \geq Z/2$ sequences actually observed from $\{= 0, \neq 0\}^{L'}$ with all initial segments of d, saving only the matches, where under ideal conditions, only a single match should occur. These matches are then extended, compared to the observations, and so forth, until recovering the full secret RSA exponent d.

Certainly, there will be noise in our measurements, so quite some more than $Z/2$ ciphertexts will be needed under realistic conditions.

And that breaks RSA!

7 How to Repair Reduce-by-Feedback to Avoid the DPA Attack on RSA

In this section, we suggest modifications to strengthen Reduce-by-Feedback against Differential Power Analysis.

As we have seen, the initial all-zero phase is exploitable by DPA. We can neither avoid $\mu_0 = 0$ in the first step, nor $\alpha_0 = 0$ once in a while — if using directly the z bits of A, and M_H, respectively.

We can, however, avoid $M = 0 \mapsto M^+ = 0$ in this cases, by using the same "1-off" trick as in property (iv) of Shift-and-add and Reduce-by-Feedback:

$$0 = 1 + (-1)$$

We just never add a zeroth multiple, but instead add B once, and subtract it (Z-fold) in the next step. This brings us back to zero every second step. Assuming B to have 50% 1's, the effect is flipping back-and-forth half of the register bits.

To be explicit, we use the case $z = 3, Z = 8$ in the sequel. The columns "old" show the regular case [15,16,20], applying properties (iii) and (iv), including a multiple 0. We also adjust the treatment of values $\Sigma = -1, 1, 2$, and 3 to minimize the information flow (bias) from $\overline{\alpha}, \overline{\mu}$ to C, A, M_H, see columns "new".

Note that we still use the "1-off trick", however in an irregular way, so that the required multiples are no longer just the even ones. In any case, all required multiples can still be obtained by shifting from only $Z/4$ values, e.g. 6, and 8.

Table 4. Old and new multiples $\overline{\alpha}_k, \overline{\mu}_k$

C_α, α_k, / C_μ M_H	Σ	$\overline{\alpha}_k$, C^+ / $\overline{\mu}_k$(old)	$\overline{\alpha}_k$, C^+ / $\overline{\mu}_k$(new)		C_α, α_k, / C_μ M_H	Σ	$\overline{\alpha}_k$, C^+ / $\overline{\mu}_k$(old)	$\overline{\alpha}_k$, C^+ / $\overline{\mu}_k$(new)
0 -1	-1	0 1	-1 0		1 -1	-9	-8 1	-8 1
0 000	0	0 0	1 1		1 000	-8	-8 0	-8 0
0 001	1	2 1	1 0		1 001	-7	-6 1	-6 1
0 010	2	2 0	3 1		1 010	-6	-6 0	-6 0
0 011	3	4 1	3 0		1 011	-5	-4 1	-4 1
0 100	4	4 0	4 0		1 100	-4	-4 0	-3 1
0 101	5	6 1	6 1		1 101	-3	-2 1	-3 0
0 110	6	6 0	6 0		1 110	-2	-2 0	-1 1
0 111	7	8 1	8 1		1 111	-1	0 1	-1 0
0 1000	8	8 0	8 0		1 1000	0	0 0	1 1

Description of Table 4, Multiples $\overline{\alpha}_k, \overline{\mu}_k$ from A, M_H

The original α_k (bits from A), may vary from 0 to $Z - 1 = 7$, M_H (upper part of M) may vary from -1 to $Z = 8$. Applying property (iv), a previous odd value was adjusted by $+1$, hence we may have to adjust now $(C_\alpha, C_\mu = 1)$ by $-Z = -8$, giving an overall sum Σ between -9 and $+8$. Σ is now split into a multiple actually added, $\overline{\alpha}_k, \overline{\mu}_k$, minus a possible new carry $C_\alpha^+, C_\mu^+ = 1$. In the original scheme, the multiples were $0, \pm 2, \pm 4, \pm 6$, and ± 8, while we now have $\pm 1, \pm 3, \pm 4, \pm 6$, and ± 8, avoiding zero.

Observe that in both cases, all multiples are shifts and negatives of just the two multiples 6 and 8. Hence, even after the modification, only these 2 multiples have actually to be stored (and computed).

Description of Table 5, Bias

There is now less bias between Σ, C and the bits of α_k, μ_k. We define bias as $\mathrm{pr}(1) - \mathrm{pr}(0)$ (not as $\mathrm{pr}(1) - \frac{1}{2}$).

We assume probability $1/8$ each for $\alpha = 0, \ldots, 7$. For, μ, by folding 3 equidistributions over the intervals $[0, 8[, [-1/2, 1/2[,$ and $[-1/2, 1/2[,$ we obtain probability $1/8$ each for $\mu = 1, \ldots, 6$, probability $5/48$ for $\mu = 0$ and 7, and probability $1/48$ for $\mu = -1$ and 8, each comprising the interval $M_H \in [\mu, \mu + 1[$.

$C = 0$ and $C = 1$ are each assigned probability $1/2$.

We consider the bias of the bits of C and Σ (internal values revealing information about the actual contents of A and M), conditional on certain value sets for $\overline{\alpha}, \overline{\mu}$, namely zero, positive, shifts of 8, and shifts of 6 (potentially observable by DPA).

We now have probability zero for $\overline{\alpha} = 0$, which was $1/8$ before. Neither can we infer anything on observing a shift of 8 (1,2,4,8) vs. a shift of 6 (3,6).

What remains is a bias from $\overline{\alpha}$ positive to $C = 0$ (which is almost a tautology). The fact $\overline{\alpha} > 0$, however, is a mix of the cases $\overline{\alpha} = 1, 2, 3, 4, 6, 8$, far more difficult to analyze by DPA than the distinction $\alpha = 0$ vs. $\alpha \neq 0$, now ruled out.

Table 5. Bias of C, Σ, conditional on $\overline{\alpha}, \overline{\mu}$

	$C\vert\alpha$	$\Sigma_2\vert\alpha$	$\Sigma_1\vert\alpha$	$\Sigma_0\vert\alpha$	$C\vert\mu$	$\Sigma_2\vert\mu$	$\Sigma_1\vert\mu$	$\Sigma_0\vert\mu$
$\overline{\alpha}, \overline{\mu} = 0$ new=old	0	0	0	0	0	0	0	0
$\overline{\alpha}, \overline{\mu} > 0$ new	-1	0	0	0	$-23/24$	$1/24$	$1/24$	$1/24$
$\overline{\alpha}, \overline{\mu} > 0$ old	-1	$1/7$	$1/7$	$1/7$	-1	$-2/21$	$-2/21$	$-2/21$
$\overline{\alpha}, \overline{\mu} \in \{\pm1, \pm2, \pm4, \pm8\}$ new=old	0	0	0	0	0	0	0	0
$\overline{\alpha}, \overline{\mu} \in \{\pm3, \pm6\}$ new=old	0	0	0	0	0	0	0	0

We now give the complete Shift-and-Add-with-Reduce-by-Feedback algorithm for $z = 3$, including the mentioned modifications, and the final adjustment from delayed-carry to a single register.

Algorithm 3. *Shift-and-Add-with-Reduce-by-Feedback*
IN *A,B,N* // each at most l bits long, N odd
OUT M // the product $M = A \cdot B \mod N, 0 \leq M < 2^l$ (not necessarily $M < N$)
// M is actually stored in a delayed-carry register (c, s).   Table 2 :

```
const mult[-9..8] = (-8,-8,-6,-6,-4,-3,-3,-1,-1,1,1,3,3,4,6,6,8,8)
const C[-9..8] = (1,0,1,0,1,1,0,1,0,1,0,1,0,0,1,0,1,0)
```

$M := 0, C_\alpha := 0, C_\mu := 0$
FOR $k := 0$ TO $\lceil l/z \rceil - 1$
  $\alpha := 4 \cdot a_{3k+2} + 2 \cdot a_{3k+1} + 1 \cdot a_{3k} - 8 \cdot C_\alpha$
  $\overline{\alpha} := \mathtt{mult}[\alpha], C_\alpha := \mathtt{C}[\alpha]$
  $\mu := \lfloor M/2^{l+z+1} \rfloor - 8 \cdot C_\mu$
  $\overline{\mu} := \mathtt{mult}[\mu], C_\mu := \mathtt{C}[\mu]$
  $M := ((M \bmod 2^{l+z+1}) << z) + \overline{\alpha} \cdot B + \overline{\mu} \cdot K$
ENDFOR

// Multiply by 2^9
FOR $k := 1$ TO 3
  $\overline{\alpha} := -8 \cdot C_\alpha, C_\alpha := 0$
  $\mu := \lfloor M/2^{l+z+1} \rfloor - 8 \cdot C_\mu$
  $\overline{\mu} := \mathtt{mult}[\mu], C_\mu := \mathtt{C}[\mu]$
  $M := ((M \bmod 2^{l+z+1}) << z) + \overline{\alpha} \cdot B + \overline{\mu} \cdot K$
ENDFOR
// Divide by 2^9, leaving $M < 2^l$
FOR $k := 1$ TO 9
  IF M is odd $N' := N$ else $N' := 0$
  $M := (M + N') >> 1$
ENDFOR
$M := C + S$ *// the final carry, using e.g. carry-look-ahead or interrupts*

Although N' is either N or zero in the last 9 steps, the result $(M + N') >> 1$ will differ from M in about half of the bits in both cases, making DPA based on flip-flop recharges extremely difficult.

Unfortunately (or luckily, if we want to promote Reduce-by-Feedback), we see no way to implement this modification with Montgomery multiplication:

The two properties (*iii*) and (*iv*) of Shift-and-Add-with-Reduce-by-Feedback can be mapped to Montgomery multiplication as

(*iii*) use shifted multiples (of N) to compensate results terminating in ...0, and
(*iv*) use the 2's complement of multiples of N terminating in ...01 to account for those terminating in ...11.

Again, we have a total of $Z/4$ multiples physically to be stored, those multiples of N terminating in ...01. However, there seems to be no workaround to replace the do-nothing (subtract $0 \cdot N$) in the case ...000 by anything else.

Conclusion

We have (re-)introduced the Reduce-by-Feedback algorithm, which can be seen as "Montgomery on the high end", but was inspired by LFSR feedback.

Reduce-by-Feedback is immune against timing attacks (as is Montgomery multiplication with dummy subtraction), with the possible exception of the final carry run.

We recalled how to avoid physically storing multiples, by providing shifted multiples, and using the "1-off trick", saving 75%.

RSA can be broken by DPA, when executed with Montgomery multiplication, or the unmodified Reduce-by-Feedback.

We proposed modifications for the choice of multiples of both the second factor B and the feedback value $K \equiv 2^{l+2z+1} \mod N$. These modifications diminish bias, avoid the multiple zero, and thereby avoid the accumulator being zero in consecutive time steps. These effects of the modification will diminish the susceptibility of Reduce-by-Feedback to Differential Power Analysis considerably. In particular, the DPA attack of Section 6 on RSA, exploiting the partial multiplier zero, is no longer possible.

Replacing a multiple zero with "$1 + (-1)$" by the "1-off trick" is not possible for Montgomery multiplication. Therefore, the DPA attack against RSA with Montgomery multiplication is still possible.

We have therefore shown that *Reduce-by-Feedback-with-Shift-and-Add* is the method of choice, to implement a timing-resistant and DPA-aware modular multiplication.

References

1. Benaloh, J., Dai, W.: Fast Modular Reduction. In: CRYPTO 1995 Rump Session (1995)
2. Beth, T., Gollmann, D.: Algorithm engineering for public key algorithms. IEEE J. SAC 7(4), 458–466 (1989)
3. Brickell, E.F.: A Fast Modular Multiplication Algorithm with Applications to Two Key Cryptography. In: Proc. CRYPTO 1982, pp. 51–60. Plenum Press (1983)
4. Chari, S., Rao, J.R., Rohatgi, P.: Template Attacks. In: Kaliski Jr., B.S., Koç, Ç.K., Paar, C. (eds.) CHES 2002. LNCS, vol. 2523, pp. 13–28. Springer, Heidelberg (2003)
5. Elbirt, A.J., Paar, C.: Towards an FPGA architecture optimized for public-key algorithms. In: The SPIE's Symposium on Voice, Video, and Communications, Boston (1999)
6. Guajardo, J., Kumar, S.S., Paar, C., Pelzl, J.: Efficient Software–Implementation of Finite Fields with Applications to Cryptography. Acta Appl. Math. 39, 75–118 (2006)
7. Jeong, Y.-J., Burleson, W.P.: VLSI array algorithms and architectures for RSA modular multiplication. IEEE Trans. VLSI Systems 5(2), 211–217 (1997)
8. Joye, M.: Highly Regular Right-to-Left Algorithms for Scalar Multiplication. In: Paillier, P., Verbauwhede, I. (eds.) CHES 2007. LNCS, vol. 4727, pp. 135–147. Springer, Heidelberg (2007)
9. Kocher, P.C.: Timing Attacks on Implementations of Diffie-Hellman, RSA, DSS, and Other Systems. In: Koblitz, N. (ed.) CRYPTO 1996. LNCS, vol. 1109, pp. 104–113. Springer, Heidelberg (1996)
10. Kocher, P.C., Jaffe, J., Jun, B.: Differential Power Analysis. In: Wiener, M. (ed.) CRYPTO 1999. LNCS, vol. 1666, pp. 388–397. Springer, Heidelberg (1999)
11. Miyamoto, A., Homma, N., Aoki, T., Satoh, A.: Enhanced Power Analysis Attack Using Chosen Message against RSA Hardware Implementations. In: ISCAS 2008, Seattle, pp. 3282–3285 (2008)
12. Montgomery, P.L.: Modular Multiplication without trial Division. Math. Comp. 44, 519–521 (1985)

13. Schindler, W.: A Timing Attack against RSA with the Chinese Remainder The-
    orem. In: Paar, C., Koç, Ç.K. (eds.) CHES 2000. LNCS, vol. 1965, pp. 109–124.
    Springer, Heidelberg (2000)
14. Sedlak, H., Golze, U.: An RSA Cryptography processor. In: Proc. Euromicro 1986,
    Microprocessing and Microprogramming, vol. 18, pp. 583–590 (1986)
15. Vielhaber, M.: Entwurf und Layout eines RSA-Koprozessors für Chipkarten.
    Diploma Thesis, TH Karlsruhe (KIT) (1987)
16. Vielhaber, M.: The Karlsruhe RSA Co-processor: ISDN Network Security by RSA
    encryption, E.I.S.S. Report 89/14a, European Institute for System Security, Karl-
    sruhe (1990)
17. Vielhaber, M.: Der Karlsruher RSA Koprozessor: Verschlüsseln mit RSA im ISDN-
    Netz, E.I.S.S. Report 89/14, European Institute for System Security, Karlsruhe
    (1990)
18. Yen, S.-M., Lien, W.-C., Moon, S.-J., Ha, J.C.: Power Analysis by Exploiting
    Chosen Message and Internal Collisions – Vulnerability of Checking Mechanism
    for RSA-Decryption. In: Dawson, E., Vaudenay, S. (eds.) Mycrypt 2005. LNCS,
    vol. 3715, pp. 183–195. Springer, Heidelberg (2005)
19. USPTO Patent US5724279: Computer-implemented method and computer for per-
    forming modular reduction, Applicants: Josh Benaloh, Wei Dai
20. Deutsches Patentamt, DE P 3924344 Multiplikations-/Reduktionseinrichtung.
    Vielhaber, Michael Johannes, Anmelder (1992)

Side Channel Attack to Actual Cryptanalysis: Breaking CRT-RSA with Low Weight Decryption Exponents

Santanu Sarkar and Subhamoy Maitra

Applied Statistics Unit, Indian Statistical Institute,
203 B. T. Road, Kolkata 700 108, India
sarkar.santanu.bir@gmail.com, subho@isical.ac.in

Abstract. Towards the cold boot attack (a kind of side channel attack), the problems of reconstructing RSA parameters when (i) certain bits are unknown (Heninger and Shacham, Crypto 2009) and (ii) the bits are available but with some error probability (Henecka, May and Meurer, Crypto 2010) have been considered very recently. In this paper we exploit the error correction heuristic proposed by Henecka et al to show that CRT-RSA schemes having low Hamming weight decryption exponents are insecure given small encryption exponents (e.g., $e = 2^{16} + 1$). In particular, we show that the CRT-RSA schemes presented by Lim and Lee (SAC 1996) and Galbraith, Heneghan and McKee (ACISP 2005) with low weight decryption exponents can be broken in a few minutes in certain cases. Further, the scheme of Maitra and Sarkar (CT-RSA 2010), where the decryption exponents are not of low weight but they have large low weight factors, can also be cryptanalysed. We also identify a few modifications of the error correction strategy that provides significantly improved experimental outcome towards the cold boot attack.

Keywords: Cold Boot Attack, CRT-RSA, Cryptanalysis, Error Correction, Exponents, Hamming Weight, RSA.

1 Introduction

Side Channel Attack. Side channel cryptanalysis is now a quite popular technique for evaluating cryptographic schemes and this method usually considers additional information available from the physical implementation of a cryptosystem, rather than exploiting the theoretical weaknesses of the algorithm itself. The additional information may be obtained from timing information, power consumption, electromagnetic leaks etc. and the attack may very well exploit technical knowledge of the internal operation of the system on which the algorithm is implemented. The initial research in this area is pioneered by Kocher [18].

Recently, the idea of cold-boot attack has been presented in [12] that shows it is possible to exploit degraded data from the computer memory to attack cryptosystems such as DES, AES, RSA etc. This idea has been studied in more

E. Prouff and P. Schaumont (Eds.): CHES 2012, LNCS 7428, pp. 476–493, 2012.

detail in [14] that shows that if certain percentage of bits of the RSA secret key are available, then it is possible to reconstruct the complete secret key (or in other words, it is possible to factorize the RSA modulus). Subsequently, in [13], a model has been considered, where the bits of the secret key are available with some probability of error. In this paper we study the work of [13] in more detail.

In general, the side channel attacks use the existing cryptanalytic techniques with additional (side channel) information. In contrast, in this paper we exploit the algorithm developed for side channel attacks [13], that is applied for a direct attack on certain versions of CRT-RSA with no extra hints or other information. Further, we also provide certain modifications on the algorithm of [13] to (heuristically) improve the results of [13]. This improved strategy can immediately be used for the side channel cryptanalysis presented in [13] which is related to cold-boot attack [12].

We also like to refer the recent paper [6] related to noisy factoring where new attacks have been proposed whose running time is essentially the "square root" of exhaustive search. In this case [6, Section 4], that attack considers that the noisy version of one of the RSA primes is available. However, the cold-boot attack model that we consider here, is different from [6] as the noisy versions of more than one secret parameters of RSA variants are in the hand of cryptanalyst.

RSA. In the seventies, the path-breaking idea of public key cryptosystem has been introduced by Diffe and Hellman [10] and as an outstanding follow-up, RSA public key cryptosystem [27] has been proposed by Rivest, Shamir and Adleman. RSA is undoubtedly the most attractive research area in cryptology with immediate applications in practice.

The RSA cryptosystem and several variants of it are in use for applications related to secure data exchange mechanisms. The encryption as well as the decryption process in RSA use modular exponentiation. As square and multiply is the most popular method for modular exponentiation, it is immediate to note that the cost is low for small exponents.

Before proceeding further, let us briefly explain the RSA public key cryptosystem. In RSA, a large integer N is generated such that $N = pq$, where p, q are primes of same bit lengths. The encryption and decryption exponents are denoted by e, d respectively and they are chosen in such a manner that $ed \equiv 1 \bmod \phi(N)$, where $\phi(N) = \phi(pq) = (p-1)(q-1)$, the Euler's totient function. The parameters e, N are distributed as the public key and the part d is kept secret. In the encryption process, we have $C = M^e \bmod N$, whereas, the decryption is performed as $M = C^d \bmod N$.

It is clear that the cost of modular exponentiation can be reduced if one can reduce the exponents e, d. However, $ed > \phi(N)$ provides the constraint that one cannot make both e, d small. For any integer x, let us denote its bit-length as ℓ_x and thus $\ell_e + \ell_d \geq \ell_N$. Towards making the decryption process faster, the secret decryption exponent d has to be made small. In this direction, using the idea of continued fraction, Wiener [28] showed that when $d < \frac{1}{3}N^{\frac{1}{4}}$, one can factor N efficiently. Later, using lattice based techniques, this result has been improved by Boneh and Durfee [3,2] till the upper bound $N^{0.292}$. To achieve further efficiency during encryption process, small e is considered. Coppersmith [7] has shown that RSA with very small e, e.g., $e = 3$ is not secure. For practical purposes, little larger encryption exponents are used. For example, it is a common practice to use $e = 2^{16} + 1$ and it is believed to be quite secure. Given small e, d becomes of the order of N, and the decryption process will be much less efficient than the encryption.

CRT-RSA. To achieve further efficiency during decryption, Wiener [28] prescribed use of Chinese Remainder Theorem (CRT) that has earlier been studied by Quisquater and Couvreur [26]. This is known as CRT-RSA. In CRT-RSA, one uses $d_p = d \bmod (p - 1)$ and $d_q = d \bmod (q - 1)$, instead of d, for the decryption process. This is the most widely used variant of RSA in practice, and decryption becomes more efficient if one pre-calculates the value of $q^{-1} \bmod p$. Thus, in PKCS [24] standard for the RSA cryptosystem, it is recommended to store the RSA secret parameters as a tuple $(p, q, d, d_p, d_q, q^{-1} \bmod p)$. For all the cryptanalytic strategy we mention here, the term $q^{-1} \bmod p$ could not be exploited. Thus, in this paper, we will refer the *secret key* of RSA as a tuple $SK = (p, q, d, d_p, d_q)$. Let us now discuss the cryptanalytic results related to CRT-RSA. The birthday attack has been pointed out by Pinch (as referred in [25]) in case of very small d_p, d_q (one may also have a look at [23]). Further, if $d_p, d_q < N^{0.073}$, one can factor N in polynomial time [17]. In [16], it has been shown that CRT-RSA is weak if $d_p - d_q$ is known and d_p, d_q are smaller than $N^{0.099}$. Broadly speaking, it is easy to see that CRT-RSA can be broken in $O(e)$ time if $d_p - d_q$ can be obtained with small effort.

There are also some important results related to RSA variants under the fault attack. Boneh et al [4] showed that CRT-RSA implementations are vulnerable in this regard. Later Coron et al [8,9] extended the results of [4]. Recently, Brier et al [5] have presented alternative key-recovery attacks on CRT-RSA signatures under fault model.

RSA and CRT-RSA Variants. There are several proposals on RSA and CRT-RSA key generation algorithms such that e is small and the secret parameters have certain special structures. For example, Lenstra [19] pointed out that by taking N with half of most significant bits to be zero, one would obtain around 30% advantage in encryption and decryption process. Similar idea for using large number of zeros in the binary representation of decryption exponent has also been used in several papers. In such a case, the multiplication effort will be reduced a lot in square and multiply algorithm. Initially Lim and Lee [21] considered the RSA keys with relatively low Hamming weight of the decryption exponent d. Later, Galbraith et al [11] proposed a key generation algorithm for CRT-RSA. Using that idea, one can generate CRT-RSA modulus N which allows the cost of encryption and decryption to be balanced according to the requirements of the applications. For faster decryption, one can choose d_p, d_q with low Hamming weight. In this regard, Galbraith et al [11] mentioned

> "In some settings we may also want to choose the d_i to have low Hamming weight. This is easily done if the k_i are small."

Towards the security analysis, for small e, the estimated time complexity to attack such a scheme [11] has been presented as $O\left(\sqrt{w_{d_p}}\binom{\ell_{d_p}/2}{w_{d_p}/2}\right)$, where w_x is the Hamming weight of the binary representation of the integer x. In line of the work of [11], another efficient scheme has been proposed in [22] that also relied on large low weight factors in the decryption exponent d_p, d_q. The security analysis of [22] show that the exhaustive search for the low Hamming weight factors in the decryption exponents is an approach to attack such a scheme. Note that the schemes of [21,11,22] are motivated towards implementing on low end devices with limited computational power (such as smart card).

Our Contribution. To the best of our knowledge, there has been no cryptanalytic result on the security of schemes [21,11,22] so far. For the first time, in contrary to the claims in [21,11,22], we show that the ideas exploiting low weight integers in the secret decryption exponents, can be broken much faster. The basic technique we use in this paper is the work of [13] related to error correction of RSA secret key. In Crypto 2010, Henecka et al [13] studied the case when the bits of SK were known with some error probability for each bit. We refer a noisy version of SK as $\tilde{SK}$, i.e., $\tilde{SK} = (\tilde{p}, \tilde{q}, \tilde{d}, \tilde{d}_p, \tilde{d}_q)$. That is, each bit of the parameters in SK is considered to be flipped with some probability $\delta \in [0, \frac{1}{2})$. The authors [13] could show that one can correct the errors in the secret key (i.e. recover the secret key) in polynomial time (for small e) when the error rate δ is less than $0.237, 0.160, 0.084$ when noisy versions of (p, q, d, d_p, d_q) or (p, q, d) or (p, q) are available.

The algorithm presented in [13] guesses the bits of one of the primes and then uses the reconstruction technique for cold-boot attack in [14] as to get approximations of the other parameters in SK. The verification of each guess is achieved by comparing the Hamming distance of the guess with the erroneous version of SK obtained through side-channel attacks. This is equivalent to pruning the search space towards the correct solution, and hence higher bit-error can be corrected if one uses more parameters from SK during the pruning phase.

In CRT-RSA situation, we have $ed_p \equiv 1 \bmod (p - 1)$. Thus, one can write $ed_p = 1 + k_p(p - 1)$ where $k_p < e$. Similarly we have $ed_q = 1 + k_q(q - 1)$ where $k_q < e$. For small values of e, one may assume k_p, k_q are known to the attacker in $O(e)$ time complexity as we explain in Section 2. In general, for a randomly chosen integer x, we have $w_x \approx \frac{\ell_x}{2}$. However, for efficient decryption, sometimes w_{d_p}, w_{d_q} are taken significantly smaller than the random case. For example, consider that $\ell_{d_p} = \ell_{d_q} = 512$ and $w_{d_p}, w_{d_q} \approx 50$. In such a scenario, one can take the all zero bit string as error-incorporated (noisy) presentation of d_p, d_q, where the error rate is around $\frac{50}{512} \approx 10\%$. As the error rate is significantly small, one can apply the error correcting algorithm of [13] to recover the secret key. Denoting the time complexity of the error-correction algorithm [13] as τ, our strategy attacks the schemes [21,11] in $\tau O(e)$ time, and the scheme [22] in $\tau O(e^3)$ time.

While attacking the schemes [21,11,22], one can attempt to recover all the bits of p as it is done for the error correcting algorithm in [13]. However, one can also try to construct only the least significant half of p using the same strategy and then use the lattice based result of [1] to get the complete p. While describing the experimental results, we present separate data for constructing all the bits of p and only least significant half of p.

While applying the heuristic [13], we noted a few modifications that can improve the performance significantly and the central idea is as follows. Instead of a single fixed threshold related to bit-matching in [13], we use multiple thresholds towards the motivation that we involve several constraints on the secret parameters in our case whereas a single constraint has been taken into consideration in [13].

Table 1. Experimental results of [13] with maximum possible δ as available from [13, Section 6, Tables 2, 3, 4].

Parameters	Upper bound of δ [13]		Success probability (expt.)		upper bound of δ
	theoretical	experimental	[13]	our	achieved in our expt.
(p, q)	0.084	0.08	0.22	0.61	0.12
(p, q, d)	0.160	0.14	0.15	0.52	0.17
(p, q, d, d_p, d_q)	0.237	0.20	0.21	0.50	0.25

To present a glimpse of our improvement, let us provide a brief comparison of our results with that of [13] in Table 1. For the experiments, we only refer to the results at maximum value of δ in [13] (and show that our success probability is better at that point) because our main contribution is to show that we can go significantly beyond the bound of δ in [13] with our heuristic strategy in Algorithm 2 (Section 3). See Section 4 for the detailed experimental results. For any other results describing lower error rates presented in [13], we always obtain improved success probability and those are not explicitly mentioned here.

We like to point out that apart from the specific attack on CRT-RSA with certain parameters, our improved heuristic can correct more noise than the existing strategy [13] for cold-boot attack in general.

Roadmap. In Section 2, we efficiently exploit the error correction strategy of [13] to show that CRT-RSA schemes that involve low weight secret parameters are not secure. We point out that the CRT-RSA based schemes of [21,11] with certain parameters can be broken in a few minutes (Section 2.1) and also present cryptanalytic results on the scheme of [22] (Section 2.2). Further, in the process, we provide modifications to the error correction heuristic of [13] that provides significantly improved experimental results. This is presented in Section 3. Detailed experimental results are presented in Section 4. Conclusion of this paper is presented in Section 5.

2 The Idea of Cryptanalysis

We start with the basic relations of CRT-RSA, such as $ed_p = 1 + k_p(p-1) \Rightarrow k_p - 1 \equiv k_p p \bmod e$, and $ed_q = 1 + k_p(q-1) \Rightarrow k_q - 1 \equiv k_q q \bmod e$. From these we get $k_q \equiv (k_p - 1)(k_p(1 - N) - 1)^{-1} \bmod e$. Since both $k_p, k_q < e$ and from the above equation a choice of k_p fixes k_q, there are $O(e)$ possible choices for the pairs (k_p, k_q). As we have assumed, the Hamming weight of d_p, d_q are considered to be significantly lower than the random case. Further, the presence of 1's in the binary representation of d_p, d_q are considered to be i.i.d. For better explanation, from now on, we will assume that k_p, k_q are known to the attacker and finally the complexity of the attack will be obtained by multiplying an $O(e)$ factor, unless mentioned otherwise.

Our idea is to guess a few bits (say a many bits) of p from the least significant side and let the corresponding integer be p'. From p', we get an approximation q' of q. From p', q' and using the knowledge of e, k_p, k_q, we obtain the approximations of d_p, d_q, that we denote by d'_p, d'_q respectively. If the Hamming weights of d'_p, d'_q are less than some predefined threshold, then p' would be a possible choice of p. This process will be repeated until we have obtained a set A of possible guesses p' for p. Then we extend the solutions by adding a more bits in the more significant side with the possible partial solutions in A. The process continues till we get a set of possible solutions for p itself.

Input: N, e, k_p, k_q and a, C
Output: Set A, containing possible guesses for p.

1   Initialize $b = 0, A = \emptyset, A_{-1} = \{\lambda\}, i = 1$;

2   **while** $b < \frac{\ell_N}{2}$ **do**

3      $A = \{0, 1\}^a \| A_{-1}$;

4   For each possible options $p' \in A$, calculate $q' = (p')^{-1} N \bmod 2^{b+a}$;

5   For each p', q', calculate
   $$d'_p = (1 + k_p(p' - 1)) e^{-1} \bmod 2^{b+a}, d'_q = (1 + k_q(q' - 1)) e^{-1} \bmod 2^{b+a};$$

6   If the number of 0's taking together the binary patterns of d'_p, d'_q in the positions b to $b + a - 1$ from the least significant side is less than C, then delete p' from A;

7   If $b \neq 0$ and $A = \emptyset$, then terminate the algorithm and report failure;

8      $A_{-1} = A; b = b + a; i = i + 1$;

   end

9   Report A;

Algorithm 1. Reconstruction algorithm for p.

In Algorithm 1, we present the algorithm formally. For that we need to use certain notations. Given two binary strings u_1, u_2, by $u_1 \| u_2$ we mean the concatenation of the strings. With abuse of notation, for two integers x, y, by $x \| y$ we denote the integer formed by the concatenation of the binary representations of x, y. We also consider the notation $X \| Y = \{x \| y : x \in X, y \in Y\}$. By λ we mean an empty or null string. Steps 4 and 5 in Algorithm 1 can be calculated efficiently using the relations [13, Equations (8), (10), (11)].

2.1 Cryptanalysis of [21,11]

In [21,11], CRT-RSA secret keys are generated in a manner such that the weights of d_p, d_q are small. By δ we denote the probability that a bit of d_p or d_q is 1. Thus, δ can be estimated as $\frac{w_{d_p}}{\ell_{d_p}}$ or $\frac{w_{d_q}}{\ell_{d_q}}$. Following the theoretical results of [13], we immediately get the result as below. For the sake of completeness, a detailed analysis regarding this is available in Appendix A.

Theorem 1. *Let* $a = \lceil \frac{\ln \ell_N}{4\epsilon^2} \rceil, \gamma_0 = \sqrt{(1 + \frac{1}{a}) \frac{\ln 2}{4}}$ *and* $C = a + 2a\gamma_0$. *We also consider that the parameters* k_p, k_q *of CRT-RSA are known. Then one can obtain* p *in time* $O(l_N^{2 + \frac{\ln 2}{2\epsilon^2}})$ *with success probability greater than* $1 - \frac{2\epsilon^2}{\ln \ell_N} - \frac{1}{\ell_N}$ *if* $\delta \leq \frac{1}{2} - \gamma_0 - \epsilon$.

To maximize δ, we need that ϵ should converge to zero and in such a case a tends to infinity. Then the value of γ_0 converges to 0.416. Thus, asymptotically Algorithm 1 works when δ is less than $0.5 - 0.416 = 0.084$. However since in this case a becomes very large, the algorithm will not be efficient and may not be implemented in practice. This is the reason, experimental results could not reach the theoretical bounds in [13].

Generally, d_p, d_q are taken to be of same bit size which is equal to $\frac{\ell_N}{2}$. Thus, following the idea of Theorem 1 above, one can cryptanalyze CRT-RSA having $w_{d_p}, w_{d_q} \leq 0.04\ell_N$ in $O(e \cdot \text{poly}(\ell_N))$ time. For each possible option of k_p, k_q (this requires $O(e)$ time), one needs to apply Algorithm 1 to obtain p. It is indeed clear that for small e the attack remains efficient.

In [21, Page 9, end of paragraph 3], example parameters have been proposed, where $\ell_N = 768$, $\ell_{d_p} = 384$ and $w_{d_p} = 30$. This falls under the condition mentioned above and we could cryptanalyze all the CRT-RSA keys with such parameters in a few minutes in practice. In another example [21, Table 2, Section 7], it has been considered that $\ell_N = 768$, $\ell_{d_p} = 377$ and $w_{d_p} = 45$ and $e = 257$. In this case $\delta = \frac{w_{d_p}}{\ell_{d_p}} \approx 0.12 > 0.08$, and thus it is not in the bound given in Theorem 1 and so Algorithm 1 would not work as it is. However, in the next section (Section 3) we will present some modifications over Algorithm 1 to get Algorithm 2 that provides significantly improved results experimentally than what presented in Theorem 1. That helps us in easily breaking CRT-RSA with the above mentioned parameters in a few minutes again.

In [11, Figure 1], parameters are proposed as $(\ell_e, \ell_{d_p}, \ell_{k_p}) = (176, 338, 2)$ with $w_{d_p} = 38$. Note that in this situation, one could easily obtain k_p and k_q, even without trying $O(e)$ steps. Here $\delta = \frac{38}{338} \approx 0.11$. Using Algorithm 2 discussed in Section 3, one can break the CRT-RSA scheme with such parameters mentioned in [11] within a few minutes.

2.2 Cryptanalysis of [22]

In [22], the CRT-RSA decryption exponents d_p, d_q have been chosen in a slightly different manner. Here the weight of d_p, d_q are not small, but they are of the form

$d_p = d_{p_1} d_{p_2}$ and $d_q = d_{q_1} d_{q_2}$. The factors d_{p_1} and d_{q_1} are of size $O(e)$ and the other factors d_{p_2} of d_p and d_{q_2} of d_q are of significantly small Hamming weight. So, in this case we have $e d_{p_1} d_{p_2} = 1 + k_p(p-1)$ and $e d_{q_1} d_{q_2} = 1 + k_q(q-1)$. There are $O(e^2)$ choices of d_{p_1}, d_{q_1}. Since $k_q \equiv (k_p-1)(k_p(1-N)-1)^{-1} \bmod e$, within $O(e^3)$ many attempts one can get the correct choice of $(d_{p_1}, d_{q_1}, k_p, k_q)$. Hence one can consider that the attacker knows d_{p_1}, d_{q_1}, k_p and k_q, perform the attack and then multiply the effort by $O(e^3)$ to get the total time complexity of the cryptanalysis. Let a many LSBs of p be known and p' be the corresponding integer. Then a many LSBs of d_{p_2}, d_{q_2} can be obtained through the following identities:

$$d_{p_2} \equiv (e d_{p_1})^{-1}(1 + k_p(p'-1)) \bmod 2^a,$$
$$d_{q_2} \equiv (e d_{q_1})^{-1}\left(1 + k_q\left(N(p')^{-1} \bmod 2^a - 1\right)\right) \bmod 2^a. \qquad (1)$$

We use the Equation (1) in step 5 of the Algorithm 1 instead of what is given there towards the cryptanalysis of [22]. Here $\ell_{d_{p_2}} \approx \frac{\ell_N}{2} - \ell_e$. When $w_{d_{p_2}}, w_{d_{q_2}} \leq 0.08\left(\frac{\ell_N}{2} - \ell_e\right)$, one could cryptanalyze the CRT-RSA scheme with the parameters proposed in [22] in time $O(e^3 \ell_N^{2 + \frac{\ln 2}{2e^2}})$.

Five challenges have been presented in [22], where $e = 2^{16} + 1$, $\ell_N = 1024$ and $w_{d_{p_2}} = w_{d_{q_2}} = 40$. As both d_{p_1} and d_{q_1} are of $O(e)$, $\ell_{d_{p_2}} \approx \ell_{d_{q_2}} \approx 512 - 16 = 496$. Hence $\delta = \frac{40}{496} \approx 0.08$. Thus, the proposal of [22] can be cryptanalysed using Algorithm 1 with a modification in step 5 as described above.

Let us now explain the efficiency of our cryptanalysis on the proposal of [22] as we could not break it in real time as had been done on the examples of [21,11]. Note that for e as described in [22], e^3 is around 2^{48} and that many runs of Algorithm 1 or Algorithm 2 are required.

The parameters of [22] are so chosen that if one tries to go for an exhaustive search, then it will require around 2^{94} effort. Based on this, it has been claimed [22] that such a scheme is secure as the best possible factorization strategy using NFS [20] requires around 2^{86} time complexity and one cannot attack the scheme in a lower complexity than that.

Now we show that one can indeed attack the scheme of [22] in a time complexity much less than 2^{86}. One can implement the attack with all possible values of $d_{p_1}, d_{q_1}, k_p, k_q$ that requires 2^{48} many invocations of Algorithm 1 or Algorithm 2 when e is 16-bit integer. As Algorithm 2 works significantly better than Algorithm 1, we estimate the time complexity of each invocation of Algorithm 2. Given a block length $a = 10$, to get p, we need $\lceil \frac{512}{10} \rceil = 52$ many generations of set A, where we bound the number of solutions in the set A by 1000 in the experiments. We can estimate the time complexity for each invocation of Algorithm 2 with the above parameters as $52 \cdot 1000 < 2^{16}$. Hence, the total complexity of the attack is around $2^{48+16} = 2^{64}$, which is significantly smaller than 2^{86}.

While the idea presented in [22] can be cryptanalyzed, it cannot be broken in a few minutes experimentally as it could be done for [21,11]. One may actually explore the ideas of countermeasure from [22]. To protect/blind the secret exponents, one needs to use the product of one large integer (of small weight) and one very small integer such that the weight of the product is not small (please see Section 2.2). Proper choice of parameters (in particular bit length of the smaller and larger factors) will make the attack less effective. As an example, instead of choosing 16-bit k_p, k_q, one may try larger ones (say, 32 bits). This will increase d_{p_1}, d_{q_1} to 32 bits. In this case, the attacker needs to know d_{p_1}, d_{q_1}, k_p and the attack complexity will this increase.

3   Heuristics for Further Improvement of the Error Correction Algorithm [13]

The theoretical bounds on the noise rate δ in [13] for which SK can be recovered from a noisy version of it are as follows: (i) $\delta < 0.237$, when $\tilde{p}, \tilde{q}, \tilde{d}, \tilde{d}_p, \tilde{d}_q$ are available, (ii) $\delta < 0.160$, when $\tilde{p}, \tilde{q}, \tilde{d}$ are available, and (iii) $\delta < 0.084$ when $\tilde{p}, \tilde{q}$ are available. However one cannot achieve these bounds due to high length of the block size. The experimental bounds achieved in [13, Section 6, Tables 2, 3, 4] are presented as 0.20, 0.14 and 0.08 for cases 1, 2 and 3 respectively with success probability less than 0.25. In this section we explain certain heuristics to improve these experimental results significantly. We present experimental results having error rates higher than the theoretical upper bound of [13]. This we get for the parameter $a = 10$ which is much lower than the values used in the experiments of [13] and increasing a in our strategy improves our results further. Let us now present the broad ideas behind our improvements.

Different Values of the Threshold C. Instead of one fixed threshold C, we take different thresholds in different steps, that depend on the value $b+a$. During the pruning, we count the number of bits at which the noisy parameters and the possible solutions match for the positions 0 to $b + a - 1$. Thus the number of comparison for each parameter is $(b + a)$ at each step and then based on that we decide whether we will accept or reject a solution. Thus we consider a cumulative measure, where for the initial steps, the bit strings compared are of the lesser size and as the solution grows, the bit strings in comparison are of larger size. The threshold parameters are chosen based on the error rate and length of the-then solutions (which are actually $b + a$ at that point of time).

Multiple Constraints on Each Round. Instead of considering the total number of mismatched bits for each component of the secret key, all possible constraints are considered at the same time in our strategy. Suppose we want to factor $N = pq$ where p, q are available with some noise. Consider an instance of the algorithm where we have reached up to the bit position i.

For proceeding further, let us have a few notations. For an integer x, by $x[i]$, we denote the i-th least significant bit of x. Further, by $x_{[i]}$, we mean the bit-string $x[i], x[i-1], \ldots, x[1], x[0]$ and this will also be interpreted as an integer. As an example, for a prime, say, $p = 23$, we have the binary representation 10111 and thus, $p_{[3]} = 0111$, which is 7 as an integer.

Let the number of matched bits between the partial solution p' and the corresponding bits of the noisy version $\tilde{p}$, i.e., $\tilde{p}_{[i]}$ be μ_1 and for q' and $\tilde{q}_{[i]}$ be μ_2. We impose constraints on both the values of μ_1, μ_2 along with $\mu_3 = \mu_1 + \mu_2$ to achieve better pruning. In [13], the pruning was applied on the sum of the individual values only. Here, in case of m many components of the secret key, total number of constraints in each iteration would be $\sum_{i=1}^{m} \binom{m}{i} = 2^m - 1$. When we work with the five parameters p, q, d, d_p, d_q, we use a total of 31 constraints instead of 1 as mentioned in [13].

Suppose that the secret key has m components. For the l-th component ($1 \leq l \leq m$) of the key, let $\mu_{2^{l-1}}$ denote the number of matched bits between the partial solution and the noisy version of the component at the corresponding bits. Then for each individual component, numbers of matched bits are given by $\{\mu_1, \mu_2, \mu_4, \ldots, \mu_{2^{m-1}}\}$ respectively. Now, for any general $k \in [1, 2^m - 1]$, we define the term μ_k as follows: $\mu_k = \sum_{i=1}^{m} k_i \mu_{2^{i-1}}$, where k_i are the bits of k for $1 \leq i \leq m$. Thus the total number of matched bits in all components of the secret key is given by $\mu_{2^m - 1} = \sum_{i=1}^{m} \mu_{2^{i-1}}$. In practice, when we work with $m = 5$ parameters p, q, d, d_p, d_q of the secret key, $\mu_{31} = \mu_1 + \mu_2 + \mu_4 + \mu_8 + \mu_{16}$ represents the cumulative sum of matched bits between the partial solution and corresponding bits of the noisy version for all parameters.

Value of Threshold Parameters and Its Run-Time Modifications. For each μ_i, we choose the value of the threshold C_i^{a+b} depending upon the noise rate δ and the value of $a + b$ at that stage. For a fixed noise rate δ, we choose the minimum C_i^{a+b} such that

$$\sum_{j=1}^{C_i^{a+b}} \binom{w_i(a+b)}{j} \delta^j (1-\delta)^{a+b-j} > \nu, \qquad (2)$$

where w_i is the Hamming weight of i as we have mentioned earlier. In the experiments we take $\nu = 0.99$ for $a + b < 150$ and $\nu = 0.98$ for $a + b \geq 150$. We choose the thresholds like this so that the possibility of rejecting a correct partial solution remains very low. It is clear that one cannot allow the size of A to increase exponentially. Thus one needs to keep some upper bound on $|A|$ while running the algorithm. Let $|A|$ be restricted by a constant upper bound B. Consider m secret parameters and take certain threshold for each μ_i. While creating the set A from A_{-1} in a loop, if $|A| > B$, we reduce $C_{2^m-1}^{a+b}$ by 1. One may ask, why do we reduce only the threshold corresponding to $\mu_{2^m-1} = \mu_1 + \mu_2 + \ldots + \mu_{2^m-1}$. We have tried in manipulating other thresholds as well, but found that this is quite an effective idea to obtain good experimental results. The study of such thresholds and their modifications during the run of the algorithm is an interesting question and requires serious attention that is not in the scope of this work.

The Modified Algorithm. Based on the above discussion, our improved error correction strategy is presented in Algorithm 2. We have presented the algorithm with all the five parameters $\tilde{p}, \tilde{q}, \tilde{d}, \tilde{d}_p, \tilde{d}_q$, though one can easily modify it for less number of parameters.

Input: N, e, k, k_p, k_q
Input: $\tilde{p}, \tilde{q}, \tilde{d}, \tilde{d}_p, \tilde{d}_q$
Input: a, B and threshold parameters as described in (2).
Output: Set A, containing possible guesses for p.

1 Initialize $b = 0, A = \emptyset, A_{-1} = \emptyset$;

2 **while** $b < \frac{\ell_N}{2}$ **do**

3     $A = \{0, 1\}^a \| A_{-1}$;

4   For each possible options $p' \in A$, calculate $q' = (p')^{-1} N \bmod 2^{b+a}$;

5   Calculate $d' = (1 + k(N + 1 - p' - q')) e^{-1}) \bmod 2^{b+a}$;

6   Calculate

    $d'_p = (1 + k_p(p' - 1)) e^{-1} \bmod 2^{b+a}, d'_q = (1 + k_q(q' - 1)) e^{-1} \bmod 2^{b+a}$;

7   Calculate μ_i's for $i = 1$ to 31 comparing least significant $b + a$ bits of the noisy strings and the corresponding possible partial solution strings of length $b + a$, i.e., through the positions 0 to $b + a - 1$;

8   If $\mu_i < C_i^{a+b}$ for any $i \in [1, \ldots, 31]$, delete the solution from A;

9   If $|A| > B$, reduce C_{31}^{a+b} by 1 and go to Step 8;

10   If $b \neq 0$ and $A = \emptyset$, then terminate the algorithm and report failure;

11     $A_{-1} = A$; $b = b + a$;

    end

12 Report A;

Algorithm 2. Improved Error Correction algorithm.

In the next section we present experimental results to highlight the significant improvements over [13]. Theoretical analysis of Algorithm 2 (possibly exploiting statistical techniques) is indeed of interest, though it is not attempted in this initiative.

4 Experimental Results

We have implemented Algorithm 2 using C programming language (with GMP library for processing large integers) on Linux Ubuntu 2.6. The hardware platform is an HP Z800 workstation with 3GHz Intel(R) Xeon(R) CPU. Our implementation is not optimized and for each run the time required varied from a few seconds to a few minutes depending on the error rates. The time estimations presented in our tables are the averages of the time required in the successful runs only. In all the experiments, we take 1024-bit RSA with public exponent $e = 2^{16} + 1$. We consider $a = 10$ and $B = 1000$. For each experiment, we generate 20 different RSA secret keys and for each secret key, we generate 20 many noisy versions by incorporating independently and uniformly distributed errors with noise rate δ. So, we have a total of 400 samples.

First we go for only two parameters. This can be interpreted in two ways: (i) the noisy versions of p, q are available with error rate δ or (ii) p, q are completely unknown, but the weights of p, q are small, i.e., $\frac{w_p}{\ell_p} \approx \frac{w_q}{\ell_q} \approx \delta$.

Table 2. Experimental results with Algorithm 2 with two parameters p, q

δ	0.08	0.09	0.10	0.11	0.12	0.13
Success probability	0.61	0.36	0.19	0.06	0.02	-
Time (in seconds)	255.97	249.56	252.23	235.34	230.13	-
Success probability (half)	0.71	0.55	0.41	0.23	0.13	0.08
Time (in seconds)	68.82	66.24	66.23	66.00	60.04	61.67

Table 3. Experimental results with Algorithm 2 with two parameters d_p, d_q

δ	0.08	0.09	0.10	0.11	0.12	0.13
Success probability	0.59	0.27	0.14	0.04	-	-
Time (in seconds)	307.00	294.81	272.72	265.66	-	-
Success probability (half)	0.68	0.49	0.25	0.18	0.08	0.02
Time (in seconds)	87.41	84.47	80.18	74.57	79.33	76.04

Note that we run Algorithm 2 till we obtain all the bits of p. However, it is known that if one obtains the least significant half of p, then it is possible to obtain the factorization of N efficiently [1]. In this case, as we need to reconstruct half of the bits of p instead of the full binary string, the success probability will increase. Keeping this in mind, in the experimental results we provide the success probability to obtain the complete bit pattern of p to compare our results with [13] as well as the success probability to obtain least significant half of p, that is actually required for the attack. We refer the second one as success probability (half) in the tables. The error rate of the order of 0.08 could be achieved with success probability 0.22 in [13], and one may see in Table 2 that our results are significantly improved.

In Table 3, we present experimental results related to our attack in Section 2. Taking $e = 2^{16} + 1$, we have generated CRT-RSA secret exponents d_p, d_q having small Hamming weights using the idea of [11]. In Table 3, we present experimental results taking 5-bit k_p, k_q. The results in Table 3 is slightly worse than Table 2. This is because in the least ℓ_{k_p} many bits of d_p, d_q, the error rate is not small due to the key generation algorithm of [11]. We obtained similar kinds of results for cryptanalysis of [22] too, with similar parameters as in the benchmark examples of [22, Appendix A]. However, in these cases, for practical experiments in a few minutes, we need to consider that $d_{p_1}, d_{q_1}, k_p, k_q$ are available. That is, the time need to be multiplied by 2^{48} for actual attack, when $e = 2^{16} + 1$, say.

Next, in Table 4, we consider the case with three parameters p, q, d. When $\delta = 0.14$, the success probability in [13, Table 3] has been reported as 0.15. The success probability using our modification is 0.52 in this scenario. Further we could demonstrate experimental results till $\delta = 0.17$ which is better than the theoretical bound of 0.16 in [13].

Next we present the results will all five parameters. It is evident from Table 5 that we obtained significant improvement over the results of [13].

Our current implementation is only towards proof-of-the-concept. The results are expected to improve further with optimized implementation. While we can work with higher error rates and the success probability of our modified version

Table 4. Experimental results with Algorithm 2 with three parameters p, q, d

δ	0.14	0.15	0.16	0.17	0.18
Experiment	0.52	0.21	0.11	0.03	-
Time (in seconds)	461.24	430.36	412.08	407.8	-
Experiment(half)	0.69	0.48	0.23	0.14	0.07
Time (in seconds)	142.99	131.69	127.12	123.00	124.00

Table 5. Experimental results with Algorithm 2 with five parameters p, q, d, d_p, d_q

δ	0.20	0.21	0.22	0.23	0.24	0.25
Experiment	0.50	0.44	0.33	0.14	0.02	0.005
Time (in seconds)	699.82	639.16	607.23	580.45	540.10	502.00
Experiment(half)	0.70	0.58	0.55	0.31	0.12	0.065
Time (in seconds)	221.00	192.35	190.94	173.96	168.00	169.61

are much better than [13], we require little more running time than [13]. However, we like to point out that our improvement is not achieved at the cost of a higher running time as our results cannot be reached taking $a = 10$ by the techniques of [13]. Towards the higher error rates, the value of a in [13] varies from 20 to 29, whereas we work with a as small as 10 for all the cases. Still we achieve better success rate. Larger the size of a, larger is the set of partial solutions A. This clearly shows that our strategy performs significantly better with the new ideas of pruning where the correct solution is retained with good success rate.

We have also explored a few other implementation strategies for Algorithm 2. Let $D(x)$ be the largest integer that divides x. From the knowledge of k, k_p and k_q, one can easily calculate $D(k), D(k_p), D(k_q)$. Thus, in the steps 5 and 6 of the Algorithm 2, one can calculate

$$d' = (1 + k(N + 1 - p' - q')) e^{-1}) \bmod 2^{b+a+D(k)},$$
$$d'_p = (1 + k_p(p' - 1)) e^{-1} \bmod 2^{b+a+D(k_p)} \text{ and}$$
$$d'_q = (1 + k_q(q' - 1)) e^{-1} \bmod 2^{b+a+D(k_q)}.$$

Since $D(k), D(k_p)$ and $D(k_q)$ are very small in general, the improvement in terms of time complexity will not be significant and we have checked that experimentally too. In the course of optimizing the algorithm, one may note that the solution sets for these equations can be evolved rapidly by Hensel lifting. These are actually used in the time of inverse calculations in our strategy. In this case also, we did not obtain major improvements in running time.

5   Conclusion

In this paper, first we apply the recently proposed error correction strategy (motivated from cold-boot attack) for RSA secret keys [13] to actual cryptanalysis of CRT-RSA under certain conditions. We studied two kinds of schemes. The first one considers the CRT-RSA decryption keys of low weight as in [21,11]. In these cases, we demonstrate complete break in a few minutes for 1024 bit RSA moduli. The next one considers the scenario when the decryption exponents are not of low weight, but they contain large low weight factors [22]. Though this scheme seems more resistant to our method than the ones in [21,11], it [22] is also prone to cryptanalysis with much lower complexity than what claimed in the paper [22].

Further, we had a detailed look at the actual error correction algorithm of [13] and provided significant improvements as evident from experimental results. The experimental results are significantly better than the ones presented in [13] and more importantly, we could demonstrate that the theoretical bound of [13] can also be crossed using our heuristic. These results can directly be applied to cold-boot attack, in general, on RSA and its variants.

Acknowledgments. The authors like to thank the Centre of Excellence in Cryptology, Indian Statistical Institute for relevant support towards this research.

References

1. Boneh, D., Durfee, G., Frankel, Y.: An Attack on RSA Given a Small Fraction of the Private Key Bits. In: Ohta, K., Pei, D. (eds.) ASIACRYPT 1998. LNCS, vol. 1514, pp. 25–34. Springer, Heidelberg (1998)
2. Boneh, D., Durfee, G.: Cryptanalysis of RSA with Private Key d Less Than $N^{0.292}$. In: Stern, J. (ed.) EUROCRYPT 1999. LNCS, vol. 1592, pp. 1–11. Springer, Heidelberg (1999)
3. Boneh, D., Durfee, G.: Cryptanalysis of RSA with Private Key d Less Than $N^{0.292}$. IEEE Transactions on Information Theory 46(4), 1339–1349 (2000)
4. Boneh, D., DeMillo, R.A., Lipton, R.J.: On the importance of checking cryptographic protocols for faults. Journal of Cryptology 14(2), 101–119 (2001)
5. Brier, É., Naccache, D., Nguyen, P.Q., Tibouchi, M.: Modulus Fault Attacks against RSA-CRT Signatures. In: Preneel, B., Takagi, T. (eds.) CHES 2011. LNCS, vol. 6917, pp. 192–206. Springer, Heidelberg (2011)
6. Chen, Y., Nguyen, P.Q.: Faster Algorithms for Approximate Common Divisors: Breaking Fully-Homomorphic-Encryption Challenges over the Integers. In: Pointcheval, D., Johansson, T. (eds.) EUROCRYPT 2012. LNCS, vol. 7237, pp. 502–519. Springer, Heidelberg (2012)
7. Coppersmith, D.: Small Solutions to Polynomial Equations and Low Exponent Vulnerabilities. Journal of Cryptology 10(4), 223–260 (1997)
8. Coron, J.-S., Joux, A., Kizhvatov, I., Naccache, D., Paillier, P.: Fault Attacks on RSA Signatures with Partially Unknown Messages. In: Clavier, C., Gaj, K. (eds.) CHES 2009. LNCS, vol. 5747, pp. 444–456. Springer, Heidelberg (2009)
9. Coron, J.-S., Naccache, D., Tibouchi, M.: Fault Attacks Against EMV Signatures. In: Pieprzyk, J. (ed.) CT-RSA 2010. LNCS, vol. 5985, pp. 208–220. Springer, Heidelberg (2010)
10. Diffie, W., Hellman, M.E.: New directions in cryptography. IEEE Transactions on Information Theory IT-22(6), 644–654 (1976)
11. Galbraith, S.D., Heneghan, C., McKee, J.F.: Tunable balancing of RSA. In: Boyd, C., González Nieto, J.M. (eds.) ACISP 2005. LNCS, vol. 3574, pp. 280–292. Springer, Heidelberg (2005),
http://www.isg.rhul.ac.uk/~sdg/full-tunable-rsa.pdf (last accessed December 8, 2011)

12. Halderman, J.A., Schoen, S., Heninger, N., Clarkson, W., Paul, W., Calandrino, J., Feldman, A., Appelbaum, J., Felten, E.: Lest we remember: Cold boot attacks on encryption keys. In: Proceedings of USENIX Security 2008, pp. 45–60. USENIX (July 2008)

13. Henecka, W., May, A., Meurer, A.: Correcting Errors in RSA Private Keys. In: Rabin, T. (ed.) CRYPTO 2010. LNCS, vol. 6223, pp. 351–369. Springer, Heidelberg (2010)

14. Heninger, N., Shacham, H.: Reconstructing RSA Private Keys from Random Key Bits. In: Halevi, S. (ed.) CRYPTO 2009. LNCS, vol. 5677, pp. 1–17. Springer, Heidelberg (2009)

15. Hoeffding, W.: Probability inequalities for sums of bounded random variables. Journal of the American Statistical Association 58(301), 13–30 (1963)

16. Jochemsz, E., May, A.: A Strategy for Finding Roots of Multivariate Polynomials with New Applications in Attacking RSA Variants. In: Lai, X., Chen, K. (eds.) ASIACRYPT 2006. LNCS, vol. 4284, pp. 267–282. Springer, Heidelberg (2006)

17. Jochemsz, E., May, A.: A Polynomial Time Attack on RSA with Private CRT-Exponents Smaller Than $N^{0.073}$. In: Menezes, A. (ed.) CRYPTO 2007. LNCS, vol. 4622, pp. 395–411. Springer, Heidelberg (2007)

18. Kocher, P.C.: Timing Attacks on Implementations of Diffie-Hellman, RSA, DSS, and Other Systems. In: Koblitz, N. (ed.) CRYPTO 1996. LNCS, vol. 1109, pp. 104–113. Springer, Heidelberg (1996)

19. Lenstra, A.: Generating RSA Moduli with a Predetermined Portion. In: Ohta, K., Pei, D. (eds.) ASIACRYPT 1998. LNCS, vol. 1514, pp. 1–10. Springer, Heidelberg (1998)

20. Lenstra, A.K., Lenstra Jr., H.W.: The Development of the Number Field Sieve. Springer (1993)

21. Lim, C.H., Lee, P.J.: Sparse RSA Secret Keys and Their Generation. In: Proceedings of SAC, pp. 117–131 (1996), http://dasan.sejong.ac.kr/~chlim/english_pub.html (last accessed December 8, 2011)

22. Maitra, S., Sarkar, S.: Efficient CRT-RSA Decryption for Small Encryption Exponents. In: Pieprzyk, J. (ed.) CT-RSA 2010. LNCS, vol. 5985, pp. 26–40. Springer, Heidelberg (2010)

23. May, A.: New RSA Vulnerabilities Using Lattice Reduction Methods. PhD thesis, University of Paderborn, Germany (2003)

24. Public-Key Cryptography Standards (PKCS) #1 v2.1: RSA Cryptography Standard. RSA Security Inc. (2002), http://www.rsa.com/rsalabs/node.asp?id=2125

25. Qiao, G., Lam, K.-Y.: RSA Signature Algorithm for Microcontroller Implementation. In: Schneier, B., Quisquater, J.-J. (eds.) CARDIS 1998. LNCS, vol. 1820, pp. 353–356. Springer, Heidelberg (2000)

26. Quisquater, J.-J., Couvreur, C.: Fast decipherment algorithm for RSA public-key cryptosystem. Electronic Letters 18, 905–907 (1982)

27. Rivest, R.L., Shamir, A., Adleman, L.: A Method for Obtaining Digital Signatures and Public Key Cryptosystems. Communications of ACM 21(2), 158–164 (1978)

28. Wiener, M.: Cryptanalysis of Short RSA Secret Exponents. IEEE Transactions on Information Theory 36, 553–558 (1990)

Appendix A

In Section 2, we have described Algorithm 1 towards cryptanalysis of CRT-RSA schemes with low weight decryption exponents when the encryption exponent is small. The theoretical result related to the algorithm has been presented in Theorem 1. Theorem 1 follows directly from the analysis of [13]. Still we explain this in detail for completeness.

We refer to Algorithm 1. Let us define a random variable X_c for the total number of 0's in the binary representation of d'_p and d'_q from the position b to $b + a - 1$, where d'_p, d'_q are correct partial solution of d_p, d_q. Clearly X_c follows binomial distribution with parameters $2a$ and probability $1 - \delta$. Hence

$$P(X_c = \gamma) = \binom{2a}{\gamma}(1 - \delta)^\gamma \delta^{2a-\gamma},$$

for $\gamma = 0, \ldots, 2a$.

Now assume one can expand some incorrect partial solutions (p', q') and obtain d'_p, d'_q. Let X_b be the number of 0's in the expanded $2a$ many bits in d'_p, d'_q. To study the distribution of X_b, we consider the following heuristic assumption that every solution generated from incorrect partial solution consists of randomly chosen bits. Thus,

$$P(X_b = \gamma) = \binom{2a}{\gamma} 2^{-2a}.$$

We have to choose a threshold C such that the two distributions X_c and X_b are sufficiently separated.

Take $C = a + 2a\gamma_0$ where $\gamma_0 = \sqrt{(1 + \frac{1}{a})\frac{\log 2}{4}}$. Let the random variable Y_i represent the number of incorrect partial solutions that pass the threshold bound C at i-th stage. Then one can get the following result.

Lemma 1. *The expectation $E[Y_i]$ of Y_i is less than 2^{a+1}.*

Proof. Let us denote Z_g as the number of incorrect candidates from the partially correct solution and let Z_b give the count of the number of incorrect candidates from each partially incorrect solution. Thus, we have

$$E[Y_1] = E[Z_g],$$
$$E[Y_2] = E[Z_g] + E[Z_b]E[Y_1],$$

$$\vdots$$

$$\begin{aligned}
E[Y_i] &= E[Z_g] + E[Z_b]E[Y_{i-1}] \\
&= E[Z_g] + E[Z_b]\left(E[Z_g] + E[Z_b]E[Y_{i-2}]\right) \\
&= \cdots \\
&= E[Z_g]\sum_{k=0}^{i-1} E[Z_b]^k \\
&= E[Z_g]\frac{1 - E[Z_b]^i}{1 - E[Z_b]}.
\end{aligned}$$

Now define 2^a random variables corresponding to $i = 1, \cdots, 2^a$ such that

$$Z_b^i = \begin{cases} 1 \text{ if } i\text{-th bad candidate passes the threshold} \\ 0 \text{ otherwise} \end{cases}$$

Clearly, $Z_b = \sum_{i=1}^{2^a} Z_b^i$. So,

$$\begin{aligned} E[Z_b] &= 2^a E\left[Z_b^i\right] \\ &= 2^a Pr\left[Z_b^i = 1\right] \\ &= 2^a Pr\left[X_b \geq C\right] \\ &= 2^a Pr\left[X_b \geq 2a\left(\frac{1}{2} + \gamma_0\right)\right] \end{aligned}$$

From Hoeffding's inequality [15] we know,

$$Pr\left[X_b \geq 2a\left(\frac{1}{2} + \gamma_0\right)\right] \leq e^{-4a\gamma_0^2} = e^{-4a\left(1 + \frac{1}{a}\right)\frac{\log 2}{4}}$$
$$= 2^{-a-1}.$$

Hence $E[Z_b] \leq 2^a \cdot 2^{-a-1} = \frac{1}{2} < 1$. So,

$$E[Y_i] < \frac{E[Z_g]}{1 - E[Z_b]} \leq 2E[Z_g] \leq 2(2^a - 1) \text{ as } E[Z_g] \leq 2^a - 1$$
$$< 2^{a+1}.$$

$\square$

To have the time complexity of Algorithm 1 polynomial in ℓ_N, one can take, for example, $a = \lceil \frac{\ln \ell_N}{4\epsilon^2} \rceil$ and $\delta \leq \frac{1}{2} - \gamma_0 - \epsilon$. Then we get the following result.

Lemma 2. *Algorithm 1 succeeds with probability greater than $1 - \frac{\epsilon^2}{\ell_N} - \frac{1}{\ell_N}$.*

Proof. The probability of pruning the correct partial solution at one step is given by $Pr[X_c < C]$. Now

$$Pr[X_c < C] = Pr[X_c < 2a(\frac{1}{2} + \gamma_0)] \leq Pr[X_c < 2a(1 - \delta - \epsilon)] \leq e^{-4a\epsilon^2} \leq \frac{1}{\ell_N}.$$

Thus,

$$\begin{aligned} Pr[success] &= (1 - Pr[X_c < C])^{\lceil \frac{\ell_N}{2a} \rceil} \geq (1 - Pr[X_c < C])^{\frac{\ell_N}{2a}+1} \\ &\geq (1 - \frac{1}{\ell_N})^{\frac{\ell_N}{2a}+1} \geq 1 - \frac{\frac{\ell_N}{2a}+1}{\ell_N} \geq 1 - \frac{1}{2a} - \frac{1}{\ell_N} \geq 1 - \frac{2\epsilon^2}{\ln \ell_N} - \frac{1}{\ell_N}. \end{aligned}$$

$\square$

Using same idea of [13], it can be shown that the time complexity of our idea is $O(\ell_N^{2+\frac{\log 2}{2\epsilon^2}})$. Hence from Lemma 1 and Lemma 2, we get Theorem 1 (Section 2.1) as follows.

Theorem 1. *Let* $a = \lceil \frac{\ln \ell_N}{4\epsilon^2} \rceil$, $\gamma_0 = \sqrt{(1 + \frac{1}{a})\frac{\ln 2}{4}}$ *and* $C = a + 2a\gamma_0$. *We also consider that the parameters* k_p, k_q *of CRT-RSA are known. Then one can obtain* p *in time* $O(l_N^{2+\frac{\ln 2}{2\epsilon^2}})$ *with success probability greater than* $1 - \frac{2\epsilon^2}{\ln \ell_N} - \frac{1}{\ell_N}$ *if* $\delta \leq \frac{1}{2} - \gamma_0 - \epsilon$.

Pushing the Limits of High-Speed $GF(2^m)$ Elliptic Curve Scalar Multiplication on FPGAs

Chester Rebeiro, Sujoy Sinha Roy, and Debdeep Mukhopadhyay

Department of Computer Science and Engineering
Indian Institute of Technology Kharagpur, India
{chester,sujoyetc,debdeep}@cse.iitkgp.ernet.in

Abstract. In this paper we present an FPGA implementation of a high-speed elliptic curve scalar multiplier for binary finite fields. High speeds are achieved by boosting the operating clock frequency while at the same time reducing the number of clock cycles required to do a scalar multiplication. To increase clock frequency, the design uses optimized implementations of the underlying field primitives and a mathematically analyzed pipeline design. To reduce clock cycles, a new scheduling scheme is presented that allows overlapped processing of scalar bits. The resulting scalar multiplier is the fastest reported implementation for generic curves over binary finite fields. Additionally, the optimized primitives leads to area requirements that is significantly lesser compared to other high-speed implementations. Detailed implementation results are furnished in order to support the claims.

Keywords: Elliptic curve scalar multiplication, FPGA, high-speed implementation, Montgomery ladder.

1 Introduction

Elliptic curve cryptography (ECC) is an asymmetric key cipher adopted by the IEEE [21] and NIST [22] as it offers more security per key bit compared to other contemporary ciphers. Security in ECC based cryptosystems is achieved through elliptic curve scalar multiplication. The complex finite field operations involved in ECC often mandates dedicated accelerators for cryptographic and cryptanalytic applications. Field programmable gate arrays (FPGAs) are a popular platform for accelerating curve scalar multiplication due to features such as in-house programmability, shorter time to market, reconfigurability, low non-recurring costs, and simpler design cycles [23]. However, the constrained resources, large granularity, and high costs of routing makes the development of high-speed hardware on FPGAs difficult. The challenges involved in development with FPGAs have led to several published articles on high-speed designs of elliptic curve scalar multiplication for FPGA platforms [1,2,3,5,6,9,10,17]. For binary finite fields over generic curves, most notable works are by Chelton and Benaissa in [5] and more recently Azarderakhsh and Reyhani-Masoleh in [2]. Chelton and Benaissa are capable of doing a scalar multiplication in 19.5μsec, while Azarderakhsh and

E. Prouff and P. Schaumont (Eds.): CHES 2012, LNCS 7428, pp. 494–511, 2012.

Reyhani-Masoleh's implementation requires $17.2\mu sec$. In this paper, we propose an elliptic curve multiplier (ECM) capable of doing scalar multiplications in $10.7\mu sec$ in the same finite field and FPGA family as [5] and [2].

The speed of a hardware design is dictated by 2 parameters: the frequency of the clock and the number of clock cycles required to perform the computation. One method to boost the maximum operable clock frequency is by reducing area. Smaller area generally implies lesser routing delay, which in turn implies higher operable clock frequencies. Another method to increase clock frequency is by pipelining. Chelton and Benaissa [5] extensively rely on this in order to achieve high-speeds. However extensive pipelining in the design is likely to increase the clock cycles required for the computation. Clock cycles can be reduced by parallelization, efficient scheduling, and advanced pipeline techniques such as data-forwarding. Parallelization by replication of computing units was used in [2] to achieve high speeds. The drawback of parallelization however is the large area requirements. Our ECM achieves high-speeds by (1) reducing area, (2) appropriate usage of FPGA hardware resources, (3) optimal pipelining enhanced with data-forwarding, (4) and efficient scheduling mechanisms.

The area requirements of the ECM is primarily due to the finite field arithmetic primitives, in particular multiplication and inversion. In [17], it was shown that an ECM developed with highly optimized field primitives is capable of achieving high computation speeds in-spite of using a naïve scalar multiplication algorithm, no pipelining, or parallelization. Our choice of finite field primitives is based on [17], and has an area requirement which is 50% lesser than [5] and 37% lesser than [2]. The reduced area results in better routing thus leading to increased operating frequencies. Besides the finite field primitives, the registers used in the ECM contribute significantly to the area. Each register in the ECM stores a field element, which can be large. Besides, there are several such registers present. We argue that the area as well as delay can be reduced by placing the registers efficiently in the FPGA.

Ideally, an L stage pipeline can boost the clock frequency up to L times. In order to achieve the maximum effectiveness of the pipelines, the design should be partitioned into L equal stages. That is, each stage of the pipeline should have the same delay. However to date the only means of achieving this is by trial-and-error. In this paper, we show that a theoretical model for FPGA designs, when applied for the ECM, can be used to first estimate the delay in the critical path and there by find the ideal pipelining. As L increases, there is likely to be more data dependencies in the computations, thus resulting in more stalls (*bubbles*) in the pipeline. The paper investigates scheduling strategies for the Montgomery scalar multiplication algorithm, which is an efficient method for pipelining the ECM [13] Compared to [5], which also uses the Montgomery ladder, our scheduling techniques require $3m$ clock cycles lesser for scalar multiplication in the field $GF(2^m)$.

The structure of the paper is as follows: Section 2 has the brief mathematical background required to understand this paper. The organization of the ECM is discussed in Section 3. Section 4 formally analyzes pipelining the ECM while

Section 5 discusses the scheduling of instructions in the pipeline. Section 6 determines the right pipeline for the ECM and Section 7 presents the architecture of the ECM with the *right pipeline*. Implementation results are presented and compared the state-of-the-art in Section 8, while the final section has the conclusion for the paper.

2   Background

An elliptic curve is either represented by 2 point *affine coordinates* or 3 point *projective coordinates*. The smaller number of finite field inversions required by projective coordinates makes it the preferred coordinate system. For the field $GF(2^m)$, the equation for an elliptic curve in projective coordinates is $Y^2 + XYZ = X^3Z + aX^2Z^2 + bZ^4$, where the curve constants a and $b \in GF(2^m)$ and $b \neq 0$. The points on the elliptic curve together with the point at infinity ($\mathcal{O}$) form an Abelian group under addition with group operations *point addition* and *point doubling*. For a given point P on the curve (called the *base point*)

Algorithm 1. *Montgomery Point Multiplication*

Input: Base point P and scalar $s = \{s_{t-1}s_{t-2}\cdots s_0\}_2$ with $s_{t-1} = 1$
Output: Point on the curve $Q = sP$
1 **begin**
2     $P_1(X_1, Z_1) \leftarrow P(X, Z)$; $P_2(X_2, Z_2) \leftarrow 2P(X, Z)$
3   **for** $k = t - 2$ *to* 0 **do**
4   **if** $s_k = 1$ **then**
5             $P_1 \leftarrow P_1 + P_2$
6             $P_2 \leftarrow 2P_2$
7   **end**
8   **else**
9             $P_2 \leftarrow P_1 + P_2$
10        $P_1 \leftarrow 2P_1$
11   **end**
12   **end**
13   **return** $Q \leftarrow Projective2Affine(P_1, P_2)$
14 **end**

---

and a scalar s, *scalar multiplication* is the computation of the scalar product sP. Algorithm 1 depicts the Montgomery algorithm [11,14] for computing sP. For each bit in s, a point addition followed by a point doubling is done (lines 5,6 and 9,10). In these operations (listed in Equation 1) only the X and Z coordinates of the points are used.

$$X_i \leftarrow X_i \cdot Z_j \;;\; Z_i \leftarrow X_j \cdot Z_i \;;\; T \leftarrow X_j \;;\; X_j \leftarrow X_j^4 + b \cdot Z_j^4$$
$$Z_j \leftarrow (T \cdot Z_j)^2 \;;\; T \leftarrow X_i \cdot Z_i \;;\; Z_i \leftarrow (X_i + Z_i)^2 \;;\; X_i \leftarrow x \cdot Z_i + T \tag{1}$$

Depending on the value of the bit s_k, operand and destination registers for the point operations vary. When $s_k = 1$ then $i = 1$ and $j = 2$, and when $s_k = 0$ then $i = 2$ and $j = 1$. The final step in the algorithm, $Projective2Affine(\cdot)$, converts the 3 coordinate scalar product in to the acceptable 2 coordinate affine form. This step involves a finite field inversion along with 9 other multiplications [13].

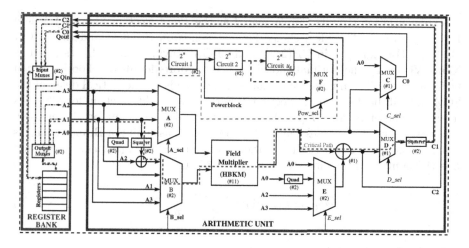

Fig. 1. Block Diagram of the Processor Organization

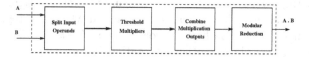

Fig. 2. Different Stages in the HBKM

3   The Processor Organization

The functionality of the ECM is to execute Algorithm 1. It comprises of 2 units: the register bank and the arithmetic unit as seen in Figure 1. In each clock cycle, control signals are generated according to the value of the bit s_k, which reads operands from the register bank, performs the computation in the arithmetic unit, and finally write back the results. In this section we present the architecture for the register bank and arithmetic units.

3.1   Arithmetic Unit

The *field multiplier* is the central part of the arithmetic unit. We choose to use a *hybrid bit-parallel Karatsuba field multiplier* (HBKM), which was first introduced in [18] and then used in [17]. The advantage of the HBKM is the sub-quadratic complexity of the Karatsuba algorithm coupled with efficient utilization of the FPGA's LUT resources. Further, the bit-parallel scheme requires lesser clock cycles compared to digit level multipliers used in [2]. The HBKM recursively splits the input operands until a threshold (τ) is reached, then threshold (school-book) multipliers are applied. The outputs of the threshold multipliers are combined and then reduced (Figure 2).

Field inversion is performed by a generalization of the Itoh-Tsujii inversion algorithm for FPGA platforms [19]. The generalization requires a cascade of

2^n exponentiation circuits (implemented as the *Powerblock* in Figure 1), where $1 \leq n \leq m - 1$. The ideal choice for n depends on the field and the FPGA platform. For example, in $GF(2^{163})$ and FPGAs having 4 input LUTs (such as Xilinx Virtex 4), the optimal choice for n is 2. More details on choosing n can be found in [8]. The number of cascades, u_s, depends on the critical delay of the ECM and will be discussed in Section 4. Further, an addition chain for $\lfloor \frac{m-1}{n} \rfloor$ is required. Therefore, for $GF(2^{163})$ and $n = 2$, an addition chain for 81 is needed. The number of clock cycles required for inversion, assuming a Brauer chain, is given by Equation 2, where the addition chain has the form $(u_1, u_2, \cdots, u_l)$, and L is the number of pipeline stages in the ECM [4].

$$cc_{ita} = L(l + 1) + \sum_{i=2}^{l} \left\lceil \frac{u_i - u_{i-1}}{u_s} \right\rceil \tag{2}$$

3.2 Register Bank

There are six registers in the register bank, each capable of storing a field element. Five of the registers are used for the computations in Equation 1, while one is used for field inversion. There are 3 ways in which the registers can be implemented in FPGAs. The first approach, using *block RAM*, is slow due to constraints in routing. The two other alternatives are *distributed RAM* and flip-flops. Distributed RAM allows the FPGA's LUTs to be configured as RAM. Each bit of the 6 registers will share the same LUT. However each register is used for a different purpose therefore the centralization effect of distributed RAM will cause long routes, leading to lowering of clock frequencies. Additionally, there is an impact on the area requirements. Flip-flops on the other hand allow de-centralization of the registers, there by allowing registers to be placed in locations close to their usage, thus routing is easier. Further, each slice in the FPGA has equal number of LUTs and flip-flops. The ECM is an LUT intensive design, due to which several of the flip-flops in the slice remain unutilized. By configuring the registers to make use of these flip-flops, no additional area (in terms of the number of slices) is required.

4 Pipelining the ECM

All combinational data paths in the ECM start from the register bank output and end at the register bank input. The maximum operable frequency of the ECM is dictated by the longest combinational path, known as the *critical path*. There can be several critical paths, one such example is highlighted through the (red) dashed line in Figure 1.

Estimating Delays in the ECM : Let t_{cp}^* be the delay of the critical paths and $f_1^* = \frac{1}{t_{cp}^*}$ the maximum operable frequency of the ECM prior to pipelining. Consider the case of pipelining the ECM into L stages, then the maximum operable frequency can be increased to at-most $f_L^* = L \times f_1^*$. This *ideal frequency* can be achieved if and only if the following two conditions are satisfied.

Table 1. LUT delays of Various Combinational Circuit Components

Component	$k - LUT$ Delay for k $\geq$ 4	$m = 163, k = 4$
m bit field adder	1	1
m bit $n : 1$ Mux $(D_{n:1}(m))$	$\lceil log_k(n + log_2n) \rceil$	2 (for $n = 4$) 1 (for $n = 2$)
Exponentiation Circuit $(D_{2^n}(m))$	$max(LUTDelay(d_i))$, where d_i is the ith output bit of the exponentiation circuit	2 (for $n = 1$) 2 (for $n = 2$)
Powerblock $(D_{powerblk}(m))$	$u_s \times D_{2^n}(m) + D_{u_s:1}(m)$	4 (for $u_s = 2$)
Modular Reduction (D_{mod})	1 for irreducible trinomials 2 for pentanomials	2 (for pentanomials)
HBKM $(D_{HBKM}(m))$	As seen in Figure 2, this can be written as $D_{split} + D_{threshold} + D_{combine} + D_{mod}$ $= \lceil log_k(\frac{m}{\tau}) \rceil + \lceil log_k(2\tau) \rceil$ $+ \lceil log_2(\frac{m}{\tau}) \rceil + D_{mod}$	11 (for $\tau = 11$)

1. Every critical path in the design should be split into L stages with each stage having a delay of exactly $\frac{t^*_{cp}}{L}$.
2. All other paths in the design should be split so that any stage in these paths should have a delay which is less than or equal to $\frac{t^*_{cp}}{L}$.

While it is not always possible to exactly obtain f^*_L, we can achieve close to the ideal clock frequency by making a theoretical estimation of t^*_{cp} and then identifying the locations in the architecture where the pipeline stages have to be inserted. We denote this theoretical estimate of delay by $t^{\#}_{cp}$. The theoretical analysis is based on the following prepositions. These propositions were first stated in [19] and used to design high-speed inversion circuits. Their correctness have been extensively validated in [19] for 4 and 6 input LUT based FPGAs.

Proposition 1. *[19] For circuits which are implemented using LUTs, the delay of a path in the circuit is proportional to the number of LUTs in the path.*

Proposition 2. *[19] The number of LUTs in the critical path of an n variable Boolean function having the form $y = g_n(x_1, x_2, \cdots, x_n)$ is given by $\lceil log_k(n) \rceil$, where k is the number of inputs to the LUTs (k $-$ LUT).*

Using these two propositions it is possible to analyze the delay of various combinational circuit components in terms of LUTs. The *LUT delays* of relevant combinational components are summarized in Table 1. The reader is referred to [19] for detailed analysis of the LUT delays. The LUT delays of all components in Figure 1 are shown in parenthesis for k = 4. Note that the analysis also considers optimizations by the synthesis tool (such as the merging of the squarer and adder before Mux B (Figure 1), which reduces the delay from 3 to 2).

Pipelining Paths in the ECM : Table 1 can be used to determine the LUT delays of any path in the ECM. For the example critical path, (the red dashed

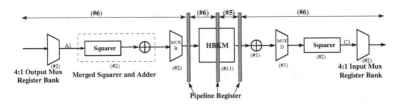

Fig. 3. Example Critical Path (with $L = 4$ and $k = 4$)

line in Figure 1), the estimate for t_{cp}^* is the sum of the LUT delays of each component in the path. This evaluates to $t_{cp}^\# = 23$. Figure 3 gives a detailed view of this path. Pipelining the paths in the ECM require pipeline registers to be introduced in between the LUTs. The following proposition determines how the pipeline registers have to be inserted in a path in order to achieve the maximum operable frequency $(f_L^\#)$ as close to the ideal (f_L^*) as possible (Note that $f_L^\# \leq f_L^*$).

Proposition 3. *If $t_{cp}^\#$ is the LUT delay of the critical paths, and L is the desired number of stages in the pipeline, then the best clock frequency $(f_L^\#)$ is achieved only if no path has delay more than $\lceil \frac{t_{cp}^\#}{L} \rceil$.*

For example for $L = 4$, no path should have a LUT delay more than $\lceil \frac{23}{4} \rceil$. This identifies the exact locations in the paths where pipeline registers have to be inserted. Figure 3 shows the positions of the pipeline register for $L = 4$ for the critical path.

On the Pipelining of the Powerblock :   The powerblock is used only once during the computation; at the end of the scalar multiplication. There are two choices with regard to implementing the powerblock, either pipeline the powerblock as per Proposition 3 or reduce the number of 2^n circuits in the cascade so that the following LUT delay condition is satisfied (refer Table 1),

$$D_{powerblock}(m) \leq \lceil \frac{t_{cp}^\#}{L} \rceil - 1 \tag{3}$$

, where -1 is due to the output mux in the register bank. However the sequential nature of the Itoh-Tsujii algorithm [7] ensures that the result of one step is used in the next. Due to the data dependencies which arise the algorithm is not suited for pipelining and hence the latter strategy is favored. For $k = 4$ and $m = 163$, the optimal exponentiation circuit is $n = 2$ having an LUT delay of 2 [19]. Thus a cascade of two 2^2 circuits would best satisfy the inequality in (3).

5   Scheduling for the ECM

In this section we discuss the scheduling of the addition-doubling loop in Algorithm 1. For each bit in the scalar (s_k), the eight operations in Equation 1 are

Table 2. Scheduling Instructions for the ECM

$c_1^k: X_i \leftarrow X_i \cdot Z_j$	$e_4^k: Z_j \leftarrow (T \cdot Z_j)^2$
$e_2^k: Z_i \leftarrow X_j \cdot Z_i$	$e_5^k: T \leftarrow X_i \cdot Z_i; \; Z_i \leftarrow (X_i + Z_i)^2$
$e_3^k: T \leftarrow X_j; \; X_j \leftarrow X_j^4 + b \cdot Z_j^4$	$e_6^k: X_i \leftarrow x \cdot Z_i + T$

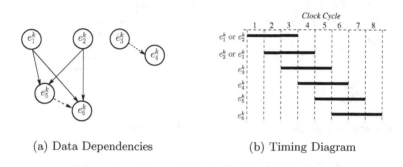

(a) Data Dependencies   (b) Timing Diagram

Fig. 4. Scheduling the Scalar Bit s_k

computed. Unlike [2], where 2 finite field multipliers are used in the architecture, we, like [5], use a single field multiplier. This restriction makes the field multiplier the most critical resource in the ECM as Equation 1 involves six field multiplications, which have to be done sequentially. The remaining operations comprise of additions, squarings, and data transfers can be done in parallel with the multiplications. Equation 1 can be rewritten as in Table 5 using 6 instructions, with each instruction capable of executing simultaneously in the ECM.

Proper scheduling of the 6 instructions is required to minimize the impact of data dependencies, thus reducing pipeline stalls. The dependencies between the instructions e_1^k to e_6^k are shown in Figure 4(a). In the figure a *solid arrow* implies that the subsequent instruction cannot be started unless the previous instruction has completed, while a *dashed arrow* implies that the subsequent instruction cannot be started unless the previous instruction has started. For example e_6^k uses Z_i, which is updated in e_5^k. Since the update does not require a multiplication (an addition followed by a squaring here), it is completed in one clock cycle. Thus e_5^k to e_6^k has a dashed arrow, and e_6^k can start one clock cycle after e_5^k. On the other hand, dependencies depicted with the solid arrow involve the multiplier output in the former instruction. This will take L clock cycles, therefore a longer wait.

The dependency diagram shows that in the longest dependency chain, e_5^k and e_6^k has dependency on e_1^k and e_2^k. Thus e_1^k and e_2^k are scheduled before e_5^k and e_4^k. Since the addition in e_5^k has a dependency on e_2^k, operation e_5^k is triggered just after completion of e_1^k and e_2^k; and operation e_6^k is triggered in the next clock cycle. When $L \geq 3$, the interval between starting and completion of e_1^k and e_2^k can be utilized by scheduling e_3^k and e_4^k. Thus, the possible scheduling schemes for the 6 instructions is

$$(\{e_1^k, e_2^k\}, e_3^k, e_4^k, e_5^k, e_6^k) \tag{4}$$

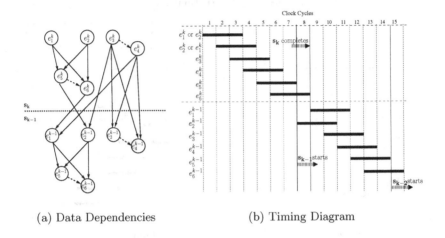

(a) Data Dependencies   (b) Timing Diagram

Fig. 5. Schedule for Two Scalar Bits when $s_{k-1} = s_k$

Where $\{\}$ implies that there is no strict order in the scheduling (either e_1 or e_2 can be scheduled first). An example of a scheduling for $L = 3$ is shown in Figure 4(b). For $L \geq 3^1$, the number of clock cycles required for each bit in the scalar is $2L + 2$. In the next part of this section we show that the clock cycles can be reduced to $2L + 1$ (and in some cases $2L$) if two consecutive bits of the scalar are considered.

5.1   Scheduling for Two Consecutive Bits of the Scalar

Consider the scheduling of operations for two bits of the scalar, s_k and s_{k-1} (Algorithm 1). We assume that the computation of bit s_k is completed and the next bit s_{k-1} is to be scheduled. Two cases arise: $s_{k-1} = s_k$ and $s_{k-1} \neq s_k$. We consider each case separately.

When the Consecutive Key Bits Are Equal :   Figure 5(a) shows the data dependencies when the two bits are equal. The last two instructions to complete for the s_k bit are e_5^k and e_6^k. For the subsequent bit (s_{k-1}), either e_1^{k-1} or e_2^{k-1} has to be scheduled first according to the sequence in (4). We see from Figure 5(a) that e_1^{k-1} depends on e_6^k, while e_2^{k-1} depends on e_5^k. Further, since e_5^k completes earlier than e_6^k, we schedule e_2^{k-1} before e_1^{k-1}. Thus the scheduling for 2 consecutive equal bits is

$$(\{e_1^k, e_2^k\}, e_3^k, e_4^k, e_5^k, e_6^k, e_2^{k-1}, e_1^{k-1}, e_3^{k-1}, e_4^{k-1}, e_5^{k-1}, e_6^{k-1})$$

An example is shown in Figure 5(b).

---

[1] The special case of $L <= 2$ can trivially be analyzed. The clock cycles required in this case is six.

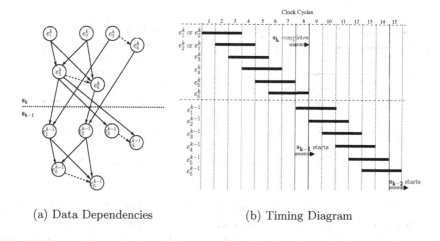

(a) Data Dependencies   (b) Timing Diagram

Fig. 6. Schedule for Two Scalar Bits when $s_{k-1} \neq s_k$

When the Consecutive Key Bits Are Not Equal : Figure 6(a) shows the data dependency for two consecutive scalar bits that are not equal. Here it can be seen that e_1^{k-1} and e_2^{k-1} depend on e_6^k and e_5^k respectively. Since, e_5^k completes before e_6^k, we schedule e_1^{k-1} before e_2^{k-1}. The scheduling for two consecutive bits is as follows

$$(\{e_1^k , e_2^k\} , e_3^k , e_4^k , e_5^k , e_6^k , e_1^{k-1}, e_2^{k-1}, e_3^{k-1}, e_4^{k-1}, e_5^{k-1}, e_6^{k-1})$$

An example is shown in Figure 6(b).

Effective Clock Cycle Requirement : Starting from e_1^k (or e_2^k), completion of e_2^k (or e_1^k) takes $L + 1$ clock cycles, for an L stage pipelined ECM. After completion of e_1^k and e_2^k, e_5^k starts. This is followed by e_6^k in the next clock cycle. So in all $2L + 2$ clock cycles are required. The last clock cycle however is also used for the next bit of the scalar. So effectively the clock cycles required per bit is $2L + 1$. Compared to the work in [5], our scheduling strategy saves two clock cycles for each bit of the scalar. For an m bit scalar, the saving in clock cycles compared to [5] is $2m$. Certain values of L allow data forwarding to take place. In such cases the clock cycles per bit reduces to $2L$, thus saving $3m$ clock cycles compared to [5].

5.2   Data Forwarding to Reduce Clock Cycles

For a given value of L, Proposition 3 specifies where the pipeline registers have to be placed in the ECM. If the value of L is such that a pipeline register is placed at the output of the field multiplier, then data forwarding can be applied to save one clock cycle per scalar bit. For example, consider $L = 4$. This has pipeline registers placed immediately after the multiplier as shown in Figure 3. This register can

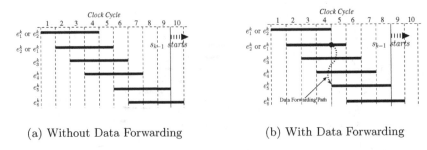

(a) Without Data Forwarding   (b) With Data Forwarding

Fig. 7. Effect of Data Forwarding in the ECM for $L = 4$

be used to start the instruction e_5^k one clock cycle earlier. Figure 5.1 compares the execution of a single bit with and without data forwarding. Though e_2^k (or e_1^k) finishes in the fifth clock cycle, the result of the multiplication is latched into the pipeline register after the fourth clock cycle. With data forwarding from this register, we start e_5^k from the fifth clock cycle, thus reducing clock cycle requirement by one to $2L$.

6   Finding the Right Pipeline

The time taken for a scalar multiplication in the ECM is the product of the number of clock cycles required and the time period of the clock. For an L stage pipeline, Section 4 determines the best time period for the clock. In this section we would first estimate the number of clock cycles required and then analyze the effect of L on the computation time.

6.1   Number of Clock Cycles

There are two parts in Algorithm 1. First the scalar multiplication in projective coordinates and then the conversion to affine coordinates. The conversion comprises of finding an inverse and 9 multiplications. The clock cycles required is given by $cc_{2scm} = cc_{3scm} + cc_{ita} + cc_{conv}$.

cc_{3scm} is the clock cycles required for the scalar multiplication in projective coordinates. From the analysis in Section 5 this can be written as $2mL$ if data forwarding is possible and $m(2L + 1)$ otherwise. For the conversion to affine coordinates, finding the inverse requires cc_{ita} clock cycles (from Equation 2), while the 9 multiplications following the inverse requires cc_{conv} clock cycles. The value of cc_{conv} for the ECM was found to be $7 + 9L$. Thus,

$$cc_{2scm} = \left[cc_{3scm}\right] + \left[L(l+1) + \sum_{i=2}^{l}\left\lceil\frac{u_i - u_{i-1}}{u_s}\right\rceil\right] + \left[7 + 9L\right] \qquad (5)$$

Table 3. Computation Time Estimates for Various Values of L for an ECM over $GF(2^{163}$ and FPGA with 4 input LUTs

L	u_s	DataForwarding Feasible	cc_{3scm}	cc_{ita}	cc_{conv}	cc_{2scm}	ct
1	9	No	978	25	16	1019	$1019t_{cp}^{\#}$
2	4	No	978	44	25	1047	$524t_{cp}^{\#}$
3	3	No	1141	61	34	1236	$412t_{cp}^{\#}$
4	2	Yes	1304	82	43	1429	$357t_{cp}^{\#}$
5	1	No	1793	130	52	1975	$395t_{cp}^{\#}$
6	1	Yes	1956	140	61	2157	$360t_{cp}^{\#}$
7	1	Yes	2282	150	70	2502	$358t_{cp}^{\#}$

6.2 Analyzing Computation Time

The procedure involved in analyzing the computation time for an L stage pipeline is as follows.

1. Determine $t_{cp}^{\#}$ (the LUT delay of the critical path of the combinational circuit) using Table 1.

2. Compute the maximum operable frequency ($\lceil \frac{t_{cp}^{\#}}{L} \rceil$) and determine the locations of the pipeline registers. Therefore determine if data forwarding is possible.

3. Determine u_s, the number of cascades in the power block, using Equation 3 and the delay of a single 2^n block (Table 1).

4. Compute cc_{2scm}, using Equation 5.

5. The computation time ct is given by $cc_{2scm} \times \lceil \frac{t_{cp}^{\#}}{L} \rceil$.

For an ECM over $GF(2^{163})$, the threshold for the HBKM set as 11, an addition chain of $(1, 2, 4, 5, 10, 20, 40, 80, 81)$, and 2^2 exponentiation circuits in the power block, the $t_{cp}^{\#}$ is 23. The estimated computation time for various values of L are given in Table 3. The cases $L = 1$ and $L = 2$ are special as for these $cc_{3scm} = 6m$. The table clearly shows that the least computation time is obtained when $L = 4$.

7 Detailed Architecture of the ECM

Figure 8 shows the detailed architecture for $L = 4$. The input to the architecture is the scalar, reset signal, and the clock. At reset, the curve constants and base point are loaded from ROM. At every clock cycle, the control unit generates signals for the register bank and the arithmetic unit. Registers are selected through multiplexers in the register bank and fed to the arithmetic unit through the buses $A0$, $A1$, $A2$, $A3$, and Qin. Multiplexers again channel the data into the multiplier. The results are written back into the registers through buses $C0$, $C1$, $C2$, $Qout$. Note the placement of the pipeline registers dividing the circuit in 4 stages and ensuring that each stage has an LUT delay which is less than or equal to $\lceil \frac{23}{4} \rceil = 6$. Note also the pipeline register present immediately after the field multiplier (HBKM) used for data forwarding.

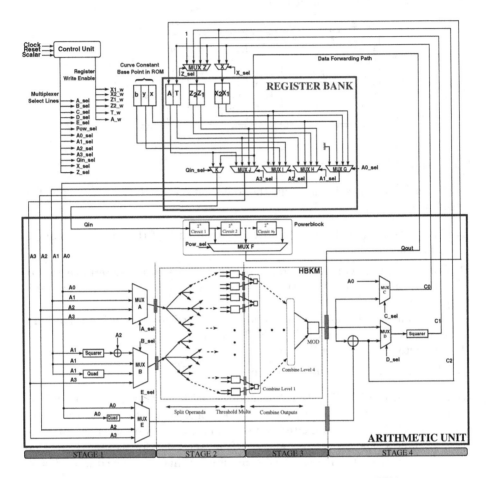

Fig. 8. Detailed Architecture for a 4 Stage Pipelined ECM

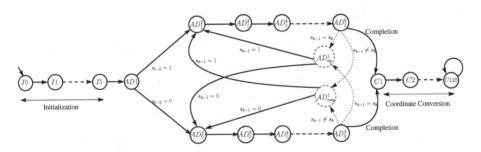

Fig. 9. Finite State Machine for 4-Stage ECM

Table 4. Comparison of the Proposed ECM with FPGA based Published Results

Work	Platform	Field (m)	Slices	LUTs	Freq (MHz)	Comp. Time (μs)
Orlando [15]	XCV400E	163	-	3002	76.7	210
Bednara [3]	XCV1000	191	-	48300	36	270
Gura [6]	XCV2000E	163	-	19508	66.5	140
Lutz [12]	XCV2000E	163	-	10017	66	233
Saqib [20]	XCV3200	191	18314	-	10	56
Pu [16]	XC2V1000	193	-	3601	115	167
Ansari [1]	XC2V2000	163	-	8300	100	42
Rebeiro [17]	XC4V140	233	19674	37073	60	31
Järvinen[1] [9]	Stratix II	163	(11800ALMs)	-	-	48.9
Kim [2] [10]	XC4VLX80	163	24363	-	143	10.1
Chelton [5]	XCV2600E	163	15368	26390	91	33
	XC4V200	163	16209	26364	153.9	19.5
Azarderakhsh[3] [2]	XC4CLX100	163	12834	22815	196	17.2
	XC5VLX110	163	6536	17305	262	12.9
Our Result (Virtex 4 FPGA)	XC4VLX80	163	8070	14265	147	9.7
	XC4V200	163	8095	14507	132	10.7
	XC4VLX100	233	13620	23147	154	12.5
Our Result (Virtex 5 FPGA)	XC5VLX85t	163	3446	10176	167	8.6
	XC5VSX240	163	3513	10195	148	9.5
	XC5VLX85t	233	5644	18097	156	12.3

1. uses 4 field multipliers; 2. uses 3 field multipliers; 3. uses 2 field multipliers

Figure 9 shows the finite state machine for $L = 4$. The states $I0$ to $I5$ are used for initialization (line 2 in Algorithm 1). State $AD1$ represents the first clock cycle for the scalar bit s_{t-2}. States AD_1^2 to AD_1^9 represent the computations when $s_k = 1$, while AD_0^2 to AD_0^9 are for $s_k = 0$. Each state corresponds to a clock cycle in Figure 7(b). Processing for the next scalar bit (s_{k-1}) begins in the same clock cycle as AD_0^9 and AD_1^9 in states AD_{eq}^1 and AD_{neq}^1. The states AD_{eq}^1 or AD_{neq}^1 are entered depending on the equality of s_k and s_{k-1}. If $s_k = s_{k-1}$ then AD_{eq}^1 is entered, else AD_{neq}^1 is entered. After processing of all scalar bits is complete, the conversion to affine coordinates ($cc_{ita} + cc_{conv}$) takes place in states C_1 to C_{125}.

8 Implementation Results and Comparisons

We evaluated the ECM using Xilinx Virtex 4 and Virtex 5 platforms. Table 4 shows the place and route results using the Xilinx ISE tool. There have been several implementations of elliptic curve processors on different fields, curves, platforms, and for different applications. Due to the vast variety of implementations available, we restrict comparisons with FPGA implementations for generic elliptic curves over binary finite fields (Table 4). In this section we analyze recent high-speed implementations.

The implementation in [17] is over the field $GF(2^{233})$ and does a scalar multiplication in $31\mu s$. The implementation relied heavily on optimized finite-field primitives and was not pipelined or parallelized. Our implementation on the same field uses enhanced primitives from [17], and therefore has smaller area requirements. Additionally higher speeds are achieved due to efficient pipelining and scheduling of instructions.

The implementation in [5] uses a 7 stage pipeline, thus achieves high operating clock frequency. However, the un-optimized pipeline and large clock cycle requirement limits performance. In comparison, the ECM uses better scheduling there by saving around 1,600 clock cycles and a better pipeline, there by obtaining frequencies close to [5], in-spite of having only 4 pipeline stages. Further, efficient field primitives and the sub-quadratic Karatsuba multiplier instead of the quadratic Mastrovito multiplier result in 50% reduction in area on Virtex 4.

In [2], two highly optimized digit field multipliers were used. This enabled parallelization of the instructions and higher clock frequency. However, the use of digit field multipliers resulted in large clock cycle requirement for scalar multiplication (estimated at 3,380). We use a single fully parallel field multiplier requiring only 1,429 clock cycles and an area which is 37% lesser in Virtex 4.

In [10], a computation time of $10.1\mu s$ was achieved while on the same platform our ECM achieves a computation time of $9.7\mu s$. Although the speed gained is minimal, it should be noted that [10] uses 3 digit-level finite field multipliers compared to one in ours, thus has an area requirement which is about 3 times ours. The compact area is useful especially for cryptanalytic applications where our ECM can test thrice as many keys compared to [10].

9   Conclusion

The papers presents techniques to reduce the computation time for scalar multiplications on elliptic curves. The techniques involve the use of highly optimized finite field primitives and efficient utilization of FPGA resources in order to reduce the area requirements, which in turn leads to better routing, hence higher clock frequencies. Additionally, a theoretical analysis of the data paths, help pipeline the multiplier. Further, efficient scheduling of elliptic curve operations, supported with data-forwarding mechanisms, reduce the number of clock cycles required to execute a scalar multiplication. These mechanisms result in a scalar multiplier that is faster than any other reported implementations, in-spite of having just a single finite field multiplier. The presence of a single optimized field multiplier additionally leads to area requirements, which is considerably lesser than contemporary implementations. Results are presented for generic curves over the field $GF(2^{163})$, however these mechanisms can be applied for other curves and fields as well.

References

1. Ansari, B., Hasan, M.: High-performance architecture of elliptic curve scalar multiplication. IEEE Transactions on Computers 57(11), 1443–1453 (2008)
2. Azarderakhsh, R., Reyhani-Masoleh, A.: Efficient FPGA Implementations of Point Multiplication on Binary Edwards and Generalized Hessian Curves Using Gaussian Normal Basis. IEEE Transactions on Very Large Scale Integration (VLSI) Systems PP(99), 1 (2011)
3. Bednara, M., Daldrup, M., von zur Gathen, J., Shokrollahi, J., Teich, J.: Reconfigurable Implementation of Elliptic Curve Crypto Algorithms. In: Proceedings of the International Parallel and Distributed Processing Symposium, IPDPS 2002, Abstracts and CD-ROM, pp. 157–164 (2002)

4. Rebeiro, C., Roy, S.S., Reddy, D.S., Mukhopadhyay, D.: Revisiting the Itoh Tsujii Inversion Algorithm for FPGA Platforms. IEEE Transactions on VLSI Systems 19(8), 1508–1512 (2011)
5. Chelton, W.N., Benaissa, M.: Fast Elliptic Curve Cryptography on FPGA. IEEE Transactions on Very Large Scale Integration (VLSI) Systems 16(2), 198–205 (2008)
6. Gura, N., Shantz, S.C., Eberle, H., Gupta, S., Gupta, V., Finchelstein, D., Goupy, E., Stebila, D.: An End-to-End Systems Approach to Elliptic Curve Cryptography. In: Kaliski Jr., B.S., Koç, Ç.K., Paar, C. (eds.) CHES 2002. LNCS, vol. 2523, pp. 349–365. Springer, Heidelberg (2003)
7. Itoh, T., Tsujii, S.: A Fast Algorithm For Computing Multiplicative Inverses in $GF(2^m)$ Using Normal Bases. Inf. Comput. 78(3), 171–177 (1988)
8. Järvinen, K.U.: On Repeated Squarings in Binary Fields. In: Jacobson Jr., M.J., Rijmen, V., Safavi-Naini, R. (eds.) SAC 2009. LNCS, vol. 5867, pp. 331–349. Springer, Heidelberg (2009)
9. Järvinen, K., Skytta, J.: On parallelization of high-speed processors for elliptic curve cryptography. IEEE Transactions on Very Large Scale Integration (VLSI) Systems 16(9), 1162–1175 (2008)
10. Kim, C.H., Kwon, S., Hong, C.P.: FPGA Implementation of High Performance Elliptic Curve Cryptographic processor over $GF(2^{163})$. Journal of Systems Architecture - Embedded Systems Design 54(10), 893–900 (2008)
11. López, J., Dahab, R.: Fast Multiplication on Elliptic Curves over $GF(2^m)$ without Precomputation. In: Koç, Ç.K., Paar, C. (eds.) CHES 1999. LNCS, vol. 1717, pp. 316–327. Springer, Heidelberg (1999)
12. Lutz, J., Hasan, A.: High Performance FPGA based Elliptic Curve Cryptographic Co-Processor. In: ITCC 2004: Proceedings of the International Conference on Information Technology: Coding and Computing (ITCC 2004), vol. 2, p. 486. IEEE Computer Society, Washington, DC (2004)
13. Menezes, A.J., van Oorschot, P.C., Vanstone, S.A.: Handbook of Applied Cryptography. CRC Press (2001)
14. Montgomery, P.L.: Speeding the pollard and elliptic curve methods of factorization. Mathematics of Computation 48, 243–264 (1987)
15. Orlando, G., Paar, C.: A High-Performance Reconfigurable Elliptic Curve Processor for $GF(2^m)$. In: Paar, C., Koç, Ç.K. (eds.) CHES 2000. LNCS, vol. 1965, pp. 41–56. Springer, Heidelberg (2000)
16. Pu, Q., Huang, J.: A Microcoded Elliptic Curve Processor for $GF(2^m)$ Using FPGA Technology. In: 2006 International Conference on Communications, Circuits and Systems Proceedings, vol. 4, pp. 2771–2775 (June 2006)
17. Rebeiro, C., Mukhopadhyay, D.: High Speed Compact Elliptic Curve Cryptoprocessor for FPGA Platforms. In: Chowdhury, D.R., Rijmen, V., Das, A. (eds.) INDOCRYPT 2008. LNCS, vol. 5365, pp. 376–388. Springer, Heidelberg (2008)
18. Rebeiro, C., Mukhopadhyay, D.: Power Attack Resistant Efficient FPGA Architecture for Karatsuba Multiplier. In: VLSID 2008: Proceedings of the 21st International Conference on VLSI Design, pp. 706–711. IEEE Computer Society, Washington, DC (2008)
19. Roy, S.S., Rebeiro, C., Mukhopadhyay, D.: Theoretical Modeling of the Itoh-Tsujii Inversion Algorithm for Enhanced Performance on k-LUT based FPGAs. In: Design, Automation, and Test in Europe, DATE 2011 (2011)

20. Saqib, N.A., Rodríiguez-Henríquez, F., Diaz-Perez, A.: A Parallel Architecture for Fast Computation of Elliptic Curve Scalar Multiplication Over $GF(2^m)$. In: Proceedings of the 18th International Parallel and Distributed Processing Symposium (April 2004)
21. I. C. Society. IEEE Standard Specifications for Public-key Cryptography (2000)
22. U.S. Department of Commerce, National Institute of Standards and Technology. Digital signature standard (DSS) (2000)
23. Wollinger, T., Guajardo, J., Paar, C.: Security on FPGAs: State-of-the-art Implementations and Attacks. Trans. on Embedded Computing Sys. 3(3), 534–574 (2004)

Appendix A

In this appendix we summarize the ideal design parameters for k = 4 (Xilinx Virtex 4 FPGA) and k = 6 (Xilinx Virtex 5) for the field $GF(2^{163})$.

Table 5. Summary of Design Parameters for $GF(2^{163})$ for k = 4 and k = 6

Parameter	k = 4	k = 6
Threshold used in HBKM (τ)	11	11
Exponentiation Circuit in Powerblock	2^2 circuit (Quad)	2^4 circuit
Addition Chain	(1, 2, 4, 5, 10, 20, 40, 80, 81)	(1, 2, 4, 5, 10, 20, 40)
Number of Cascades in Powerblock (u_s)	2	1
LUT Delay ($t_{cp}^{\#}$)	23	17
Ideal Number of Pipeline Stages (L)	4	4

Table 6. LUT requirement for different Primitives in $GF(2^{163})$

Primitives	No. of Instances	LUTs in Virtex 4 per unit	total	LUTs in Virtex 5 per unit	total
Adder	1	163	163	163	163
Squarer	1	163	163	163	163
Adder merged with Squarer[1]	1	163	163	163	163
Quad Circuit[2]	4	315	1260	249	996
Mux 2:1[3]	5, 4	163	815	163	652
Mux 4:1	8	326	2608	163	1304
Multiplier	1	9092	9092	6313	6313
Total	-	-	14264	-	9754

1. This is present before Mux B
2. Two of these are present in the Powerblock
3. On Virtex 4, Mux F is 2 : 1. On Virtex 5, this is not required as there is single 2^4 circuit

Appendix B

In this appendix we present more details about the implementation. In order to understand how the FPGA's LUTs have been utilized, we have synthesized each module individually. Table 6 gives the details according to Figure 8. It may be noted that these results may not exactly match the results in Table 4 because (1) they have been synthesized individually (2) and it does not have the top module which contains the control unit.

For the Virtex 5 FPGA, the Powerblock should ideally have a single 2^4 circuit as seen in Table 5. This we have implemented using a cascade of two quad circuits.

The critical path for the design (both in Virtex 4 and Virtex 5) obtained from the Xilinx tool, was through Mux H (in the register bank), the quad circuit, and then the Mux B (refer Figure 8). This path is present in the first stage of the pipeline and corresponds to the maximum operating clock frequency specified in Table 4.

On the Design of Hardware Building Blocks for Modern Lattice-Based Encryption Schemes

Norman Göttert, Thomas Feller, Michael Schneider,
Johannes Buchmann, and Sorin Huss

CASED - Center for Advanced Security Research Darmstadt
Technische Universität Darmstadt, Germany
{norman.gottert,thomas.feller,michael.schneider,
johannes.buchmann,sorin.huss}@cased.de

Abstract. We present both a hardware and a software implementation variant of the *learning with errors* (LWE) based cryptosystem presented by Lindner and Peikert. This work helps in assessing the practicality of lattice-based encryption. For the software implementation, we give a comparison between a matrix and polynomial based variant of the LWE scheme. This module includes multiplication in polynomial rings using Fast Fourier Transform (FFT). In order to implement lattice-based cryptography in an efficient way, it is crucial to apply the systems over polynomial rings. FFT speeds up multiplication in polynomial rings, which is the most critical operation in lattice-based cryptography, from quadratic to quasi-linear runtime. For the hardware variant, we show how this fundamental building block of lattice-based cryptography can be implemented and evaluated in terms of performance. A second important component for lattice-based cryptosystems is the sampling from discrete Gaussian distributions. We examine three different variants for sampling Gaussian distributed integers, namely rejection sampling, a rounding based approach, and a look-up table based approach in hardware.

Keywords: LWE, Lattice-Based Encryption, Hardware, FPGA.

1 Introduction

Lattice-based cryptography is currently enjoying high attention in the cryptographic community. Related systems offer an alternative security background to factoring and discrete logarithm based schemes. Moreover, while the latter two may be broken using quantum computers, so far there is no quantum computer algorithm known that solves hard lattice problems faster than classical algorithms. Unlike factoring and discrete logarithms, there are even no subexponential time attacks known against lattice systems on classical computers. Last, but not least, lattice-based cryptosystems only apply simple and fast arithmetic operations and asymptotically allow for quasi-linear runtimes, which is nearly optimal. Lattice-based schemes are usually accompanied by very strong security proofs, which relate breaking the system to solving worst-case problems in lattices (compared to basing the security on average-case problems only, like we

E. Prouff and P. Schaumont (Eds.): CHES 2012, LNCS 7428, pp. 512–529, 2012.

know from other areas of cryptography). All these facts distinguish lattice-based cryptosystems as promising candidates to replace systems based on number theoretic problems, like factoring and computing discrete logarithms.

Originally, most lattice systems require the storage of huge matrices over integer rings and are quite inefficient both in runtime and storage space. The idea of replacing matrices by polynomials over ideals in integer rings allows to reduce both. Hence, replacing lattices by ideal lattices results in very efficient systems. Instead of storing huge matrices of space $\mathcal{O}(n^2)$, where n is larger than 128, it is sufficient to store just $\mathcal{O}(n \log n)$ elements. Moreover, the multiplication of elements of ideal lattices can be performed efficiently using the Fast Fourier Transform (FFT) [CT65] in time $\mathcal{O}(n \log n)$ for a serial and in $\mathcal{O}(\log n)$ for a parallel implementation, instead of $\mathcal{O}(n^2)$ for straightforward multiplication.

Based on lattice problems, many cryptographic primitives were already developed in theory. Among others, there are a hash function [ADL+08], digital signatures [Lyu09], encryption schemes [SS11, LP11], fully homomorphic encrpytion [Gen09], and many more. The security of most encryption schemes is based on the *learning with errors* problem (LWE). Regev [Reg05] and Peikert [Pei09] proved that the LWE problem is at least as hard as solving certain lattice problems in the worst case, which is the background of the strong security of LWE-based cryptosystems. What is missing for nearly all lattice-based encryption systems, however, are implementations. To the best of our knowledge, there is no publicly available implementation of any provably secure lattice-based cryptosystem available yet. The NTRUEncrypt system [HPS98] is a special case, where implementations are provided, but they are protected by patents. Further, the original NTRUEncrypt scheme lacks a security proof. In order to show that lattice-based cryptography is ready for practical real-world applications, the schemes have to be implemented first. The asymptotic advantage gained by FFT is well-known, however, it lacks a practical evaluation for this application.

For sampling of Gaussian distributed integers, the situation is similar. The theoretical evaluation of rejection sampling is known meanwhile, but the practical efficiency of this approach is still unclear. We are also not aware of any comparison to the rounding-based approach as presented by Devroye in [Dev86].

When lattice-based cryptography is to be used in practice, efficient hardware components are required as well. Therefore, it is necessary to investigate into design optimizations of current cryptosystems as hardware modules. Hence, we present efficient hardware modules for the fundamental building blocks of lattice-based encryption schemes, such as the FFT-based polynomial multiplication and a Gaussian sampler.

In addition to providing a reference software implementation of the matrix and polynomial variants of the encryption scheme, we detail a fully engineered hardware implementation using FFT for polynomial multiplication. Furthermore, an evaluation of all these implementation variants is given.

1.1   Related Work

Regev introduced the first worst-case hardness proof for the LWE problem together with the first LWE-based encryption scheme in 2005 [Reg05]. Various improvements of this scheme appeared later, such as [ACPS09, LPR10, Mic10]. In 2011, Lindner and Peikert proposed in [LP11] an adaption of the system of [Mic10] and various efficiency improvements. This is the most recent and at the same time the most promising LWE-based encryption scheme. These authors detail an improved security analysis and multiple parameter sets for different security levels. They introduce a matrix-based variant as well as an instantiation based on polynomials over residue rings.

The encryption scheme of [SS11] is a variant of the NTRUEncrypt system equipped with a quantum security reduction. Its security is based on the LWE problem in polynomial rings in the standard model. To our knowledge there is no practical investigation of this scheme available so far. We expect its efficiency to be comparable to the LWE scheme of [LP11].

1.2   Our Contribution

To the best of our knowledge we present the first practical evaluation of an LWE-based cryptosystem. This paper details a performance comparison of two realization variants based on matrix and polynomial operations, whereas the latter one is using FFT for fast multiplication. Further, real-world implications are evaluated by a comparison of the measured error-rate to theoretical expectations. Moreover, we present an optimal set of parameters for hardware implementations, which allow for optimizations of the largest hardware modules. In our hardware implementation we apply the FFT approach of [CLRS09]. Additionally, well-known improvements related to the inverse FFT allow for the removal of half of the residue class multipliers, together with the reduction of the critical path and hence provide a higher performance.

We propose efficient hardware modules for evaluation of the FFT as well as for the discrete Gaussian sampler. Both modules can be used for numerous lattice-based encryption [LPR10, LP11] and signature schemes as in, e.g. [GPV08, Lyu12].

Our experiments illustrate the sizes of the keys, the message expansion factor, and the timing results for the hardware and software implementations. The polynomial variant of LWE performes as expected: The size of private and public keys grows linearly in the security parameter. For the medium security parameters given in [LP11], the size of the secret key is 0.5 KB, whereas the public key is 1 KB in size. In software, key generation for the same security level takes $3.1ms$, whereas encryption and decryption take $1.5ms$ and $0.6ms$, respectively. In contrast, dedicated hardware modules speed-up encryption by a factor of nearly 200 and decryption by nearly 70 for this level of security. The message expansion factor, i.e., the size of a ciphertext divided by the size of a plaintext, is about 50. While the polynomial variant is superior in all other measured characteristics, the matrix variant features smaller message expansion factors.

2   Preliminaries

The authors of [LP11] propose two implementation variants, which exploit either
a matrix-based or a polynomial-based representation. For this reason we refer to
these representations as LWE-Matrix and LWE-Polynomial, respectively.

2.1   The LWE Problem

The security background of the cryptosystem under examination is the LWE
problem, which was introduced by Regev in 2005.

Consider a dimension $n > 1$, an integer module $p \geq 2$ and an error distribution
χ. The distribution χ will be the discrete Gaussian error distribution. Given a
vector $s \in \mathbb{Z}_q^n$, a vector $a \in \mathbb{Z}_q^n$ is chosen uniformly at random. Further, an
error term $e \leftarrow \chi$ is chosen and the pair $(a, t = \langle a|s \rangle + e \mod p)$ is computed.
The search version of LWE asks to find s given an arbitrary number of sample
pairs (a_i, t_i). The decision version of LWE asks to distinguish between arbitrary
numbers of sample pairs (a_i, t_i) and uniformly drawn samples from $\mathbb{Z}_q^n \times \mathbb{Z}_q$.

For hardness results on LWE we refer the reader to the work of [Reg05, Pei09].
Practical attacks on the LWE-based cryptosystems were described in [LP11]. The
ring LWE problem defined in [LPR10] is the adaption of LWE to polynomial
rings. It is important as security background for the LWE-Polynomial scheme.
An attacker breaking the LWE-Polynomial encryption system is able to solve the
ring LWE problem instance, and thus is able to solve certain lattice problems in
all lattices of a certain smaller dimension (the so-called *worst-case hardness*).

2.2   LWE-Based Encryption

Here we recall the more efficient polynomial variant, for the matrix variant we
refer to [Mic10, LP11]. Define the polynomial rings $R = \mathbb{Z}[X]/\langle f(x) \rangle$ and $R_q = \mathbb{Z}_q[X]/\langle f(x) \rangle$ for a polynomial $f(x)$ that is monic and irreducible. Example
choices are $f(x) = x^n + 1$ for n being a power of 2. Further, χ_k and χ_e are
error distributions over R for key generation and encryption, respectively. Useful
parameters for different levels of security were presented in [LPR10, LP11].

The LWE-Polynomial encryption is denoted as (KeyGen, Enc, Dec), where

- KeyGen(a): choose $r_1, r_2 \leftarrow \chi_k$ and let $p = r_1 - a \cdot r_2$. Output public key p
  and secret key r_2.
- Enc($a, p, m \in \Sigma^n$): choose $e_1, e_2, e_3 \leftarrow \chi_e$. Let $\bar{m} = \text{encode}(m) \in R_q$. The
  ciphertext is then $(c_1 = a \cdot e_1 + e_2, c_2 = p \cdot e_1 + e_3 + \bar{m}) \in R_q^2$.
- Dec($(c_1, c_2), r_2$): output $\text{decode}(c_1 \cdot r_2 + c_2)$.

Decoding fails if $|e_1 \cdot r_1 + e_2 \cdot r_2 + e_3|$ is bigger than the threshold $t = \lfloor q/4 \rfloor$. This
per-symbol error probability is denoted δ. It is depending on the error distribu-
tions χ_k and χ_e. More exactly, δ is an upper bound on the error probability per
symbol. Following the proposal of [LP11], we choose $\chi_k = \chi_e = \chi$. The Gaussian

standard deviation s for the distribution χ is selected depending on the dimension n, threshold t, parameter c, and the error probability δ by means of

$$s^2 = \frac{\sqrt{2\pi}}{c} \cdot \frac{t}{\sqrt{2n \cdot ln(2/\delta)}}. \qquad (1)$$

The variant LWE-Matrix is denoted in a similar manner, it uses matrices over $\mathbb{Z}_q$ instead of polynomials over the ring R. Since the arithmetic in polynomial rings can be performed more efficiently, the polynomial variant seems to be more appropriate in practice. A possible disadvantage of LWE-Polynomial concerns security. The system is provably secure as long as the decision ring LWE problem is hard. LWE-Matrix only requires the decision LWE problem to be hard, which is a weaker assumption, since it is unknown if the additional ring structure influences the hardness of the LWE problem.

Message Encoding. Error-tolerant encoder and decoder functions are required by the presented encryption system. In the following the message encoding of LWE-Polynomial is detailed, which can analogously be applied to LWE-Matrix.

A message m, represented as a bit-vector $m \in \Sigma^n = \{0,1\}^n$, is transformed into a vector $\bar{m} \in R_q$. Therefore, the encoding and decoding are functions encode:$\Sigma \to R_q$ and decode:$R_q \to \Sigma$, respectively. The equation $encode(decode(m) + e \ mod \ q) = m$ is satisfied as long as all coefficients of the error-polynomial $e \in R_q$ are within the threshold t, for which we selected $[-t, t) = [-\lfloor \frac{q}{4} \rfloor, \lfloor \frac{q}{4} \rfloor)$.

2.3   Fast Fourier Transform

The FFT is used to convert the coefficient representation of the polynomial to a point-value representation. The multiplication using the coefficients of two polynomials with degree n takes $\mathcal{O}(n^2)$ time. In point-value representation, the multiplication is performed in $\mathcal{O}(n \log n)$ as serial implementation and in $\mathcal{O}(\log n)$ if implemented in parallel, which we selected for hardware implementation.

Algorithm 1 characterizes the polynomial multiplication, which applies the FFT to convert the polynomials from coefficient to point-value representation.

To exploit the full potential of the FFT and speed-up the polynomial reduction (cf. Appendix A), we restrict the choice of the irreducible polynomial $f(x)$ to a cyclotomic one with the form $f(x) = x^n + 1$, where n is a power of 2. This allows for further improvements which are detailed in Section 4.3.

There are multiple reasons why we favour FFT over other polynomial multiplication approaches (e.g., Toom-Cook[Coo66]). FFT is easily parallelizable, and the asymptotic runtime of the parallel FFT is $\mathcal{O}(\log n)$ compared to $\mathcal{O}(n^{1+\epsilon})$, where $0 < \epsilon < 1$ for Toom-Cook. Additionally, the FFT hardware implementation greatly benefits from the utilized polynomial $f(x) = x^n + 1$ with n a power of 2, as this saves a lot of hardware resources. However, the Toom-Cook approach might be faster for practical parameters, but the comparison of polynomial multiplication algorithms is out of scope of this work.

Algorithm 1. Polynomial Multiplication using FFT

Input: $a, b \in \mathbb{Z}_q^{2n}$, ω_m
Output: $c \in \mathbb{Z}_q^{2n}$

1 $A = FFT(a, \omega_m)$
2 $B = FFT(b, \omega_m)$
3 **for** $i = 0$ *to* $2n - 1$ **do**
4 $\quad | \quad C[i] = A[i] \cdot B[i]$
5 **end**
6 $c = FFT^{-1}(C, \omega_m)$
7 **return** c

---

3   Software Implementation

The general purpose of the software implementation presented herein was to provide a reference implementation of the LWE based encryption scheme and to assess its real-world properties. We are using C++ on a Linux-based operating system, with the GCC 4.6.1 compiler. We integrated the NTL library [Sho] in version 5.5.2, which comprises data types for matrices and vectors over residue classes $\mathbb{Z}_q$ as well as elements in polynomial rings $\mathbb{Z}_q[X]$ and in factor rings $\mathbb{Z}_q[X]/\langle f(x) \rangle$. NTL applies FFT for its polynomial multiplication routines. Our outlined software implementation contains both variants – LWE-Matrix as well as LWE-Polynomial, which are available online.[1] The tested parameter sets of n, q, c, and s, for different values of δ are denoted in Table 1. The first column is taken from [LP11] and the values for the standard deviation of Gaussian sampling s are computed according to (1). It should be noted that the toy parameter set for $n = 128$ can not be considered secure in practice. For $n = 256$ and $\delta = 10^{-2}$ the estimated runtime/advantage ratio of the strongest (so-called decoding) attack is 2^{120} seconds, which is compared to the security of AES-128 in [LP11]. Unfortunately, an approach to compute "real" security estimates (bit-security) for lattice-based cryptosystems is not known so far.

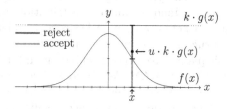

Fig. 1. Rejection sampling for
$f(x) = \frac{1}{s} \cdot e^{\frac{-\pi \cdot x^2}{s^2}}$, $g(x) = \frac{1}{n}$ and $k = \lceil \frac{n}{s} \rceil$.

Algorithm 2.
Rejection Sampling

1 **repeat**
2 $\quad | \quad x \xleftarrow{\$} \mathbb{Z} \cap [-t, t]$
3 $\quad | \quad u \xleftarrow{\$} \mathbb{R} \cap [0, 1]$
4 **until** $u \cdot k \cdot g(x) < f(x)$
5 **return** $x \in \mathbb{Z}_q$ following a
Gaussian distribution

---

[1] https://www.cdc.informatik.tu-darmstadt.de/de/cdc/
personen/michael-schneider/

Table 1. Computed values for parameter s in dependence of combinations of all used tuples (n, c, q) and δ, as proposed in [LP11]. For $\delta = 10^{-2}$ this reflects the classifications toy, low, medium, and high for n set to 128, 192, 256, and 320, respectively

(n, c, q)	$s_{\delta=10^{-2}}$	class[LP11]	$s_{\delta=10^{-3}}$	$s_{\delta=10^{-4}}$	$s_{\delta=10^{-5}}$	$s_{\delta=10^{-6}}$
$(128, 1.35, 2053)$	6.77	toy	6.19	5.79	5.5	5.26
$(192, 1.28, 4093)$	8.87	low	8.11	7.59	7.2	6.9
$(256, 1.25, 4093)$	8.35	medium	7.63	7.15	6.78	6.5
$(320, 1.22, 4093)$	8.0	high	7.31	6.84	6.5	6.22
$(384, 1.2, 4451)$	8.04	–	7.34	6.87	6.52	6.25
$(448, 1.184, 4723)$	8.02	–	7.33	6.86	6.51	6.23
$(512, 1.172, 4987)$	8.01	–	7.32	6.85	6.5	6.23

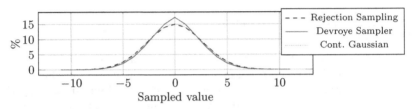

Fig. 2. Histogram of 10^8 samples of rejection sampling and the sampler of Devroye

We denote the expression $x \xleftarrow{\$} S$ for a value of x that is being sampled uniformly at random from the set S. For sampling Gaussian distributed integers, we apply the *rejection sampling* approach. This method is, among others, exploited in [GPV08] already. Algorithm 2 and Fig. 1 illustrate the rejection sampling approach. An exercise of this method revealed that for the chosen parameters the sampling success rate is approx. 20%. This is due to the fact that in the second sampling step in Algorithm 2, values far from the origin being accepted with a very small probability. Tests with the Gaussian Sampler of Devroye [Dev86, Chap. 3, Exercise 3] with standard deviation of $\sigma = \frac{1}{\sqrt{2\pi}} \cdot s$ showed a success rate of 85.2% for this sampler. The generation of 10^8 samples on our test platform took $19s$ compared to $75s$ for the rejection sampler. Therefore, using Devroye's sampler would allow for faster sampling in the encryption system. Unfortunately, the output of this sampler differs from the continuous Gaussian distribution, as shown in Fig. 2. Further, the performance benefits by using a Devroye sampler were negligible in our tests, although the success rates differ significantly.

The performance tests presented in this paper have been executed on an Intel Core 2 Duo CPU running at 3.00 GHz and 4Gb of RAM. As clearly visible from Fig. 3, LWE-Polynomial benefits from the fewer coefficients and outperforms LWE-Matrix by at least a factor of 4. The superior performance results, the smaller memory footprint, and less key data were the reasons to consider only LWE-Polynomial for hardware implementation. To be more specific, the analysis of the memory footprint revealed that LWE-Polynomial utilized 3.8 to 23 times less memory during key generation, 2.6 to 17.2 times while encrypting, and decryption took 2 to 5 times less memory resources than LWE-Matrix.

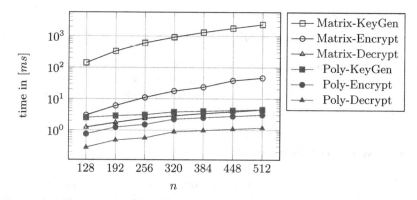

Fig. 3. Computation times of LWE software variants for all basic functions KeyGen, Encrypt, and Decrypt (cf. Table 2).

Table 2. LWE-Matrix vs. LWE-Polynomial: Time in milliseconds for key generation, encryption, and decryption of a 16 byte plaintext.

	KeyGen			Encrypt			Decrypt		
n	t_{Matrix}	t_{Poly}	t_{Matrix}/t_{Poly}	t_{Matrix}	t_{Poly}	t_{Matrix}/t_{Poly}	t_{Matrix}	t_{Poly}	t_{Matrix}/t_{Poly}
128	141.3	2.51	56.2	3.01	0.76	3.98	1.24	0.28	4.40
256	604.9	3.10	195.3	11.01	1.52	7.23	2.37	0.57	4.15
384	1311.2	4.05	323.6	23.41	2.51	9.34	3.41	0.98	3.46
512	2338.5	4.53	516.5	46.05	3.06	15.04	4.52	1.18	3.84

Table 3. Filesizes in bytes of LWE public and private keys as well as size of a ciphertext for a 16 byte plaintext. It is remarkable that the ciphertext of the matrix variant is smaller than that of the polynomial variant.

	Public Key			Private Key			Cyphertext		
n	Matrix	Poly	Matrix/Poly	Matrix	Poly	Matrix/Poly	Matrix	Poly	Matrix/Poly
128	146811	1154	127.22	53602	394	136.05	1142	1143	1.00
256	465851	2435	191.31	108298	883	122.65	1816	2423	0.75
384	935676	3659	255.72	162232	1271	127.64	2433	3650	0.67
512	1567516	4912	319.12	216624	1665	130.10	3058	4893	0.62

The filesize of the generated key material is depicted in Table 3. We directly used the NTL [Sho] output, which is on the one hand clearly not an optimal representation for the data, but allowed for interoperability with other tools. The theoretical key and ciphertext sizes for the software implementation are given in (2) and (3), respectively. Here, n denotes the dimension and l the message length in bits. For software based approaches a short integer (16 bit) length has been selected instead of $\lceil log_2(q) \rceil$ for the required bit width, which has been chosen for the hardware implementation. The ratios between the filesizes of LWE-Matrix and LWE-Polynomial are on the other hand good estimations for the real values.

$$Size_{Matrix,Public} = (n \cdot l + n^2) \cdot \lceil log_2(q) \rceil \qquad Size_{Matrix,Private} = n \cdot l \cdot \lceil log_2(q) \rceil$$
$$Size_{Poly,Public} = 2 \cdot n \cdot \lceil log_2(q) \rceil \qquad Size_{Poly,Private} = n \cdot \lceil log_2(q) \rceil \qquad (2)$$

$$Size_{Matrix,Cipher} = (n + l) \cdot \lceil log_2(q) \rceil \qquad Size_{Poly,Cipher} = 2n \cdot \lceil log_2(q) \rceil \qquad (3)$$

Another noteable result of our software evaluation revealed that the ciphertext for higher dimensions is smaller in LWE-Matrix compared to LWE-Polynomial, as denoted in Table 3. Further on, we noticed that different values of the error probability δ have only a marginal impact on the related runtimes.

4 Hardware Implementation

As aforementioned, only the LWE-Polynomial variant has been selected for hardware implementation, as the software evaluation indicated the performance benefits of this representation (see Fig. 3). The evaluation platform of the hardware implementation is a Xilinx ML-605 evaluation board providing a Virtex-6 LX240T FPGA. Hardware modules have been designed in a vendor independent manner, which allows the utilization of FPGAs from other vendors as well as a representation as an Application Specific Integrated Circuit (ASIC) implementation. Additionally, we also present related synthesis results using a Virtex-7 device in Table 6.

The parameters exploited for the hardware implementation are given in Table 4 and we assigned $\delta = 10^{-2}$. Recall that the parameter set with $n = 128$ does not supply sufficient security guarantees ("toy" parameters). The FFT requires for all roots of unity that $q - 1$ is a multiple of $2n$ (see Corollary 30.4 [CLRS09]).

The dataflow for the three primitive operations of the LWE-based scheme are depicted in Fig. 4. In contrast to the rejection sampling approach applied in the software variant, which would require floating point arithmetic, the Gaussian sampler has been implemented by means of a look-up table. Integers in the range of $[-\lceil 2 \cdot s \rceil, \lceil 2 \cdot s \rceil]$ are selected by applying the random output of a linear feedback shift register (LFSR) as an address to an array of Gaussian distributed values. In order to save resources, this array has been embodied using only start and end addresses of the values. An unoptimized Gaussian array would require (resolution $\cdot \lceil log_2(q) \rceil$) bits, whereas the optimized version requires $3 \cdot (2 \cdot \lceil 2s \rceil + 1) \cdot \lceil log_2(q) \rceil$ only. For example, with $s = 6.67$, $q = 3329$ and a resolution of 1023 the optimized version requires 1044 bits (c.f. Table 5) whereas a straight forward array would require 12288 bits, which saves in this case approx. 92% of the memory. Additionally, the uniform sampler, required during key generation is realized by an LFSR (cf. Sect. 4.1). For the look-up-table with interval size of $4s$, the probability of a sample outside of this interval is $4.6 \cdot 10^{-6}$. This probability can be lowered further by choosing a larger interval.

Table 4. LWE parameters for hardware tests using $\delta = 10^{-2}$ and the bit width ($\lceil log_2(q) \rceil$) for representing coefficient values.

n	q	s	ω	$\lceil log_2(q) \rceil$
128	3329	8.62	17	12
256	7681	11.31	62	13
512	12289	12.18	49	14

Table 5. Implemented approach of the Gaussian array

Start/End-address	...	...	157	237	238	337	338	451	452	570	571	684	685	784	785	865	...	...
Gaussian value	...		-3		-2		-1		0		1		2		3		...	

The message encoding, as outlined in Sect. 2.2, is performed by the encode and decode modules depicted in Fig. 4(a) and Fig. 4(c), respectively. The encryption datapath, as displayed in Fig. 4(a), shows how Gaussian distributed random errors are introduced within the cipertext by multiplication and addition. As a result of the encryption, the two vectors c_1 and c_2 contain the ciphertext. Decryption of the ciphertext (c_1, c_2) is performed by a multiplication of the private key r_2 with c_1 followed by an addition of c_2 as detailed in Fig. 4(c).

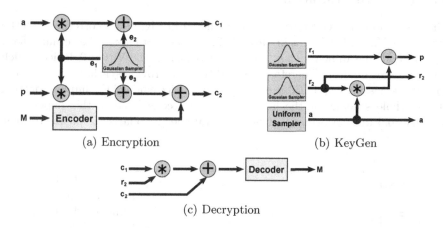

(a) Encryption   (b) KeyGen

(c) Decryption

Fig. 4. Overview of LWE Encryption Scheme Datapaths

The modular multiplication of polynomials is by far the most expensive operation in this scheme. Therefore, we apply the FFT for polynomial reduction and a Montgomery multiplier [Mon85] to realize the modular multiplication of the polynomial coefficients as depicted in Fig. 5.

4.1   Random Numbers for Keys and Errors

Uniformly distributed random numbers are required by the Gaussian sampler. In this work we emphasize on the implementation of the Gaussian sampler and therfore the quality of the LFSR-based RNG is not in the scope of this paper. For the practical use of the herein proposed scheme, attacks on the random number generators, such as frequency injection [MM09], have to be considered. Novel concepts for RNGs addressing reconfigurable hardware have been presented in previous works, such as [KG04], [Gün10], [VD10] and [MKD11].

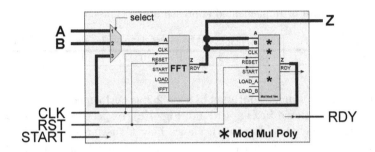

Fig. 5. Arithmetic unit for polynomial multiplication

4.2   Resource Utilization

The utilization of device resources are depicted in Fig. 6 and correspond to the data of the Virtex-6 columns in Table 6. Taking a closer look at the actual values, one can find that dimensions $n > 128$ did not fit into the Virtex-6 device, which we used for evaluation. Therefore, we additionally provide synthesis results for a Virtex-7 series device (cf. Table 6). Using this device enabled the implementation of the whole scheme for the largest dimension considered in this paper. Our primary goal was performance, which naturally leads to larger implementations.

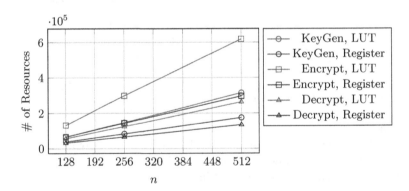

Fig. 6. Hardware resource utilization of the top-modules LWE-KeyGen, LWE-Encrypt, and LWE-Decrypt for $n = \{128, 256, 512\}$. Detailed data is denoted in Table 6.

4.3   Design Improvements

The choice of parameters allows for some optimizations of certain modules. As an example, the inverse FFT can be considerably improved in the case that if the reduction polynomial for the residue ring follows the structure of $f(x) = x^n + 1$ for which n is a power of 2, i.e., it is cyclotomic. The *root of unity* $\omega \in \mathbb{Z}_q$ is selected such that $\omega^{2n} = 1 \bmod q$ with q is prime and $q - 1$ is a multiple of $2n$. Based on Collary 30.4 in [CLRS09], ω is determined by $\omega^n = -1 \bmod q$

Table 6. Top-level module resource utilization of KeyGen, Encrypt, and Decrypt on a Xilinx XC6VLX240T FPGA and XC7V2000T for $n = \{128, 256, 512\}$

(a) LWE-KeyGen

	Virtex-6 LX240T						Virtex-7 2000T					
	n = 128	%	n = 256	%	n = 512	%	n = 128	%	n = 256	%	n = 512	%
# Registers	37918	12	82463	27	174757	57	37918	1	85472	3	174757	7
# LUTs	64804	42	146718	97	314635	208	69140	5	163209	13	348204	28

(b) LWE-Encrypt

	Virtex-6 LX240T						Virtex-7 2000T					
	n = 128	%	n = 256	%	n = 512	%	n = 128	%	n = 256	%	n = 512	%
# Registers	65680	21	143396	47	296207	98	65680	2	143396	5	296207	12
# LUTs	131254	87	298016	197	618934	410	131187	10	320816	26	634893	51

(c) LWE-Decrypt

	Virtex-6 LX240T						Virtex-7 2000T					
	n = 128	%	n = 256	%	n = 512	%	n = 128	%	n = 256	%	n = 512	%
# Registers	31884	10	65174	21	134036	44	31884	1	65174	2	134036	5
# LUTs	56311	37	124158	82	263083	174	56313	4	124265	10	260772	21

throughout this paper and it is used as a common parameter of all butterfly modules. Generally speaking, this eliminates the final polynomial reduction step, resulting in a cut of half the residue class multipliers.

Assume that the inputs of a butterfly module are denoted by x_i and x_{i+n} and the outputs are denoted by y_i and y_{i+n}. Then the inner calculation of the butterfly module is given by

$$y_i = x_i + \omega_j x_{i+n}$$
$$y_{i+n} = x_i - \omega_j x_{i+n} . \tag{4}$$

In the last step of the inverse FFT each coefficient is multiplied by the inverse element of $2n$ in the residue class ring $\mathbb{Z}_q$. Applying this multiplication to every output of a butterfly module results in

$$y_i \cdot (2n)^{-1} = x_i \cdot (2n)^{-1} + \omega_j x_{i+n} \cdot (2n)^{-1}$$
$$y_{i+n} \cdot (2n)^{-1} = x_i \cdot (2n)^{-1} - \omega_j x_{i+n} \cdot (2n)^{-1} . \tag{5}$$

Applying the reduction step of cyclotomic polynomials (cf. (10)), both outputs of a butterfly module are subtracted as follows

$$y_i \cdot (2n)^{-1} - y_{i+n} \cdot (2n)^{-1} = x_i \cdot (2n)^{-1} + \omega_j x_{i+n} \cdot (2n)^{-1}$$
$$- x_i \cdot (2n)^{-1} + \omega_j x_{i+n} \cdot (2n)^{-1} \tag{6}$$
$$= 2\omega_j (2n)^{-1} \cdot x_{i+n} .$$

An important consequence of (6) is that not every input x_i is required to calculate the inverse FFT. If this consequence is considered for the inputs of the inverse FFT, only those inputs characterized by an odd index are used. So, each input with an odd index is connected to a wire that is placed in the lower half of the parallel FFT depicted as in Fig. 7. The term $2\omega_j(2n)^{-1}$ in (6) is precomputed in order to further reduce the ammount of utilized resources.

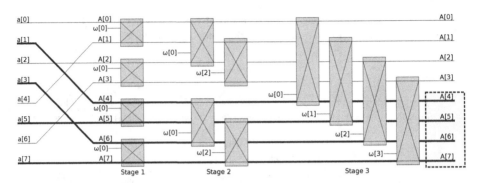

Fig. 7. Optimization of the inverse FFT followed by a polynomial reduction (cf. (6))

5   Evaluation and Practical Implications

An evaluation of the presented results shows that for the LWE-Polynomial implementation, encryption and decryption differ only by a factor of roughly 2.6 for the software version. The hardware implementation as presented in this paper is, because of its full parallel structure, able to perform encryption and decryption at the same speed. The LWE-Matrix does not share this property as the size of the matrix grows quadratically with the dimension n.

To compare the performance of each implementation, the achieved throughputs of both, the hardware and software implementation variants are depicted in Fig. 8. Due to the increased parallelism and the structure of the encryption scheme, the hardware outperforms the software by a factor up to 316 for encryption and of 122 for decryption. For the key generation the results are even better and show a performance gain of roughly 400 in all three dimensions ($n = \{128, 256, 512\}$). The reason for this considerable gain is the difference between the rejection sampling approach in software and the look-up table method in hardware. Additionally, the hardware benefits from a full parallel implementation for sampling values, in contrast to the serial software implementation.

5.1   Message Expansion Factors

The estimated message expansion factor for the dimension $n = 128$ and the parameters $q = 2053$ and $s = 6.77$ is 22. For the software-based implementations of LWE-Matrix and LWE-Polynomial a short integer (16 bit) has been chosen to represent the coefficient values, resulting in the difference in Table 7. As aforementioned, the parameters for the hardware variant have been selected to improve the FFT, hence the message expansion factors are also different (as q is set to 3329). A comparison of all implementation variants is detailed in Table 7. Obviously, using the optimal representation for the polynomial coefficients, the message expansion factor of 24 is still very close to the expected factor of 22.

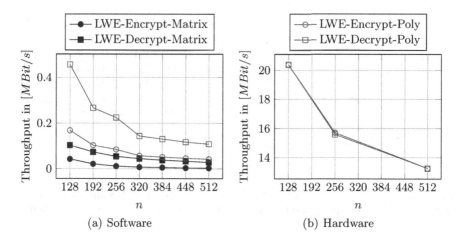

Fig. 8. Throughput values of evaluated implementation variants

Table 7. Ciphertext in bytes for 16 byte plaintext and corresponding expansion factors

n	LWE-Matrix Cipher	LWE-Matrix Cipher/Plain	LWE-Polynomial Cipher	LWE-Polynomial Cipher/Plain	LWE-Hardware Cipher	LWE-Hardware Cipher/Plain
128	512	32	512	32	384	24
192	640	40	768	48	–	–
256	768	48	1024	64	832	52
320	896	56	1280	80	–	–
384	1024	64	1536	96	–	–
448	1152	72	1792	112	–	–
512	1280	80	2048	128	1792	112

5.2 Error Rates

An interesting result of the software implementaion tests, is the practical error rate that can be observed during decryption, as depicted in Figure 9. As a part of the evaluation we assessed the error rates, which represent the error probability per symbol, against the upper bound δ outlined in Sect. 2.2. We state that the practical error rate is more than a factor 100 smaller than its upper bound δ, whereas the error rate increases with the dimension n.

The exact measurement values for the encountered bit errors for LWE-Matrix and LWE-Polynomial are given in Table 8 and Table 9, respectively.

A reduction of the error probability may be achieved by changing the parameters which determine s (cf. (1)) q and n. The Gaussian standard deviation s decreases when δ decreases as well. For the security guarantee to hold, it is necessary that $s \cdot q > 2\sqrt{n}$ holds. This implies that, for the same security level n, q has to be increased when smaller δ (and with this smaller s) is required. A second way to deal with decryption errors is the application of error correcting codes. This approach allows to keep the system parameters unchanged, but enlarges the message expansion factor.

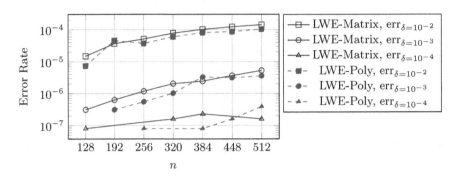

Fig. 9. Rate of bit-errors for LWE-Matrix and LWE-Polynomial for $\delta \in \{10^{-2}, 10^{-3}, 10^{-4}\}$ for $1,600,000$ byte plaintext, as detailed in Table 8 and Table 9

Table 8. LWE-Matrix: Bit errors and error rate for a $1,600,000$ byte plaintext

n	$\delta = 10^{-2}$ Errors	%	$\delta = 10^{-3}$ Errors	%	$\delta = 10^{-4}$ Errors	%	$\delta = 10^{-5}$ Errors	%	$\delta = 10^{-6}$ Errors	%
128	189	0.001477	4	0.000031	1	0.000008	0	0.00	0	0.00
192	467	0.003648	8	0.000063	0	0.00	0	0.00	0	0.00
256	650	0.005078	15	0.000117	0	0.00	0	0.00	0	0.00
320	994	0.007766	26	0.000203	2	0.000016	0	0.00	0	0.00
384	1304	0.010188	31	0.000242	3	0.000023	0	0.00	0	0.00
448	1567	0.012242	47	0.000367	0	0.00	0	0.00	0	0.00
512	1820	0.014219	68	0.000531	2	0.000016	0	0.00	0	0.00

Table 9. LWE-Polynomial: Bit errors and error rate for a $1,600,000$ byte plaintext

n	$\delta = 10^{-2}$ Errors	%	$\delta = 10^{-3}$ Errors	%	$\delta = 10^{-4}$ Errors	%	$\delta = 10^{-5}$ Errors	%	$\delta = 10^{-6}$ Errors	%
128	94	0.000734	0	0.00	0	0.00	0	0.00	0	0.00
192	586	0.004578	4	0.000031	0	0.00	0	0.00	0	0.00
256	474	0.003703	7	0.000055	1	0.000008	0	0.00	0	0.00
320	766	0.005984	13	0.000102	0	0.00	0	0.00	0	0.00
384	1031	0.008055	42	0.000328	1	0.000008	0	0.00	0	0.00
448	1108	0.008656	40	0.000313	2	0.000016	0	0.00	0	0.00
512	1329	0.010383	46	0.000359	5	0.000039	0	0.00	0	0.00

6   Future Work

Usage of error correcting codes, such as Viterbi[Vit67], in order to overcome decryption errors will be addressed in future work. Another not yet addressed aspect is that error detection itself is not sufficient since it allows for a correlation with the private key if decryption fails; error corection however may interact with the security guarantees of the whole scheme. Applying the central limit theorem enables the precomputation of expected error rates instead of upper bounds, which leads to a better estimation than the upper bound δ.

We consider as additional goals for the hardware implementation to investigate into an architecture which exploits resource sharing in order to reduce the amount of required resources. Further, a hardware version without the use of an FFT is envisaged to quantify the tradeoff between resource utilization and throughput. An investigation on the benefits of constant multipliers, to further improve the design, is part of future work.

Acknowledgements. This work was supported by CASED (`www.cased.de`). We thank Richard Linder for his contributions and fruitful discussions. We also thank the anonymous reviewers for their helpful comments.

References

[ACPS09] Applebaum, B., Cash, D., Peikert, C., Sahai, A.: Fast Cryptographic Primitives and Circular-Secure Encryption Based on Hard Learning Problems. In: Halevi, S. (ed.) CRYPTO 2009. LNCS, vol. 5677, pp. 595–618. Springer, Heidelberg (2009)

[ADL+08] Arbitman, Y., Dogon, G., Lyubashevsky, V., Micciancio, D., Peikert, C., Rosen, A.: SWIFFTX: A proposal for the SHA-3 standard. In: The First SHA-3 Candidate Conference (2008)

[CLRS09] Cormen, T.H., Leiserson, C.E., Rivest, R.L., Stein, C.: Introduction to Algorithms, 3rd edn. The MIT Press (2009)

[Coo66] Cook, S.A.: On the minimum computation time of functions. PhD thesis. Harvard Univ., Cambridge (1966)

[CT65] Cooley, J.W., Tukey, J.W.: An algorithm for the machine calculation of complex fourier series. Mathematics of Computation 19(90) (1965)

[Dev86] Devroye, L.: Non-uniform random variate generation. Springer-Verlag New York (1986)

[Gen09] Gentry, C.: Fully homomorphic encryption using ideal lattices. In: STOC. ACM (2009)

[GPV08] Gentry, C., Peikert, C., Vaikuntanathan, V.: Trapdoors for hard lattices and new cryptographic constructions. In: STOC. ACM (2008)

[Gün10] Güneysu, T.: True random number generation in block memories of reconfigurable devices. In: FPT. IEEE (2010)

[HPS98] Hoffstein, J., Pipher, J., Silverman, J.H.: NTRU: A Ring-Based Public Key Cryptosystem. In: Buhler, J.P. (ed.) ANTS 1998. LNCS, vol. 1423, pp. 267–288. Springer, Heidelberg (1998)

[KG04] Kohlbrenner, P., Gaj, K.: An embedded true random number generator for FPGAs. In: ACM/SIGDA FPGA (2004)

[LP11] Lindner, R., Peikert, C.: Better Key Sizes (and Attacks) for LWE-Based Encryption. In: Kiayias, A. (ed.) CT-RSA 2011. LNCS, vol. 6558, pp. 319–339. Springer, Heidelberg (2011)

[LPR10] Lyubashevsky, V., Peikert, C., Regev, O.: On Ideal Lattices and Learning with Errors over Rings. In: Gilbert, H. (ed.) EUROCRYPT 2010. LNCS, vol. 6110, pp. 1–23. Springer, Heidelberg (2010)

[Lyu09] Lyubashevsky, V.: Fiat-Shamir with Aborts: Applications to Lattice and Factoring-Based Signatures. In: Matsui, M. (ed.) ASIACRYPT 2009. LNCS, vol. 5912, pp. 598–616. Springer, Heidelberg (2009)

[Lyu12]   Lyubashevsky, V.: Lattice Signatures without Trapdoors. In: Pointcheval, D., Johansson, T. (eds.) EUROCRYPT 2012. LNCS, vol. 7237, pp. 738–755. Springer, Heidelberg (2012)

[Mic10]   Micciancio, D.: Duality in lattice cryptography (2010) (Invited talk)

[MKD11]   Majzoobi, M., Koushanfar, F., Devadas, S.: FPGA-Based True Random Number Generation Using Circuit Metastability with Adaptive Feedback Control. In: Preneel, B., Takagi, T. (eds.) CHES 2011. LNCS, vol. 6917, pp. 17–32. Springer, Heidelberg (2011)

[MM09]   Markettos, A.T., Moore, S.W.: The Frequency Injection Attack on Ring-Oscillator-Based True Random Number Generators. In: Clavier, C., Gaj, K. (eds.) CHES 2009. LNCS, vol. 5747, pp. 317–331. Springer, Heidelberg (2009)

[Mon85]   Montgomery, P.L.: Modular multiplication without trial division. Mathematics of Computation 44 (1985)

[Paa94]   Paar, C.: Efficient VLSI Architectures for Bit-Parallel Computation in Galois Fields. PhD thesis, Universität Essen (1994)

[Pei09]   Peikert, C.: Public-key cryptosystems from the worst-case shortest vector problem: extended abstract. In: STOC. ACM (2009)

[Reg05]   Regev, O.: On lattices, learning with errors, random linear codes, and cryptography. In: STOC. ACM (2005)

[Sho]   Shoup, V.: Number theory library (NTL), http://www.shoup.net/ntl/

[SS11]   Stehlé, D., Steinfeld, R.: Making NTRU as Secure as Worst-Case Problems over Ideal Lattices. In: Paterson, K.G. (ed.) EUROCRYPT 2011. LNCS, vol. 6632, pp. 27–47. Springer, Heidelberg (2011)

[VD10]   Varchola, M., Drutarovsky, M.: New High Entropy Element for FPGA Based True Random Number Generators. In: Mangard, S., Standaert, F.-X. (eds.) CHES 2010. LNCS, vol. 6225, pp. 351–365. Springer, Heidelberg (2010)

[Vit67]   Viterbi, A.J.: Error bounds for convolutional codes and an asymptotically optimum decoding algorithm. IEEE Transactions on Information Theory 13(2) (1967)

A   Mathematical Background

The polynomial reduction can be written as $r(x) = g(x) \bmod f(x)$ and represented in matrix notation [Paa94] as a multiplication of $g(x)$ with the reduction matrix $\mathbf{M}$ as follows:

$$
\begin{pmatrix} r_0 \\ r_1 \\ \vdots \\ r_{n-1} \end{pmatrix} = \begin{pmatrix} 1 & 0 & \cdots & 0 & \mu_{0,0} & \cdots & \mu_{0,n-2} \\ 0 & 1 & \cdots & 0 & \mu_{1,0} & \cdots & \mu_{1,n-2} \\ \vdots & \vdots & \ddots & \vdots & \vdots & \ddots & \vdots \\ 0 & 0 & \cdots & 1 & \mu_{n-1,0} & \cdots & \mu_{n-1,n-2} \end{pmatrix} \cdot \begin{pmatrix} g_0 \\ g_1 \\ \vdots \\ g_{n-1} \\ g_n \\ \vdots \\ g_{2n-2} \end{pmatrix} . \tag{7}
$$

The elements $\mu_{j,i}$ of matrix $\mathbf{M}$ are calculated from

$$
\mu_{j,i} = \begin{cases} -f_j & , \text{ for } j = 0, \ldots, n-1; \ i = 0 \\ \mu_{j-1,i-1} + \mu_{n-1,i-1} \cdot \mu_{j,0} & , \text{ for } j = 0, \ldots, n-1; i = 1, .., n-2; \end{cases} \tag{8}
$$

where $\mu_{j-1,i-1} = 0$, if $j = 0$. This procedure gets very simple if $f(x)$ is a cyclotomic polynomial of the form $x^n + 1$, with n is a power of 2

$$
\mathbf{M} = \begin{pmatrix} 1 & 0 & \cdots & 0 & -1 & 0 & \cdots & 0 \\ 0 & 1 & \cdots & 0 & 0 & -1 & \cdots & 0 \\ \vdots & \vdots & \ddots & \vdots & \vdots & & \ddots & \vdots \\ \vdots & \vdots & & \vdots & 0 & \cdots & 0 & -1 \\ 0 & 0 & \cdots & 1 & 0 & \cdots & 0 & 0 \end{pmatrix}. \tag{9}
$$

By means of this simplified matrix, the calculation of the matrix-vector multiplication can be reduced to:

$$
\begin{pmatrix} r_0 \\ r_1 \\ \vdots \\ r_{n-1} \end{pmatrix} = \begin{pmatrix} g_0 - g_n \\ g_1 - g_{n+1} \\ \vdots \\ g_{n-2} - g_{2n-2} \\ g_{n-1} \end{pmatrix}. \tag{10}
$$

This simplification (10) leads to the fact that the polynomial reduction takes only linear time.

Practical Lattice-Based Cryptography: A Signature Scheme for Embedded Systems

Tim Güneysu[1,*], Vadim Lyubashevsky[2,**], and Thomas Pöppelmann[1,*]

[1] Horst Görtz Institute for IT-Security, Ruhr-University Bochum, Germany
[2] INRIA / ENS, Paris

Abstract. Nearly all of the currently used and well-tested signature schemes (e.g. RSA or DSA) are based either on the factoring assumption or the presumed intractability of the discrete logarithm problem. Further algorithmic advances on these problems may lead to the unpleasant situation that a large number of schemes have to be replaced with alternatives. In this work we present such an alternative – a signature scheme whose security is derived from the hardness of lattice problems. It is based on recent theoretical advances in lattice-based cryptography and is highly optimized for practicability and use in embedded systems. The public and secret keys are roughly 12000 and 2000 bits long, while the signature size is approximately 9000 bits for a security level of around 100 bits. The implementation results on reconfigurable hardware (Spartan/Virtex 6) are very promising and show that the scheme is scalable, has low area consumption, and even outperforms some classical schemes.

Keywords: Post-Quantum Cryptography, Lattice-Based Cryptography, Ideal Lattices, Signature Scheme Implementation, FPGA.

1 Introduction

Due to the yet unpredictable but possibly imminent threat of the construction of a quantum computer, a number of alternative cryptosystems to RSA and ECC have gained significant attention during the last years. In particular, it has been widely accepted that relying solely on asymmetric cryptography based on the hardness of factoring or the (elliptic curve) discrete logarithm problem is certainly not sufficient in the long term [7]. This has been mainly due to the work of Shor [34], who demonstrated that both classes of problems can be efficiently attacked with quantum computers. As a consequence, first steps towards the required diversification and investigation of alternative fundamental problems and schemes have been taken. This has already led to efficient implementations of various schemes based on multivariate quadratic systems [5,3] and the code-based McEliece cryptosystem [10,35].

---

* This work was partially supported by European Commission through the ICT programme under contract ICT-2007-216676 ECRYPT II.
** Work supported in part by the European Research Council.

E. Prouff and P. Schaumont (Eds.): CHES 2012, LNCS 7428, pp. 530–547, 2012.

Another promising alternative to number-theoretic constructions are lattice-based cryptosystems because they admit security proofs based on well-studied problems that currently cannot be solved by quantum algorithms. For a long time, however, lattice constructions have only been considered secure for inefficiently large parameters that are well beyond practicability[1] or were, like GGH [14] and NTRUSign [17], broken due to flaws in the ad-hoc design approach [30]. This has changed since the introduction of cyclic and ideal lattices [26] and related computationally hard problems like RING-SIS [31,22,24] and RING-LWE [25] which enabled the constructions of a great variety of theoretically elegant and efficient cryptographic primitives.

In this work we try to further close the gap between the advances in theoretical lattice-based cryptography and real-world implementation issues by constructing and implementing a provably-secure digital signature scheme based on ideal lattices. While maintaining the connection to hard ideal lattice problems we apply several performance optimizations for practicability that result in moderate signature and key sizes as well as performance suitable for embedded and hardware systems.

Digital Signatures and Related Work. Digital signatures are arguably the most used public-key cryptographic primitive in practical applications, and a lot of effort has gone into trying to construct such schemes from lattice assumptions. Due to the success of the NTRU encryption scheme, it was natural to try to design a signature scheme based on the same principles. Unlike the encryption scheme, however, the proposed NTRU signature scheme [18,16] has been completely broken by Nguyen and Regev [30]. Provably-secure digital signatures were finally constructed in 2008, by Gentry, Peikert, and Vaikuntanathan [13], and, using different techniques, by Lyubashevsky and Micciancio [23]. The scheme in [13] was very inefficient in practice, with outputs and keys being megabytes long, while the scheme in [23] was only a one-time signature that required the use of Merkle trees to become a full signature scheme. The work of [23] was extended by Lyubashevsky [20,21], who gave a construction of a full-fledged signature scheme whose keys and outputs are currently on the order of 15000 bits each, for an 80-bit security level. The work of [13] was also recently extended by Micciancio and Peikert [27], where the size of the signatures and keys is roughly 100, 000 bits.

Our Contribution. The main contribution of this work is the implementation of a digital signature scheme from [20,21] optimized for embedded systems. In addition, we propose an improvement to the above-mentioned scheme which preserves the security proof, while lowering the signature size by approximately a factor of two. We demonstrate the practicability of our scheme by implementing a scalable and efficient signing and verification engine. For example, on the low-cost Xilinx Spartan-6 we are 1.5 times faster and use only half of the resources

---

[1] One notable exception is the NTRU public-key encryption scheme [17], which has essentially remained unbroken since its introduction.

of the optimized RSA implementation of Suzuki [38]. With more than 12000 signatures and over 14000 signature verifications per second, we can satisfy even high-speed demands using a Virtex-6 device.

Outline. The paper is structured as follows. First we give a short overview on our hardness assumption in Section 2 and then introduce the highly efficient and practical signature scheme in Section 3. Based on this description, we introduce our implementation and the hardware architecture of the signing and signature verification engine in Section 4 and analyze its performance on different FPGAs in Section 5. In Section 6 we summarize our contribution and present an outlook for future work.

2   Preliminaries

2.1   Notation

Throughout the paper, we will assume that n is an integer that is a power of 2, p is a prime number congruent to 1 modulo $2n$, and $\mathcal{R}^{p^n}$ is the ring $\mathbb{Z}_p[\mathbf{x}]/(\mathbf{x}^n + 1)$. Elements in $\mathcal{R}^{p^n}$ can be represented by polynomials of degree $n - 1$ with coefficients in the range $[-(p-1)/2, (p-1)/2]$, and we will write $\mathcal{R}_k^{p^n}$ to be a subset of the ring $\mathcal{R}^{p^n}$ that consists of all polynomials with coefficients in the range $[-k, k]$. For a set S, we write $s \xleftarrow{\$} S$ to indicate that s is being chosen uniformly at random from S.

2.2   Hardness Assumption

In a particular version of the RING-SIS problem, one is given an ordered pair of polynomials $(\mathbf{a}, \mathbf{t}) \in \mathcal{R}^{p^n} \times \mathcal{R}^{p^n}$ where $\mathbf{a}$ is chosen uniformly from $\mathcal{R}^{p^n}$ and $\mathbf{t} = \mathbf{a}\mathbf{s}_1 + \mathbf{s}_2$, where $\mathbf{s}_1$ and $\mathbf{s}_2$ are chosen uniformly from $\mathcal{R}_k^{p^n}$, and is asked to find an ordered pair $(\mathbf{s}_1', \mathbf{s}_2')$ such that $\mathbf{a}\mathbf{s}_1' + \mathbf{s}_2' = \mathbf{t}$. It can be shown that when $k > \sqrt{p}$, the solution is not unique and finding any one of them, for $\sqrt{p} < k \ll p$, was proven in [31,22] to be as hard as solving worst-case lattice problems in ideal lattices. On the other hand, when $k < \sqrt{p}$, it can be shown that the only solution is $(\mathbf{s}_1, \mathbf{s}_2)$ with high probability, and there is no classical reduction known from worst-case lattice problems to finding this solution. In fact, this latter problem is a particular instance of the RING-LWE problem. It was recently shown in [25] that if one chooses the $\mathbf{s}_i$ from a slightly different distribution (i.e., a Gaussian distribution instead of a uniform one), then solving the RING-LWE problem (i.e., recovering the $\mathbf{s}_i$ when given $(\mathbf{a}, \mathbf{t})$) is as hard as solving worst-case lattice problems using a quantum algorithm. Furthermore, it was shown that solving the decision version of RING-LWE, that is distinguishing ordered pairs $(\mathbf{a}, \mathbf{a}\mathbf{s}_1 + \mathbf{s}_2)$ from uniformly random ones in $\mathcal{R}^{p^n} \times \mathcal{R}^{p^n}$, is still as hard as solving worst-case lattice problems.

In this paper, we implement our signature scheme based on the presumed hardness of the decision RING-LWE problem with particularly "aggressive" parameters. We define the $\mathbf{DCK}_{p,n}$ problem (Decisional Compact Knapsack problem) to be the problem of distinguishing between the uniform distribution over

$\mathcal{R}^{p^n} \times \mathcal{R}^{p^n}$ and the distribution $(\mathbf{a}, \mathbf{as}_1 + \mathbf{s}_2)$ where $\mathbf{a}$ is uniformly random in $\mathcal{R}^{p^n}$ and $\mathbf{s}_i$ are uniformly random in $\mathcal{R}_1^{p^n}$. As of now, there are no known algorithms that take advantage of the fact that the distribution of $\mathbf{s}_i$ is uniform (i.e., not Gaussian) and consists of only $-1/0/1$ coefficients[2], and so it is very reasonable to conjecture that this problem is still hard. In fact, this is essentially the assumption that the NTRU encryption scheme is based on. Due to lack of space, we direct the interested reader to Section 3 of the full version of [21] for a more in-depth discussion of the hardness of the different variants of the SIS and LWE problems.

2.3 Cryptographic Hash Function H with Range D_{32}^n

Our signature scheme uses a hash function, and it is quite important for us that the output of this function is of a particular form. The range of this function, D_{32}^n, for $n \geq 512$ consists of all polynomials of degree $n - 1$ that have all zero coefficients except for at most 32 coefficients that are ± 1.

We denote by H the hash function that first maps $\{0, 1\}^*$ to a 160-bit string and then *injectively* maps the resulting 160-bit string r to D_{32}^n via an efficient procedure we now describe. To map a 160-bit string into the range D_{32}^n for $n \geq 512$, we look at 5 bits of r at a time, and transforms them into a 16-digit string with at most one non-zero coefficient as follows: let $r_1 r_2 r_3 r_4 r_5$ be the five bits we are currently looking at. If r_1 is 0, then put a -1 in position number $r_2 r_3 r_4 r_5$ (where we read the 4-digit string as a number between 0 and 15) of the 16-digit string. If r_1 is 1, then put a 1 in position $r_2 r_3 r_4 r_5$. This converts a 160-bit string into a 512-digit string with at most 32 ± 1's.[3] We then convert the 512-bit string into a polynomial of degree at least 512 in the natural way by assigning the ith coefficient of the polynomial the ith bit of the bit-string. If the polynomial is of degree greater than 512, then all of its higher-order terms will be 0.

3   The Signature Scheme

In this section, we will present the lattice-based signature scheme whose hardware implementation we describe in Section 4. This scheme is a combination of the schemes from [20] and [21] as well as an additional optimization that allows us to reduce the signature length by almost a factor of two. In [20], Lyubashevsky constructed a lattice-based signature scheme based on the hardness of the RING-SIS problem, and this scheme was later improved in two ways [21].

---

[2] For readers familiar with the Arora-Ge algorithm for solving LWE with small noise [2], we would like to point out that it is does not apply to our problem because this algorithm requires polynomially-many samples of the form $(\mathbf{a}_i, \mathbf{a}_i\mathbf{s} + \mathbf{e}_i)$, whereas in our problem, only one such sample is given.

[3] There is a more "compact" way to do it (see for example [11] for an algorithm that can convert a 160-bit string into a 512-digit one with at most 24 ± 1 coefficients), but the resulting transformation algorithm is quadratic rather than linear.

The first improvement results in signatures that are asymptotically shorter, but unfortunately involves a somewhat more complicated rejection sampling algorithm during the singing procedure, involving sampling from the normal distribution and computing quotients to a very high precision, which would not be very well supported in hardware. We do not know whether the actual savings achieved in the signature length would justify the major slowdown incurred, and we do leave the possibility of efficiently implementing this rejection sampling algorithm to future work. The second improvement from [21], which we do use, shows how the size of the keys and the signature can be made significantly smaller by changing the assumption from RING-SIS to RING-LWE.

3.1 The Basic Signature Scheme

For ease of exposition, we first present the basic combination scheme of [20] and [21] in Figure 1, and sketch its security proof. Full security proofs are available in [20] and [21]. We then present our optimization in Sections 3.2 and 3.3.

Signing Key: $s_1, s_2 \overset{\$}{\leftarrow} \mathcal{R}_1^{p^n}$

Verification Key: $a \overset{\$}{\leftarrow} \mathcal{R}^{p^n}, t \leftarrow as_1 + s_2$

Cryptographic Hash Function: $H : \{0,1\}^* \rightarrow D_{32}^n$

$\text{Sign}(\mu, a, s_1, s_2)$
1: $y_1, y_2 \overset{\$}{\leftarrow} \mathcal{R}_k^{p^n}$
2: $c \leftarrow H(ay_1 + y_2, \mu)$
3: $z_1 \leftarrow s_1 c + y_1, z_2 \leftarrow s_2 c + y_2$
4: if z_1 or $z_2 \notin \mathcal{R}_{k-32}^{p^n}$, then goto step 1
5: output (z_1, z_2, c)

$\text{Verify}(\mu, z_1, z_2, c, a, t)$
1: Accept iff
   $z_1, z_2 \in \mathcal{R}_{k-32}^{p^n}$ and
   $c = H(az_1 + z_2 - tc, \mu)$

Fig. 1. The Basic Signature Scheme

The secret keys are random polynomials $s_1, s_2 \overset{\$}{\leftarrow} \mathcal{R}_1^{p^n}$ and the public key is (a, t), where $a \overset{\$}{\leftarrow} \mathcal{R}^{p^n}$ and $t \leftarrow as_1 + s_2$. The parameter k in our scheme which first appears in line 1 of the signing algorithm controls the trade-off between the security and the runtime of our scheme. The smaller we take k, the more secure the scheme becomes (and the shorter the signatures get), but the time to sign will increase. We explain this as well as the choice of parameters below.

To sign a message μ, we pick two "masking" polynomials $y_1, y_2 \overset{\$}{\leftarrow} \mathcal{R}_k^{p^n}$ and compute $c \leftarrow H(ay_1 + y_2, \mu)$ and the potential signature (z_1, z_2, c) where $z_1 \leftarrow s_1 c + y_1, z_2 \leftarrow s_2 c + y_2$[4]. But before sending the signature, we must perform a rejection-sampling step where we only send if z_1, z_2 are both in $\mathcal{R}_{k-32}^{p^n}$. This part is crucial for security and it is also where the size of k matters. If k is too small, then z_1, z_2 will almost never be in $\mathcal{R}_{k-32}^{p^n}$, whereas if its too big, it will

---

[4] We would like to draw the reader's attention to the fact that in step 3, reduction modulo p is not performed since all the polynomials involved have small coefficients.

be easy for the adversary to forge messages[5]. To verify the signature $(\mathbf{z}_1, \mathbf{z}_2, \mathbf{c})$, the verifier simply checks that $\mathbf{z}_1, \mathbf{z}_2 \in \mathcal{R}_{k-32}^{p^n}$ and that $\mathbf{c} = \mathrm{H}(\mathbf{az}_1 + \mathbf{z}_2 - \mathbf{tc}, \mu)$.

Our security proof follows that in [21] except that it uses the rejection sampling algorithm from [20]. Given a random polynomial $\mathbf{a} \in \mathcal{R}^{p^n}$, we pick two polynomials $\mathbf{s}_1, \mathbf{s}_2 \xleftarrow{\$} \mathcal{R}_{k'}^{p^n}$ for a sufficiently large k' and return $(\mathbf{a} \in \mathcal{R}^{p^n}, \mathbf{t} = \mathbf{as}_1' + \mathbf{s}_2')$ as the public key. By the $\mathbf{DCK}_{p,n}$ assumption (and a standard hybrid argument), this looks like a valid public key (i.e., the adversary cannot tell that the $\mathbf{s}_i$ are chosen from $\mathcal{R}_{k'}^{p^n}$ rather than from $\mathcal{R}_1^{p^n}$). When the adversary gives us signature queries, we appropriately program the hash function outputs so that our signatures are valid even though we do not know a valid secret key (in fact, a valid secret key does not even exist). When the adversary successfully forges a new signature, we then use the "forking lemma" [33] to produce two signatures of the message μ, $(\mathbf{z}_1, \mathbf{z}_2, \mathbf{c})$ and $(\mathbf{z}_1', \mathbf{z}_2', \mathbf{c}')$, such that

$$\mathrm{H}(\mathbf{az}_1 + \mathbf{z}_2 - \mathbf{tc}, \mu) = \mathrm{H}(\mathbf{az}_1' + \mathbf{z}_2' - \mathbf{tc}', \mu), \tag{1}$$

which implies that

$$\mathbf{az}_1 + \mathbf{z}_2 - \mathbf{tc} = \mathbf{az}_1' + \mathbf{z}_2' - \mathbf{tc}' \tag{2}$$

and because we know that $\mathbf{t} = \mathbf{as}_1 + \mathbf{s}_2$, we can obtain

$$\mathbf{a}(\mathbf{z}_1 - \mathbf{cs}_1 - \mathbf{z}_1' + \mathbf{c}'\mathbf{s}_1) + (\mathbf{z}_2 - \mathbf{cs}_2 - \mathbf{z}_2' + \mathbf{c}'\mathbf{s}_2) = \mathbf{0}.$$

Because $\mathbf{z}_i, \mathbf{s}_i, \mathbf{c}$, and $\mathbf{c}'$ have small coefficients, we found two polynomials $\mathbf{u}_1, \mathbf{u}_2$ with small coefficients such that $\mathbf{au}_1 + \mathbf{u}_2 = 0$[6] By [21, Lemma 3.7], knowing such small $\mathbf{u}_i$ allows us to solve the $\mathbf{DCK}_{p,n}$ problem.

We now explain the trick that we use to lower the size of the signature as returned by the optimized scheme presented in Section 3.3. Notice that if Equation (2) does not hold exactly, but only approximately (i.e., $\mathbf{az}_1 + \mathbf{z}_2 - \mathbf{tc} - (\mathbf{az}_1' + \mathbf{z}_2' - \mathbf{tc}') = \mathbf{w}$ for some small polynomial $\mathbf{w}$), then we can still obtain small $\mathbf{u}_1, \mathbf{u}_2$ such that $\mathbf{au}_1 + \mathbf{u}_2 = 0$, except that the value of $\mathbf{u}_2$ will be larger by at most the norm of $\mathbf{w}$. Thus if $\mathbf{az}_1 + \mathbf{z}_2 - \mathbf{tc} \approx \mathbf{az}_1' + \mathbf{z}_2' - \mathbf{tc}'$, we will still be able to produce small $\mathbf{u}_1, \mathbf{u}_2$ such that $\mathbf{au}_1 + \mathbf{u}_2 = 0$. This could make us consider only sending $(\mathbf{z}_1, \mathbf{c})$ as a signature rather than $(\mathbf{z}_1, \mathbf{z}_2, \mathbf{c})$, and the proof will go through fine. The problem with this approach is that the verification algorithm will no longer work, because even though $\mathbf{az}_1 + \mathbf{z}_2 - \mathbf{tc} \approx \mathbf{az}_1 - \mathbf{tc}$, the output of the hash function H will be different. A way to go around the problem is to only evaluate H on the "high order bits" of the coefficients comprising the polynomial $\mathbf{az}_1 + \mathbf{z}_2 - \mathbf{tc}$ which we could hope to be the same as those of the polynomial $\mathbf{az}_1 - \mathbf{tc}$. But in practice, too many bits would be different (because of the carries caused by $\mathbf{z}_2$) for this to be a useful trick. What we do instead is send $(\mathbf{z}_1, \mathbf{z}_2', \mathbf{c})$ as the signature where $\mathbf{z}_2'$ only tells us the carries that $\mathbf{z}_2$ would have created in the high order bits in the sum of $\mathbf{az}_1 + \mathbf{z}_2 - \mathbf{tc}$, and so $\mathbf{z}_2'$ can be represented with much fewer bits than $\mathbf{z}_2$. In the next subsection, we explain

---

[5] The exact probability that $\mathbf{z}_1, \mathbf{z}_2$ will be in $\mathcal{R}_{k-32}^{p^n}$ is $\left(1 - \frac{64}{2k+1}\right)^{2n}$.

[6] It is also important that these polynomials are non-zero.

exactly what we mean by "high-order bits" and give an algorithm that produces a z_2' from z_2, and then provide an optimized version of the scheme in this section that uses the compression idea.

3.2 The Compression Algorithm

For every integer y in the range $\left[-\frac{p-1}{2}, \frac{p-1}{2}\right]$ and any positive integer k, y can be uniquely written as $y = y^{(1)}(2k+1) + y^{(0)}$ where $y^{(0)}$ is an integer in the range $[-k, k]$ and $y^{(1)} = \frac{y - y^{(0)}}{2k+1}$. Thus $y^{(0)}$ are the "lower-order" bits of y, and $y^{(1)}$ are the "higher-order" ones[7]. For a polynomial $\mathbf{y} = \mathbf{y}[0] + \mathbf{y}[1]\mathbf{x} + \ldots + \mathbf{y}[n-1]\mathbf{x}^{n-1} \in \mathcal{R}^{p^n}$, we define $\mathbf{y}^{(1)} = \mathbf{y}[0]^{(1)} + \mathbf{y}[1]^{(1)}\mathbf{x} + \ldots + \mathbf{y}[n-1]^{(1)}\mathbf{x}^{n-1}$ and $\mathbf{y}^{(0)} = \mathbf{y}[0]^{(0)} + \mathbf{y}[1]^{(0)}\mathbf{x} + \ldots + \mathbf{y}[n-1]^{(0)}\mathbf{x}^{n-1}$.

The Lemma below states that given two vectors $\mathbf{y}, \mathbf{z} \in \mathcal{R}^{p^n}$ where the coefficients of $\mathbf{z}$ are small, we can replace $\mathbf{z}$ by a much more compressed vector $\mathbf{z}'$ while keeping the higher order bits of $\mathbf{y} + \mathbf{z}$ and $\mathbf{y} + \mathbf{z}'$ the same. The algorithm that satisfies this lemma is presented in Figure 5 in Appendix A.

Lemma 3.1. *There exists a linear-time algorithm* Compress$(\mathbf{y}, \mathbf{z}, p, k)$ *that for any p, n, k where $2nk/p > 1$ takes as inputs $\mathbf{y} \xleftarrow{\$} \mathcal{R}^{p^n}, \mathbf{z} \in \mathcal{R}_k^{p^n}$, and with probability at least .98 (over the choices of.$\mathbf{y} \in \mathcal{R}^{p^n}$), outputs a $\mathbf{z}' \in \mathcal{R}_k^{p^n}$ such that*

1. $(\mathbf{y} + \mathbf{z})^{(1)} = (\mathbf{y} + \mathbf{z}')^{(1)}$
2. $\mathbf{z}'$ *can be represented with only $2n + \lceil \log(2k+1) \rceil \cdot \frac{6kn}{p}$ bits.*

3.3 A Signature Scheme for Embedded Systems

We now present the version of the signature scheme that incorporates the compression idea from Section 3.2 (see Figure 2). We will use the following notation that is similar to the notation in Section 3.2: every polynomial $\mathbf{Y} \in \mathcal{R}^{p^n}$ can be written as

$$\mathbf{Y} = \mathbf{Y}^{(1)}(2(k-32)+1) + \mathbf{Y}^{(0)}$$

where $\mathbf{Y}^{(0)} \in \mathcal{R}_{k-32}^{p^n}$ and k corresponds to the k in the signature scheme in Figure 2. Notice that there is a bijection between polynomials $\mathbf{Y}$ and this representation $(\mathbf{Y}^{(1)}, \mathbf{Y}^{(0)})$ where

$$\mathbf{Y}^{(0)} = \mathbf{Y} \bmod (2(k-32)+1),$$

and

$$\mathbf{Y}^{(1)} = \frac{\mathbf{Y} - \mathbf{Y}^{(0)}}{2(k-32)+1}.$$

Intuitively, $\mathbf{Y}^{(1)}$ is comprised of the higher order bits of $\mathbf{Y}$.

The secret key in our scheme consists of two polynomials $\mathbf{s}_1, \mathbf{s}_2$ sampled uniformly from $\mathcal{R}_1^{p^n}$ and the public key consists of two polynomials $\mathbf{a} \xleftarrow{\$} \mathcal{R}^{p^n}$ and

---

[7] Note that these only roughly correspond to the notion of most and least significant bits.

Signing Key: $\mathbf{s}_1, \mathbf{s}_2 \xleftarrow{\$} \mathcal{R}_1^{p^n}$

Verification Key: $\mathbf{a} \xleftarrow{\$} \mathcal{R}^{p^n}, \mathbf{t} \leftarrow \mathbf{a}\mathbf{s}_1 + \mathbf{s}_2$

Cryptographic Hash Function: $H : \{0,1\}^* \to D_{32}^n$

Sign$(\mu, \mathbf{a}, \mathbf{s}_1, \mathbf{s}_2)$

  1: $\mathbf{y}_1, \mathbf{y}_2 \xleftarrow{\$} \mathcal{R}_k^{p^n}$

  2: $\mathbf{c} \leftarrow H\left((\mathbf{a}\mathbf{y}_1 + \mathbf{y}_2)^{(1)}, \mu\right)$

  3: $\mathbf{z}_1 \leftarrow \mathbf{s}_1\mathbf{c} + \mathbf{y}_1, \mathbf{z}_2 \leftarrow \mathbf{s}_2\mathbf{c} + \mathbf{y}_2$

  4: if $\mathbf{z}_1$ or $\mathbf{z}_2 \notin \mathcal{R}_{k-32}^{p^n}$, then goto step 1

  5: $\mathbf{z}_2' \leftarrow$ Compress $(\mathbf{a}\mathbf{z}_1 - \mathbf{t}\mathbf{c}, \mathbf{z}_2, p, k - 32)$

  6: if $\mathbf{z}_2' = \perp$, then goto step 1

  7: output $(\mathbf{z}_1, \mathbf{z}_2', \mathbf{c})$

Verify$(\mu, \mathbf{z}_1, \mathbf{z}_2', \mathbf{c}, \mathbf{a}, \mathbf{t})$

  1: Accept iff

    $\mathbf{z}_1, \mathbf{z}_2' \in \mathcal{R}_{k-32}^{p^n}$ and

    $\mathbf{c} = H\left((\mathbf{a}\mathbf{z}_1 + \mathbf{z}_2' - \mathbf{t}\mathbf{c})^{(1)}, \mu\right)$

Fig. 2. Optimized Signature Scheme

$\mathbf{t} = \mathbf{a}\mathbf{s}_1 + \mathbf{s}_2$. In step 1 of the signing algorithm, we choose the "masking polynomials" $\mathbf{y}_1, \mathbf{y}_2$ from $\mathcal{R}_k^{p^n}$. In step 2, we let $\mathbf{c}$ be the hash function value of the high order bits of $\mathbf{a}\mathbf{y}_1 + \mathbf{y}_2$ and the message μ. In step 3, we compute $\mathbf{z}_1, \mathbf{z}_2$ and proceed only if they fall into a certain range. In step 5, we compress the value $\mathbf{z}_2$ using the compression algorithm implied in Lemma 3.1, and obtain a value $\mathbf{z}_2'$ such that $(\mathbf{a}\mathbf{z}_1 - \mathbf{t}\mathbf{c} + \mathbf{z}_2)^{(1)} = (\mathbf{a}\mathbf{z}_1 - \mathbf{t}\mathbf{c} + \mathbf{z}_2')^{(1)}$ and send $(\mathbf{z}_1, \mathbf{z}_2', \mathbf{c})$ as the signature of μ. The verification algorithm checks whether $\mathbf{z}_1, \mathbf{z}_2'$ are in $\mathcal{R}_{k-32}^{p^n}$ and that $\mathbf{c} = H\left((\mathbf{a}\mathbf{z}_1 + \mathbf{z}_2' - \mathbf{t}\mathbf{c})^{(1)}, \mu\right)$.

The running time of the signature algorithm depends on the relationship of the parameter k with the parameter p. The larger the k, the more chance that $\mathbf{z}_1$ and $\mathbf{z}_2$ will be in $\mathcal{R}_{k-32}^{p^n}$ in step 4 of the signing algorithm, but the easier the signature will be to forge. Thus it is prudent to set k as small as possible while keeping the running time reasonable.

3.4 Concrete Instantiation

We now give some concrete instantiations of our signature scheme from Figure 2. The security of the scheme depends on two things: the hardness of the underlying $\mathbf{DCK}_{p,n}$ problem and the hardness of finding pre-images in the random oracle H[8]. For simplicity, we fixed the output of the random oracle to 160 bits and so finding pre-images is 160 bits hard. Judging the security of the lattice problem, on the other hand, is notoriously more difficult. For this part, we rely on the extensive experiments performed by Gama and Nguyen [12] and Chen and Nguyen [8] to determine the hardness of lattice reductions for certain classes of lattices. The lattices that were used in the experiments of [12] were a little different than ours, but we believe that barring some unforeseen weakness due to the

---

[8] It is generally considered folklore that for obtaining signatures with λ bits of security using the Fiat-Shamir transform, one only needs random oracles that output λ bits (i.e., collision-resistance is not a requirement). While finding collisions in the random oracle does allow the *valid* signer to produce two distinct messages that have the same signature, this does not constitute a break.

Table 1. Signature Scheme Parameters

Aspect	Set I	Set II
n	512	1024
p	8383489	16760833
k	2^{14}	2^{15}
Approximate signature bit size	8,950	18,800
Approximate secret key bit size	1,620	3,250
Approximate public key bit size	11,800	25,000
Expected number of repetitions	7	7
Approximate root Hermite factor	1.0066	1.0035
Equivalent symmetric security in bits	≈ 100	> 256

added algebraic structure of our lattices and the parameters, the results should be quite similar. We consider it somewhat unlikely that the algebraic structure causes any weaknesses since for certain parameters, our signature scheme is as hard as RING-LWE (which has a quantum reduction from worst-case lattice problems [25]), but we do encourage cryptanalysis for our particular parameters because they are somewhat smaller than what is required for the worst-case to average-case reduction in [37,25] to go through.

The methodology for choosing our parameters is the same as in [21], and so we direct the interested reader to that paper for a more thorough discussion. In short, one needs to make sure that the length of the secret key $[\mathbf{s}_1|\mathbf{s}_2]$ as a vector is not too much smaller than $\sqrt{p}$ and that the allowable length of the signature vector, which depends on k, is not much larger than $\sqrt{p}$. Using these quantities, one can perform the now-standard calculation of the "root Hermite factor" that lattice reduction algorithms must achieve in order to break the scheme (see [12,28,21] for examples of how this is done). According to experiments in [12,8] a factor of 1.01 is achievable now, a factor of 1.007 seems to have around 80 bits of security, and a factor of 1.005 has more than 256-bit security. In Figure 1, we present two sets of parameters. According to the aforementioned methodology, the first has somewhere around 100 bits of security, while the second has more than 256.

We will now explain how the signature, secret key, and public key sizes are calculated. We will use the concrete numbers from set I as example. The signature size is calculated by summing the bit lengths of $\mathbf{z}_1$, $\mathbf{z}_2'$, and $\mathbf{c}$. Since $\mathbf{z}_1$ is in $\mathcal{R}_{k-32}^{p^n}$, it can be represented by $n\lceil\log(2(k-32)+1)\rceil \leq n\log k + n = 7680$ bits. From Lemma 3.1, we know that $\mathbf{z}_2'$ can be represented with $2n + \lceil\log(2(k-32)+1)\rceil \cdot \frac{6(k-32)n}{p} \leq 2n + 6\log(2k) = 1114$ bits. And $\mathbf{c}$ can be represented with 160 bits, for a total signature size of 8954 bits. The secret key consists of polynomials $\mathbf{s}_1, \mathbf{s}_2 \in \mathcal{R}_1^{p^n}$, and so they can be represented with $2\lceil n\log(3)\rceil = 1624$ bits, but a simpler representation can be used that requires 2048 bits. The public key consists of the polynomials $(\mathbf{a}, \mathbf{t})$, but the polynomial $\mathbf{a}$ does not need to be unique for every secret key, and can in fact be some randomness that is agreed upon by everyone who uses the scheme. Thus the public key can be just $\mathbf{t}$, which can be represented using $\lceil n \log p \rceil = 11776$ bits.

We point out that even though the signature and key sizes are larger than in some number theory based schemes, the signature scheme in Figure 2 is quite efficient, (in software and in hardware), with all operations taking quasi-linear time, as opposed to at least quadratic time for number-theory based schemes. The most expensive operation of the signing algorithm is in step 2 where we need to compute $\mathbf{a}\mathbf{y}_1 + \mathbf{y}_2$, which also could be done in quasilinear time using FFT. In step 3, we also need to perform polynomial multiplication, but because $\mathbf{c}$ is a very sparse polynomial with only 32 non-zero entries, this can be performed with just 32 vector additions. And there is no multiplication needed in step 5 because $\mathbf{a}\mathbf{z}_1 - \mathbf{t}\mathbf{c} = \mathbf{a}\mathbf{y}_1 + \mathbf{y}_2 - \mathbf{z}_2$.

4 Implementation

In this section we provide a detailed description of our FPGA implementation of the signature scheme's signing and verification procedures for parameter set $\mathbf{I}$ with about 100 bits of equivalent symmetric security. In order to improve the speed and resource consumption on the FPGA, we utilize internal block memories (BRAM) and DSP hardcores spanning over three clock domains. We designed dedicated implementations of the signing and verification operation that work with externally generated keys.

Roughly speaking, the signing engine is composed out of a scalable amount of area-efficient polynomial multipliers to compute $\mathbf{a}\mathbf{y}_1 + \mathbf{y}_2$. Fresh randomness for $\mathbf{y}_1, \mathbf{y}_2$ is supplied each run by a random number generator (in this prototype implementation an LFSR). To ensure a steady supply of fresh polynomials from the multiplier for the subsequent parts of the design and the actual signing operation, we have included a buffer of a configurable size that pre-stores pairs $(\mathbf{a}\mathbf{y}_1 + \mathbf{y}_2, \mathbf{y}_1 \| \mathbf{y}_2)$. The hash function H saves its state after the message has been hashed and thus prevents rehashing of the (presumably long) message in each new rejection-sampling step. The sparse multiplication of $\mathbf{s}\mathbf{c}$ works coefficient-wise and thus allows immediate testing for the rejection condition. If an out-of-bound coefficient occurs (line 4 and 6 of Figure 2), the multiplication and compression is immediately interrupted and a new polynomial pair is retrieved from the buffer. For the verification engine, we rely on the polynomial multiplier used to compute $\mathbf{a}\mathbf{y}_1 + \mathbf{y}_2$ twice as we compute $\mathbf{a}\mathbf{z}_1 + \mathbf{z}'_2$ first, maintain the internal state and therefore add $\mathbf{t}(-\mathbf{c})$ in a second round to produce the input for the hash function. When signatures are fed into or returned by both engines, they are encoded in order to meet the signature size (see Lemma A.2 for a detailed algorithm).

4.1 Message Signing

The detailed top-level design of the signing engine is depicted in Figure 3. The computation of $\mathbf{a}\mathbf{y}_1 + \mathbf{y}_2$ is implemented in clock domain (1) and carried out by a number of `PolyMul` units (three units are shown in the depicted setup). The BRAMs storing the initial parameters $\mathbf{y} = \mathbf{y}_1 \| \mathbf{y}_2$ are refilled by a random number generator (RNG) running independently in clock domain (3) and the constant polynomial $\mathbf{a}$ is loaded during device initialization. When a `PolyMul` unit has finished the computation of $\mathbf{r} = \mathbf{a}\mathbf{y}_1 + \mathbf{y}_2$, it requests exclusive access to the `Buffer` and stores $\mathbf{r}$ and $\mathbf{y}$ when free space is available. Internally the `Buffer` consists of the two configurable FIFOs `FIFO(r)` and `FIFO(y)`. As all operations in clock domain (1) and (3) are independent of the secret key or message, they are triggered when space in the `Buffer` becomes available. As described in Section 3.4, the polynomial $\mathbf{r} = \mathbf{a}\mathbf{y}_1 + \mathbf{y}_2$ is needed as input to the hashing as well as for the compression components and is thus stored in `BRAM_BUF(r)` while the coefficients of $\mathbf{y}_1, \mathbf{y}_2$ are only needed once and therefore taken directly out of the FIFOs.

When a signature for a message stored in `FIFO(m)` is requested, the sampling-rejection is repeated in clock domain (2) until a valid signature has been written into `FIFO(σ)`. The message to be signed is first hashed and its internal state saved. Therefore, it is only necessary to rehash $\mathbf{r}$ in case the computed signature is rejected (but not the message again). When the hash $\mathbf{c}$ is ready, the `Compression` component is started. In this component, the values $\mathbf{z}_1 = \mathbf{s}_1\mathbf{c} + \mathbf{y}_1$ and $\mathbf{z}_2 = \mathbf{s}_2\mathbf{c} + \mathbf{y}_2$ are computed column/coefficient-wise with a Comba-style sparse multiplier [9] followed by an addition so that coefficients of $\mathbf{z}_1$ or $\mathbf{z}_2$ are sequentially generated. Rejection-sampling is directly performed on these coefficients and the whole pair $(\mathbf{r}, \mathbf{y})$ is rejected once a coefficient is encountered that is not in the desired range. The secret key $\mathbf{s} = \mathbf{s}_1\|\mathbf{s}_2$ is stored in the block RAM `BRAM(s)` which can be initialized during device initialization or set from the outside during runtime. The whole signature $\sigma = (\mathbf{z}_1, \mathbf{z}_2', \mathbf{c})$ is encoded by the `Encoder` component in order to meet the desired signature size (max. 8954 bits) and then written into the FIFO `FIFO(σ)`. The usage of FIFOs and BRAMs as I/O port allows easy integration of our engine into other designs and provides the ability for clock domain separation.

Polynomial Multiplication. The most time-consuming operation of the signature scheme is the polynomial multiplication $\mathbf{a} \cdot \mathbf{y}_1$ (with the addition of $\mathbf{y}_2$ being rather simple). Recall that $\mathbf{a} \in \mathcal{R}^{p^n}$ has 512 23-bit wide coefficients and that $\mathbf{y}_1 \in \mathcal{R}_k^{p^n}$ consists of 512 16-bit wide coefficients. We are aware that the selected schoolbook algorithm (complexity of $\mathcal{O}(n^2)$) is theoretically inferior compared to Karatsuba [19] ($\mathcal{O}(n^{\log 3})$) or the FFT [29] ($\mathcal{O}(n \log n)$). However, its regular structure and iterative nature allows very high clock frequencies and an area efficient implementation on very small and cheap devices. The polynomial reduction with $f = x^n + 1$ is performed in place which leads to the negacyclic convolution

$$\mathbf{r} = \sum_{i=0}^{511} \sum_{j=0}^{511} (-1)^{\lfloor \frac{i+j}{n} \rfloor} \mathbf{a}_i \mathbf{y}_j x^{i+j} \quad \bmod 512$$

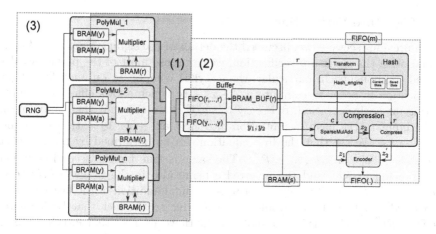

Fig. 3. Block structure of the implemented signing engine. The three different clock domains are denoted by (1), (2), (3).

of $\mathbf{a}$ and $\mathbf{y}_1$. The data path for the arithmetic is depicted in Figure 4(a). The computation of $\mathbf{a}_i\mathbf{y}_j$ is realized in a multiplication core. We avoid dealing with signed values by determining the sign of the value added to the intermediate coefficient from the MSB sign bit of $\mathbf{y}_j$ and if a reduction modular $x^n + 1$ is necessary. As all coefficients of $\mathbf{a}$ are stored in the range $[0, p-1]$ they do not affect the sign of the result. Modular reduction (see Figure 4(b)) by $p = 8383489$ is implemented based on the idea of Solinas [36] as 2^{23} mod $8383489 = 5119$ is very small. For the modular addition of y_2 the multiplier's arithmetic pipeline is reused in a final round in which the output of BRAM(a) is being set to 1 and the coefficients of y_2 are being fed into the BRAM(y) port. Each PolyMul unit also acts as an additional buffer as it can hold one complete result of $\mathbf{r}$ in its internal temporary BRAM and thus reduces latency further in a scenario with precomputation. All in all, one PolyMul unit requires 204 slices, 3 BRAMs, 4 DSPs and is able to generate approx. 1130 pairs of (r,y) per second at a clock frequency of 300 MHz on a Spartan-6.

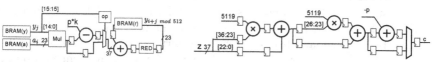

(a) Pipelined data-path of PolyMul. (b) DSP based modular reduction with $p = 8383489$.

Fig. 4. Implementation of PolyMul

4.2 Signature Verification

In the previous sections we discussed the details of the signing algorithm. When dealing with the signature verification, we can reuse most of the previously described components. In particular, the PolyMul component only needs a slight modification in order to compute $az_1 + z_2' - tc$ which allows efficient resource sharing for both operation. It is easy to see that we can split the computation of the input to the hash instantiation into $t_1 = az_1 + z_2'$, $t_2 = t(-c) + 0$ and $t = t_1 + t_2$. We see that the first equation can be performed by the PolyMul core as $a \in \mathcal{R}^{p^n}$ and $z_1, z_2' \in \mathcal{R}_k^{p^n}$. The same is true for the second equation with t being in $\mathcal{R}^{p^n}$ and the inverted c being also in the range $[-k, k]$ (c is even much smaller). The only problem is the final addition of the last equation as a third call to PolyMul would not work due to the fact that both inputs are from $\mathcal{R}^{p^n}$ which PolyMul cannot handle. However, note that PolyMul stores the intermediate state of the schoolbook multiplication in BRAM(r) but initializes the block RAM with zero coefficients prior to the next computation of a new $ay_1 + y_2$. As a consequence, PolyMul supports a special flag that triggers a multiply-accumulate behavior in which the content of BRAM(r) is preserved after a full run of the schoolbook multiplication (ay_1) and an addition of y_2. Therefore, the intermediate values t_1 and t_2 are summed up in BRAM(r) and we do not need the final addition. This enabled us to design a verification engine that performs its arithmetic operations with just two runs of the PolyMul core.

5 Results and Comparison

All presented results below were obtained after post-place-and-route (PAR) and were generated with Xilinx ISE 13.3. We have implemented the signing and verification engine (parameter set **I**, buffer of size one) on two devices of the low-cost Spartan-6 device family and on one high-speed Virtex-6 (all speed grade -3). Detailed information regarding performance and resources consumption is given in Table 2 and Table 3, respectively. For the larger devices we instantiate multiple distinct engines as the Compression and Hash components become the bottleneck when a certain amount of PolyMul components are instantiated. Note also that our implementation is small enough to fit the signing (two PolyMul units) or verification engine on the second-smallest Spartan-6 LX9.

When comparing our results to other work as given in Table 4, we conservatively assume that RSA signatures (one modular exponentiation) with a key size of 1024 bit and ECDSA signatures (one point multiplication) with a key size of 160 bit are comparable to our scheme in terms of security (see Section 3.4 for details on the parameters). In comparison with RSA, our implementation on the low-cost Spartan-6 is 1.5 times faster than the high-speed implementation of Suzuki [38] – that still needs twice as many device resources and runs on the more expensive Virtex-4 device. Note however, that ECC over binary curves is very well suited for hardware and even implementations on old FPGAs like the Virtex-2 [1] are faster than our lattice-based scheme. For the NTRUSign lattice-based signature scheme (introduced in [17] and broken by Nguyen [30]) and the

Table 2. Performance of signing and verification for different design targets

	Aspect	Spartan-6 LX16	Spartan-6 LX100	Virtex-6 LX130
Signing	Engines/Multiplier	1/7	4/9	9/8
	Total Multipliers	7	36	72
	Max. freq. domain (1)	270 MHz	250 MHz	416 MHz
	Max. freq. domain (2)	162 MHz	154 MHz	204 MHz
	Throughput Ops/s	931	4284	12627
Verification	Independent engines	2	14	20
	Max. frequency domain (1)	272 MHz	273 MHz	402 MHz
	Max. frequency domain (2)	158 MHz	103 MHz	156 MHz
	Throughput Ops/s	998	7015	14580

Table 3. Resource consumption of signing and verification for different design targets

	Aspect	Spartan 6 LX16	Spartan 6 LX100	Virtex 6 LX130
Signing	Slices	2273	11006	19896
	LUT/FF	7465/8993	30854/34108	67027/95511
	18K BRAM	29.5	138	234
	DPS48A1	28	144	216
Verification	Slices	2263	14649	18998
	LUT/FF	6225/6663	44727/45094	61360/57903
	18K BRAM	15	90	120
	DPS48A1	8	56	60

XMSS [6] hash-based signature scheme we are not aware of any implementation results for FPGAs. Hardware implementations of Multivariate Quadratic (MQ) cryptosystems [5,3] show that these schemes are faster (factor 2-50) than ECC but also suffer from impractical key sizes for the private and public key (e.g., 80 Kb for Unbalanced Oil and Vinegar (UOV)) [32]. While implementations of the McEliece encryption scheme offer good performance [10,35] the only implementation of a code based signature scheme [4] is extremely slow with a runtime of 830 ms for signing.

6   Conclusion

In this paper we presented a provably secure lattice based digital signature scheme and its implementation on a wide scale of reconfigurable hardware. With moderate resource requirements and more than 12,000 and 14,000 signing and verification operations per second on a Virtex-6 FPGA, our prototype implementation even outperforms classical and alternative cryptosystems in terms of signature size and performance.

Table 4. Implementation results for comparable signature schemes (signing)

Operation	Algorithm	Device	Resources	Ops/s
RSA Signature [38]	RSA-1024; private key	XC4VFX12-10	3937 LS/ 17 DSPs	548
ECDSA [15]	NIST-P224; point mult.	XC4VFX12-12	1580 LS/ 26 DSPs	2,739
ECDSA [1]	NIST-B163; point mult.	XC2V2000	8300 LUTs/ 7 BRAMs	24,390
UOV-Signature [5]	UOV(60,20)	XC5VLX50-3	13437 LUTs	170,940

Future work consists of optimization of the rejection-sampling steps as well as evaluation of different polynomial multiplication methods like the FFT. We also plan to investigate practicability of the signature scheme on other platforms like microcontrollers or graphic cards.

References

1. Ansari, B., Hasan, M.: High performance architecture of elliptic curve scalar multiplication. CACR Research Report 1, 2006 (2006)
2. Arora, S., Ge, R.: New Algorithms for Learning in Presence of Errors. In: Aceto, L., Henzinger, M., Sgall, J. (eds.) ICALP 2011, Part I. LNCS, vol. 6755, pp. 403–415. Springer, Heidelberg (2011)
3. Balasubramanian, S., Carter, H., Bogdanov, A., Rupp, A., Ding, J.: Fast multivariate signature generation in hardware: The case of rainbow. In: Application-Specific Systems, Architectures and Processors, ASAP 2008, pp. 25–30. IEEE (2008)
4. Beuchat, J., Sendrier, N., Tisserand, A., Villard, G., et al.: FPGA implementation of a recently published signature scheme. Rapport de Recherche RR LIP 2004-14 (2004)
5. Bogdanov, A., Eisenbarth, T., Rupp, A., Wolf, C.: Time-Area Optimized Public-Key Engines: $\mathcal{MQ}$-Cryptosystems as Replacement for Elliptic Curves? In: Oswald, E., Rohatgi, P. (eds.) CHES 2008. LNCS, vol. 5154, pp. 45–61. Springer, Heidelberg (2008)
6. Buchmann, J., Dahmen, E., Hülsing, A.: XMSS - A Practical Forward Secure Signature Scheme Based on Minimal Security Assumptions. In: Yang, B.-Y. (ed.) PQCrypto 2011. LNCS, vol. 7071, pp. 117–129. Springer, Heidelberg (2011)
7. Buchmann, J., May, A., Vollmer, U.: Perspectives for cryptographic long-term security. Commun. ACM 49, 50–55 (2006)
8. Chen, Y., Nguyen, P.Q.: BKZ 2.0: Better Lattice Security Estimates. In: Lee, D.H., Wang, X. (eds.) ASIACRYPT 2011. LNCS, vol. 7073, pp. 1–20. Springer, Heidelberg (2011)
9. Comba, P.G.: Exponentiation cryptosystems on the IBM PC. IBM Syst. J. 29, 526–538 (1990)
10. Eisenbarth, T., Güneysu, T., Heyse, S., Paar, C.: MicroEliece: McEliece for Embedded Devices. In: Clavier, C., Gaj, K. (eds.) CHES 2009. LNCS, vol. 5747, pp. 49–64. Springer, Heidelberg (2009)
11. Fischer, J.-B., Stern, J.: An Efficient Pseudo-random Generator Provably as Secure as Syndrome Decoding. In: Maurer, U.M. (ed.) EUROCRYPT 1996. LNCS, vol. 1070, pp. 245–255. Springer, Heidelberg (1996)
12. Gama, N., Nguyen, P.Q.: Predicting Lattice Reduction. In: Smart, N.P. (ed.) EUROCRYPT 2008. LNCS, vol. 4965, pp. 31–51. Springer, Heidelberg (2008)

13. Gentry, C., Peikert, C., Vaikuntanathan, V.: Trapdoors for hard lattices and new cryptographic constructions. In: STOC, pp. 197–206 (2008)

14. Goldreich, O., Goldwasser, S., Halevi, S.: Public-Key Cryptosystems from Lattice Reduction Problems. In: Kaliski Jr., B.S. (ed.) CRYPTO 1997. LNCS, vol. 1294, pp. 112–131. Springer, Heidelberg (1997)

15. Güneysu, T., Paar, C.: Ultra High Performance ECC over NIST Primes on Commercial FPGAs. In: Oswald, E., Rohatgi, P. (eds.) CHES 2008. LNCS, vol. 5154, pp. 62–78. Springer, Heidelberg (2008)

16. Hoffstein, J., Howgrave-Graham, N., Pipher, J., Silverman, J.H., Whyte, W.: NTRUSign: Digital Signatures Using the NTRU Lattice. In: Joye, M. (ed.) CT-RSA 2003. LNCS, vol. 2612, pp. 122–140. Springer, Heidelberg (2003)

17. Hoffstein, J., Pipher, J., Silverman, J.H.: NTRU: A Ring-Based Public Key Cryptosystem. In: Buhler, J.P. (ed.) ANTS 1998. LNCS, vol. 1423, pp. 267–288. Springer, Heidelberg (1998)

18. Hoffstein, J., Pipher, J., Silverman, J.H.: NSS: An NTRU Lattice-Based Signature Scheme. In: Pfitzmann, B. (ed.) EUROCRYPT 2001. LNCS, vol. 2045, pp. 211–228. Springer, Heidelberg (2001)

19. Karatsuba, A., Ofman, Y.: Multiplication of multidigit numbers on automata. Soviet Physics Doklady 7, 595 (1963)

20. Lyubashevsky, V.: Fiat-Shamir with Aborts: Applications to Lattice and Factoring-Based Signatures. In: Matsui, M. (ed.) ASIACRYPT 2009. LNCS, vol. 5912, pp. 598–616. Springer, Heidelberg (2009)

21. Lyubashevsky, V.: Lattice Signatures without Trapdoors. In: Pointcheval, D., Johansson, T. (eds.) EUROCRYPT 2012. LNCS, vol. 7237, pp. 738–755. Springer, Heidelberg (2012), Full version at http://eprint.iacr.org/2011/537

22. Lyubashevsky, V., Micciancio, D.: Generalized Compact Knapsacks Are Collision Resistant. In: Bugliesi, M., Preneel, B., Sassone, V., Wegener, I. (eds.) ICALP 2006, Part II. LNCS, vol. 4052, pp. 144–155. Springer, Heidelberg (2006)

23. Lyubashevsky, V., Micciancio, D.: Asymptotically Efficient Lattice-Based Digital Signatures. In: Canetti, R. (ed.) TCC 2008. LNCS, vol. 4948, pp. 37–54. Springer, Heidelberg (2008)

24. Lyubashevsky, V., Micciancio, D., Peikert, C., Rosen, A.: SWIFFT: A Modest Proposal for FFT Hashing. In: Nyberg, K. (ed.) FSE 2008. LNCS, vol. 5086, pp. 54–72. Springer, Heidelberg (2008)

25. Lyubashevsky, V., Peikert, C., Regev, O.: On Ideal Lattices and Learning with Errors over Rings. In: Gilbert, H. (ed.) EUROCRYPT 2010. LNCS, vol. 6110, pp. 1–23. Springer, Heidelberg (2010)

26. Micciancio, D.: Generalized compact knapsacks, cyclic lattices, and efficient one-way functions. Computational Complexity 16(4), 365–411 (2007)

27. Micciancio, D., Peikert, C.: Trapdoors for Lattices: Simpler, Tighter, Faster, Smaller. In: Pointcheval, D., Johansson, T. (eds.) EUROCRYPT 2012. LNCS, vol. 7237, pp. 700–718. Springer, Heidelberg (2012), Full version at http://eprint.iacr.org/2011/501

28. Bernstein, D.J., Buchmann, J., Dahmen, E.: Post-Quantum Cryptography. Springer (2009) ISBN: 978-3-540-88701-0

29. Moenck, R.T.: Practical fast polynomial multiplication. In: Proceedings of the Third ACM Symposium on Symbolic and Algebraic Computation, SYMSAC 1976, pp. 136–148. ACM, New York (1976)

30. Nguyen, P., Regev, O.: Learning a parallelepiped: Cryptanalysis of GGH and NTRU signatures. Journal of Cryptology 22, 139–160 (2009)

31. Peikert, C., Rosen, A.: Efficient Collision-Resistant Hashing from Worst-Case Assumptions on Cyclic Lattices. In: Halevi, S., Rabin, T. (eds.) TCC 2006. LNCS, vol. 3876, pp. 145–166. Springer, Heidelberg (2006)
32. Petzoldt, A., Thomae, E., Bulygin, S., Wolf, C.: Small Public Keys and Fast Verification for Multivariate Quadratic Public Key Systems. In: Preneel, B., Takagi, T. (eds.) CHES 2011. LNCS, vol. 6917, pp. 475–490. Springer, Heidelberg (2011)
33. Pointcheval, D., Stern, J.: Security arguments for digital signatures and blind signatures. J. Cryptology 13(3), 361–396 (2000)
34. Shor, P.: Algorithms for quantum computation: discrete logarithms and factoring. In: 1994 Proceedings of the 35th Annual Symposium on Foundations of Computer Science, pp. 124–134. IEEE (1994)
35. Shoufan, A., Wink, T., Molter, H., Huss, S., Kohnert, E.: A novel cryptoprocessor architecture for the McEliece public-key cryptosystem. IEEE Transactions on Computers 59(11), 1533–1546 (2010)
36. Solinas, J.: Generalized mersenne numbers. Faculty of Mathematics, University of Waterloo (1999)
37. Stehlé, D., Steinfeld, R., Tanaka, K., Xagawa, K.: Efficient Public Key Encryption Based on Ideal Lattices. In: Matsui, M. (ed.) ASIACRYPT 2009. LNCS, vol. 5912, pp. 617–635. Springer, Heidelberg (2009)
38. Suzuki, D.: How to Maximize the Potential of FPGA Resources for Modular Exponentiation. In: Paillier, P., Verbauwhede, I. (eds.) CHES 2007. LNCS, vol. 4727, pp. 272–288. Springer, Heidelberg (2007)

A   Compression Algorithm

In this section we present our compression algorithm. For two vectors $\mathbf{y}, \mathbf{z}$, the algorithm first checks whether the coefficient $\mathbf{y}[i]$ of $\mathbf{y}$ is greater than $(p-1)/2 - k$ in absolute value. If it is, then there is a possibility that $\mathbf{y}[i] + \mathbf{z}[i]$ will need to be reduced modulo p and in this case we do not compress $\mathbf{z}[i]$. Ideally there should not be many such elements, and we can show that for the parameters used in the signature scheme, there will be at most 6 (out of n) with high probability. It's possible to set the parameters so that there are no such elements, but this decreases the efficiency and is not worth the very slight savings in the compression.

Assuming that $\mathbf{y}[i]$ is in the range where $\mathbf{z}[i]$ can be compressed, we assign the value of k to $\mathbf{z}'[i]$ if $\mathbf{y}[i]^{(0)} + \mathbf{z}[i] > k$, assign $-k$ if $\mathbf{y}[i]^{(0)} + \mathbf{z}[i] < -k$, and 0 otherwise. We now move on to proving that the algorithm satisfies Lemma 3.1.

Lemma A.1. *Item 1 of Lemma 3.1 holds.*

Proof. Given in the full version of this paper.

Lemma A.2. *Item 2 of Lemma 3.1 holds.*

Proof. If $\mathbf{z}[i]' = 0$, we represent it with the bit string '00'. If $\mathbf{z}[i]' = k$, we represent it with the bit string '01'. $\mathbf{z}[i]' = -k$, we represent it with the bit string '10'. If $\mathbf{z}[i]' = \mathbf{z}[i]$ (in other words, it is uncompressed), we represent it with the string '11$\mathbf{z}[i]$' where $\mathbf{z}[i]$ can be represented by $2 \log k$ bits (the '11' is necessary to signify that the following $\log 2k$ bits represent an uncompressed

Compress($\mathbf{y}, \mathbf{z}, p, k$)
1: *uncompressed* $\leftarrow 0$
2: **for** i=1 to n **do**
3:     **if** $|\mathbf{y}[i]| > \frac{p-1}{2} - k$ **then**
4:         $\mathbf{z}'[i] \leftarrow \mathbf{z}[i]$
5:         *uncompressed* $\leftarrow$ *uncompressed* $+ 1$
6:     **else**
7:         write $\mathbf{y}[i] = \mathbf{y}[i]^{(1)}(2k+1) + \mathbf{y}[i]^{(0)}$ where $-k \leq \mathbf{y}[i]^{(0)} \leq k$
8:         **if** $\mathbf{y}[i]^{(0)} + \mathbf{z}[i] > k$ **then**
9:             $\mathbf{z}'[i] \leftarrow k$
10:         **else if** $\mathbf{y}[i]^{(0)} + \mathbf{z}[i] < -k$ **then**
11:             $\mathbf{z}'[i] \leftarrow -k$
12:         **else**
13:             $\mathbf{z}'[i] \leftarrow 0$
14:         **end if**
15:     **end if**
16: **end for**
17: **if** *uncompressed* $\leq \frac{6kn}{p}$ **then**
18:     **return** $\mathbf{z}'$
19: **else**
20:     **return** $\perp$
21: **end if**

Fig. 5. The Compression Algorithm

value). Thus uncompressed values use $2 + \log 2k$ bits and the other values use just 2 bits. Since there are at most $6kn/p$ uncompressed values, the maximum number of bits that are needed is

$$(2+\log 2k)\cdot\frac{6kn}{p} + 2\left(n - \frac{6kn}{p}\right) = 2n + \lceil\log(2k+1)\rceil\cdot\frac{6kn}{p}. \qquad \square$$

Finally, we show that if $\mathbf{y}$ is uniformly distributed in $\mathcal{R}^{p^n}$, then with probability at least .98, the algorithm will not have more than 6 uncompressed elements.

Lemma A.3. *If $\mathbf{y}$ is uniformly distributed modulo p and $2nk/p \geq 1$, then the compression algorithm outputs $\perp$ with probability less than 2%.*

Proof. The probability that the inequality in line 3 will be true is exactly $2k/p$. Thus the value of the *"uncompressed"* variable follows the binomial distribution with n samples each being 1 with probability $2k/p$. Since we will always set $n >> 2k/p$, this distribution can be approximated by the Poisson distribution with $\lambda = 2nk/p$. If $\lambda \geq 1$ then the probability that the number of occurrences is greater than 3λ is at most 2% (this occurs for $\lambda = 1$). Since we output $\perp$ when *uncompressed* $> 6kn/p = 3\lambda$, it is output with probability at most 2%. $\qquad \square$

An Efficient Countermeasure against Correlation Power-Analysis Attacks with Randomized Montgomery Operations for DF-ECC Processor

Jen-Wei Lee, Szu-Chi Chung, Hsie-Chia Chang, and Chen-Yi Lee

Department of Electronics Engineering and Institute of Electronics,
National Chiao Tung University, Hsinchu, Taiwan
jenweilee@gmail.com, {phonchi,hcchang,cylee}@si2lab.org

Abstract. Correlation power-analysis (CPA) attacks are a serious threat for cryptographic device because the key can be disclosed from data-dependent power consumption. *Hiding* power consumption of encryption circuit can increase the security against CPA attacks, but it results in a large overhead for cost, speed, and energy dissipation. *Masking* processed data such as randomized scalar or primary base point on elliptic curve is another approach to prevent CPA attacks. However, these methods requiring pre-computed data are not suitable for hardware implementation of real-time applications. In this paper, a new CPA countermeasure performing all field operations in a randomized Montgomery domain is proposed to eliminate the correlation between target and reference power traces. After implemented in 90-nm CMOS process, our protected 521-bit dual-field elliptic curve cryptographic (DF-ECC) processor can perform one elliptic curve scalar multiplication (ECSM) in 4.57ms over $GF(p^{521})$ and 2.77ms over $GF(2^{409})$ with 3.6% area and 3.8% power overhead. Experiments from an FPGA evaluation board demonstrate that the private key of unprotected device will be revealed within 10^3 power traces, whereas the same attacks on our proposal cannot successfully extract the key value even after 10^6 measurements.

Keywords: Elliptic curve cryptography (ECC), side-channel attacks, power-analysis attacks, Montgomery algorithm.

1 Introduction

Elliptic curve cryptography (ECC) independently introduced by Koblitz [1] and Miller [2] has been widely applied to provide a confident scheme for information exchange. For the past several years, many previous works [3], [4], [5], [6] have been published for ECC hardware implementation aiming at the performance improvement. However, even the ECC is secure at cryptanalysis, the private data of a unprotected hardware device can be extracted by the physical attacks due to side-channel leakage. The power-analysis attacks, initially presented by Kocher [7], can reveal the key value by analyzing the power information of a cryptographic implementation such as on an ASIC, FPGA or microprocessor.

E. Prouff and P. Schaumont (Eds.): CHES 2012, LNCS 7428, pp. 548–564, 2012.

During the device processing, simple power-analysis (SPA) attacks can distinguish the key value through visual inspection because of the specifically active circuit with direct hardware scheduling. The unified elliptic curve (EC) point calculation [8], [9] is usually used to avoid the variation of power consumption over time. However, the correlation power-analysis (CPA) attacks [10] computing the correlation between target power traces and power model by statistical approach can reveal the key value due to the existence of key-dependent operations in every round of calculation. For ECC primitives specified in IEEE P1363 [11], the CPA attacks can be applied to EC integrated encryption system, single pass EC Diffie-Hellman or single pass EC Menezes-Qu-Vanstone key agreement because the private key is kept invariant for a long time duration.

Hiding technique with algorithm-independent dedicated circuit is a common approach to protect cryptographic processors from attackers collecting the key-dependent characteristics of power traces. In [12], wave dynamic differential logic circuit with regular routing algorithm is exploited to equalize the current between rising and falling transitions. However, at least double hardware latency, area cost, and energy for unprotected encryption engines are required due to precharging for half cycle, and generating complementary logic outputs from divided single ended modules with equivalent power consumption. Switched capacitor [13] is able to isolate the encryption core from the external power supplies, but this approach results in 50% speed loss for replenishing charge every cycle. In order to avoid the throughput degradation, a countermeasure circuit using digital controlled ring oscillators [14] is designed outside of the critical path. The concept is to generate random noise power to dominate the power consumption of arithmetic unit, and then the correlation peak would not be found even matching the correct key value. But this demands extra 100% power overhead for the key-dependent processing element.

At the algorithm level, *masking* the processed data independent of power consumption is another approach to avoid the CPA attacks. Since the scalar K of EC point calculation is periodic with the point order $\#E$, a randomized scalar technique proposed by Coron [15] can be adopted to change the key value by adding $\alpha \cdot \#E$ for every elliptic curve scalar multiplication (ECSM) such as $KP = (K + \alpha \cdot \#E)P$, where α is a random integer and P is a primary base point on EC. However, with this method, the throughput overhead is inevitable due to extending the key length. In [9], the ECSM of 521-bit key extended with a 32-bit random value needs 10% more execution time to be carried out than that of unprotected approach. Another CPA countermeasure also presented in [15] is to mask the primary base point with pre-computed random points R and $S = KR$. Then the ECSM is achieved by computing $K(P + R) = KP'$ and subtracting S before returning such that $KP' - S = KP$. For every next ECSM calculation, the random points R and S are refreshed by performing $R \leftarrow (-1)^{\beta}2R$ and $S \leftarrow (-1)^{\beta}2S$ with a single random bit β. But the time-cost random point generation is not suitable for real-time applications as the EC parameters are various with different users. In [16], the EC isomorphism method can randomize the primary base point by simple finite field operations without pre-computing random points. However, it is limited to be applied in single finite field $GF(p)$.

In this brief, we propose a new efficient countermeasure to overcome the CPA attacks by computing overall dual-field ECC functions in a *randomized Montgomery domain*. The feature of our approach is to mask the intermediate values in not only the arithmetic but also the temporary register. Thus it is unnecessary to extend the key length, customize circuit and modify the routing algorithm in ASIC or FPGA design flow. Since our proposed design adopts simple logic circuit to counteract CPA attacks, the hardware cost overhead could be significantly reduced, and the maximum operating frequency of protected design is the same as that of unprotected design using conventional Montgomery algorithm. Additionally, by reducing the iteration time of the division, which dominates other field operations in computation time, the speed can be improved further.

The remainder of the paper is outlined as follows. CPA attacks applied on the ECC device are introduced in Section 2. The proposed countermeasure method and design architecture are given in Section 3 and Section 4, respectively. Section 5 shows the FPGA power measurement and ASIC implementation results. Section 6 concludes this work.

2   CPA Attacks on ECC Device

Algorithm 1 presented in [8] is a usually adopted approach to counteract SPA attacks by regularly performing the ECSM $KP = P + \cdots + P$, where K is the m-bit private key and P is a point on elliptic curves (ECs). But the intermediate values of elliptic curve point doubling in Step 3 and Step 4 still have dependence on the zero and non-zero bit of the key value. Hence, with a chosen point P, the key value can be distinguished by matching the power trace segment of accessing the memory storage for point coordinates P_1 or P_2.

Algorithm 1. Montgomery ladder ECSM algorithm

Input: K and P
Output: KP
1. Let $P_1 \leftarrow P, P_2 \leftarrow 2P$
2. **For** i from $m - 2$ to 0 **do**
3.   **If** $K_i = 1$ **then** $P_1 \leftarrow P_1 + P_2, P_2 \leftarrow 2P_2$
4.   **else** $P_2 \leftarrow P_1 + P_2, P_1 \leftarrow 2P_1$
5. **Return** P_1

Fig. 1 illustrates the scenario of CPA attacks. For ECC primitives, the primary base point is commonly public. Thus the power model can be characterized from the hamming distance of memory storage for key-dependent point coordinates by measuring the device sample before the statistical analysis, which computes the correlation between the measured target power traces and the power model. The correlation value of correct hypothesis will be larger than that of the others due to the same hamming distance of processed data. Through this approach, the overall binary key can be extracted after $m - 1$ rounds in linear time.

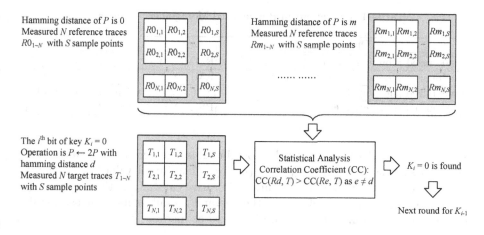

Fig. 1. CPA attacks on an ECC device operating in a specific domain

3   Proposed Algorithm against CPA Attacks

The fundamental concept of CPA countermeasure is to break the dependency between intermediate values and power traces. For achieving the EC point calculation, the well-known Montgomery algorithm [17] is usually adopted to perform the field arithmetic in a specific domain such that $A \equiv a \cdot r \pmod{p}$, where a is in the integer domain and $r \equiv 2^m \pmod{p}$ is the Montgomery constant with m-bit field length. In this work, we introduce an approach to resist the CPA attacks at modular algorithm by calculating the operands in a randomized Montgomery domain $A \equiv a \cdot 2^\lambda \pmod{p}$, where the domain value λ equals the hamming weight (HW) of an n-bit random value α. Note that n is the maximum field length and the bit values of $(\alpha_{n-1}, \alpha_{n-2}, \ldots, \alpha_m)$ are set to zero for preventing λ from exceeding m. By exploiting this approach, the intermediate values can be masked because they are various with different domain values such as $2^g \pmod{p} \neq 2^h \pmod{p}$ when $0 \leq g \neq h < m$. Since the proposed method is to randomize intermediate values in basic modular operations, the SPA resistant ECSM algorithm shown in Algorithm 1 can still be applied without computation overhead from extended scalar length, and there is no need for pre-computed EC points. The overall randomized Montgomery operations for input operands $X \equiv x \cdot 2^\lambda \pmod{p}$ and $Y \equiv y \cdot 2^\lambda \pmod{p}$ are summarized in Table 1.

3.1   Randomized Montgomery Multiplication

Algorithm 2 shows our proposed randomized Montgomery multiplication which contains two operating steps in every iteration to change the intermediate domain value λ', and these steps are determined by the ith bit of random value α. If $\alpha_i = 1$, the domain value of output operand R decreases by one in Step 4 such

Table 1. Operations in Randomized Montgomery Domain

Operation	Arithmetic
Randomized Montgomery multiplication (RMM)	$\mathrm{RMM}(X,Y) \equiv x \cdot y \cdot 2^\lambda \pmod p$
Randomized Montgomery division (RMD)	$\mathrm{RMD}(X,Y) \equiv x \cdot y^{-1} \cdot 2^\lambda \pmod p$
Randomized addition (RA)	$\mathrm{RA}(X,Y) \equiv (x+y) \cdot 2^\lambda \pmod p$
Randomized subtraction (RS)	$\mathrm{RS}(X,Y) \equiv (x-y) \cdot 2^\lambda \pmod p$

as $R = (R + V_0 \cdot S)/2 \pmod p$; the domain value remains the same as $\alpha_i = 0$ in Step 5 such as $R = (R + V_0 \cdot S) \pmod p$. The initial values of operands (V, R, S) are set to be $(X, 0, Y)$. In further iterative calculation, the bit value V_0 is equal to the ith bit value of X, and the operand S doubles its value as $\alpha_i = 0$. Base on these, the functionality can be derived as follows:

- For 1st iteration, the intermediate result of R is $(X_0 \cdot Y) \cdot 2^{-\alpha_0} \pmod p$.
- For 2nd iteration, R becomes $((X_0 \cdot Y) \cdot 2^{-\alpha_0} \pmod p + X_1 \cdot (2^{1-\mathrm{HW}(\alpha_0)} \cdot Y)) \cdot 2^{-\alpha_1} \pmod p$.
- Until m^{th} iteration, the final result of R is $(\cdots(((X_0 \cdot Y) \cdot 2^{-\alpha_0} \pmod p + X_1 \cdot (2^{1-\mathrm{HW}(\alpha_0)} \cdot Y)) \cdot 2^{-\alpha_1} \pmod p + X_2 \cdot (2^{2-\mathrm{HW}(\alpha_1, \alpha_0)} \cdot Y)) \cdot 2^{-\alpha_2} \pmod p +$
$\cdots + X_{m-1} \cdot (2^{m-1-\mathrm{HW}(\alpha_{m-2}, \cdots, \alpha_1, \alpha_0)} \cdot Y)) \cdot 2^{-\alpha_{m-1}} \pmod p$
$\equiv (X_0 \cdot Y \cdot 2^{-\mathrm{HW}(\alpha_{m-1}, \ldots, \alpha_0)}) \pmod p + (X_1 \cdot Y \cdot 2^{-\mathrm{HW}(\alpha_{m-1}, \ldots, \alpha_0)+1}) \pmod p +$
$\cdots + (X_{m-1} \cdot Y \cdot 2^{-\mathrm{HW}(\alpha_{m-1}, \ldots, \alpha_0)+m-1}) \pmod p$
$\equiv X \cdot Y \cdot 2^{-\mathrm{HW}(\alpha_{m-1}, \ldots, \alpha_0)} \pmod p$
$\equiv X \cdot Y \cdot 2^{-\lambda} \pmod p$.

Hence, the randomized Montgomery multiplication in Algorithm 2 can be performed in m iterations, the same as those in conventional radix-2 Montgomery multiplication.

Algorithm 2. Radix-2 randomized Montgomery multiplication

Input: X, Y, p, and α
Output: $R = \mathrm{RMM}(X,Y)$
1. Let $V = X$, $R = 0$, $S = Y$
2. **For** i from 0 to $m-1$ **do**
3. $R \equiv R + V_0 \cdot S \pmod p$, $V = V/2$
4.   **If** $\alpha_i = 1$ **then** $R \equiv R/2 \pmod p$
5.   **else** $S \equiv 2S \pmod p$
6. **Return** R

Algorithm 3 shows a radix-4 approach to Algorithm 2 for almost 50% iteration reduction. The domain value of R is determined by the HW of two continuous bits of random value α in Steps 5, 6, and 7. For the case of HW = 2, it is reduced by two through performing quartering operation such as $R \equiv R/4 \pmod p$. While halving R and doubling S operations are performed as HW = 1, these are deduced by computing one iteration of radix-2 Montgomery reduction and one iteration of radix-2 modular reduction in single period. For the rest case of HW $= 0$, the operand $S \equiv 4S \pmod{p}$ is performed due to the unchanged domain value of R.

Algorithm 3. Radix-4 randomized Montgomery multiplication

Input: X, Y, p, and α

Output: $R = \mathrm{RMM}(X, Y)$

1. Let $V = X$, $R = 0$, $S = Y$
2. **For** i from 0 to $\lceil \frac{m}{2} \rceil - 1$ **do**
3.   **If** $m \pmod{2} \equiv 1$ and $i = \lceil \frac{m}{2} \rceil - 1$ **then**
           $R \equiv R + V_0 \cdot S \pmod{p}$, $V = \frac{V}{2}$
4.   **else**
           $R \equiv R + V_0 \cdot S + V_1 \cdot 2S \pmod{p}$, $V = \frac{V}{4}$
5.   **If** $(\alpha_{2i+1}, \alpha_{2i}) = (1, 1)$ **then**
           $R \equiv \frac{R}{4} \pmod{p}$
6.   **else if** $(\alpha_{2i+1}, \alpha_{2i}) = (1, 0)$ or $(0, 1)$ **then**
           $R \equiv \frac{R}{2} \pmod{p}$, $S \equiv 2S \pmod{p}$
7.   **else**
           $S \equiv 4S \pmod{p}$
8. **Return** R

3.2   Randomized Montgomery Division

To achieve the division in Montgomery domain, Kaliski [18] first proposed an iterative algorithm which needs $m \sim 2m$ iterations of successive reduction, $0 \sim m$ iterations for degree recovery (reduce intermediate domain value λ' to be m as $\lambda' > m$), and two additional Montgomery multiplications with a final modular reduction $p - R$. The algorithm presented in [18] is formulated from the identical equations as follows:

$$\begin{cases} Y \cdot R \equiv -U \cdot 2^{\lambda'} \pmod{p} \\ Y \cdot S \equiv V \cdot 2^{\lambda'} \pmod{p}. \end{cases}$$

Based on Kaliski's method, we derive a new randomized Montgomery division which is described in Algorithm 4. To directly achieve the division operation without additional multiplication and final modular reduction, our method is to modify the initial values of (U, V, R, S) to be $(p, Y, 0, X)$ in Step 1 and the RS data path with modular subtraction in Steps 10, 11, 13, 14. Then the identities become

$$\begin{cases} X^{-1} \cdot Y \cdot R \equiv U \cdot 2^{\lambda'} \pmod{p} \\ X^{-1} \cdot Y \cdot S \equiv V \cdot 2^{\lambda'} \pmod{p}. \end{cases}$$

Similar to RMM, the RS data path between the Montgomery domain and integer domain is determined by the ith bit value of α. The domain value of operands R and S increases by one as $\alpha_i = 1$ and remains the same as $\alpha_i = 0$.

For further reducing the degree recovery phase, the RS data path turns into dividing values by two in Steps 5, 8, 11, 14 to keep the intermediate domain

Algorithm 4. Radix-2 randomized Montgomery division

Input: $X, Y, p,$ and α

Output: $R = \text{RMD}(X, Y)$

1. Let $U = p, V = Y, R = 0, S = X$
2. **While** $(V > 0)$ **do**
3.   **If** U is even **then** $U = U/2$
4.   **If** $\alpha_i = 1$ **then** $S \equiv 2S \pmod{p}$
5.   **else** $R \equiv R/2 \pmod{p}$
6.   **else if** V is even **then** $V = V/2$
7.   **If** $\alpha_i = 1$ **then** $R \equiv 2R \pmod{p}$
8.   **else** $S \equiv S/2 \pmod{p}$
9.   **else if** $U > V$ **then** $U = (U - V)/2$
10.   **If** $\alpha_i = 1$ **then** $R \equiv R - S \pmod{p}, S \equiv 2S \pmod{p}$
11.   **else** $R \equiv (R - S)/2 \pmod{p}$
12.   **else** $V = (V - U)/2$
13.   **If** $\alpha_i = 1$ **then** $S \equiv S - R \pmod{p}, R \equiv 2R \pmod{p}$
14.   **else** $S \equiv (S - R)/2 \pmod{p}$
15.   **If** $i < m$ **then** $i = i + 1$
16. **Return** R

---

value in $\lambda = \text{HW}(\alpha)$ as $i = m$. Thus the identities in Algorithm 4 are given as follows:

$$\text{If } i < m, \text{then} \begin{cases} X^{-1} \cdot Y \cdot R \equiv U \cdot 2^{\lambda'} \pmod{p} \\ X^{-1} \cdot Y \cdot S \equiv V \cdot 2^{\lambda'} \pmod{p} \end{cases}$$

$$\text{else} \begin{cases} X^{-1} \cdot Y \cdot R \equiv U \cdot 2^{\lambda} \pmod{p} \\ X^{-1} \cdot Y \cdot S \equiv V \cdot 2^{\lambda} \pmod{p}. \end{cases}$$

Before the last iteration, both U and V are 1 because the initial values of U and V are relatively prime. Then after finishing the iterative operations in Step 2, the values of (U, V, R, S) become $(1, 0, X \cdot Y^{-1} \cdot 2^{\lambda} \pmod{p}, 0)$. As a result, the proposed randomized division algorithm requires at most $2m$ iterations of successive reduction. Table 2 shows the expected operation time and the comparison with related works on modifying radix-2 Montgomery division algorithm. With randomization capability, Algorithm 4 will also benefit the hardware design owing to the low latency.

Table 2. Analysis of Various Division Algorithms

	Algorithm 4	TCAS-I'06 [3]	ESSCIRC'10 [9]
Iteration Time	$m \sim 2m$	$m \sim 2m$	$m \sim 3m$
Multiplication	0	$2 \sim 3$	0
Domain	Random $2^{\lambda}, 0 \leq \lambda \leq m$	Fixed 2^m	Fixed 2^m

Algorithm 5 shows the radix-4 randomized Montgomery division derived from Algorithm 4, and there are more branches in the algorithm as the radix becomes lager. To remain the domain value of R unpredictable in the flexible range of

$[0, m-1)$, it is determined by the HW of random value α_i or (α_{i+1}, α_i). The values of UV is reduced to at least $UV/4$ except $U \equiv 1 \pmod 4$, $V \equiv 3 \pmod 4$ or $U \equiv 3 \pmod 4$, $V \equiv 1 \pmod 4$ in Steps 17 and 18. With this approach and a radix-4 RMM given in Algorithm 3, the EC point calculation can be carried out faster in affine coordinates than that in projective coordinates [19], where the iteration time ratio RMD/RMM $\cong 1.32$ over $GF(p)$ and 1.44 over $GF(2^m)$.

4   Hardware Architecture of DF-ECC Processor

Fig. 2 shows the block diagram of the proposed dual-field ECC (DF-ECC) processor. For the CPA resistance, all field operations over $GF(p)$ and $GF(2^m)$ are performed by the Galois field arithmetic unit (GFAU) in a randomized Montgomery domain. The operating domain is determined by the value in domain shift register, which is sourced from a 1-bit random number generator (RNG) and refreshed before the next ECSM calculation. For the flexibility, we use an all-digital RNG utilizing the cycle-to-cycle time jitter in free-running oscillators with a synchronous feedback post-processor [20]. The overall architecture of CPA countermeasure circuit is shown in Fig. 3. Besides, to efficiently store the long bit length operands including EC parameters and points, a block memory of register file is exploited.

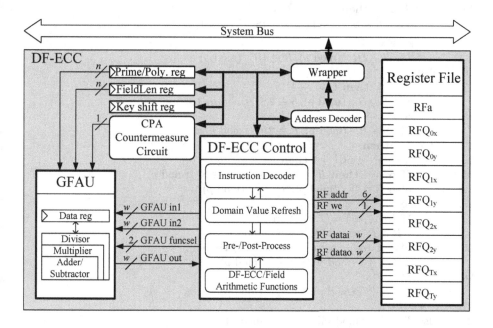

Fig. 2. Overall diagram for the DF-ECC processor

Algorithm 5. Radix-4 randomized Montgomery division

Input: X, Y, p, and α

Output: $R = \text{RMD}(X, Y)$

1. Let $U = p, V = Y, R = 0, S = X, i = 0$
2. **While** $(V > 0)$ **do**
3. $c \equiv U \pmod 4, d \equiv V \pmod 4, t = 2$
4.   **If** $i = m - 1$ **then**
           $R \equiv 2R \pmod p, S \equiv 2S \pmod p, t = 1$
5.   **else if** $c = 0$ **then** $U = \frac{U}{4}, S \equiv 4S \pmod p$
6.   **else if** $d = 0$ **then** $V = \frac{V}{4}, R \equiv 4R \pmod p$
7.   **else if** $c = d$ **then**
8.   **If** $U > V$ **then** $U = \frac{U-V}{4}$,
             $R \equiv R - S \pmod p, S \equiv 4S \pmod p$
9.   **else** $V = \frac{V-U}{4}$,
             $S \equiv S - R \pmod p, R \equiv 4R \pmod p$
10.   **else if** $c = 2$ **then**
11.   **If** $\frac{U}{2} > V$ **then** $U = \frac{\frac{U}{2}-V}{2}$,
              $R \equiv R - 2S \pmod p, S \equiv 4S \pmod p$
12.   **else** $V = \frac{V-\frac{U}{2}}{2}, U = \frac{U}{2}$,
              $S \equiv 2S - R \pmod p, R \equiv 2R \pmod p$
13.   **else if** $d = 2$ **then**
14.   **If** $U > \frac{V}{2}$ **then** $U = \frac{U-\frac{V}{2}}{2}, V = \frac{V}{2}$,
              $R \equiv 2R - S \pmod p, S \equiv 2S \pmod p$
15.   **else** $V = \frac{\frac{V}{2}-U}{2}$,
              $S \equiv S - 2R \pmod p, R \equiv 4R \pmod p$
16.   **else**
17.   **If** $U > V$ **then** $U = \frac{U-V}{2}$,
              $R \equiv R - S \pmod p, S \equiv 2S \pmod p, t = 1$
18.   **else** $V = \frac{V-U}{2}$,
              $S \equiv S - R \pmod p, R \equiv 2R \pmod p, t = 1$
19.   **If** $i < m$ **then**
20.   **If** $i = m - 1$ or $t = 1$ **then**
21.   **If** $\alpha_i = 1$ **then** $R \equiv R \pmod p, S \equiv S \pmod p$
22.   **else** $R \equiv \frac{R}{2} \pmod p, S \equiv \frac{S}{2} \pmod p$
23.   **else**
24.   **If** $(\alpha_{i+1}, \alpha_i) = (1, 1)$ **then**
                  $R \equiv R \pmod p, S \equiv S \pmod p$
25.   **else if** $(\alpha_{i+1}, \alpha_i) = (1, 0)$ or $(0, 1)$ **then**
                  $R \equiv \frac{R}{2} \pmod p, S \equiv \frac{S}{2} \pmod p$
26.   **else**
                  $R \equiv \frac{R}{4} \pmod p, S \equiv \frac{S}{4} \pmod p$
27. $i = i + t$
28.   **else** $R \equiv \frac{R}{2^t} \pmod p, S \equiv \frac{S}{2^t} \pmod p$
29. **Return** R

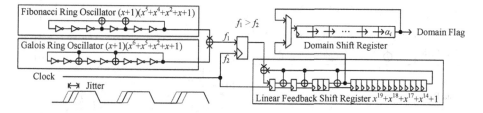

Fig. 3. The domain flag is to randomly assign operating domain for GFAU

As the iterative operations in Algorithm 4 and Algorithm 5 are performed in one cycle, the critical path is to calculate the results of R or S consisting of the UV comparison with modular operations. For the modular division by 2 or 4 in Steps 5, 8, 11, 14 of Algorithm 4 and Steps 22, 25, 26, 28 of Algorithm 5, multiples of the prime p are added to enable the lowest part of R or S is zero so that they can be carried out by simple shift logic operator. Further, since the results of R, S are irrelevant to the results of operands U or V, a fully-pipelined stage can be inserted between the UV and RS data path to moderate the critical path. As the UV data path is determined, then the next cycle is to set the values of the operands R, S and simultaneously determine the next case until $V = 0$. Although one additional cycle is needed after pipelining, this is negligible as the operation takes hundreds or thousands of cycles. The timing flow of pipelined scheme is shown in Fig. 4. Besides, to reduce the hardware cost, symmetric modular operations such as $R \equiv (R - S)/2 \pmod{p}$ and $S \equiv (S-R)/2 \pmod{p}$ in Algorithm 4, $R \equiv (R-S)/4 \pmod{p}$ and $S \equiv (S-R)/4 \pmod{p}$ in Algorithm 5 can be executed by the same computational unit with a swap logic circuit, which is to switch the input operands of RS data path. In Algorithm 4, the RS data path can be classified into two groups: the first group includes Steps 4, 5 and Steps 10, 11; the second one consists of Steps 7, 8 and Steps 13, 14. In Algorithm 5, the two groups of RS data path are classified as follows: Steps 6, 9, 12, 15, and 18 belong in the first group; the second one consists of the others. The data flows of R and S are switched as the processing group is different from the group in previous cycle. Moreover, since the EC point calculation is a serial field operation, both of the temporary registers and modular operations can be shared for the operands V, S, R in Algorithm 2 and Algorithm 4 (or Algorithm 3 and Algorithm 5). These multiple modular operations in the iterative calculation can be effectively implemented by using a programmable data path of bit-level architecture, which consists of the carry-save adders with a carry-lookahead adder at last stage. The detailed radix-2 and radix-4 GFAU architecture is shown in Fig. 5 and Fig. 6, respectively.

5   Power Measurement and Implementation Results

Based on our proposed architecture using Montgomery ladder ECSM method, four different 160-bit DF-ECC processors with radix-2 and radix-4 algorithms are

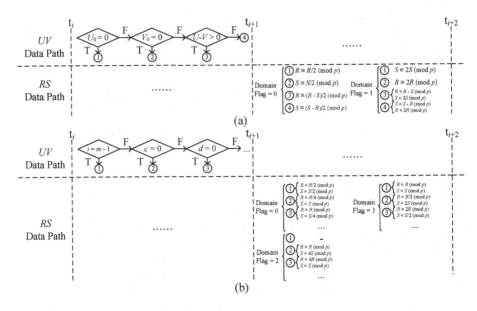

Fig. 4. Fully-pipelined scheme for the (a) radix-2 (b) radix-4 randomized Montgomery division

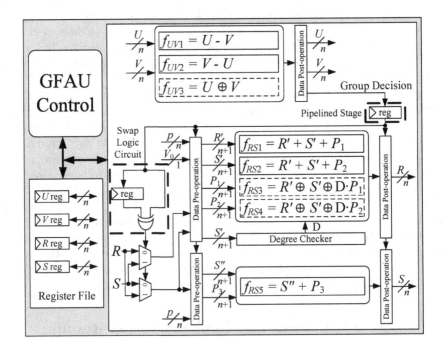

Fig. 5. Radix-2 GFAU

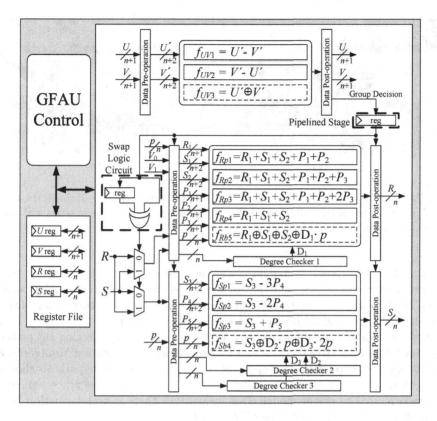

Fig. 6. Radix-4 GFAU

independently designed on an FPGA platform to evaluate the CPA resistance. The performance results are given in Table 3, and the verification environment is shown in Fig. 7.

Table 3. FPGA Implementation Results

Design	Area (Slices)	f_{max} (MHz)	Field Arithmetic
I	7,573 (32%)	27.7	Radix-2 Montgomery
II	8,158 (34%)	27.7	Radix-2 Randomized Montgomery
III	9,828 (41%)	20.2	Radix-4 Montgomery
IV	10,460 (43%)	20.2	Radix-4 Randomized Montgomery

As shown in Algorithm 1, the point coordinate value P_2 is dependent on the bit value of the key in every iteration. Fig. 8(a) and Fig. 8(b) illustrate the CPA attacks on the unprotected Design-I and Design-III, respectively, using conventional Montgomery algorithm [21] to reveal the key value. The correlation

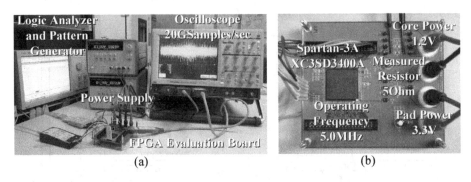

Fig. 7. (a) Environment of power measurement. (b) Current running through the DF-ECC processor recorded by measuring the voltage drop via a resistor in series with the board power pin and FPGA power pin.

coefficients for all possible hamming distances of the point coordinate P_2 are plotted over power traces, and that of the correct key hypothesis is plotted in black. In this case, as more than 10^3 power traces are used, the correlation of the correct key is the highest one among that of all the other key hypotheses, and then the key value can be found easily. However, even after collecting 10^6 power measurements from the Design-II and Design-IV using randomized Montgomery operations, the correlation coefficients of correct and incorrect hypothesis shown in Fig. 9 cannot be scattered, and they are near zero because the processed data are uncorrelated to power model. This means that there is no information bias of the key value extracted by the CPA attacks.

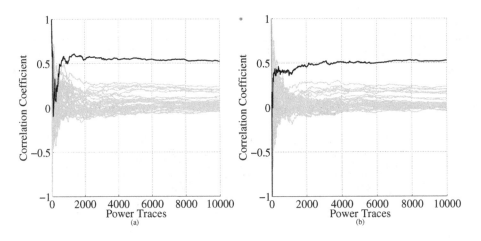

Fig. 8. Correlation coefficients of the target traces and power model over power traces obtained from the (a) Design-I (b) Design-III performing arithmetic in a fixed domain

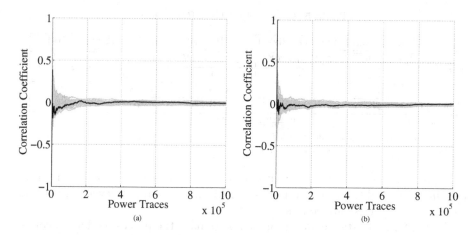

Fig. 9. Correlation coefficients of the target traces and power model over power traces obtained from the (a) Design-II (b) Design-IV performing arithmetic in a randomized domain.

Table 4. Implementation Results Compared with Related Works

	Technology	Field Length	Area (mm²)	KGates	Galois Field	f_{max} (MHz)	Time (ms/ECSM)	Energy (µJ/ECSM)	AT Product
Ours (Radix-2)	90-nm	160	0.21	61.3	$GF(p_{160})$	277	0.71	11.9	1
					$GF(2^{160})$	277	0.61	9.6	1
Ours (Radix-4)	90-nm	160	0.29	83.2	$GF(p_{160})$	238	0.43	11.2	0.82
					$GF(2^{160})$	238	0.39	8.97	0.87
TCAS-II'09 [5]	0.13-µm	160	1.44	169	$GF(p_{160})$	121	0.61	42.6	1.63*
					$GF(2^{160})$	146	0.37	30.5	1.16*
Ours (Radix-2)	90-nm	521	0.58	168	$GF(p_{521})$	250	8.08	452	1
					$GF(2^{409})$	263	4.65	246	1
Ours (Radix-4)	90-nm	521	0.93	265	$GF(p_{521})$	232	4.57	435	0.89
					$GF(2^{409})$	238	2.77	238	0.94
ESSCIRC'10 [9]	90-nm	521	0.55	170	$GF(p_{521})$	132	19.2	1,123	2.40
					$GF(2^{409})$	166	8.2	480	1.78

* Technology scaled area-time product = Gates × (Time × t), where t = 90-nm/0.13-µm.

Our proposed DF-ECC processor was also implemented by UMC 90-nm CMOS technology, and the post-layout simulations for ASIC implementation with comparisons are given in Table 4. In terms of area-time product, our DF-ECC processor outperforms other approaches. By reducing the division iteration time and randomizing intermediate values in field arithmetic without increasing the key size, our work using radix-2 approach is at least 44% faster than the previous 521-bit design [9] with comparable hardware complexity. Compared with a four

Table 5. Overhead for CPA Resistance

	Ours (Radix-2)	Ours (Radix-4)	ESSCIRC'10 [9]	JSSC'06 [12]	JSSC'10 [13]
Design	521 DF-ECC	521 DF-ECC	521 DF-ECC	128 AES	128 AES
Area	4.3%	3.6%	10%	210%	7.2%
Time	0	0	14.0%[a]	288%	100%
Energy	5.2%	3.8%	20.8%[b]	270%	33%

Overhead = $\frac{\text{Result differences between protected and unprotected circuit}}{\text{Results of unprotected circuit}}$ ×100%.

a. Estimated by cycle count×clock period.

b. Estimated by operation time×average power.

multipliers based ECC processor without power-analysis protection [5], our fully-pipelined and highly-integrated radix-4 GFAU architecture achieves competitive speed with 51% less gate counts.

For the CPA resistance, our approach is to mask the processed data uncorrelated with power traces without lengthening the hardware latency and without dominating the power consumption of key-dependent operations. From the comparison given in Table 5, our proposed countermeasure is superior to others not only in operation time but also in energy dissipation.

6   Conclusion

In this paper, we introduced a randomized dual-field Montgomery algorithm which is suitable for ECC hardware implementation against the CPA attacks. Without modifying logic circuit and without pre-computing data from host system, the relationship between target power traces and power model can be broken by performing the field arithmetic in a unpredictable operating domain. The proposed CPA countermeasure approach has been analyzed on an FPGA platform. Attacks on the unprotected designs reveal the private key within one thousand power traces, while the key value of the protected core cannot be extracted after one million power traces. Circuit overhead for randomly determining the operating domain can be integrated into the system without speed degradation. Implemented by an UMC 90-nm technology, our protected 521-bit DF-ECC processor using radix-4 randomized Montgomery operations, with 3.6% area and 3.8% average power overhead, can perform one $GF(p_{521})$ ECSM in 4.57ms and one $GF(2^{409})$ ECSM in 2.77ms. We believe that both high performance and efficient CPA countermeasure are achieved in our proposed DF-ECC processor.

Acknowledgments. The work described in this paper was supported by Taiwan's National Science Council (NSC) and Ministry of Economic Affairs (MOEA) under Grants NSC 100-2220-E-009-016, NSC 101-2220-E-009-060, and MOEA 100-EC-17-A-01-S1-180. The authors are grateful to United Microelectronics Corporation (UMC), Taiwan, for technology support. The authors would also like to thank Po-Ming Tu for his suggestion of FPGA power measurement, and anonymous reviewers for their constructive comments to improve the quality of this paper.

References

1. Koblitz, N.: Elliptic Curve Cryptosystems. Math. Comp. 48, 203–209 (2001)
2. Miller, V.S.: Use of Elliptic Curves in Cryptography. In: Williams, H.C. (ed.) CRYPTO 1985. LNCS, vol. 218, pp. 417–426. Springer, Heidelberg (1986)
3. McIvor, C.J., McLoone, M., McCanny, J.V.: Hardware Elliptic Curve Cryptographic Processor over $GF(p)$. IEEE Trans. Circuits Syst. I 53(9), 1946–1957 (2006)
4. Sakiyama, K., Batina, L., Preneel, B., Verbauwhede, I.: Multicore Curve-Based Cryptoprocessor With Reconfigurable Modular Arithmetic Logic Units over $GF(2^n)$. IEEE Trans. Comput. 56(9), 1269–1282 (2007)
5. Lai, J.-Y., Huang, C.-T.: A Highly Efficient Cipher Processor for Dual-Field Elliptic Curve Cryptography. IEEE Trans. Circuits Syst. II 56(5), 394–398 (2009)
6. Chen, J.-H., Shieh, M.-D., Lin, W.-C.: A High-Performance Unified-Field Reconfigurable Cryptographic Processor. IEEE Trans. VLSI Syst. 18(8), 1145–1158 (2010)
7. Kocher, P., Jaffe, J., Jun, B.: Differential Power Analysis. In: Wiener, M. (ed.) CRYPTO 1999. LNCS, vol. 1666, pp. 388–397. Springer, Heidelberg (1999)
8. Montgomery, P.: Speeding the Pollard and Elliptic Curve Methods of Factorization. Math. Comp. 48, 243–264 (1987)
9. Lee, J.-W., Chen, Y.-L., Tseng, C.-Y., Chang, H.-C., Lee, C.-Y.: A 521-bit Dual-Field Elliptic Curve Cryptographic Processor With Power Analysis Resistance. In: European Solid-State Circuits Conference (ESSCIRC 2010), pp. 206–209 (2010)
10. Brier, E., Clavier, C., Olivier, F.: Correlation Power Analysis with a Leakage Model. In: Joye, M., Quisquater, J.-J. (eds.) CHES 2004. LNCS, vol. 3156, pp. 16–29. Springer, Heidelberg (2004)
11. IEEE: Standard Specifications or Public-Key Cryptography. IEEE Std. 1363 (2000)
12. Hwang, D., Tiri, K., Hodjat, A., Lai, B.-C., Yang, S., Schaumont, P., Verbauwhede, I.: AES-Based Security Coprocessor IC in 0.18-μm CMOS With Resistance to Differential Power Analysis Side-Channel Attacks. IEEE J. Solid-State Circuits 41(4), 781–792 (2006)
13. Tokunaga, C., Blaauw, D.: Securing Encryption Systems With a Switched Capacitor Current Equalizer. IEEE J. Solid-State Circuits 45(1), 23–31 (2010)
14. Liu, P.-C., Chang, H.-C., Lee, C.-Y.: A True Random-Based Differential Power Analysis Countermeasure Circuit for an AES Engine. IEEE Trans. Circuits Syst. II 59(2), 103–107 (2012)
15. Coron, J.-S.: Resistance against Differential Power Analysis for Elliptic Curve Cryptosystems. In: Koç, Ç.K., Paar, C. (eds.) CHES 1999. LNCS, vol. 1717, pp. 292–302. Springer, Heidelberg (1999)
16. Joye, M., Tymen, C.: Protections against Differential Analysis for Elliptic Curve Cryptography. In: Koç, Ç.K., Naccache, D., Paar, C. (eds.) CHES 2001. LNCS, vol. 2162, pp. 377–390. Springer, Heidelberg (2001)
17. Montgomery, P.: Modular Multiplication Without Trial Division. Math. Comp. 44, 519–521 (1985)
18. Kaliski, B.: The Montgomery Inverse and Its Applications. IEEE Trans. Comput. 44(8), 1064–1065 (1995)

19. Cohen, H., Miyaji, A., Ono, T.: Efficient Elliptic Curve Exponentiation Using Mixed Coordinates. In: Ohta, K., Pei, D. (eds.) ASIACRYPT 1998. LNCS, vol. 1514, pp. 51–65. Springer, Heidelberg (1998)
20. Golic, J.D.: New Methods for Digital Generation and Postprocessing of Random Data. IEEE Trans. Comp. 55, 1217–1229 (2006)
21. Chen, Y.-L., Lee, J.-W., Liu, P.-C., Chang, H.-C., Lee, C.-Y.: A Dual-Field Elliptic Curve Cryptographic Processor With a Radix-4 Unified Division Unit. In: IEEE Int. Symp. on Circuits Syst. (ISCAS 2011), pp. 713–716 (2011)

Author Index